THEORY AND APPLICATIONS OF
FRACTIONAL DIFFERENTIAL EQUATIONS

NORTH-HOLLAND MATHEMATICS STUDIES 204

(Continuation of the Notas de Matemática)

Editor: Jan van Mill
Faculteit der Exacte Wetenschappen
Amsterdam, The Netherlands

ELSEVIER

2006

Amsterdam – Boston – Heidelberg – London – New York – Oxford
Paris – San Diego – San Francisco – Singapore – Sydney – Tokyo

THEORY AND APPLICATIONS OF FRACTIONAL DIFFERENTIAL EQUATIONS

ANATOLY A. KILBAS
Belarusian State University
Minsk, Belarus

HARI M. SRIVASTAVA
University of Victoria
Victoria, British Columbia, Canada

JUAN J. TRUJILLO
Universidad de La Laguna
La Laguna (Tenerife)
Canary Islands, Spain

ELSEVIER

Amsterdam – Boston – Heidelberg – London – New York – Oxford
Paris – San Diego – San Francisco – Singapore – Sydney – Tokyo

ELSEVIER B.V.
Radarweg 29
P.O. Box 211, 1000 AE Amsterdam
The Netherlands

ELSEVIER Inc.
525 B Street, Suite 1900
San Diego, CA 92101-4495
USA

ELSEVIER Ltd.
The Boulevard, Langford Lane
Kidlington, Oxford OX5 1GB
UK

ELSEVIER Ltd.
84 Theobalds Road
London WC1X 8RR
UK

First edition 2006

Library of Congress Cataloging in Publication Data
A catalog record is available from the Library of Congress.

British Library Cataloguing in Publication Data
A catalogue record is available from the British Library.

ISBN-13: 978-0-444-51832-3
ISBN-10: 0-444-51832-0
ISSN (Series): 0304-0208

∞ The paper used in this publication meets the requirements of ANSI/NISO Z39.48-1992 (Permanence of Paper).

Transferred to digital printing in 2006.

To

Tamara, Rekha, and Margarita

Preface

The subject of fractional calculus (that is, calculus of integrals and derivatives of any arbitrary real or complex order) has gained considerable popularity and importance during the past three decades or so, due mainly to its demonstrated applications in numerous seemingly diverse and widespread fields of science and engineering. It does indeed provide several potentially useful tools for solving differential and integral equations, and various other problems involving special functions of mathematical physics as well as their extensions and generalizations in one and more variables.

The concept of fractional calculus is popularly believed to have stemmed from a question raised in the year 1695 by Marquis de L'Hôpital (1661-1704) to Gottfried Wilhelm Leibniz (1646-1716), which sought the meaning of Leibniz's (currently popular) notation $\frac{d^n y}{dx^n}$ for the derivative of order $n \in \mathbb{N}_0 := \{0, 1, 2, \cdots\}$ when $n = \frac{1}{2}$ (What if $n = \frac{1}{2}$?). In his reply, dated 30 September 1695, Leibniz wrote to L'Hôpital as follows: "... *This is an apparent paradox from which, one day, useful consequences will be drawn.* ..."

Subsequent mention of fractional derivatives was made, in some context or the other, by (for example) Euler in 1730, Lagrange in 1772, Laplace in 1812, Lacroix in 1819, Fourier in 1822, Liouville in 1832, Riemann in 1847, Greer in 1859, Holmgren in 1865, Grünwald in 1867, Letnikov in 1868, Sonin in 1869, Laurent in 1884, Nekrassov in 1888, Krug in 1890, and Weyl in 1917. In fact, in his 700-page textbook, entitled "*Traité du Calcul Différentiel et du Calcul Intégral*" (Second edition; Courcier, Paris, 1819), S. F. Lacroix devoted two pages (pp. 409-410) to fractional calculus, showing *eventually* that

$$\frac{d^{\frac{1}{2}}}{dv^{\frac{1}{2}}} v = \frac{2\sqrt{v}}{\sqrt{\pi}}.$$

In addition, of course, to the theories of differential, integral, and integro-differential equations, and special functions of mathematical physics as well as their extensions and generalizations in one and more variables, some of the areas of present-day applications of fractional calculus include Fluid Flow, Rheology, Dynamical Processes in Self-Similar and Porous Structures, Diffusive Transport Akin to Diffusion, Electrical Networks, Probability and Statistics, Control Theory of Dynamical Systems, Viscoelasticity, Electrochemistry of Corrosion, Chemical Physics, Optics and Signal Processing, and so on.

The first work, devoted exclusively to the subject of fractional calculus, is the book by Oldham and Spanier [643] published in 1974. One of the most recent works on the subject of fractional calculus is the book of Podlubny [682] published in 1999, which deals principally with fractional differential equations. Some of the latest (but certainly not the last) works especially on fractional models of anomalous kinetics of complex processes are the volumes edited by Carpinteri and Mainardi [132] in 1997 and by Hilfer [340] in 2000, and the book by Zaslavsky [915] published in 2005. Indeed, in the meantime, numerous other works (books, edited volumes, and conference proceedings) have also appeared. These include (for example) the remarkably comprehensive encyclopedic-type monograph by Samko, Kilbas and Marichev [729], which was published in Russian in 1987 and in English in 1993, and the book devoted substantially to fractional differential equations by Miller and Ross [603], which was published in 1993. And today there exist at least two international journals which are devoted almost entirely to the subject of fractional calculus: (i) *Journal of Fractional Calculus* and (ii) *Fractional Calculus and Applied Analysis.*

The main objective of this book is to complement the contents of the other books mentioned above. Many new results, obtained recently in the theory of ordinary and partial differential equations, are not specifically reflected in the book. We aim at presenting, in a systematic manner, results including the existence and uniqueness of solutions for the Cauchy Type and Cauchy problems involving nonlinear ordinary fractional differential equations, explicit solutions of linear differential equations and of the corresponding initial-value problems by their reduction to Volterra integral equations and by using operational and compositional methods, applications of the one- and multi-dimensional Laplace, Mellin, and Fourier integral transforms in deriving closed-form solutions of ordinary and partial differential equations, and a theory of the so-called sequential linear fractional differential equations including a generalization of the classical Frobenius method.

This book consists of a total of eight chapters. Chapter 1 (Preliminaries) provides some basic definitions and properties from such topics of Mathematical Analysis as functional spaces, special functions, integral transforms, generalized functions, and so on. The extensive modern-day usages of such special functions as the classical Mittag-Leffler functions and its various extensions, the Wright (or, more precisely, the Fox-Wright) generalization of the relatively more familiar hypergeometric ${}_pF_q$ function, and the Fox H-function in the solutions of ordinary and partial fractional differential equations have indeed motivated a major part of Chapter 1. Chapter 2 (Fractional Integrals and Fractional Derivatives) contains the definitions and some potentially useful properties of several different families of fractional integrals and fractional derivatives. Chapter 1 and Chapter 2, together, are meant to prepare the reader for the understanding of the various mathematical tools and techniques which are developed in the later chapters of this book.

The fundamental existence and uniqueness theorems for *ordinary* fractional differential equations are presented in Chapter 3 with special reference to the Cauchy Type problems. Here, in Chapter 3, we also consider nonlinear and linear fractional differential equations in one-dimensional and vectorial cases. Chapter

4 is devoted to explicit and numerical solutions of fractional differential equations and boundary-value problems associated with them. Our approaches in this chapter are based mainly upon the reduction to Volterra integral equations, upon compositional relations, and upon operational calculus.

In Chapter 5, we investigate the applications of the Laplace, Mellin, and Fourier integral transforms with a view to constructing explicit solutions of linear differential equations involving the Liouville, Caputo, and Riesz fractional derivatives with constant coefficients. Chapter 6 is devoted to a survey of the developments and results in the fields of *partial* fractional differential equations and to the applications of the Laplace and Fourier integral transforms in order to obtain closed-form solutions of the Cauchy Type and Cauchy problems for the fractional diffusion-wave and evolution equations.

Linear differential equations of *sequential* and *non-sequential* fractional order, as well as systems of linear fractional differential equations associated with the Riemann-Liouville and Caputo derivatives, are investigated in Chapter 7, which incidentally also develops an interesting generalization of the classical Frobenius Method for solving fractional differential equations with *variable* coefficients and a direct way to obtain explicit solutions of such types of differential equations with *constant* coefficients. And, while a survey of a variety of applications of fractional differential equations are treated briefly in many chapters of this book (especially in Chapter 7), a review of some important applications involving fractional models is presented systematically in the last chapter of the book (Chapter 8).

At the end of this book, for the convenience of the readers interested in *further* investigations on these and other closely-related topics, we include a rather large and up-to-date Bibliography. We also include a Subject Index.

Operators of fractional integrals and fractional derivatives, which are based essentially upon the familiar Cauchy-Goursat Integral Formula, were considered by (among others) Sonin in 1869, Letnikov in 1868 onwards, and Laurent in 1884. In recent years, many authors have demonstrated the usefulness of such types of fractional calculus operators in obtaining particular solutions of numerous families of homogeneous (as well as nonhomogeneous) linear ordinary and partial differential equations which are associated, for example, with many of the *celebrated* equations of mathematical physics such as (among others) the *Gauss hypergeometric equation*:

$$z(1-z)\frac{d^2w}{dz^2} + [\gamma - (\alpha+\beta+1)z]\frac{dw}{dz} - \alpha\beta w = 0$$

and the relatively more familiar *Bessel equation*:

$$z^2\frac{d^2w}{dz^2} + z\frac{dw}{dz} + (z^2-\nu^2)w = 0.$$

In the cases of (ordinary as well as partial) differential equations of *higher* orders, which have stemmed naturally from the Gauss hypergeometric equation, the Bessel equation, and their many relatives and extensions, there have been several seemingly independent attempts to present a remarkably large number of scattered results in a unified manner. For developments dealing extensively with such applications of fractional calculus operators in the solution of ordinary and partial differential equations, the interested reader is referred to the numerous recent works cited in the Bibliography.

This book is written primarily for the graduate students and researchers in many different disciplines in the mathematical, physical, and engineering sciences, who are interested not only in learning about the various mathematical tools and techniques used in the theory and widespread applications of fractional differential equations, but also in *further* investigations which emerge naturally from (or which are motivated substantially by) the physical situations modelled mathematically in the book.

Right from the conceptualization of this book project until the finalization of the typescript in this present form, the authors have gratefully received invaluable suggestions and comments from colleagues at many different academic institutions and research centers around the world. Special mention ought to be made of the help and assistance so generously and meticulously provided by Dr. Margarita Rivero, Dr. Blanca Bonilla, Dr. José R. Franco, Ldo. Luís Rodríguez-Germá, all at the Universidad de La Laguna. Special mention should be made also of Engr. Sergio Ortiz-Villajos Eirín and of Dr. Luís Vázquez at the Universidad Complutense de Madrid.

The first- and the second-named authors would like to express their appreciation for the kind hospitality and support by the Universidad de La Laguna during their many short-term visits to the Departamento de Análisis Matemático (Universidad de La Laguna) throughout the tenure of this book project. All three authors, and particularly the second-named author, are especially thankful to the second-named author's wife and colleague, Professor Rekha Srivastava, for her constant support, encouragement, and understanding. The first-named author would also like to thank his wife, Mrs. Tamara Sapova, for her support and encouragement.

We are immensely grateful to Professor Nacere Hayek Calil for his constant inspiration and encouragement throughout the tenure of this book project.

Finally, the authors (particularly the third-named author) would like to thankfully acknowledge the financial grants and support for this book project, which were awarded by the Ministry of Education and Sciences (MTM2004/00327), FEDER and the Autonomous Government of Canary Islands (PI 2003/133). This book was prepared in the frame of joint collaboration of the first- and the third-named authors supported by Belarusian Fundamental Research Fund under Projects F03MC-008 and F05MC-050. This work was also supported, in part, by the Universidad de La Laguna, the Academia Canaria de Ciencias, and the Natural Sciences and Engineering Research Council of Canada.

September 2005

Anatoly A. Kilbas
Belarusian State University

Hari M. Srivastava
University of Victoria

Juan J. Trujillo
Universidad de La Laguna

Contents

Chapter 1

PRELIMINARIES

This chapter is preliminary in character and contains definitions and properties from such topics of *Analysis* as functional spaces, special functions, and integral transforms.

1.1 Spaces of Integrable, Absolutely Continuous, and Continuous Functions

In this section we present definitions of spaces of p-integrable, absolutely continuous, and continuous functions and their weighted modifications. We also give characterizations of those modified spaces which will be used later.

Let $\Omega = [a, b]$ $(-\infty \leqq a < b \leqq \infty)$ be a finite or infinite interval of the real axis $\mathbb{R} = (-\infty, \infty)$. We denote by $L_p(a, b)$ $(1 \leqq p \leqq \infty)$ the set of those Lebesgue complex-valued measurable functions f on Ω for which $\|f\|_p < \infty$, where

$$\|f\|_p = \left(\int_a^b |f(t)|^p dt\right)^{1/p} \quad (1 \leqq p < \infty) \tag{1.1.1}$$

and

$$\|f\|_\infty = \text{ess sup}_{a \leqq x \leqq b} |f(x)|. \tag{1.1.2}$$

Here ess sup$|f(x)|$ is the essential maximum of the function $|f(x)|$ [see, for example, Nikol'skii [628], pp. 12-13)].

We also need the weighted L^p-space with the power weight. Such a space, which we denote by $X_c^p(a, b)$ $(c \in \mathbb{R};\ 1 \leqq p \leqq \infty)$, consists of those complex-valued Lebesgue measurable functions f on (a, b) for which $\|f\|_{X_c^p} < \infty$, with

$$\|f\|_{X_c^p} = \left(\int_a^b |t^c f(t)|^p \frac{dt}{t}\right)^{1/p} \quad (1 \leqq p < \infty) \tag{1.1.3}$$

and

$$\|f\|_{X_c^\infty} = \text{ess sup}_{a \leqq x \leqq b}[x^c|f(x)|]. \tag{1.1.4}$$

In particular, when $c = 1/p$, the space $X_c^p(a,b)$ coincides with the $L_p(a,b)$-space: $X_{1/p}^p(a,b) = L_p(a,b)$.

Let now $[a,b]$ $(-\infty < a < b < \infty)$ be a finite interval and let $AC[a,b]$ be the space of functions f which are absolutely continuous on $[a,b]$. It is known [see Kolmogorov and Fomin ([434], p. 338)] that $AC[a,b]$ coincides with the space of primitives of Lebesgue summable functions:

$$f(x) \in AC[a,b] \Leftrightarrow f(x) = c + \int_a^x \varphi(t)dt \quad (\varphi(t) \in L(a,b)), \tag{1.1.5}$$

and therefore an absolutely continuous function $f(x)$ has a summable derivative $f'(x) = \varphi(x)$ almost everywhere on $[a,b]$. Thus (1.1.5) yields

$$\varphi(t) = f'(t) \text{ and } c = f(a). \tag{1.1.6}$$

For $n \in \mathbb{N} := \{1,2,3,\cdots\}$ we denote by $AC^n[a,b]$ the space of complex-valued functions $f(x)$ which have continuous derivatives up to order $n-1$ on $[a,b]$ such that $f^{(n-1)}(x) \in AC[a,b]$:

$$AC^n[a,b] = \left\{f : [a,b] \to \mathbb{C} \text{ and } (D^{n-1}f)(x)] \in AC[a,b] \quad (D = \frac{d}{dx})\right\}, \tag{1.1.7}$$

$\mathbb{C}$ being the set of complex numbers. In particular, $AC^1[a,b] = AC[a,b]$.

This space is characterized by the following assertion [see Samko et al. ([729], Lemma 2.4)].

Lemma 1.1 *The space $AC^n[a,b]$ consists of those and only those functions $f(x)$ which can be represented in the form*

$$f(x) = (I_{a+}^n \varphi)(x) + \sum_{k=0}^{n-1} c_k(x-a)^k, \tag{1.1.8}$$

where $\varphi(t) \in L(a,b)$, c_k $(k = 0,1,\cdots, n-1)$ are arbitrary constants, and

$$(I_{a+}^n \varphi)(x) = \frac{1}{(n-1)!} \int_a^x (x-t)^{n-1} \varphi(t)dt. \tag{1.1.9}$$

It follows from (1.1.8) that

$$\varphi(t) = f^{(n)}(t), \quad c_k = \frac{f^{(k)}(a)}{k!} \quad (k = 0,1,\cdots, n-1). \tag{1.1.10}$$

We also use a weighted modification of the space $AC^n[a,b]$ $(n \in \mathbb{N})$, in which the usual derivative $D = d/dx$ is replaced by the so-called δ-derivative, defined by

$$\delta = xD \quad (D = \frac{d}{dx}). \tag{1.1.11}$$

Such a modification, which we denote by $AC^n_{\delta,\mu}[a,b]$ ($n \in \mathbb{N}$; $\mu \in \mathbb{R}$), involves the complex-valued Lebesgue measurable functions g on (a,b) such that $x^\mu g(x)$ has δ-derivatives up to order $n-1$ on $[a,b]$ and $\delta^{n-1}[x^\mu g(x)]$ is absolutely continuous on $[a,b]$:

$$AC^n_{\delta,\mu}[a,b] = \left\{ g : [a,b] \to \mathbb{C} : \ \delta^{n-1}[x^\mu g(x)] \in AC[a,b], \quad \mu \in \mathbb{R}, \ \delta = x\frac{d}{dx} \right\}. \tag{1.1.12}$$

In particular, when $\mu = 0$, the space $AC^n_\delta[a,b] := AC^n_{\delta,0}[a,b]$ is defined by

$$AC^n_\delta[a,b] = \left\{ g : [a,b] \to \mathbb{C} : \ \delta^{n-1}[g(x)] \in AC[a,b], \quad \delta = x\frac{d}{dx} \right\}. \tag{1.1.13}$$

If $\mu = 0$ and $n = 1$, the space $AC^1_\delta[a,b]$ coincides with $AC[a,b]$.

When $a > 0$, the space $AC^n_{\delta,\mu}[a,b]$ is characterized by the following result [see Kilbas ([370], Theorem 3.1)].

Lemma 1.2 *Let $0 < a < b < \infty$, $n \in \mathbb{N}$ and $\mu \in \mathbb{R}$. The space $AC^n_{\delta,\mu}[a,b]$ consists of those and only those functions $g(x)$ which are represented in the form*

$$g(x) = x^\mu \left[\frac{1}{(n-1)!} \int_a^x \left(\log\frac{x}{t}\right)^{n-1} \varphi(t)\frac{dt}{t} + \sum_{k=0}^{n-1} d_k \left(\log\frac{x}{a}\right)^k \right], \tag{1.1.14}$$

where $\varphi(t) \in L(a,b)$ and d_k ($k = 0, 1, \cdots, n-1$) are arbitrary constants.

In particular, $g(x) \in AC^n_\delta[a,b]$ if, and only if,

$$g(x) = \frac{1}{(n-1)!} \int_a^x \left(\log\frac{x}{t}\right)^{n-1} \varphi(t)dt + \sum_{k=0}^{n-1} d_k \left(\log\frac{x}{a}\right)^k. \tag{1.1.15}$$

We note that $\varphi(t)$ and d_k in (1.1.14) are given by

$$\varphi(t) = g'_{n-1}(t), \quad d_k = \frac{g_k(a)}{k!} \quad (k = 0, 1, \cdots, n-1) \tag{1.1.16}$$

with

$$g_k(x) = \delta^k[x^\mu g(x)] \ (k = 1, \cdots, n-1), \quad g_0(x) = x^\mu g(x), \tag{1.1.17}$$

while in (1.1.15) they take the more simple forms

$$\varphi(t) = \frac{d}{dt}[\delta^{n-1}[g(t)], \ d_k = \frac{\delta^k g(a)}{k!} \quad (k = 0, 1, \cdots, n-1). \tag{1.1.18}$$

Let $\Omega = [a,b]$ ($-\infty \leqq a < b \leqq \infty$) and $m \in \mathbb{N}_0 = \{0, 1, \cdots\}$. We denote by $C^m(\Omega)$ a space of functions f which are m times continuously differentiable on Ω with the norm

$$\|f\|_{C^m} = \sum_{k=0}^m \|f^{(k)}\|_C = \sum_{k=0}^m \max_{x\in\Omega} |f^{(k)}(x)|, \quad m \in \mathbb{N}_0. \tag{1.1.19}$$

In particular, for $m = 0$, $C^0(\Omega) \equiv C(\Omega)$ is the space of continuous functions f on Ω with the norm

$$\|f\|_C = \max_{x\in\Omega} |f(x)|. \tag{1.1.20}$$

When $\Omega = [a,b]$ is a finite interval and $\gamma \in \mathbb{C}$ $(0 \leqq \Re(\gamma) < 1)$, we introduce the weighted space $C_\gamma[a,b]$ of functions f given on $(a,b]$, such that the function $(x-a)^\gamma f(x) \in C[a,b]$, and

$$\|f\|_{C_\gamma} = \|(x-a)^\gamma f(x)\|_C, \quad C_0[a,b] = C[a,b]. \tag{1.1.21}$$

For $n \in \mathbb{N}$ we denote by $C_\gamma^n[a,b]$ the Banach space of functions $f(x)$ which are continuously differentiable on $[a,b]$ up to order $n-1$ and have the derivative $f^{(n)}(x)$ of order n on $(a,b]$ such that $f^{(n)}(x) \in C_\gamma[a,b]$:

$$C_\gamma^n[a,b] = \left\{ f: \ \|f\|_{C_\gamma^n} = \sum_{k=0}^{n-1} \|f^{(k)}\|_C + \|f^{(n)}\|_{C_\gamma} \right\}, \quad C_\gamma^0[a,b] = C_\gamma[a,b]. \tag{1.1.22}$$

From this definition we have the following characterization of the space $C_\gamma^n[a,b]$ [see Kilbas et al. ([375], Lemma 1)].

Lemma 1.3 *Let $n \in \mathbb{N}_0 = \{0, 1, \cdots\}$ and $\gamma \in \mathbb{C}$ $(0 \leqq \Re(\gamma) < 1)$. The space $C_\gamma^n[a,b]$ consists of those and only those functions f which are represented in the form*

$$f(x) = \frac{1}{(n-1)!} \int_a^x (x-t)^{n-1} \varphi(t)dt + \sum_{k=0}^{n-1} c_k (x-a)^k, \tag{1.1.23}$$

where $\varphi(t) \in C_\gamma[a,b]$ and c_k $(k = 0, 1, \cdots, n-1)$ are arbitrary constants.

Moreover,

$$\varphi(t) = f^{(n)}(t), \quad c_k = \frac{f^{(k)}(a)}{k!} \quad (k = 0, 1, \cdots, n-1). \tag{1.1.24}$$

In particular, when $\gamma = 0$, the space $C^n[a,b]$ consists of those and only those functions f which are represented in the form (1.1.23), *where $\varphi(t) \in C[a,b]$ and c_k $(k = 0, 1, \cdots, n-1)$ are arbitrary constants. Moreover, the relations in* (1.1.24) *hold.*

We also introduce two subspaces $C_a^0[a,b]$ and $C_b^0[a,b]$ of the space $C[a,b]$ defined by

$$C_a^0[a,b] = \{f(x) \in C_a^0[a,b]; \ f(a) = 0, \ \|f\|_{C_a} = \|f\|_C\} \tag{1.1.25}$$

and

$$C_b^0[a,b] = \{f(x) \in C_a^0[a,b]; \ f(b) = 0, \ \|f\|_{C_a} = \|f\|_C\}. \tag{1.1.26}$$

When $\Omega = [a,b]$ $(0 < a < b < \infty)$ is a finite interval and $\gamma \in \mathbb{C}$ $(0 \leqq \Re(\gamma) < 1)$, we introduce the weighted space $C_{\gamma,\log}[a,b]$ of functions g, given on $(a,b]$ and such that $[\log(x/a)]^\gamma g(x) \in C[a,b]$:

$$\|g\|_{C_{\gamma,\log}} = \left\| \left(\log \frac{x}{a}\right)^\gamma f(x) \right\|_C, \quad C_{0,\log}[a,b] = C[a,b]. \tag{1.1.27}$$

For $n \in \mathbb{N}$ we denote by $C^n_{\delta,\gamma}[a,b]$ the Banach space of functions $g(x)$ which have continuous δ-derivatives on $[a,b]$ up to order $n-1$ and derivative $(\delta^n g)(x)$ on $(a,b]$ of order n such that $(\delta^n g)(x) \in C_{\gamma,\log}[a,b]$:

$$C^n_{\delta,\gamma}[a,b] = \left\{ g : \ ||g||_{C^n_{\delta,\gamma}} = \sum_{k=0}^{n-1} ||\delta^k g||_C + ||\delta^n g||_{C_{\gamma,\log}} \right\}, \quad C^0_{\delta,\gamma}[a,b] = C_{\gamma,\log}[a,b]. \tag{1.1.28}$$

From this definition we have the following characterization of the space $C^n_{\delta,\gamma}[a,b]$.

Lemma 1.4 *Let $0 < a < b < \infty$, $n \in \mathbb{N}_0$ and $\gamma \in \mathbb{C}$ $(0 \leqq \Re(\gamma) < 1)$. The space $C^n_{\delta,\gamma}[a,b]$ consists of those and only those functions g which are represented in the form*

$$g(x) = \frac{1}{(n-1)!} \int_a^x \left(\log\frac{x}{t}\right)^{n-1} \varphi(t)\frac{dt}{t} + \sum_{k=0}^{n-1} d_k \left(\log\frac{x}{a}\right)^k, \tag{1.1.29}$$

where $\varphi(t) \in C_{\gamma,\log}[a,b]$ and d_k $(k = 0,1,\cdots, n-1)$ are arbitrary constants.

Moreover,

$$\varphi(t) = (\delta^n g)(t), \ d_k = \frac{(\delta^k g)(a)}{k!} \quad (k = 0,1,\cdots, n-1). \tag{1.1.30}$$

In particular, when $\gamma = 0$, the space $C^n_{\delta,0}[a,b] = C^n_\delta[a,b]$ consists of those and only those functions g which are represented in the form (1.1.29), *where $\varphi(t) \in C[a,b]$ and d_k $(k = 0,1,\cdots, n-1)$ are arbitrary constants. Moreover, the relations in* (1.1.30) *hold.*

For Banach spaces X and Y, we denote by $X \to Y$ the continuous embedding:

$$(a) \text{ if } f \in X, \text{ then } f \in Y; \quad (b) \ ||f||_Y \leqq K||f||_X, \tag{1.1.31}$$

where the constant $K > 0$ does not depend on f.

From the definitions (1.1.22) and (1.1.28) we derive the property of the spaces $C^n_\gamma[a,b]$ and $C^n_{\delta,\gamma}[a,b]$.

Property 1.1 *Let $n \in \mathbf{N}_0$, and let γ_1 and γ_2 be complex numbers such that*

$$0 \leqq \Re(\gamma_1) \leqq \Re(\gamma_2) < 1. \tag{1.1.32}$$

The following embedding hold:

$$C^n[a,b] \to C^n_{\gamma_1}[a,b] \to C^n_{\gamma_2}[a,b], \tag{1.1.33}$$

with

$$||f||_{C^n_{\gamma_2}} \leqq K||f||_{C^n_{\gamma_1}}, \ K = \min\left[1, (b-a)^{\Re(\gamma_2-\gamma_1)}\right]; \tag{1.1.34}$$

$$C^n_\delta[a,b] \to C^n_{\delta,\gamma_1}[a,b] \to C^n_{\delta,\gamma_2}[a,b], \tag{1.1.35}$$

with

$$\|f\|_{C^n_{\delta;\gamma_2}} \leqq K_\delta \|f\|_{C^n_{\delta;\gamma_1}}, \quad K_\delta = \min\left[1, \left(\log\frac{b}{a}\right)^{\Re(\gamma_2-\gamma_1)}\right]. \tag{1.1.36}$$

In particular,

$$C[a,b] \to C_{\gamma_1}[a,b] \to C_{\gamma_2}[a,b], \quad \text{with} \quad \|f\|_{C_{\gamma_2}} \leqq (b-a)^{\Re(\gamma_2-\gamma_1)}\|f\|_{C_{\gamma_1}}; \tag{1.1.37}$$

$$C[a,b] \to C_{\gamma_1,\log}[a,b] \to C_{\gamma_2,\log}[a,b], \quad \text{with} \quad \|f\|_{C_{\delta;\gamma_2}} \leqq \left(\log\frac{b}{a}\right)^{\Re(\gamma_2-\gamma_1)} \|f\|_{C_{\delta;\gamma_1}}. \tag{1.1.38}$$

1.2 Generalized Functions

In this section we present definitions and some properties of certain spaces of test and generalized functions. More detailed information for such classical spaces can be found in the books by Gel'fand and Shilov [276], Vladimirov [859] and Zemanian [921], and for other ones in the books by Samko et al. [729] and McBride [566].

Let $\mathbb{R}^n$ be the n-dimensional Euclidean space, let $\mathbf{x} = (x_1, \cdots, x_n)$ and $\mathbf{t} = (t_1, \cdots, t_n)$ be points in $\mathbb{R}^n$, $\mathbf{x} \cdot \mathbf{t} = x_1t_1 + \cdots + x_nt_n$ be their scalar product, and $|\mathbf{t}| = \left(t_1^2 + \cdots + t_n^2\right)^{1/2}$, $d\mathbf{t} = dt_1 \cdots dt_n$. Let

$$\mathbb{N}^n = \{\mathbf{k} = (k_1, \cdots, k_n), \; k_j \in \mathbb{N}; \; j = 1, \cdots, n\}, \tag{1.2.1}$$

$$\mathbb{N}_0^n = \{\mathbf{k} = (k_1, \cdots, k_n), \; k_j \in \mathbb{N}_0; \; j = 1, \cdots, n\} \tag{1.2.2}$$

The elements $\mathbf{k} \in \mathbb{N}_0^n$ are called *multi-indices.* For $\mathbf{x} \in \mathbb{R}^n$ and $\mathbf{k} \in \mathbb{N}_0^n$ we shall use the notations

$$\mathbf{x}^{\mathbf{k}} = x_1^{k_1} \cdots x_n^{k_n}, \quad |\mathbf{k}| = k_1 + \cdots + k_n, \tag{1.2.3}$$

and

$$\mathbf{D} = \left(\frac{\partial}{\partial x_1}, \cdots, \frac{\partial}{\partial x_n}\right), \quad \mathbf{D}^{\mathbf{k}} = \frac{\partial^{|k|}}{\partial x_1 \cdots \partial x_n}. \tag{1.2.4}$$

We denote by $\mathbb{C}^n$ $(n \in \mathbb{N})$ the n-dimensional space of n complex numbers $z = (z_1, \cdots, z_n)$ $(z_j \in \mathbb{C}; \; j = 1, \cdots, n)$.

We shall consider generalized functions over Ω, where Ω is a domain in $\mathbb{R}^n$. We choose test functions on Ω as the infinitely differentiable functions at the interior points of Ω with prescribed behavior at the boundary points of Ω. We denote by $< f, \varphi >$ the value of the generalized function f as a functional on the test function φ. A generalized function f is called *regular* if it is a locally integrable function such that $\int_\Omega f(\mathbf{x})\varphi(\mathbf{x})d\mathbf{x}$ exists for each test function $\varphi(\mathbf{x})$ and

$$< f, \varphi > = \int_\Omega f(\mathbf{x})\varphi(\mathbf{x})d\mathbf{x}. \tag{1.2.5}$$

It is assumed that the bilinear form $< f, \varphi >$ is chosen in such a way that it coincides with (1.2.5) in the case of a regular generalized function.

We assume that the space of test functions $X = X(\Omega)$ is a topological space and denote by $X' = X'(\Omega)$ the space of continuous linear functional space on X which is called *the topological dual of* X.

First of all, we recall the notion of a generalized function concentrated at a point. A generalized function $f \in X'$ is said to be zero on open set G, if $< f, \varphi >= 0$ for each test function $\varphi \in X$ which is zero beyond G. The union $\mathcal{O}_f$ of all open sets where $f = 0$ is called *a null set of the function* f. The complement of the null set with respect to Ω is called *the support of the generalized function* and such a complement is denoted by $\text{supp}(f) = \Omega \setminus \mathcal{O}_f$. The generalized function is said to be concentrated at the point $\mathbf{t}_0$, if $\text{supp}(f)$ is this point $\mathbf{t}_0$.

The classical *Dirac function* $\delta(\mathbf{t} - \mathbf{t}_0)$, $\mathbf{t}_0 \in \Omega$, defined by

$$\langle \delta(\mathbf{t} - \mathbf{t}_0), \varphi \rangle = \varphi(\mathbf{t}_0) \tag{1.2.6}$$

and its derivatives defined as

$$\left\langle (\mathbf{D}^{\mathbf{k}}\delta)(\mathbf{t} - \mathbf{t}_0), \varphi \right\rangle = (-1)^{|\mathbf{k}|}(\mathbf{D}^{\mathbf{k}}\varphi)(\mathbf{t}_0) \quad (\mathbf{k} \in \mathbb{N}^n), \tag{1.2.7}$$

provide simple examples of generalized functions concentrated at the point $\mathbf{t}_0$.

Any function with domain Ω in $\mathbb{R}^n$ or $\mathbb{C}^n$ and range in $\mathbb{R} \equiv \mathbb{R}^1$ or $\mathbb{C} \equiv \mathbb{C}^1$ is called *an ordinary function.* An ordinary function is called *smooth* (or $\mathbb{C}^\infty$) if it is infinitely differentiable. An ordinary function φ on $\mathbb{R}^n$ is said to be of *rapid descent* if $\varphi(\mathbf{t}) = o\left(|\mathbf{t}|^{-m}\right)$ for every integer $m \in \mathbb{Z} = \{0, \pm 1, \pm 2, \cdots\}$. An ordinary function φ is said to be of *slow growth* if there exists an integer $m \in \mathbb{Z}$ such that $\varphi(\mathbf{t}) = o\left(|\mathbf{t}|^{m}\right)$. If an ordinary function $\varphi(\mathbf{t})$ is defined and continuous on some open set Ω in $\mathbb{R}^n$, then the *support* of φ (denoted by $\text{supp}(\varphi)$) is the closure with respect to Ω of the set of points where $\varphi(\mathbf{t}) \neq 0$. The generalized function $f \in X'$ is said to be zero on an open set Ω if $< f, \varphi >= 0$ for every $\varphi \in X(\Omega)$ with $\text{supp}(\varphi) \subset \Omega$. If Ω_f is the largest open set where $f = 0$, then the set $\Omega \setminus \Omega_f$ is called *the support of the generalized function* f and is denoted by $\text{supp}(f)$. If $\text{supp}(f) \subset G \subset \Omega$, then f is said to be *concentrated on* G. If G is bounded, then f is called a *generalized function with compact support.*

When $\Omega = \mathbb{R}$, the support of an ordinary or generalized function f is said to be *bounded on the left or on the right* if there exists a real number $T_f \in \mathbb{R}$ such that $f(t) = 0$ for $t < T_f$ or $t > T_f$, respectively.

Now we give definitions and certain properties of some spaces of test and generalized functions. First we present the *Schwartz test-function space* $\mathcal{S} \equiv \mathcal{S}(\mathbb{R}^n)$, defined on the whole space ($\Omega = \mathbb{R}^n$), and its dual $\mathcal{S}' \equiv \mathcal{S}'(\mathbb{R}^n)$. $\mathcal{S}$ is the linear space of all complex-valued smooth functions $\varphi(\mathbf{t})$ on $\mathbb{R}^n$ such that

$$\gamma_{m,\mathbf{k}} = \sup_{\mathbf{t} \in \mathbb{R}^n} \left[(1 + |\mathbf{t}|^{2m}\right] |\mathbf{D}^{\mathbf{k}}\varphi(\mathbf{t})| < \infty \tag{1.2.8}$$

for all nonnegative integers $m \in \mathbb{N}_0$ and $\mathbf{k} \in \mathbb{N}_0^n$. The topology of $\mathcal{S}$ is generated by the countable collection of seminorms $\{\gamma_{m,\mathbf{k}}\}_{m,|\mathbf{k}|=0}^{\infty}$. It follows from (1.2.8) that every $\varphi(\mathbf{t}) \in \mathcal{S}$ is a function of rapid descent, and therefore $\mathcal{S}$ is called *the space of smooth functions of rapid descent.*

The *Schwartz space of generalized* functions $\mathcal{S}'$ is the dual space of $\mathcal{S}$. Every generalized function $f \in \mathcal{S}'$ can be represented in the form

$$f(\mathbf{t}) = \mathbf{D}^{\mathbf{m}} F(\mathbf{t}), \tag{1.2.9}$$

where $\mathbf{m} \in \mathbb{N}_0^n$ and $F(\mathbf{t})$ is a continuous function of slow growth as $|\mathbf{t}| \to \infty$. Therefore, $\mathcal{S}'$ is called *the space of generalized functions of slow growth* or *the space of tempered distributions.*

When $n = 1$ and $\Omega = [0, \infty)$, the space $\mathcal{S}_+ \equiv \mathcal{S}_+([0, \infty))$ is defined as the test-function space consisting of all smooth functions $\varphi(t)$ on $[0, \infty)$ which are of rapid descent as $t \to +\infty$. The corresponding space of generalized functions $\mathcal{S}'_+$ is the dual space of $\mathcal{S}_+$.

Next we consider the test-spaces $\mathcal{D}_K$, $\mathcal{D}$ and their dual spaces. Let $\Omega = K$ be a compact (i.e. bounded and closed) set in $\mathbb{R}^n$. We denote by $\mathcal{D}_K = \mathcal{D}_K(\mathbb{R}^n)$ the linear space of all complex-valued smooth functions $\varphi(\mathbf{t})$ that are equal to zero outside K. The topology of $\mathcal{D}_K$ is generated by the countable collection of seminorms $\{\gamma_{\mathbf{k}}(\varphi)\}_{|\mathbf{k}|=0}^{\infty}$, where

$$\gamma_{\mathbf{k}}(\varphi) = \sup_{\mathbf{t} \in \mathbb{R}^n} |\mathbf{D}^{\mathbf{k}} \varphi(\mathbf{t})|, \quad \varphi(\mathbf{t}) \in \mathcal{D}_K, \tag{1.2.10}$$

and $\mathbf{k} \in \mathbb{N}_0^n$. The space of generalized functions $\mathcal{D}'_K$ is the dual space of $\mathcal{D}_K$. Every function $f \in \mathcal{D}'_K$ has a representation of the form

$$f(\mathbf{t}) = \mathbf{D}^{\mathbf{m}} F(\mathbf{t}), \tag{1.2.11}$$

where $F(\mathbf{t})$ is some ordinary function on K and $\mathbf{m} \in \mathbb{N}_0^n$.

Let $\{K_m\}_{m=1}^{\infty}$ be a sequence of compact sets K_m in $\mathbb{R}^n$ such that

$$K_1 \subset K_2 \subset \cdots \subset K_m \subset \cdots, \qquad \mathbb{R}^n = \cup_{m=1}^{\infty} K_m. \tag{1.2.12}$$

The test space $\mathcal{D} = \mathcal{D}(\mathbb{R}^n)$ is defined as the countable-union space of all $\mathcal{D}_{K_m}$: $\mathcal{D} = \cup_{m=1}^{\infty} \mathcal{D}_{K_m}$. This space consists of all smooth functions with compact supports. The space of generalized functions $\mathcal{D}'$ is the dual space of $\mathcal{D}$. Every generalized function $f \in \mathcal{D}'$ can be represented in the form

$$f(\mathbf{t}) = \sum_{|\mathbf{m}|=1}^{\infty} \mathbf{D}^{\mathbf{p}_{\mathbf{m}}} F_{\mathbf{m}}(\mathbf{t}), \tag{1.2.13}$$

where $\mathbf{m}$, $\mathbf{p}_{\mathbf{m}} \in \mathbb{N}_0^n$, and $F_{\mathbf{m}}(\mathbf{t})$ are continuous functions with compact supports such that the supports of $F_{\mathbf{m}}$ are separated from the origin as $|\mathbf{m}| \to \infty$ and are contained in an arbitrary neighborhood of $\mathrm{supp}(f)$.

The elements of the spaces $\mathcal{D}'_K$ and $\mathcal{D}'$ are called *distributions.*

The notation of (1.2.13) for every generalized function $f \in \mathcal{D}'$ with compact support can be simplified to

$$f(\mathbf{t}) = \sum_{|\mathbf{m}|=0}^{M} \mathbf{D}^{\mathbf{m}} F_{\mathbf{m}}(\mathbf{t}) \quad (M \in \mathbb{N}_0), \tag{1.2.14}$$

where $\mathbf{m} \in \mathbb{N}_0^n$ and $F_{\mathbf{m}}(\mathbf{t})$ are continuous functions concentrated in an arbitrary neighborhood of supp(f). In particular, every generalized function $f \in \mathcal{D}'$ concentrated at the point $\mathbf{t}_0$ is expressed in terms of the Dirac delta function (1.2.6) and its derivatives (1.2.7) by

$$f(\mathbf{t}) = \sum_{|\mathbf{m}|=0}^{M} a_{\mathbf{m}} \mathbf{D}^{\mathbf{m}} \delta(\mathbf{t} - \mathbf{t}_0), \tag{1.2.15}$$

where $\mathbf{m} \in \mathbb{N}_0^n$ and $a_{\mathbf{m}}$ are given constants.

When $n = 1$ and $\Omega = [0, \infty)$, we denote by $\mathcal{D}_+ \equiv \mathcal{D}_+([0, \infty))$ the subspace of $\mathcal{D}(\mathbb{R})$ consisting of all $\varphi(t) \in \mathcal{D}(\mathbb{R})$ concentrated on $[0, \infty)$, and the corresponding space of generalized functions $\mathcal{D}'_+$ is the dual space of $\mathcal{D}_+$. We denote by $\mathcal{D}'_R$ and $\mathcal{D}'_L$ the subspaces of $\mathcal{D}'(\mathbb{R})$ consisting of all generalized functions with supports bounded on the left and on the right, respectively.

The test-function space $\mathcal{E} = \mathcal{E}(\mathbb{R}^n)$ is the linear space of all complex-valued smooth functions on $\mathbb{R}^n$. Its dual $\mathcal{E}' = \mathcal{E}'(\mathbb{R}^n)$ consists of all generalized functions with compact supports. When $n = 1$ and $\Omega = [0, \infty)$, $\mathcal{E}_+ = \mathcal{E}_+([0, \infty))$ is defined as the space of all complex-valued smooth functions on $[0, \infty)$, and $\mathcal{E}'_+$ is the dual space of $\mathcal{E}_+$.

The test-function space $\mathcal{L} = \mathcal{L}(\mathbb{R}^n)$ is the linear space of all entire functions $\varphi(\mathbf{t})$ such that

$$\xi_{\mathbf{k},\mathbf{m}}(\varphi) = \sup_{\mathbf{t} \in \mathbb{C}^n} \left(|\Re(\mathbf{t}|^{\mathbf{m}} |\mathbf{D}^{\mathbf{k}} \varphi(\mathbf{t})| \exp\left[-\sum_{j} a_j |\Im(\mathbf{t})| \right] \right) < \infty, \tag{1.2.16}$$

where $\mathbf{k}, \ \mathbf{m} \in \mathbb{N}_0^n$ and the constants a_j $(j = 1, \cdots, n)$ depend on $\varphi(\mathbf{t})$. The topology of $\mathcal{L}$ is generated by the countable set of seminorms $\{\xi_{\mathbf{k},\mathbf{m}}(\varphi)\}_{|\mathbf{k}|,|\mathbf{m}|=0}^{\infty}$. The space of generalized functions $\mathcal{L}' = \mathcal{L}'(\mathbb{R}^n)$ dual to $\mathcal{L}$ is called the space of *ultradistributions*.

Other properties of the above spaces of test and generalized functions may be found in the books by Gel'fand and Shilov [276], Vladimirov [859], and Zemanian [921]. Now we indicate some spaces of test and generalized functions suitable for fractional differentiation and integration operators.

First we present the *Lizorkin spaces* Ψ, Φ and Ψ', Φ' of test and generalized functions on $\mathbb{R}^n$; see Section 25.1 in the book by Samko et al. [729]. The former test-function space Ψ is a linear space of all complex-valued infinitely differentiable functions $\varphi(\mathbf{t})$ whose derivatives vanish at the origin:

$$\Psi = \{\varphi(\mathbf{t}) : \quad \varphi \in \mathcal{S}(\mathbb{R}^n), \quad (\mathbf{D}^{\mathbf{k}}\varphi)(\mathbf{0}) = 0 \ \ (|\mathbf{k}| \in \mathbb{N}_0). \tag{1.2.17}$$

Lastly, we introduce the test-function space Φ which is a subspace of the Schwartz space $\mathcal{S}(\mathbb{R}^n)$. It consists of those Schwartzian functions $\varphi \in \mathbf{S}$ which are orthogonal to polynomials, that is,

$$\int_{\mathbb{R}^n} \mathbf{t}^{\mathbf{k}} \varphi(\mathbf{t}) d\mathbf{t} = 0 \quad (\mathbf{k} \in \mathbb{N}_0^n). \tag{1.2.18}$$

The spaces of generalized functions Ψ' and Φ' are the spaces dual to Ψ and Φ, respectively.

Finally, we mention *McBride spaces* $\mathcal{F}_{p,\mu}$ and $\mathcal{F}'_{p,\mu}$ of test and generalized functions defined for $1 \leqq p \leqq \infty$ and $\mu \in \mathbb{R}$ [see McBride [566]]. The test-function space $\mathcal{F}_{p,\mu}$ is a linear space of all complex-valued functions $\varphi(x)$ on $\mathbb{R}^+$ such that

$$\gamma_k^{p,\mu}(\varphi) = ||t^k[t^{-\mu}\varphi(t)]||_{L^p} < \infty, \tag{1.2.19}$$

where $k \in \mathbb{N}_0$. The topology of $\mathcal{F}_{p,\mu}$ is generated by the countable collection of seminorms $\{\gamma_k^{p,\mu}(\varphi)\}_{k=0}^{\infty}$. The space of generalized functions $\mathcal{F}'_{p,\mu}$ is the dual space of $\mathcal{F}_{p,\mu}$.

1.3 Fourier Transforms

In this section we present definitions and some properties of one- and multi-dimensional Fourier transforms in spaces of p-summable and generalized functions. More detailed information may be found in the books by Sneddon [773] and Titchmarsh [819] (one-dimensional case), Kolmogorov and Fomin [435], Nikol'skii [628] and Stein and Weiss [802] (multidimensional case), and Brychkov and Prudnikov [108] (generalized functions).

We begin with the one-dimensional case. *The Fourier transform* of a function $\varphi(t)$ of real variable $t \in \mathbb{R} = (-\infty, \infty)$ is defined by

$$(\mathcal{F}\varphi)(x) = \mathcal{F}[\varphi(t)](x) = \hat{\varphi}(x) := \int_{-\infty}^{\infty} e^{ixt}\varphi(t)dt \quad (x \in \mathbb{R}). \tag{1.3.1}$$

The inverse Fourier transform is given by the formula

$$(\mathcal{F}^{-1}g)(x) = \mathcal{F}^{-1}[g(t)](x) = \frac{1}{2\pi}\hat{g}(-x) := \frac{1}{2\pi}\int_{-\infty}^{\infty} e^{-ixt}g(t)dt \quad (x \in \mathbb{R}). \tag{1.3.2}$$

The integrals in (1.3.1) and (1.3.2) converge absolutely for functions $\varphi, g \in L^1(\mathbb{R})$ and in the norm of the space $L^2(\mathbb{R})$ for $\varphi, g \in L^2(\mathbb{R})$. The L^1-theory and L^2-theory of the above Fourier integrals is described in the books by Sneddon [773] and Titchmarsh [819].

In particular, each of these transforms is inverse to the other one for "sufficiently good" functions φ, g:

$$\mathcal{F}^{-1}\mathcal{F}\varphi = \varphi, \quad \mathcal{F}\mathcal{F}^{-1}g = g, \tag{1.3.3}$$

and the following simple relation holds:

$$(\mathcal{F}\mathcal{F}\varphi)(x) = \varphi(-x). \tag{1.3.4}$$

The next properties of the Fourier transform give the connection between it and the *translation* τ_h *and dilation* Π_λ *operators* defined by

$$(\tau_h\varphi)(x) = \varphi(x-h) \quad (x, h \in \mathbb{R}) \tag{1.3.5}$$

and

$$(\Pi_\lambda\varphi)(x) = \varphi(\lambda x) \quad (x \in \mathbb{R};\ \ \lambda > 0), \tag{1.3.6}$$

respectively. The following relations are valid for $\varphi \in L^j(\mathbb{R})$ $(j = 1, 2)$:

$$(\mathcal{F}\tau_h\varphi)(x) = e^{-ixh}(\mathcal{F}\varphi)(x) \quad (x, h \in \mathbb{R}), \tag{1.3.7}$$

$$(\mathcal{F}\Pi_\lambda\varphi)(x) = \frac{1}{\lambda}\mathcal{F}\left(\frac{x}{\lambda}\right) \quad (x \in \mathbb{R};\ \ \lambda \in \mathbb{R}^+), \tag{1.3.8}$$

$$\mathcal{F}[e^{iat}\varphi(t)](x) = (\tau_{-a}\mathcal{F})[\varphi](x) \equiv \mathcal{F}(x+a) \quad (x, a \in \mathbb{R}). \tag{1.3.9}$$

If $\varphi(t) \in L_1(\mathbb{R})$, then $(\mathcal{F}\varphi)(x)$ is a bounded continuous function and, in accordance with the Riemann-Liouville theorem, $(\mathcal{F}\varphi)(x)$ tends to zero as $|x| \to \infty$:

$$\lim_{|x|\to\infty} \int_{-\infty}^{\infty} e^{ixt}\varphi(t)dt = 0, \quad \varphi(t) \in L_1(\mathbb{R}). \tag{1.3.10}$$

The rate of decrease of $(\mathcal{F}\varphi)(x)$ at infinity is connected with the smoothness of the function $\varphi(t)$. This connection is given by the following relations:

$$\mathcal{F}[D^k\varphi(t)](x) = (-ix)^k(\mathcal{F}\varphi)(x) \quad (k \in \mathbb{N}) \tag{1.3.11}$$

and

$$D^k(\mathcal{F}\varphi)(x) = (it)^k\mathcal{F}[\varphi(t)](x) \quad (k \in \mathbb{N}). \tag{1.3.12}$$

These equations are valid for "sufficiently good" functions φ; for example, for functions $\varphi(t) \in C^k(\mathbb{R})$ such that $\varphi^{(j)}(t) \in L^1(\mathbb{R})$ $(j = 0, 1, \cdots, k)$.

The Fourier convolution operator of two functions h and φ is defined by the integral

$$h * \varphi := (h * \varphi)(x) = \int_{-\infty}^{\infty} h(x-t)\varphi(t)dt \quad (x \in \mathbb{R}), \tag{1.3.13}$$

which has the commutative property

$$h * \varphi = \varphi * h. \tag{1.3.14}$$

The boundedness of the convolution operator in the space $L^p(\mathbb{R})$ is given by the *Young theorem.*

Theorem 1.1 *If $h(t) \in L_1(\mathbb{R})$ and $\varphi(t) \in L_p(\mathbb{R})$, then $(h * \varphi)(x) \in L_p(\mathbb{R})$ $(1 \leqq p \leqq \infty)$ and the following inequality holds:*

$$\|h * \varphi\|_p \leqq \|h\|_1 \|\varphi\|_p. \tag{1.3.15}$$

In particular, if $h(t) \in L_1(\mathbb{R})$ and $\varphi(t) \in L_2(\mathbb{R})$, then $(h * \varphi)(x) \in L_2(\mathbb{R})$ and

$$\|h * \varphi\|_2 \leqq \|h\|_1 \|\varphi\|_2. \tag{1.3.16}$$

When $h(t) \in L_2(\mathbb{R})$ and $\varphi(t) \in L_2(\mathbb{R})$, then the convolution (1.3.13) has the following property: *the function $(h * \varphi)(x)$ is continuous, bounded, and vanishes at infinity.*

The Fourier transform of the convolution (1.3.13) is given by the *Fourier convolution theorem.*

Theorem 1.2 *Let either $h(t) \in L_1(\mathbb{R})$ and $\varphi(t) \in L_1(\mathbb{R})$, or $h(t) \in L_1(\mathbb{R})$ and $\varphi(t) \in L_2(\mathbb{R})$, or $h(t) \in L_2(\mathbb{R})$ and $\varphi(t) \in L_2(\mathbb{R})$.*

Then the Fourier transform of the convolution $h * \varphi$ is given by

$$\left(\mathcal{F}(h * \varphi)\right)(x) = (\mathcal{F}h)(x)(\mathcal{F}\varphi)(x). \tag{1.3.17}$$

We also indicate that the *cosine-* and *sine-Fourier transforms* of a function $\varphi(t)$ $(t \in \mathbb{R}^+)$ are defined by

$$(\mathcal{F}_c\varphi)(x) = \int_0^\infty \cos(xt)\varphi(t)dt \quad (x \in \mathbb{R}^+) \tag{1.3.18}$$

and

$$(\mathcal{F}_s\varphi)(x) = \int_0^\infty \sin(xt)\varphi(t)dt \quad (x \in \mathbb{R}^+), \tag{1.3.19}$$

respectively, while the corresponding inverse transforms have the forms

$$(\mathcal{F}_c^{-1}g)(x) = \frac{2}{\pi}\int_0^\infty \cos(xt)g(t)dt \quad (x \in \mathbb{R}^+) \tag{1.3.20}$$

and

$$(\mathcal{F}_s^{-1}g)(x) = \frac{2}{\pi}\int_0^\infty \sin(xt)g(t)dt \quad (x \in \mathbb{R}^+). \tag{1.3.21}$$

Next we consider the multidimensional case. We shall use the notation given in Section 1.2.

The *n-dimensional Fourier transform* of a function $\varphi(\mathbf{t})$ of $\mathbf{t} \in \mathbb{R}^n$ is defined by

$$(\mathcal{F}\varphi)(\mathbf{x}) = \mathcal{F}[\varphi(\mathbf{t})](\mathbf{x}) = \hat{\varphi}(\mathbf{x}) := \int_{\mathbb{R}^n} e^{i\mathbf{x}\cdot\mathbf{t}}\varphi(\mathbf{t})d\mathbf{t} \quad (\mathbf{x} \in \mathbb{R}^n), \tag{1.3.22}$$

while the corresponding *inverse Fourier transform* is given by the formula

$$(\mathcal{F}^{-1}g)(\mathbf{x}) = \mathcal{F}^{-1}[g(\mathbf{t})](\mathbf{x}) = \frac{1}{2\pi}\hat{g}(-\mathbf{x}) := \frac{1}{(2\pi)^n}\int_{\mathbb{R}^n} e^{-i\mathbf{x}\cdot\mathbf{t}}g(\mathbf{t})dt \quad (\mathbf{x} \in \mathbb{R}^n). \tag{1.3.23}$$

The integrals in (1.3.22) and (1.3.23) have the same properties as those of the one-dimensional ones in (1.3.1) and (1.3.2). They converge absolutely for functions $\varphi, g \in L_1(\mathbb{R}^n)$ and in the norm of the space $L_2(\mathbb{R}^n)$ for $\varphi, g \in L_2(\mathbb{R}^n)$. The L_1-theory and the L_2-theory of the above Fourier integrals can be found in the books by Kolmogorov and Fomin [434], Nikol'skii [628], and Stein and Weiss [802].

All of the above properties of the one-dimensional Fourier transforms are extended to the multidimensional Fourier transforms. In particular, the relations (1.3.7)-(1.3.9) take the following forms:

$$(\mathcal{F}\tau_{\mathbf{h}}\varphi)(\mathbf{x}) = e^{-i\mathbf{x}\cdot\mathbf{h}}(\mathcal{F}\varphi)(\mathbf{x}) \quad (\mathbf{x}, \mathbf{h} \in \mathbb{R}^n), \tag{1.3.24}$$

$$(\mathcal{F}\Pi_\lambda\varphi)(\mathbf{x}) = \frac{1}{\lambda^n}\mathcal{F}\varphi\left(\frac{\mathbf{x}}{\lambda}\right) \quad (\mathbf{x} \in \mathbb{R}^n;\ \lambda > 0), \tag{1.3.25}$$

$$\mathcal{F}[e^{i\mathbf{a}\cdot\mathbf{t}}\varphi(\mathbf{t})](x) = (\tau_{-\mathbf{a}}\mathcal{F})[\varphi](\mathbf{x}) \equiv \mathcal{F}(\mathbf{x}+\mathbf{a}) \quad (\mathbf{x}, \mathbf{a} \in \mathbb{R}^n); \tag{1.3.26}$$

$\tau_{\mathbf{h}}$ and Π_λ being defined by

$$(\tau_{\mathbf{h}}\varphi)(\mathbf{x}) = \varphi(\mathbf{x}-\mathbf{h}), \quad (\Pi_\lambda\varphi)(\mathbf{x}) = \varphi(\lambda\mathbf{x}) \quad (\mathbf{x}, \mathbf{h} \in \mathbb{R}^n;\ \lambda \in \mathbb{R}^+); \tag{1.3.27}$$

while (1.3.10)-(1.3.12) are extended as follows:

$$\lim_{|\mathbf{x}|\to\infty}\int_{\mathbb{R}^n} e^{i\mathbf{x}\cdot\mathbf{t}}\varphi(\mathbf{t})dt = 0, \quad \varphi(\mathbf{t}) \in L_1(\mathbb{R}^n), \tag{1.3.28}$$

$$\mathcal{F}[\mathbf{D}^{\mathbf{k}}\varphi(\mathbf{t})](\mathbf{x}) = (-i\mathbf{x})^{\mathbf{k}}(\mathcal{F}\varphi)(\mathbf{x}) \quad (\mathbf{x} \in \mathbb{R}^n;\ k \in \mathbb{N}^n) \tag{1.3.29}$$

and

$$\mathbf{D}^{\mathbf{k}}(\mathcal{F}\varphi)(\mathbf{x}) = \mathcal{F}[(i\mathbf{t})^{\mathbf{k}}\varphi(\mathbf{t})](\mathbf{x}) \quad (\mathbf{x} \in \mathbb{R}^n;\ \mathbf{k} \in \mathbb{N}^n), \tag{1.3.30}$$

respectively. If Δ is the n-dimensional Laplace operator

$$\Delta = \frac{\partial^2}{\partial x_1^2} + \cdots + \frac{\partial^2}{\partial x_n^2}. \tag{1.3.31}$$

then (1.3.29) yields

$$(\mathcal{F}\Delta\varphi(\mathbf{t})](\mathbf{x}) = -|x|^2(\mathcal{F}\varphi)(\mathbf{x}) \quad (\mathbf{x} \in \mathbb{R}^n). \tag{1.3.32}$$

Analogous to (1.3.13), the *Fourier convolution operator* of two functions h and φ is defined by

$$h * \varphi := (h * \varphi)(\mathbf{x}) = \int_{\mathbb{R}^n} h(\mathbf{x}-\mathbf{t})\varphi(\mathbf{t})dt \quad (\mathbf{x} \in \mathbb{R}^n), \tag{1.3.33}$$

which has the commutative property (1.3.14). For such a convolution operator, the Young theorem (Theorem 1.1) and the Fourier convolution theorem (Theorem 1.2) still apply.

Theorem 1.3 *If $h(t) \in L_1(\mathbb{R}.)$ and $\varphi(t) \in L_p(\mathbb{R}^n)$ with $1 \leqq p \leqq \infty$, then $(h * \varphi)(x) \in L_p(\mathbb{R}^n)$ and the inequality* (1.3.15) *holds.*

In particular, if $h(t) \in L_1(\mathbb{R}^n)$ and $\varphi(t) \in L_2(\mathbb{R}^n)$, then $(h * \varphi)(x) \in L_2(\mathbb{R}^n)$ and the formula (1.3.16) *is valid.*

Theorem 1.4 *Let either $h(t) \in L_1(\mathbb{R}^n)$ and $\varphi(t) \in L_1(\mathbb{R}^n)$, or $h(t) \in L_1(\mathbb{R}^n)$ and $\varphi(t) \in L_2(\mathbb{R}^n)$, or $h(t) \in L_2(\mathbb{R}^n)$ and $\varphi(t) \in L_2(\mathbb{R}^n)$.*

Then the Fourier transform of the convolution $h * \varphi$ is given by the formula (1.3.17).

Now we present Fourier transforms of generalized functions defined in Section 1.2; see Section 2.2 in the book by Brychkov and Prudnikov [108]. First we present the definitions by L. Schwartz of the Fourier transform in the space of generalized functions $\mathcal{S}'(\mathbb{R}^n)$. Such a *direct Fourier transform* is defined by

$$< \mathcal{F}[f], \varphi > = < f, \mathcal{F}[\varphi] > \quad (\varphi \in \mathcal{S}), \tag{1.3.34}$$

while the *inverse Fourier transform* has the form

$$\mathcal{F}^{-1}[f(\mathbf{t})] = \frac{1}{(2\pi)^n} \mathcal{F}[f(-\mathbf{t})] \quad (f \in \mathcal{S}'). \tag{1.3.35}$$

When $f(\mathbf{t}) \in L_1(\mathbb{R}^n))$, it is easily seen that $\mathcal{F}[f]$ and $\mathcal{F}^{-1}[f]$ coincide with (1.3.22) and (1.3.23), respectively.

$\mathcal{F}$ and $\mathcal{F}^{-1}$ specify an automorphism of the generalized function space $\mathcal{S}'$ and they are inverse to each other:

$$\mathcal{F}^{-1}\left[\mathcal{F}[f]\right] = f, \quad \mathcal{F}\left[\mathcal{F}^{-1}[f]\right] = f \quad (f \in \mathcal{S}'). \tag{1.3.36}$$

The formulas (1.3.24), (1.3.26), (1.3.29), and (1.3.30) are valid for $f \in \mathcal{S}'$:

$$\mathcal{F}[\tau_{\mathbf{h}} f(\mathbf{t})](\mathbf{x}) = e^{-i\mathbf{x}\cdot\mathbf{h}} \mathcal{F}[f](\mathbf{x}) \quad (\mathbf{x}, \mathbf{h} \in \mathbb{R}^n), \tag{1.3.37}$$

$$\mathcal{F}[\Pi_\lambda f(\mathbf{t}](\mathbf{x}) = \frac{1}{\lambda^n} \mathcal{F}[f]\left(\frac{\mathbf{x}}{\lambda}\right) \quad \left(\mathbf{x} \in \mathbb{R}^n; \ \lambda \in \mathbb{R}^+\right), \tag{1.3.38}$$

$$\mathcal{F}[e^{i\mathbf{a}\cdot\mathbf{t}} \varphi(\mathbf{t})](x) = (\tau_{-\mathbf{a}} \mathcal{F})[\varphi](\mathbf{x}) \equiv \mathcal{F}(\mathbf{x} + \mathbf{a}) \quad (\mathbf{x}, \mathbf{a} \in \mathbb{R}^n); \tag{1.3.39}$$

$$\mathcal{F}[\mathbf{D}^{\mathbf{k}} f(\mathbf{t})](\mathbf{x}) = (-i\mathbf{x})^{\mathbf{k}} \mathcal{F}[f](\mathbf{x}) \quad (\mathbf{x} \in \mathbb{R}^n; \ k \in \mathbb{N}^n) \tag{1.3.40}$$

$$D^{\mathbf{k}} \mathcal{F}[f](\mathbf{x}) = \mathcal{F}[(i\mathbf{t})^{\mathbf{k}} \varphi(\mathbf{t})](\mathbf{x}) \quad (\mathbf{x} \in \mathbb{R}^n; \ \mathbf{k} \in \mathbb{N}^n). \tag{1.3.41}$$

Here $\tau_{\mathbf{h}} f$, $\Pi_\delta f$ and $\mathbf{D}^{\mathbf{k}} f$ are defined for $f \in \mathcal{S}'$ by

$$< \tau_{\mathbf{h}} f, \varphi > = < f, \tau_{-\mathbf{h}} \varphi > \quad (\varphi \in \mathbf{S}), \tag{1.3.42}$$

$$< \Pi_\lambda f, \varphi > = \frac{1}{\lambda^n} < f, \Pi_{1/\lambda} \varphi > \quad (\varphi \in \mathbf{S}) \tag{1.3.43}$$

and

$$< D^{\mathbf{k}} f, \varphi > = < f, (-1)^{|\mathbf{k}|} D^{\mathbf{k}} \varphi > \quad (\varphi \in \mathbf{S}). \tag{1.3.44}$$

To obtain a convolution formula of the form (1.3.33), we need to introduce the Fourier transform in the space of generalized functions $\mathcal{E}'$. For $f \in \mathcal{E}'$, the Fourier transform $\mathcal{F}[f]$ can be represented in the form

$$\mathcal{F}[f](\mathbf{x}) = < f(\mathbf{t}), \eta(\mathbf{t})e^{i\mathbf{x}\cdot\mathbf{t}} >, \tag{1.3.45}$$

where $\eta(\mathbf{t})$ is any function in the space $\mathcal{D}$ of test functions that is equal to unity in the neighborhood of supp$f$. Moreover, $\mathcal{F}[f](\mathbf{x})$ is an entire function of $\mathbf{x}$, and for any $\mathbf{k} \in \mathbb{N}_0^n$, there exist numbers $A_\mathbf{k} \geqq 0$ and $m \geqq 0$ such that

$$|\mathbf{D}^\mathbf{k}\mathcal{F}[f](\mathbf{x})| \leqq A_\mathbf{k}\left(1 + |\mathbf{x}|^2\right)^m. \tag{1.3.46}$$

Now the convolution $f * g$ of two generalized functions $f \in \mathcal{S}'$ and $g \in \mathcal{E}'$ is defined as follows:

$$< f * g, \varphi > = \langle f(\mathbf{t}), < g(\mathbf{u}), \varphi(\mathbf{t} + \mathbf{u}) > \rangle \quad (\varphi \in \mathcal{S}). \tag{1.3.47}$$

Here the right-hand side makes sense because $< g(\mathbf{u}), \varphi(\mathbf{t} + \mathbf{u}) > \in \mathcal{S}$. Such a convolution $f * g$ is a member of $\mathcal{S}'$ and the formula of the form (1.3.17) holds:

$$\mathcal{F}[f * g] = \mathcal{F}[f]\mathcal{F}[g] \quad (f \in \mathcal{S}'; \;\; g \in \mathcal{E}'). \tag{1.3.48}$$

The Fourier transform (1.3.22) is an isomorphism from the test space $\mathcal{D}$ onto the other one $\mathcal{L}$. A function $\varphi(\mathbf{t}) \in \mathcal{D}$ has the property that $\varphi(\mathbf{t})$ is equal to zero outside the domain

$$G = \{\mathbf{t} = (t_1, \cdots, t_n) \in \mathbb{R}^n : \;\; |t_j| \leqq a_j, \;\; a_j > 0 \;\; (j = 1, \cdots n)\}, \tag{1.3.49}$$

$a_j > 0$ $(j = 1, \cdots n)$ being positive numbers, if, and only if, its Fourier transform $\mathcal{F}[\varphi](z)$ is an entire function of $z = x + iy$ satisfying the inequalities

$$|z^\mathbf{k}\mathcal{F}[\varphi](z)| \leqq B_\mathbf{k}e^{\mathbf{a}\cdot\mathbf{x}} \quad (\mathbf{a} = (a_1, \cdots, a_n) \in \mathbb{R}^n) \tag{1.3.50}$$

for every nonnegative integer $\mathbf{k} \in \mathbb{N}_0^n$ and for some $B_\mathbf{k} \geqq 0$.

The above properties of functions $\varphi \in \mathcal{D}$ were used by Gelfand and Shilov [276] to define the direct and inverse Fourier transforms in the space of generalized functions $\mathcal{D}'$ by

$$< \mathcal{F}_\mathcal{D}[f], \varphi > = < f, \mathcal{F}[\varphi] > \quad (\varphi \in \mathcal{L}) \tag{1.3.51}$$

and

$$< \mathcal{F}_\mathcal{D}^{-1}[f], \varphi > = < f, \mathcal{F}^{-1}[\varphi] > \quad (\varphi \in \mathcal{L}), \tag{1.3.52}$$

respectively. Analogous to (1.3.51)-(1.3.52), the Fourier transforms are defined in the space $\mathcal{D}'$ as follows:

$$< \mathcal{F}_\mathcal{L}[f], \varphi > = < f, \mathcal{F}[\varphi] > \quad (\varphi \in \mathcal{D}) \tag{1.3.53}$$

and

$$< \mathcal{F}_\mathcal{L}^{-1}[f], \varphi > = < f, \mathcal{F}^{-1}[\varphi] > \quad (\varphi \in \mathcal{D}). \tag{1.3.54}$$

These Fourier transforms have the properties (1.3.37)-(1.3.41) in the spaces $\mathcal{D}'$ and $\mathcal{L}'$. The convolution $f * g$ is defined by (1.3.47) for $f \in \mathcal{D}'$ and $g \in \mathcal{L}'$, and the formula (1.3.48) holds for such a convolution.

To conclude this section, we present formulas for the Fourier transforms (1.3.1) and (1.3.22) of some elementary and generalized functions. First we present the one-dimensional case. The Fourier transform (1.3.1) of power functions $|t|^\lambda$ and $|t|^\lambda \text{sgn}(t)$ are given for $x \in \mathbb{R}$ $(x \neq 0)$ by

$$\mathcal{F}\left[|t|^\lambda\right](x) = 2^{\lambda+1}\pi^{1/2}\frac{\Gamma\left(\frac{\lambda+1}{2}\right)}{\Gamma\left(-\frac{\lambda}{2}\right)}|x|^{-\lambda-1}$$

$$= -2\sin\left(\frac{\lambda\pi}{2}\right)\Gamma(\lambda+1)|x|^{-\lambda-1} \quad (\lambda \in \mathbb{C};\ \lambda \neq 0;\ \lambda \neq -1-2k;\ k \in \mathbb{N}_0), \tag{1.3.55}$$

$$\mathcal{F}\left[|t|^{-2k-1}\right](x) = \frac{2(-1)^k}{(2k)!}x^{2k}\left[\psi(2k+1) - \log(|x|)\right] \quad (k \in \mathbb{N}_0), \tag{1.3.56}$$

and

$$\mathcal{F}\left[|t|^\lambda \text{sgn}(t)\right](x) = 2i\cos\left(\frac{\lambda\pi}{2}\right)\Gamma(\lambda+1)|x|^{-\lambda-1}\text{sgn}(x) \quad (\lambda \neq -2k;\ k \in \mathbb{N}), \tag{1.3.57}$$

$$\mathcal{F}\left[t^{-2k}\text{sgn}(t)\right](x) = i\frac{2(-1)^{k+1}}{(2k-1)!}x^{2k-1}\left[\psi(2k) - \log(|x|)\right] \quad (k \in \mathbb{N}), \tag{1.3.58}$$

$\psi(z)$ being the Euler psi function (1.5.17) [see Brychkov and Prudnikov ([108], formulas 10.1.11 and 10.1.12)]. The Fourier transform of generalized power functions t^k $(k \in \mathbb{N}_0)$ and of the polynomial $P_m(t)$ of degree $m \in \mathbb{N}$ are expressed via the Dirac delta function (1.2.6) and its derivatives (1.2.7):

$$\mathcal{F}[1](x) = 2\pi\delta(x), \quad \mathcal{F}\left[t^k\right](x) = 2\pi(-i)^k\delta^{(k)}(x) \quad (k \in \mathbb{N}), \tag{1.3.59}$$

and

$$\mathcal{F}[P_m(t)](x) = 2\pi P_m\left(-i\frac{d}{dx}\right)\delta(x) \quad \left(x \in \mathbb{R};\ P_m(x) = \sum_{j=0}^m a_j x^j,\ a_m \neq 0;\ m \in \mathbb{N}\right), \tag{1.3.60}$$

with $a_j \in \mathbb{C}$ $(j = 0, 1, \cdots, m;\ a_m \neq 0)$ [see Brychkov and Prudnikov ([108], formulas 10.1.1 and 10.1.5) and Gel'fand and Shilov ([276], formula (4) in Table 1)]. The following relations hold for the one-dimensional Fourier transform (1.3.1) of the Dirac delta function (1.2.6) and of its derivatives (1.2.7):

$$\mathcal{F}[\delta(t)](x) = 1 \quad (x \in \mathbb{R}), \tag{1.3.61}$$

$$\mathcal{F}[\delta(t-a)](x) = e^{-iax} \quad (x, a \in \mathbb{R}), \tag{1.3.62}$$

$$\mathcal{F}[\delta^{(k)}(t)](x) = (-ix)^k \quad (x \in \mathbb{R};\ k \in \mathbb{N}), \tag{1.3.63}$$

$$\mathcal{F}[\delta^k(t-a)](x) = e^{-iax}(-ix)^k \quad (x, a \in \mathbb{R};\ k \in \mathbb{N}). \tag{1.3.64}$$

The next formulas are extensions of the relations given in (1.3.55) and (1.3.59)-(1.3.62) to the multidimensional Fourier transform (1.3.22) in $\mathbb{R}^n$ ($n \in \mathbb{N}$):

$$\mathcal{F}\left[|\mathbf{t}|^{\lambda}\right](\mathbf{x}) = 2^{\lambda+n}\pi^{n/2}\frac{\Gamma\left(\frac{\lambda+n)}{2}\right)}{\Gamma\left(-\frac{\lambda}{2}\right)}|\mathbf{x}|^{-\lambda-n}$$

$$= -2^{\lambda+n}\pi^{(n-2)/2}\sin\left(\frac{\lambda\pi}{2}\right)\Gamma\left(\frac{\lambda+n}{2}\right)\Gamma\left(1+\frac{\lambda}{2}\right)|x|^{-\lambda-n} \tag{1.3.65}$$

$$\left(|\mathbf{t}| = \left(t_1^2+\cdots+t_n^2\right)^{1/2}\right)$$

$$(\lambda \in \mathbb{C};\ \lambda \neq 0;\ \lambda \neq -1-2k,\ k \in \mathbb{N}_0;\quad \mathbf{x} \in \mathbb{R}^n;\ n \in \mathbb{N};\ \mathbf{x} \neq \mathbf{0} := (0,\cdots,0)),$$

$$\mathcal{F}[1,1,\cdots,1](\mathbf{x}) = 2\pi\delta(\mathbf{x})\quad (\mathbf{x} \in \mathbb{R}^n,\ n \in \mathbb{N}), \tag{1.3.66}$$

$$\mathcal{F}\left[t_1^k\cdots t_n^k\right](\mathbf{x}) = (2\pi)^n(-i)^{k_1+\cdots+k_n}\delta^{(k_1)}(x_1)\cdots\delta^{(k_n)}(x_n) \tag{1.3.67}$$

$$(\mathbf{x} \in \mathbb{R}^n,\ n \in \mathbb{N};\ k_j \in \mathbb{N},\ j = 1,\cdots,n),$$

$$\mathcal{F}[P_{|m|}(t_1,\cdots,t_n)](x) = (2\pi)^n P_{|m|}\left(-i\frac{d}{dx_1},\cdots,-i\frac{d}{dx_n}\right)\delta(\mathbf{x}) \tag{1.3.68}$$

$$(\mathbf{x} \in \mathbb{R}^n,\ n \in \mathbb{N}),$$

$P_{|m|}(\mathbf{x})$ being a polynomial of degree $|\mathbf{m}| = m_1+\cdots+m_n$ ($m_j \in \mathbb{N};\ j = 1,\cdots,n$)

$$P(x_1,\cdots,x_n) = \sum_{j_1=0}^{m_1}\cdots\sum_{j_n=0}^{m_n} a_{j_1\cdots j_m}x_1^{j_1}\cdots x_n^{j_n} \tag{1.3.69}$$

with constant coefficients $a_{j_1\cdots j_m} \in \mathbb{C}$,

$$\mathcal{F}[\delta(t_1,\cdots,t_n)](\mathbf{x}) = \mathbf{1} := (1,\cdots,1)\quad (\mathbf{x} \in \mathbb{R}^n;\ n \in \mathbb{N}); \tag{1.3.70}$$

$$\mathcal{F}[\delta(|\mathbf{t}|-a)](\mathbf{x}) = \frac{(2\pi)^{(n-1)/2}}{\Gamma[(n-2)/2]}a^{n/2}|\mathbf{x}|^{(2-n)/2}J_{(n-2)/2}(a|\mathbf{x}|); \tag{1.3.71}$$

$$(\mathbf{x} \in \mathbb{R}^n,\ n \in \mathbb{N}\backslash\{1\};\ a \in \mathbb{R}\backslash\{0\})$$

$J_{(n-2)/2}(z)$ being the Bessel function of the first kind (1.7.1); in particular, when $n = 3$,

$$\mathcal{F}[\delta(|\mathbf{t}|-a)](\mathbf{x}) = 4\pi a\frac{\sin(a|\mathbf{x}|)}{|\mathbf{x}|}\quad \left(\mathbf{x} \in \mathbb{R}^3;\ a \in \mathbb{R};\ a \neq 0\right). \tag{1.3.72}$$

[see Brychkov and Prudnikov ([108], formulas 707-709, 711, 713, 714)].

We also note that the relations in (1.3.55) and (1.3.65) for $\lambda = k \in \mathbb{N}$ take the following forms:

$$\mathcal{F}\left[|t|^{2k}\right](x) = 0\quad (x \in \mathbb{R};\ k \in \mathbb{N}), \tag{1.3.73}$$

$$\mathcal{F}\left[|t|^{2k+1}\right](x) = 2(-1)^{k+1}(2k+1)!|x|^{-2k-2}\quad (x \in \mathbb{R};\ k \in \mathbb{N}_0), \tag{1.3.74}$$

and

$$\mathcal{F}\left[|\mathbf{t}|^{2k}\right](\mathbf{x}) = \mathbf{0} \quad (\mathbf{x} \in \mathbb{R}^n;\ n, k \in \mathbb{N}), \tag{1.3.75}$$

$$\mathcal{F}\left[|\mathbf{t}|^{2k+1}\right](\mathbf{x}) = 2^n(-1)^{k+1}\pi^{(n-1)/2}\frac{(2k+1)!}{k!}\Gamma\left(k+\frac{n+1}{2}\right)|x|^{-2k-n-1}$$
$$(\mathbf{x} \in \mathbb{R}^n;\ n \in \mathbb{N};\ k \in \mathbb{N}_0), \tag{1.3.76}$$

respectively.

1.4 Laplace and Mellin Transforms

In this section we present definitions and some properties of one- and multidimensional Laplace and Mellin transforms. More detailed information may be found in the books by Ditkin and Prudnikov [195], Doetsch [198], [199],Sneddon [773] and Titchmarsh [819] (for the one-dimensional case), and Brychkov et al. [107] (for the multidimensional case).

The *Laplace transform* of a function $\varphi(t)$ of a real variable $t \in \mathbb{R}^+ = (0, \infty)$ is defined by

$$(\mathcal{L}\varphi)(s) = \mathcal{L}[\varphi(t)](s) = \tilde{\varphi}(s) := \int_0^\infty e^{-st}\varphi(t)dt \quad (s \in \mathbb{C}). \tag{1.4.1}$$

If the integral (1.4.1) is convergent at the point $s_0 \in \mathbb{C}$, then it converges absolutely for $s \in \mathbb{C}$ such that $\Re(s) > \Re(s_0)$. The infimum σ_φ of values s for which the Laplace integral (1.4.1) converges is called *the abscissa of convergence.* Thus the Laplace integral (1.4.1) converges for $\Re(s) > \sigma_\varphi$ and diverges for $\Re(s) < \sigma_\varphi$.

The *inverse Laplace transform* is given for $x \in \mathbb{R}^+$ by the formula

$$(\mathcal{L}^{-1}g)(x) = \mathcal{L}^{-1}[g(s)](x) := \frac{1}{2\pi i}\int_{\gamma-i\infty}^{\gamma+i\infty} e^{sx}g(s)ds \quad (\gamma = \Re(s) > \sigma_\varphi). \tag{1.4.2}$$

The direct and inverse Laplace transforms are inverse to each other for "sufficiently good" functions φ and g:

$$\mathcal{L}^{-1}\mathcal{L}\varphi = \varphi \quad \text{and} \quad \mathcal{L}\mathcal{L}^{-1}g = g. \tag{1.4.3}$$

First we present some simple properties of the Laplace transform analogous to those given in Section 1.3 by (1.3.7)-(1.3.9) and (1.3.11)-(1.3.12) for the Fourier transform:

$$(\mathcal{L}\tau_h\varphi)(p) = e^{-ph}(\mathcal{L}\varphi)(p) \quad (h \in \mathbb{R}), \tag{1.4.4}$$

$$(\mathcal{L}\Pi_\lambda\varphi)(p) = \frac{1}{\lambda}\mathcal{L}\left(\frac{p}{\lambda}\right) \quad (\lambda \in \mathbb{R}^+), \tag{1.4.5}$$

$$\mathcal{L}[e^{-at}\varphi(t)](p) = (\tau_{-a}\mathcal{L})(p) \equiv \mathcal{L}(p+a) \quad (a \in \mathbb{C}), \tag{1.4.6}$$

$$\mathcal{L}[D^k\varphi(t)](p) = p^k(\mathcal{L}\varphi)(p) \quad (k \in \mathbb{N}) \tag{1.4.7}$$

and

$$D^k(\mathcal{L}\varphi)(s) = (-1)^k\mathcal{L}[t^k\varphi(t)](s) \quad (k \in \mathbb{N}). \tag{1.4.8}$$

These equations are valid for "sufficiently good" functions φ.

When $\varphi \in C^k(\mathbb{R}^+)$, the Laplace transforms $(\mathcal{L}\varphi)(s)$ and $\mathcal{L}[D^k\varphi(t)](s)$ exist and $\lim_{t\to+\infty}(D\varphi)^{(j)}(t) = 0$ for $j = 0, 1, \cdots, k-1$, then the relation in (1.4.7) is replaced by the more general one

$$\mathcal{L}[D^k\varphi(t)](s) = s^k(\mathcal{L}\varphi)(s) - \sum_{j=0}^{k-1} s^{k-j-1}(D^j\varphi)(0) \quad (k \in \mathbb{N}). \tag{1.4.9}$$

The *Laplace convolution operator* of two functions $h(t)$ and $\varphi(t)$, given on $\mathbb{R}^+$, is defined for $x \in \mathbb{R}^+$ by the integral

$$h * \varphi = (h * \varphi)(x) := \int_0^x h(x-t)\varphi(t)dt, \tag{1.4.10}$$

which has the commutative property

$$h * \varphi = \varphi * h. \tag{1.4.11}$$

The Convolution Theorem 1.2 applied to (1.4.10) yields the form

$$(\mathcal{L}(h * \varphi)(p) = (\mathcal{L}h)(p)(\mathcal{L}\varphi)(p), \tag{1.4.12}$$

which holds for "sufficiently good" functions h and φ.

We denote by $\mathbb{R}^n_{+\cdots+}$ the region in $\mathbb{R}^n$ with positive coordinates:

$$\mathbb{R}^n_{+\cdots+} - \{\mathbf{t} - (t_1, \cdots, t_n) \in \mathbb{R}^n : \ t_j > 0 \ (j = 1, \cdots, n)\}. \tag{1.4.13}$$

The *n-dimensional Laplace transform* of a function $\varphi(\mathbf{t})$ of $\mathbf{t} \in \mathbb{R}^n_{+\cdots+}$ is defined by

$$(\mathcal{L}\varphi)(\mathbf{s}) = \mathcal{L}[\varphi(\mathbf{t})](\mathbf{s}) = \tilde{\varphi}(\mathbf{s}) := \int_{\mathbb{R}^n_+} e^{-\mathbf{s}\cdot\mathbf{t}}\varphi(\mathbf{t})d\mathbf{t} \quad (\mathbf{s} \in \mathbb{C}^n), \tag{1.4.14}$$

while the *inverse Laplace transform* is given for $\mathbf{x} \in \mathbb{R}^n_{+\cdots+}$ by the formula

$$(\mathcal{L}^{-1}g)(\mathbf{x}) = \mathcal{L}^{-1}[g(\mathbf{s})](\mathbf{x}) := \frac{1}{(2\pi i)^n}\int_{\gamma_1-i\infty}^{\gamma_1+i\infty} \cdots \int_{\gamma_n-i\infty}^{\gamma_n+i\infty} e^{\mathbf{x}\cdot\mathbf{s}}g(\mathbf{s})d\mathbf{s} \tag{1.4.15}$$

where $\gamma_j = \Re(s_j) > \sigma_{\varphi,j}$ $(\sigma_{\varphi,j} \in \mathbb{R};\ j = 1, \cdots, n)$. The theory of these multidimensional Laplace transforms is found in the book by Brychkov et al. [107].

The integrals in (1.4.14) and (1.4.15) have some properties analogous to those for the one-dimensional Laplace transforms given by (1.4.1) and (1.4.2). In particular, they satisfy the mutually invertible relations in (1.4.3) for "sufficiently good" functions φ and g, and for such functions the above properties (1.4.4)-(1.4.8) of the one-dimensional Laplace transforms are extended to the multidimensional case as follows:

$$(\mathcal{L}\tau_{\mathbf{h}}\varphi)(\mathbf{p}) = e^{-\mathbf{p}\cdot\mathbf{h}}(\mathcal{L}\varphi)(\mathbf{p}) \quad (\mathbf{h} \in \mathbb{R}^n), \tag{1.4.16}$$

$$(\mathcal{L}\Pi_\lambda\varphi)(\mathbf{p}) = \frac{1}{\lambda^n}\mathcal{L}\varphi\left(\frac{\mathbf{x}}{\lambda}\right) \quad (\lambda \in \mathbb{R}^+), \tag{1.4.17}$$

$$\mathcal{L}[e^{-\mathbf{a}\cdot\mathbf{t}}\varphi(\mathbf{t})](p) = (\tau_{-\mathbf{a}}\mathcal{L}[\varphi](\mathbf{p}) \equiv \mathcal{L}(\mathbf{p}+\mathbf{a}) \quad (\mathbf{a} \in \mathbb{R}^n); \tag{1.4.18}$$

$$\mathcal{L}[\mathbf{D}^{\mathbf{k}}\varphi(\mathbf{t})](\mathbf{x}) = \mathbf{p}^{\mathbf{k}}(\mathcal{L}\varphi)(\mathbf{p}) \quad (k \in \mathbb{N}^n) \tag{1.4.19}$$

and

$$\mathbf{D}^{\mathbf{k}}(\mathcal{L}\varphi)(\mathbf{x}) = (-1)^{|\mathbf{k}|}\mathcal{L}[\mathbf{t}^{\mathbf{k}}\varphi(\mathbf{t})](\mathbf{p}) \quad (\mathbf{k} \in \mathbb{N}^n), \tag{1.4.20}$$

respectively.

Analogous to (1.4.10), the *Laplace convolution operator* of two functions h and φ is defined by

$$h * \varphi \equiv (h * \varphi)(\mathbf{x}) = \int_{\mathbf{0}}^{\mathbf{x}} h(\mathbf{x}-\mathbf{t})\varphi(\mathbf{t})dt \quad \left(\mathbf{x} \in \mathbb{R}^n_{+\cdots+}\right), \tag{1.4.21}$$

where

$$\int_{\mathbf{0}}^{\mathbf{x}} := \int_0^{x_1} \cdots \int_0^{x_1} \text{ and } \quad (h(\mathbf{x}-\mathbf{t}) := h(x_1-t_1,\cdots,\ x_n-t_n)). \tag{1.4.22}$$

The operation in (1.4.21) has the commutative property (1.4.11), and for such a convolution the Laplace convolution theorem (1.4.12) also holds.

The *Mellin transform* of a function $\varphi(t)$ of a real variable $t \in \mathbb{R}^+ = (0,\infty)$ is defined by

$$(\mathcal{M}\varphi)(p) = \mathcal{M}[\varphi(t)](p) = \varphi^*(s) := \int_0^\infty t^{s-1}\varphi(t)dt \quad (s \in \mathbb{C}) \tag{1.4.23}$$

and the *inverse Mellin transform* is given for $x \in \mathbb{R}^+$ by the formula

$$(\mathcal{M}^{-1}g)(x) = \mathcal{M}^{-1}[g(s)](x) := \frac{1}{2\pi i}\int_{\gamma-i\infty}^{\gamma+i\infty} x^{-s}g(s)ds \quad (\gamma = \Re(s)). \tag{1.4.24}$$

These relations can be derived from (1.3.1) and (1.3.2) if we replace $\varphi(t)$ by $\varphi(e^t)$ and ix by s. Thus the conditions for the existence of the integrals in (1.4.23) and (1.4.24) can be derived from the corresponding conditions for the direct and inverse Fourier transforms.

The direct and inverse Mellin transforms are inverse to each other for "sufficiently good" functions φ and g:

$$\mathcal{M}^{-1}\mathcal{M}\varphi = \varphi \quad \text{and} \quad \mathcal{M}\mathcal{M}^{-1}g = g. \tag{1.4.25}$$

Two simple properties of the Mellin transform are connected with the elementary operators M_a and N_a defined by

$$(M_a\varphi)(x) = x^a\varphi(x) \quad (x \in \mathbb{R};\ a \in \mathbb{C}) \tag{1.4.26}$$

and

$$(N_a\varphi)(x) = \varphi(x^a) \quad (x \in \mathbb{R};\ \ a \in \mathbb{R}\backslash\{0\}), \tag{1.4.27}$$

respectively. The following relations hold for "sufficiently good" functions φ:

$$(\mathcal{M}M_a\varphi)(s) = \mathcal{M}_{-a}(s) := \mathcal{M}(s+a) \quad (a \in \mathbb{C}) \tag{1.4.28}$$

and

$$(\mathcal{M}N_a\varphi)(s) = \frac{1}{|a|}\mathcal{M}\left(\frac{s}{a}\right) \quad (a \in \mathbb{R}\backslash\{0\}) \tag{1.4.29}$$

In particular,

$$(\mathcal{M}N_{-1}\varphi)(s) = \mathcal{M}(-s). \tag{1.4.30}$$

We also indicate some other properties for the Mellin transform which are valid for "sufficiently good" functions φ:

$$(\mathcal{M}\Pi_\lambda\varphi)(s) = \lambda^{-s}\mathcal{M}(s) \quad (\lambda \in \mathbb{R}^+) \tag{1.4.31}$$

with the Π_λ-dilation operator (1.3.6);

$$\mathcal{M}[D^m\varphi(t)](s) = (-1)^m(s-1)\cdots(s-m)(\mathcal{M}\varphi)(s-m) \tag{1.4.32}$$

$$= \frac{\Gamma(1+m-s)}{\Gamma(1-s)}(\mathcal{M}\varphi)(s-m) \tag{1.4.33}$$

$$= (-1)^m\frac{\Gamma(s)}{\Gamma(s-m)}(\mathcal{M}\varphi)(s-m), \tag{1.4.34}$$

$$\mathcal{M}[\delta^m\varphi(t)](s) = (-s)^m(\mathcal{M}\varphi)(s), \tag{1.4.35}$$

and

$$D^m(\mathcal{M}\varphi)(s) = \mathcal{M}[(\log t)^m\varphi(t)](s), \tag{1.4.36}$$

with $m \in \mathbb{N}$ and the δ-derivative (1.1.11).

When $\varphi \in C^m(\mathbb{R}^+)$, the Mellin transforms $(\mathcal{M}[\varphi(t)](s-m)$ and $\mathcal{M}[D^m\varphi(t)](s)$ exist, and $\lim_{t\to 0+}[t^{s-k-1}\varphi^{(m-k-1)}(t)]$ and $\lim_{t\to+\infty}[t^{s-k-1}\varphi^{(m-k-1)}(t)]$ are finite for $k = 0, 1, \cdots, m-1$, then the relations in (1.4.33) and (1.4.34) are replaced by the following more general ones:

$$\mathcal{M}[D^m\varphi(t)](s) = \frac{\Gamma(1+m-s)}{\Gamma(1-s)}(\mathcal{M}\varphi)(s-m)$$

$$+\sum_{k=0}^{m-1}\frac{\Gamma(1+k-s)}{\Gamma(1-s)}[x^{s-k-1}\varphi^{(m-k-1)}(x)]_0^\infty \tag{1.4.37}$$

and

$$\mathcal{M}[D^m\varphi(t)](s) = (-1)^m\frac{\Gamma(s)}{\Gamma(s-m)}(\mathcal{M}\varphi)(s-m)$$

$$+\sum_{k=0}^{m-1}(-1)^k\frac{\Gamma(s)}{\Gamma(s-k)}[x^{s-k-1}\varphi^{(m-k-1)}(x)]_0^\infty, \tag{1.4.38}$$

with $m \in \mathbb{N}$, respectively.

The *Mellin convolution operator* of two functions $h(t)$ and $\varphi(t)$, given on $\mathbb{R}^+$, is defined for $x \in \mathbb{R}^+$ by the integral

$$h * \varphi = (h * \varphi)(x) := \int_0^x h\left(\frac{x}{t}\right) \varphi(t) \frac{dt}{t}, \tag{1.4.39}$$

which has the commutative property

$$h * \varphi = \varphi * h. \tag{1.4.40}$$

The Convolution Theorem 1.2 with respect to (1.4.39) yields the form

$$(\mathcal{M}(h * \varphi))(s) = (\mathcal{M}h)(s)(\mathcal{M}\varphi)(s), \tag{1.4.41}$$

which holds for "sufficiently good" functions h and φ.

The *n-dimensional Mellin transform* of a function $\varphi(\mathbf{t})$ of $\mathbf{t} \in \mathbb{R}^n_{+\cdots+}$ is defined by

$$(\mathcal{M}\varphi)(\mathbf{s}) = \mathcal{M}[\varphi(\mathbf{t})](\mathbf{s}) = \varphi^*(\mathbf{s}) := \int_{\mathbb{R}^n_+} e^{-\mathbf{s}\cdot\mathbf{t}} \varphi(\mathbf{t}) d\mathbf{t} \quad (\mathbf{s} \in \mathbb{C}^n), \tag{1.4.42}$$

while the *inverse Mellin transform* is given for $\mathbf{x} \in \mathbb{R}^n_{+\cdots+}$ by the formula

$$(\mathcal{M}^{-1}g)(\mathbf{x}) \equiv \mathcal{M}^{-1}[g(\mathbf{p})](\mathbf{x}) = \frac{1}{(2\pi i)^n} \int_{\gamma_1 - i\infty}^{\gamma_1 + i\infty} \cdots \int_{\gamma_n - i\infty}^{\gamma_n + i\infty} \mathbf{x}^{-\mathbf{s}} g(\mathbf{s}) d\mathbf{s} \tag{1.4.43}$$

with $\gamma_j = \Re(s_j)$ $(j = 1, \cdots, n)$. The theory for these multidimensional Mellin transforms appears in the book by Brychkov et al. [107].

The integrals in (1.4.42) and (1.4.43) have some properties analogous to those of the one-dimensional Mellin transforms given by (1.4.23) and (1.4.24). In particular, the mutually invertible relations in (1.4.25) are valid for "sufficiently good" functions φ and g, and for such functions the above properties (1.4.28)-(1.4.36) of the one-dimensional Mellin transforms are extended to the multidimensional case as follows:

$$(\mathcal{M}M_{\mathbf{a}}\varphi)(\mathbf{s}) = \mathcal{M}_{-\mathbf{a}}(\mathbf{s}) \equiv \mathcal{M}(\mathbf{s} + \mathbf{a}) \quad (a \in \mathbb{C}^n); \tag{1.4.44}$$

$$(\mathcal{M}N_{\mathbf{a}}\varphi)(\mathbf{s}) = \frac{1}{|a|} \mathcal{M}\left(\frac{\mathbf{s}}{\mathbf{a}}\right) \quad (\mathbf{a} \in \mathbb{R}^n, \ \mathbf{a} \neq \mathbf{0}); \tag{1.4.45}$$

$$(\mathcal{M}N_{-1}\varphi)(\mathbf{s}) = \mathcal{M}(-\mathbf{s}); \tag{1.4.46}$$

$$(\mathcal{M}\Pi_\lambda\varphi)(\mathbf{s}) = \lambda^{-|\mathbf{s}|}\mathcal{M}(\mathbf{s}) \quad (\lambda \in R_+); \tag{1.4.47}$$

$$\mathcal{M}[\mathbf{D}^{\mathbf{m}}\varphi(\mathbf{t})](\mathbf{s}) = (-1)^{|\mathbf{m}|}(\mathbf{s} - \mathbf{1}) \cdots (\mathbf{s} - \mathbf{m})(\mathcal{M}\varphi)(\mathbf{s} - \mathbf{m})$$

$$= \frac{\Gamma(\mathbf{1} + \mathbf{m} - \mathbf{s})}{\Gamma(\mathbf{1} - \mathbf{s})}(\mathcal{M}\varphi)(\mathbf{s} - \mathbf{m}) \tag{1.4.48}$$

$$= (-1)^{|\mathbf{m}|} \frac{\Gamma(\mathbf{s})}{\Gamma(\mathbf{s} - \mathbf{m})}(\mathcal{M}\varphi)(\mathbf{s} - \mathbf{m}) \tag{1.4.49}$$

$$(\mathbf{m} = (m_1, \cdots, m_n) \in \mathbb{N}^n := \mathbb{N} \times \cdots \times \mathbb{N}; \quad |\mathbf{m}| = m_1 + \cdots + m_n);$$

$$\mathcal{M}[\delta^{\mathbf{m}}\varphi(\mathbf{t})](\mathbf{s}) = (-\mathbf{s})^{\mathbf{m}}(\mathcal{M}\varphi)(\mathbf{s}) \quad (\mathbf{m} \in \mathbb{N}^n) \tag{1.4.50}$$

and

$$\mathbf{D}^{\mathbf{m}}(\mathcal{M}\varphi)(\mathbf{s}) = \mathcal{M}[(\log \mathbf{t})^{\mathbf{m}}\varphi(\mathbf{t})](\mathbf{s}) \quad (\mathbf{m} \in \mathbb{N}^n). \tag{1.4.51}$$

Here the operators $M_{\mathbf{a}}$ and $N_{\mathbf{a}}$ are defined by

$$(M_{\mathbf{a}}\varphi)(\mathbf{x}) = \mathbf{x}^{\mathbf{a}}\varphi(\mathbf{x}) \quad (\mathbf{x} \in \mathbb{R}^n; \ \mathbf{a} \in \mathbb{C}^n) \tag{1.4.52}$$

and

$$(N_{\mathbf{a}}\varphi)(\mathbf{x}) = \varphi(\mathbf{x}^{\mathbf{a}}) \quad (\mathbf{x} \in \mathbb{R}^n, \ \ \mathbf{a} \in \mathbb{R}^n; \ \mathbf{a} \neq \mathbf{0}), \tag{1.4.53}$$

respectively, while

$$(-\mathbf{s}) \cdots (\mathbf{s} - \mathbf{m}) = \prod_{j=1}^{n}(s_j - 1) \cdots (s_j - m_j), \ \ (-\mathbf{s})^{\mathbf{m}} = (-s_1)^{m_1} \cdots (-s_n)^{m_n}, \tag{1.4.54}$$

$$\delta^{-\mathbf{s}} = \delta_1^{-s_1} \cdots \delta_n^{-s_n}; \quad \delta^{\mathbf{m}} = \left(\frac{1}{t_1}\frac{\partial}{\partial t_1}\right)^{m_1} \cdots \left(\frac{1}{t_n}\frac{\partial}{\partial t_n}\right)^{m_n}. \tag{1.4.55}$$

Analogous to (1.4.39), the *Mellin convolution operator* of two functions h and φ is defined by

$$h * \varphi = (h * \varphi)(\mathbf{x}) := \int_{\mathbf{0}}^{\infty} h\left(\frac{\mathbf{x}}{\mathbf{t}}\right)\varphi(\mathbf{t})\frac{d\mathbf{t}}{\mathbf{t}} \quad \left(\mathbf{x} \in \mathbb{R}^n_{+\cdots+}\right), \tag{1.4.56}$$

where

$$\int_{\mathbf{0}}^{\infty} := \int_0^{\infty} \cdots \int_0^{\infty}, \quad h\left(\frac{\mathbf{x}}{\mathbf{t}}\right) := h\left(\frac{x_1}{t_1}, \cdots, \frac{x_n}{t_n}\right) \text{ and } \frac{d\mathbf{t}}{\mathbf{t}} := \frac{dt_1}{t_1} \cdots \frac{dt_n}{t_n}. \tag{1.4.57}$$

The commutative property (1.4.40) holds for such a convolution, as does the Mellin convolution theorem (1.4.41).

To conclude this section, we present formulas for the Laplace transform (1.4.1) and the Mellin transform (1.4.23) of some elementary and generalized functions. First we present the former. The Laplace transform (1.4.1) of power function t^{λ} is given by

$$\mathcal{L}\left[t^{\lambda}\right](p) = \frac{\Gamma(\lambda+1)}{p^{\lambda+1}} \quad (\Re(\lambda) > -1; \ \Re(p) > 0), \tag{1.4.58}$$

$$\mathcal{L}\left[t^{k}\right](p) = \frac{k!}{p^{k+1}} \quad (k \in \mathbb{N}_0; \ \Re(p) > 0), \tag{1.4.59}$$

$$\mathcal{L}\left[t^{-1-k}\right](p) = (-1)^{k+1}\frac{p^k}{k!}[\log(p) - \psi(k+1)] \quad (k \in \mathbb{N}_0; \ \Re(p) > 0), \tag{1.4.60}$$

where $\psi(z)$ is the Euler psi function (1.5.17) [see Brychkov and Prudnikov ([108], formulas 739 and 742)]. The Laplace transform (1.4.1) of the Dirac delta function (1.2.6) and its derivatives (1.2.7) have the following forms:

$$\mathcal{L}[\delta(t)](s) = 1 \quad (s \in \mathbb{C}), \tag{1.4.61}$$

$$\mathcal{L}\left[\delta^{(k)}(t)\right](s) = s^k \quad (s \in \mathbb{C};\ k \in \mathbb{N}), \tag{1.4.62}$$

$$\mathcal{L}\left[\delta^{(k)}(t-a)\right](s) = e^{-as}s^k \quad (a \in \mathbb{C};\ s \in \mathbb{C};\ k \in \mathbb{N}); \tag{1.4.63}$$

[see Brychkov and Prudnikov ([108], formulas 732 and 733)].

The Mellin transform of the exponential function $e^{-\lambda t}$ and the power function $(1+t)^{-\sigma}$ is expressed in terms of the Euler Gamma function (1.5.1) by

$$\mathcal{M}\left[e^{-\lambda t}\right](s) = \frac{\Gamma(s)}{\lambda^s} \quad (\Re(\lambda) > 0;\ \Re(s) > 0) \tag{1.4.64}$$

and

$$\mathcal{M}\left[(t+1)^{-\sigma}\right](s) = \frac{\Gamma(s)\Gamma(\sigma - s)}{\Gamma(\sigma)} \quad (0 < \Re(s) < \Re(\sigma)), \tag{1.4.65}$$

respectively. In particular, when $\lambda = 0$, (1.4.64) yields

$$\mathcal{M}\left[e^{-t}\right](s) = \Gamma(s) \quad (\Re(s) > 0). \tag{1.4.66}$$

1.5 The Gamma Function and Related Special Functions

In this section we present the definitions and some properties of the Euler gamma function and of some special functions connected with this function. More detailed information may be found in the book by Erdélyi et al. ([249], Vol. 1, Chapter I).

The *Euler gamma function* $\Gamma(z)$ is defined by the so-called *Euler integral of the second kind*:

$$\Gamma(z) = \int_0^\infty t^{z-1}e^{-t}dt \quad (\Re(z) > 0), \tag{1.5.1}$$

where $t^{z-1} = e^{(z-1)\log(t)}$. This integral is convergent for all complex $z \in \mathbb{C}$ $(\Re(z) > 0)$. It follows from (1.4.66) that the gamma function is the Mellin transform of the exponential function:

$$\mathcal{M}[e^{-t}](z) = \Gamma(z) \quad (\Re(z) > 0). \tag{1.5.2}$$

For this function the *reduction formula*

$$\Gamma(z+1) = z\Gamma(z) \qquad (\Re(z) > 0) \tag{1.5.3}$$

holds; it is obtained from (1.5.1) by integration by parts. Using this relation, the Euler gamma function is extended to the half-plane $\Re(z) \leqq 0$ by

$$\Gamma(z) = \frac{\Gamma(z+n)}{(z)_n} \quad (\Re(z) > -n;\ \ n \in \mathbb{N};\ \ z \notin \mathbb{Z}_0^- := \{0, -1, -2, \cdots\}). \tag{1.5.4}$$

Here $(z)_n$ is the *Pochhammer symbol*, defined for complex $z \in \mathbb{C}$ and non-negative integer $n \in \mathbb{N}_0$ by

$$(z)_0 = 1 \ \text{ and } \ (z)_n = z(z+1)\cdots(z+n-1) \ \ (n \in \mathbb{N}). \tag{1.5.5}$$

Equations (1.5.3) and (1.5.5) yield

$$\Gamma(n+1) = (1)_n = n! \quad (n \in \mathbb{N}_0) \tag{1.5.6}$$

with (as usual) $0! = 1$.

It follows from (1.5.4) that the gamma function is analytic everywhere in the complex plane $\mathbb{C}$ except at $z = 0, -1, -2, \cdots$, where $\Gamma(z)$ has simple poles and is represented by the asymptotic formula

$$\Gamma(z) = \frac{(-1)^k}{z+k}\,[1 + O(z+k)] \quad (z \to -k; \quad k \in \mathbb{N}_0). \tag{1.5.7}$$

We also indicate some other properties of the gamma function such as the *functional equation:*

$$\Gamma(z)\Gamma(1-z) = \frac{\pi}{\sin(\pi z)} \quad (z \notin \mathbb{Z}_0;\ 0 < \Re(z) < 1); \quad \Gamma\left(\frac{1}{2}\right) = \sqrt{\pi}; \tag{1.5.8}$$

the *Legendre duplication formula:*

$$\Gamma(2z) = \frac{2^{2z-1}}{\sqrt{\pi}}\Gamma(z)\Gamma\left(z+\frac{1}{2}\right) \quad (z \in \mathbb{C}); \tag{1.5.9}$$

and the more general *Gauss-Legendre multiplication theorem:*

$$\Gamma(mz) = \frac{2^{mz-1}}{(2\pi)^{(m-1)/2}}\prod_{k=0}^{m-1}\Gamma\left(z+\frac{k}{m}\right) \quad (z \in \mathbb{C}; \quad m \in \mathbb{N}\backslash\{1\}); \tag{1.5.10}$$

the *Stirling asymptotic formula:*

$$\Gamma(z) = (2\pi)^{1/2} z^{z-1/2} e^{-z}\left[1 + O\left(\frac{1}{z}\right)\right] \quad (|\arg(z)| < \pi; \ |z| \to \infty); \tag{1.5.11}$$

and its corollary for $|\Gamma(x+iy)|$ $(x, y \in \mathbb{R})$:

$$|\Gamma(x+iy)| = (2\pi)^{1/2}|x|^{x-1/2} e^{-x-\pi[1-\mathrm{sign}(x)y]/2}\left[1 + O\left(\frac{1}{x}\right)\right] \quad (x \to \infty). \tag{1.5.12}$$

In particular,

$$n! = (2\pi n)^{1/2}\left(\frac{n}{e}\right)^n\left[1 + O\left(\frac{1}{n}\right)\right] \quad (n \in \mathbb{N}; \quad n \to \infty) \tag{1.5.13}$$

and

$$|\Gamma(x+iy)| = (2\pi)^{1/2}|y|^{x-1/2} e^{-x-\pi|y|/2}\left[1 + O\left(\frac{1}{y}\right)\right] \quad (y \to \infty); \tag{1.5.14}$$

the quotient expansion of two gamma functions at infinity:

$$\frac{\Gamma(z+a)}{\Gamma(z+b)} = z^{a-b}\left[1 + O\left(\frac{1}{z}\right)\right] \quad |(\arg(z+a)| < \pi; \ |z| \to \infty); \tag{1.5.15}$$

and the equality:

$$\Gamma\left(n+\frac{1}{2}\right)=\frac{(2n-1)!!}{2^n}\sqrt{\pi}, \quad (2n-1)!!:=1\cdot3\cdots(2n-1) \quad (n\in\mathbb{N}), \tag{1.5.16}$$

which follows from (1.5.3) and the second relation in (1.5.8).

The *Euler psi function* is defined as the logarithmic derivative of the gamma-function:

$$\psi(z)=\frac{d}{dz}\log\Gamma(z)=\frac{\Gamma'(z)}{\Gamma(z)} \quad (z\in\mathbb{C}). \tag{1.5.17}$$

This function has the following property:

$$\psi(z+m)=\psi(z)+\sum_{k=0}^{m-1}\frac{1}{z+k} \quad (z\in\mathbb{C};\ m\in\mathbb{N}), \tag{1.5.18}$$

which, for $m=1$, yields

$$\psi(z+1)=\psi(z)+\frac{1}{z} \quad (z\in\mathbb{C}). \tag{1.5.19}$$

The *beta function* is defined by the Euler integral of the first kind:

$$B(z,w)=\int_0^1 t^{z-1}(1-t)^{w-1}dt \quad (\Re(z)>0;\ \Re(w)>0), \tag{1.5.20}$$

This function is connected with the gamma functions by the relation

$$B(z,w)=\frac{\Gamma(z)\Gamma(w)}{\Gamma(z+w)} \quad (z,w\notin\mathbb{Z}_0^-). \tag{1.5.21}$$

The *binomial coefficients* are defined for $\alpha\in\mathbb{C}$ and $n\in\mathbb{N}_0:=\{0,1,\cdots\}$ by the formula

$$\binom{\alpha}{0}=1,\quad \binom{\alpha}{n}=\frac{\alpha(\alpha-1)\cdots(\alpha-n+1)}{n!}=\frac{(-1)^n(-\alpha)_n}{n!} \quad (n\in\mathbb{N}). \tag{1.5.22}$$

In particular, when $\alpha=m$ $(m\in\mathbb{N}_0)$, we have

$$\binom{m}{n}=\frac{m!}{n!(m-n)!} \quad (m,n\in\mathbb{N}_0;\ m\geqq n) \tag{1.5.23}$$

and

$$\binom{m}{n}=0 \quad (m,n\in\mathbb{N}_0;\ 0\leqq m<n) \tag{1.5.24}$$

If $\alpha\notin\mathbb{Z}^-:=\{-1,-2,-3,\cdots\}=:\mathbb{Z}_0^-\backslash\{0\}$, (1.5.22) is represented via the gamma function by

$$\binom{\alpha}{n}=\frac{\Gamma(\alpha+1)}{n!\Gamma(\alpha-n+1)} \quad (\alpha\in\mathbb{C};\ \alpha\notin\mathbb{Z}^-;\ n\in\mathbb{N}_0). \tag{1.5.25}$$

Such a relation can be extended from $n \in \mathbb{N}_0$ to arbitrary complex $\beta \in \mathbb{C}$ by

$$\binom{\alpha}{\beta} = \frac{\Gamma(\alpha+1)}{\Gamma(\alpha-\beta+1)\Gamma(\beta+1)} \quad (\alpha, \beta \in \mathbb{C}; \quad \alpha \notin \mathbb{Z}^-). \tag{1.5.26}$$

The *incomplete gamma functions* $\gamma(z, w)$ and $\Gamma(z, w)$ are defined for $z, w \in \mathbb{C}$ by

$$\gamma(z, w) = \int_0^w t^{z-1} e^{-t} dt \quad (\Re(z) > 0) \tag{1.5.27}$$

and

$$\Gamma(z, w) = \int_w^\infty t^{z-1} e^{-t} dt, \tag{1.5.28}$$

respectively. The following relation is evident:

$$\gamma(z, \infty) = \Gamma(z, 0) = \Gamma(z) = \gamma(z, w) + \Gamma(z, w) \quad (\Re(z) > 0). \tag{1.5.29}$$

1.6 Hypergeometric Functions

In this section we present the definitions and some properties of Gauss, Kummer, and generalized hypergeometric functions. More detailed information may be found in the book by Erdélyi et al. ([249] Vol. 1, Chapters II, IV, and VI).

The *Gauss hypergeometric function* ${}_2F_1(a, b; c; z)$ is defined in the unit disk as the sum of the hypergeometric series

$$ {}_2F_1(a, b; c; z) = \sum_{k=0}^{\infty} \frac{(a)_k (b)_k}{(c)_k} \frac{z^k}{k!}, \tag{1.6.1}$$

where $|z| < 1$; $a, b \in \mathbb{C}$; $c \in \mathbb{C} \backslash \mathbb{Z}_0^-$, and $(a)_k$ is the Pochhammer symbol (1.5.5). The series in (1.6.1) is absolutely convergent for $|z| < 1$ and for $|z| = 1$, when $\Re(c - a - b) > 0$, while it is conditionally convergent for $|z| = 1$ $(z \neq 1)$ if $-1 < \Re(c - a - b) \leqq 0$. For other values of z, the Gauss hypergeometric function is defined as an analytic continuation of the series (1.6.1). One such analytic continuation is given by the *Euler integral representation*:

$$ {}_2F_1(a, b; c; z) = \frac{\Gamma(c)}{\Gamma(b)\Gamma(c-b)} \int_0^1 t^{b-1} (1-t)^{c-b-1} (1-zt)^{-a} dt \tag{1.6.2}$$

$$(0 < \Re(b) < \Re(c); \ |\arg(1-z)| < \pi).$$

If $c \notin \mathbb{Z}_0^-$, then ${}_2F_1(a, b; c; z)$ has another integral representation in terms of the Mellin-Barnes contour integral.

$$ {}_2F_1(a, b; c; z) = \frac{1}{2\pi i} \frac{\Gamma(c)}{\Gamma(a)\Gamma(b)} \int_{\gamma - i\infty}^{\gamma + i\infty} \frac{\Gamma(a-s)\Gamma(b-s)}{\Gamma(c-s)} (-z)^{-s} \Gamma(s) ds. \tag{1.6.3}$$

Here $|\arg(-z)| < \pi$ and the path of integration starts at the point $\gamma - i\infty$ $(\gamma \in \mathbb{R})$ and terminates at the point $\gamma + i\infty$, separating all the poles $s = -k$ $(k = 0, 1, 2 \cdots)$

to the left and all the poles $s = a + n$ ($n \in \mathbb{N}_0$) and $s = b + m$ ($m \in \mathbb{N}_0$) to the right.

The asymptotic behavior of ${}_2F_1(a, b; c; z)$ at infinity is given by

$$ {}_2F_1(a,b;c;z) = A(-z)^{-a}\left[1 + O\left(\frac{1}{z}\right)\right] + B(-z)^{-b}\left[1 + O\left(\frac{1}{z}\right)\right] \tag{1.6.4} $$

$$ (|z| \to \infty;\ |\arg(-\mathrm{z})| < \pi) $$

when $a - b \notin \mathbb{Z} := \{0, \pm 1, \pm 2, \cdots\}$, and by

$$ {}_2F_1(a,b;c;z) = C(-z)^{-a}\log(z)\left[1 + O\left(\frac{1}{z}\right)\right] \tag{1.6.5} $$

$$ (|z| \to \infty;\ |\arg(-\mathrm{z})| < \pi) $$

when $a - b \in \mathbb{Z}$, where $A, B, C \in \mathbb{C}$ are constants.

The Gauss hypergeometric function has the following simple properties:

$$ {}_2F_1(b,a;c;z) = {}_2F_1(a,b;c;z), \tag{1.6.6} $$

$$ {}_2F_1(a,b;c;0) = {}_2F_1(0,b;c;z) = 1, \tag{1.6.7} $$

$$ {}_2F_1(a,b;b;z) = (1-z)^{-a}, \tag{1.6.8} $$

$$ {}_2F_1(a,b;c;1) = \frac{\Gamma(c)\Gamma(c-a-b)}{\Gamma(c-a)\Gamma(c-b)} \quad (\Re(c-a-b) > 0), \tag{1.6.9} $$

the *Euler transformation formula:*

$$ {}_2F_1(a,b;c;z) = (1-z)^{c-a-b}\ {}_2F_1(c-a,c-b;c;z), \tag{1.6.10} $$

and the following differentiation relations:

$$ \left(\frac{d}{dz}\right)^n {}_2F_1(a,b;c;z) = \frac{(a)_n(b)_n}{(c)_n}\ {}_2F_1(a+n,b+n;c+;z) \quad (n \in \mathbb{N}) \tag{1.6.11} $$

and

$$ \left(\frac{d}{dz}\right)^n \left[z^{a+n-1}\ {}_2F_1(a,b;c;z)\right] = (a)_n z^{a-1}\ {}_2F_1(a+n,b;c;z) \quad (n \in \mathbb{N}). \tag{1.6.12} $$

We also note that, if c is not an integer ($c \notin \mathbb{Z}$), then the Gauss hypergeometric function

$$ u_1(z) = {}_2F_1(a,b;c;z) $$

and the function

$$ u_2(z) = z^{1-c}\ {}_2F_1(a-c+1, b-c+1; 2-c; z) $$

are two linearly independent solutions of the *hypergeometric differential equation*

$$ z(1-z)\frac{d^2u}{dz^2} + [c - (a+b+1)z]\frac{du}{dz} - abu(z) = 0. \tag{1.6.13} $$

The *confluent hypergeometric Kummer function* is also defined by the series

$$\Phi(a;c;z) = {}_1F_1(a;c;z) := \sum_{k=0}^{\infty} \frac{(a)_k}{(c)_k} \frac{z^k}{k!}, \tag{1.6.14}$$

where $z, a \in \mathbb{C};, c \in \mathbb{C}\backslash\mathbb{Z}_0^-$; but, in contrast to the hypergeometric series in (1.6.1), this series is convergent for any $z \in \mathbb{C}$. It has an integral representations similar to (1.6.2) and (1.6.3):

$$\Phi(a;c;z) = \frac{\Gamma(c)}{\Gamma(a)\Gamma(c-a)} \int_0^1 t^{a-1}(1-t)^{c-a-1}e^{zt}dt \quad (0 < \Re(a) < \Re(c)) \tag{1.6.15}$$

and

$$\Phi(a;c;z) = \frac{1}{2\pi i} \frac{\Gamma(c)}{\Gamma(a)} \int_{\gamma-i\infty}^{\gamma+i\infty} \frac{\Gamma(a-s)}{\Gamma(c-s)}(-z)^{-s}\Gamma(s)ds, \tag{1.6.16}$$

where $|\arg(-z)| < \pi$ and the path of integration separates all the poles $s = -k$ $(k = 0, 1, 2\cdots)$ to the left and all the poles $s = a + n$ $(n \in \mathbb{N}_0)$ to the right.

The asymptotic behavior of $\Phi(a;c;z)$ at infinity has the form

$$\Phi(a;c;z) = \frac{\Gamma(c)}{\Gamma(c-a)} \left(\frac{e^{i\pi\varepsilon}}{z}\right)^a \sum_{k=0}^{M} \frac{(a)_n(a-c+1)_n}{n!} \left(-\frac{1}{z}\right)^n \left[1 + O\left(\frac{1}{z}\right)\right]$$

$$+\frac{\Gamma(c)}{\Gamma(a)} e^z z^{a-c} \sum_{k=0}^{N} \frac{(c-a)_n(1-a)_n}{n!} \frac{1}{z^n} \left[1 + O\left(\frac{1}{z}\right)\right], \tag{1.6.17}$$

when $|z| \to \infty$ and $|\arg(z)| < \pi$, and where $\varepsilon = 1$ for $\Im(z) > 0$, while $\varepsilon = -1$ for $\Im(z) < 0$.

In particular,

$$\Phi(a;c;z) = \frac{\Gamma(c)}{\Gamma(a)} e^z z^{a-c} \left[1 + O\left(\frac{1}{z}\right)\right] \quad (\Re(z) \to \infty) \tag{1.6.18}$$

and

$$\Phi(a;c;z) = \frac{\Gamma(c)e^{-i\pi a}}{\Gamma(c-a)} z^{-a} \left[1 + O\left(\frac{1}{z}\right)\right] \quad (\Re(z) \to -\infty). \tag{1.6.19}$$

The following differentiation relations hold for Kummer's confluent hypergeometric function:

$$\left(\frac{d}{dz}\right)^n \Phi(a;c;z) = \frac{(a)_n}{(c)_n} \Phi(a+n;c+n;z) \quad (n \in \mathbb{N}), \tag{1.6.20}$$

$$\left(\frac{d}{dz}\right)^n [z^{c-1}\Phi(a;c;z)] = (-1)^n(1-c)_n z^{c-n-1}\Phi(a;c-n;z) \quad (n \in \mathbb{N}), \tag{1.6.21}$$

$$\left(\frac{d}{dz}\right)^n [e^{-z}\Phi(a;c;z)] = (-1)^n \frac{(c-a)_n}{(c)_n} e^{-z}\Phi(a;c+n;z) \quad (n \in \mathbb{N}), \tag{1.6.22}$$

and

$$\left(\frac{d}{dz}\right)^n \left[e^{-z}z^{c-a+n-1}\Phi(a;c;z)\right] = (c-a)_n e^{-z}z^{c-a-1}\Phi(a-n;c;z) \quad (n \in \mathbb{N}). \tag{1.6.23}$$

The Kummer hypergeometric function $u_1(z) = \Phi(a;c;z)$ is the solution to the *confluent hypergeometric equation*

$$z\frac{d^2u}{dz^2} + (c-z)\frac{du}{dz} - au(z) = 0. \tag{1.6.24}$$

The other solution to this equation is given by *Tricomi's confluent hypergeometric function* $\Psi(a;c;z)$ defined by

$$\Psi(a;c;z) = \frac{1}{\Gamma(a)}\int_0^\infty t^{a-1}(1+t)^{c-a-1}e^{-zt}dt \quad (\Re(z) > 0;\ \Re(a) > 0). \tag{1.6.25}$$

Using the δ-derivative defined in (1.1.11), equations (1.6.13) and (1.6.24) can be rewritten as

$$[\delta(\delta+c-1) - z(\delta+a)(\delta+b)]u(z) = 0 \ (\delta = z\frac{d}{dz}) \tag{1.6.26}$$

and

$$[\delta(\delta+c-1) - z(\delta+a))]u(z) = 0 \ (\delta = z\frac{d}{dz}), \tag{1.6.27}$$

respectively.

The Gauss hypergeometric series (1.6.1) and the Kummer hypergeometric series (1.6.14) are extended to the *generalized hypergeometric series* defined by

$${}_pF_q(a_1,\cdots,\ a_p;b_1,\cdots,\ b_q;z) = \sum_{k=0}^\infty \frac{(a_1)_k\cdots(a_p)_k}{(b_1)_k\cdots(b_q)_k}\frac{z^k}{k!}, \tag{1.6.28}$$

where $a_l, b_j \in \mathbb{C}$, $b_j \neq 0, -1, -2, \cdots$ $(l = 1, \cdots,\ p;\ j = 1, \cdots,\ q)$. This series is absolutely convergent for all values of $z \in \mathbb{C}$ if $p \leqq q$. When $p = q+1$, the series in (1.3.28) is absolutely convergent for $|z| < 1$ and for $|z| = 1$ when $\Re\left(\sum_{j=1}^q b_j - \sum_{l=1}^p a_l\right) > 0$, while it is conditionally convergent for $|z| = 1$ $(z \neq 1)$ if $-1 < \Re\left(\sum_{j=1}^q b_j - \sum_{l=1}^p a_l\right) \leqq 0$.

If $b_j \notin \mathbb{Z}_0^-$ $(j = 1, \cdots,\ q)$, this function has an integral representation in terms of the Mellin-Barnes contour integral, generalizing those in (1.6.3) and (1.6.16), given by

$$\begin{aligned}&{}_pF_q(a_1,\cdots,\ a_p;b_1,\cdots,\ b_q;z)\\ &= \frac{1}{2\pi i}\frac{\prod_{j=1}^q \Gamma(b_j)}{\prod_{l=1}^p \Gamma(a_l)}\int_{\gamma-i\infty}^{\gamma+i\infty}\frac{\prod_{l=1}^p\Gamma(a_l-s)}{\prod_{j=1}^q\Gamma(b_j-s)}(-z)^{-s}\Gamma(s)ds.\end{aligned} \tag{1.6.29}$$

Here $|\arg(-z)| < \pi$ and the path of integration separates all the poles $s = -k$ $(k = 0, 1, 2, \cdots)$ to the left and all the poles $s = a_j + n$ $(n \in \mathbb{N}_0;\ j = 1, \cdots,\ p)$ to the right.

If $a_j - a_l \notin \mathbb{Z}$ are not integers for all $j, l = 1, \cdots, p$, then the asymptotic behavior of (1.6.28) at infinity, generalizing the one in (1.6.4), is given by

$$ {}_pF_q(a_1, \cdots, a_p; b_1, \cdots, b_q; z) = \sum_{j=1}^{p} A_j(-z)^{-a_j} \left[1 + O\left(\frac{1}{z}\right)\right] \tag{1.6.30} $$

$$ (|z| \to \infty;\ |\arg(-z)| < \pi), $$

where $A_j \in \mathbb{C}$ $(j = 1, \cdots, p)$ are constants. When $a_j - a_l \in \mathbb{Z}$ for some of $j, l = 1, \cdots, p$, then, analogous to (1.6.5), the corresponding $(-z)^{-a_j}$ has to be multiplied by $\log(z)$.

For ${}_pF_q(a_1, \cdots, a_p; b_1, \cdots, b_q; z)$ the following differentiation relation holds, generalizing those in (1.6.11) and (1.6.20):

$$ \left(\frac{d}{dz}\right)^n {}_pF_q(a_1, \cdots, a_p; b_1, \cdots, b_q; z) $$

$$ = \frac{(a_1)_n \cdots (a_p)_n}{(b_1)_n \cdots (b_q)_n} {}_pF_q(a_1 + n, \cdots, a_p + n; b_1 + n, \cdots, b_q + n; z) \quad (n \in \mathbb{N}). \tag{1.6.31} $$

The function $u(z) = {}_pF_q(a_1, \cdots, a_p; b_1, \cdots, b_q; z)$ is a solution to the *generalized hypergeometric equation*

$$ \left[\delta \prod_{j=1}^{q}(\delta + b_j - 1) - z \prod_{l=1}^{p}(\delta + a_l)\right] u(z) = 0 \quad (\delta = z\frac{d}{dz}). \tag{1.6.32} $$

We note that (1.6.32) yields (1.6.26) and (1.6.27) in the cases where $p = 2$, $q = 1$, $a_1 = a$, $a_2 = b$, $b_1 = c$ and $p = 1$, $q = 1$, $a_1 = a$, $b_1 = c$, respectively.

To conclude this section, we present formulas for the Laplace transform (1.4.1) and the Mellin transform (1.4.23) of the functions considered here. First we consider the former. Applying the relation (1.4.59) to (1.6.1), (1.6.14), (1.6.25), and (1.6.28), direct evaluation leads to the following relations for the Laplace transform of the above hypergeometric functions:

$$ \mathcal{L}[t^{\lambda-1}\, {}_2F_1(a, b; c; t])(s) = \frac{\Gamma(\lambda)}{s^\lambda}\, {}_3F_1\left(\lambda, a, b; c; \frac{1}{s}\right) \tag{1.6.33} $$

$$ (a, b \in \mathbb{C};\ c \in \mathbb{C}\backslash\mathbb{Z}_0^-;\ \Re(s) > 0;\ \Re(\lambda) > 0), $$

where $G^{3,1}_{2,3}[s\,|\ldots]$ denotes the Meijer G-function given by (1.12.51);

$$ \mathcal{L}[t^{\lambda-1}\Phi(a; c; t)](s) = \frac{\Gamma(\lambda)}{s^\lambda}\, {}_2F_1\left(\lambda, a; c; \frac{1}{s}\right) \tag{1.6.34} $$

$$ (a \in \mathbb{C};\ c \in \mathbb{C}\backslash\mathbb{Z}_0^-;\ \Re(s) > 0;\ \Re(\lambda) > 0) $$

$$ \mathcal{L}[t^{\lambda-1}\Psi(a; c; t)](s) = \frac{\Gamma(\lambda)\Gamma(\lambda - c + 1)}{\Gamma(a - c + \lambda + 1)}\, {}_2F_1(\lambda, \lambda - c + 1; a - c + \lambda + 1; 1 - \frac{1}{s}) \tag{1.6.35} $$

$$(\Re(\lambda) > 0;\ \Re(c) < 1 + \Re(\lambda);\ |1 - s| < 1);$$

$$\mathcal{L}[t^{\lambda-1}{}_pF_q[a_1, \cdots, a_p; b_1, \cdots, b_q; t)](s) = \frac{\Gamma(\lambda)}{s^\lambda}\,{}_{p+1}F_q\left[\lambda, a_1, \cdots, a_p; b_1, \cdots, b_q; \frac{1}{s}\right] \tag{1.6.36}$$

$$\left(a_j \in \mathbb{C},\ (j = 1, \cdots, p);\ b_j \in \mathbb{C}\backslash\mathbb{Z}_0^-,\ (j = 1, \cdots, q);\ p \leqq q\right)$$

$$(\Re(\lambda) > 0;\ \Re(s) > 0 \text{ when } p < q;\ \Re(s) > 1 \text{ when } p = q).$$

According to (1.4.23) and (1.4.24), from relations (1.6.3), (1.6.16), and (1.6.29) we derive the following formulas for the Mellin transform (1.4.23) of the Gauss, Kummer, and generalized hypergeometric functions (1.6.1), (1.6.14), and (1.6.28):

$$\mathcal{M}[{}_2F_1(a, b; c; -t)](s) = \frac{\Gamma(c)}{\Gamma(a)\Gamma(b)}\frac{\Gamma(s)\Gamma(a-s)\Gamma(b-s)}{\Gamma(c-s)} \tag{1.6.37}$$

$$\left(a, b, c, s \in \mathbb{C},\ c \notin \mathbb{Z}_0^-;\ 0 < \Re(s) < \min[\Re(a), \Re(b)]\right),$$

$$\mathcal{M}[\Phi(a; c; -t)](s) = \frac{\Gamma(c)}{\Gamma(a)}\frac{\Gamma(a-s)\Gamma(s)}{\Gamma(c-s)} \tag{1.6.38}$$

$$\left(a, b, c, s \in \mathbb{C};\ c \notin \mathbb{Z}_0^-;\ 0 < \Re(s) < \Re(a)\right),$$

$$\mathcal{M}[{}_pF_q[a_1, \cdots, a_p; b_1, \cdots, b_q; -t)](s) = \frac{\Gamma(s)\prod_{j=1}^q \Gamma(b_j)}{\prod_{l=1}^p \Gamma(a_l)}\frac{\prod_{l=1}^p \Gamma(a_l - s)}{\prod_{j=1}^q \Gamma(b_j - s)} \tag{1.6.39}$$

$$\left(a_l, b_j \in \mathbb{C};\ b_j \notin \mathbb{Z}_0^-;\ (l = 1, \cdots, p;\ j = 1, \cdots, q);\ \ 0 < \Re(s) < \min_{1\leqq l\leqq p}[\Re(a_l)]\right).$$

1.7 Bessel Functions

In this section we present the definitions and some properties of classical Bessel functions and some of their modifications. More detailed information about the Bessel functions may be found in the book by Erdélyi et al. ([249], Vol. 2).

The *Bessel function of the first kind* $J_\nu(z)$ is defined by

$$J_\nu(z) = \sum_{k=0}^\infty \frac{(-1)^k (z/2)^{2k+\nu}}{k!\Gamma(\nu + k + 1)}, \tag{1.7.1}$$

where $z \in \mathbb{C}\backslash(-\infty, 0]$ and $\nu \in \mathbb{C}$. This function can be represented in terms of the hypergeometric function

$${}_0F_1(c; z) = \sum_{k=0}^\infty \frac{1}{(c)_k}\frac{z^k}{k!} \quad (z \in \mathbb{C};\ c \in \mathbb{C}\backslash\mathbb{Z}_0^-). \tag{1.7.2}$$

by

$$J_\nu(z) = \frac{1}{\Gamma(\nu + 1)}\left(\frac{z}{2}\right)^\nu {}_0F_1\left(\nu + 1; -\frac{z^2}{4}\right), \tag{1.7.3}$$

and so the series in (1.7.1) is convergent for all $z \in \mathbb{C}$. Thus $J_\nu(z)$ is analytic in z. In particular, when $\nu = -1/2$ and $\nu = 1/2$, we have

$$J_{-1/2}(z) = \left(\frac{2}{\pi z}\right)^{1/2} \cos(z) \text{ and } J_{1/2}(z) = \left(\frac{2}{\pi z}\right)^{1/2} \sin(z). \tag{1.7.4}$$

The Bessel function $J_\nu(z)$ is given by the *Poisson integral representation:*

$$J_\nu(z) = \frac{(z/2)^\nu}{\sqrt{\pi}\Gamma(\nu + 1/2)} \int_{-1}^{1} (1 - t^2)^{\nu - 1/2} \cos(zt)dt \quad \left(\Re(\nu) > -\frac{1}{2}\right). \tag{1.7.5}$$

The following lemma yields the other integral representation of $J_\nu(z)$ by the Mellin-Barnes contour integral.

Lemma 1.5 *For $z \in \mathbb{C}$ ($|\arg(z)| < \pi$) there holds the relation*

$$J_\nu(z) = \frac{2^{-1-\nu}}{i\sqrt{\pi}} \int_{\gamma - i\infty}^{\gamma + i\infty} \frac{\Gamma(\nu + s)}{\Gamma([1 + \nu + s]/2)\Gamma(1 + [\nu - s]/2)} z^{-s} ds \tag{1.7.6}$$

where the path of integration separates all the poles $s = -\nu - n$ ($n \in \mathbb{N}_0$) to the left.

Proof. The usual technique for evaluating the Mellin-Barnes contour integral in (1.7.6) and thus prove Lemma 1.5 can be found (for example) in Erdélyi et al. ([249], Vol. 1, Section 1.19).

The main terms of the asymptotic behavior of $J_\nu(z)$ near zero and infinity are given by

$$J_\nu(z) = \frac{1}{\Gamma(\nu + 1)} \left(\frac{z}{2}\right)^\nu [1 + O(z)] \quad (z \to 0;\ \nu \in \mathbb{C} \backslash \mathbb{Z}^-) \tag{1.7.7}$$

and

$$J_\nu(z) = \left(\frac{2}{\pi z}\right)^{1/2} \left[\cos\left(z - \frac{\nu\pi}{2} - \frac{\pi}{4}\right) + O\left(\frac{1}{z}\right)\right] \quad (|z| \to \infty), \tag{1.7.8}$$

respectively.

The function $u(z) = J_\nu(z)$ is a solution to the *Bessel differential equation*

$$z^2 \frac{d^2 u}{dz^2} + z \frac{du}{dz} + (z^2 - \nu^2) u(z) = 0. \tag{1.7.9}$$

The Bessel function of the first kind satisfies the differential recurrence relations which, in terms of the δ-operator, have the forms

$$\delta^n \left[z^\nu J_\nu(z)\right] = z^{\nu - n} J_{\nu - n}(z) \text{ and } \delta^n \left[z^{-\nu} J_\nu(z)\right] = (-1)^n z^{-\nu - n} J_{\nu + n}(z), \tag{1.7.10}$$

where $\delta = z\frac{d}{dz}$ and $n \in \mathbb{N}$.

The *Bessel function of the second kind*, or the *Neumann function* $Y_\nu(z)$, is defined via the Bessel function of the first kind by

$$Y_\nu(z) = \frac{1}{\sin(\nu\pi)}[\cos(\nu\pi)J_\nu(z) - J_{-\nu}(z)] \quad (\nu \in \mathbb{C}\backslash\mathbb{Z}). \tag{1.7.11}$$

The integral representation of $Y_\nu(z)$ in terms of the Mellin-Barnes contour integral is given by the following lemma.

Lemma 1.6 *For $z \in \mathbb{C}$ ($|\arg(z)| < \pi$) the following relation holds:*

$$Y_\nu(z) = \frac{2^{-1-\nu}}{i\sqrt{\pi}} \int_{\gamma-i\infty}^{\gamma+i\infty} \frac{\Gamma(\nu+s)\Gamma([\nu-s]/2)\, z^{-s}}{\Gamma([\nu+1+s]/2)\Gamma([\nu+1-s]/2)\Gamma(1+[\nu-s]/2)} ds, \tag{1.7.12}$$

where the path of integration separates all the poles $s = \nu + 2m$ ($m \in \mathbb{N}_0$) to the right and all the poles $s = -\nu - n$ ($n \in \mathbb{N}_0$) to the left.

Proof. We can easily prove Lemma 1.6 by using the method based on the direct and inverse Mellin transforms (see Section 1.4).

The Neumann function $Y_\nu(z)$ has the following asymptotic behavior near zero and infinity:

$$Y_\nu(z) = -\frac{\Gamma(\nu)}{\pi}\left(\frac{2}{z}\right)^\nu [1 + O(z)] \quad (|z| \to 0) \tag{1.7.13}$$

and

$$Y_\nu(z) = \left(\frac{2}{\pi z}\right)^{1/2} \left[\sin\left(z - \frac{\nu\pi}{2} - \frac{\pi}{4}\right) + O\left(\frac{1}{z}\right)\right] \quad (|z| \to \infty) \tag{1.7.14}$$

The functions $J_\nu(z)$ and $Y_\nu(z)$ are linearly independent solutions to the Bessel differential equation (1.7.9). The Neumann function satisfies the relations similar to those in (1.7.10):

$$\delta^n [z^\nu Y_\nu(z)] = z^{\nu-n} Y_{\nu-n}(z) \text{ and } \delta^n [z^{-\nu} Y_\nu(z)] = (-1)^n z^{-\nu-n} Y_{\nu+n}(z), \tag{1.7.15}$$

where $\delta = z\frac{d}{dz}$ and $n \in \mathbb{N}$.

The *modified Bessel function* $I_\nu(z)$ is defined by

$$I_\nu(z) = \sum_{k=0}^{\infty} \frac{(z/2)^{2k+\nu}}{k!\Gamma(\nu+k+1)} \quad (z \in \mathbb{C}\backslash(-\infty, 0]). \tag{1.7.16}$$

Such a function is expressed via the Bessel function of the first kind as follows:

$$I_\nu(z) = e^{-\nu\pi i/2} J_\nu\left(e^{\pi i/2} z\right). \tag{1.7.17}$$

Analogous to (1.7.5), $I_\nu(z)$ has integral representations of the *Poisson* type:

$$I_\nu(z) = \frac{(z/2)^\nu}{\sqrt{\pi}\Gamma(\nu+1/2)} \int_{-1}^{1} (1-t^2)^{\nu-1/2} \cosh(zt)dt \quad \left(\Re(\nu) > -\frac{1}{2}\right), \tag{1.7.18}$$

and by the contour integral

$$I_\nu(z) = \frac{1}{2\pi i}\int_{-i\pi+\infty}^{i\pi+\infty} e^{z\cosh(t)-\nu t}dt \quad \left(\nu \in \mathbb{C};\ |\arg z| < \frac{\pi}{2}\right). \tag{1.7.19}$$

The main terms of the asymptotic behavior of $I_\nu(z)$ near zero and infinity are given by

$$I_\nu(z) = \frac{1}{\Gamma(\nu+1)}\left(\frac{z}{2}\right)^\nu [1+O(z)], \quad (z \to 0;\ \nu \in \mathbb{C}\backslash\mathbb{Z}^-) \tag{1.7.20}$$

and

$$I_\nu(z) = \left(\frac{\pi z}{2}\right)^{1/2} e^z \left[1+O\left(\frac{1}{z}\right)\right] + ie^{-z+\nu\pi i}\left[1+O\left(\frac{1}{z}\right)\right] \tag{1.7.21}$$

$$\left(|z| \to \infty;\ -\frac{\pi}{2} < \arg(z) < \frac{3\pi}{2}\right). \tag{1.7.22}$$

The function $u(z) = I_\nu(z)$ is a solution to the *modified Bessel differential equation*

$$z^2\frac{d^2u}{dz^2} + z\frac{du}{dz} - (z^2+\nu^2)u(z) = 0. \tag{1.7.23}$$

The modified Bessel function satisfies the differential recurrence relations of the form (1.7.10):

$$\delta^n\left[z^\nu I_\nu(z)\right] = z^{\nu-n}I_{\nu-n}(z) \text{ and } \delta^n\left[z^{-\nu}I_\nu(z)\right] = z^{-\nu-n}I_{\nu+n}(z), \tag{1.7.24}$$

where $\delta = z\frac{d}{dz}$ and $n \in \mathbb{N}$.

The *modified Bessel function of the third kind*, or the *Macdonald function* $K_\nu(z)$, is defined via the modified Bessel function by

$$K_\nu(z) = \frac{\pi}{2\sin(\nu\pi)}[\cos(\nu\pi)I_{-\nu}(z) - I_\nu(z)] \quad (\nu \in \mathbb{C}\backslash\mathbb{Z}). \tag{1.7.25}$$

This function has the following integral representations:

$$K_\nu(z) = \frac{1}{2}\int_0^\infty t^{-\nu-1}\exp\left[-\frac{z}{2}\left(t+\frac{1}{t}\right)\right]dt, \quad (\Re(z) > 0), \tag{1.7.26}$$

but, when $0 < \Re(z) < \frac{1}{2}$, then

$$K_\nu(z) = \frac{\sqrt{\pi}}{\Gamma(-\nu+1/2)}\left(\frac{2}{z}\right)^\nu \int_1^\infty (t^2-1)^{-\nu-1/2}e^{-zt}dt \tag{1.7.27}$$

The following lemma presents the integral representation of $K_\nu(z)$ by means of the Mellin-Barnes contour integral.

Lemma 1.7 *For $z \in \mathbb{C}$ ($|\arg(z)| < \pi$) the following relation holds:*

$$K_\nu(z) = \frac{2^{-2-\nu}}{i\sqrt{\pi}} \int_{\gamma-i\infty}^{\gamma+i\infty} \frac{\Gamma(\nu+s)\Gamma([s-\nu]/2)}{\Gamma([\nu+1+s]/2)} z^{-s} ds, \tag{1.7.28}$$

where the path of integration separates all the poles $s = -\nu - n$ ($n \in \mathbb{N}_0$) and $s = \nu - 2m$ ($m \in \mathbb{N}_0$) to the left.

Proof. This lemma can be proved just as Lemma 1.6 by using the known Mellin transform of $K_\nu(x)$, properties of the gamma function, and the inverse Mellin transform.

The asymptotic representations of $K_\nu(z)$ near zero and infinity are given by

$$K_\nu(z) = \frac{\Gamma(\nu)}{2}\left(\frac{2}{z}\right)^\nu [1 + O(z)] \quad (z \to 0;\ \Re(\nu) > 0) \tag{1.7.29}$$

and

$$K_\nu(z) = \left(\frac{\pi}{2z}\right)^{1/2} e^{-z}\left[1 + O\left(\frac{1}{z}\right)\right] \quad \left(|z| \to \infty;\ |\arg(z)| < \frac{3\pi}{2}\right), \tag{1.7.30}$$

respectively.

The functions $I_\nu(z)$ and $K_\nu(z)$ are linearly independent solutions to the Bessel differential equation (1.7.23). The Macdonald function $K_\nu(z)$ satisfies the relations similar to those in (1.7.24):

$$\delta^n [z^\nu K_\nu(z)] = (-1)^n z^{\nu-n} K_{\nu-n}(z) \quad \left(\delta = z\frac{d}{dz};\ n \in \mathbb{N}\right) \tag{1.7.31}$$

and

$$\delta^n \left[z^{-\nu} K_\nu(z)\right] = (-1)^n z^{-\nu-n} K_{\nu+n}(z) \quad \left(\delta = z\frac{d}{dz};\ n \in \mathbb{N}\right). \tag{1.7.32}$$

We also note that the functions $Y_n(z)$ and $K_\nu(z)$ have the symmetric property with respect to indices:

$$Y_{-n}(z) = Y_n(z) \quad (n \in \mathbb{N}) \text{ and } K_{-\nu}(z) = K_\nu(z) \quad (\nu \in \mathbb{R}). \tag{1.7.33}$$

Now we present formulas for the Laplace transform (1.4.1) and the Mellin transform (1.4.23) of the functions considered in this section. The Laplace transforms of Bessel functions (1.7.1), (1.7.11), (1.7.16), and (1.7.25) are given by

$$\mathcal{L}[J_\nu(t)](s) = (s^2+1)^{-1/2}\left[s + (s^2+1)^{1/2}\right]^{-\nu}, \tag{1.7.34}$$

when $\Re(\nu) > -1$ and $\Re(s) > 0$.

$$\mathcal{L}[Y_\nu(t)](s) = (s^2+1)^{-1/2}\left\{\frac{\cot(\nu\pi) - \csc(\nu\pi)\left[s + (s^2+1)^{1/2}\right]^{2\nu}}{\left[s + (s^2+1)^{1/2}\right]^\nu}\right\}, \tag{1.7.35}$$

when $|\Re(\nu)| < 1$, $\nu \neq 0$, and $\Re(s) > 0$.

$$\mathcal{L}[I_\nu(t)](s) = (s^2-1)^{-1/2}\left[s+(s^2-1)^{1/2}\right]^{-\nu} \tag{1.7.36}$$

when $\Re(\nu) > -1$, and $\Re(s) > 1$, and

$$\mathcal{L}[K_\nu(t)](s) = \pi\csc(\nu\pi)(s^2-1)^{-1/2}\left\{\left[s+(s^2+1)^{1/2}\right]^{\nu} - \left[s-(s^2-1)^{1/2}\right]^{-\nu}\right\} \tag{1.7.37}$$

when $|\Re(\nu)| < 1$, $\nu \neq 0$, and $\Re(s) > -1$.

The Mellin transforms of the Bessel functions are given by

$$\mathcal{M}[J_\nu(t)](s) = 2^{-\nu}\sqrt{\pi}\frac{\Gamma(\nu+s)}{\Gamma\left(\frac{1+\nu+s}{2}\right)\Gamma\left(1+\frac{\nu-s}{2}\right)} = 2^{s-1}\frac{\Gamma\left(\frac{\nu+s}{2}\right)}{\Gamma\left(1+\frac{\nu-s}{2}\right)} \tag{1.7.38}$$

when $-\Re(\nu) < \Re(s) < \frac{3}{2}$.

$$\mathcal{M}[Y_\nu(ct)](s) = 2^{-\nu}\sqrt{\pi}\frac{\Gamma(\nu+s)\Gamma(\frac{\nu-s}{2})\Gamma(1-\frac{\nu-s}{2})}{\Gamma(\frac{1+\nu+s}{2})\Gamma(\frac{1+\nu-s}{2})\Gamma(1+\frac{\nu-s}{2})\Gamma(\frac{1-\nu+s}{2})} \tag{1.7.39}$$

$$= 2^{s-1}\cot\left[\frac{(\nu-s)\pi}{2}\right]\frac{\Gamma\left(\frac{\nu+s}{2}\right)}{\Gamma\left(1+\frac{\nu-s}{2}\right)} \tag{1.7.40}$$

when $|\Re(\nu)| < \Re(s) < \frac{3}{2}$, and

$$\mathcal{M}[K_\nu(t)](s) = 2^{-1-\nu}\sqrt{\pi}\,\frac{\Gamma(\nu+s)\Gamma(\frac{s-\nu}{2})}{\Gamma(\frac{1+\nu+s}{2})} = 2^{s-2}\Gamma\left(\frac{s+\nu}{2}\right)\Gamma\left(\frac{s-\nu}{2}\right) \tag{1.7.41}$$

when $|\Re(\nu)| < \Re(s)$.

We note that the formulas (1.7.38), (1.7.40), and (1.7.41) are well known; [see, for example, Prudnikov et al. ([689], Vol. 2, formulas 2.12.8.2, 2.13.2.2, and 2.16.6.1)].

Next we consider a function $Z_\rho^\nu(z)$ defined by

$$Z_\rho^\nu(z) = \int_0^\infty t^{\nu-1}\exp\left(-t^\rho - \frac{z}{t}\right)dt \quad (\nu \in \mathbb{C};\ \rho \in \mathbb{R}), \tag{1.7.42}$$

with an additional condition $\Re(\nu) \leqq 0$ for $\rho < 0$. This function was introduced by Kratzel [451] for $\rho \geqq 1$ and investigated (for any real $\rho \in R$) by Kilbas et al. [401]. When $\rho = 1$ and z is replaced by $z^2/4$, then in accordance with (1.7.26), it coincides with the Macdonald function $K_{-\nu}(z)$ with the exactness of the multiplier $2(z/2)^\nu$:

$$Z_1^\nu\left(\frac{z^2}{4}\right) = 2\left(\frac{z}{2}\right)^\nu K_{-\nu}(z) \quad (\Re(z) > 0)\,. \tag{1.7.43}$$

The function $Z_\rho^\nu(z)$ is analytic with respect to $z \in \mathbb{C}$ ($\Re(z) > 0$). It has different asymptotic estimates at zero and infinity for $\rho > 0$ and $\rho < 0$. Thus we have

$$Z_\rho^\nu(z) = O\left(z^{\min[0,\Re(\nu)]}\right) \quad (\rho > 0;\quad z \to 0), \tag{1.7.44}$$

if the poles of the gamma functions $\Gamma(s)$ and $\Gamma([\nu+s]/\rho)$ do not coincide, while if these poles coincide, the logarithmic multiplier $\log(z)$ must be included in the asymptotic estimate (1.7.36);

$$Z_\rho^\nu(z) \sim \gamma z^{(2\nu-\rho)/(2\rho+2)} \exp\left[-\beta z^{\rho/(\rho+1)}\right] \tag{1.7.45}$$

where $\rho > 0$, $|z| \to \infty$, $|\arg(z)| < \frac{(\rho+1)\pi)}{2\rho} - \epsilon$, and

$$\gamma = \left(\frac{2\pi}{\rho+1}\right)\rho^{-(2\nu+1)/(2\rho+2)}, \quad \beta = \left(1+\frac{1}{\rho}\right)^{1/(\rho+1)}, \quad 0 < \epsilon < \frac{(\rho+1)\pi)}{2\rho}. \tag{1.7.46}$$

When $\rho < 0$ and $|\arg(z)| < [(\rho-1)\pi/(2\rho)]$, then the asymptotic estimates of $Z_\rho^\nu(z)$ at zero and infinity are given by

$$Z_\rho^\nu(z) = O(1) \quad (\rho < 0; \;\; z \to 0) \tag{1.7.47}$$

and

$$Z_\rho^\nu(z) = O\left(z^{\Re(\eta)}\right) \quad (\rho < 0; \; |z| \to \infty), \tag{1.7.48}$$

respectively. The results in (1.7.44)-(1.7.48) were established in Kilbas et al. [401]. We only note that the relations in (1.7.44) and (1.7.45) for $\rho > 1$ and $x > 0$ were given by Kratzel [451].

The integral representation of $Z_\rho^\nu(z)$ in terms of the Mellin-Barnes contour integral is given by the following lemma.

Lemma 1.8 *For $z \in \mathbb{C}$ ($|\arg(z)| < \pi$) the following relation holds:*

$$Z_\rho^\nu(z) = \frac{1}{2\pi|\rho|}\int_{\gamma-i\infty}^{\gamma+i\infty} \Gamma(s)\Gamma\left(\frac{\nu+s}{\rho}\right) z^{-s} ds, \tag{1.7.49}$$

where the path of integration separates all the poles at $s_n = -n$ ($n \in \mathbb{N}_0$) and at $s_m = -m\rho - \nu$ ($m \in \mathbb{N}_0$), to the left when $\rho > 0$, while for $\rho < 0$ it separates all the poles at $s_n = -n$ ($n \in \mathbb{N}_0$) to the left and at $s_m = -m\rho - \nu$ ($m \in \mathbb{N}_0$) to the right.

Proof. This lemma is proved just as Lemma 1.6 by using the following known Mellin transform $\mathcal{M}$ of $Z_\rho^\nu(x)$:

$$\left(\mathcal{M}Z_\rho^\nu\right)(s) = \frac{1}{|\rho|}\Gamma(s)\Gamma\left(\frac{\nu+s}{\rho}\right) \quad (\Re(s) > \min[0, -\Re(\nu)]). \tag{1.7.50}$$

Such a relation for $\rho > 0$ was proved by Kilbas and Shlapakov [403], and for the general case when $\rho \in \mathbb{R}$ ($\rho \neq 0$) by Kilbas et al. [401].

Finally, we consider a function $\lambda_{\gamma,\sigma}^{(\beta)}(z)$ defined by

$$\lambda_{\nu,\sigma}^{(\beta)}(z) = \frac{\beta}{\Gamma(\nu+1-1/\beta)}\int_1^\infty (t^\beta - 1)^{\nu-1/\beta} t^\sigma e^{-zt} dt \tag{1.7.51}$$

with $\beta > 0$, $\Re(\nu) > \frac{1}{\beta} - 1$, $\sigma \in \mathbb{C}$, and $\Re(z) > 0$.

This function was introduced by Glaeske et al. [283]. It is an analytic function with respect to $z \in \mathbb{C}$. When $z = x > 0$, the asymptotic behavior of $\lambda_{\nu,\sigma}^{(\beta)}(x)$ at zero and infinity is given by

$$\lambda_{\nu,\sigma}^{(\beta)}(x) \sim \frac{\Gamma[-\nu - \sigma/\beta]}{\Gamma[1 - (\sigma+1)/\beta]} \quad \left(\Re(\nu) < -\frac{\sigma}{\beta};\ x \to 0+\right) \tag{1.7.52}$$

and

$$\lambda_{\nu,\sigma}^{(\beta)}(x) \sim \beta^{\nu+1-1/\beta} e^{-x} x^{(1/\beta)-\nu-1} \quad \left(\Re(\nu) > 1 - \frac{1}{\beta};\ x \to \infty\right). \tag{1.7.53}$$

The following lemma gives the integral representation of $\lambda_{\nu,\sigma}^{(\beta)}(z)$ as a Mellin-Barnes contour integral.

Lemma 1.9 *For $z \in \mathbb{C}$ ($|\arg(z)| < \pi$) the following relation holds:*

$$\lambda_{\nu,\sigma}^{(\beta)}(z) = \frac{1}{2\pi i} \int_{\gamma-i\infty}^{\gamma+i\infty} \frac{\Gamma(s)\Gamma(-\nu + [-\sigma+s]/\beta)}{\Gamma(1 - [1+\sigma-s]/\beta)} z^{-s} ds, \tag{1.7.54}$$

where the path of integration separates all the poles at $s = -n$ ($n \in \mathbb{N}_0$) and at $s = -m\beta + \nu\beta + \sigma$ ($m \in \mathbb{N}_0$) to the left.

Proof. This lemma is proved just as Lemma 1.6 by using the following known Mellin transform $\mathcal{M}$ of $\lambda_{\nu,\sigma}^{(\beta)}(x)$:

$$\left(\mathcal{M}\lambda_{\nu,\sigma}^{(\beta)}\right)(s) = \frac{\Gamma(s)\Gamma(-\nu + [-\sigma+s]/\beta)}{\Gamma(1 - [1+\sigma-s]/\beta)} \quad (\Re(s) > \min[0, 1-\beta+\Re(\sigma)]), \tag{1.7.55}$$

which was established in Glaeske et al. [283].

To conclude this section, we indicate that the function $\lambda_{\nu,\sigma}^{(\beta)}(z)$ is a generalization of the function $\lambda_\gamma^{(n)}(z)$ introduced by Kratzel [447] and defined by

$$\lambda_\gamma^{(n)}(z) = \frac{(2\pi)^{(n-1)/2}\sqrt{n}}{\Gamma(\gamma+1-1/n)} \left(\frac{z}{n}\right)^{\gamma n} \int_1^\infty (t^n - 1)^{\gamma-1/n} e^{-zt} dt \tag{1.7.56}$$

$$\left(n \in \mathbb{N};\ \Re(\gamma) > \frac{1}{n} - 1;\ \Re(z) > 0\right).$$

When $n = 2$, this function reduces to the Macdonald function $K_{-\gamma}(z)$ as follows:

$$\lambda_\gamma^{(2)}(z) = 2\left(\frac{z}{2}\right)^\gamma K_{-\gamma}(z) \quad \left(\Re(z) > 0;\ \Re(\gamma) > -\frac{1}{2}\right). \tag{1.7.57}$$

Therefore, in accordance with (1.7.43) and (1.7.57), functions such as $Z_\rho^\nu(x)$ and $\lambda_{\nu,\sigma}^{(\beta)}(x)$ are called Bessel-type functions.

1.8 Classical Mittag-Leffler Functions

In this section we present the definitions and some properties of two classical *Mittag-Leffler functions.* More detailed information may be found in the books by Erdélyi et al. ([249], Vol. 3, Section 18.1) and Dzhrbashyan ([205], Chapter III and in [208]).

The function $E_\alpha(z)$ defined by

$$E_\alpha(z) := \sum_{k=0}^{\infty} \frac{z^k}{\Gamma(\alpha k+1)} \quad (z \in \mathbb{C};\ \Re(\alpha) > 0), \tag{1.8.1}$$

was introduced by Mittag-Leffler [604] and is, therefore, known as the Mittag-Leffler function. The basic properties of this function were studied by Mittag-Leffler ([604], [605] and [606]) and by Wiman ([890]). We present some properties of this function. It is an entire function of z with order $[\Re(\alpha)]^{-1}$ and type 1. In particular, when $\alpha = 1$ and $\alpha = 2$, we have

$$E_1(z) = e^z \text{ and } E_2(z) = \cosh(\sqrt{z}). \tag{1.8.2}$$

When $\alpha = n \in \mathbb{N}$, the following differentiation formulas hold for the function $E_n(\lambda z^n)$:

$$\left(\frac{d}{dz}\right)^n E_n(\lambda z^n) = \lambda E_n(\lambda z^n) \quad (n \in \mathbb{N};\ \lambda \in \mathbb{C}) \tag{1.8.3}$$

and

$$\left(\frac{d}{dz}\right)^n \left[z^{n-1} E_n\left(\frac{\lambda}{z^n}\right)\right] = \frac{(-1)^n \lambda}{z^{n+1}} E_n\left(\frac{\lambda}{z^n}\right) \quad (z \neq 0;\ n \in \mathbb{N};\ \lambda \in \mathbb{C}). \tag{1.8.4}$$

When $\alpha = 1/n$ ($n \in \mathbb{N}\backslash\{1\}$), the function $E_{1/n}(z)$ has the following representation:

$$E_{1/n}(z) = e^{z^n}\left[1 + n\int_0^z e^{-t^n}\left(\sum_{k=1}^{n-1} \frac{t^{k-1}}{\Gamma(k/n)}\right) dt\right] \quad (n \in \mathbb{N}\backslash\{1\}). \tag{1.8.5}$$

In particular, for $n = 2$, we have

$$E_{1/2}(z) = e^{z^2}\left[1 + \frac{2}{\sqrt{\pi}}\int_0^z e^{-t^2} dt\right], \tag{1.8.6}$$

which yields the asymptotic estimate

$$E_{1/2}(z) \sim 2e^{z^2} \quad \left(|z| \to \infty;\ |\arg(z)| < \frac{\pi}{4}\right). \tag{1.8.7}$$

The asymptotic behavior of $E_\alpha(z)$ for any α is more complicated. It is based on the integral representation of $E_\alpha(z)$ in the form

$$E_\alpha(z) = \frac{1}{2\pi}\int_C \frac{t^{\alpha-1}e^t}{t^\alpha - z} dt. \tag{1.8.8}$$

Here the path of integration $\mathcal{C}$ is a loop which starts and ends at $-\infty$ and encircles the circular disk $|t| \leqq |z|^{1/\alpha}$ in the positive sense: $|\arg(t)| \leqq \pi$ on $\mathcal{C}$. The integrand in (1.8.8) has a branch point at $t = 0$. The complex t-plane is cut along the negative real axis, and in the cut plane the integrand is single-valued: the principal branch of t^α is taken in the cut plane.

When $\alpha > 0$, $E_\alpha(z)$ has different asymptotic behavior at infinity for $0 < \alpha < 2$ and $\alpha \geqq 2$. If $0 < \alpha < 2$ and μ is a real number such that

$$\frac{\pi\alpha}{2} < \mu < \min[\pi, \pi\alpha], \tag{1.8.9}$$

then, for $N \in \mathbb{N}\backslash\{1\}$, the following asymptotic expansions are valid:

$$E_\alpha(z) = \frac{1}{\alpha} z^{(1-\beta)/\alpha} \exp\left(z^{1/\alpha}\right) - \sum_{k=1}^{N} \frac{1}{\Gamma(1-\alpha k)} \frac{1}{z^k} + O\left(\frac{1}{z^{N+1}}\right) \tag{1.8.10}$$

with $|z| \to \infty$, $|\arg(z)| \leqq \mu$; and

$$E_\alpha(z) = -\sum_{k=1}^{N} \frac{1}{\Gamma(1-\alpha k)} \frac{1}{z^k} + O\left(\frac{1}{z^{N+1}}\right) \tag{1.8.11}$$

with $|z| \to \infty$, $\mu \leqq |\arg(z)| \leqq \pi$.

When $\alpha \geqq 2$, then (for $N \in \mathbb{N}\backslash\{1\}$) the following asymptotic estimate holds:

$$E_\alpha(z) = \frac{1}{\alpha} \sum_n z^{1/\alpha} \exp\left[\exp\left(\frac{2n\pi i}{\alpha}\right) z^{1/\alpha}\right] - \sum_{k=1}^{N} \frac{1}{\Gamma(1-\alpha k)} \frac{1}{z^k} + O\left(\frac{1}{z^{N+1}}\right) \tag{1.8.12}$$

with $|z| \to \infty$, $|\arg(z)| \leqq \frac{\alpha\pi}{2}$, and where the first sum is taken over all integers n such that

$$|\arg(z) + 2\pi n| \leqq \frac{\alpha\pi}{2}. \tag{1.8.13}$$

When $\alpha > 0$, the following lemma yields the other integral representation of $E_\alpha(z)$ as a Mellin-Barnes contour integral.

Lemma 1.10 *For $\alpha > 0$ and $z \in \mathbb{C}$ ($|\arg(z)| < \pi$), the following relation holds:*

$$E_\alpha(z) = \frac{1}{2\pi i} \int_{\gamma-i\infty}^{\gamma+i\infty} \frac{\Gamma(s)\Gamma(1-s)}{\Gamma(1-\alpha s)} (-z)^{-s} ds, \tag{1.8.14}$$

where the path of integration separates all the poles at $s = -k$ ($k \in \mathbb{N}_0$) to the left and all the poles at $s = n+1$ ($n \in \mathbb{N}_0$) to the right.

Proof. This lemma is proved just as Lemma 1.5.

According to (1.4.23) and (1.4.24), from (1.8.14) we derive the Mellin transform of the Mittag-Leffler function (1.8.1):

$$\mathcal{M}[E_\alpha(-t)](s) = \frac{\Gamma(s)\Gamma(1-s)}{\Gamma(1-\alpha s)} \quad (0 < \Re(s) < 1). \tag{1.8.15}$$

The Laplace transform (1.4.1) of $E_\alpha(t)$ has a more complicated form. It is expressed in terms of the Wright function (1.11.1) with $s=2$ and $q=1$ by

$$\mathcal{L}[E_\alpha(t)](s) = \frac{1}{s}\, {}_2\Psi_1\left[\begin{matrix}(1,1),(1,1)\\ \\(\alpha,1)\end{matrix}\,\middle|\,\frac{1}{s}\right] \quad (\Re(s)>0). \tag{1.8.16}$$

The Mittag-Leffler function $E_{\alpha,\beta}(z)$, generalizing the one in (1.8.1), is defined by

$$E_{\alpha,\beta}(z) := \sum_{k=0}^{\infty} \frac{z^k}{\Gamma(\alpha k+\beta)} \quad (z,\ \beta\in\mathbb{C};\ \ \Re(\alpha)>0). \tag{1.8.17}$$

This function, sometimes called a Mittag-Leffler type function, first appeared in a paper by Wiman [890], and its basic properties were investigated (almost five decades later) by Agarwal [5], Humbert [353], Humbert and Agarwal [354], and Dhrbashyan[205]. When $\beta=1$, $E_{\alpha,\beta}(z)$ coincides with the Mittag-Leffler function (1.8.1):

$$E_{\alpha,1}(z) = E_\alpha(z) \quad (z\in\mathbb{C};\ \ \Re(\alpha)>0). \tag{1.8.18}$$

We also recall two other particular cases of (1.8.17):

$$E_{1,2}(z) = \frac{e^z-1}{z}, \quad \text{and} \quad E_{2,2}(z) = \frac{\sinh(\sqrt{z})}{\sqrt{z}}. \tag{1.8.19}$$

Like the Mittag-Leffler function $E_\alpha(z)$, $E_{\alpha,\beta}(z)$ is an entire function of z with order $[\Re(\alpha)]^{-1}$ and type 1, and satisfies the following differentiation formulas generalizing those in (1.8.3) and (1.8.4):

$$\left(\frac{d}{dz}\right)^n \left[z^{\beta-1}E_{n,\beta}(\lambda z^n)\right] = z^{\beta-n-1}E_{n,\beta-n}(\lambda z^n) \quad (n\in\mathbb{N};\ \ \lambda\in\mathbb{C}) \tag{1.8.20}$$

and

$$\left(\frac{d}{dz}\right)^n \left[z^{n-\beta}E_{n,\beta}\left(\frac{\lambda}{z^n}\right)\right] = \frac{(-1)^n\lambda}{z^{n+\beta}}E_{n,\beta}\left(\frac{\lambda}{z^n}\right) \quad (z\neq 0;\ \ n\in\mathbb{N};\ \ \lambda\in\mathbb{C}). \tag{1.8.21}$$

It may be directly proved that the usual derivatives of $E_{\alpha,\beta}(z)$ are expressed in terms of the generalized Mittag-Leffler functions (1.9.1) by

$$\left(\frac{d}{dz}\right)^n [E_{\alpha,\beta}(z)] = n!E^{n+1}_{\alpha,\beta+\alpha n}(z) \quad (n\in\mathbb{N}). \tag{1.8.22}$$

In particular, we have

$$\left(\frac{d}{dz}\right)^n [E_{\alpha}(z)] = n!E^{n+1}_{\alpha,1+\alpha n}(z) \quad (n\in\mathbb{N}). \tag{1.8.23}$$

When $\alpha=1/n$ $(n\in\mathbb{N})$, the function $E_{1/n,\beta}(z)$ has a more general representation than the one in (1.8.5):

$$E_{1/n,\beta}(z) = z^{(1-\beta)n}e^{z^n}\left[z_0^{(\beta-1)n}e^{-z_0^n}E_{1/n,\beta}(z_0)\right. \tag{1.8.24}$$

$$+ n \int_0^z e^{-t^n} \left(\sum_{k=1}^{n} \frac{t^{\beta n-k-1}}{\Gamma(\beta - [k/n])} \right) dt \Bigg] \tag{1.8.25}$$

for any $z_0 \in \mathbb{C}\backslash\{0\}$.

The function $E_{\alpha,\beta}(z)$ has the integral representation

$$E_{\alpha,\beta}(z) = \frac{1}{2\pi} \int_{\mathcal{C}} \frac{t^{\alpha-\beta} e^t}{t^\alpha - z} dt, \tag{1.8.26}$$

which generalizes the one in (1.8.8) with the same path $\mathcal{C}$.

The representation (1.8.26) can be applied to obtain the asymptotic behavior of $E_{\alpha,\beta}(z)$ at infinity, which is different for the cases $0 < \alpha < 2$ and $\alpha \geqq 2$. When $0 < \alpha < 2$ and for a real number μ satisfying (1.8.9), then (for $N \in \mathbb{N}$)

$$E_{\alpha,\beta}(z) = \frac{1}{\alpha} z^{(1-\beta)/\alpha} \exp\left(z^{1/\alpha}\right) - \sum_{k=1}^{N} \frac{1}{\Gamma(\beta - \alpha k)} \frac{1}{z^k} + O\left(\frac{1}{z^{N+1}}\right) \tag{1.8.27}$$

with $|z| \to \infty$, $|\arg(z)| \leqq \mu$, and

$$E_{\alpha,\beta}(z) = - \sum_{k=1}^{N} \frac{1}{\Gamma(\beta - \alpha k)} \frac{1}{z^k} + O\left(\frac{1}{z^{N+1}}\right) \tag{1.8.28}$$

with $|z| \to \infty$ and $\mu \leqq |\arg(z)| \leqq \pi$

If $\alpha \geqq 2$, then

$$E_{\alpha,\beta}(z) = \frac{1}{\alpha} \sum_{n} \left(z^{1/\alpha} \exp\left[\frac{2n\pi i}{\alpha}\right]\right)^{1-\beta} \exp\left[\exp\left(\frac{2n\pi i}{\alpha}\right) z^{1/\alpha}\right]$$

$$- \sum_{k=1}^{N} \frac{1}{\Gamma(\beta - \alpha k)} \frac{1}{z^k} + O\left(\frac{1}{z^{N+1}}\right) \quad \left(|z| \to \infty;\ |\arg(z)| \leqq \frac{\alpha\pi}{2}\right), \tag{1.8.29}$$

where the first sum is taken over all integers n satisfying the condition (1.8.13).

In particular, when $\alpha = 2$, we have

$$E_{2,\beta}(z) = \frac{1}{2} z^{(1-\beta)/2} \left[e^{\sqrt{z}} + e^{-\pi i(1-\beta)\mathrm{sign}(\arg z)} e^{-\sqrt{z}} \right]$$

$$- \sum_{k=1}^{N} \frac{1}{\Gamma(\beta - 2k)} \frac{1}{z^k} + O\left(\frac{1}{z^{N+1}}\right) \quad (|z| \to \infty;\ |\arg(z)| \leqq \pi), \tag{1.8.30}$$

and, for $z = x > 0$, we have

$$E_{2,\beta}(-x) = x^{(1-\beta)/2} \cos\left(\sqrt{x} + \frac{\pi(1-\beta)}{2}\right)$$

$$- \sum_{k=1}^{N} \frac{(-1)^k}{\Gamma(\beta - 2k)} \frac{1}{x^k} + O\left(\frac{1}{x^{N+1}}\right) \quad (x > 0;\ x \to \infty). \tag{1.8.31}$$

The following result, generalizing Lemma 1.10, provides the integral representation for $E_{\alpha,\beta}(z)$ $(\alpha > 0)$ in terms of a Mellin-Barnes contour integral.

Lemma 1.11 *For $\alpha > 0$ and $z \in \mathbb{C}$ ($|\arg(-z)| < \pi$) the following relation holds:*

$$E_{\alpha,\beta}(z) = \frac{1}{2\pi i}\int_{\gamma-i\infty}^{\gamma+i\infty} \frac{\Gamma(s)\Gamma(1-s)}{\Gamma(\beta-\alpha s)}(-z)^{-s}ds, \tag{1.8.32}$$

where the path of integration separates all the poles at $s = -k$ ($k \in \mathbb{N}_0$) to the left and all the poles at $s = n+1$ ($n \in \mathbb{N}_0$) to the right.

From (1.8.32) we arrive at the Mellin transform of the Mittag-Leffler function $E_{\alpha,\beta}(Z)$ in the form

$$\mathcal{M}[E_{\alpha,\beta}(-t)](s) = \frac{\Gamma(s)\Gamma(1-s)}{\Gamma(\beta-\alpha s)} \quad (0 < \Re(s) < 1). \tag{1.8.33}$$

It can also be directly shown that the Laplace transform of $E_{\alpha,\beta}(t)$ is given in terms of the Wright function (1.11.1) by

$$\mathcal{L}[E_{\alpha,\beta}(t)](s) = \frac{1}{s}\, {}_2\Psi_1\left[\begin{array}{c|c} (1,1),(1,1) & \frac{1}{s} \\ (\alpha,\beta) & \end{array}\right] \quad (\Re(s) > 0). \tag{1.8.34}$$

When $\beta = 1$, the relations (1.8.33) and (1.8.34) coincide with the ones in (1.8.15) and (1.8.16).

To conclude this section, we consider the Fourier transform (1.3.1) of the Mittag-Leffler function $E_{\alpha,\beta}(|t|)$ in (1.8.17) with $\alpha > 1$. Making a term-by-term integration of the series and using the formula (1.3.55), we have

$$\mathcal{F}\left[E_{\alpha,\beta}(|t|)\right](x) = \sum_{k=0}^{\infty} \frac{1}{\Gamma(\alpha k+\beta)}\mathcal{F}\left[|t|^{\gamma+k}\right](x)$$

$$= \frac{2\pi\delta(x)}{\Gamma(\beta)} - 2\sum_{k=0}^{\infty}\frac{(-1)^{k+1}(2k+1)!|x|^{-2k-2}}{\Gamma(2\alpha k+\alpha+\beta)}$$

$$= \frac{2\pi\delta(x)}{\Gamma(\beta)} - \frac{2}{x^2}\sum_{k=0}^{\infty}\frac{\Gamma(2k+2)\Gamma(k+1)}{\Gamma(2\alpha k+\alpha+\beta)}\left(-\frac{1}{x^2}\right)^k\frac{1}{k!}, \tag{1.8.35}$$

where $\delta(x)$ is the Dirac delta function (1.2.6). Hence, in accordance with (1.11.10), the result is expressed in terms of the generalized Wright function ${}_2\Psi_1(z)$ by

$$\mathcal{F}\left[E_{\alpha,\beta}(|t|)\right](x) = \frac{2\pi\delta(x)}{\Gamma(\beta)} - \frac{2}{x^2}\, {}_2\Psi_1\left[\begin{array}{c|c} (2,2),(1,1) & -\frac{1}{x^2} \\ (\alpha+\beta,2\alpha) & \end{array}\right] \quad (\alpha > 1;\ \beta \in \mathbb{C}). \tag{1.8.36}$$

According to Theorem 1.5 in Section 1.11, the Wright function is defined for $\alpha > 1$. In particular, when $\beta = 1$, in accordance with (1.8.18), we have

$$\mathcal{F}\left[E_{\alpha}(|t|)\right](x) = 2\pi\delta(x) - \frac{2}{x^2}\, {}_2\Psi_1\left[\begin{array}{c|c} (2,2),(1,1) & -\frac{1}{x^2} \\ (\alpha+1,2\alpha) & \end{array}\right] \quad (\alpha > 1). \tag{1.8.37}$$

Since

$$E_{\alpha,\beta}(|t|) - \frac{1}{\Gamma(\beta)} = |t|E_{\alpha,\alpha+\beta}(|t|), \tag{1.8.38}$$

if $\alpha > 1$, we find from (1.8.36) and (1.8.37) that

$$\mathcal{F}\left[|t|E_{\alpha,\alpha+\beta}(|t|)\right](x) = \frac{2}{x^2}\, {}_2\Psi_1\left[\begin{array}{c|c} (2,2),(1,1) & \\ & -\frac{1}{x^2} \\ (\alpha+\beta,2\alpha) & \end{array}\right] \quad (\beta \in \mathbb{C}), \tag{1.8.39}$$

and

$$\mathcal{F}\left[|t|E_{\alpha,\alpha+1}(|t|)\right](x) = \frac{2}{x^2}\, {}_2\Psi_1\left[\begin{array}{c|c} (2,2),(1,1) & \\ & -\frac{1}{x^2} \\ (\alpha+1,2\alpha) & \end{array}\right], \tag{1.8.40}$$

respectively.

1.9 Generalized Mittag-Leffler Functions

In this section we present definitions and properties of some functions which generalize Mittag-Leffler functions (1.8.1) and (1.8.17).

First we consider the generalized Mittag-Leffler function defined for complex $z \in \mathbb{C}$, α, β, $\rho \in \mathbb{C}$, and $\Re(\alpha) > 0$ by

$$E^{\rho}_{\alpha,\beta}(z) := \sum_{k=0}^{\infty} \frac{(\rho)_k}{\Gamma(\alpha k+\beta)} \frac{z^k}{k!} = \frac{1}{\Gamma(\rho)}\, {}_1\Psi_1\left[\begin{array}{c|c} (\rho,1) & \\ & z \\ (\beta,\alpha) & \end{array}\right], \tag{1.9.1}$$

where $(\rho)_k$ is the Pochhammer symbol (1.5.5). This function, introduced by Prabhakar [687], is an entire function of z of order $[\Re(\alpha)]^{-1}$. Its properties were investigated by Prabhakar [687] and Kilbas et al. [400].

In particular, when $\rho = 1$, it coincides with the Mittag-Leffler function (1.8.17):

$$E^{1}_{\alpha,\beta}(z) = E_{\alpha,\beta}(z) \quad (z \in \mathbb{C}). \tag{1.9.2}$$

When $\alpha = 1$, $E^{\rho}_{1,\beta}(z)$ coincides with the Kummer confluent hypergeometric function $\Phi(\rho;\beta;z)$ (see (1.6.14)), apart from a constant factor $[\Gamma(\beta)]^{-1}$:

$$E^{\rho}_{1,\beta}(z) = \frac{1}{\Gamma(\beta)}\Phi(\rho;\beta;z), \tag{1.9.3}$$

When $\alpha = m \in \mathbb{N}$ is a positive integer, $E^{\rho}_{m,\beta}(z)$ coincides with the generalized hypergeometric function (1.6.28) with $p = 1$ and $q = m$, apart from a constant multiplier factor:

$$E^{\rho}_{m,\beta}(z) = \frac{1}{\Gamma(\beta)}\, {}_1F_m\left(\rho; \frac{\beta}{m}, \frac{\beta+1}{m}, \cdots, \frac{\beta+m-1}{m}; \frac{z}{m^m}\right). \tag{1.9.4}$$

For the generalized Mittag-Leffler function (1.9.1) the following differentiation formulas hold:

$$\left(\frac{d}{dz}\right)^n \left[E^{\rho}_{\alpha,\beta}(z)\right] = (\rho)_n E^{\rho+n}_{\alpha,\beta+\alpha n}(z) \quad (n \in \mathbb{N}) \tag{1.9.5}$$

and

$$\left(\frac{d}{dz}\right)^n \left[z^{\beta-1} E^{\rho}_{\alpha,\beta}\left(\lambda z^{\alpha}\right)\right] = z^{\beta-n-1} E^{\rho}_{\alpha,\beta-n}\left(\lambda z^{\alpha}\right) \quad (\lambda \in \mathbb{C};\ \ n \in \mathbb{N}), \tag{1.9.6}$$

When $\rho = 1$, (1.9.5) coincides with (1.8.22); while for $\rho = 1$ and for $\alpha = 1$, $\rho = a$, $\beta = c$, (1.9.6) coincides with (1.8.20) and (1.6.21), respectively.

The following relation is valid for the δ-derivative (1.1.11):

$$\left(\prod_{j=1}^{n} \delta z^{\beta-j}\right) E^{\rho}_{\alpha,\beta}(z) = z^{\beta-n-1} E^{\rho}_{\alpha,\beta-n}(z) \quad \left(\delta = z\frac{d}{dz};\ \ n \in \mathbb{N}\right). \tag{1.9.7}$$

When $\rho = 1$, (1.9.7) yields the corresponding formulas for the Mittag-Leffler function (1.8.17):

$$\left(\prod_{j=1}^{n} \delta z^{\beta-j}\right) E_{\alpha,\beta}(z) = z^{\beta-n-1} E_{\alpha,\beta-n}(z); \tag{1.9.8}$$

and, in particular,

$$\left(\prod_{j=1}^{n} \delta z^{1-j}\right) E_{\alpha}(z) = z^{-n} E_{\alpha,\beta-n}(z); \tag{1.9.9}$$

where $\delta = z\frac{d}{dz}$ and $n \in \mathbb{N}$.

For $\alpha > 0$, the integral representation for $E^{\rho}_{\alpha,\beta}(z)$ can be expressed in terms of a Mellin-Barnes contour integral. Thus, just as we proved Lemma 1.5, we arrive at the following result.

Lemma 1.12 *For $\alpha > 0$ and $z \in \mathbb{C}$ $(|\arg(-z)| < \pi)$, the following relation holds:*

$$E^{\rho}_{\alpha,\beta}(z) = \frac{1}{2\pi i \Gamma(\rho)} \int_{\gamma-i\infty}^{\gamma+i\infty} \frac{\Gamma(s)\Gamma(\rho-s)}{\Gamma(\beta-\alpha s)} (-z)^{-s} ds, \tag{1.9.10}$$

where the path of integration separates all the poles at $s = -k$ $(k \in \mathbb{N}_0)$ to the left and all the poles at $s = n + \rho$ $(n \in \mathbb{N}_0)$ to the right.

In accordance with (1.4.23) and (1.4.24), (1.9.10) yields the Mellin transform of the generalized Mittag-Leffler function (1.9.1) as follows:

$$\mathcal{M}[E^{\rho}_{\alpha,\beta}(-t)](s) = \frac{1}{\Gamma(\rho)} \frac{\Gamma(s)\Gamma(\rho-s)}{\Gamma(\beta-\alpha s)} \quad (0 < \Re(s) < \Re(\rho)). \tag{1.9.11}$$

It may be directly verified that the Laplace transform (1.4.1) of $E^{\rho}_{\alpha,\beta}(t)$ is expressed in terms of the Wright function (1.11.1) with $p = 2$ and $q = 1$ as follows:

$$\mathcal{L}[E^{\rho}_{\alpha,\beta}(t)](s) = \frac{1}{s}\, {}_2\Psi_1 \left[\begin{array}{c|c} (\rho,1),(1,1) & \frac{1}{s} \\ (\alpha,\beta) & \end{array} \right] \quad (\Re(s) > 0). \tag{1.9.12}$$

When $\rho = 1$, (1.9.11) and (1.9.12) coincide with (1.8.33) and (1.8.34), respectively.

The following formula also holds for the Laplace transform of the function $t^{\beta-1}E^{\rho}_{\alpha,\beta}(\lambda t^{\alpha})$:

$$\mathcal{L}\left[t^{\beta-1}E^{\rho}_{\alpha,\beta}(\lambda t^{\alpha})\right](s) = \frac{s^{\alpha\rho-\beta}}{(s^{\alpha}-\lambda)^{\rho}} \tag{1.9.13}$$

where $\Re(s) > 0$, $\Re(\beta) > 0$, $\lambda \in \mathbb{C}$, and $|\lambda s^{-\alpha}| < 1$.

The next function generalizing the one in (1.9.1) is defined by

$$E_{\rho}((\alpha_j,\beta_j)_{1,m}; z) := \sum_{k=0}^{\infty} \frac{(\rho)_k}{\prod_{j=1}^{m}\Gamma(\alpha_j k + \beta_j)} \frac{z^k}{k!}$$

$$= \frac{1}{\Gamma(\rho)}\, {}_1\Psi_m \left[\begin{array}{c|c} (\rho,1) & z \\ (\beta_1,\alpha_1),\cdots,(\beta_m,\alpha_m) & \end{array} \right] \tag{1.9.14}$$

where $z, \rho, \beta_j \in \mathbb{C}$, $\Re(\alpha_j) > 0$, $j = 1, \cdots, n$, and $m \in \mathbb{N}$.

When $m = 1$, (1.9.14) coincides with (1.9.1), that is, $E_{\rho}((\alpha_1,\beta_1); z) = E^{\rho}_{\alpha,\beta}(z)$.

For this function the differentiation formula, generalizing the one in (1.9.6), is given by

$$\prod_{j=1}^{m}\left[\left(\frac{d}{dz}\right)^{n_j} z^{\beta_j-1}\right][E_{\rho}((\alpha_j,\beta_j)_{1,m}; z)]$$

$$= z^{\beta_1+\cdots+\beta_m-n_1-\cdots-n_m-1}E_{\rho}((\alpha_j,\beta_j-n)_{1,m}; z) \quad (n_j \in \mathbb{N};\ j = 1,\cdots,n). \tag{1.9.15}$$

The following lemma, generalizing Lemma 1.12, presents the integral representation of $E_{\rho}((\alpha,\beta)_n; z)$ with $\alpha_j > 0$ $(j = 1,\cdots,m)$ via the Mellin-Barnes contour integral.

Lemma 1.13 *If $\alpha_j > 0$ $(j = 1,\cdots,m)$, then, for $z \in \mathbb{C}$ $(|\arg(-z)| < \pi)$, the following relation holds:*

$$E_{\rho}((\alpha_j,\beta_j)_{1,m}; z) = \frac{1}{2\pi i\Gamma(\rho)} \int_{\gamma-i\infty}^{\gamma+i\infty} \frac{\Gamma(s)\Gamma(\rho-s)}{\prod_{j=1}^{m}\Gamma(\beta_j-\alpha_j s)}(-z)^{-s}ds, \tag{1.9.16}$$

where the path of integration separates all the poles at $s = -k$ $(k \in \mathbb{N}_0)$ to the left and all the poles at $s = n + \rho$ $(n \in \mathbb{N}_0)$ to the right.

From (1.9.16) we derive the Mellin transform of the generalized Mittag-Leffler function (1.9.14):

$$\mathcal{M}[E_\rho\left((\alpha_j,\beta_j)_{1,m};-t\right)](s)=\frac{1}{\Gamma(\rho)}\frac{\Gamma(s)\Gamma(\rho-s)}{\prod_{j=1}^m\Gamma(\beta_j-\alpha_j s))}\quad(0<\Re(s)<\Re(\rho)).\tag{1.9.17}$$

The Laplace transform of $E_\rho\left((\alpha_j,\beta_j)_{1,m};t\right)$ is given in terms of the Wright function (1.11.1) with $p=2$ and $q=m$ by

$$\mathcal{L}[E_\rho\left((\alpha_j,\beta_j)_{1,m};t\right)](s)=\frac{1}{s}\,{}_2\Psi_m\left[\begin{array}{c|c}(\rho,1),(1,1) & \frac{1}{s}\\ (\alpha_j,\beta_j)_{1,m} & \end{array}\right]\quad(\Re(s)>0).\tag{1.9.18}$$

When $m=1$, (1.9.18) yields (1.9.12).

The above functions (1.9.1) and (1.9.14), generalizing the classical Mittag-Leffler function $E_\alpha(z)$, extend such properties of $E_{\alpha,\beta}(z)$ as the differentiation formula of the form (1.8.20) and the integral representation (1.8.32) via Mellin-Barnes contour integrals. Now we consider the generalized Mittag-Leffler function which extends the differentiation properties (1.8.20) and (1.8.21). Such a function $E_{\alpha,m,l}(z)$, introduced by Kilbas and Saigo [392], is defined by the following series:

$$E_{\alpha,m,l}(z):=\sum_{k=0}^{\infty}c_k z^k\tag{1.9.19}$$

with

$$c_0=1\text{ and }\quad c_k=\prod_{j=0}^{k-1}\frac{\Gamma[\alpha(jm+l)+1]}{\Gamma[\alpha(jm+l+1)+1]}\quad(k\in\mathbb{N}).\tag{1.9.20}$$

Here an empty product is taken to be one, and α, $l\in\mathbb{C}$ are complex numbers, and $m\in\mathbb{R}$ such that

$$\Re(\alpha)>0,\quad m>0,\text{ and }\quad\alpha(jm+l)\notin\mathbb{Z}^-\ (j\in\mathbb{N}_0).\tag{1.9.21}$$

$E_{\alpha,m,l}(z)$ is an entire function of z of order $[\Re(\alpha)]^{-1}$ and type m [see Gorenflo et al. [294], [295]].

In particular, if $m=1$, the conditions in (1.9.21) take the form

$$\Re(\alpha)>0\text{ and }\alpha(j+l)\notin\mathbb{Z}^-\quad(j\in\mathbb{N}_0),\tag{1.9.22}$$

and (1.9.19) is reduced to the Mittag-Leffler type function given in (1.8.17), apart from a constant factor $\Gamma(\alpha l+1)$:

$$E_{\alpha,1,l}(z)=\Gamma(\alpha l+1)E_{\alpha,\alpha l+1}(z).\tag{1.9.23}$$

When $\alpha=n\in\mathbb{N}$, (1.9.18) takes the form

$$E_{n,m,l}(z)=1+\sum_{k=1}^{\infty}\left(\prod_{q=0}^{k-1}\prod_{j=1}^{n}\frac{1}{n(qm+l)+j}\right)z^k\tag{1.9.24}$$

$$(n \in \mathbb{N}; \quad m > 0; \quad n(qm+l) \notin \mathbb{Z}^-(q \in \mathbb{N}_0))$$

The last function has the following differentiation properties:

$$\frac{d^n}{dz^n}\left[z^{n(l-m+1)}E_{n,m,l}\left(\lambda z^{nm}\right)\right] = \lambda z^{nl}E_{n,m,l}\left(\lambda z^{nm}\right) \tag{1.9.25}$$

and

$$\frac{d^n}{dz^n}\left[z^{n(m-l)-1}E_{n,m,l}\left(\frac{\lambda}{z^{nm}}\right)\right] = \frac{(-1)^n\lambda}{z^{n(l+1)+1}}E_{n,m,l}\left(\frac{\lambda}{z^{nm}}\right) \quad (z \neq 0), \tag{1.9.26}$$

with $n \in \mathbb{N}$ and $\lambda \in \mathbb{C}$. When $m = 1$ and $nl + 1 = \beta$, then, in accordance with (1.9.23), the relations (1.9.25) and (1.9.26) coincide with (1.8.20) and (1.8.21), respectively.

The so-called multivariate Mittag-Leffler function $E_{(a_1,\cdots,\,a_n),b}(z_1,\cdots,\,z_n)$ of n complex variables $z_1,\cdots,\,z_n \in \mathbb{C}$ with complex parameters $a_1,\cdots,\,a_n,b \in \mathbb{C}$ is defined by

$$E_{(a_1,\cdots,\,a_n),b}(z_1,\cdots,\,z_n) = \sum_{k=0}^{\infty}\sum_{l_1,\cdots,\,l_n \geqq 0}^{l_1+\cdots+l_n=k}\begin{pmatrix} k \\ l_1,\cdots,l_n \end{pmatrix}\frac{\prod_{j=1}^n z_j^{l_j}}{\Gamma(b+\sum_{j=1}^n a_j l_j)}, \tag{1.9.27}$$

in terms of the multinomial coefficients:

$$\begin{pmatrix} k \\ l_1,\cdots,l_n \end{pmatrix} = \frac{k!}{l_1!\cdots l_n!} \quad (k,l_1,\cdots,l_n \in \mathbb{N}_0). \tag{1.9.28}$$

1.10 Functions of the Mittag-Leffler Type

In this section we present the definitions and properties of special functions defined in terms of the Mittag-Leffler function $E_{\alpha,\alpha}(z)$ and the power function $z^{\alpha-1}$ and its matrix analog. These functions will play the main role in investigating the so-called subsequential differential equations of fractional order.

First we consider a function defined for $z \in \mathbb{C}\backslash\{0\}$ and $\alpha, \lambda \in \mathbb{C}$ in terms of the Mittag-Leffler function (1.8.1) by

$$E_\alpha\left(\lambda z^\alpha\right) \quad (z \in \mathbb{C}\backslash\{0\};\ \lambda \in \mathbb{C};\ \Re(\alpha) > 0). \tag{1.10.1}$$

The following differentiation formulas hold for this function with respect to z:

$$\left(\frac{\partial}{\partial z}\right)^n\left[E_\alpha\left(\lambda z^\alpha\right)\right] = z^{-n}E_{\alpha,1-n}\left(\lambda z^\alpha\right) \tag{1.10.2}$$

and with respect to λ:

$$\left(\frac{\partial}{\partial \lambda}\right)^n\left[E_\alpha\left(\lambda z^\alpha\right)\right] = n!z^{\alpha n}E_{\alpha,\alpha n+1}^{n+1}\left(\lambda z^\alpha\right), \tag{1.10.3}$$

where $E_\alpha(\lambda z^\alpha)$ is the generalized Mittag-Leffler function (1.9.1) and $n \in \mathbb{N}$.

Putting $\rho = \beta = 1$ in (1.9.13) and taking (1.9.2) and (1.8.18) into account, we obtain the Laplace transform of the function (1.10.1):

$$\mathcal{L}\left[E_\alpha(\lambda t^\alpha)\right](s) = \frac{s^{\alpha-1}}{s^\alpha - \lambda} \quad (\Re(s) > 0;\ \lambda \in \mathbb{C};\ |\lambda s^{-\alpha}| < 1), \tag{1.10.4}$$

and differentiating (1.10.4) n times with respect to λ leads to the relation

$$\mathcal{L}\left[t^{\alpha n}\left(\frac{\partial}{\partial\lambda}\right)^n E_\alpha(\lambda t^\alpha)\right](s) = \frac{n!s^{\alpha-1}}{(s^\alpha-\lambda)^{n+1}} \quad (n \in \mathbb{N}). \tag{1.10.5}$$

Next we consider a function, more general than that in (1.10.1), defined by

$$z^{\beta-1}E_{\alpha,\beta}(\lambda z^\alpha) \quad (z \in \mathbb{C}\backslash\{0\};\ \alpha,\beta,\lambda \in \mathbb{C};\ \Re(\alpha) > 0). \tag{1.10.6}$$

The following relations, analogous to those in (1.10.2)-(1.10.5), are valid for the function in (1.10.6) ($n \in \mathbb{N}$):

$$\left(\frac{\partial}{\partial z}\right)^n \left[z^{\beta-1}E_{\alpha,\beta}(\lambda z^\alpha)\right] = z^{\beta-n-1}E_{\alpha,\beta-n}(\lambda z^\alpha), \tag{1.10.7}$$

$$\left(\frac{\partial}{\partial\lambda}\right)^n \left[z^{\beta-1}E_{\alpha,\beta}(\lambda z^\alpha)\right] = n!z^{\alpha n+\beta-1}E^{n+1}_{\alpha,\alpha n+\beta}(\lambda z^\alpha), \tag{1.10.8}$$

$$\mathcal{L}\left[t^{\beta-1}E_{\alpha,\beta}(\lambda t^\alpha)\right](s) = \frac{s^{\alpha-\beta}}{s^\alpha-\lambda} \quad (\Re(s) > 0;\ \lambda \in \mathbb{C};\ |\lambda s^{-\alpha}| < 1), \tag{1.10.9}$$

and

$$\mathcal{L}\left[t^{\alpha n+\beta-1}\left(\frac{\partial}{\partial\lambda}\right)^n E_{\alpha,\beta}(\lambda t^\alpha)\right](s) = \frac{n!s^{\alpha-\beta}}{(s^\alpha-\lambda)^{n+1}} \quad (|\lambda s^{-\alpha}| < 1). \tag{1.10.10}$$

Now we consider the special case of the function (1.10.6) when $\beta = \alpha$. This special function, which we denote by $e_\alpha^{\lambda z}$, and it is called α-Exponential functions, is defined by

$$e_\alpha^{\lambda z} := z^{\alpha-1}E_{\alpha,\alpha}(\lambda z^\alpha) \tag{1.10.11}$$

where $z \in \mathbb{C}\backslash\{0\}$, $\Re(\alpha) > 0$, and $\lambda \in \mathbb{C}$.

According to (1.8.17), $e_\alpha^{\lambda z}$ is a series of the form

$$e_\alpha^{\lambda z} = z^{\alpha-1}\sum_{k=0}^{\infty}\lambda^k\frac{z^{\alpha k}}{\Gamma[(k+1)\alpha]} \quad (\Re(\alpha) > 0). \tag{1.10.12}$$

Hence $e_\alpha^{\lambda z}$ is analytic with respect to $z \in \mathbb{C}\backslash\{0\}$ and

$$\lim_{z\to 0}\left[z^{1-\alpha}e_\alpha^{\lambda z}\right] = \frac{1}{\Gamma(\alpha)} \quad (\Re(\alpha) > 0). \tag{1.10.13}$$

In particular, we have

$$\lim_{x\to a+}\left[(x-a)^{1-\alpha}e_{\alpha}^{\lambda(x-a)}\right]=\frac{1}{\Gamma(\alpha)}\quad(\Re(\alpha)>0).\tag{1.10.14}$$

When $\alpha = 1$, then in accordance with (1.8.18) and (1.8.2), $e_1^{\lambda z}$ coincides with the exponential function $e^{\lambda z}$:

$$e_1^{\lambda z}=e^{\lambda z}\quad(z,\lambda\in\mathbb{C}),\tag{1.10.15}$$

Therefore, the function in (1.10.11) can be considered as a generalization of the exponential function, and as such we can call $e_\alpha^{\lambda z}$ the *α-exponential function.* The semigroup property for the exponential functions is well known:

$$e^{\lambda z}e^{\mu z}=e^{(\lambda+\mu)z}\quad(z,\lambda,\mu\in\mathbb{C}).\tag{1.10.16}$$

But the α-exponential function does not possess such a property:

$$e_\alpha^{\lambda z}e_\alpha^{\mu z}\neq e_\alpha^{(\lambda+\mu)z}\quad(\alpha\neq 1).\tag{1.10.17}$$

For example, if $\alpha = 2$ and $z = 1$, then in accordance with the second relation in (1.8.19), we have

$$e_2^{\lambda}=\sqrt{\lambda}\sinh(\sqrt{\lambda}),\tag{1.10.18}$$

but

$$[\sqrt{\lambda}\sinh(\sqrt{\lambda})][\sqrt{\mu}\sinh(\sqrt{\mu})]\neq\sqrt{(\lambda+\mu)}\sinh(\sqrt{(\lambda+\mu)}).\tag{1.10.19}$$

From (1.8.27)-(1.8.29) we derive the asymptotic estimates for (1.10.11) at infinity. If $0 < \alpha < 2$ and for a real number μ satisfying (1.8.9), then

$$e_\alpha^{\lambda z}=\frac{\lambda^{(1-\alpha)/\alpha}}{\alpha}\exp\left(\lambda^{1/\alpha}z\right)-\sum_{k=1}^{N-1}\frac{\lambda^{-k-1}}{\Gamma(-\alpha k)}\frac{1}{z^{\alpha k+1}}+O\left(\frac{1}{z^{\alpha N+1}}\right)\tag{1.10.20}$$

where $|z|\to\infty$, $N\in\mathbb{N}\backslash\{1\}$, and $|\arg(\lambda z^\alpha)|\leqq\mu$; and

$$e_\alpha^{\lambda z}=-\sum_{k=1}^{N-1}\frac{\lambda^{-k-1}}{\Gamma(-\alpha k)}\frac{1}{z^{\alpha k+1}}+O\left(\frac{1}{z^{\alpha N+1}}\right)\tag{1.10.21}$$

where $|z|\to\infty$, $N\in\mathbb{N}\backslash\{1\}$, and $\mu\leqq|\arg(\lambda z^\alpha)|\leqq\pi$;

If $\alpha\geqq 2$, then

$$e_\alpha^{\lambda z}=\frac{1}{\alpha}\sum_n\left(\lambda^{(1-\alpha)/\alpha}\exp\left[\frac{2\pi ni}{\alpha}\right]\right)^{1-\alpha}\exp\left[\exp\left(\frac{2n\pi i}{\alpha}\right)\lambda^{1/\alpha}z\right]$$

$$-\sum_{k=1}^{N-1}\frac{\lambda^{-k-1}}{\Gamma(-\alpha k)}\frac{1}{z^{\alpha k+1}}+O\left(\frac{1}{z^{\alpha N+1}}\right)\quad\left(|z|\to\infty;\ N\in\mathbb{N}\backslash\{1\};\ |\arg(\lambda z^\alpha)|\leqq\frac{\alpha\pi}{2}\right),\tag{1.10.22}$$

where $|z| \to \infty$, $N \in \mathbb{N}\backslash\{1\}$, $|\arg(\lambda z^\alpha)| \leqq \frac{\alpha\pi}{2}$, and the first sum is taken over all integers n satisfying the condition

$$|\arg(\lambda z^\alpha) + 2\pi n| \leqq \frac{\alpha\pi}{2}. \tag{1.10.23}$$

Relations (1.10.7)-(1.10.10) yield the corresponding formulas for the function in (1.10.11) as follows:

$$\left(\frac{d}{dz}\right)^n \left[e_\alpha^{\lambda z}\right] = z^{\alpha-n-1} E_{\alpha,\alpha-n}(\lambda z^\alpha) \quad (n \in \mathbb{N};\ \lambda \in \mathbb{C}), \tag{1.10.24}$$

$$\left(\frac{\partial}{\partial\lambda}\right)^n \left[e_\alpha^{\lambda z}\right] = n! z^{\alpha n+\alpha-1} E_{\alpha,(n+1)\alpha}^{n+1}(\lambda z^\alpha) \quad (n \in \mathbb{N}), \tag{1.10.25}$$

$$\mathcal{L}\left[e_\alpha^{\lambda t}\right](s) = \frac{1}{s^\alpha - \lambda} \quad \left(\Re(s) > 0;\ \lambda \in \mathbb{C};\ |\lambda s^{-\alpha}| < 1\right), \tag{1.10.26}$$

and

$$\mathcal{L}\left[\left(\frac{\partial}{\partial\lambda}\right)^n e_\alpha^{\lambda t}\right](s) = \frac{n!}{(s^\alpha - \lambda)^{n+1}} \quad (n \in \mathbb{N}). \tag{1.10.27}$$

Formulas (1.10.25) and (1.10.27) above allow us to generalize the function in (1.10.11) to the form

$$e_{\alpha,n}^{\lambda z} = \frac{1}{n!}\left(\frac{\partial}{\partial\lambda}\right)^n \left[e_\alpha^{\lambda z}\right] \tag{1.10.28}$$

where $z \in \mathbb{C} - \{0\}$, $\Re(\alpha) > 0$, $\lambda \in \mathbb{C}$, and $n \in \mathbb{N}_0$.

When $n = 0$, (1.10.28) coincides with function (1.10.11):

$$e_{\alpha,0}^{\lambda z} = e_\alpha^{\lambda z}. \tag{1.10.29}$$

By (1.10.25), the following relation holds:

$$e_{\alpha,n}^{\lambda z} = n! z^{\alpha-1} E_{\alpha,(n+1)\alpha}^{n+1}(\lambda z^\alpha) \quad (n \in \mathbb{N}), \tag{1.10.30}$$

and hence, in accordance with (1.9.1), $e_{\alpha,n}^{\lambda z}$ has the series representation

$$e_{\alpha,n}^{\lambda z} = z^{\alpha-1}\sum_{k=0}^{\infty} \frac{(k+n)!}{\Gamma[(k+n+1)\alpha]}\frac{(\lambda z^\alpha)^k}{k!} = z^{\alpha-1}\,{}_1\Psi_1\left[\begin{array}{c} (n+1,1) \\ \\ \left(\frac{n+1}{\alpha}, \frac{1}{\alpha}\right) \end{array}\middle|\ \lambda z^\alpha\right]. \tag{1.10.31}$$

Formulas (1.10.26) and (1.10.27) yield the Laplace transform of $t^{\alpha n} e_{\alpha,n}^{\lambda t}$:

$$\mathcal{L}\left[t^{\alpha n} e_{\alpha,n}^{\lambda z}\right](s) = \frac{n!}{(s^\alpha - \lambda)^{n+1}} \quad \left(n \in \mathbb{N}_0;\ \Re(s) > 0;\ |\lambda s^{-\alpha}| < 1\right). \tag{1.10.32}$$

If $x > 0$ and $\lambda = b + ic$ $(b, c \in \mathbb{R})$, then it directly follows that the real and imaginary parts of $e_{\alpha,n}^{\lambda x}$ are given by

$$\Re\left[e_{\alpha,n}^{\lambda x}\right] = \sum_{j=0}^{\infty} (-1)^j c^{2j} \frac{x^{2j\alpha}}{(2j)!} e_{\alpha,n+2j}^{bx} \tag{1.10.33}$$

and

$$\mathfrak{I}\left[e_{\alpha,n}^{\lambda x}\right]=\sum_{j=0}^{\infty}(-1)^{j}c^{2j+1}\frac{x^{(2j+1)\alpha}}{(2j+1)!}e_{\alpha,n+2j+1}^{bx}, \tag{1.10.34}$$

respectively.

We also need a matrix analog of the function $e_\alpha^{\lambda z}$. Let $M_n(\mathbb{R})$ $(n \in \mathbb{N})$ be a set of all matrices $\mathbf{A} = [a_{jk}]$ of order $n \times n$ with $a_{jk} \in \mathbb{R}$ $(j = 1, \cdots, n)$. By analogy with (1.10.12) for $z \in \mathbb{C}\backslash\{0\}$ $(\Re(\alpha) > 0)$, and $\mathbf{A} \in M_n(\mathbb{R})$, here we introduce a *matrix α-exponential function* defined by

$$e_\alpha^{\mathbf{A}z} := z^{\alpha-1}\sum_{k=0}^{\infty}\mathbf{A}^k\frac{z^{\alpha k}}{\Gamma[(k+1)\alpha]}. \tag{1.10.35}$$

When $\alpha = 1$, then $e_1^{\mathbf{A}z}$ coincides with the matrix exponential function $e^{\mathbf{A}z}$:

$$e_1^{\mathbf{A}z} = e^{\mathbf{A}z} \quad (z \in \mathbb{C}). \tag{1.10.36}$$

In general, the semigroup property of the form

$$e^{\mathbf{A}z}e^{\mathbf{B}z} = e^{(\mathbf{A}+\mathbf{B})z} \quad (z \in \mathbb{C};\ \mathbf{A},\mathbf{B} \in M_n(\mathbb{R})), \tag{1.10.37}$$

does not hold for the matrix exponential functions $e^{\mathbf{A}z}$ and $e^{\mathbf{B}z}$. So neither does such a property hold for the matrix α-exponential functions $e_\alpha^{\mathbf{A}z}$ and $e_\alpha^{\mathbf{B}z}$:

$$e_\alpha^{\mathbf{A}z}e_\alpha^{\mathbf{B}z} \neq e_\alpha^{(\mathbf{A}+\mathbf{B})z} \quad (z, \alpha \in \mathbb{C};\ \mathbf{A},\mathbf{B} \in M_n(\mathbb{R})). \tag{1.10.38}$$

Similarly, the inversion formula

$$(e^{\mathbf{A}z})^{-1} = e^{-\mathbf{A}z}, \tag{1.10.39}$$

valid for the matrix exponential function $e^{\mathbf{A}z}$, is not true for matrix α-exponential function $e_\alpha^{\mathbf{A}z}$:

$$(e_\alpha^{\mathbf{A}z})^{-1} \neq e_\alpha^{-\mathbf{A}z}. \tag{1.10.40}$$

We define the norm $||\mathbf{A}||$ of the matrix $\mathbf{A}$ with elements $a_{jk} \in \mathbb{R}$ $(j, k = 1, \cdots, n)$ by

$$||\mathbf{A}|| = \max_{j,k\in\mathbb{N}} |a_{jk}|. \tag{1.10.41}$$

Then, from (1.10.35), we derive the estimate for the norm of $e_\alpha^{\mathbf{A}z}$. For any fixed $z \in \mathbb{C}$, the following relation holds:

$$||e_\alpha^{\mathbf{A}z}|| \leqq \sum_{k=0}^{\infty}||\mathbf{A}||^k\frac{|z|^{\Re(\alpha)k}}{|\Gamma[(k+1)\alpha]|}. \tag{1.10.42}$$

When $z = x > 0$ and $\alpha > 0$, the above formula takes the more simple form

$$||e_\alpha^{\mathbf{A}x}|| \leqq \sum_{k=0}^{\infty}||\mathbf{A}||^k\frac{x^{\alpha k}}{\Gamma[(k+1)\alpha]}. \tag{1.10.43}$$

1.11 Wright Functions

In this section we present definitions and properties of the so-called Wright (or, more appropriately, the Fox-Wright) functions given in Erdélyi et al. ([249], Vol. 1, Section 4.1), ([249], Vol. 3, Section 18.1), Kilbas et al. [402], and in Srivastava and Karlsson [790].

The simplest Wright function $\phi(\alpha, \beta; z)$ is defined (for $z, \alpha, \beta \in \mathbb{C}$) by the series

$$\phi(\alpha,\beta;z) = {}_0\Psi_1\left[\begin{array}{c|c} \text{———} & \\ (\beta,\alpha) & z \end{array}\right] := \sum_{k=0}^{\infty} \frac{1}{\Gamma(\alpha k+\beta)} \frac{z^k}{k!}. \tag{1.11.1}$$

If $\alpha > -1$, the series in (1.11.1) is absolutely convergent for all $z \in \mathbb{C}$, while for $\alpha = -1$ this series is absolutely convergent for $|z| < 1$ and for $|z| = 1$ and $\Re(\beta) > -1$ [see Kilbas et al. ([402], Corollary 1.2)]. Moreover, for $\alpha > -1$, $\phi(\alpha, \beta; z)$ is an entire function of z. The function $\phi(\alpha, \beta; z)$ was introduced by Wright [891], who in [892] and [896] investigated its asymptotic behavior at infinity provided that $\alpha > -1$.

When $\alpha = 1$ and $\beta = \nu + 1$, the function $\phi(1, \nu + 1; \pm z^2/4)$ is expressed in terms of the Bessel functions $J_\nu(z)$ and $I_\nu(z)$, given by (1.7.1) and (1.7.16), as follows:

$$\phi\left(1, \nu+1; -\frac{z^2}{4}\right) = \left(\frac{2}{z}\right)^{\nu} J_\nu(z), \quad \phi\left(1, \nu+1; \frac{z^2}{4}\right) = \left(\frac{2}{z}\right)^{\nu} I_\nu(z). \tag{1.11.2}$$

The integral representation of (1.11.1) in terms of the Mellin-Barnes contour integral is given by

$$\phi(\alpha,\beta;z) = \frac{1}{2\pi i} \int_{\mathcal{C}} \frac{\Gamma(s)}{\Gamma(\beta-\alpha s)} (-z)^{-s} ds, \tag{1.11.3}$$

where the path of integration $\mathcal{C}$ separates all the poles at $s = -k$ ($k \in \mathbb{N}_0$) to the left. If $\mathcal{C} = (\gamma - i\infty, \gamma + i\infty)$ ($\gamma \in \mathbb{R}$), then the representation (1.11.3) is valid if either of the following conditions holds:

$$0 < \alpha < 1, \ |\arg(-z)| < \frac{(1-\alpha)\pi}{2}, \quad z \neq 0 \tag{1.11.4}$$

or

$$\alpha = 1, \ \Re(\beta) > 1 + 2\gamma, \ \arg(-z) = 0, \ z \neq 0. \tag{1.11.5}$$

This result was established by Kilbas et al. ([402], Corollary 4.1). In their paper, other conditions for the representation (1.11.3) were also given for the case when $\mathcal{C} = \mathcal{L}_{-\infty}$ ($\mathcal{L}_{\infty}$) is a loop situated in a horizontal strip starting at the point $-\infty + i\varphi_1$ ($\infty + i\varphi_1$) and terminating at the point $-\infty + i\varphi_2$ ($\infty + i\varphi_2$) with $-\infty < \varphi_1 < \varphi_2 < \infty$.

In accordance with (1.4.23) and (1.4.24), the relation (1.11.3) means that the Mellin transform of (1.11.1) is given by the formula

$$\mathcal{M}[\phi(\alpha,\beta;t)](s) = \frac{\Gamma(s)}{\Gamma(\beta-\alpha s)} \quad (\Re(s) > 0). \tag{1.11.6}$$

The Laplace transform (1.4.1) of (1.11.1) is expressed in terms of the Mittag-Leffler function (1.8.17):

$$\mathcal{L}[\phi(\alpha,\beta;t)](s) = \frac{1}{s}E_{\alpha,\beta}\left(\frac{1}{s}\right) \quad (\alpha > -1;\ \ \beta \in \mathbb{C};\ \ \Re(s) > 0); \tag{1.11.7}$$

[see Erdélyi et al. ([249], Vol. 3, Section 18.2)].

If $z \in \mathbb{C}$ and $|\arg(z)| \leqq \pi - \epsilon$ $(0 < \epsilon < \pi)$, then the asymptotic behavior of $\phi(\alpha,\beta;z)$ at infinity is given by

$$\phi(\alpha,\beta;z)$$
$$= a_0(\alpha z)^{(1-\beta)/(1+\alpha)} \exp\left[\left(1+\frac{1}{\alpha}\right)(\alpha z)^{1/(1+\alpha)}\right]\left[1+O\left(\left(\frac{1}{z}\right)^{1/(1+\alpha)}\right)\right], \tag{1.11.8}$$

where $a_0 = [2\pi(\alpha+1)]^{-1/2}$ $(z \to \infty)$.

The following differentiation formula for $\phi(\alpha,\beta;z)$ is a direct consequence of the definition (1.11.1):

$$\left(\frac{d}{dz}\right)^n \phi(\alpha,\beta;z) = \phi(\alpha,\alpha+n\beta;z) \quad (n \in \mathbb{N}). \tag{1.11.9}$$

When $\alpha = \mu$, $\beta = \nu + 1$, and z is replaced by $-z$, the function $\phi(\alpha,\beta;z)$ is denoted by $J_\nu^\mu(z)$:

$$J_\nu^\mu(z) := \psi(\mu,\nu+1;-z) = \sum_{k=0}^{\infty} \frac{1}{\Gamma(\mu k+\nu+1)}\frac{(-z)^k}{k!}, \tag{1.11.10}$$

and this function is known as the *Bessel-Wright* function, or the *Wright generalized Bessel function* [see Prudnikov et al. ([689], Vol.3), and Kiryakova([417], p. 352)]. When $\mu = 1$, the Bessel function of the first kind (1.7.1) is connected with (1.11.10) by

$$J_\nu(z) = \left(\frac{z}{2}\right)^\nu J_\nu^1\left(\frac{z^2}{4}\right). \tag{1.11.11}$$

Using this definition, one may derive from (1.11.3), (1.11.6)-(1.11.8), and (1.11.9) the representation of $J_\nu(z)$ in terms of a Mellin-Barnes contour integral, its Mellin and Laplace transforms, and the differentiation formula. We only note that if $\mu = p/q$ is a rational expression, the function $u(z) = J_\nu^{p/q}(z)$ is the solution to the differential equation of order $p+q$:

$$\left[\frac{(-z)^q}{q^q}p^{-p} - \prod_{l=1}^{q-1}\left(\frac{1}{q}\delta - \frac{l}{q}\right)\prod_{j=q}^{p+q-1}\left(\frac{1}{q}\delta - 1 + \frac{\nu-q+1-j}{p}\right)\right]u(z) = 0 \tag{1.11.12}$$

where $\delta = z\frac{d}{dz}$. This equation can be rewritten in the alternative form as follows:

$$\left[(-1)^q z^{\nu q/p} - \left(\frac{p}{q}z^{1-q/p}\frac{d}{dz}\right)^p z^{q+\nu q/p}\left(\frac{d}{dz}\right)^q\right]u(z) = 0 \tag{1.11.13}$$

[see Kiryakova ([417], formula (E.37))].

The more general function ${}_p\Psi_q(z)$ is defined for $z \in \mathbb{C}$, complex $a_l, b_j \in \mathbb{C}$, and real α_l, $\beta_j \in \mathbb{R}$ ($l = 1, \cdots, p$; $j = 1, \cdots, q$) by the series

$$ {}_p\Psi_q(z) = {}_p\Psi_q \left[\begin{array}{c|c} (a_l, \alpha_l)_{1,p} & \\ & z \\ (b_l, \beta_l)_{1,q} & \end{array} \right] := \sum_{k=0}^{\infty} \frac{\prod_{l=1}^{p} \Gamma(a_l + \alpha_l k)}{\prod_{j=1}^{q} \Gamma(b_j + \beta_j k)} \frac{z^k}{k!}, \tag{1.11.14} $$

where $z, a_l, b_j \in \mathbb{C}$, $\alpha_l, \beta_j \in \mathbb{R}$, $l = 1, \cdots, p$, and $j = 1, \cdots, q$. This general Wright (or, more appropriately, Fox-Wright) function was investigated by Fox ([263] and [264]) and Wright ([893], [894] and [895]), who presented its asymptotic expansion for large values of the argument z under the condition

$$ \sum_{j=1}^{q} \beta_j - \sum_{l=1}^{p} \alpha_l > -1. \tag{1.11.15} $$

If these conditions are satisfied, the series in (1.11.14) is convergent for any $z \in \mathbb{C}$. This result follows from the following assertion, which yields the conditions for the convergence of the series (1.11.14) [see Kilbas et al. ([402], Theorem 1)].

Theorem 1.5 *Let $a_l, b_j \in \mathbb{C}$ and α_l, $\beta_j \in \mathbb{R}$ ($l = 1, \cdots p$; $j = 1, \cdots, q$) and let*

$$ \Delta = \sum_{j=1}^{q} \beta_j - \sum_{l=1}^{p} \alpha_l, \tag{1.11.16} $$

$$ \delta = \prod_{l=1}^{p} |\alpha_l|^{-\alpha_l} \prod_{j=1}^{q} |\beta_j|^{\beta_j}, \tag{1.11.17} $$

and

$$ \mu = \sum_{j=1}^{q} b_j - \sum_{l=1}^{p} a_l + \frac{p-q}{2}. \tag{1.11.18} $$

(a) *If $\Delta > -1$, then the series in* (1.11.14) *is absolutely convergent for all $z \in \mathbb{C}$.*

(b) *If $\Delta = -1$, then the series in* (1.11.14) *is absolutely convergent for $|z| < \delta$ and for $|z| = \delta$ and $\Re(\mu) > 1/2$.*

Proof. (1.11.14) is a power series of the form

$$ {}_p\Psi_q(z) = \sum_{k=0}^{\infty} c_k z^k, \quad c_k = \frac{\prod_{l=1}^{p} \Gamma(a_l + \alpha_l k)}{\prod_{j=1}^{q} \Gamma(b_j + \beta_j k)} \frac{1}{k!} \quad (k \in \mathbb{N}_0). $$

We investigate the asymptotic behavior of $|c_k|$ as $k \to \infty$. In accordance with (1.5.12), we have

$$ |\Gamma(a_l + \alpha_l k)| \sim A_l \left(\frac{k}{e}\right)^{\alpha_l k} |\alpha_l|^{\alpha_l k} k^{\Re(a_l) - 1/2}, \quad A_l = (2\pi)^{1/2} |\alpha_l|^{\Re(a_l) - 1/2}, $$

with $k \to \infty$, and $l = 1, \cdots, p$; and

$$|\Gamma(b_j + \beta_j k)| \sim B_j \left(\frac{k}{e}\right)^{\beta_j k} |\beta_j|^{\beta_j k} k^{\Re(b_j)-1/2}, \quad B_j = (2\pi)^{1/2} |\beta_j|^{\Re(b_j)-1/2},$$

with $k \to \infty$ and $j = 1, \cdots, q$, while, by (1.5.13), we have

$$k! \sim (2\pi)^{1/2} \left(\frac{k}{e}\right)^k k^{1/2} \quad (k \to \infty).$$

Using these estimates and taking (1.11.16)-(1.11.18) into account, we obtain an estimate for $|c_k|$, as $k \to \infty$, of the form

$$|c_k| \sim A \left(\frac{k}{e}\right)^{-(\Delta+1)k} \delta^{-k} k^{-[\Re(\mu)+1/2]} \quad (k \to \infty), \tag{1.11.19}$$

where

$$A = (2\pi)^{(p-q-1)/2} \frac{\prod_{l=1}^{p} |\alpha_l|^{\Re(a_l)-1/2}}{\prod_{j=1}^{q} |\beta_j|^{\Re(b_j)-1/2}}. \tag{1.11.20}$$

Now the results in (a) and (b) follow from the known convergence principles for power series, which completes the proof of Theorem 1.5.

When $\alpha_l, \beta_j \in \mathbb{R}_+$, $(l = 1, \cdots, p;\ j = 1, \cdots, q)$, the generalized Wright function ${}_p\Psi_q(z)$ has the following integral representation as a Mellin-Barnes contour integral.

$${}_p\Psi_q\left[\begin{array}{c|c} (a_l, \alpha_l)_{1,p} & \\ & z \\ (b_l, \beta_l)_{1,q} & \end{array}\right] = \frac{1}{2\pi i} \int_{\mathcal{C}} \frac{\Gamma(s) \prod_{l=1}^{p} \Gamma(a_l - \alpha_l s)}{\prod_{j=1}^{q} \Gamma(b_j - \beta_j s)} (-z)^{-s} ds, \tag{1.11.21}$$

where the path of integration $\mathcal{C}$ separates all the poles at $s = -k$ $(k \in \mathbb{N}_0)$ to the left and all the poles at $s = (a_l + n_l)/\alpha_l$ $(l = 1, \cdots, p;\ n_l \in \mathbb{N})$ to the right. If $\mathcal{C} = (\gamma - i\infty, \gamma + i\infty)$ $(\gamma \in \mathbb{R})$, then the representation (1.11.21) is valid if either of the following conditions holds:

$$\Delta < 1, \ |\arg(-z)| < \frac{(1-\Delta)\pi}{2}, \quad z \neq 0 \tag{1.11.22}$$

or

$$\Delta = 1, \ (\Delta + 1)\gamma + \frac{1}{2} < \Re(\mu), \ \ \arg(-z) = 0, \ \ z \neq 0. \tag{1.11.23}$$

This result was established by Kilbas et al. ([402], Theorem 4). In their paper, other conditions for the representation (1.11.21) were also given for the case where $\mathcal{C} = \mathcal{L}_{-\infty}$ $(\mathcal{L}_{\infty})$ is a loop situated in a horizontal strip starting at the point $-\infty + i\varphi_1$ $(\infty + i\varphi_1)$ and terminating at the point $-\infty + i\varphi_2$ $(\infty + i\varphi_2)$ with $-\infty < \varphi_1 < \varphi_2 < \infty$.

By (1.4.23) and (1.4.24), (1.11.21) yields the Mellin transform of the generalized Wright function (1.11.14):

$$\mathcal{M}\left[{}_p\Psi_q\left[\begin{array}{c|c} (a_l, \alpha_l)_{1,p} & \\ & -t \\ (b_j, \beta_j)_{1,q} & \end{array}\right]\right](s) = \frac{\Gamma(s) \prod_{l=1}^{p} \Gamma(a_l - \alpha_l s)}{\prod_{j=1}^{q} \Gamma(b_j - \beta_j s)} \tag{1.11.24}$$

with $\alpha_l, \beta_j \in \mathbb{R}^+$; $l = 1, \cdots, p$; $j = 1, \cdots, q$; $0 < \Re(s) < \min_{1 \leqq l \leqq p} \left[\frac{\Re(a_l)}{\alpha_l}\right]$.

It can be directly verified by using (1.11.14) and (1.4.59) that the Laplace transform of the generalized Wright function (1.11.14) is given by

$$\mathcal{L}\left[{}_p\Psi_q\left[\begin{array}{c|c} (a_l,\alpha_l)_{1,p} & \\ & -t \\ (b_j,\beta_j)_{1,q} & \end{array}\right]\right](s) = \frac{1}{s}\,{}_{p+1}\Psi_q\left[\begin{array}{c|c} (1,1),(a_l,\alpha_l)_{1,p} & \\ & -\frac{1}{s} \\ (b_j,\beta_j)_{1,q} & \end{array}\right] \quad (\Re(s) > 0). \tag{1.11.25}$$

To conclude this section, we note that when $\alpha_l = \beta_j = 1$, with $l = 1, \cdots, p$; $j = 1, \cdots, q$, and $b_j \notin \mathbb{Z}_0^-$, with $j = 1, \cdots, q$; then, in accordance with (1.5.4), the generalized Wright function ${}_p\Psi_q(z)$ in (1.11.10) coincides with the generalized hypergeometric function ${}_pF_q(z)$ in (1.6.28) as follows:

$${}_p\Psi_q\left[\begin{array}{c|c} (a_l,1)_{1,p} & \\ & z \\ (b_j,1)_{1,q} & \end{array}\right] = \frac{\prod_{l=1}^{p}\Gamma(a_l)}{\prod_{j=1}^{q}\Gamma(b_j)}\,{}_pF_q[a_1,\cdots,a_p;b_1,\cdots,b_q;z] \tag{1.11.26}$$

In particular, the representation (1.11.21) yields the one in (1.6.29).

1.12 The H-Function

In this section we present the definitions and properties of the so-called H-function of Fox [264]. More detailed information may be found in the books by Mathai and Saxena ([560] Chapter 1), Srivastava et al. ([789], Chapter 1), Prudnikov et al. ([689], Vol. 3, Section 8.2), and Kilbas and Saigo [398].

For integers m, n, p, q such that $0 \leqq m \leqq q$ and $0 \leqq n \leqq p$, for $a_l, b_j \in \mathbb{C}$, and for $\alpha_l, \beta_j \in \mathbb{R}^+$ $(l = 1, \cdots, p;\ j = 1, \cdots, q)$, the H-function $H_{p,q}^{mn}(z)$ is defined via a Mellin-Barnes contour integral of the form

$$H_{p,q}^{m,n}(z) = H_{p,q}^{m,n}\left[z\left|\begin{array}{c} (a_l,\alpha_l)_{1,p} \\ \\ (b_j,\beta_j)_{1,q} \end{array}\right.\right]$$

$$= H_{p,q}^{m,n}\left[z\left|\begin{array}{c} (a_1,\alpha_1),\cdots,(a_p,\alpha_p) \\ \\ (b_1,\beta_1),\cdots,(b_q,\beta_q) \end{array}\right.\right] := \frac{1}{2\pi i}\int_{\mathcal{C}} \mathcal{H}_{p,q}^{m,n}(s)z^{-s}ds \tag{1.12.1}$$

with

$$\mathcal{H}_{p,q}^{m,n}(s) := \frac{\prod_{j=1}^{m}\Gamma(b_j+\beta_j s)\prod_{l=1}^{n}\Gamma(1-a_l-\alpha_l s)}{\prod_{l=n+1}^{r}\Gamma(a_l+\alpha_l s)\prod_{j=m+1}^{q}\Gamma(1-b_j-\beta_j s)}. \tag{1.12.2}$$

An empty product in (1.12.2), if it occurs, is taken to be one, and the poles

$$b_{jl} = -\frac{b_j + l}{\beta_j} \quad (j = 1, \cdots, m;\ l \in \mathbb{N}_0) \tag{1.12.3}$$

and the poles

$$a_{rk} = \frac{1 - a_r + k}{\alpha_r} \quad (r = 1, \cdots, n; \quad k \in \mathbb{N}_0) \tag{1.12.4}$$

do not coincide:

$$\alpha_r(b_j + l) \neq \beta_j(a_r - k - 1) \quad (r = 1, \cdots, n; \ j = 1, \cdots, m; \ k, l \in \mathbb{N}_0). \tag{1.12.5}$$

$\mathcal{C}$ in (1.12.1) is an infinite contour which separates all the poles at $s = b_{jl}$ in (1.12.3) to the left and all the poles at $s = a_{rk}$ in (1.12.4) to the right of $\mathcal{C}$, and $\mathcal{C}$ has one of the following forms:

(a) $\mathcal{C} = \mathcal{L}_{-\infty}$ is a left loop situated in a horizontal strip starting at the point $-\infty + i\varphi_1$ and terminating at the point $-\infty + i\varphi_2$ with $-\infty < \varphi_1 < \varphi_2 < \infty$;

(b) $\mathcal{C} = \mathcal{L}_{+\infty}$ is a left loop situated in a horizontal strip starting at the point $+\infty + i\varphi_1$ and terminating at the point $+\infty + i\varphi_2$ with $-\infty < \varphi_1 < \varphi_2 < \infty$;

(c)) $\mathcal{C} = (\gamma - i\infty, \gamma + i\infty)$ $(\gamma \in \mathbb{R})$ is a contour starting at the point $\gamma - i\infty$ and terminating at the point $\gamma + i\infty$.

The properties of the H-function (1.12.1) depend on the numbers Δ, δ, and μ, which are expressed via m, n, p, q, a_l, α_l $(l = 1, \cdots, p)$ and b_j, β_j $(j = 1, \cdots, q)$ by the relations (1.11.12)-(1.11.14), and on the number a^*:

$$a^* = \sum_{l=1}^{n} \alpha_l - \sum_{l=n+1}^{p} \alpha_l + \sum_{j=1}^{m} \beta_j - \sum_{j=m+1}^{q} \beta_j. \tag{1.12.6}$$

The conditions for the existence of the H-function are given by the following result [see Kilbas and Saigo ([398]; Theorem 1.1)].

Theorem 1.6 *The H-function $H^{m,n}_{p,q}(z)$ defined by* (1.12.1) *makes sense in the following cases:*

$$\mathcal{L} = \mathcal{L}_{-\infty}, \quad \Delta > 0, \quad z \neq 0; \tag{1.12.7}$$

$$\mathcal{L} = \mathcal{L}_{-\infty}, \quad \Delta = 0, \quad 0 < |z| < \delta; \tag{1.12.8}$$

$$\mathcal{L} = \mathcal{L}_{-\infty}, \quad \Delta = 0, \quad |z| = \delta, \quad \Re(\mu) < -1; \tag{1.12.9}$$

$$\mathcal{L} = \mathcal{L}_{+\infty}, \quad \Delta < 0, \quad z \neq 0; \tag{1.12.10}$$

$$\mathcal{L} = \mathcal{L}_{+\infty}, \quad \Delta = 0, \quad |z| > \delta; \tag{1.12.11}$$

$$\mathcal{L} = \mathcal{L}_{+\infty}, \quad \Delta = 0, \quad |z| = \delta, \quad \Re(\mu) < -1; \tag{1.12.12}$$

$$\mathcal{L} = (\gamma - i\infty, \gamma + i\infty), \quad a^* > 0, \quad |\arg(z)| < \frac{a^*\pi}{2}, \quad z \neq 0; \tag{1.12.13}$$

$$\mathcal{L} = (\gamma - i\infty, \gamma + i\infty), \quad a^* = 0, \quad \gamma\Delta + \Re(\mu) < -1, \quad \arg(z) = 0, \quad z \neq 0. \tag{1.12.14}$$

Let the conditions in (1.12.5) be satisfied. If all poles at $s = b_{jl}$ in (1.12.3) are simple, i.e., if

$$\beta_j(b_r + k) \neq \beta_r(b_r + l) \quad (r \neq j; \ r, j = 1, \cdots, m; \ k, l \in \mathbb{N}_0), \tag{1.12.15}$$

and either $\Delta > 0$, and $z \neq 0$ or $\Delta = 0$ and $0 < |z| < \delta$, then the H-function (1.12.1) has the following power series expansion:

$$H_{p,q}^{m,n}(z) = \sum_{j=1}^{m} \sum_{l=0}^{\infty} h_{jl}^{*} z^{(b_j+l)/\beta_j}, \tag{1.12.16}$$

where

$$h_{jl}^{*} = \frac{(-1)^l}{l!\beta_j} \frac{\prod_{k=1,\ k\neq j}^{m} \Gamma\left(b_k - [b_j+l]\frac{\beta_k}{\beta_j}\right) \prod_{k=1}^{n} \Gamma\left(1-a_k+[b_j+l]\frac{\alpha_k}{\beta_j}\right)}{\prod_{k=n+1}^{p} \Gamma\left(a_k - [b_j+l]\frac{\alpha_k}{\beta_j}\right) \prod_{k=m+1}^{q} \Gamma\left(1-b_k+[b_j+l]\frac{\beta_k}{\beta_j}\right)}. \tag{1.12.17}$$

If all poles at $s = a_{rk}$ in (1.12.4) are simple, i.e., if

$$\alpha_j(1-a_r+k) \neq \alpha_r(1-a_j+l) \quad (r \neq j;\ r, j = 1, \cdots,\ n;\ k, l \in \mathbb{N}_0), \tag{1.12.18}$$

and either $\Delta < 0$ and $z \neq 0$ or $\Delta = 0$ and $|z| > \delta$, then the H-function (1.12.1) has a power series expansion of the form

$$H_{p,q}^{m,n}(z) = \sum_{r=1}^{n} \sum_{k=0}^{\infty} h_{rk} z^{(a_r-1-k)/\alpha_r}, \tag{1.12.19}$$

where

$$h_{rk} = \frac{(-1)^k}{k!\alpha_r} \frac{\prod_{j=1}^{m} \Gamma\left(b_j + [1-a_r+k]\frac{\beta_j}{\alpha_r}\right) \prod_{j=1, j\neq r}^{n} \Gamma\left(1-a_j-[1-a_r+k]\frac{\alpha_j}{\alpha_r}\right)}{\prod_{j=n+1}^{p} \Gamma\left(a_j - [1-a_r+k]\frac{\alpha_j}{\alpha_r}\right) \prod_{j=m+1}^{q} \Gamma\left(1-b_j-[1-a_r+k]\frac{\beta_j}{\alpha_r}\right)} \tag{1.12.20}$$

Remark 1.1 When the conditions in (1.12.5) are satisfied, but some of the poles at $s = b_{jl}$ in (1.12.3) or at $s = a_{rk}$ in (1.12.4) coincide, the power series expansions (1.12.16) and (1.12.19) are replaced by expansions containing two terms, one of which has power series expansions of the form (1.12.17) and (1.12.19), while the other has power-logarithmic expansions [see Kilbas and Saigo ([398], Section 1.4)].

The relation (1.12.16) yields the power asymptotic expansion of the H-function (1.12.1) near zero. If the conditions in (1.12.5) and (1.12.15) are satisfied and either $\Delta \geqq 0$ or $\Delta < 0$ and $a^* > 0$, then the principal terms of the asymptotic estimate of $H_{p,q}^{m,n}(z)$ have the form

$$H_{p,q}^{m,n}(z) = \sum_{j=1}^{m} \left[h_j^* z^{b_j/\beta_j} + o\left(z^{b_j/\beta_j}\right)\right] \quad (z \to 0) \tag{1.12.21}$$

with the additional condition $|\arg(z)| < a^*\pi/2$ when $\Delta < 0$ and $a^* > 0$, where

$$h_j^* := h_{j0}^* = \frac{1}{\beta_j} \frac{\prod_{r=1,r\neq j}^{m} \Gamma\left(b_r - b_j\frac{\beta_r}{\beta_j}\right) \prod_{r=1}^{n} \Gamma\left(1-a_r+b_j\frac{\alpha_r}{\beta_j}\right)}{\prod_{r=n+1}^{p} \Gamma\left(a_r - b_j\frac{\alpha_r}{\beta_j}\right) \prod_{r=m+1}^{q} \Gamma\left(1-b_r+b_j\frac{\beta_r}{\beta_j}\right)}. \tag{1.12.22}$$

The main term of this asymptotic estimate is given by

$$H_{p,q}^{m,n}(z) = O\left(z^{\varrho^*}\right) \quad \left(\varrho^* := \min_{1\leqq j\leqq m}\left[\frac{\Re(b_j)}{\beta_j}\right]\right). \tag{1.12.23}$$

For the case where $\Delta < 0$ and $a^* = 0$, the H-function (1.12.1) has exponential asymptotic behavior near zero. If the conditions in (1.12.5) and (1.12.15) are satisfied, the main part of such an expansion has the form

$$H_{p,q}^{m,n}(z) = \sum_{j=1}^{m}\left[h_j^* z^{b_j/\beta_j} + o\left(z^{b_j/\beta_j}\right)\right]$$

$$+A^* z^{-(\mu+1/2)/|\Delta|}\left(c_0^* \exp\left[-(B^* + C^* z^{-1/|\Delta|}\right) i\right] - d_0^* \exp\left[\left(B^* + C^* z^{-1/|\Delta|}\right) i\right]$$

$$+o\left(z^{-(\mu+1/2)/|\Delta|}\right) \quad (z \to 0; \ |\arg(z)| \leqq \epsilon^*). \tag{1.12.24}$$

Here h_j^* $(1 \leqq j \leqq m)$ are given by (1.12.22), ϵ^* is a constant such that

$$0 < \epsilon^* < \frac{\pi}{2} \min_{1\leqq r\leqq n;\ m+1\leqq j\leqq q} [\alpha_r, \beta_j], \tag{1.12.25}$$

$$c_0^* = (-2\pi i)^{n+m-q} \exp\left[-\left(\sum_{j=m+1}^{q} b_j - \sum_{r=1}^{n} a_r\right)\pi i\right], \tag{1.12.26}$$

$$d_0^* = (2\pi i)^{n+m-q} \exp\left[\left(\sum_{j=m+1}^{q} b_j - \sum_{r=1}^{n} a_r\right)\pi i\right], \tag{1.12.27}$$

$$A^* = \frac{1}{2\pi i|\Delta|}(2\pi)^{(q-p+1)/2}|\Delta|^{-\mu}\prod_{r=1}^{p}\alpha_r^{a_r-1/2}\prod_{j=1}^{q}\beta_j^{-b_r+1/2}\left(|\Delta|^{|\Delta|}\delta\right)^{(\mu+1/2)/|\Delta|}, \tag{1.12.28}$$

$$B^* = \frac{(2\mu+1)\pi}{4}, \quad C^* = \left(|\Delta|^{|\Delta|}\delta\right)^{1/|\Delta|}. \tag{1.12.29}$$

In particular, when

$$\min_{1\leqq j\leqq m}\left[\frac{\Re(b_j}{\beta_j}\right] = -\frac{\Re(\mu)+1/2}{|\Delta|}, \tag{1.12.30}$$

the main term of the asymptotic estimate (1.12.24) has the form

$$H_{p,q}^{m,n}(z) = O\left(z^{\varrho^*}\right) \quad \left(\varrho^* = \min_{1\leqq j\leqq m}\left[\frac{\Re(b_j)}{\beta_j}, -\frac{\Re(\mu)+1/2}{|\Delta|}\right]\right) \quad (z \to 0). \tag{1.12.31}$$

Formula (1.12.19) yields the power asymptotic expansion of the H-function (1.12.1) near infinity. If the conditions in (1.12.5) and (1.12.18) are satisfied and either $\Delta \leqq 0$ or $\Delta > 0$ and $a^* > 0$, then the principal terms of the asymptotic estimate of $H^{m,n}_{p,q}(z)$ have the form

$$H^{m,n}_{p,q}(z) = \sum_{r=1}^{n} \left[h_r z^{(a_r-1)/\alpha_r} + o\left(z^{(a_r-1/\alpha_r)}\right)\right] \quad (|z| \to \infty), \tag{1.12.32}$$

with the additional condition $|\arg(z)| < a^*\pi/2$ when $\Delta > 0$ and $a^* > 0$, where

$$h_r := h_{r0} = \frac{1}{\alpha_r} \frac{\prod_{j=1}^{m} \Gamma\left(b_j + [1-a_r]\frac{\beta_j}{\alpha_r}\right) \prod_{j=1, j\neq r}^{n} \Gamma\left(1 - a_j - [1-a_r]\frac{\alpha_j}{\alpha_r}\right)}{\prod_{j=n+1}^{p} \Gamma\left(a_j - [1-a_r]\frac{\alpha_j}{\alpha_r}\right) \prod_{j=m+1}^{q} \Gamma\left(1 - b_j - [1-a_r]\frac{\beta_j}{\alpha_r}\right)}. \tag{1.12.33}$$

The main term of this asymptotic estimate is given by

$$H^{m,n}_{p,q}(z) = O\left(z^{\varrho}\right) \quad \left(\varrho^* := \min_{1\leqq r\leqq n} \left[\frac{\Re(a_r) - 1}{\alpha_r}\right]\right) \quad (|z| \to \infty). \tag{1.12.34}$$

For the case where $\Delta > 0$ and $a^* = 0$, the H-function (1.12.1) has exponential asymptotic behavior near infinity. If the conditions in (1.12.5) and (1.12.18) are satisfied, the main part of such an expansion has the form

$$H^{m,n}_{p,q}(z) = \sum_{r=1}^{n} \left[h_r z^{(a_r-1)/\alpha_r} + o\left(z^{(a_- -1)/\alpha_r}\right)\right]$$

$$+ A z^{(\mu+1/2)/\Delta} \left(c_0 \exp\left[\left(B + C z^{1/\Delta}\right) i\right] - d_0 \exp\left[-\left(B + C z^{1/\Delta}\right) i\right]\right.$$

$$+ o\left(z^{(\mu+1/2)/|\Delta|}\right) \quad (z \to \infty;\ |\arg(z)| \leqq \epsilon). \tag{1.12.35}$$

Here h_r $(1 \leqq r \leqq n)$ are given by (1.12.33), ϵ is a constant such that

$$0 < \epsilon < \frac{\pi}{2} \min_{1\leqq r\leqq n;\ m+1\leqq j\leqq q} [\alpha_r, \beta_j], \tag{1.12.36}$$

$$c_0 = (2\pi i)^{m+n-p} \exp\left[\left(\sum_{r=n+1}^{p} a_r - \sum_{j=1}^{m} b_j\right) \pi i\right], \tag{1.12.37}$$

$$d_0 = (-2\pi i)^{m+n-p} \exp\left[-\left(\sum_{r=n+1}^{p} a_r - \sum_{j=1}^{m} b_j\right) \pi i\right], \tag{1.12.38}$$

$$A = \frac{1}{2\pi i \Delta} (2\pi)^{(p-q+1)/2} \Delta^{-\mu} \prod_{r=1}^{p} \alpha_r^{-a_r+1/2} \prod_{j=1}^{q} \beta_j^{b_r-1/2} \left(\frac{\Delta^{\Delta}}{\delta}\right)^{(\mu+1/2)/\Delta}, \tag{1.12.39}$$

$$B = \frac{(2\mu+1)\pi}{4}, \quad C = \left(\frac{\Delta^{\Delta}}{\delta}\right)^{1/\Delta}. \tag{1.12.40}$$

In particular, when

$$\min_{1 \leqq r \leqq n} \left[\frac{\Re(a_r) - 1}{\alpha_r} \right] = \frac{\Re(\mu) + 1/2}{\Delta}, \tag{1.12.41}$$

the main term of the asymptotic estimate (1.12.35) has the form

$$H^{m,n}_{p,q}(z) = O\left(z^{\varrho}\right) \quad \left(\varrho := \min_{1 \leqq r \leqq n} \left[\frac{\Re(a_r) - 1}{\alpha_r}, \frac{\Re(\mu) + 1/2}{\Delta} \right] \right) \quad (|z| \to \infty). \tag{1.12.42}$$

Remark 1.2 When conditions in (1.12.5) are satisfied, but some of the poles at $s = b_{jl}$ in (1.12.3) or at $s = a_{rk}$ in (1.12.4) coincide, the logarithmic multipliers of the form $[\log(z)]^n$ ($n \in \mathbb{N}$) must be included in the asymptotic estimates (1.12.21), (1.12.23), (1.12.24), (1.12.31) and (1.12.32), (1.12.34), (1.12.35), (1.12.42) [see Kilbas and Saigo ([398], Sections 1.8, 1.9 and 1.5, 1.6)].

Now we present some simple properties of the H-function (1.12.1).

The $H^{m,n}_{p,q}(z)$ in (1.12.1) is symmetric in the set of pairs $(a_1, \alpha_1), \cdots, (a_n, \alpha_n)$, in $(a_{n+1}, \alpha_{n+1}), \cdots, (a_p, \alpha_p)$, in $(b_1, \beta_1), \cdots, (b_m, \beta_m)$, and in $(b_{m+1}, \beta_{m+1}), \cdots, (b_q, \beta_q)$.

If one of (a_l, α_l) $(l = 1, \cdots, n)$ is equal to one of (b_j, β_j) $(l = m+1, \cdots, q)$, or if one of (a_l, α_l) $(l = n+1, \cdots, p)$ is equal to one of (b_j, β_j) $(j = 1, \cdots, m)$, then the H-function reduces to a lower order H-function, that is, p, q, and n (or m) decrease by one. For example, the following two *reduction formulas* hold:

$$H^{m,n}_{p,q}\left[z \,\middle|\, \begin{matrix} (a_l, \alpha_l)_{1,p} \\ (b_j, \beta_j)_{1,q-1}, (a_1, \alpha_1) \end{matrix} \right] = H^{m,n-1}_{p-1,q-1}\left[z \,\middle|\, \begin{matrix} (a_l, \alpha_l)_{2,p} \\ (b_j, \beta_j)_{1,q-1} \end{matrix} \right] \tag{1.12.43}$$

and

$$H^{m,n}_{p,q}\left[z \,\middle|\, \begin{matrix} (a_l, \alpha_l)_{1,p-1}, (b_1, \beta_1) \\ (b_j, \beta_j)_{1,q} \end{matrix} \right] = H^{m-1,n}_{p-1,q-1}\left[z \,\middle|\, \begin{matrix} (a_l, \alpha_l)_{1,p-1} \\ (b_j, \beta_j)_{2,q} \end{matrix} \right]. \tag{1.12.44}$$

The following relations are valid:

$$z^{\sigma} H^{m,n}_{p,q}\left[z \,\middle|\, \begin{matrix} (a_l, \alpha_l)_{1,p} \\ (b_j, \beta_j)_{1,q} \end{matrix} \right] = H^{m,n}_{p,q}\left[z \,\middle|\, \begin{matrix} (a_l + \sigma\alpha_l, \alpha_l)_{1,p} \\ (b_j + \sigma\beta_j, \beta_j)_{1,q} \end{matrix} \right] \quad (z \in \mathbb{C}) \tag{1.12.45}$$

and

$$H^{m,n+1}_{p+1,q+1}\left[z \,\middle|\, \begin{matrix} (c, \alpha), (a_l, \alpha_l)_{1,p} \\ (b_j, \beta_j)_{1,q}, (c+k, \alpha) \end{matrix} \right]$$

$$= (-1)^k H^{m+1,n}_{p+1,q+1}\left[z \,\middle|\, \begin{matrix} (a_l, \alpha_l)_{1,p}, (c, \alpha) \\ (c+k, \alpha), (b_j, \beta_j)_{1,q} \end{matrix} \right] \tag{1.12.46}$$

where $c \in \mathbb{C};\ \alpha \in \mathbb{R}^+;\ k \in \mathbb{Z}$.

The following *translation formula* holds:

$$H^{m,n}_{p,q}\left[z \,\middle|\, \begin{matrix} (a_l,\alpha_l)_{1,p} \\ (b_j,\beta_j)_{1,q} \end{matrix}\right] = H^{n,m}_{q,p}\left[\frac{1}{z} \,\middle|\, \begin{matrix} (1-b_j,\beta_j)_{1,q} \\ (1-a_l,\alpha_l)_{1,p} \end{matrix}\right]. \tag{1.12.47}$$

Next we present the *differentiation formulas* for the H-function (1.12.1):

$$\left(\frac{d}{dz}\right)^k \left\{z^\omega H^{m,n}_{p,q}\left[cz^\sigma \,\middle|\, \begin{matrix} (a_l,\alpha_l)_{1,p} \\ (b_j,\beta_j)_{1,q} \end{matrix}\right]\right\}$$

$$= z^{\omega-k} H^{m,n+1}_{p+1,q+1}\left[cz^\sigma \,\middle|\, \begin{matrix} (-\omega,\sigma),(a_l,\alpha_l)_{1,p} \\ (b_j,\beta_j)_{1,q},(k-\omega,\sigma) \end{matrix}\right] \quad (\omega, c \in \mathbb{C};\ \sigma \in \mathbb{R}^+); \tag{1.12.48}$$

$$\prod_{j=1}^k \left(z\frac{d}{dz} - c_j\right)\left[cz^\sigma \,\middle|\, \begin{matrix} (a_l,\alpha_l)_{1,p} \\ (b_j,\beta_j)_{1,q} \end{matrix}\right]$$

$$= z^\omega H^{m,n+k}_{p+k,q+k}\left[cz^\sigma \,\middle|\, \begin{matrix} (c_j-\omega,\sigma)_{1,k},(a_l,\alpha_l)_{1,p} \\ (b_j,\beta_j)_{1,q},(c_j+1-\omega,\sigma)_{1,k} \end{matrix}\right] \tag{1.12.49}$$

where $\omega,\ c,\ c_j \in \mathbb{C};\ j = 1,\cdots,\ k;\ \sigma \in \mathbb{R}^+$;

$$\left(\frac{d}{dz}\right)^k H^{m,n}_{p,q}\left[(cz+d)^\sigma \,\middle|\, \begin{matrix} (a_l,\alpha_l)_{1,p} \\ (b_j,\beta_j)_{1,q} \end{matrix}\right]$$

$$= \left(\frac{c}{cz+d}\right)^k H^{m,n+1}_{p+1,q+1}\left[(cz+d)^\sigma \,\middle|\, \begin{matrix} (0,\sigma),(a_l,\alpha_l)_{1,p} \\ (b_j,\beta_j)_{1,q},(k,\sigma) \end{matrix}\right] \tag{1.12.50}$$

where $(c, d \in \mathbb{C};\ \sigma \in \mathbb{R}^+)$.

Many special functions are a particular case of the H-function (1.12.1). When $\alpha_l = \beta_j = 1$ $(l = 1,\cdots p;\ j = 1,\cdots,\ q)$, it reduces to the Meijer G-function $G^{m,n}_{p,q}(z)$:

$$H^{m,n}_{p,q}\left[z \,\middle|\, \begin{matrix} (a_l,1)_{1,p} \\ (b_j,1)_{1,q} \end{matrix}\right] = G^{m,n}_{p,q}\left[z \,\middle|\, \begin{matrix} (a_l)_{1,p} \\ (b_j)_{1,q} \end{matrix}\right] = G^{m,n}_{p,q}\left[z \,\middle|\, \begin{matrix} a_1,\cdots,\ a_p \\ b_1,\cdots,\ b_q \end{matrix}\right]$$

$$= \frac{1}{2\pi i}\int_{\mathcal{L}} \frac{\prod_{j=1}^m \Gamma(b_j+s)\prod_{l=1}^n \Gamma(1-a_l-s)}{\prod_{l=n+1}^r \Gamma(a_l+s)\prod_{j=m+1}^q \Gamma(1-b_j-s)} z^{-s}ds. \tag{1.12.51}$$

Most of the functions presented in Sections 6-9 and 11 can be expressed in terms of the H-function (1.12.1). It follows from (1.6.3) and (1.6.16) that the Gauss and Kummer hypergeometric functions are represented by

$$ {}_2F_1(a,b;c;z) = \frac{\Gamma(c)}{\Gamma(a)\Gamma(b)} H_{2,2}^{1,2}\left[-z \middle| \begin{matrix} (1-a,1), 1-b, 1 \\ (0,1),(1-c,1) \end{matrix} \right] $$

$$ = \frac{\Gamma(c)}{\Gamma(a)\Gamma(b)} G_{2,2}^{1,2}\left[-z \middle| \begin{matrix} 1-a, 1-b \\ 0, 1-c \end{matrix} \right] \tag{1.12.52} $$

and

$$ \Phi(a;c;z) = \frac{\Gamma(c)}{\Gamma(a)} H_{1,2}^{1,1}\left[-z \middle| \begin{matrix} (1-a,1) \\ (0,1),(1-c,1) \end{matrix} \right] = \frac{\Gamma(c)}{\Gamma(a)} G_{1,2}^{1,1}\left[-z \middle| \begin{matrix} 1-a \\ 0, 1-c \end{matrix} \right], \tag{1.12.53} $$

respectively.

From (1.6.29) we have the following representation for the generalized hypergeometric function:

$$ {}_pF_q(a_1,\cdots,a_p;b_1,\cdots,b_q;z) = \frac{\prod_{j=1}^{q}\Gamma(b_j)}{\prod_{l=1}^{p}\Gamma(a_l)} H_{p,q+1}^{1,p}\left[-z \middle| \begin{matrix} (1-a_l,1)_{1,p} \\ (0,1),(1-b_j,1)_{1,q} \end{matrix} \right] $$

$$ = \frac{\prod_{j=1}^{q}\Gamma(b_j)}{\prod_{l=1}^{p}\Gamma(a_l)} G_{p,q+1}^{1,p}\left[-z \middle| \begin{matrix} (1-a_l)_{1,p} \\ 0,(1-b_j)_{1,q} \end{matrix} \right]. \tag{1.12.54} $$

Formulas (1.7.6), (1.7.12), and (1.7.28) yield the following results for the Bessel function of the first kind, the modified Bessel function, and the Macdonald function:

$$ J_\nu(z) = 2^{-\nu}\sqrt{\pi} H_{1,2}^{1,0}\left[z \middle| \begin{matrix} ([1+\nu]/2, 1/2) \\ (\nu,1),(-\nu/2,1/2) \end{matrix} \right], \tag{1.12.55} $$

$$ Y_\nu(z) = 2^{-\nu}\sqrt{\pi} H_{2,3}^{1,1}\left[z \middle| \begin{matrix} (1-\nu/2,1/2),([1+\nu]/2,1/2) \\ (\nu,1),(-\nu/2,1/2),((1-\nu)/2,1/2) \end{matrix} \right] \tag{1.12.56} $$

and

$$ K_\nu(z) = 2^{-\nu-1}\sqrt{\pi} H_{1,2}^{2,0}\left[z \middle| \begin{matrix} ([1+\nu]/2, 1/2) \\ (\nu,1),(-\nu/2,1/2) \end{matrix} \right], \tag{1.12.57} $$

respectively.

Remark 1.3 There are other known representations for the Bessel function of the first kind, the modified Bessel function, and the Macdonald function given by

$$J_\nu(z) = \left(\frac{2}{z}\right)^a H^{1,0}_{0,2}\left[\frac{z^2}{4}\,\middle|\, \begin{matrix} \overline{\qquad\qquad} \\ ([a+\nu]/2,1),([a-\nu]/2,1) \end{matrix}\right] \quad (a \in \mathbb{C}) \tag{1.12.58}$$

$$Y_\nu(z) = \left(\frac{2}{z}\right)^a H^{2,0}_{1,3}\left[\frac{z^2}{4}\,\middle|\, \begin{matrix} (\frac{a-\nu-1}{2},1) \\ ([a+\nu]/2,1),([a-\nu]/2,1),([a-\nu-1]/2,1) \end{matrix}\right] \quad (a \in \mathbb{C}) \tag{1.12.59}$$

and

$$K_\nu(z) = \frac{1}{2}\left(\frac{2}{z}\right)^a H^{2,0}_{0,2}\left[\frac{z^2}{4}\,\middle|\, \begin{matrix} \overline{\qquad\qquad} \\ ([a-\nu]/2,1),([a+\nu]/2,1) \end{matrix}\right] \quad (a \in \mathbb{C}), \tag{1.12.60}$$

[see Kilbas and Saigo ([398], (2.9.18)-(2.9.20))].

The following relations for the Bessel-type functions follow from (1.7.49) and (1.7.54):

$$Z^\nu_\rho(z) = \begin{cases} \frac{1}{\rho} H^{2,0}_{0,2}\left[z\,\middle|\, \begin{matrix} \overline{\qquad\qquad} \\ (0,1),(\nu/\rho,1/\rho) \end{matrix}\right] & (\rho > 0), \\ -\frac{1}{\rho} H^{1,1}_{0,2}\left[z\,\middle|\, \begin{matrix} (1-\nu/\rho,-1/\rho) \\ (0,1) \end{matrix}\right] & (\rho < 0;\ \Re(\nu) \leqq 0) \end{cases} \tag{1.12.61}$$

and

$$\lambda^{(\beta)}_{\nu,\sigma}(x) = H^{2,0}_{1,2}\left[z\,\middle|\, \begin{matrix} (1-[\sigma+1]/\beta,1/\beta) \\ (0,1),(-\nu-\sigma/\beta,1/\beta) \end{matrix}\right]. \tag{1.12.62}$$

From (1.8.14) and (1.8.32) we obtain the following representations for the Mittag-Leffler functions (1.8.1) and (1.8.17):

$$E_\alpha(z) = H^{1,1}_{1,2}\left[-z\,\middle|\, \begin{matrix} (0,1) \\ (0,1),(0,\alpha) \end{matrix}\right] \tag{1.12.63}$$

and

$$E_{\alpha,\beta}(z) = H^{1,1}_{1,2}\left[-z\,\middle|\, \begin{matrix} (0,1) \\ (0,1),(1-\beta,\alpha) \end{matrix}\right], \tag{1.12.64}$$

while (1.9.10) and (1.9.16) yield the corresponding formulas for the generalized Mittag-Leffler functions (1.9.1) and (1.9.14):

$$E_{\alpha,\beta}^{\rho}(z) = \frac{1}{\Gamma(\rho)} H_{1,2}^{1,1}\left[-z \left| \begin{array}{c} (1-\rho,1) \\ (0,1),(1-\beta,\alpha) \end{array} \right.\right] \tag{1.12.65}$$

and

$$E_{\rho}\left((\alpha_j,\beta_j)_{1,m};z\right) = \frac{1}{\Gamma(\rho)} H_{1,m+1}^{1,1}\left[-z \left| \begin{array}{c} (1-\rho,1) \\ (0,1),(1-\beta_j,\alpha_j)_{1,m} \end{array} \right.\right]. \tag{1.12.66}$$

Finally, from (1.11.3) we derive the representation for the Wright function $\phi(\alpha,\beta;z)$ defined by (1.11.1):

$$\phi(\alpha,\beta;z) = H_{0,2}^{1,0}\left[-z \left| \begin{array}{c} \overline{} \\ (0,1),(1-\beta,\alpha) \end{array} \right.\right], \tag{1.12.67}$$

and (1.11.21) gives us the corresponding result for the generalized Wright function (1.11.14):

$${}_p\Psi_q\left[\begin{array}{c} (a_l,\alpha_l)_{1,p} \\ (b_j,\beta_j)_{1,q} \end{array} \right| z \left.\vphantom{\begin{array}{c} a \\ b \end{array}}\right] = H_{p,q+1}^{1,p}\left[-z \left| \begin{array}{c} (1-a_l,\alpha_l)_{1,p} \\ (0,1),(1-b_j,\beta_j)_{1,q} \end{array} \right.\right]. \tag{1.12.68}$$

1.13 Fixed Point Theorems

In this section we present various existence and uniqueness theorems based on classical theorems which assert the existence and uniqueness of fixed points of certain operators. We shall use known definitions and notions from functional analysis which can be found, for example, in the book by Kolmogorov and Fomin [434].

First we present *Schauder's fixed point theorem* which yields only the existence of a fixed point without its uniqueness.

Theorem 1.7 *Let (D,d) be a complete metric space, let U be a closed convex subset of D, and let $T : U \to U$ be the map such that the set $Tu : u \in U$ is relatively compact in D.*

Then the operator T has at least one fixed point $u^* \in U$:

$$Tu^* = u^*. \tag{1.13.1}$$

The necessary and sufficient conditions required for the set U to be relatively compact in the space of continuous functions $C[a,b]$ on a finite closed interval $[a,b]$ of the real axis $\mathbb{R}$ are given by the Arzela-Ascoli theorem. To formulate this theorem, we recall the definitions of equicontinuous and uniformly bounded sets. A set G is called *equicontinuous* if, for every $\epsilon > 0$, there exists some $\delta > 0$ such that, for all $g \in G$ and all $x_1, x_2 \in [a,b]$ with $|x_1 - x_2| < \delta$, we have $|g(x_1) - g(x_2)| < \epsilon$. A set G is called *uniformly bounded* if there exists a constant $M > 0$ such that $||g||_\infty \leqq M$ for every $g \in G$. The following *Arzela-Ascoli theorem* can now be recalled.

Theorem 1.8 *Let G be a subset of $C[a,b]$ equipped with the Chebyshev norm. Then G is relatively compact in $C[a,b]$ if, and only if, G is equicontinuous and uniformly bounded.*

Next we present the classical *Banach fixed point theorem* in a complete metric space.

Theorem 1.9 *Let (U,d) be a nonempty complete metric space, let $0 \leqq \omega < 1$, and let $T: U \to U$ be the map such that, for every $u, v \in U$, the relation*

$$d(Tu,Tv) \leqq \omega d(u,v) \quad (0 \leqq \omega < 1) \tag{1.13.2}$$

holds. Then the operator T has a unique fixed point $u^* \in U$.

Furthermore, if T^k $(k \in \mathbb{N})$ is the sequence of operators defined by

$$T^1 = T \quad \text{and} \quad T^k = TT^{k-1} \; (k \in \mathbb{N}\backslash\{1\}), \tag{1.13.3}$$

then, for any $u_0 \in U$, the sequence $\{T^k u_0\}_{k=1}^{\infty}$ converges to the above fixed point u^*.

We note that the map $T : U \to U$ satisfying the condition (1.13.2) is called a *contractive map* or a *contractive mapping.*

We also indicate a generalization of the above Banach fixed point theorem given by Weissinger [see Diethelm ([173], Theorem C.5)].

Theorem 1.10 *Let (U,d) be a nonempty complete metric space, and let $\omega_k \geqq 0$ for any $k \in \mathbb{N}_0$ be such that the series $\sum_{k=0}^{\infty} \omega_k$ converges. Further, let $T: U \to U$ be such a map that, for every $k \in \mathbb{N}$ and for every $u, v \in U$, the relation*

$$d(T^k u, Tk^v) \leqq \omega_k d(u,v) \quad (k \in \mathbb{N}) \tag{1.13.4}$$

holds. Then the operator T has a unique fixed point $u^* \in U$. Furthermore, for any $u_0 \in U$, the sequence $\{T^k u_0\}_{k=1}^{\infty}$ converges to this fixed point u^*.

Chapter 2

FRACTIONAL INTEGRALS AND FRACTIONAL DERIVATIVES

This chapter contains the definitions and some properties of fractional integrals and fractional derivatives of different types.

2.1 Riemann-Liouville Fractional Integrals and Fractional Derivatives

In this section we give the definitions of the Riemann-Liouville fractional integrals and fractional derivatives on a finite interval of the real line and present some of their properties in spaces of summable and continuous functions. More detailed information may be found in the book by Samko et al. ([729], Section 2).

Let $\Omega = [a, b]$ $(-\infty < a < b < \infty)$ be a finite interval on the real axis $\mathbb{R}$. The *Riemann-Liouville fractional integrals* $I_{a+}^{\alpha} f$ and $I_{b-}^{\alpha} f$ of order $\alpha \in \mathbb{C}$ $(\Re(\alpha) > 0)$ are defined by

$$(I_{a+}^{\alpha} f)(x) := \frac{1}{\Gamma(\alpha)} \int_a^x \frac{f(t)dt}{(x-t)^{1-\alpha}} \quad (x > a;\ \Re(\alpha) > 0) \tag{2.1.1}$$

and

$$(I_{b-}^{\alpha} f)(x) := \frac{1}{\Gamma(\alpha)} \int_x^b \frac{f(t)dt}{(t-x)^{1-\alpha}} \quad (x < b;\ \Re(\alpha) > 0), \tag{2.1.2}$$

respectively. Here $\Gamma(\alpha)$ is the Gamma function (1.5.1). These integrals are called the *left-sided and the right-sided fractional integrals.* When $\alpha = n \in \mathbb{N}$, the definitions (2.1.1) and (2.1.2) coincide with the nth integrals of the form

$$(I^n_{a+}f)(x) = \int_a^x dt_1 \int_a^{t_1} dt_2 \cdots \int_a^{t_{n-1}} f(t_n)dt_n$$

$$= \frac{1}{(n-1)!}\int_a^x (x-t)^{n-1}f(t)dt \quad (n \in \mathbb{N}) \tag{2.1.3}$$

and

$$(I^n_{b-}f)(x) = \int_x^b dt_1 \int_{t_1}^{b} dt_2 \cdots \int_{t_{n-1}}^{b} f(t_n)dt_n$$

$$= \frac{1}{(n-1)!}\int_x^b (t-x)^{n-1}f(t)dt \quad (n \in \mathbb{N}). \tag{2.1.4}$$

The *Riemann-Liouville fractional derivatives* $D^\alpha_{a+}y$ and $D^\alpha_{b-}y$ of order $\alpha \in \mathbb{C}$ ($\Re(\alpha) \geqq 0$) are defined by

$$(D^\alpha_{a+}y)(x) := \left(\frac{d}{dx}\right)^n (I^{n-\alpha}_{a+}y)(x)$$

$$= \frac{1}{\Gamma(n-\alpha)}\left(\frac{d}{dx}\right)^n \int_a^x \frac{y(t)dt}{(x-t)^{\alpha-n+1}} \quad (n = [\Re(\alpha)]+1;\ x > a) \tag{2.1.5}$$

and

$$(D^\alpha_{b-}y)(x) := \left(-\frac{d}{dx}\right)^n (I^{n-\alpha}_{b-}y)(x)$$

$$= \frac{1}{\Gamma(n-\alpha)}\left(-\frac{d}{dx}\right)^n \int_x^b \frac{y(t)dt}{(t-x)^{\alpha-n+1}} \quad (n = [\Re(\alpha)]+1;\ x < b), \tag{2.1.6}$$

respectively, where $[\Re(\alpha)]$ means the integral part of $\Re(\alpha)$. In particular, when $\alpha = n \in \mathbb{N}_0$, then

$$(D^0_{a+}y)(x) = (D^0_{b-}y)(x) = y(x);\ (D^n_{a+}y)(x) = y^{(n)}(x),$$

$$\text{and } (D^n_{b-}y)(x) = (-1)^n y^{(n)}(x) \quad (n \in \mathbb{N}), \tag{2.1.7}$$

where $y^{(n)}(x)$ is the usual derivative of $y(x)$ of order n. If $0 < \Re(\alpha) < 1$, then

$$(D^\alpha_{a+}y)(x) = \frac{1}{\Gamma(1-\alpha)}\frac{d}{dx}\int_a^x \frac{y(t)dt}{(x-t)^{\alpha-[\Re(\alpha)]}} \quad (0 < \Re(\alpha) < 1;\ x > a), \tag{2.1.8}$$

$$(D^\alpha_{b-}y)(x) = -\frac{1}{\Gamma(1-\alpha)}\frac{d}{dx}\int_x^b \frac{y(t)dt}{(t-x)^{\alpha-[\Re(\alpha)]}} \quad (0 < \Re(\alpha) < 1;\ x < b). \tag{2.1.9}$$

When $\alpha \in \mathbb{R}^+$, then (2.1.5) and (2.1.6) take the following forms:

$$(D^\alpha_{a+}y)(x) = \frac{1}{\Gamma(n-\alpha)}\left(\frac{d}{dx}\right)^n \int_a^x \frac{y(t)dt}{(x-t)^{\alpha-n+1}} \quad (n = [\alpha]+1;\ x > a) \tag{2.1.10}$$

and

$$(D_{b-}^{\alpha}y)(x)=\frac{1}{\Gamma(n-\alpha)}\left(-\frac{d}{dx}\right)^{n}\int_{x}^{b}\frac{y(t)dt}{(t-x)^{\alpha-n+1}}\quad(n=[\alpha]+1;\ x<b),\tag{2.1.11}$$

while (2.1.8) and (2.1.9) are given by

$$(D_{a+}^{\alpha}y)(x)=\frac{1}{\Gamma(1-\alpha)}\frac{d}{dx}\int_{a}^{x}\frac{y(t)dt}{(x-t)^{\alpha}}\quad(0<\alpha<1;\ x>a)\tag{2.1.12}$$

and

$$(D_{b-}^{\alpha}y)(x)=-\frac{1}{\Gamma(1-\alpha)}\frac{d}{dx}\frac{1}{\Gamma(1-\alpha)}\int_{x}^{b}\frac{y(t)dt}{(t-x)^{\alpha}}\quad(0<\alpha<1;\ x<b),\tag{2.1.13}$$

respectively.

If $\Re(\alpha)=0$ $(\alpha\neq 0)$, then (2.1.5) and (2.1.6) yield fractional derivatives of a purely imaginary order:

$$(D_{a+}^{i\theta}y)(x)=\frac{1}{\Gamma(1-i\theta)}\frac{d}{dx}\int_{a}^{x}\frac{y(t)dt}{(x-t)^{i\theta}}\quad(\theta\in\mathbb{R}\backslash\{0\};\ x>a),\tag{2.1.14}$$

$$(D_{b-}^{i\theta}y)(x)=-\frac{1}{\Gamma(1-i\theta)}\frac{d}{dx}\int_{x}^{b}\frac{y(t)dt}{(t-x)^{i\theta}}\quad(\theta\in\mathbb{R}\backslash\{0\};\ x<b).\tag{2.1.15}$$

It can be directly verified that the Riemann-Liouville fractional integration and fractional differentiation operators (2.1.1), (2.1.5) and (2.1.2), (2.1.6) of the power functions $(x-a)^{\beta-1}$ and $(b-x)^{\beta-1}$ yield power functions of the same form.

Property 2.1 *If $\Re(\alpha)\geqq 0$ and $\beta\in\mathbb{C}$ $(\Re(\beta)>0)$, then*

$$\left(I_{a+}^{\alpha}(t-a)^{\beta-1}\right)(x)=\frac{\Gamma(\beta)}{\Gamma(\beta+\alpha)}(x-a)^{\beta+\alpha-1}\quad(\Re(\alpha)>0),\tag{2.1.16}$$

$$\left(D_{a+}^{\alpha}(t-a)^{\beta-1}\right)(x)=\frac{\Gamma(\beta)}{\Gamma(\beta-\alpha)}(x-a)^{\beta-\alpha-1}\quad(\Re(\alpha)\geqq 0)\tag{2.1.17}$$

and

$$\left(I_{b-}^{\alpha}(b-t)^{\beta-1}\right)(x)=\frac{\Gamma(\beta)}{\Gamma(\beta+\alpha)}(b-x)^{\beta+\alpha-1}\quad(\Re(\alpha)>0),\tag{2.1.18}$$

$$\left(D_{b-}^{\alpha}(b-t)^{\beta-1}\right)(x)=\frac{\Gamma(\beta)}{\Gamma(\beta-\alpha)}(b-x)^{\beta-\alpha-1}\quad(\Re(\alpha)\geqq 0).\tag{2.1.19}$$

In particular, if $\beta=1$ and $\Re(\alpha)\geqq 0$, then the Riemann-Liouville fractional derivatives of a constant are, in general, not equal to zero:

$$\left(D_{a+}^{\alpha}1\right)(x)=\frac{(x-a)^{-\alpha}}{\Gamma(1-\alpha)},\quad\left(D_{b-}^{\alpha}1\right)(x)=\frac{(b-x)^{-\alpha}}{\Gamma(1-\alpha)}\quad(0<\Re(\alpha)<1).\tag{2.1.20}$$

On the other hand, for $j=1,2,\cdots,[\Re(\alpha)]+1$,

$$\left(D_{a+}^{\alpha}(t-a)^{\alpha-j}\right)(x)=0,\quad\left(D_{b-}^{\alpha}(b-t)^{\alpha-j}\right)(x)=0.\tag{2.1.21}$$

From (2.1.21) we derive the following result.

Corollary 2.1 *Let $\Re(\alpha) > 0$ and $n = [\Re(\alpha)] + 1$.*

(a) *The equality $(D^{\alpha}_{a+}y)(x) = 0$ is valid if, and only if,*

$$y(x) = \sum_{j=1}^{n} c_j (x-a)^{\alpha-j},$$

where $c_j \in \mathbb{R}$ $(j = 1, \cdots, n)$ are arbitrary constants.

In particular, when $0 < \Re(\alpha) \leqq 1$, the relation $(D^{\alpha}_{a+}y)(x) = 0$ holds if, and only if, $y(x) = c(x-a)^{\alpha-1}$ with any $c \in \mathbb{R}$.

(b) *The equality $(D^{\alpha}_{b-}y)(x) = 0$ is valid if, and only if,*

$$y(x) = \sum_{j=1}^{n} d_j (b-x)^{\alpha-j},$$

where $d_j \in \mathbb{R}$ $(j = 1, \cdots, n)$ are arbitrary constants.

In particular, when $0 < \Re(\alpha) \leqq 1$, the relation $(D^{\alpha}_{b-}y)(x) = 0$ holds if, and only if, $y(x) = d(b-x)^{\alpha-1}$ with any $d \in \mathbb{R}$.

We indicate some properties of the left- and right-sided fractional calculus operators I^{α}_{a+}, D^{α}_{a+} and I^{α}_{b-}, D^{α}_{b-} basically presented in Samko et al. ([729], Section 2). The first result yields the boundedness of the fractional integration operators $I^{\alpha}_{a+}f$ and $I^{\alpha}_{b-}f$ from the space $L_p(a,b)$ $(1 \leqq p \leqq \infty)$ with the norm $\|f\|_p$, defined according to (1.1.1) and (1.1.2) with $c = 1/p$, by

$$\|f\|_p := \left(\int_a^b |f(x)|^p dx \right)^{1/p} \quad (1 \leqq p < \infty), \quad \|f\|_p := \text{esssup}_{a \leqq x \leqq b} |f(x)| \ (p = \infty). \tag{2.1.22}$$

Lemma 2.1 **(a)** *The fractional integration operators I^{α}_{a+} and I^{α}_{b-} with $\Re(\alpha) > 0$ are bounded in $L_p(a,b)$ $(1 \leqq p \leqq \infty)$:*

$$\|I^{\alpha}_{a+}f\|_p \leqq K\|f\|_p, \quad \|I^{\alpha}_{b-}f\|_p \leqq K\|f|_p \quad \left(K = \frac{(b-a)^{\Re(\alpha)}}{\Re(\alpha)|\Gamma(\alpha)|} \right). \tag{2.1.23}$$

(b) *If $0 < \alpha < 1$ and $1 < p < 1/\alpha$, then the operators I^{α}_{a+} and I^{α}_{b-} are bounded from $L_p(a,b)$ into $L_q(a,b)$, where $q = p/(1-\alpha p)$.*

Remark 2.1 Lemma 2.1(a) was proved in Samko et al. ([729] Theorem 2.6), while Lemma 2.1(b) is known as the Hardy-Littlewood theorem [see Samko et al. ([729], Theorem 3.5)].

The next result characterizes the conditions for the existence of the fractional derivatives D^{α}_{a+} and D^{α}_{b-} in the space $AC^n[a,b]$ defined in (1.1.7).

Lemma 2.2 *Let $\Re(\alpha) \geqq 0$, and $n = [\Re(\alpha)] + 1$. If $y(x) \in AC^n[a,b]$, then the fractional derivatives $D^{\alpha}_{a+}y$ and $D^{\alpha}_{b-}y$ exist almost everywhere on $[a,b]$ and can be represented in the forms*

$$(D^{\alpha}_{a+}y)(x) = \sum_{k=0}^{n-1} \frac{y^{(k)}(a)}{\Gamma(1+k-\alpha)}(x-a)^{k-\alpha} + \frac{1}{\Gamma(n-\alpha)} \int_a^x \frac{y^{(n)}(t)dt}{(x-t)^{\alpha-n+1}} \tag{2.1.24}$$

and

$$(D^{\alpha}_{b-}y)(x) = \sum_{k=0}^{n-1} \frac{(-1)^k y^{(k)}(b)}{\Gamma(1+k-\alpha)}(b-x)^{k-\alpha} + \frac{(-1)^n}{\Gamma(n-\alpha)} \int_x^b \frac{y^{(n)}(t)dt}{(t-x)^{\alpha-n+1}}, \tag{2.1.25}$$

respectively.

Proof. Relation (2.1.24) was established in Samko et al. ([729], Theorem 2.2) on the basis of Definition (2.1.5) and Lemma 1.1. Formula (2.1.25) is proved similarly by using Definition (2.1.6) and the representation for the function $g(x) \in AC^n[a,b]$ of the form (1.1.8):

$$g(x) = \frac{(-1)^n}{(n-1)!} \int_x^b (t-x)^{n-1}\phi(t)dt + \sum_{k=0}^{n-1} d_k(-1)^k(b-x)^k, \tag{2.1.26}$$

where

$$\phi(t) = g^{(n)}(t) \text{ and } d_k = \frac{g^{(k)}(b)}{k!}. \tag{2.1.27}$$

Corollary 2.2 *If $0 \leqq \Re(\alpha) < 1$ $(\alpha \neq 0)$ and $y(x) \in AC[a,b]$, then*

$$(D^{\alpha}_{a+}y)(x) = \frac{1}{\Gamma(1-\alpha)} \left[\frac{y(a)}{(x-a)^{\alpha}} + \int_a^x \frac{y'(t)dt}{(x-t)^{\alpha}}\right] \tag{2.1.28}$$

and

$$(D^{\alpha}_{b-}y)(x) = \frac{1}{\Gamma(1-\alpha)} \left[\frac{y(b)}{(b-x)^{\alpha}} - \int_x^b \frac{y'(t)dt}{(t-x)^{\alpha}}\right]. \tag{2.1.29}$$

Remark 2.2 Relations (2.1.28) and (2.1.29) were proved in Samko et al. ([729], Lemmas 2.2 and 2.3).

The *semigroup property* of the fractional integration operators I^{α}_{a+} and I^{α}_{b-} are given by the following result [see Samko et al. ([729], Sections 2.3 and 2.5)].

Lemma 2.3 *If $\Re(\alpha) > 0$ and $\Re(\beta) > 0$, then the equations*

$$(I^{\alpha}_{a+}I^{\beta}_{a+}f)(x) = (I^{\alpha+\beta}_{a+}f)(x), \text{ and } (I^{\alpha}_{b-}I^{\beta}_{b-}f)(x) = (I^{\alpha+\beta}_{b-}f)(x) \tag{2.1.30}$$

are satisfied at almost every point $x \in [a,b]$ for $f(x) \in L_p(a,b)$ $(1 \leqq p \leqq \infty)$. If $\alpha + \beta > 1$, then the relations in (2.1.30) hold at any point of $[a,b]$.

The following assertion shows that the fractional differentiation is an operation inverse to the fractional integration from the left.

Lemma 2.4 *If $\Re(\alpha) > 0$ and $f(x) \in L_p(a,b)$ $(1 \leqq p \leqq \infty)$, then the following equalities*

$$(D^\alpha_{a+} I^\alpha_{a+} f)(x) = f(x) \text{ and } (D^\alpha_{b-} I^\alpha_{b-} f)(x) = f(x) \quad (\Re(\alpha) > 0) \tag{2.1.31}$$

hold almost everywhere on $[a,b]$.

Remark 2.3 The first relation in (2.1.31) for $f(x) \in L(a,b)$ was established in Samko et al. ([729], Theorem 2.4). The second one can be proved similarly.

From Lemmas 2.2-2.4 we derive the following composition relations between fractional differentiation and fractional integration operators.

Property 2.2 *If $\Re(\alpha) > \Re(\beta) > 0$, then, for $f(x) \in L_p(a,b)$ $(1 \leqq p \leqq \infty)$, the relations*

$$(D^\beta_{a+} I^\alpha_{a+} f)(x) = I^{\alpha-\beta}_{a+} f(x) \text{ and } (D^\beta_{b-} I^\alpha_{b-} f)(x) = I^{\alpha-\beta}_{b-} f(x) \tag{2.1.32}$$

hold almost everywhere on $[a,b]$.

In particular, when $\beta = k \in \mathbb{N}$ and $\Re(\alpha) > k$, then

$$(D^k I^\alpha_{a+} f)(x) = I^{\alpha-k}_{a+} f(x) \text{ and } (D^k I^\alpha_{b-} f)(x) = (-1)^k I^{\alpha-k}_{b-} f(x). \tag{2.1.33}$$

Property 2.3 *Let $\Re(\alpha) \geqq 0$, $m \in \mathbb{N}$ and $D = d/dx$.*

(a) *If the fractional derivatives $(D^\alpha_{a+} y)(x)$ and $(D^{\alpha+m}_{a+} y)(x)$ exist, then*

$$(D^m D^\alpha_{a+} y)(x) = (D^{\alpha+m}_{a+} y)(x). \tag{2.1.34}$$

(b) *If the fractional derivatives $(D^\alpha_{b-} y)(x)$ and $(D^{\alpha+m}_{b-} y)(x)$ exist, then*

$$(D^m D^\alpha_{b-} y)(x) = (-1)^m (D^{\alpha+m}_{b-} y)(x). \tag{2.1.35}$$

To present the next property, we use the spaces of functions $I^\alpha_{a+}(L_p)$ and $I^\alpha_{b-}(L_p)$ defined for $\Re(\alpha) > 0$ and $1 \leqq p \leqq \infty$ by

$$I^\alpha_{a+}(L_p) := \left\{ f : \ f = I^\alpha_{a+} \varphi, \ \ \varphi \in L_p(a,b) \right\} \tag{2.1.36}$$

and

$$I^\alpha_{b-}(L_p) := \left\{ f : \ f = I^\alpha_{b-} \phi, \ \ \phi \in L_p(a,b) \right\}, \tag{2.1.37}$$

respectively. The composition of the fractional integration operator I^α_{a+} with the fractional differentiation operator D^α_{a+} is given by the following result.

Lemma 2.5 *Let $\Re(\alpha) > 0$, $n = [\Re(\alpha)] + 1$ and let $f_{n-\alpha}(x) = (I^{n-\alpha}_{a+} f)(x)$ be the fractional integral* (2.1.1) *of order $n - \alpha$.*

(a) *If $1 \leqq p \leqq \infty$ and $f(x) \in I^\alpha_{a+}(L_p)$, then*

$$\left(I^\alpha_{a+} D^\alpha_{a+} f \right)(x) = f(x). \tag{2.1.38}$$

(b) *If* $f(x) \in L_1(a,b)$ *and* $f_{n-\alpha}(x) \in AC^n[a,b]$, *then the equality*

$$(I^\alpha_{a+} D^\alpha_{a+} f)(x) = f(x) - \sum_{j=1}^{n} \frac{f^{(n-j)}_{n-\alpha}(a)}{\Gamma(\alpha - j + 1)}(x-a)^{\alpha-j}, \tag{2.1.39}$$

holds almost everywhere on $[a,b]$.

In particular, if $0 < \Re(\alpha) < 1$, *then*

$$(I^\alpha_{a+} D^\alpha_{a+} f)(x) = f(x) - \frac{f_{1-\alpha}(a)}{\Gamma(\alpha)}(x-a)^{\alpha-1}, \tag{2.1.40}$$

where $f_{1-\alpha}(x) = (I^{1-\alpha}_{a+} f)(x)$, *while for* $\alpha = n \in \mathbb{N}$, *the following equality holds:*

$$(I^n_{a+} D^n_{a+} f)(x) = f(x) - \sum_{k=0}^{n-1} \frac{f^{(k)}(a)}{k!}(x-a)^k. \tag{2.1.41}$$

The following index rule Property is also easy to prove.

Property 2.4 *Let* $\alpha > 0$ *and* $\beta > 0$ *be such that* $n - 1 < \alpha \leq n$, $m - 1 < \beta \leq m$ $(n, m \in \mathbf{N})$ *and* $\alpha + \beta < n$, *and let* $f \in L_1(a,b)$ *and* $f_{m-\alpha} \in AC^m([a,b])$. *Then we have the following index rule:*

$$(D^\alpha_{a+} D^\beta_{a+} f)(x) = (D^{\alpha+\beta}_{a+} f)(x) - \sum_{j=1}^{m} (D^{\beta-j}_{a+} f)(a+) \frac{(x-a)^{-j-\alpha}}{\Gamma(1-j-\alpha)}. \tag{2.1.42}$$

Proof. Since $n > \alpha + \beta$, then using (2.1.5) and the semigroup property (2.1.30), we have

$$(D^\alpha_{a+} D^\beta_{a+} f)(x) = \left(\frac{d}{dx}\right)^n (I^{n-\alpha}_{a+} D^\beta_{a+} f)(x) = \left(\frac{d}{dx}\right)^n (I^{n-\alpha-\beta}_{a+} [I^\beta_{a+} D^\beta_{a+} f])(x) \tag{2.1.43}$$

Since $f \in L_1(a,b)$ and $f_{m-\alpha} \in AC^m([a,b])$, then by Lemma 2.4 (with α replaced by β) we have

$$(I^\beta_{a+} D^\beta_{a+} f)(t) = f(t) - \sum_{j=1}^{m} \frac{(I^{m-\beta}_{a+} f)^{(m-j)}(a+)}{\Gamma(\beta - j + 1)}(x-a)^{\beta-j}. \tag{2.1.44}$$

According to (2.1.5), $[(I^{m-\beta}_{a+} f)]^{(m-j)}(x) = (D^{\beta-j}_{a+} f)(x)$, and thus substituting (2.1.44) into (2.1.44) using the relation (2.1.7), we arrive at (2.1.42).

Lemma 2.5 was proved in Samko et al. ([729], Theorem 2.4). The following statement, which can be proved just as Lemma 2.5, characterizes the composition of the fractional integration operator I^α_{b-} with the fractional differentiation operator D^α_{b-}.

Lemma 2.6 *Let* $\Re(\alpha) > 0$ *and* $n = [\Re(\alpha)] + 1$. *Also let* $g_{n-\alpha}(x) = (I^{n-\alpha}_{b-} g)(x)$ *be the fractional integral* (2.1.2) *of order* $n - \alpha$.

(a) *If* $1 \leqq p \leqq \infty$ *and* $g(x) \in I^{\alpha}_{b-}(L_p)$, *then*

$$\left(I^{\alpha}_{b-} D^{\alpha}_{b-} g\right)(x) = g(x). \tag{2.1.45}$$

(b) *If* $g(x) \in L_1(a,b)$ *and* $g_{n-\alpha}(x) \in AC^n[a,b]$, *then the formula*

$$(I^{\alpha}_{b-} D^{\alpha}_{b-} g)(x) = g(x) - \sum_{j=1}^{n} \frac{(-1)^{n-j} g^{(n-j)}_{n-\alpha}(a)}{\Gamma(\alpha - j + 1)} (b-x)^{\alpha - j} \tag{2.1.46}$$

holds almost everywhere on $[a,b]$.

In particular, if $0 < \Re(\alpha) < 1$, *then*

$$(I^{\alpha}_{b-} D^{\alpha}_{b-} g)(x) = g(x) - \frac{g_{1-\alpha}(a)}{\Gamma(\alpha)} (b-x)^{\alpha-1}, \tag{2.1.47}$$

where $g_{1-\alpha}(x) = (I^{1-\alpha}_{b-} g)(x)$, *while for* $\alpha = n \in \mathbb{N}$, *the following equality holds:*

$$(I^{n}_{b-} D^{n}_{b-} g)(x) = g(x) - \sum_{k=0}^{n-1} \frac{(-1)^k g^{(k)}(b)}{k!} (b-x)^k. \tag{2.1.48}$$

Now we present the rules for *fractional integration by parts*, which were proved in Samko et al. ([729], Corollary of Theorem 3.5 and Corollary 2 of Theorem 2.4).

Lemma 2.7 *Let* $\alpha > 0$, $p \geqq 1$, $q \geqq 1$, *and* $(1/p) + (1/q) \leqq 1 + \alpha$ $(p \neq 1$ *and* $q \neq 1$ *in the case when* $(1/p) + (1/q) = 1 + \alpha.)$

(a) If $\varphi(x) \in L_p(a,b)$ *and* $\psi(x) \in L_q(a,b)$, *then*

$$\int_a^b \varphi(x) \left(I^{\alpha}_{a+} \psi\right)(x) dx = \int_a^b \psi(x) \left(I^{\alpha}_{b-} \varphi\right)(x) dx. \tag{2.1.49}$$

(b) If $f(x) \in I^{\alpha}_{-}(L_p)$ *and* $g(x) \in I^{\alpha}_{a+}(L_q)$, *then*

$$\int_a^b f(x) \left(D^{\alpha}_{a+} g\right)(x) dx = \int_a^b g(x) \left(I^{\alpha}_{b-} f\right)(x) dx. \tag{2.1.50}$$

Now we consider the properties of (2.1.1) and (2.1.2) and fractional derivatives (2.1.5) and (2.1.6) in the spaces $C_{\gamma}[a,b]$ and $C^n_{\gamma}[a,b]$ defined in (1.1.21) and (1.1.22), respectively. The existence of the fractional integrals $I^{\alpha}_{a+} f$ and $I^{\alpha}_{b-} f$ in the space $C_{\gamma}[a,b]$ and the fractional derivatives $D^{\alpha}_{a+} y$ and $D^{\alpha}_{b-} y$ in the space $C^n_{\gamma}[a,b]$ are given by the following lemma.

Lemma 2.8 *Let* $\Re(\alpha) \geqq 0$ *and* $\gamma \in \mathbb{C}$.

(a) *Let* $\Re(\alpha) > 0$ *and* $0 \leqq \Re(\gamma) < 1$.

If $\Re(\gamma) > \Re(\alpha)$, *then the fractional integration operators* I^{α}_{a+} *and* I^{α}_{b-} *are bounded from* $C_{\gamma}[a,b]$ *into* $C_{\gamma-\alpha}[a,b]$:

$$\|I^{\alpha}_{a+} f\|_{C_{\gamma-\alpha}} \leqq k_1 \|f\|_{C_{\gamma}} \text{ and } \|I^{\alpha}_{b-} f\|_{C_{\gamma-\alpha}} \leqq k_1 \|f|_{C_{\gamma}}, \tag{2.1.51}$$

$$k_1 = \frac{\Gamma[\Re(\alpha)]|\Gamma(1-\Re(\gamma))|}{|\Gamma(\alpha)|\Gamma[1+\Re(\alpha-\gamma)]}.$$

In particular I^α_{a+} *and* I^α_{b-} *are bounded in* $C_\gamma[a,b]$.

If $\Re(\gamma) \leqq \Re(\alpha)$, *then the fractional integration operators* I^α_{a+} *and* I^α_{b-} *are bounded from* $C_\gamma[a,b]$ *into* $C[a,b]$:

$$||I^\alpha_{a+}f||_C \leqq k_2||f||_{C_\gamma}, \quad and \quad ||I^\alpha_{b-}f||_C \leqq k_2||f|_{C_\gamma}, \tag{2.1.52}$$

$$k_2 = (b-a)^{\Re(\alpha-\gamma)}\frac{\Gamma[\Re(\alpha)]|\Gamma(1-\Re(\gamma))|}{|\Gamma(\alpha)|\Gamma[1+\Re(\alpha-\gamma)]}.$$

In particular, I^α_{a+} *and* I^α_{b-} *are bounded in* $C_\gamma[a,b]$.

(b) *If* $\Re(\alpha) \geqq 0$, $n = [\Re(\alpha)] + 1$ *and* $y(x) \in C^n_\gamma[a,b]$, *then the fractional derivatives* $D^\alpha_{a+}y$ *and* $D^\alpha_{b-}y$ *exist on* $(a,b]$ *and can be represented by* (2.1.24) *and* (2.1.25), *respectively. In particular,* $D^\alpha_{a+}y$ *and* $D^\alpha_{b-}y$ *are given by* (2.1.28) *and* (2.1.29), *respectively, when* $0 \leqq \Re(\alpha) < 1$ $(\alpha \neq 0)$ *and* $y(x) \in C_\gamma[a,b]$.

The following assertion for fractional calculus operators (2.1.1), and (2.1.2), and (2.1.5) and (2.1.6), is analogous to those in Lemmas 2.3-2.6 and Property 2.2.

Lemma 2.9 *Let* $\Re(\alpha) > 0$, $\Re(\beta) > 0$ *and* $0 \leqq \Re(\gamma) < 1$. *The following assertions are then true:*

(a) *If* $f(x) \in C_\gamma[a,b]$, *then the first and second relations in* (2.1.30) *hold at any point* $x \in (a,b]$ *and* $x \in [a,b)$, *respectively. When* $f(x) \in C[a,b]$, *these relations are valid at any point* $x \in [a,b]$.

(b) *If* $f(x) \in C_\gamma[a,b]$, *then the first and second equalities in* (2.1.31) *hold at any point* $x \in (a,b]$ *and* $x \in [a,b)$, *respectively. When* $f(x) \in C[a,b]$, *then these equalities are valid at any point* $x \in [a,b]$.

(c) *Let* $\Re(\alpha) > \Re(\beta) > 0$. *If* $f(x) \in C_\gamma[a,b]$, *then the first and second relations in* (2.1.32) *hold at any point* $x \in (a,b]$ *and* $x \in [a,b)$, *respectively. When* $f(x) \in C[a,b]$, *then these relations are valid at any point* $x \in [a,b]$. *In particular, when* $\beta = k \in \mathbb{N}$ *and* $\Re(\alpha) > k$, *the relations in* (2.1.33) *are valid in their respective cases.*

(d) *Let* $n = [\Re(\alpha)] + 1$. *Also let* $f_{n-\alpha}(x) = (I^{n-\alpha}_{a+}f)(x)$ *the fractional integral* (2.1.1) *and* $g_{n-\alpha}(x) = (I^{n-\alpha}_{b-}g)(x)$ *the fractional integral* (2.1.2) *be, of order* $n-\alpha$.

If $f(x) \in C_\gamma[a,b]$ *and* $f_{n-\alpha}(x) \in C^n_\gamma[a,b]$, *then the relation* (2.1.39) *holds at any point* $x \in (a,b]$. *In particular, when* $0 < \Re(\alpha) < 1$ *and* $f_{1-\alpha}(x) \in C^1_\gamma[a,b]$, *the equality* (2.1.40) *is valid.*

If $g(x) \in C_\gamma[a,b]$ *and* $g_{n-\alpha}(x) \in C^n_\gamma[a,b]$, *then the equality* (2.1.46) *holds at any point* $x \in [a,b)$. *In particular, when* $0 < \Re(\alpha) < 1$ *and* $f_{1-\alpha}(x) \in C^1_\gamma[a,b]$, *the equality* (2.1.47) *is valid.*

If $f(x) \in C[a,b]$ *and* $f_{n-\alpha}(x) \in C^n[a,b]$, *then* (2.1.39) *and* (2.1.46) *hold at any point* $x \in [a,b]$. *In particular, if* $f(x) \in C^n[a,b]$, *the relations* (2.1.41) *and* (2.1.48) *are valid at any point* $x \in [a,b]$.

Remark 2.4 The assertions of Lemmas 2.8 and 2.9 for the left-sided fractional calculus operators $I_{a+}^{\alpha} f$ and $D_{a+}^{\alpha} y$ were given in Kilbas et al. [375]. The results of Lemmas 2.8 and 2.9 for the corresponding right-sided fractional calculus operators $I_{b-}^{\alpha} f$ and $D_{b-}^{\alpha} y$ are proved similarly.

To conclude this section, we give the formula which shows that the Riemann-Liouville fractional integral (2.1.1) of the Mittag-Leffler function (1.8.17) with special parameters also yields a function of the same kind.

$$\left(I_{a+}^{\alpha}(t-a)^{\beta-1}E_{\mu,\beta}\left[\lambda(t-a)^{\mu}\right]\right)(x) = (x-a)^{\alpha+\beta-1}E_{\mu,\alpha+\beta}\left[\lambda(x-a)^{\mu}\right] \quad (2.1.53)$$

with $\lambda \in \mathbb{C}$, $\Re(\alpha) > 0$, $\Re(\beta) > 0$, *and* $\Re(\mu) \geqq 0$, [see, for example, ([729], Table 9.1, formula 23)]. A similar relation for the Riemann-Liouville fractional derivative (2.1.5) directly follows:

$$\left(D_{a+}^{\alpha}(t-a)^{\beta-1}E_{\mu,\beta}\left[\lambda(t-a)^{\mu}\right]\right)(x) = (x-a)^{\beta-\alpha-1}E_{\mu,\beta-\alpha}\left[\lambda(x-a)^{\mu}\right] \quad (2.1.54)$$

with $\lambda \in \mathbb{C}$, $\Re(\alpha) > 0$, $\Re(\beta) > 0$, *and* $\Re(\mu) \geqq 0$.

In particular, when $\beta = \mu = \alpha$, we can apply the well-known limit formula

$$\lim_{z\to 0} \frac{1}{\Gamma(z)} = 0 \quad (2.1.55)$$

to derive from (2.1.54) the following relation for the function $e_{\alpha}^{\lambda(x-a)}$ defined in (1.10.12):

$$\left(D_{0+}^{\alpha} e_{\alpha}^{\lambda(t-a)}\right)(x) = \lambda e_{\alpha}^{\lambda(x-a)} \quad (\Re(\alpha) > 0;\ \ \lambda \in \mathbb{C}). \quad (2.1.56)$$

When $\beta = 1$ and $\mu = \alpha$, then (2.1.54), in accordance with (1.8.17) and (1.8.18), yields the following formula for the Mittag-Leffler function (1.8.1):

$$\left(D_{a+}^{\alpha} E_{\alpha}\left[\lambda(t-a)^{\alpha}\right]\right)(x) = \lambda E_{\alpha}\left[\lambda(x-a)^{\alpha}\right] \quad (\Re(\alpha) > 0;\ \ \lambda \in \mathbb{C}). \quad (2.1.57)$$

Finally, we give an analog of the well-known property from analysis

$$\frac{d}{dx}\int_a^x K(x,t)dt = \int_a^x \frac{\partial}{\partial x}K(x,t)dt + \lim_{t\to x-0} K(x,t), \quad (2.1.58)$$

which is valid provided that $K(x,t)$ is continuous on $[a,b]\times[a,b]$ and that $K(x,t)$ has a continuous partial derivative $(\partial/\partial x)K(x,t)$ with respect to $x \in [a,b]$ for any fixed $t \in [a,b]$, for the Riemann-Liouville fractional derivative $(D_{a+}^{\alpha}y)(x)$ of order α $(0 < \alpha < 1)$ in the case when $K(x,t) = k(x-t)f(t)$. For this we need the so-called *partial Riemann-Liouville fractional derivative of order* α $(0 < \alpha < 1)$ with respect to x of a function $y(x,t)$ of two variables $(x,t) \in [a,b]\times[a,b]$ defined by

$$(D_{a+,x}^{\alpha}y)(x,t) = \frac{1}{\Gamma(1-\alpha)}\frac{\partial}{\partial x}\int_a^x \frac{y(u,t)du}{(x-u)^{\alpha}}, \quad (2.1.59)$$

with $0 < \alpha < 1$; $x > a$; $t \in [a,b]$, (see Section 2.9 in this regard). The following notation is also suitable for the Riemann-Liouville fractional derivative $(D_{a+}^{\alpha}y)(x)$

in (2.1.8) and for the Riemann-Liouville fractional integral $(I_{a+}^{1-\alpha}f)(x)$ defined by (2.1.1):

$$(D_{a+}^{\alpha}y)(x) = D_{a+}^{\alpha}[y(t)](x) \textit{ and } (I_{a+}^{1-\alpha}f)(x) = I_{a+}^{1-\alpha}[f(t)](x) \quad (0 < \alpha < 1). \tag{2.1.60}$$

The following lemma can be proved fairly easily.

Lemma 2.10 *Let $a \in \mathbb{R}$ and $0 < \alpha < 1$. Also let the function $f(x)$ and $k(x)$ be defined on $[a, b]$ such that*

$$f(x) \in C[a, b] \textit{ and } L(x) = \int_0^x \tau^{-\alpha}k(x-\tau)d\tau \in C^1[a, b]. \tag{2.1.61}$$

Then, for any $x \in [a, b]$,

$$D_{a+}^{\alpha}\left[\int_a^t k(t-u)f(u)du\right](x)$$

$$= \int_a^x D_{a+}^{\alpha}[k(t-a)](u)f(x+a-u)du + f(x)\lim_{x\to a+} I_{a+}^{1-\alpha}[k(t-a)](x). \tag{2.1.62}$$

Remark 2.5 Lemma 2.10 is stated here for $0 < \alpha < 1$, provided that the conditions of (2.1.61) are satisfied. In fact, this lemma also holds for $\alpha \in \mathbb{C}$ $(0 < \Re(\alpha) < 1)$ under conditions for functions $f(x)$ and $k(x)$ different from those in (2.1.61). For example, that $f(x)$ can be Lebesgue measurable on $[a, b]$ and that $k(x)$ can have a measurable derivative $k'(x)$ almost everywhere on $[a, b]$.

Remark 2.6 For $a = 0$, the relation of the form (2.1.62) was formally proposed by Podlubny ([682], 2.213), together with its generalization of the form (2.1.58) in which the derivative d/dx is replaced by the Riemann-Liouville fractional derivative D_{0+}^{α}.

2.2 Liouville Fractional Integrals and Fractional Derivatives on the Half-Axis

In this section we present the definitions and some properties of the Liouville fractional integrals and fractional derivatives on the half-axis $\mathbb{R}^+$. More detailed information may be found in the book by Samko et al. ([729], Section 5]).

The Riemann-Liouville fractional integrals (2.1.1) and (2.1.2) and fractional derivative (2.1.5) and (2.1.6), defined on a finite interval $[a, b]$ of the real line $\mathbb{R}$, are naturally extended to the half-axis $\mathbb{R}^+$. The fractional integration constructions, corresponding to the ones in (2.1.1) and (2.1.2), have the following forms:

$$(I_{0+}^{\alpha}f)(x) := \frac{1}{\Gamma(\alpha)}\int_0^x \frac{f(t)dt}{(x-t)^{1-\alpha}} \quad (x > 0;\ \Re(\alpha) > 0) \tag{2.2.1}$$

and

$$(I_{-}^{\alpha}f)(x) := \frac{1}{\Gamma(\alpha)}\int_x^{\infty} \frac{f(t)dt}{(t-x)^{1-\alpha}} \quad (x > 0;\ \Re(\alpha) > 0), \tag{2.2.2}$$

respectively, while the fractional differentiation constructions, corresponding to those in (2.1.5) and (2.1.6), are defined by

$$(D_{0+}^{\alpha}y)(x) := \left(\frac{d}{dx}\right)^n (I_{0+}^{n-\alpha}y)(x) = \frac{1}{\Gamma(n-\alpha)}\left(\frac{d}{dx}\right)^n \int_0^x \frac{y(t)dt}{(x-t)^{\alpha-n+1}} \tag{2.2.3}$$

and

$$(D_{-}^{\alpha}y)(x) := \left(-\frac{d}{dx}\right)^n (I_{-}^{n-\alpha}y)(x) = \frac{1}{\Gamma(n-\alpha)}\left(-\frac{d}{dx}\right)^n \int_x^{\infty} \frac{y(t)dt}{(t-x)^{\alpha-n+1}}, \tag{2.2.4}$$

with $n = [\mathfrak{R}(\alpha)] + 1$; $\mathfrak{R}(\alpha) \geqq 0$; $x > 0$.

The expressions for $I_{0+}^{\alpha}f$ and $I_{-}^{\alpha}f$ in (2.2.1) and (2.2.2), and $D_{0+}^{\alpha}y$ and $D_{-}^{\alpha}y$ in (2.2.3) and (2.2.4), are called the *Liouville left- and right-sided fractional integrals and fractional derivatives on the half-axis* $\mathbb{R}^+$. In particular, when $\alpha = n \in \mathbb{N}_0$, then

$$(D_{+}^{0}y)(x) = (D_{-}^{0}y)(x) = y(x); \quad (D_{+}^{n}y)(x) = y^{(n)}(x),$$

$$(D_{-}^{n}y)(x) = (-1)^n y^{(n)}(x) \quad (n \in \mathbb{N}), \tag{2.2.5}$$

where $y^{(n)}(x)$ is the usual derivative of $y(x)$ of order n.

If $0 < \mathfrak{R}(\alpha) < 1$ and $x > 0$, then

$$(D_{0+}^{\alpha}y)(x) = \frac{1}{\Gamma(1-\alpha)}\frac{d}{dx}\int_0^x \frac{y(t)dt}{(x-t)^{\alpha-[\mathfrak{R}(\alpha)]}} \tag{2.2.6}$$

and

$$(D_{-}^{\alpha}y)(x) = -\frac{1}{\Gamma(1-\alpha)}\frac{d}{dx}\int_x^{\infty} \frac{y(t)dt}{(t-x)^{\alpha-[\mathfrak{R}(\alpha)]}}. \tag{2.2.7}$$

If $\mathfrak{R}(\alpha) = 0$ $(\alpha \neq 0)$, then the Liouville fractional derivatives (2.2.6) and (2.2.7) are of a purely imaginary order and have the following forms:

$$(D_{0+}^{i\theta}y)(x) = \frac{1}{\Gamma(1-i\theta)}\frac{d}{dx}\int_0^x \frac{y(t)dt}{(x-t)^{i\theta}} \quad (\theta \in \mathbb{R}\backslash\{0\};\ x > 0) \tag{2.2.8}$$

and

$$(D_{-}^{i\theta}y)(x) = -\frac{1}{\Gamma(1-i\theta)}\frac{d}{dx}\int_x^{\infty} \frac{y(t)dt}{(t-x)^{i\theta}} \quad (\theta \in \mathbb{R}\backslash\{0\};\ x > 0), \tag{2.2.9}$$

respectively.

The Liouville fractional calculus operators I_{0+}^{α} and D_{0+}^{α} satisfy the relations (2.1.16) and (2.1.17) with $a = 0$, while the Liouville fractional calculus operators I_{-}^{α} and D_{-}^{α} of the power function $x^{\beta-1}$ and the exponential function $e^{-\lambda x}$ yield a power function of the same form and the same exponential function, respectively, both apart from a constant multiplication factor.

Property 2.5 *Let* $\Re(\alpha) \geqq 0$.

(a) *If* $\Re(\beta) > 0$, *then*

$$\left(I_{0+}^{\alpha} t^{\beta-1}\right)(x) = \frac{\Gamma(\beta)}{\Gamma(\beta+\alpha)} x^{\beta+\alpha-1} \quad (\Re(\alpha) > 0;\ \Re(\beta) > 0), \tag{2.2.10}$$

$$\left(D_{0+}^{\alpha} t^{\beta-1}\right)(x) = \frac{\Gamma(\beta)}{\Gamma(\beta-\alpha)} x^{\beta-\alpha-1} \quad (\Re(\alpha) \geqq 0;\ \Re(\beta) > 0). \tag{2.2.11}$$

(b) *If* $\beta \in \mathbb{C}$, *then*

$$\left(I_{-}^{\alpha} t^{\beta-1}\right)(x) = \frac{\Gamma(1-\alpha-\beta)}{\Gamma(1-\beta)} x^{\beta+\alpha-1} \quad (\Re(\alpha) > 0;\ \Re(\alpha+\beta) < 1), \tag{2.2.12}$$

$$\left(D_{-}^{\alpha} t^{\beta-1}\right)(x) = \frac{\Gamma(1+\alpha-\beta)}{\Gamma(1-\beta)} x^{\beta-\alpha-1} \quad (\Re(\alpha) \geqq 0;\ \Re(\alpha+\beta-[\Re(\alpha)]) < 1). \tag{2.2.13}$$

(c) *If* $\Re(\lambda) > 0$, *then*

$$(I_{-}^{\alpha} e^{-\lambda t})(x) = \lambda^{-\alpha} e^{-\lambda x} \quad (\Re(\alpha) > 0), \tag{2.2.14}$$

$$(D_{-}^{\alpha} e^{-\lambda t})(x) = \lambda^{\alpha} e^{-\lambda x} \quad (\Re(\alpha) \geqq 0). \tag{2.2.15}$$

When $0 < \alpha < 1$ *and* $1 \leqq p < 1/\alpha$, *the integrals* $I_{0+}^{\alpha} f$ *and* $I_{-}^{\alpha} f$ *are defined for a function* $f(x) \in L_p(\mathbb{R}^+)$.

Proof. Formulas (2.2.10) and (2.2.11) follow from (2.1.16) and (2.1.17) for $\beta = 0$. (2.2.12) is known [Samko et al. ([729], Table 9.3.1)]. (2.2.13) is proved by using the definitions (2.2.4) and (2.2.12) with α being replaced by $n-\alpha$, where $n = [\Re(\alpha)]+1$. Using these formulas and differentiating the obtained relation n times, we have

$$\begin{aligned}
\left(D_{-}^{\alpha} t^{\beta-1}\right)(x) &= \left(-\frac{d}{dx}\right)^{n} \left(I_{-}^{n-\alpha} t^{\beta-1}\right)(x) \\
&= \left(-\frac{d}{dx}\right)^{n} \left[\frac{\Gamma(1-n+\alpha-\beta)}{\Gamma(1-\beta)} x^{\beta+n-\alpha-1}\right] \\
&= (-1)^{n} \frac{\Gamma(1-n+\alpha-\beta)}{\Gamma(1-\beta)} \frac{\Gamma(\beta+n-\alpha)}{\Gamma(\beta-\alpha)} x^{\beta-\alpha-1}.
\end{aligned} \tag{2.2.16}$$

Also, by (1.5.8), we have

$$\Gamma(1-n+\alpha-\beta)\Gamma(\beta+n-\alpha) = \frac{\pi}{\sin[(\beta-\alpha+n)\pi]} = \frac{(-1)^{n}\pi}{\sin[(\beta-\alpha)\pi]} \tag{2.2.17}$$

and

$$\frac{1}{\Gamma(\beta-\alpha)} = \frac{\Gamma(1+\alpha-\beta)}{\Gamma(\beta-\alpha)\Gamma(1+\alpha-\beta)} = \frac{\Gamma(1+\alpha-\beta)\sin[(\beta-\alpha)\pi]}{\pi}. \tag{2.2.18}$$

Substituting these relations in (2.2.16), we obtain (2.2.13). Properties (2.2.14) and (2.2.15) are well known [Samko et al. ([729],(5.20))].

Lemma 2.11 (Hardy-Littlewood Theorem). *Let $1 \leqq p \leqq \infty$, $1 \leqq q \leqq \infty$, and $\alpha > 0$. Then the operators I^α_{0+} and I^α_- are bounded from $L_p(\mathbb{R}^+)$ into $L_q(\mathbb{R}^+)$ if, and only if,*

$$0 < \alpha < 1, \quad 1 < p < \frac{1}{\alpha} \text{ and } q = \frac{p}{1-\alpha p}. \tag{2.2.19}$$

The weighted analog of the assertion in the space $X^p_c(\mathbb{R}^+)$, defined in (1.1.3) and (1.1.4), is given by the following result.

Lemma 2.12 [Samko et al. ([729], Theorems 5.3 and 5.4)]. *Let $1 \leqq p < \infty$, $\mu \in \mathbb{R}$ and*

$$0 < \alpha < m + \frac{1}{p}, \quad 0 \leqq m \leqq \alpha, \quad q = \frac{1}{1-(\alpha - m)p}, \quad \nu = \left(\frac{\mu}{p} - m\right) q, \tag{2.2.20}$$

with $m \neq 0$ for $p = 1$.

(a) *If $\mu < p - 1$, then the operator I^α_{0+} is bounded from $X^p_{(\mu+1)/p}(\mathbb{R}^+)$ into $X^q_{(\nu+1)/q}(\mathbb{R}^+)$:*

$$\left(\int_0^\infty x^\nu \left| \left(I^\alpha_{0+} f\right)(x)\right|^q dx\right)^{1/q} \leqq k_1 \left(\int_0^\infty x^\mu |f(x)|^p dx\right)^{1/p}, \tag{2.2.21}$$

where the constant $k_1 > 0$ does not depend on f.

(b) *If $\mu > \alpha p - 1$, then the operator I^α_- is bounded from $X^p_{(\mu+1)/p}(\mathbb{R}^+)$ into $X^q_{(\nu+1)/q}(\mathbb{R}^+)$:*

$$\left(\int_0^\infty x^\nu \left| \left(I^\alpha_- f\right)(x)\right|^q dx\right)^{1/q} \leqq k_2 \left(\int_0^\infty x^\mu |f(x)|^p dx\right)^{1/p}, \tag{2.2.22}$$

where the constant $k_2 > 0$ does not depend on f.

Lemma 2.13 *Let $1 \leqq p < \infty$.*

(a) *If $1 < p < \infty$ and $\alpha > 0$, then the operator I^α_{0+} is bounded from $L_p(\mathbb{R}^+)$ into $X^p_{1/p-\alpha}(\mathbb{R}^+)$:*

$$\left(\int_0^\infty x^{-\alpha p} \left| \left(I^\alpha_{0+} f\right)(x)\right|^p dx\right)^{1/p} \leqq \frac{\Gamma(1/p')}{\Gamma(\alpha + 1/p)} \left(\int_0^\infty |f(x)|^p dx\right)^{1/p}. \tag{2.2.23}$$

(b) *If $1 \leqq p < 1/\alpha$ and $0 < \alpha < 1$, then the operator I^α_- is bounded from $L_p(\mathbb{R}^+)$ into $X^p_{1/p-\alpha}(\mathbb{R}^+)$:*

$$\left(\int_0^\infty x^{-\alpha p} \left| \left(I^\alpha_- f\right)(x)\right|^p dx\right)^{1/p} \leqq \frac{\Gamma(1/p - \alpha)}{\Gamma(1/p)} \left(\int_0^\infty |f(x)|^p dx\right)^{1/p}. \tag{2.2.24}$$

The following assertion, similar to Lemma 2.3, also holds [see Samko et al. ([729], Section 5.1)].

Property 2.6 *Let $\alpha > 0$, $\beta > 0$, $p \geqq 1$ and $\alpha + \beta < 1/p$. If $f(x) \in L_p(\mathbb{R}^+)$, then the semigroup properties*

$$(I_{0+}^{\alpha} I_{0+}^{\beta} f)(x) = (I_{0+}^{\alpha+\beta} f)(x), \text{ and } (I_{-}^{\alpha} I_{-}^{\beta} f)(x) = (I_{-}^{\alpha+\beta} f)(x) \tag{2.2.25}$$

hold. The relations in (2.2.25) *are also valid for "sufficiently good" functions $f(x)$.*

The Liouville fractional derivatives $(D_{0+}^{\alpha} y)(x)$ and $(D_{-}^{\alpha} y)(x)$ exist for "sufficiently good" functions $y(x)$; for example, for functions $y(x)$ in the space $C_0^{\infty}(\mathbb{R}^+)$ of all infinitely differentiable functions on $\mathbb{R}^+$ with a compact support. Therefore, the following properties, similar to those in Lemma 2.4 and Properties 2.2 and 2.3, are valid.

Property 2.7 *If $\alpha > 0$, then the relations*

$$(D_{0+}^{\alpha} I_{0+}^{\alpha} f)(x) = f(x), \text{ and } (D_{-}^{\alpha} I_{-}^{\alpha} f)(x) = f(x) \tag{2.2.26}$$

are true for "sufficiently good" functions $f(x)$. In particular, these formulas hold for $f(x) \in L_1(\mathbb{R}^+)$.

Property 2.8 *If $\alpha > \beta > 0$, then the formulas*

$$(D_{0+}^{\beta} I_{0+}^{\alpha} f)(x) = (I_{0+}^{\alpha-\beta} f)(x) \text{ and } (D_{-}^{\beta} I_{-}^{\alpha} f)(x) = (I_{-}^{\alpha-\beta} f)(x) \tag{2.2.27}$$

hold for "sufficiently good" functions $f(x)$ such as (for example) $f(x) \in L_1(\mathbb{R}^+)$.

In particular, when $\beta = k \in \mathbb{N}$ and $\Re(\alpha) > k$, then

$$(D^k I_{0+}^{\alpha} f)(x) = (I_{0+}^{\alpha-k} f)(x) \text{ and } (D^k I_{-}^{\alpha} f)(x) = (-1)^k (I_{-}^{\alpha-k} f)(x). \tag{2.2.28}$$

Property 2.9 *Let $\alpha > 0$, $m \in \mathbb{N}$ and $D = d/dx$.*

(a) *If the fractional derivatives $(D_{0+}^{\alpha} y)(x)$ and $(D_{0+}^{\alpha+m} y)(x)$ exist, then*

$$(D^m D_{0+}^{\alpha} y)(x) = (D_{0+}^{\alpha+m} f)(x). \tag{2.2.29}$$

(b) *If the fractional derivatives $(D_{-}^{\alpha} y)(x)$ and $(D_{-}^{\alpha+m} y)(x)$ exist, then*

$$(D^m D_{-}^{\alpha} y)(x) = (-1)^m (D_{-}^{\alpha+m} f)(x). \tag{2.2.30}$$

The formulas for *fractional integration by parts*, analogous to those in Lemma 2.7, for the Liouville fractional integrals (2.2.1) and (2.2.2) and fractional derivatives (2.2.3) and (2.2.4) are given by the following result.

Property 2.10 *If $\alpha > 0$, then the relations*

$$\int_0^{\infty} \varphi(x) \left(I_{0+}^{\alpha} \psi\right)(x) dx = \int_0^{\infty} \psi(x) \left(I_{-}^{\alpha} \varphi\right)(x) dx \tag{2.2.31}$$

and

$$\int_0^{\infty} f(x) \left(D_{0+}^{\alpha} g\right)(x) dx = \int_0^{\infty} g(x) \left(D_{-}^{\alpha} f\right)(x) dx \tag{2.2.32}$$

are valid for "sufficiently good" functions φ, ψ and f, g.

In particular, (2.2.31) *holds for functions $\varphi(x) \in L_p(\mathbb{R}^+)$ and $\psi(x) \in L_q(\mathbb{R}^+)$, while* (2.2.32) *for $f(x) \in I_{-}^{\alpha}(L_p(\mathbb{R}^+))$ and $g(x) \in I_{0+}^{\alpha}(L_q(\mathbb{R}^+))$, provided that $p > 1$, $q > 1$, $(1/p) + (1/q) = 1 + \alpha$.*

Remark 2.7 Relation (2.2.31) was given in Samko et al. ([729], (5.16′)). The expression (2.2.32) follows from (2.2.31) by taking $I_{0+}^{\alpha}\psi = g$ and $I_{-}^{\alpha}\varphi = f$.

The next assertion yields the Laplace transform (1.4.1) of the Liouville fractional integrals and fractional derivatives $I_{0+}^{\alpha}f$ and $D_{0+}^{\alpha}y$ in (2.2.1) and (2.2.3) [see Samko et al. ([729], Theorems 7.2 and 7.3)].

Lemma 2.14 *Let $\Re(\alpha) > 0$ and $f(x) \in L_1(0,b)$ for any $b > 0$. Also let the estimate*

$$|f(x)| \leqq Ae^{p_0 x} \quad (x > b > 0) \tag{2.2.33}$$

hold, for constants $A > 0$ and $p_0 > 0$.

(a) *If $f(x) \in L_1(0,b)$ for any $b > 0$, then the relation*

$$(\mathcal{L}I_{0+}^{\alpha}f)(s) = s^{-\alpha}(\mathcal{L}f)(p) \tag{2.2.34}$$

is valid for $\Re(s) > p_0$.

(b) *If $n = [\Re(\alpha)] + 1$, $y(x) \in AC^n[0,b]$ for any $b > 0$, and the following estimate of the form (2.2.33)*

$$|y(x)| \leqq Be^{q_0 x} \quad (x > b > 0) \tag{2.2.35}$$

holds for constants $B > 0$ and $q_0 > 0$, and if $y^{(k)}(0) = 0$ $(k = 0, 1, \cdots, n-1)$, then the relation

$$(\mathcal{L}D_{0+}^{\alpha}y)(s) = s^{\alpha}(\mathcal{L}y)(s). \tag{2.2.36}$$

is valid for $\Re(s) > q_0$.

Remark 2.8 If $\Re(\alpha) > 0$, $n = [\Re(\alpha)] + 1$, $y(x) \in AC^n[0,b]$ for any $b > 0$, the condition in (2.2.35) is satisfied and there exist the finite limits

$$\lim_{x\to 0+}[D^k I_{0+}^{n-\alpha}y(x)] \text{ and } \lim_{x\to\infty}[D^k I_{0+}^{n-\alpha}y(x)] = 0 \ (D = d/dx; k = 0, 1, \cdots, n-1),$$

then from (2.2.6) and (1.4.9) we derive a relation, more general than that in (2.2.36), of the form

$$(\mathcal{L}D_{0+}^{\alpha}y)(s) = s^{\alpha}(\mathcal{L}y)(s) - \sum_{k=0}^{n-1} s^{n-k-1}D^k(I_{0+}^{n-\alpha}y)(0+) \ (\Re(s) > q_0). \tag{2.2.37}$$

In particular, when $0 < \Re(\alpha) < 1$ and $y(x) \in AC[0,b]$ for any $b > 0$, then

$$(\mathcal{L}D_{0+}^{\alpha}y)(s) = s^{\alpha}(\mathcal{L}y)(s) - (I_{0+}^{1-\alpha}y)(0+). \tag{2.2.38}$$

The Mellin transform (1.4.23) of the the Liouville fractional integrals $I_{0+}^{\alpha}\varphi$ and $I_{-}^{\alpha}\varphi$ and fractional derivatives $D_{a+}^{\alpha}f$ and $D_{-}^{\alpha}y$ are given by the following statements [Samko et al. ([729], Theorems 7.4 and 7.5)].

Lemma 2.15 *Let* $\Re(\alpha) > 0$, $s \in \mathbb{C}$ *and* $f(x) \in X^1_{s+\alpha}(\mathbb{R}^+)$.

(a) *If* $\Re(s) < 1 - \Re(\alpha)$, *then*

$$(\mathcal{M}I^\alpha_{0+}f)(s) = \frac{\Gamma(1-\alpha-s)}{\Gamma(1-s)}(\mathcal{M}f)(s+\alpha) \ (\Re(s+\alpha) < 1). \tag{2.2.39}$$

(b) *If* $\Re(s) > 0$, *then*

$$(\mathcal{M}I^\alpha_{-}f)(s) = \frac{\Gamma(s)}{\Gamma(s+\alpha)}(\mathcal{M}f)(s+\alpha) \ (\Re(s) > 0). \tag{2.2.40}$$

Lemma 2.16 *Let* $\Re(\alpha) > 0$, $n = [\Re(\alpha)] + 1$, $s \in \mathbb{C}$ *and* $y(x) \in X^1_{s-\alpha}(\mathbb{R}^+)$.

(a) *If* $\Re(s) < 1 + \Re(\alpha)$ *and the conditions*

$$\lim_{x\to 0+} \left[x^{s-k-1}(I^{n-\alpha}_{0+}y)(x)\right] = 0 \ (k = 0, 1, \cdots, n-1) \tag{2.2.41}$$

and

$$\lim_{x\to \infty} \left[x^{s-k-1}(I^{n-\alpha}_{0+}y)(x)\right] = 0 \ (k = 0, 1, \cdots, n-1) \tag{2.2.42}$$

hold, then

$$(\mathcal{M}D^\alpha_{0+}y)(s) = \frac{\Gamma(1+\alpha-s)}{\Gamma(1-s)}(\mathcal{M}y)(s-\alpha) \ (\Re(s-\alpha) < 1). \tag{2.2.43}$$

(b) *If* $\Re(s) > 0$ *and the conditions:*

$$\lim_{x\to 0+} \left[x^{s-k-1}(I^{n-\alpha}_{-}y)(x)\right] = 0 \ (k = 0, 1, \cdots, n-1) \tag{2.2.44}$$

and

$$\lim_{x\to \infty} \left[x^{s-k-1}(I^{n-\alpha}y)(x)\right] = 0 \ (k = 0, 1, \cdots, n-1) \tag{2.2.45}$$

are valid, then

$$(\mathcal{M}D^\alpha_{-}y)(s) = \frac{\Gamma(s)}{\Gamma(s-\alpha)}(\mathcal{M}y)(s-\alpha) \ (\Re(s) > 0). \tag{2.2.46}$$

Remark 2.9 If $\Re(\alpha) > 0$, $n = [\Re(\alpha)] + 1$, and the conditions in (2.2.41), (2.2.42) and (2.2.44), (2.2.45) are not satisfied, then from (2.2.3), (1.4.37), (2.2.39) and (2.2.4), (1.4.38), (2.2.40) we derive the following formulas, more general than those in (2.2.43) and (2.2.46):

$$(\mathcal{M}D^\alpha_{0+}y)(s) = \frac{\Gamma(1+\alpha-s)}{\Gamma(1-s)}(\mathcal{M}y)(s-\alpha)$$

$$+ \sum_{k=0}^{n-1} \frac{\Gamma(1+k-s)}{\Gamma(1-s)} \left[x^{s-k-1}(I^{n-\alpha}_{0+}y)(x)\right]_0^\infty \tag{2.2.47}$$

and

$$(\mathcal{M}D_{-}^{\alpha}y)(s) = \frac{\Gamma(s)}{\Gamma(s-\alpha)}(\mathcal{M}y)(s-\alpha)$$
$$+\sum_{k=0}^{n-1}(-1)^{n-k}\frac{\Gamma(s)}{\Gamma(s-k)}\left[x^{s-k-1}(I_{-}^{n-\alpha}y)(x)\right]_0^{\infty}, \qquad (2.2.48)$$

respectively. In particular, when $0 < \Re(\alpha) < 1$, then

$$(\mathcal{M}D_{0+}^{\alpha}y)(s) = \frac{\Gamma(1+\alpha-s)}{\Gamma(1-s)}(\mathcal{M}y)(s-\alpha) + \left[x^{s-1}(I_{0+}^{1-\alpha}y)(x)\right]_0^{\infty} \qquad (2.2.49)$$

and

$$(\mathcal{M}D_{-}^{\alpha}y)(s) = \frac{\Gamma(s)}{\Gamma(s-\alpha)}(\mathcal{M}y)(s-\alpha) + \left[x^{s-k-1}(I_{-}^{n-\alpha}y)(x)\right]_0^{\infty}. \qquad (2.2.50)$$

Formulas (2.1.53) and (2.1.54), involving the Mittag-Leffler function (1.8.17), are also valid for the Liouville fractional integrals (2.2.1) and fractional derivatives (2.2.3):

$$\left(I_{0+}^{\alpha}t^{\beta-1}E_{\mu,\beta}\left(\lambda t^{\mu}\right)\right)(x) = x^{\alpha+\beta-1}E_{\mu,\alpha+\beta}\left(\lambda x^{\mu}\right) \qquad (2.2.51)$$

and

$$\left(D_{0+}^{\alpha}t^{\beta-1}E_{\mu,\beta}\left(\lambda t^{\mu}\right)\right)(x) = x^{\beta-\alpha-1}E_{\mu,\beta-\alpha}\left(\lambda x^{\mu}\right) \qquad (2.2.52)$$

where $\lambda \in \mathbb{C}$; $\Re(\alpha) > 0$, $\Re(\beta) > 0$, and $\Re(\mu) > 0$.

When $\mu = \alpha$, the formula (2.2.52) can be rewritten as follows:

$$\left(D_{0+}^{\alpha}t^{\beta-1}E_{\alpha,\beta}\left[\lambda t^{\alpha}\right]\right)(x) = \frac{x^{\beta-\alpha-1}}{\Gamma(\beta-\alpha)} + \lambda x^{\beta-1}E_{\alpha,\beta}\left(\lambda x^{\alpha}\right) \quad (\lambda \in \mathbb{C}). \qquad (2.2.53)$$

A relation similar to (2.2.53) is valid for the left-sided Liouville fractional derivative (2.2.4) with $\Re(\alpha) > 0$ and $\Re(\beta) > [\Re(\alpha)] + 1$:

$$\left(D_{-}^{\alpha}t^{\alpha-\beta}E_{\alpha,\beta}\left[\lambda t^{-\alpha}\right]\right)(x) = \frac{x^{-\beta}}{\Gamma(\beta-\alpha)} + \lambda x^{-\alpha-\beta}E_{\alpha,\beta}\left(\lambda x^{-\alpha}\right) \quad (\lambda \in \mathbb{C}). \quad (2.2.54)$$

The results in (2.2.53) and (2.2.54) for $\alpha > 0$ and $\beta > [\alpha] + 1$ were proved by Kilbas and Saigo in [391] and presented in Kilbas and Saigo ([397] in 1997, [724] in 1998). They can be extended to a complex α and β by analytic continuation.

In conclusion we indicate the composition properties of the Liouville fractional integration operators (2.2.1) and (2.2.2) with the operator of translation τ_h and the operator of dilation Π_{λ}, defined in (1.3.5) and (1.3.6), respectively. For "sufficiently good" functions $f(x)$, the following formulas hold:

$$\tau_h I_{0+}^{\alpha}f = I_{h+}^{\alpha}\tau_h f, \quad \textit{and } \tau_h I_{-}^{\alpha}f = I_{-}^{\alpha}\tau_h f \quad (\alpha > 0; \quad h \in \mathbb{R}), \qquad (2.2.55)$$

and

$$\Pi_{\lambda}I_{0+}^{\alpha}f = \lambda^{\alpha}I_{0+}^{\alpha}\Pi_{\lambda}f, \ \textit{and } \Pi_{\lambda}I_{-}^{\alpha}f = \lambda^{\alpha}I_{-}^{\alpha}\Pi_{\lambda}f \quad (\alpha > 0; \ \lambda > 0) \qquad (2.2.56)$$

[see Samko et al. ([729], (5.12)-(5.14))].

2.3 Liouville Fractional Integrals and Fractional Derivatives on the Real Axis

In this section we present the definitions and some properties of the Liouville fractional integrals and fractional derivatives on the whole axis $\mathbb{R} = (-\infty, \infty)$. More detailed information may be found in the book by Samko et al. ([729], Section 5).

The Liouville fractional integrals and fractional derivatives on $\mathbb{R}$ are defined similarly to those on the half-axis in Section 2.2. The fractional integrals have the form

$$(I_+^\alpha f)(x) := \frac{1}{\Gamma(\alpha)} \int_{-\infty}^x \frac{f(t)dt}{(x-t)^{1-\alpha}} \tag{2.3.1}$$

and

$$(I_-^\alpha f)(x) := \frac{1}{\Gamma(\alpha)} \int_x^\infty \frac{f(t)dt}{(t-x)^{1-\alpha}}, \tag{2.3.2}$$

where $x \in \mathbb{R}$ and $\Re(\alpha) > 0$, while the fractional derivatives corresponding to those in (2.2.3) and (2.2.4) are defined by

$$(D_+^\alpha y)(x) := \left(\frac{d}{dx}\right)^n (I_+^{n-\alpha} y)(x) = \frac{1}{\Gamma(n-\alpha)} \left(\frac{d}{dx}\right)^n \int_{-\infty}^x \frac{y(t)dt}{(x-t)^{\alpha-n+1}} \tag{2.3.3}$$

and

$$(D_-^\alpha y)(x) := \left(-\frac{d}{dx}\right)^n (I_-^{n-\alpha} y)(x) = \frac{1}{\Gamma(n-\alpha)} \left(-\frac{d}{dx}\right)^n \int_x^\infty \frac{y(t)dt}{(t-x)^{\alpha-n+1}}, \tag{2.3.4}$$

where $n = [\Re(\alpha)] + 1$, $\Re(\alpha) \geqq 0$ and $x \in \mathbb{R}$, respectively.

The expressions for $I_+^\alpha f$ and $I_-^\alpha f$ in (2.3.1) and (2.3.2), and for $D_+^\alpha y$ and $D_-^\alpha y$ in (2.3.3) and (2.3.4), are called *Liouville left- and right-sided fractional integrals and fractional derivatives* on the whole axis $\mathbb{R}$. In particular, when $\alpha = n \in \mathbb{N}_0$, then

$$(D_+^0 y)(x) = (D_-^0 y)(x) = y(x);$$

$$(D_+^n y)(x) = y^{(n)}(x), \quad (D_-^n y)(x) = (-1)^n y^{(n)}(x) \quad (n \in \mathbb{N}) \tag{2.3.5}$$

where $y^{(n)}(x)$ is the usual derivative of $y(x)$ of order n.

If $0 < \Re(\alpha) < 1$ and $x \in \mathbb{R}$, then

$$(D_+^\alpha y)(x) = \frac{1}{\Gamma(1-\alpha)} \frac{d}{dx} \int_{-\infty}^x \frac{y(t)dt}{(x-t)^{\alpha-[\Re(\alpha)]}} \tag{2.3.6}$$

and

$$(D_-^\alpha y)(x) = -\frac{1}{\Gamma(1-\alpha)} \frac{d}{dx} \int_x^\infty \frac{y(t)dt}{(t-x)^{\alpha-[\Re(\alpha)]}}. \tag{2.3.7}$$

When $\Re(\alpha) = 0$ $(\alpha \neq 0)$, the Liouville fractional derivatives (2.2.38) and (2.3.7) of a purely imaginary order have the forms

$$(D_+^{i\theta} y)(x) = \frac{1}{\Gamma(1-i\theta)} \frac{d}{dx} \int_{-\infty}^x \frac{y(t)dt}{(x-t)^{i\theta}} \quad (\theta \in \mathbb{R}\backslash\{0\};\ x \in \mathbb{R}) \tag{2.3.8}$$

and

$$(D_{-}^{i\theta}y)(x) = -\frac{1}{\Gamma(1-i\theta)}\frac{d}{dx}\int_x^\infty \frac{y(t)dt}{(t-x)^{i\theta}} \quad (\theta \in \mathbb{R}\backslash\{0\};\ x \in \mathbb{R}), \tag{2.3.9}$$

respectively.

Analogous to Property 2.5(b), the Liouville fractional calculus operators I_+^α and D_+^α of the exponential function $e^{\lambda x}$ yield the same exponential function, both apart from a constant multiplication factor.

Property 2.11 *Let* $\Re(\lambda) > 0$.

(a) *If* $\Re(\alpha) \geqq 0$, *then*

$$(I_+^\alpha e^{\lambda t})(x) = \lambda^{-\alpha} e^{\lambda x}. \tag{2.3.10}$$

(b) *If* $\Re(\alpha) \geqq 0$, *then*

$$(D_+^\alpha e^{\lambda t})(x) = \lambda^{\alpha} e^{\lambda x}. \tag{2.3.11}$$

When $0 < \alpha < 1$ and $1 \leqq p < 1/\alpha$, the integrals $I_+^\alpha f$ and $I_-^\alpha f$ are defined for a function $f(x) \in L_p(\mathbb{R})$. Indeed, the following *Hardy-Littlewood theorem* holds [see Samko et al. ([729], Theorem 5.3)].

Lemma 2.17 *Let* $1 \leqq p \leqq \infty$, $1 \leqq q \leqq \infty$ *and* $\alpha > 0$. *The operators* I_+^α *and* I_-^α *are bounded from* $L_p(\mathbb{R})$ *into* $L_q(\mathbb{R}^+)$ *if, and only if, the conditions in* (2.2.14) *are satisfied.*

Let $\dot{\mathbb{R}}$ be the axis $\mathbb{R}$ supplemented by the one infinite point: $\dot{\mathbb{R}} = \mathbb{R}\bigcup\{\infty\}$. We denote by $L_{p,\omega}$ $(1 \leqq p < \infty)$ a space of functions $f(x)$ with an exponential weight $\omega \in \mathbb{R}$ on $\dot{\mathbb{R}}$ by defining the norm as follows:

$$L_{p,\omega}(\dot{\mathbb{R}}) := \left\{ f :\ \ ||f||_{p,\omega} = \left(\int_{-\infty}^{\infty} e^{-\omega t}|f(t)|^p dt \right)^{1/p} < \infty \right\}, \tag{2.3.12}$$

with $1 \leqq p < \infty$, while by $L_{\infty,\omega} = C_\omega$ the space of functions $f(x)$ such that $e^{-\omega x}f(x) \in C(\dot{\mathbb{R}})$ with the norm

$$L_{\infty,\omega}(\dot{\mathbb{R}}) = C_\omega(\dot{\mathbb{R}}) := \left\{ f :\ \ ||f||_\omega = \max_{t\in\dot{\mathbb{R}}} e^{-\omega t}|f(t)| < \infty \right\}. \tag{2.3.13}$$

Spaces (2.3.12) and (2.3.13) are invariant with respect to the Liouville fractional integration I_+^α and I_-^α [see Samko et al. ([729], Theorem 5.7)].

Lemma 2.18 *Let* $1 \leqq p \leqq \infty$, $\alpha > 0$ *and* $\omega > 0$.

(a) *The operator* I_+^α *is bounded in the space* $L_{p,\omega}$:

$$||I_+^\alpha f||_{p,\omega} \leqq k||f||_{p,\omega}, \quad k = \left(\frac{p}{\omega}\right)^\alpha \ (1 \leqq p < \infty), \ k = \omega^\alpha \ (p = \infty). \tag{2.3.14}$$

(b) *The operator* I_-^α *is bounded in the space* $L_{p,-\omega}$:

$$||I_-^\alpha f||_{p,-\omega} \leqq k||f||_{p,-\omega}, \quad k = \left(\frac{p}{|\omega|}\right)^\alpha \ (1 \leqq p < \infty), \ k = |\omega|^\alpha \ (p = \infty). \tag{2.3.15}$$

The following assertions, similar to Property 2.6, hold [see Samko et al. ([729], Section 5.1)].

Lemma 2.19 *Let $\alpha > 0$, $\beta > 0$, and $p \geqq 1$ be such that $\alpha + \beta < 1/p$. If $f(x) \in L_p(\mathbb{R})$, then*

$$(I^{\alpha}_{+} I^{\beta}_{+} f)(x) = (I^{\alpha+\beta}_{+} f)(x) \text{ and } (I^{\alpha}_{-} I^{\beta}_{-} f)(x) = (I^{\alpha+\beta}_{-} f)(x). \tag{2.3.16}$$

The Liouville fractional derivatives$(D^{\alpha}_{+} y)(x)$ and $(D^{\alpha}_{-} y)(x)$ exist for "sufficiently good" functions $y(x)$; for example, for functions $y(x)$ in the space $C^{\infty}_{0}(\mathbb{R})$ of all infinitely differentiable functions on $\mathbb{R}$ with compact support. Therefore, the following properties, similar to Properties 2.7 and 2.10, are valid.

Lemma 2.20 *If $\alpha > 0$, then, for "sufficiently good" functions $f(x)$, the relations*

$$(D^{\alpha}_{+} I^{\alpha}_{+} f)(x) = f(x), \text{ and } (D^{\alpha}_{-} I^{\alpha}_{-} f)(x) = f(x) \tag{2.3.17}$$

are true. In particular, these formulas hold for $f(x) \in L_1(\mathbb{R})$.

Property 2.12 *If $\alpha > \beta > 0$, then the formulas*

$$(D^{\beta}_{+} I^{\alpha}_{+} f)(x) = (I^{\alpha-\beta}_{0+} f)(x), \text{ and } (D^{\beta}_{-} I^{\alpha}_{-} f)(x) = (I^{\alpha-\beta}_{-} f)(x) \tag{2.3.18}$$

hold for "sufficiently good" functions $f(x)$. In particular, these relations hold for $f(x) \in L_1(\mathbb{R})$.

Furthermore, when $\beta = k \in \mathbb{N}$ and $\mathfrak{R}(\alpha) > k$, then

$$(D^{k} I^{\alpha}_{+} f)(x) = I^{\alpha-k}_{+} f(x), \text{ and } (D^{k} I^{\alpha}_{-} f)(x) = (-1)^{k} I^{\alpha-k}_{-} f(x). \tag{2.3.19}$$

Property 2.13 *Let $\alpha > 0$, $m \in \mathbb{N}$ and $D = d/dx$.*

(a) *If the fractional derivatives $(D^{\alpha}_{+} y)(x)$ and $(D^{\alpha+m}_{+} y)(x)$ exist, then*

$$(D^{m} D^{\alpha}_{+} y)(x) = (D^{\alpha+m}_{+} f)(x). \tag{2.3.20}$$

(b) *If the fractional derivatives $(D^{\alpha}_{-} y)(x)$ and $(D^{\alpha+m}_{-} y)(x)$ exist, then*

$$(D^{m} D^{\alpha}_{-} y)(x) = (-1)^{m} (D^{\alpha+m}_{-} f)(x). \tag{2.3.21}$$

Property 2.14 *If $\alpha > 0$, then the relations*

$$\int_{-\infty}^{\infty} \varphi(x) \left(I^{\alpha}_{+} \psi\right)(x) dx = \int_{-\infty}^{\infty} \psi(x) \left(I^{\alpha}_{-} \varphi\right)(x) dx \tag{2.3.22}$$

and

$$\int_{-\infty}^{\infty} f(x) \left(D^{\alpha}_{+} g\right)(x) dx = \int_{-\infty}^{\infty} g(x) \left(D^{\alpha}_{-} f\right)(x) dx \tag{2.3.23}$$

are valid for "sufficiently good" functions φ, ψ and f, g.

In particular, (2.3.22) *holds for functions $\varphi(x) \in L_p(\mathbb{R})$ and $\psi(x) \in L_q(\mathbb{R})$, while* (2.3.23) *holds for $f(x) \in I^{\alpha}_{-}(L_p(\mathbb{R}^{+}))$ and $g(x) \in I^{\alpha}_{+}(L_q(\mathbb{R}^{+}))$, provided that $p > 1$, $q > 1$, and $(1/p) + (1/q) = 1 + \alpha$.*

Remark 2.10 Property 2.14 was given in Samko et al. ([729], (5.16) and (5.17)).

The Fourier transform (1.3.1) of the Liouville fractional integrals $I_+^\alpha f$ and $I_-^\alpha f$ is given by the following result [see Samko et al. ([729], Theorem 7.1)].

Property 2.15 *Let* $0 < \Re(\alpha) < 1$ *and* $f(x) \in L_1(\mathbb{R})$. *Then the following relations hold:*

$$(\mathcal{F} I_+^\alpha f)(x) = \frac{(\mathcal{F}f)(x)}{(-ix)^\alpha} \tag{2.3.24}$$

and

$$(\mathcal{F} I_-^\alpha f)(x) = \frac{(\mathcal{F}f)(x)}{(ix)^\alpha}. \tag{2.3.25}$$

Here $(\mp ix)^\alpha$ *means*

$$(\mp ix)^\alpha = |x|^\alpha e^{\mp \alpha\pi i \,\mathrm{sgn}(x)/2}. \tag{2.3.26}$$

Remark 2.11 When $\Re(\alpha) > 0$, then, for "sufficiently good" functions $f(x)$, the equations (2.3.24) and (2.3.25) are valid as well as the following corresponding relations for the Liouville fractional derivatives $D_+^\alpha f$ and $D_-^\alpha f$:

$$(\mathcal{F} D_+^\alpha f)(x) = (-ix)^\alpha (\mathcal{F}f)(x) \tag{2.3.27}$$

and

$$(\mathcal{F} D_-^\alpha f)(x) = (ix)^\alpha (\mathcal{F}f)(x), \tag{2.3.28}$$

where $(\mp ix)^\alpha$ is defined by (2.3.26).

In conclusion, we give the composition properties of the Liouville fractional integration operators (2.3.1) and (2.3.2), similar to those given in (2.2.55) and (2.2.56) for the Liouville fractional integration operators on $\mathbb{R}^+$. The formulas

$$\tau_h I_+^\alpha f = I_+^\alpha \tau_h f, \quad \tau_h I_-^\alpha f = I_-^\alpha \tau_h f \quad (\alpha > 0; \quad h \in \mathbb{R}), \tag{2.3.29}$$

and

$$\Pi_\lambda I_+^\alpha f = \lambda^\alpha I_+^\alpha \Pi_\lambda f, \quad \Pi_\lambda I_-^\alpha f = \lambda^\alpha I_-^\alpha \Pi_\lambda f \quad (\alpha > 0; \ \lambda > 0), \tag{2.3.30}$$

hold for "sufficiently good" functions $f(x)$. In particular, they are true for $f(x) \in L_p(\mathbb{R})$, $1 \leqq p < 1/\alpha$, when $0 < \alpha < 1$ [see Samko et al. ([729], (5.12)-(5.13))].

2.4 Caputo Fractional Derivatives

In this section we present the definitions and some properties of the *Caputo fractional derivatives*. Let $[a, b]$ be a finite interval of the real line $\mathbb{R}$, and let $D_{a+}^\alpha[y(t)](x) \equiv (D_{a+}^\alpha y)(x)$ and $D_{b-}^\alpha[y(t)](x) \equiv (D_{b-}^\alpha y)(x)$ be the Riemann-Liouville fractional derivatives of order $\alpha \in \mathbb{C}$ ($\Re(\alpha) \geqq 0$) defined by (2.1.5) and (2.1.6), respectively. The fractional derivatives $({}^C D_{a+}^\alpha y)(x)$ and $({}^C D_{b-}^\alpha y)(x)$ of order $\alpha \in \mathbb{C}$

($\Re(\alpha) \geqq 0$) on $[a, b]$ are defined via the above Riemann-Liouville fractional derivatives by

$$({}^C D_{a+}^{\alpha} y)(x) := \left(D_{a+}^{\alpha} \left[y(t) - \sum_{k=0}^{n-1} \frac{y^{(k)}(a)}{k!}(t-a)^k \right] \right)(x) \tag{2.4.1}$$

and

$$({}^C D_{b-}^{\alpha} y)(x) := \left(D_{b-}^{\alpha} \left[y(t) - \sum_{k=0}^{n-1} \frac{y^{(k)}(b)}{k!}(b-t)^k \right] \right)(x), \tag{2.4.2}$$

respectively, where

$$n = [\Re(\alpha)] + 1 \ for \ \alpha \notin \mathbb{N}_0; \quad n = \alpha \ for \ \alpha \in \mathbb{N}_0. \tag{2.4.3}$$

These derivatives are called *left-sided and right-sided Caputo fractional derivatives of order* α.

In particular, when $0 < \Re(\alpha) < 1$, the relations (2.4.1) and (2.4.2) take the following forms:

$$({}^C D_{a+}^{\alpha} y)(x) = \left(D_{a+}^{\alpha} [y(t) - y(a)] \right)(x), \tag{2.4.4}$$

$$({}^C D_{b-}^{\alpha} y)(x) = \left(D_{b-}^{\alpha} [y(t) - y(b)] \right)(x). \tag{2.4.5}$$

If $\alpha \notin \mathbb{N}_0$ and $y(x)$ is a function for which the Caputo fractional derivatives $({}^C D_{a+}^{\alpha} y)(x)$ and $({}^C D_{b-}^{\alpha} y)(x)$ of order $\alpha \in \mathbb{C}$ ($\Re(\alpha) \geqq 0$) exist together with the Riemann-Liouville fractional derivatives $(D_{a+}^{\alpha} y)(x)$ and $(D_{b-}^{\alpha} y)(x)$, then, in accordance with (2.1.7) and (2.1.9), they are connected with each other by the following relations:

$$({}^C D_{a+}^{\alpha} y)(x) = (D_{a+}^{\alpha} y)(x) - \sum_{k=0}^{n-1} \frac{y^{(k)}(a)}{\Gamma(k-\alpha+1)}(x-a)^{k-\alpha} \quad (n = [\Re(\alpha)] + 1) \tag{2.4.6}$$

and

$$({}^C D_{b-}^{\alpha} y)(x) = (D_{b-}^{\alpha} y)(x) - \sum_{k=0}^{n-1} \frac{y^{(k)}(b)}{\Gamma(k-\alpha+1)}(b-x)^{k-\alpha} \quad (n = [\Re(\alpha)] + 1). \tag{2.4.7}$$

In particular, when $0 < \Re(\alpha) < 1$, we have

$$({}^C D_{a+}^{\alpha} y)(x) = (D_{a+}^{\alpha} y)(x) - \frac{y(a)}{\Gamma(1-\alpha)}(x-a)^{-\alpha}, \tag{2.4.8}$$

$$({}^C D_{b-}^{\alpha} y)(x) = (D_{b-}^{\alpha} y)(x) - \frac{y(b)}{\Gamma(1-\alpha)}(b-x)^{-\alpha}. \tag{2.4.9}$$

If $\alpha \notin \mathbb{N}_0$, then the Caputo fractional derivatives (2.4.1) and (2.4.2) coincide with the Riemann-Liouville fractional derivatives (2.1.5) and (2.1.6) in the following cases:

$$({}^C D_{a+}^{\alpha} y)(x) = (D_{a+}^{\alpha} y)(x), \tag{2.4.10}$$

if $y(a) = y'(a) = \cdots = y^{(n-1)}(a) = 0 (n = [\Re(\alpha)] + 1)$; and

$$({}^C D_{b-}^{\alpha} y)(x) = (D_{b-}^{\alpha} y)(x), \tag{2.4.11}$$

if $y(b) = y'(b) = \cdots = y^{(n-1)}(b) = 0 (n = [\Re(\alpha)] + 1)$.

In particular, when $0 < \Re(\alpha) < 1$, we have

$$({}^C D_{a+}^{\alpha} y)(x) = (D_{a+}^{\alpha} y)(x), \quad \text{when} \quad y(a) = 0, \tag{2.4.12}$$

$$({}^C D_{b-}^{\alpha} y)(x) = (D_{b-}^{\alpha} y)(x), \quad \text{when} \quad y(b) = 0. \tag{2.4.13}$$

If $\alpha = n \in \mathbb{N}_0$ and the usual derivative $y^{(n)}(x)$ of order n exists, then $({}^C D_{a+}^{n} y)(x)$ coincides with $y^{(n)}(x)$, while $(C D_{b-}^{n} y)(x)$ coincides with $y^{(n)}(x)$ with exactness to the constant multiplier $(-1)^n$:

$$({}^C D_{a+}^{n} y)(x) = y^{(n)}(x) \text{ and } ({}^C D_{b-}^{n} y)(x) = (-1)^n y^{(n)}(x) \quad (n \in \mathbb{N}). \tag{2.4.14}$$

The Caputo fractional derivatives $({}^C D_{a+}^{\alpha} y)(x)$ and $({}^C D_{b-}^{\alpha} y)(x)$ are defined for functions $y(x)$ for which the Riemann-Liouville fractional derivatives of the right-hand sides of (2.4.1) and (2.4.2) exist. In particular, they are defined for $y(x)$ belonging to the space $AC^n[a, b]$ of absolutely continuous functions defined in (1.1.7). Thus the following statement holds.

Theorem 2.1 *Let $\Re(\alpha) \geqq 0$ and let n be given by (2.4.3). If $y(x) \in AC^n[a, b]$, then the Caputo fractional derivatives $({}^C D_{a+}^{\alpha} y)(x)$ and $({}^C D_{b-}^{\alpha} y)(x)$ exist almost everywhere on $[a, b]$.*

(a) *If $\alpha \notin \mathbb{N}_0$, $({}^C D_{a+}^{\alpha} y)(x)$ and $({}^C D_{b-}^{\alpha} y)(x)$ are represented by*

$$({}^C D_{a+}^{\alpha} y)(x) = \frac{1}{\Gamma(n-\alpha)} \int_a^x \frac{y^{(n)}(t)dt}{(x-t)^{\alpha-n+1}} =: (I_{a+}^{n-\alpha} D^n y)(x) \tag{2.4.15}$$

and

$$({}^C D_{b-}^{\alpha} y)(x) = \frac{(-1)^n}{\Gamma(n-\alpha)} \int_x^b \frac{y^{(n)}(t)dt}{(t-x)^{\alpha-n+1}} =: (-1)^n (I_{b-}^{n-\alpha} D^n y)(x), \tag{2.4.16}$$

respectively, where $D = d/dx$ and $n = [\Re(\alpha)] + 1$.

In particular, when $0 < \Re(\alpha) < 1$ and $y(x) \in AC[a, b]$,

$$({}^C D_{a+}^{\alpha} y)(x) = \frac{1}{\Gamma(1-\alpha)} \int_a^x \frac{y'(t)dt}{(x-t)^{\alpha}} =: (I_{a+}^{1-\alpha} D y)(x) \tag{2.4.17}$$

and

$$({}^C D_{b-}^{\alpha} y)(x) = -\frac{1}{\Gamma(1-\alpha)} \int_x^b \frac{y'(t)dt}{(t-x)^{\alpha}} =: -(I_{b-}^{1-\alpha} D y)(x). \tag{2.4.18}$$

(b) *If $\alpha = n \in \mathbb{N}_0$, then $({}^C D_{a+}^{n} y)(x)$ and $({}^C D_{b-}^{n} y)(x)$ are represented by (2.4.14). In particular,*

$$({}^C D_{a+}^{0} y)(x) = ({}^C D_{b-}^{0} y)(x) = y(x). \tag{2.4.19}$$

Proof. Let $\alpha \notin \mathbb{N}_0$. Using (2.4.1) and (2.1.5), integrating by parts the inner integral and differentiating (which is possible by the conditions of the theorem), we have

$$({}^C D_{a+}^{\alpha} y)(x)$$

$$= \frac{1}{\Gamma(n-\alpha)} \left(\frac{d}{dx}\right)^n \left\{ -\frac{(x-t)^{n-\alpha}}{n-\alpha} \left[y(t) - \sum_{k=0}^{n-1} \frac{y^{(k)}(a)}{k!}(t-a)^k \right] \Big|_{t=a}^{x} + \right.$$

$$\left. + \int_a^x \frac{(x-t)^{n-\alpha}}{n-\alpha} \frac{d}{dt} \left[y(t) - \sum_{k=0}^{n-1} \frac{y^{(k)}(a)}{k!}(t-a)^k \right] dt \right\}$$

$$= \frac{1}{\Gamma(n-\alpha)} \left(\frac{d}{dx}\right)^{n-1} \int_a^x (x-t)^{n-\alpha-1} \left[y'(t) - \sum_{k=1}^{n-1} \frac{y^{(k)}(a)}{(k-1)!}(t-a)^{k-1} \right] dt$$

$$= \cdots = \frac{1}{\Gamma(n-\alpha)} \frac{d}{dx} \int_a^x (x-t)^{n-\alpha-1} \left[y^{(n-1)}(t) - y^{(n-1)}(a) \right] dt.$$

Using the above argument again, we derive that

$$({}^C D_{a+}^{\alpha} y)(x) = \frac{1}{\Gamma(n-\alpha)} \int_a^x (x-t)^{n-\alpha-1} y^{(n)}(t) dt,$$

which yields the result in (2.4.15). Relation (2.4.16) is proved similarly.

If $\alpha = n \in \mathbb{N}_0$, then (2.4.1), in accordance with the first formula in (2.1.7), takes the form:

$$({}^C D_{a+}^{n} y)(x) = \left(\frac{d}{dx}\right)^n \left[y(x) - \sum_{k=0}^{n-1} \frac{y^{(k)}(a)}{k!}(x-a)^k \right], \tag{2.4.20}$$

and from Lemma 1.1 we derive $({}^C D_{a+}^{n} y)(x) = y^{(n)}(x)$, which yields the first result in (2.4.14). The second one is proved similarly.

The following statement is analogous to that of Theorem 2.1 for functions $y(x)$ in the space $C^n[a,b]$ of continuously differentiable functions on $[a,b]$ up to order n.

Theorem 2.2 *Let $\Re(\alpha) \geqq 0$ and let n be given by (2.4.3). Also let $y(x) \in C^n[a,b]$. Then the Caputo fractional derivatives $({}^C D_{a+}^{\alpha} y)(x)$ and $({}^C D_{b-}^{\alpha} y)(x)$ are continuous on $[a,b]$: $({}^C D_{a+}^{\alpha} y)(x) \in C[a,b]$ and $({}^C D_{b-}^{\alpha} y)(x) \in C[a,b]$.*

(a) *If $\alpha \notin \mathbb{N}_0$, then $({}^C D_{a+}^{\alpha} y)(x)$ and $({}^C D_{b-}^{\alpha} y)(x)$ are represented by (2.4.15) and (2.4.16), respectively. Moreover,*

$$({}^C D_{a+}^{\alpha} y)(a) = ({}^C D_{b-}^{\alpha} y)(b) = 0. \tag{2.4.21}$$

In particular, they have, respectively, the forms (2.4.17) and (2.4.18) for $0 < \Re(\alpha) < 1$.

(b) *If $\alpha = n \in \mathbb{N}_0$, then the fractional derivatives $({}^C D_{a+}^{n} y)(x)$ and $({}^C D_{b-}^{n} y)(x)$ have representations given in (2.4.14). In particular, the relations in (2.4.19) hold.*

Proof. Let $\alpha \notin \mathbb{N}_0$. Formulas (2.4.15) and (2.4.16) are proved as in Theorem 2.1. The continuity of the functions $({}^C D_{a+}^{\alpha} y)(x)$ and $({}^C D_{b-}^{\alpha} y)(x)$ on $[a,b]$ follows from the representations (2.4.15) and (2.4.16), according to Lemma 2.8(a) with $f(t) = y^{(n)}(t) \in C[a,b]$ and $\gamma = 0$. The relations in (2.4.21) follow from the following inequalities:

$$\left|(I_{a+}^{n-\alpha} D^n y)(x)\right| \leqq \frac{\|y^{(n)}\|_C}{|\Gamma(n-\alpha)|[n-\Re(\alpha)+1]}(x-a)^{n-\Re(\alpha)} \tag{2.4.22}$$

and

$$\left|(I_{b-}^{n-\alpha} D^n y)(x)\right| \leqq \frac{\|y^{(n)}\|_C}{|\Gamma(n-\alpha)|[n-\Re(\alpha)+1]}(b-x)^{n-\Re(\alpha)}, \tag{2.4.23}$$

which are valid for any $x \in [a,b]$ and proved directly by using (2.1.1) and (2.1.2) (with n replaced by $n-\alpha$), and (2.1.6) and (2.1.16) with $\beta = 1$, and (1.1.19).

When $\alpha \in \mathbb{N}_0$, the first relation in (2.4.14) follows from (2.4.21) and Lemma 1.3 with $\gamma = 0$. The second one is proved similarly.

Corollary 2.3 *Let $\Re(\alpha) \geqq 0$ and let n be given by* (2.4.3).

(a) *If $\alpha \notin \mathbb{N}_0$, then the Caputo fractional differentiation operators ${}^C D_{a+}^{\alpha}$ and ${}^C D_{b-}^{\alpha}$ are bounded from the space $C^n[a,b]$ to the spaces $C_a[a,b]$ and $C_b[a,b]$ respectively, defined by* (1.1.25) *and* (1.1.26). *Moreover,*

$$\| {}^C D_{a+}^{\alpha} y\|_{C_a} \leqq k_\alpha \| y\|_{C^n} \text{ and } \| {}^C D_{b-}^{\alpha} y\|_{C_b} \leqq k_\alpha \| y\|_{C^n}, \tag{2.4.24}$$

where

$$k_\alpha = \frac{(b-a)^{n-\Re(\alpha)}}{|\Gamma(n-\alpha)|[n-\Re(\alpha)+1]}. \tag{2.4.25}$$

(b) *If $\alpha = n \in \mathbb{N}_0$, then the operators ${}^C D_{a+}^{n}$ and ${}^C D_{b-}^{n}$ are bounded from $C^n[a,b]$ to $C[a,b]$. Moreover,*

$$\| {}^C D_{a+}^{n} y\|_C = \| y\|_{C^n} \text{ and } \| {}^C D_{b-}^{n} y\|_C = \| y\|_{C^n}. \tag{2.4.26}$$

Proof. Assertion (a) follows from Theorem 2.2(a) if we take into account the estimates (4.2.22) and (4.2.23), definitions (1.1.25) and (1.1.26) of the spaces $C_a[a,b]$ and $C_b[a,b]$, definitions (1.1.19) and (1.1.20), and the estimate $\|y^{(n)}\|_C \leqq \|y\|_{C^n}$. Assertion (b) follows from Theorem 2.2(b) and from the relations in (2.4.14), in accordance with (1.1.19) and (1.1.20).

Remark 2.12 For $n-1 < \alpha < n$ $(n \in \mathbb{N})$, the derivative $({}^C D_{a+}^{\alpha} y)(x)$ in the form (2.4.15) was defined by Caputo [121] and presented in his book [122], and therefore, the derivatives $({}^C D_{a+}^{\alpha} y)(x)$ and $({}^C D_{b-}^{\alpha} y)(x)$ are called Caputo derivatives. In this regard, see also Mainardi [520] and Podlubny ([682], Section 2.4.1).

Remark 2.13 For $\alpha \geqq 0$ and $y(x) \in AC^n[a,b]$, the relation of the form (2.4.15) was proved by Diethelm ([173], Theorem 3.1) for the Caputo derivative $(D_{*a}^{\alpha} y)(x)$ defined by (2.4.1) with $n = [\alpha]$:

$$(D_{*a}^{\alpha} y)(x) = \left(D_{a+}^{\alpha} \left[y(t) - \sum_{k=0}^{n-1} \frac{y^{(k)}(a)}{k!}(t-a)^k \right] \right)(x). \tag{2.4.27}$$

The Caputo derivatives $({}^C D^\alpha_{a+}y)(x)$ and $({}^C D^\alpha_{b-}y)(x)$ have properties similar to those of the Riemann-Liouville fractional derivatives $(D^\alpha_{a+}y)(x)$ and $(D^\alpha_{b-}y)(x)$ given in (2.1.17) and (2.1.19), but different from those in (2.1.20).

Property 2.16 *Let $\Re(\alpha) > 0$ and let n be given by* (2.4.3). *Also let $\Re(\beta) > 0$. Then the following relations hold:*

$$\left({}^C D^\alpha_{a+}(t-a)^{\beta-1}\right)(x) = \frac{\Gamma(\beta)}{\Gamma(\beta-\alpha)}(x-a)^{\beta-1} \quad (\Re(\beta) > n), \tag{2.4.28}$$

$$\left({}^C D^\alpha_{b-}(b-t)^{\beta-1}\right)(x) = \frac{\Gamma(\beta)}{\Gamma(\beta-\alpha)}(b-x)^{\beta-1} \quad (\Re(\beta) > n) \tag{2.4.29}$$

and

$$({}^C D^\alpha_{a+}(t-a)^k)(x) = 0 \text{ and } ({}^C D^\alpha_{b-}(t-a)^k)(x) = 0 \quad (k = 0, 1, \cdots, n-1). \tag{2.4.30}$$

In particular,

$$({}^C D^\alpha_{a+}1)(x) = 0 \text{ and } ({}^C D^\alpha_{b-}1)(x) = 0. \tag{2.4.31}$$

When $\Re(\alpha) \notin \mathbb{N}_0$ and when $\alpha \in \mathbb{N}$, the Caputo fractional differentiation operators ${}^C D^\alpha_{a+}$ and ${}^C D^\alpha_{b-}$ provide operations inverse to the Riemann-Liouville fractional integration operators I^α_{a+} and I^α_{b-}, given in (2.1.1) and (2.1.2), from the left. But they do not have such a property when $\Re(\alpha) \in \mathbb{N}_0$ and $\Im(\alpha) \neq 0$. The following analog of Lemma 2.4 and Lemma 2.9(b) is valid.

Lemma 2.21 *Let $\Re(\alpha) > 0$ and let $y(x) \in L_\infty(a,b)$ or $y(x) \in C[a,b]$.*

(a) *If $\Re(\alpha) \notin \mathbb{N}$ or $\alpha \in \mathbb{N}$, then*

$$({}^C D^\alpha_{a+}I^\alpha_{a+}y)(x) = y(x) \text{ and } ({}^C D^\alpha_{b-}I^\alpha_{b-}y)(x) = y(x). \tag{2.4.32}$$

(b) *If $\Re(\alpha) \in \mathbb{N}$ and $\Im(\alpha) \neq 0$, then*

$$({}^C D^\alpha_{a+}I^\alpha_{a+}y)(x) = y(x) - \frac{(I^{\alpha+1-n}_{a+}y)(a+)}{\Gamma(n-\alpha)}(x-a)^{n-\alpha} \tag{2.4.33}$$

and

$$^C D^\alpha_{b-}I^\alpha_{a+}y)(x) = y(x) - \frac{(I^{\alpha+1-n}_{b-}y)(b-)}{\Gamma(n-\alpha)}(b-x)^{n-\alpha}. \tag{2.4.34}$$

Proof. Let $y(x) \in L_\infty(a,b)$ $(y(x) \in C[a,b])$, and let $\Re(\alpha) \notin \mathbb{N}$, $n = [\Re(\alpha)] + 1$ or $n \in \mathbb{N}$ and $k = 0, 1, \cdots, n-1$. By Property 2.2 (Lemma 2.9(c)), we have

$$(I^\alpha_{a+}y)^{(k)}(x) = (I^{\alpha-k}_{a+}y)(x) \quad (k = 0, 1, \cdots n-1). \tag{2.4.35}$$

Since $y(x) \in L_\infty(a,b)$ $(y(x) \in C[a,b])$, then for a.e. (for any) $x \in [a,b]$, analogous to (2.4.22), we get

$$\left|(I^{\alpha-k}_{a+}y)(x)\right| \leqq \frac{K}{|\Gamma(\alpha-k)|[\Re(\alpha)-k]}(x-a)^{\Re(\alpha)-k} \quad (K = \|y\|_{L_\infty}(= \|y\|_C)), \tag{2.4.36}$$

for any $k = 0, 1, \cdots, n-1 = [\Re(\alpha)]$, and hence

$$(I_{a+}^{\alpha}y)^{(k)}(a+) = 0 \quad (k = 0, 1, \cdots, n-1). \tag{2.4.37}$$

Thus, using (2.4.10) for $\Re(\alpha) \notin \mathbb{N}$ and (2.4.20) for $n \in \mathbb{N}$, with $y(x)$ replaced by $(I_{a+}^{\alpha}y)(x)$, and applying Lemma 2.4 (Lemma 2.9(b)), according to which the first relation in (2.1.31) holds, we derive

$$({}^{C}D_{a+}^{\alpha}I_{a+}^{\alpha}y)(x) = (D_{a+}^{\alpha}I_{a+}^{\alpha}y)(x) = y(x), \tag{2.4.38}$$

which yields the first formula in (2.4.32). The second one is proved similarly.

Let now $\alpha = m + i\theta$ ($m \in \mathbb{N}$; $\theta \neq 0$). Then $n = m + 1 \geqq 2$ and, analogous to (2.4.35) and (2.4.36), we have

$$(I_{a+}^{\alpha}y)^{(k)}(x) = (I_{a+}^{\alpha-k}y)(x) \quad (k = 0, 1, \cdots, n-2) \tag{2.4.39}$$

and

$$\left|(I_{a+}^{m+i\theta-k}D^{n}y)(x)\right| \leqq \frac{K(x-a)^{m-k}}{|\Gamma(m+i\theta-k)|(m-k)} \tag{2.4.40}$$

where $K = \|y\|_{L_\infty} (= \|y\|_C)$ $(k = 0, 1, \cdots, m-1)$ and hence

$$(I_{a+}^{\alpha}y)^{(k)}(a+) = 0 \quad (k = 0, 1, \cdots, n-2). \tag{2.4.41}$$

Using this formula, from (2.4.1) we obtain (2.4.33). (2.4.34) is proved similarly.

The next statement is an analog of Lemma 2.6(b) and Lemma 2.9(d).

Lemma 2.22 *Let $\Re(\alpha) > 0$ and let n be given by (2.4.3). If $y(x) \in AC^n[a,b]$ or $y(x) \in C^n[a,b]$, then*

$$(I_{a+}^{\alpha}\, {}^{C}D_{a+}^{\alpha}y)(x) = y(x) - \sum_{k=0}^{n-1} \frac{y^{(k)}(a)}{k!}(x-a)^k \tag{2.4.42}$$

and

$$(I_{b-}^{\alpha}\, {}^{C}D_{b-}^{\alpha}y)(x) = y(x) - \sum_{k=0}^{n-1} \frac{(-1)^k y^{(k)}(b)}{k!}(b-x)^k. \tag{2.4.43}$$

In particular, if $0 < \Re(\alpha) \leqq 1$ and $y(x) \in AC[a,b]$ or $y(x) \in C[a,b]$, then

$$(I_{a+}^{\alpha}\, {}^{C}D_{a+}^{\alpha}y)(x) = y(x) - y(a), \ \textit{and}\ (I_{b-}^{\alpha}\, {}^{C}D_{b-}^{\alpha}y)(x) = y(x) - y(b). \tag{2.4.44}$$

Proof. Let $\alpha \notin \mathbb{N}$. If $y(x) \in AC^n[a,b]$ ($y(x) \in C^n[a,b]$), then by Theorem 2.1(a) (Theorem 2.2(a)), $({}^{C}D_{a+}^{\alpha}y)(x)$ and $({}^{C}D_{-}^{\alpha}y)(x)$ are represented by (2.4.15) and (2.4.16), respectively. Then, using the semigroup property in (2.1.30), being held by Lemma 2.3 (Lemma 2.9(a)), and (2.1.8), we have

$$(I_{a+}^{\alpha}\, {}^{C}D_{a+}^{\alpha}y)(x) = (I_{a+}^{\alpha}I_{a+}^{n-\alpha}D^{n}y)(x) = (I_{a+}^{n}D_{a+}^{n}y)(x) \tag{2.4.45}$$

and

$$(I_{b-}^{\alpha}\, {}^{C}D_{b-}^{\alpha}y)(x) = (-1)^n(I_{b-}^{\alpha}I_{b-}^{n-\alpha}D^{n}y)(x) = (I_{b-}^{n}D_{b-}^{n}y)(x). \tag{2.4.46}$$

Then, by (2.1.41) and (2.1.48), from (2.4.45) and (2.4.46) we derive (2.4.42) and (2.4.43), respectively.

If $\alpha \in \mathbb{N}$, then applying Theorem 2.1(b) (Theorem 2.2(b)) and using (2.1.41) and (2.1.48), we obtain (2.4.42) and (2.4.43).

Remark 2.14 For $\alpha > 0$, the first formula in (2.4.32) and the relation (2.4.42) for the Caputo fractional derivative $(D_{*a}^{\alpha}y)(x)$ of the form (2.4.23) were obtained by Diethelm ([173], Theorems 3 and 4). We only note that these theorems are formulated for $\alpha \geqq 0$, but the case $\alpha = 0$ needs an additional explanation.

We have defined the Caputo derivatives on a finite interval $[a, b]$ by (2.4.1) and (2.4.2) and shown, in Theorems 2.1 and 2.2, that they can be represented in the forms (2.4.14) or (2.4.15) and (2.4.16), provided that $f(x) \in AC^n[a, b]$ and $f(x) \in C^n[a, b]$. Formulas (2.4.15) and (2.4.16) can be used for the definition of the Caputo fractional derivatives on the half axis $\mathbb{R}^+$ and on the whole axis $\mathbb{R}$. Thus the corresponding *Caputo fractional derivative of order* $\alpha \in \mathbb{C}$ (with $\Re(\alpha) > 0$ and $\alpha \notin \mathbb{N}$) on the half axis $\mathbb{R}^+$ and on the whole axis $\mathbb{R}$ can be defined as follows:

$$({}^C D_{0+}^{\alpha}y)(x) = \frac{1}{\Gamma(n-\alpha)} \int_0^x \frac{y^{(n)}(t)dt}{(x-t)^{\alpha+1-n}} \quad (x \in \mathbb{R}^+), \tag{2.4.47}$$

$$({}^C D_{-}^{\alpha}y)(x) = \frac{(-1)^n}{\Gamma(n-\alpha)} \int_x^{\infty} \frac{y^{(n)}(t)dt}{(t-x)^{\alpha+1-n}} \quad (x \in \mathbb{R}^+) \tag{2.4.48}$$

and

$$({}^C D_{+}^{\alpha}y)(x) = \frac{1}{\Gamma(n-\alpha)} \int_{-\infty}^x \frac{y^{(n)}(t)dt}{(x-t)^{\alpha+1-n}} \quad (x \in \mathbb{R}), \tag{2.4.49}$$

$$({}^C D_{-}^{\alpha}y)(x) = \frac{(-1)^n}{\Gamma(n-\alpha)} \int_x^{\infty} \frac{y^{(n)}(t)dt}{(t-x)^{\alpha+1-n}} \quad (x \in \mathbb{R}), \tag{2.4.50}$$

respectively.

When $0 < \Re(\alpha) < 1$, the relations (2.4.47)-(2.4.48) and (2.4.49)-(2.4.50) take the following forms:

$$({}^C D_{0+}^{\alpha}y)(x) = \frac{1}{\Gamma(1-\alpha)} \int_0^x \frac{y'(t)dt}{(x-t)^{\alpha}} \quad (x \in \mathbb{R}^+), \tag{2.4.51}$$

$$({}^C D_{-}^{\alpha}y)(x) = -\frac{1}{\Gamma(1-\alpha)} \int_x^{\infty} \frac{y'(t)dt}{(t-x)^{\alpha}} \quad (x \in \mathbb{R}^+) \tag{2.4.52}$$

and

$$({}^C D_{+}^{\alpha}y)(x) = \frac{1}{\Gamma(1-\alpha)} \int_{-\infty}^x \frac{y'(t)dt}{(x-t)^{\alpha}} \quad (x \in \mathbb{R}), \tag{2.4.53}$$

$$({}^C D_{b-}^{\alpha}y)(x) = -\frac{1}{\Gamma(1-\alpha)} \int_x^{\infty} \frac{y'(t)dt}{(t-x)^{\alpha}} \quad (x \in \mathbb{R}). \tag{2.4.54}$$

For $\alpha = n \in \mathbb{N}$, we define the Caputo derivatives $({}^C D_{0+}^n y)(x)$, $({}^C D_{-}^n y)(x)$ and $({}^C D_{+}^n y)(x)$, $({}^C D_{-}^n y)(x)$ by

$$({}^C D_{+}^n y)(x) := y^{(n)}(x),\ \ {}^C D_{-}^n y)(x) := (-1)^n y^{(n)}(x) \ \ (x \in \mathbb{R}^+) \tag{2.4.55}$$

and

$$({}^C D_{+}^\alpha y)(x) := y^{(n)}(x),\ \ ({}^C D_{-}^n y)(x) := (-1)^n y^{(n)}(x) \ \ (x \in \mathbb{R}). \tag{2.4.56}$$

The Caputo derivatives $({}^C D_{+}^\alpha y)(x)$ and $({}^C D_{-}^\alpha y)(x)$ have properties similar to those in (2.3.11) and (2.2.13).

Property 2.17 *If $\Re(\alpha) > 0$ and $\lambda > 0$, then*

$$\left({}^C D_{+}^\alpha e^{\lambda t}\right)(x) = \lambda^\alpha e^{\lambda x} \text{ and } \left({}^C D_{-}^\alpha e^{-\lambda t}\right)(x) = \lambda^\alpha e^{-\lambda x}. \tag{2.4.57}$$

The Mittag-Leffler function $E_\alpha[\lambda(x-a)^\alpha]$ is invariant with respect to the Caputo derivative ${}^C D_{a+}^\alpha$, but it is not the case for the Caputo derivative ${}^C D_{-}^\alpha$. In fact, the following assertion holds.

Lemma 2.23 *If $\alpha > 0$, $a \in \mathbb{R}$ and $\lambda \in \mathbb{C}$, then*

$$\left({}^C D_{a+}^\alpha E_\alpha[\lambda(t-a)^\alpha]\right)(x) = \lambda E_\alpha[\lambda(x-a)^\alpha] \tag{2.4.58}$$

and

$$\left({}^C D_{-}^\alpha t^{\alpha-1} E_\alpha\left(\lambda t^{-\alpha}\right)\right)(x) = \frac{1}{x} E_{\alpha,1-\alpha}\left(\lambda x^{-\alpha}\right). \tag{2.4.59}$$

In particular, when $\alpha = n \in \mathbb{N}$,

$$D^n E_n[\lambda(x-a)^n] = E_n[\lambda(x-a)^n] \tag{2.4.60}$$

and

$$D^n\left[t^{n-1} E_n\left(\lambda t^{-n}\right)\right)(x) = \frac{1}{x} E_{n,1-n}\left(\lambda x^{-n}\right) = \frac{\lambda}{x^{n+1}} E_n\left(\lambda x^{-n}\right). \tag{2.4.61}$$

Proof. Relations (2.4.58) and (2.4.59) are proved directly by using the definition (1.8.1) of the Mittag-Leffler function $E_\alpha(z)$ and by the term-by-term differentiation of the series in the left-hand sides of (2.4.58) and (2.4.59).

The following assertion, which yields the Laplace transform of the Caputo fractional derivative ${}^C D_{0+}^\alpha y$, is derived from Lemma 2.13(a) and (1.4.9).

Lemma 2.24 *Let $\alpha > 0$, $n-1 < \alpha \leqq n$ $(n \in \mathbb{N})$ be such that $y(x) \in C^n(\mathbb{R}^+)$, $y^{(n)}(x) \in L_1(0,b)$ for any $b > 0$, the estimate (2.2.35) holds for $y^{(n)}(x)$, the Laplace transforms $(\mathcal{L}y)(p)$ and $\mathcal{L}[D^n y(t)]$ exist, and $\lim_{x\to+\infty}(D^k y)(x) = 0$ for $k = 0, 1, \cdots, n-1$. Then the following relation holds:*

$$(\mathcal{L}^C D_{0+}^\alpha y)(s) = s^\alpha(\mathcal{L}y)(s) - \sum_{k=0}^{n-1} s^{\alpha-k-1}(D^k y)(0). \tag{2.4.62}$$

In particular, if $0 < \alpha \leqq 1$, then

$$(\mathcal{L}^C D_{0+}^\alpha y)(s) = s^\alpha(\mathcal{L}y)(s) - s^{\alpha-1}y(0). \tag{2.4.63}$$

Similarly, from Lemma 2.14, (1.4.37)-(1.4.38), and Remark 2.9, we derive the Mellin transform of the Caputo fractional derivatives ${}^C D_{0+}^{\alpha} y$ and ${}^C D_{-}^{\alpha} y$.

Lemma 2.25 *Let $\alpha > 0$, $n - 1 < \alpha \leqq n$ $(n \in \mathbb{N})$ be such that $y(x) \in C^n(\mathbb{R}^+)$, $y(x) \in X_{s+n-\alpha}^1(\mathbb{R}^+)$, and let there exist $(\mathcal{M}y^{(n)})(s + n - \alpha)$ and $(\mathcal{M}y)(s - \alpha)$.*

(a) *If $\Re(s) < 1 - \Re(n - \alpha)$, then*

$$(\mathcal{M}\ {}^C D_{0+}^{\alpha} y)(s) = \frac{\Gamma(1 + \alpha - s)}{\Gamma(1 - s)}(\mathcal{M}y)(s - \alpha)$$

$$+ \sum_{k=0}^{n-1} \frac{\Gamma(1 + k + \alpha - n - s)}{\Gamma(1 - s)} \left[x^{s+n-\alpha-k-1} y^{(n-k-1)}(x)\right]_0^{\infty}. \tag{2.4.64}$$

In particular, when $0 < \alpha < 1$,

$$(\mathcal{M}\ {}^C D_{0+}^{\alpha} y)(s) = \frac{\Gamma(1 + \alpha - s)}{\Gamma(1 - s)}(\mathcal{M}y)(s - \alpha) + \frac{\Gamma(\alpha - s)}{\Gamma(1 - s)} \left[x^{s-\alpha} y(x)\right]_0^{\infty}. \tag{2.4.65}$$

(b) *If $\Re(s) > 0$, then*

$$(\mathcal{M}\ {}^C D_{-}^{\alpha} y)(s) = \frac{\Gamma(s)}{\Gamma(s - \alpha)}(\mathcal{M}y)(s - \alpha)$$

$$+ \sum_{k=0}^{n-1} (-1)^{n-k} \frac{\Gamma(s)}{\Gamma(s + n - \alpha - k)} \left[x^{s+n-\alpha-k-1} y^{n-k-1}(x)\right]_0^{\infty}. \tag{2.4.66}$$

In particular, when $0 < \alpha < 1$,

$$(\mathcal{M}\ {}^C D_{-}^{\alpha} y)(s) = \frac{\Gamma(s)}{\Gamma(s - \alpha)}(\mathcal{M}y)(s - \alpha) - \frac{\Gamma(s)}{\Gamma(s + 1 - \alpha)} \left[x^{s-\alpha} y(x)\right]_0^{\infty}. \tag{2.4.67}$$

2.5 Fractional Integrals and Fractional Derivatives of a Function with Respect to Another Function

In this section we present the definitions and some properties of the fractional integrals and fractional derivatives of a function f with respect to another function g. Some of these definitions and results were given in Samko et al. ([729], Section 18.2).

Let (a, b) $(-\infty \leqq a < b \leqq \infty)$ be a finite or infinite interval of the real line $\mathbb{R}$ and $\Re(\alpha) > 0$. Also let $g(x)$ be an increasing and positive monotone function on $(a, b]$, having a continuous derivative $g'(x)$ on (a, b). The *left- and right-sided fractional integrals of a function f with respect to another function g* on $[a, b]$ are defined by

$$(I_{a+;g}^{\alpha} f)(x) := \frac{1}{\Gamma(\alpha)} \int_a^x \frac{g'(t) f(t) dt}{[g(x) - g(t)]^{1-\alpha}} \quad (x > a;\ \Re(\alpha) > 0) \tag{2.5.1}$$

and

$$(I^\alpha_{b-;g}f)(x) := \frac{1}{\Gamma(\alpha)}\int_x^b \frac{g'(t)f(t)dt}{[g(t)-g(x)]^{1-\alpha}} \quad (x < b;\ \Re(\alpha) > 0), \tag{2.5.2}$$

respectively. When $a = 0$ and $b = \infty$, we shall use the following notations:

$$(I^\alpha_{0+;g}f)(x) := \frac{1}{\Gamma(\alpha)}\int_0^x \frac{g'(t)f(t)dt}{[g(x)-g(t)]^{1-\alpha}} \quad (x > 0;\ \Re(\alpha) > 0), \tag{2.5.3}$$

$$(I^\alpha_{-;g}f)(x) := \frac{1}{\Gamma(\alpha)}\int_x^\infty \frac{g'(t)f(t)dt}{[g(t)-g(x)]^{1-\alpha}} \quad (x > 0;\ \Re(\alpha) > 0); \tag{2.5.4}$$

while, for $a = -\infty$ and $b = \infty$, we have

$$(I^\alpha_{+;g}f)(x) := \frac{1}{\Gamma(\alpha)}\int_{-\infty}^x \frac{g'(t)f(t)dt}{[g(x)-g(t)]^{1-\alpha}} \quad (x \in \mathbb{R};\ \Re(\alpha) > 0), \tag{2.5.5}$$

$$(I^\alpha_{-;g}f)(x) = \frac{1}{\Gamma(\alpha)}\int_x^\infty \frac{g'(t)f(t)dt}{[g(t)-g(x)]^{1-\alpha}} \quad (x \in \mathbb{R};\ \Re(\alpha) > 0). \tag{2.5.6}$$

Integrals (2.5.1) and (2.5.2) are called *the Riemann-Liouville fractional integrals on a finite interval* $[a, b]$, (2.5.3) and (2.5.4) *the Liouville fractional integrals on a half-axis* $\mathbb{R}^+$, while (2.5.5) and (2.5.6) are called *the Liouville fractional integrals on the whole axis* $\mathbb{R}$. If $g'(x) \neq 0$ $(-\infty \leqq a < x < b \leqq \infty)$, then the operators in (2.5.1)-(2.5.2), (2.5.3)-(2.5.4), and (2.5.5)-(2.5.6) are expressed via the Riemann-Liouville operators (2.1.1)-(2.1.2), the Liouville operators (2.2.1)-(2.2.2) and (2.3.1)-(2.3.2) by

$$(I^\alpha_{a+;g}f)(x) = (Q_g I^\alpha_{g(a)+} Q_g^{-1} f)(x), \quad (I^\alpha_{b-;g}f)(x) = (Q_g I^\alpha_{g(b)-} Q_g^{-1} f)(x), \tag{2.5.7}$$

$$(I^\alpha_{0+;g}f)(x) = (Q_g I^\alpha_{g(0)+} Q_g^{-1} f)(x), \quad (I^\alpha_{-;g}f)(x) = (Q_g I^\alpha_{g(+\infty)-} Q_g^{-1} f)(x), \tag{2.5.8}$$

and

$$(I^\alpha_{+;g}f)(x) = (Q_g I^\alpha_{g(-\infty)+} Q_g^{-1} f)(x), \quad (I^\alpha_{-;g}f)(x) = (Q_g I^\alpha_{g(+\infty)-} Q_g^{-1} f)(x), \tag{2.5.9}$$

where Q_g is the substitution operator

$$(Q_g f)(x) = f[g(x)], \tag{2.5.10}$$

and Q_g^{-1} is its inverse operator. Therefore, the properties of the integrals (2.5.1)-(2.5.6) follow from the corresponding properties of the Riemann-Liouville and Liouville fractional integrals given in Sections 2.1-2.3. For example, the following assertions hold, generalizing those in (2.1.16), (2.1.18), (2.3.10), and (2.2.14).

Property 2.18 *Let* $\Re(\alpha) > 0$ *and* $\Re(\beta) > 0$.

(a) *If* $f_+(x) = [g(x) - g(a)]^{\beta-1}$, *then*

$$(I^\alpha_{a+;g}f_+)(x) = \frac{\Gamma(\beta)}{\Gamma(\alpha+\beta)}[g(x)-g(a)]^{\alpha+\beta-1}. \tag{2.5.11}$$

(b) *If* $f_-(x) = [g(b) - g(x)]^{\beta-1}$, *then*

$$(I^{\alpha}_{b;g} f_-)(x) = \frac{\Gamma(\beta)}{\Gamma(\alpha+\beta)}[g(b) - g(x)]^{\alpha+\beta-1}. \tag{2.5.12}$$

Property 2.19 *Let* $\Re(\alpha) > 0$ *and* $\lambda > 0$. *Then*

$$(I^{\alpha}_{+;g} e^{\lambda g(t)})(x) = \lambda^{-\alpha} e^{\lambda g(x)} \tag{2.5.13}$$

and

$$(I^{\alpha}_{-;g} e^{-\lambda g(t)})(x) = \lambda^{-\alpha} e^{-\lambda g(x)}. \tag{2.5.14}$$

The semigroup property also holds.

Lemma 2.26 *Let* $\Re(\alpha) > 0$ *and* $\Re(\beta) > 0$. *Then the relations*

$$(I^{\alpha}_{a+;g} I^{\beta}_{a+;g} f)(x) = (I^{\alpha+\beta}_{a+;g} f)(x), \quad (I^{\alpha}_{b-;g} I^{\beta}_{b-;g} f)(x) = (I^{\alpha+\beta}_{b-;g} f)(x) \tag{2.5.15}$$

and

$$(I^{\alpha}_{+;g} I^{\beta}_{+;g} f)(x) = (I^{\alpha+\beta}_{+;g} f)(x), \quad (I^{\alpha}_{-;g} I^{\beta}_{-;g} f)(x) = (I^{\alpha+\beta}_{-;g} f)(x) \tag{2.5.16}$$

hold for "sufficiently good" functions $f(x)$.

Let $g'(x) \neq 0$ $(-\infty \leqq a < x < b \leqq \infty)$ and $\Re(\alpha) \geqq 0$ $(\alpha \neq 0)$. Also let $n = [\Re(\alpha)] + 1$ and $D = d/dx$. The Riemann-Liouville and Liouville fractional derivatives of a function y with respect to g of order α ($\Re(\alpha) \geqq 0$; $\alpha \neq 0$), corresponding to the Riemann-Liouville and Liouville integrals in (2.5.1)-(2.5.2), (2.5.3)-(2.5.4), and (2.5.5)-(2.5.6), are defined by

$$(D^{\alpha}_{a+;g} y)(x) := \left(\frac{1}{g'(x)} D\right)^n (I^{n-\alpha}_{a+;g} y)(x)$$

$$= \frac{1}{\Gamma(n-\alpha)} \left(\frac{1}{g'(x)} D\right)^n \int_a^x \frac{g'(t)y(t)dt}{[g(x) - g(t)]^{\alpha-n+1}} \quad (x > a), \tag{2.5.17}$$

$$(D^{\alpha}_{b-;g} y)(x) := \left(-\frac{1}{g'(x)} D\right)^n (I^{n-\alpha}_{b-;g} y)(x)$$

$$= \frac{1}{\Gamma(n-\alpha)} \left(-\frac{1}{g'(x)} D\right)^n \int_x^b \frac{g'(t)y(t)dt}{[g(t) - g(x)]^{\alpha-n+1}} \quad (x < b), \tag{2.5.18}$$

and

$$(D^{\alpha}_{0+;g} y)(x) := \left(\frac{1}{g'(x)} D\right)^n (I^{n-\alpha}_{0+;g} y)(x)$$

$$= \frac{1}{\Gamma(n-\alpha)} \left(\frac{1}{g'(x)} D\right)^n \int_0^x \frac{g'(t)y(t)dt}{[g(x) - g(t)]^{\alpha-n+1}} \quad (x > 0), \tag{2.5.19}$$

$$(D^{\alpha}_{-;g} y)(x) := \left(-\frac{1}{g'(x)} D\right)^n (I^{n-\alpha}_{-;g} y)(x)$$

$$= \frac{1}{\Gamma(n-\alpha)} \left(-\frac{1}{g'(x)}D\right)^n \int_x^\infty \frac{g'(t)y(t)dt}{[g(t)-g(x)]^{\alpha-n+1}} \quad (x>0), \tag{2.5.20}$$

and

$$(D^\alpha_{+;g}y)(x) := \left(\frac{1}{g'(x)}D\right)^n (I^{n-\alpha}_{+;g}y)(x)$$

$$= \frac{1}{\Gamma(n-\alpha)} \left(\frac{1}{g'(x)}D\right)^n \int_{-\infty}^x \frac{g'(t)y(t)dt}{[g(x)-g(t)]^{\alpha-n+1}} \quad (x\in\mathbb{R}), \tag{2.5.21}$$

$$(D^\alpha_{-;g}y)(x) = \left(-\frac{1}{g'(x)}D\right)^n (I^{n-\alpha}_{-;g}y)(x)$$

$$= \frac{1}{\Gamma(n-\alpha)} \left(-\frac{1}{g'(x)}D\right)^n \int_x^\infty \frac{g'(t)y(t)dt}{[g(t)-g(x)]^{\alpha-n+1}} \quad (x\in\mathbb{R}), \tag{2.5.22}$$

respectively.

When $g(x) = x$, (2.5.17) and (2.5.18) coincide with the Riemann-Liouville fractional derivatives (2.1.5) and (2.1.6):

$$(D^\alpha_{a+;x}y)(x) = (D^\alpha_{a+}y)(x) \text{ and } (D^\alpha_{b-;x}y)(x) = (D^\alpha_{b-}y)(x), \tag{2.5.23}$$

(2.5.19) and (2.5.20) coincide with the Liouville fractional derivatives (2.2.3) and (2.2.4):

$$(D^\alpha_{0+;x}y)(x) = (D^\alpha_{0+}y)(x) \text{ and } (D^\alpha_{-;x}y)(x) = (D^\alpha_{-}y)(x), \tag{2.5.24}$$

and (2.5.21) and (2.5.22) coincide with the Liouville fractional derivatives (2.3.3) and (2.3.4):

$$(D^\alpha_{+;x}y)(x) = (D^\alpha_{+}y)(x) \text{ and } (D^\alpha_{-;x}y)(x) = (D^\alpha_{-}y)(x). \tag{2.5.25}$$

When $\alpha = n \in \mathbb{N}$, the above general fractional derivatives (2.5.17)-(2.5.22) have the following forms:

$$(D^n_{a+;g}y)(x) = (D^n_{+;g}y)(x) = \left(\frac{1}{g'(x)}\frac{d}{dx}\right)^n y(x), \tag{2.5.26}$$

$$(D^n_{b-;g}y)(x) = (D^n_{-;g}y)(x) = \left(-\frac{1}{g'(x)}\frac{d}{dx}\right)^n y(x). \tag{2.5.27}$$

In particular, we have

$$(D^1_{a+;g}y)(x) = (D^1_{+;g}y)(x) = \frac{y'(x)}{g'(x)}, \tag{2.5.28}$$

$$(D^1_{b-;g}y)(x) = (D^1_{-;g}y)(x) = -\frac{y'(x)}{g'(x)}. \tag{2.5.29}$$

The following results, generalizing those in (2.1.17), (2.1.19) and (2.3.11), (2.2.15), are analogous to those in Properties 2.18 and 2.19:

Property 2.20 *Let* $\Re(\alpha) \geqq 0$ $(\alpha \neq 0)$ *and* $\Re(\beta) > 0$.

(a) *If* $y_+(x) = [g(x) - g(a)]^{\beta-1}$, *then*

$$(D^\alpha_{a+;g}y_+)(x) = \frac{\Gamma(\beta)}{\Gamma(\beta-\alpha)}[g(x) - g(a)]^{\beta-\alpha-1}. \tag{2.5.30}$$

(b) *If* $y_-(x) = [g(b) - g(x)]^{\beta-1}$, *then*

$$(D^\alpha_{b-;g}y_-)(x) = \frac{\Gamma(\beta)}{\Gamma(\beta-\alpha)}[g(b) - g(x)]^{\beta-\alpha-1}. \tag{2.5.31}$$

Property 2.21 *Let* $\Re(\alpha) \geqq 0$ $(\alpha \neq 0)$ *and* $\lambda > 0$. *Then*

$$(D^\alpha_{+;g}e^{\lambda g(t)})(x) = \lambda^\alpha e^{\lambda g(x)} \tag{2.5.32}$$

and

$$(D^\alpha_{-;g}e^{-\lambda g(t)})(x) = \lambda^\alpha e^{-\lambda g(x)}. \tag{2.5.33}$$

Analogous to the properties of the Riemann-Liouville and Liouville fractional derivatives, presented in Sections 2.1-2.3, there exist the corresponding properties for the general fractional derivatives in (2.5.17)-(2.5.22).

In conclusion, we consider the following special cases of the general integrals and derivatives (2.5.3) and (2.5.4) with $g(x) = x^\sigma$ $(\sigma > 0)$:

$$(I^\alpha_{0+;x^\sigma}f)(x) := \frac{\sigma}{\Gamma(\alpha)}\int_0^x \frac{t^{\sigma-1}f(t)dt}{(x^\sigma - t^\sigma)^{1-\alpha}} \quad (x > 0;\ \Re(\alpha) > 0), \tag{2.5.34}$$

$$(I^\alpha_{-;x^\sigma}f)(x) := \frac{\sigma}{\Gamma(\alpha)}\int_x^\infty \frac{t^{\sigma-1}f(t)dt}{(t^\sigma - x^\sigma)^{1-\alpha}} \quad (x > 0;\ \Re(\alpha) > 0) \tag{2.5.35}$$

and

$$(D^\alpha_{0+;x^\sigma}y)(x) := \frac{\sigma^{1-n}}{\Gamma(n-\alpha)}\left(x^{1-\sigma}D\right)^n \int_0^x \frac{t^{\sigma-1}y(t)dt}{(x^\sigma - t^\sigma)^{\alpha-n+1}} \quad (x > 0), \tag{2.5.36}$$

$$(D^\alpha_{-;x^\sigma}y)(x) := \frac{\sigma^{1-n}}{\Gamma(n-\alpha)}\left(-x^{1-\sigma}D\right)^n \int_x^\infty \frac{t^{\sigma-1}y(t)dt}{(t^\sigma - x^\sigma)^{\alpha-n+1}} \quad (x > 0) \tag{2.5.37}$$

$$(\Re(\alpha) \geqq 0\ (\alpha \neq 0);\ n = [\Re(\alpha)] + 1;\ D = \frac{d}{dx})$$

We note that $I^\alpha_{0+;x^\sigma}f$ and $I^\alpha_{-;x^\sigma}f$ are connected with the fractional integrals $I^\alpha_{0+}f$ and $I^\alpha_{-}f$ by

$$(I^\alpha_{0+;x^\sigma}f)(x) = \left(I^\alpha_{0+}f(t^{1/\sigma})\right)(x^\sigma) = (N_\sigma I^\alpha_{0+}N_{1/\sigma}f)(x) \quad (x > 0) \tag{2.5.38}$$

and

$$(I^\alpha_{-;x^\sigma}f)(x) = \left(I^\alpha_{-}f(t^{1/\sigma})\right)(x^\sigma) = (N_\sigma I^\alpha_{-}N_{1/\sigma}f)(x) \quad (x > 0), \tag{2.5.39}$$

where N_σ is an elementary operator given by (1.4.27).

As a particular case of (2.5.11), (2.5.30) and (2.5.14), (2.5.33), we have the following results.

Property 2.22 *Let $\sigma > 0$ and $\Re(\alpha) \geqq 0$ $(\alpha \neq 0)$.*

(a) *If $\Re(\alpha) > 0$ and $\Re(\beta) > 0$, then*

$$(I^{\alpha}_{0+;x^\sigma} t^{\sigma(\beta-1)})(x) = \frac{\Gamma(\beta)}{\Gamma(\alpha+\beta)} x^{\sigma(\alpha+\beta-1)} \tag{2.5.40}$$

and

$$(D^{\alpha}_{0+;x^{\sigma(\beta-1)}} t^{\sigma})(x) = \frac{\Gamma(\beta)}{\Gamma(\beta-\alpha)} x^{\sigma(\beta-\alpha-1)}. \tag{2.5.41}$$

(b) *If $\Re(\alpha) \geqq 0$ $(\alpha \neq 0)$ and $\lambda > 0$, then*

$$(I^{\alpha}_{-;x^\sigma} e^{-\lambda t^\sigma})(x) = \lambda^{-\alpha} e^{-\lambda x^\sigma} \tag{2.5.42}$$

and

$$(D^{\alpha}_{-;x^\sigma} e^{-\lambda t^\sigma})(x) = \lambda^{\alpha} e^{-\lambda x^\sigma}. \tag{2.5.43}$$

Using the representations (2.5.38) and (2.5.39) and the relation (1.4.29), and taking (2.2.36) and (2.2.37) into account, we obtain the Mellin transform of the general fractional integrals (2.5.34) and (2.5.35).

Lemma 2.27 *Let $\Re(\alpha) > 0$, $\sigma > 0$ and $f(x) \in X^1_{s+\sigma\alpha}(\mathbb{R}^+)$.*

(a) *If $\Re(s) < \sigma[1 - \Re(\alpha)]$, then*

$$(\mathcal{M} I^{\alpha}_{0+;x^\sigma} f)(s) = \frac{\Gamma(1-\alpha-s/\sigma)}{\Gamma(1-s/\sigma)} (\mathcal{M} f)(s+\sigma\alpha). \tag{2.5.44}$$

(b) *If $\Re(s) > 0$, then*

$$(\mathcal{M} I^{\alpha}_{-;x^\sigma} f)(s) = \frac{\Gamma(s/\sigma)}{\Gamma(\alpha+s/\sigma)} (\mathcal{M} f)(s+\sigma\alpha). \tag{2.5.45}$$

Remark 2.15 The formulas for the Mellin transform of the fractional derivatives (2.5.36) and (2.5.37) are more complicated than the relations in (2.5.44) and (2.5.45). For example, if $\Re(\alpha) \geqq 0$, $(\alpha \neq 0)$, $n = [\Re(\alpha)] + 1$, and $\sigma > 0$, then the Mellin transform of $(D^{\alpha}_{0+;x^\sigma} y)(x)$ for "sufficiently good" functions $y(x)$ is given by

$$(\mathcal{M} D^{\alpha}_{0+;x^\sigma} y)(s) = \frac{\Gamma(1+\alpha-s/\sigma)}{\Gamma(1-s/\sigma)} (\mathcal{M} y)(s-\sigma\alpha)$$

$$+\sigma^{-n} \left[x^{s-\sigma} \left(x^{1-\sigma} D \right)^{n-1} \left(I^{n-\alpha}_{0+;x^\sigma} y \right)(x) \right]_{x=0}^{\infty}$$

$$+ \sum_{k=1}^{n-1} \left(\prod_{j=1}^{k} (\sigma j - s) \right) \left[x^{s-(k+1)\sigma} (x^{1-\sigma} D)^{n-1-k} \left(I^{n-\alpha}_{0+;x^\sigma} y \right)(x) \right]_{x=0}^{\infty}. \tag{2.5.46}$$

In particular, when

$$\lim_{x \to 0+} \left[x^{s-(k+1)\sigma} (x^{1-\sigma} D)^{n-1-k} \left(I^{n-\alpha}_{0+;x^\sigma} y \right)(x) \right] = 0 \quad (k = 0, 1, \cdots, n-1) \tag{2.5.47}$$

and

$$\lim_{x\to\infty}\left[x^{s-(k+1)\sigma}(x^{1-\sigma}D)^{n-1-k}\left(I^{n-\alpha}_{0+;x^\sigma}y\right)(x)\right]=0 \quad (k=0,1,\cdots,\ n-1), \tag{2.5.48}$$

the formula (2.5.46) is simplified to the form

$$(\mathcal{M}D^{\alpha}_{0+;x^\sigma}y)(s)=\frac{\Gamma(1+\alpha-s/\sigma)}{\Gamma(1-s/\sigma)}(\mathcal{M}y)(s-\sigma\alpha). \tag{2.5.49}$$

When $\sigma=1$, then (2.5.49) coincides with (2.2.43), and (2.5.46) coincides with (2.2.47).

2.6 Erdélyi-Kober Type Fractional Integrals and Fractional Derivatives

In this section we present the definitions and some properties of the Erdélyi-Kober type fractional integrals and fractional derivatives and their special cases. Some of these definitions and results were presented in Samko et al. ([729], Section 18.1), Kiryakova [417] and McBride [566].

Let (a,b) $(-\infty \leqq a < b \leqq \infty)$ be a finite or infinite interval of the half-axis $\mathbb{R}^+$. Also let $\Re(\alpha)>0$, $\sigma>0$, and $\eta\in\mathbb{C}$. We consider the left- and right-sided integrals of order $\alpha\in\mathbb{C}$ $(\Re(\alpha)>0)$ defined by

$$(I^{\alpha}_{a+;\sigma,\eta}f)(x):=\frac{\sigma x^{-\sigma(\alpha+\eta)}}{\Gamma(\alpha)}\int_a^x\frac{t^{\sigma\eta+\sigma-1}f(t)dt}{(x^\sigma-t^\sigma)^{1-\alpha}} \quad (0\leqq a<x<b\leqq\infty) \tag{2.6.1}$$

and

$$(I^{\alpha}_{b-;\sigma,\eta}f)(x):=\frac{\sigma x^{\sigma\eta}}{\Gamma(\alpha)}\int_x^b\frac{t^{\sigma(1-\alpha-\eta)-1}f(t)dt}{(t^\sigma-x^\sigma)^{1-\alpha}} \quad (0\leqq a<x<b\leqq\infty), \tag{2.6.2}$$

respectively. When $a=-\infty$ and $b=\infty$, we shall use the following notations:

$$(I^{\alpha}_{+;\sigma,\eta}f)(x):=\frac{\sigma x^{-\sigma(\alpha+\eta)}}{\Gamma(\alpha)}\int_{-\infty}^x\frac{t^{\sigma\eta+\sigma-1}f(t)dt}{(x^\sigma-t^\sigma)^{1-\alpha}} \quad (x>0) \tag{2.6.3}$$

and

$$(I^{\alpha}_{-;\sigma,\eta}f)(x):=\frac{\sigma x^{\sigma\eta}}{\Gamma(\alpha)}\int_x^\infty\frac{t^{\sigma(1-\alpha-\eta)-1}f(t)dt}{(t^\sigma-x^\sigma)^{1-\alpha}} \quad (x>0). \tag{2.6.4}$$

Integrals (2.6.1) and (2.6.2), as well as (2.6.3) and (2.6.4), are called the *Erdélyi-Kober type fractional integrals.* If $\sigma=2$, $a=0$, and $b=\infty$, then integrals (2.6.1) and (2.6.4) are denoted by

$$(I_{\eta,\alpha}f)(x):=(I^{\alpha}_{0+;2,\eta}f)(x)=\frac{2x^{-2(\alpha+\eta)}}{\Gamma(\alpha)}\int_0^x\frac{t^{2\eta+1}f(t)dt}{(x^2-t^2)^{1-\alpha}} \quad (x>0) \tag{2.6.5}$$

and

$$(K_{\eta,\alpha}f)(x) := (I^{\alpha}_{-;2,\eta}f)(x) = \frac{2x^{2\eta}}{\Gamma(\alpha)}\int_x^{\infty}\frac{t^{1-2(\alpha+\eta)}f(t)dt}{(t^2-x^2)^{1-\alpha}} \quad (x>0). \tag{2.6.6}$$

These operators are called the *Erdélyi-Kober operators.* When $\sigma = 1$, $a = 0$, and $b = \infty$, the integrals in (2.6.1) and (2.6.4) take the forms

$$(I^{+}_{\eta,\alpha}f)(x) := (I^{\alpha}_{0+;1,\eta}f)(x) = \frac{x^{-\alpha-\eta}}{\Gamma(\alpha)}\int_0^{x}\frac{t^{\eta}f(t)dt}{(x-t)^{1-\alpha}} \quad (x>0) \tag{2.6.7}$$

and

$$(K^{-}_{\eta,\alpha}f)(x) := (I^{\alpha}_{-;1,\eta}f)(x) = \frac{x^{\eta}}{\Gamma(\alpha)}\int_x^{\infty}\frac{t^{-\alpha-\eta}f(t)dt}{(t-x)^{1-\alpha}} \quad (x>0). \tag{2.6.8}$$

These operators are known as the *Kober operators* or the *Kober-Erdélyi operators.*

Using the operators M_a and N_a defined in (1.4.26) and (1.4.27), we can rewrite (2.6.1) and (2.6.2) in terms of the Riemann-Liouville fractional integrals (2.1.1) and (2.1.2) in the form

$$(I^{\alpha}_{a+;\sigma,\eta}f)(x) = (N_\sigma M_{-\alpha-\eta} I^{\alpha}_{a^\sigma+} M_\eta N_{1/\sigma} f)(x) \quad (0 \leqq a < x < b \leqq \infty) \tag{2.6.9}$$

and

$$(I^{\alpha}_{b-;\sigma,\eta}f)(x) = (N_\sigma M_{\eta} I^{\alpha}_{b^\sigma-} M_{-\alpha-\eta} N_{1/\sigma} f)(x) \quad (0 \leqq a < x < b \leqq \infty), \tag{2.6.10}$$

respectively. Similarly, (2.6.3) and (2.6.4) have the following representations in terms of the Liouville fractional integrals (2.3.1) and (2.3.2):

$$(I^{\alpha}_{+;\sigma,\eta}f)(x) = (N_\sigma M_{-\alpha-\eta} I^{\alpha}_{+} M_\eta N_{1/\sigma} f)(x) \quad (x>0), \tag{2.6.11}$$

$$(I^{\alpha}_{-;\sigma,\eta}f)(x) = (N_\sigma M_{\eta} I^{\alpha}_{-} M_{-\alpha-\eta} N_{1/\sigma} f)(x) \quad (x>0). \tag{2.6.12}$$

In particular, the integrals in (2.6.5) and (2.6.6) are represented via the Liouville fractional integrals (2.2.1) and (2.2.2) as follows:

$$(I_{\eta,\alpha}f)(x) = (N_2 M_{-\alpha-\eta} I^{\alpha}_{0+} M_\eta N_{1/2} f)(x) \quad (x>0), \tag{2.6.13}$$

$$(K_{\eta,\alpha}f)(x) = (N_2 M_{\eta} I^{\alpha}_{-} M_{-\alpha-\eta} N_{1/2} f)(x) \quad (x>0), \tag{2.6.14}$$

while (2.6.7) and (2.6.8) take the forms

$$(I^{+}_{\eta,\alpha}f)(x) = (M_{-\alpha-\eta} I^{\alpha}_{0+} M_\eta f)(x) \quad (x>0), \tag{2.6.15}$$

$$(K^{-}_{\eta,\alpha}f)(x) = (M_{\eta} I^{\alpha}_{-} M_{-\alpha-\eta} f)(x) \quad (x>0). \tag{2.6.16}$$

In accordance with (2.6.9)-(2.6.10), the properties of the Erdélyi-Kober type fractional integrals (2.6.1)-(2.6.2) can be derived from the corresponding properties of the Riemann-Liouville fractional integrals. Similarly, from (2.6.11)-(2.6.12) and the properties of the Liouville fractional integrals (2.3.1)-(2.3.2), we can obtain the properties of the operators (2.6.3)-(2.6.4). In particular, the properties of the operators (2.6.5)-(2.6.6) and (2.6.7)-(2.6.8) can also be derived from the properties of Liouville fractional integrals (2.2.1) and (2.2.2). We now present some of these properties. The boundedness of the operators $I^{\alpha}_{a+;\sigma,\eta}$ and $I^{\alpha}_{b-;\sigma,\eta}$ in the space $L_p(a.b)$ is given by the following result.

Lemma 2.28 *Let $\Re(\alpha) > 0$ and $1 \leqq p < \infty$.*

(a) *If $0 < a < b < \infty$, then the operators $I^{\alpha}_{a+;\sigma,\eta}$ and $I^{\alpha}_{b-;\sigma,\eta}$ are bounded in $L_p(a,b)$.*

(b) *If $a = 0$, $b = \infty$, and $\Re(\eta) > -1 + 1/(p\sigma)$, then the operator $I^{\alpha}_{0+;\sigma,\eta}$ is bounded in $L_p(\mathbb{R}^+)$. In particular, the operators $I_{\eta,\alpha}$ and $I^{+}_{\eta,\alpha}$ are bounded in $L_p(\mathbb{R}^+)$, provided that $\Re(\eta) > -1 + 1/(2p)$ and $\Re(\eta) > -1 + 1/p$, respectively.*

(c) *If $a = 0$, $b = \infty$, and $\Re(\eta) > -1/(p\sigma)$, then the operator $I^{\alpha}_{-;\sigma,\eta}$ is bounded in $L_p(\mathbb{R}^+)$. In particular, the operators $K_{\eta,\alpha}$ and $K^{-}_{\eta,\alpha}$ are bounded in $L_p(\mathbb{R}^+)$, provided that $\Re(\eta) > -1/(2p)$ and $\Re(\eta) > -1/p$, respectively.*

The semigroup property also holds.

Lemma 2.29 *Let $\Re(\alpha) > 0$, $\Re(\beta) > 0$, and $1 \leqq p < \infty$.*

(a) *If $0 < a < b < \infty$ and $f(x) \in L_p(a,b)$, then*

$$\left(I^{\alpha}_{a+;\sigma,\eta} I^{\beta}_{a+;\sigma,\eta+\alpha} f\right)(x) = \left(I^{\alpha+\beta}_{a+;\sigma,\eta} f\right)(x) \tag{2.6.17}$$

and

$$\left(I^{\alpha}_{b-;\sigma,\eta} I^{\beta}_{b-;\sigma,\eta+\alpha} f\right)(x) = \left(I^{\alpha+\beta}_{b-;\sigma,\eta} f\right)(x). \tag{2.6.18}$$

(b) *If $a = 0$, $b = \infty$, $f(x) \in L_p(\mathbb{R}^+)$, and $\Re(\eta) > -1 + 1/(p\sigma)$, then*

$$\left(I^{\alpha}_{0+;\sigma,\eta} I^{\beta}_{0+;\sigma,\eta+\alpha} f\right)(x) = \left(I^{\alpha+\beta}_{0+;\sigma,\eta} f\right)(x). \tag{2.6.19}$$

In particular, if $\Re(\eta) > -1 + 1/(2p)$, then

$$\left(I_{\eta,\alpha} I_{\eta+\alpha,\beta} f\right)(x) = \left(I_{\eta,\alpha+\beta} f\right)(x), \tag{2.6.20}$$

while, for $\Re(\eta) > -1 + 1/p$,

$$\left(I^{+}_{\eta,\alpha} I^{+}_{\eta+\alpha,\beta} f\right)(x) = \left(I^{+}_{\eta,\alpha+\beta} f\right)(x). \tag{2.6.21}$$

(c) *If $a = 0$, $b = \infty$, $f(x) \in L_p(\mathbb{R}^+)$, and $\Re(\eta) > -1/(p\sigma)$, then*

$$\left(I^{\alpha}_{-;\sigma,\eta} I^{\beta}_{-;\sigma,\eta+\alpha} f\right)(x) = \left(I^{\alpha+\beta}_{-;\sigma,\eta} f\right)(x). \tag{2.6.22}$$

In particular, if $\Re(\eta) > -1/(2p)$, then

$$\left(K_{\eta,\alpha} K_{\eta+\alpha,\beta} f\right)(x) = \left(K_{\eta,\alpha+\beta} f\right)(x), \tag{2.6.23}$$

while, for $\Re(\eta) > -1/p$,

$$\left(K^{-}_{\eta,\alpha} K^{-}_{\eta+\alpha,\beta} f\right)(x) = \left(K^{-}_{\eta,\alpha+\beta} f\right)(x). \tag{2.6.24}$$

The weighted analog of the relation of fractional integration by parts (2.1.49) also holds.

Lemma 2.30 *Let $\Re(\alpha) > 0$ and $0 \leqq a < b \leqq \infty$. Then the following relations are valid for "sufficiently good" functions $f(x)$ and $g(x)$:*

$$\int_a^b x^{\sigma-1} f(x)(I^\alpha_{a+;\sigma,\eta} g)(x)dx = \int_a^b x^{\sigma-1} g(x)(I^\alpha_{b-;\sigma,\eta} f)(x)dx. \tag{2.6.25}$$

If $a = 0$ and $b = \infty$, then

$$\int_0^\infty x^{\sigma-1} f(x)(I^\alpha_{0+;\sigma,\eta} g)(x)dx = \int_0^\infty x^{\sigma-1} g(x)(I^\alpha_{-;\sigma,\eta} f)(x)dx. \tag{2.6.26}$$

In particular,

$$\int_0^\infty x f(x)(I_{\eta,\alpha} g)(x)dx = \int_0^\infty x g(x)(K_{\eta,\alpha} f)(x)dx \tag{2.6.27}$$

and

$$\int_0^\infty f(x)(I^+_{\eta,\alpha} g)(x)dx = \int_0^\infty g(x)(K^+_{\eta,\alpha} f)(x)dx. \tag{2.6.28}$$

Let $\Re(\alpha) \geqq 0$, $(\alpha \neq 0)$ and $n = [\Re(\alpha)] + 1$. Also let $\sigma > 0$ and $\eta \in \mathbb{C}$. *The Erdélyi-Kober type fractional derivatives*, corresponding to the Erdélyi-Kober type fractional integrals (2.6.1) and (2.6.2), are defined, for $0 \leqq a < x < b \leqq \infty$, by

$$(D^\alpha_{a+;\sigma,\eta} y)(x) := x^{-\sigma\eta}\left(\frac{1}{\sigma x^{\sigma-1}} D\right)^n x^{\sigma(n+\eta)} \left(I^{n-\alpha}_{a+;\sigma,\eta+\alpha} y\right)(x) \tag{2.6.29}$$

and

$$(D^\alpha_{b-;\sigma,\eta} y)(x) := x^{\sigma(\eta+\alpha)}\left(-\frac{1}{\sigma x^{\sigma-1}} D\right)^n x^{\sigma(n-\eta-\alpha)} \left(I^{n-\alpha}_{b-;\sigma,\eta+\alpha-n} y\right)(x), \tag{2.6.30}$$

respectively, with $D = d/dx$. When $a = -\infty$ and $b = \infty$, these definitions are written as follows:

$$(D^\alpha_{+;\sigma,\eta} y)(x) := x^{-\sigma\eta}\left(\frac{1}{\sigma x^{\sigma-1}} D\right)^n x^{\sigma(n+\eta)} \left(I^{n-\alpha}_{+;\sigma,\eta+\alpha} y\right)(x), \tag{2.6.31}$$

$$(D^\alpha_{-;\sigma,\eta} y)(x) := x^{\sigma(\eta+\alpha)}\left(-\frac{1}{\sigma x^{\sigma-1}} D\right)^n x^{\sigma(n-\eta-\alpha)} \left(I^{n-\alpha}_{-;\sigma,\eta+\alpha-n} y\right)(x). \tag{2.6.32}$$

When $\sigma = 2$, relations (2.6.29) (with $a = 0$) and (2.6.32) take the forms

$$(D^+_{\eta,\alpha} y)(x) := (D^\alpha_{0+;2,\alpha} y)(x) = x^{-2\eta}\left(\frac{1}{2x} D\right)^n x^{2(n+\eta)} \left(I_{\eta+\alpha,n-\alpha} y\right)(x), \tag{2.6.33}$$

$$(D^-_{\eta,\alpha} y)(x) := (D^\alpha_{-;2,\alpha} y)(x)$$
$$= x^{2(\eta+n)}\left(-\frac{1}{2x} D\right)^n x^{2(n-\eta-\alpha)} \left(K_{n+\alpha-n,n-\alpha} y\right)(x), \tag{2.6.34}$$

while, for $\sigma = 1$, they are given by

$$(\tilde{D}^{+}_{\eta,\alpha}y)(x) := (D^{\alpha}_{a+;1,\eta}y)(x) = x^{-\eta}D^{n}x^{n+\eta}\left(I_{\eta+\alpha,n-\alpha}y\right)(x), \tag{2.6.35}$$

$$(\tilde{D}^{-}_{\eta,\alpha}y)(x) := (D^{\alpha}_{-;1,\eta}y)(x) = x^{\eta+\alpha}(-D)^{n}x^{n-\eta-\alpha}\left(K_{\eta+\alpha-n,n-\alpha}y\right)(x). \tag{2.6.36}$$

If we suppose $\alpha = n \in \mathbb{N}$ in (2.6.29) and (2.6.30), then we obtain

$$(D^{n}_{a+;\sigma,\eta}y)(x) = (D^{n}_{+;\sigma,\eta}y)(x) = x^{-\sigma\eta}\left(\frac{1}{\sigma x^{\sigma-1}}D\right)^{n} x^{\sigma(n+\eta)}y(x), \tag{2.6.37}$$

$$(D^{n}_{b-;\sigma,\eta}y)(x) = (D^{n}_{-;\sigma,\eta}y)(x) = x^{\sigma(\eta+n)}\left(-\frac{1}{\sigma x^{\sigma-1}}D\right)^{n} x^{-\sigma\eta}y(x). \tag{2.6.38}$$

In particular, when $\sigma = 2$, $a = 0$, and $b = \infty$, we have

$$(D^{+}_{\eta,n}y)(x) = x^{-2\eta}\left(\frac{1}{2x}D\right)^{n} x^{2(n+\eta)}y(x), \tag{2.6.39}$$

$$(D^{-}_{\eta,n}y)(x) = x^{2(\eta+n)}\left(-\frac{1}{2x}D\right)^{n} x^{-2\eta}y(x), \tag{2.6.40}$$

while, for $\sigma = 1$, $a = 0$, and $b = \infty$, we have

$$(\tilde{D}^{+}_{\eta,n}y)(x) = x^{-\eta}D^{n}x^{n+\eta}y(x), \tag{2.6.41}$$

$$(\tilde{D}^{-}_{\eta,n}y)(x) = x^{\eta+n}(-D)^{n}x^{-\eta}y(x). \tag{2.6.42}$$

The Erdélyi-Kober type fractional differentiation operators in (2.6.29) and (2.6.30) provide operations inverse to the Erdélyi-Kober type operators (2.6.1) and (2.6.2) from the left.

Property 2.23 *If $\Re(\alpha) > 0$ and $0 \leqq a < b \leqq \infty$, then, for "sufficiently good" functions $f(x)$, the formulas*

$$\left(D^{\alpha}_{a+;\sigma,\eta}I^{\alpha}_{a+;\sigma,\eta}f\right)(x) = f(x), \quad \left(D^{\alpha}_{b-;\sigma,\eta}I^{\alpha}_{b-;\sigma,\eta}f\right)(x) = f(x), \tag{2.6.43}$$

$$\left(D^{\alpha}_{0+;\sigma,\eta}I^{\alpha}_{0+;\sigma,\eta}f\right)(x) = f(x), \quad \left(D^{\alpha}_{-;\sigma,\eta}I^{\alpha}_{b-;\sigma,\eta}f\right)(x) = f(x) \tag{2.6.44}$$

hold. In particular, when $\sigma = 2$,

$$\left(D^{+}_{\eta,\alpha}I_{\eta,\alpha}f\right)(x) = f(x), \text{ and } \left(D^{-}_{\eta,\alpha}K_{\eta,\alpha}f\right)(x) = f(x), \tag{2.6.45}$$

while, for $\sigma = 1$,

$$\left(\tilde{D}^{+}_{\eta,\alpha}I^{+}_{\eta,\alpha}f\right)(x) = f(x) \text{and} \quad \left(\tilde{D}^{-}_{\eta,\alpha}K^{-}_{\eta,\alpha}f\right)(x) = f(x). \tag{2.6.46}$$

The Mellin transform (1.4.23) of the Erdélyi-Kober type fractional integrals $I^{\alpha}_{0+;\sigma,\eta}f$ and $I^{\alpha}_{-;\sigma,\eta}f$ on the half-axis $\mathbb{R}^{+}$ are given by the following lemma.

Lemma 2.31 *Let $\Re(\alpha) > 0$, $\sigma > 0$ and $\eta \in \mathbb{C}$. Also let $f(x) \in L_p(\mathbb{R}^+)$ with $1 \leqq p \leqq 2$, and suppose that $s = (1/p) + i\tau$ ($\tau \in \mathbb{R}$).*

(a) *If $\Re(\eta - s/\sigma) > -1$, then*

$$\left(\mathcal{M}I^{\alpha}_{0+;\sigma,\eta}f\right)(s) = \frac{\Gamma(1+\eta-s/\sigma)}{\Gamma(1+\eta+\alpha-s/\sigma)}(\mathcal{M}f)(s). \tag{2.6.47}$$

In particular, when $\Re(\eta - s/2) > -1$, then

$$\left(\mathcal{M}I_{\eta,\alpha}f\right)(s) = \frac{\Gamma(1+\eta-s/2)}{\Gamma(1+\eta+\alpha-s/2)}(\mathcal{M}f)(s), \tag{2.6.48}$$

while, for $\Re(\eta - s) > -1$,

$$\left(\mathcal{M}I^{+}_{\eta,\alpha}f\right)(s) = \frac{\Gamma(1+\eta-s)}{\Gamma(1+\eta+\alpha-s)}(\mathcal{M}f)(s). \tag{2.6.49}$$

(b) *If $\Re(\eta + s/\sigma) > 0$, then*

$$\left(\mathcal{M}I^{\alpha}_{-;\sigma,\eta}f\right)(s) = \frac{\Gamma(\eta+s/\sigma)}{\Gamma(\eta+\alpha+s/\sigma)}(\mathcal{M}f)(s). \tag{2.6.50}$$

In particular, when $\Re(\eta + s/2) > 0$, then

$$\left(\mathcal{M}K_{\eta,\alpha}f\right)(s) = \frac{\Gamma(\eta+s/2)}{\Gamma(\eta+\alpha+s/2)}(\mathcal{M}f)(s), \tag{2.6.51}$$

while, for $\Re(\eta + s) > 0$,

$$\left(\mathcal{M}K^{-}_{\eta,\alpha}f\right)(s) = \frac{\Gamma(\eta+s)}{\Gamma(\eta+\alpha+s)}(\mathcal{M}f)(s). \tag{2.6.52}$$

2.7 Hadamard Type Fractional Integrals and Fractional Derivatives

In this section we present the definitions and some properties of the Hadamard type fractional integrals and fractional derivatives. Some of these definitions and results were presented in Samko et al. ([729], Section 18.3), Butzer et al. ([113], [112], [114]), Kilbas [370], and Kilbas and Titura [405].

Let (a, b) $(0 \leqq a < b \leqq \infty)$ be a finite or infinite interval of the half-axis $\mathbb{R}^+$, and let $\Re(\alpha) > 0$ and $\mu \in \mathbb{C}$. We consider the left-sided and right-sided integrals of fractional order $\alpha \in \mathbb{C}$ $(\Re(\alpha) > 0)$ defined by

$$(\mathcal{J}^{\alpha}_{a+}f)(x) := \frac{1}{\Gamma(\alpha)}\int_a^x \left(\log\frac{x}{t}\right)^{\alpha-1}\frac{f(t)dt}{t} \quad (a < x < b) \tag{2.7.1}$$

and

$$(\mathcal{J}^{\alpha}_{b-}f)(x) := \frac{1}{\Gamma(\alpha)}\int_x^b \left(\log\frac{t}{x}\right)^{\alpha-1}\frac{f(t)dt}{t} \quad (a < x < b), \tag{2.7.2}$$

respectively. When $a = 0$ and $b = \infty$, these relations are given by

$$(\mathcal{J}_{0+}^{\alpha} f)(x) := \frac{1}{\Gamma(\alpha)} \int_0^x \left(\log \frac{x}{t}\right)^{\alpha-1} \frac{f(t)dt}{t} \quad (x > 0) \tag{2.7.3}$$

and

$$(\mathcal{J}_{-}^{\alpha} f)(x) := \frac{1}{\Gamma(\alpha)} \int_x^{\infty} \left(\log \frac{t}{x}\right)^{\alpha-1} \frac{f(t)dt}{t} \quad (x > 0). \tag{2.7.4}$$

The fractional integrals, more general than those in (2.7.3) and (2.7.4) with $\mu \in \mathbb{C}$, are defined by

$$(\mathcal{J}_{0+,\mu}^{\alpha} f)(x) := \frac{1}{\Gamma(\alpha)} \int_0^x \left(\frac{t}{x}\right)^{\mu} \left(\log \frac{x}{t}\right)^{\alpha-1} \frac{f(t)dt}{t} \quad (x > 0) \tag{2.7.5}$$

and

$$(\mathcal{J}_{-,\mu}^{\alpha} f)(x) := \frac{1}{\Gamma(\alpha)} \int_x^{\infty} \left(\frac{x}{t}\right)^{\mu} \left(\log \frac{t}{x}\right)^{\alpha-1} \frac{f(t)dt}{t} \quad (x > 0). \tag{2.7.6}$$

The integral in (2.7.3) was introduced by Hadamard [319] [J.Math.Pure et Appl.(1892)]. Therefore, the integrals (2.7.1), (2.7.2) and (2.7.3), (2.7.4) are often referred to as the *Hadamard fractional integrals of order* α [see Samko et al. ([729], Section 18.3 and Section 23.1, notes to Section 18.3)]. The general integrals (2.7.5) and (2.7.6), introduced by Butzer et al. [113], are called the *Hadamard type fractional integrals of order* α.

Let $\delta = xD$ ($D = d/dx$) be the δ-derivative (1.1.11). The left- and right-sided *Hadamard fractional derivatives of order* $\alpha \in \mathbb{C}$ ($\Re(\alpha) \geqq 0$) on (a, b) are defined by

$$(\mathcal{D}_{a+}^{\alpha} y)(x) := \delta^n (\mathcal{J}_{a+}^{n-\alpha} y)(x)$$

$$= \left(x\frac{d}{dx}\right)^n \frac{1}{\Gamma(n-\alpha)} \int_a^x \left(\log \frac{x}{t}\right)^{n-\alpha+1} \frac{y(t)dt}{t} \quad (a < x < b) \tag{2.7.7}$$

and

$$(\mathcal{D}_{b-}^{\alpha} y)(x) := (-\delta)^n (\mathcal{J}_{b-}^{n-\alpha} y)(x)$$

$$= \left(-x\frac{d}{dx}\right)^n \frac{1}{\Gamma(n-\alpha)} \int_x^b \left(\log \frac{t}{x}\right)^{n-\alpha+1} \frac{y(t)dt}{t} \quad (a < x < b), \tag{2.7.8}$$

where $n = [\Re(\alpha)] + 1$ [see Samko et al. ([729], Section 18.3)]. When $a = 0$ and $b = \infty$, we have

$$(\mathcal{D}_{0+}^{\alpha} y)(x) := \delta^n (\mathcal{J}_{0+}^{n-\alpha} y)(x) \quad \left(x > 0;\ \delta = x\frac{d}{dx}\right), \tag{2.7.9}$$

$$(\mathcal{D}_{-}^{\alpha} y)(x) := (-\delta)^n (\mathcal{J}_{-}^{n-\alpha} y)(x) \quad \left(x > 0;\ \delta = x\frac{d}{dx}\right). \tag{2.7.10}$$

The *Hadamard type fractional derivatives of order* α, more general than those in (2.7.9) and (2.7.10) with $\mu \in \mathbb{C}$, are defined for $\Re(\alpha) \geqq 0$ by

$$(\mathcal{D}_{0+,\mu}^{\alpha}y)(x) := x^{-\mu}\delta^n x^{\mu}(\mathcal{J}_{0+,\mu}^{n-\alpha}y)(x), \tag{2.7.11}$$

$$(\mathcal{D}_{-,\mu}^{\alpha}y)(x) := x^{\mu}(-\delta)^n x^{-\mu}(\mathcal{J}_{-,\mu}^{n-\alpha}y)(x), \tag{2.7.12}$$

where $x > 0$, $\delta = x\frac{d}{dx}$, $\Re(\alpha) \geqq 0$, and $\mathcal{J}_{+,\mu}^{n-\alpha}y$, $\mathcal{J}_{-,\mu}^{n-\alpha}y$ are the Hadamard type fractional integrals (2.7.5) and (2.7.6) of order $n-\alpha$ ($n = [\Re(\alpha)]+1$) [see Butzer et al. [113, 112, 114]].

When $\alpha = m \in \mathbb{N}$, then

$$(\mathcal{D}_{a+}^{m}y)(x) = (\delta^m y)(x) \text{ and } (\mathcal{D}_{b-}^{\alpha}y)(x) = (-1)^m(\delta^m y)(x) \tag{2.7.13}$$

with $0 \leqq a < x < b \leqq \infty$, and

$$(\mathcal{D}_{0+,\mu}^{m}y)(x) = x^{\mu}\delta^m x^{-\mu}y(x) \text{ and } (\mathcal{D}_{-,\mu}^{m}y)(x) = x^{\mu}(-\delta)^m x^{-\mu}y(x) \tag{2.7.14}$$

with $x > 0$.

It can be directly verified that the Hadamard fractional integrals and fractional derivatives (2.7.1), (2.7.7) and (2.7.2), (2.7.8) of the logarithmic functions $[\log(x/a)]^{\beta-1}$ and $[\log(b/x)]^{\beta-1}$ yield logarithmic functions of the same form.

Property 2.24 *If $\Re(\alpha) > 0$, $\Re(\beta$, and $0 < a < b < \infty$, then*

$$\left(\mathcal{J}_{a+}^{\alpha}\left(\log\frac{t}{a}\right)^{\beta-1}\right)(x) = \frac{\Gamma(\beta)}{\Gamma(\beta+\alpha)}\left(\log\frac{x}{a}\right)^{\beta+\alpha-1}, \tag{2.7.15}$$

$$\left(\mathcal{D}_{a+}^{\alpha}\left(\log\frac{t}{a}\right)^{\beta-1}\right)(x) = \frac{\Gamma(\beta)}{\Gamma(\beta-\alpha)}\left(\log\frac{x}{a}\right)^{\beta-\alpha-1} \tag{2.7.16}$$

and

$$\left(\mathcal{J}_{b-}^{\alpha}\left(\log\frac{b}{t}\right)^{\beta-1}\right)(x) = \frac{\Gamma(\beta)}{\Gamma(\beta+\alpha)}\left(\log\frac{b}{x}\right)^{\beta+\alpha-1}, \tag{2.7.17}$$

$$\left(\mathcal{D}_{b-}^{\alpha}\left(\log\frac{b}{t}\right)^{\beta-1}\right)(x) = \frac{\Gamma(\beta)}{\Gamma(\beta-\alpha)}\left(\log\frac{b}{x}\right)^{\beta-\alpha-1}. \tag{2.7.18}$$

In particular, if $\beta = 1$ and $\Re(\alpha) \geqq 0$, then the Hadamard fractional derivatives of a constant, in general, are not equal to zero:

$$(\mathcal{D}_{a+}^{\alpha}1)(x) = \frac{1}{\Gamma(1-\alpha)}\left(\log\frac{x}{a}\right)^{-\alpha} \text{ and } (\mathcal{D}_{b-}^{\alpha}1)(x) = \frac{1}{\Gamma(1-\alpha)}\left(\log\frac{b}{x}\right)^{-\alpha}, \tag{2.7.19}$$

when $0 < \Re(\alpha) < 1$. On the other hand, for $j = [\Re(\alpha)]+1$,

$$\left(\mathcal{D}_{a+}^{\alpha}\left(\log\frac{t}{a}\right)^{\alpha-j}\right)(x) = 0 \text{ and } \left(\mathcal{D}_{b-}^{\alpha}\left(\log\frac{b}{t}\right)^{\alpha-j}\right)(x) = 0. \tag{2.7.20}$$

From (2.7.20) we derive the following result.

Corollary 2.4 *Let $\Re(\alpha) > 0$, $n = [\Re(\alpha)] + 1$, and $0 < a < b < \infty$.*

(a) *The equality $(\mathcal{D}_{a+}^{\alpha} y)(x) = 0$ is valid if, and only if,*

$$y(x) = \sum_{j=1}^{n} c_j \left(\log \frac{x}{a}\right)^{\alpha - j},$$

where $c_j \in \mathbb{R}$ $(j = 1, \cdots, n)$ are arbitrary constants.

In particular, when $0 < \Re(\alpha) \leqq 1$, the relation $(\mathcal{D}_{a+}^{\alpha} y)(x) = 0$ holds if, and only if, $y(x) = c\left(\log \frac{x}{a}\right)^{\alpha-1}$ for any $c \in \mathbb{R}$.

(b) *The equality $(\mathcal{D}_{b-}^{\alpha} y)(x) = 0$ is valid if, and only if,*

$$y(x) = \sum_{j=1}^{n} d_j \left(\log \frac{b}{x}\right)^{\alpha - j},$$

where $d_j \in \mathbb{R}$ $(j = 1, \cdots, n)$ are arbitrary constants.

In particular, when $0 < \Re(\alpha) \leqq 1$, the relation $(\mathcal{D}_{b-}^{\alpha} y)(x) = 0$ holds if, and only if, $y(x) = d\left(\log \frac{b}{x}\right)^{\alpha-1}$ for any $d \in \mathbb{R}$.

It can also be directly verified that the Hadamard and Hadamard type fractional integrals and derivatives (2.7.3)-(2.7.6) and (2.7.9)-(2.7.12) of the power function x^{β} yield the same function, apart from a constant multiplication factor.

Property 2.25 *Let $\Re(\alpha) > 0$, $\beta \in \mathbb{C}$, and $\mu \in \mathbb{C}$.*

(a) *If $\Re(\beta + \mu) > 0$, then*

$$\left(\mathcal{J}_{0+,\mu}^{\alpha} t^{\beta}\right)(x) = (\mu + \beta)^{-\alpha} x^{\beta} \text{ and } \left(\mathcal{D}_{0+,\mu}^{\alpha} t^{\beta}\right)(x) = (\mu + \beta)^{\alpha} x^{\beta}. \qquad (2.7.21)$$

In particular, if $\Re(\beta) > 0$, then

$$\left(\mathcal{J}_{0+}^{\alpha} t^{\beta}\right)(x) = \beta^{-\alpha} x^{\beta} \text{ and } \left(\mathcal{D}_{0+}^{\alpha} t^{\beta}\right)(x) = \beta^{\alpha} x^{\beta}. \qquad (2.7.22)$$

(b) *If $\Re(\beta - \mu) < 0$, then*

$$\left(\mathcal{J}_{-,\mu}^{\alpha} t^{\beta}\right)(x) = (\mu - \beta)^{-\alpha} x^{\beta} \text{ and } \left(\mathcal{D}_{-,\mu}^{\alpha} t^{\beta}\right)(x) = (\mu - \beta)^{\alpha} x^{\beta}. \qquad (2.7.23)$$

In particular, if $\Re(\beta) < 0$, then

$$\left(\mathcal{J}_{-}^{\alpha} t^{\beta}\right)(x) = (-\beta)^{-\alpha} x^{\beta} \text{ and } \left(\mathcal{D}_{-}^{\alpha} t^{\beta}\right)(x) = (-\beta)^{\alpha} x^{\beta}. \qquad (2.7.24)$$

The Hadamard fractional integrals (2.7.1)-(2.7.2) with $0 < a < b < \infty$ are defined on $L^p(a, b)$, and the Hadamard type fractional integrals (2.7.3)-(2.7.6) on $X_c^p(\mathbb{R}^+)$. The next assertion is proved by using the Minkowski inequality.

Lemma 2.32 *Let $\Re(\alpha) > 0$, $1 \leqq p \leqq \infty$, and $0 < a < b < \infty$.*

Then the operators $\mathcal{J}^{\alpha}_{a+}$ and $\mathcal{J}^{\alpha}_{b+}$ are bounded in $L^p(a,b)$ as follows:

$$\|\mathcal{J}^{\alpha}_{a+}f\|_p \leqq K_1\|f\|_p \text{ and } \|\mathcal{J}^{\alpha}_{b-}f\|_p \leqq K_2\|f\|_p, \tag{2.7.25}$$

where

$$K_1 = \frac{1}{|\Gamma(\alpha)|}\int_0^{\log(b/a)} t^{\Re(\alpha)-1}e^{t/p}dt \text{ and } K_2 = \frac{1}{|\Gamma(\alpha)|}\int_0^{\log(b/a)} t^{\Re(\alpha)-1}e^{-t/p}dt. \tag{2.7.26}$$

In particular,

$$\|\mathcal{J}^{\alpha}_{a+}f\|_\infty \leqq K\|f\|_\infty \text{ and } \|\mathcal{J}^{\alpha}_{b-}f\|_\infty \leqq K\|f\|_\infty, \tag{2.7.27}$$

where $K = \frac{1}{|\Gamma(\alpha)|}\left(\log\frac{b}{a}\right)^{\Re(\alpha)}$.

Lemma 2.33 *Let $\Re(\alpha) > 0$, $1 \leqq p \leqq \infty$, $c \in \mathbb{R}$, and $\mu \in \mathbb{C}$.*

(a) *If $\Re(\mu) > c$, then the operator $\mathcal{J}^{\alpha}_{0+,\mu}$ is bounded in $X^p_c(\mathbb{R}^+)$ as follows:*

$$\|\mathcal{J}^{\alpha}_{0+,\mu}f\|_{X^p_c} \leqq K_2^+\|f\|_{X^p_c} \quad (K_2^+ = [\Re(\mu) - c]^{-\Re(\alpha)}). \tag{2.7.28}$$

In particular, if $c < 0$, the operator $\mathcal{J}^{\alpha}_{0+}$ is bounded in $X^p_c(\mathbb{R}^+)$ by

$$\|\mathcal{J}^{\alpha}_{0+}f\|_{X^p_c} \leqq K_2^+\|f\|_{X^p_c} \quad (K_2^+ = [-c]^{-\Re(\alpha)}), \tag{2.7.29}$$

(b) *If $\Re(\mu) > -c$, then the operator $\mathcal{J}^{\alpha}_{-,\mu}$ is bounded in $X^p_c(\mathbb{R}^+)$ by*

$$\|\mathcal{J}^{\alpha}_{-,\mu}f\|_{X^p_c} \leqq K_2^-\|f\|_{X^p_c} \quad (K_2^- = [\Re(\mu) + c]^{-\Re(\alpha)}). \tag{2.7.30}$$

In particular, if $c > 0$, the operator $\mathcal{J}^{\alpha}_{-}$ is bounded in $X^p_c(\mathbb{R}^+)$ by

$$\|\mathcal{J}^{\alpha}_{0+}f\|_{X^p_c} \leqq K_2^-\|f\|_{X^p_c} \quad (K_2^- = c^{-\Re(\alpha)}). \tag{2.7.31}$$

Remark 2.16 Lemma 2.33 was established in Butzer et al. ([113], Theorems 4 and 6).

The Hadamard and Hadamard type fractional integrals (2.7.1)-(2.7.6) satisfy the following semigroup property.

Property 2.26 *Let $\Re(\alpha) > 0$, $\Re(\beta) > 0$, and $1 \leqq p \leqq \infty$.*

(a) *If $0 < a < b < \infty$, then, for $f \in L^p(a,b)$,*

$$\mathcal{J}^{\alpha}_{a+}\mathcal{J}^{\beta}_{a+}f = \mathcal{J}^{\alpha+\beta}_{a+}f \ (c \leqq 0) \text{ and } \mathcal{J}^{\alpha}_{b-}\mathcal{J}^{\beta}_{b-}f = \mathcal{J}^{\alpha+\beta}_{b-}f \ (c \geqq 0). \tag{2.7.32}$$

(b) *If $\mu \in \mathbb{C}$, $c \in \mathbb{R}$, $a = 0$ and $b = \infty$, then, for $f \in X^p_c(\mathbb{R}^+)$,*

$$\mathcal{J}^{\alpha}_{0+,\mu}\mathcal{J}^{\beta}_{0+,\mu}f = \mathcal{J}^{\alpha+\beta}_{0+,\mu}f \ (\Re(\mu) > c) \text{ and } \mathcal{J}^{\alpha}_{-,\mu}\mathcal{J}^{\beta}_{-,\mu}f = \mathcal{J}^{\alpha+\beta}_{-,\mu}f \ (\Re(\mu) > -c). \tag{2.7.33}$$

In particular, when $\mu = 0$,

$$\mathcal{J}^{\alpha}_{0+}\mathcal{J}^{\beta}_{0+}f = \mathcal{J}^{\alpha+\beta}_{0+}f \ (c < 0), \quad \mathcal{J}^{\alpha}_{-}\mathcal{J}^{\beta}_{-}f = \mathcal{J}^{\alpha+\beta}_{-}f \ (c > 0). \tag{2.7.34}$$

The conditions for the existence of the Hadamard fractional derivatives (2.7.7) and (2.7.8) are given by the following assertion, proved for $\alpha > 0$ in Kilbas [370]. Here the space of functions (1.1.13) is used.

Lemma 2.34 *Let $\Re(\alpha) \geqq 0$ and $n = [\Re(\alpha)] + 1$. If $y(x) \in AC^n_\delta[a,b]$ $(0 < a < b < \infty)$, then the Hadamard fractional derivatives $\mathcal{D}^\alpha_{a+}y$ and $\mathcal{D}^\alpha_{b-}y$ exist almost everywhere on $[a,b]$ and can be represented in the forms*

$$(\mathcal{D}^\alpha_{a+}y)(x) = \sum_{k=0}^{n-1} \frac{(\delta^k y)(a)}{\Gamma(1+k-\alpha)} \left(\log\frac{x}{a}\right)^{k-\alpha} + \frac{1}{\Gamma(n-\alpha)} \int_a^x \left(\log\frac{x}{t}\right)^{n-\alpha-1} (\delta^n y)(t)dt \tag{2.7.35}$$

and

$$(\mathcal{D}^\alpha_{b-}y)(x) = \sum_{k=0}^{n-1} \frac{(-1)^k(\delta^k y)(b)}{\Gamma(1+k-\alpha)} \left(\log\frac{b}{x}\right)^{k-\alpha} + \frac{(-1)^n}{\Gamma(n-\alpha)} \int_a^x \left(\log\frac{t}{x}\right)^{n-\alpha-1} (\delta^n y)(t)dt, \tag{2.7.36}$$

respectively.

In particular, when $0 < \Re(\alpha) < 1$, then, for $y(x) \in AC[a,b]$,

$$(\mathcal{D}^\alpha_{a+}y)(x) = \frac{y(a)}{\Gamma(1-\alpha)} \left(\log\frac{x}{a}\right)^{-\alpha} + \frac{1}{\Gamma(1-\alpha)} \int_a^x \left(\log\frac{x}{t}\right)^{-\alpha} y'(t)\frac{dt}{t} \tag{2.7.37}$$

and

$$(\mathcal{D}^\alpha_{b-}y)(x) = \frac{y(b)}{\Gamma(1-\alpha)} \left(\log\frac{b}{x}\right)^{-\alpha} - \frac{1}{\Gamma(1-\alpha)} \int_x^b \left(\log\frac{t}{x}\right)^{-\alpha} y'(t)\frac{dt}{t}. \tag{2.7.38}$$

When $y(x)$ is an arbitrarily-often differentiable function and $\mu \in \mathbb{R}$, the Hadamard type fractional integration (2.7.5) and the Hadamard type fractional differentiation (2.7.11) can be expressed in terms of an infinite series involving the generalized Stirling numbers [see Butzer et al. [115]].

Compositions between the operators of fractional differentiation (2.7.7)-(2.7.12) and fractional integration (2.7.1)-(2.7.6) are given by the following property.

Property 2.27 *Let $\alpha \in \mathbb{C}$ and $\beta \in \mathbb{C}$ be such that $\Re(\alpha) > \Re(\beta) > 0$.*

(a) *If $0 < a < b < \infty$ and $1 \leqq p \leqq \infty$, then, for $f \in L^p(a,b)$,*

$$\mathcal{D}^\beta_{a+}\mathcal{J}^\alpha_{a+}f = \mathcal{J}^{\alpha-\beta}_{a+}f \text{ and } \mathcal{D}^\beta_{b-}\mathcal{J}^\alpha_{b-}f = \mathcal{J}^{\alpha-\beta}_{b-}f. \tag{2.7.39}$$

In particular, if $\beta = m \in \mathbb{N}$, then

$$\mathcal{D}^m_{a+}\mathcal{J}^\alpha_{a+}f = \mathcal{J}^{\alpha-m}_{a+}f \text{ and } \mathcal{D}^m_{b-}\mathcal{J}^\alpha_{b-}f = \mathcal{J}^{\alpha-m}_{b-}f. \tag{2.7.40}$$

(b) *If $\mu \in \mathbb{C}$, $c \in \mathbb{R}$, $a = 0$ and $b = \infty$, then, for $f \in X^p_c(\mathbb{R}^+)$,*

$$\mathcal{D}^{\beta}_{0+,\mu}\mathcal{J}^{\alpha}_{0+,\mu}f = \mathcal{J}^{\alpha-\beta}_{0+,\mu}f \ (\Re(\mu) > c); \ \mathcal{D}^{\beta}_{-,\mu}\mathcal{J}^{\alpha}_{-,\mu}f = \mathcal{J}^{\alpha-\beta}_{-,\mu}f \ (\Re(\mu) > -c). \quad (2.7.41)$$

In particular, if $\beta = m \in \mathbb{N}$, *then*

$$\mathcal{D}^{m}_{0+,\mu}\mathcal{J}^{\alpha}_{0+,\mu}f = \mathcal{J}^{\alpha-m}_{0+,\mu}f \ (\Re(\mu) > c); \ \mathcal{D}^{m}_{-,\mu}\mathcal{J}^{\alpha}_{-,\mu}f = \mathcal{J}^{\alpha-m}_{-,\mu}f \ (\Re(\mu) > -c). \quad (2.7.42)$$

while, when $\mu = 0$ *and* $m \in \mathbb{N}$,

$$\mathcal{D}^{\beta}_{0+}\mathcal{J}^{\alpha}_{0+}f = \mathcal{J}^{\alpha-\beta}_{0+}f \ (c < 0); \ \ \mathcal{D}^{\beta}_{-}\mathcal{J}^{\alpha}_{-}f = \mathcal{J}^{\alpha-\beta}_{-}f \ (c > 0), \quad (2.7.43)$$

$$\mathcal{D}^{m}_{0+}\mathcal{J}^{\alpha}_{0+}f = \mathcal{J}^{\alpha-m}_{0+}f \ (c < 0) \ \text{and} \ \mathcal{D}^{m}_{-}\mathcal{J}^{\alpha}_{-}f = \mathcal{J}^{\alpha-m}_{-}f \ (c > 0). \quad (2.7.44)$$

The Hadamard and Hadamard type fractional derivatives (2.7.7), (2.7.8) and (2.7.11), (2.7.12) are operators inverse to the corresponding fractional integrals (2.7.1), (2.7.2) and (2.7.5), (2.7.6).

Property 2.28 *Let* $\Re(\alpha) > 0$.
(a) *If* $0 < a < b < \infty$ *and* $1 \leqq p \leqq \infty$, *then, for* $f \in L^p(a,b)$,

$$\mathcal{D}^{\alpha}_{a+}\mathcal{J}^{\alpha}_{a+}f = f \ (c \leqq 0) \ \text{and} \ \mathcal{D}^{\alpha}_{b-}\mathcal{J}^{\alpha}_{b-}f = f \ (c \geqq 0). \quad (2.7.45)$$

(b) *If* $\mu \in \mathbb{C}$, $c \in \mathbb{R}$, $a = 0$ *and* $b = \infty$, *then, for* $f \in X^p_c(\mathbb{R}^+)$,

$$\mathcal{D}^{\alpha}_{0+,\mu}\mathcal{J}^{\alpha}_{0+,\mu}f = f \ (\Re(\mu) > c) \ \text{and} \ \mathcal{D}^{\alpha}_{-,\mu}\mathcal{J}^{\alpha}_{-,\mu}f = f \ (\Re(\mu) > -c). \quad (2.7.46)$$

In particular, if $\mu = 0$, *then*

$$\mathcal{D}^{\alpha}_{0+}\mathcal{J}^{\alpha}_{0+}f = f \ (c < 0) \ \text{and} \ \mathcal{D}^{\alpha}_{-}\mathcal{J}^{\alpha}_{-}f = f \ (c > 0). \quad (2.7.47)$$

The following result yields the formula for the composition of the fractional differentiation operator $\mathcal{J}^{\alpha}_{a+}$ with the fractional integration operator $\mathcal{D}^{\alpha}_{a+}$.

Theorem 2.3 *Let* $\Re(\alpha) > 0$, $n = -[-\Re(\alpha)]$ *and* $0 < a < b < \infty$. *Also let* $(\mathcal{J}^{n-\alpha}_{a+}y)(x)$ *be the Hadamard type fractional integral of the form* (2.7.1). *If* $y(x) \in L(a,b)$ *and* $(\mathcal{J}^{n-\alpha}_{a+}y)(x) \in AC^n_\delta[a,b]$, *then*

$$(\mathcal{J}^{\alpha}_{a+}\mathcal{D}^{\alpha}_{a+}y)(x) = y(x) - \sum_{k=1}^{n} \frac{(\delta^{n-k}(\mathcal{J}^{n-\alpha}_{a+}y))(a)}{\Gamma(\alpha-k+1)} \left(\log\frac{x}{a}\right)^{\alpha-k}. \quad (2.7.48)$$

In particular, if $\alpha = n \in \mathbb{N}$ *and* $y(x) \in AC^n_\delta[a,b]$, *then*

$$(\mathcal{J}^{n}_{a+}\mathcal{D}^{n}_{a+}\,y)(x) = y(x) - \sum_{k=0}^{n-1} \frac{(\delta^{k}y)(a)}{k!} \left(\log\frac{x}{a}\right)^{k}. \quad (2.7.49)$$

Remark 2.17 For a real $\alpha > 0$, the relation (2.7.48) was proved in Kilbas and Titiuora ([405], Theorem 9).

We now consider the following spaces of functions of the forms (2.1.36) and (2.1.37):

$$\mathcal{J}_{a+}^{\alpha}(L^p) := \left\{g := \mathcal{J}_{a+}^{\alpha}\varphi \text{ and } \varphi \in L^p(a,b)\right\}, \tag{2.7.50}$$

$$\mathcal{J}_{b-}^{\alpha}(L^p) := \left\{g := \mathcal{J}_{b-}^{\alpha}\varphi \text{ and } \varphi \in L^p(a,b)\right\} \ (0 < a < b < \infty), \tag{2.7.51}$$

$$\mathcal{J}_{0+,\mu}^{\alpha}(X_c^p) = \left\{g := \mathcal{J}_{0+,\mu}^{\alpha}\varphi \text{ and } \varphi \in X_c^p(\mathbb{R}^+)\right\} \tag{2.7.52}$$

and

$$\mathcal{J}_{-,\mu}^{\alpha}(X_c^p) := \left\{g := \mathcal{J}_{-,\mu}^{\alpha}\varphi \text{ and } \varphi \in X_c^p(\mathbb{R}^+)\right\} \ (a = 0; b = \infty). \tag{2.7.53}$$

In particular, if

$$\mathcal{J}_{0+}^{\alpha}(X_c^p) := \left\{g := \mathcal{J}_{0+}^{\alpha}\varphi \text{ and } \varphi \in X_c^p(\mathbb{R}^+)\right\} \tag{2.7.54}$$

and

$$\mathcal{J}_{-}^{\alpha}(X_c^p) := \left\{g := \mathcal{J}_{-}^{\alpha}\varphi \text{ and } \varphi \in X_c^p(\mathbb{R}^+)\right\}, \tag{2.7.55}$$

then the relation in (2.7.48) can be simplified. From Property 2.28 we thus derive the following assertion.

Lemma 2.35 *Let* $\Re(\alpha) > 0$.

(a) *If* $0 < a < b < \infty$, *then*

$$(\mathcal{J}_{a+}^{\alpha}\mathcal{D}_{a+}^{\alpha}y)(x) = y(x) \ \left(y \in \mathcal{J}_{a+}^{\alpha}(L^p)\right), \quad (\mathcal{J}_{b-}^{\alpha}\mathcal{D}_{b-}^{\alpha}y)(x) = y(x) \ \left(y \in \mathcal{J}_{b-}^{\alpha}(L^p)\right). \tag{2.7.56}$$

(b) *If* $\mu \in \mathbb{C}$, $c \in \mathbb{R}$, $a = 0$ *and* $b = \infty$, *then*

$$(\mathcal{J}_{0+,\mu}^{\alpha}\mathcal{D}_{0+,\mu}^{\alpha}y)(x) = y(x) \ \left(y \in \mathcal{J}_{0+,\mu}^{\alpha}(X_c^p); \ \Re(\mu) > c\right) \tag{2.7.57}$$

and

$$(\mathcal{J}_{-,\mu}^{\alpha}\mathcal{D}_{-,\mu}^{\alpha}y)(x) = y(x) \ \left(y \in \mathcal{J}_{-,\mu}^{\alpha}(X_c^p); \ \Re(\mu) > -c\right). \tag{2.7.58}$$

In particular, when $\mu = 0$,

$$(\mathcal{J}_{0+}^{\alpha}\mathcal{D}_{0+,\mu}^{\alpha}y)(x) = y(x) \ \left(y \in \mathcal{J}_{0+}^{\alpha}(X_c^p); \ c < 0\right), \tag{2.7.59}$$

$$(\mathcal{J}_{-}^{\alpha}\mathcal{D}_{-}^{\alpha}y)(x) = y(x) \ \left(y \in \mathcal{J}_{-}^{\alpha}(X_c^p); \ c > 0\right). \tag{2.7.60}$$

Now we consider the properties of the Hadamard fractional integrals (2.7.1), (2.7.2) and the Hadamard fractional derivatives (2.7.7), (2.7.8) in the spaces $C_{\gamma,\log}[a,b]$ and $C_{\delta,\gamma}^{n}[a,b]$ defined in (1.1.27) and (1.1.28), respectively. The existence of the fractional integrals $\mathcal{J}_{a+}^{\alpha}f$, $\mathcal{J}_{b-}^{\alpha}f$ in the space $C_{\gamma,\log}[a,b]$ and of the fractional derivatives $\mathcal{D}_{a+}^{\alpha}y$, $\mathcal{D}_{b-}^{\alpha}y$ in the space $C_{\gamma,\log}^{n}[a,b]$ are given by the following assertion.

Lemma 2.36 *Let $0 < a < b < \infty$, $\Re(\alpha) \geqq 0$, and $0 \leqq \Re(\gamma) < 1$.*

(a) *If $\Re(\gamma) > \Re(\alpha) > 0$, then the fractional integration operators $\mathcal{J}^{\alpha}_{a+}$ and $\mathcal{J}^{\alpha}_{b-}$ are bounded from $C_{\gamma,\log}[a,b]$ into $C_{\gamma-\alpha,\log}[a,b]$:*

$$\|\mathcal{J}^{\alpha}_{a+}f\|_{C_{\gamma-\alpha,\log}} \leqq k_1\|f\|_{C_{\gamma,\log}} \text{ and } \|\mathcal{J}^{\alpha}_{b-}f\|_{C_{\gamma-\alpha,\log}} \leqq k_1\|f|_{C_{\gamma,\log}} \tag{2.7.61}$$

where

$$k_1 = \left(\log\frac{b}{a}\right)^{\Re(\alpha)} \frac{\Gamma[\Re(\alpha)]|\Gamma(1-\Re(\gamma)|}{|\Gamma(\alpha)|\Gamma[1+\Re(\alpha-\gamma)]}. \tag{2.7.62}$$

In particular, $\mathcal{J}^{\alpha}_{a+}$ and $\mathcal{J}^{\alpha}_{b-}$ are bounded in $C_{\gamma,\log}[a,b]$.

If $\Re(\gamma) \leqq \Re(\alpha)$, then the fractional integration operators $\mathcal{J}^{\alpha}_{a+}$ and $\mathcal{J}^{\alpha}_{b-}$ are bounded from $C_{\gamma,\log}[a,b]$ into $C[a,b]$:

$$\|\mathcal{J}^{\alpha}_{a+}f\|_{C} \leqq k_2\|f\|_{C_{\gamma,\log}} \text{ and} \|\mathcal{J}^{\alpha}_{b-}f\|_{C} \leqq k_2\|f|_{C_{\gamma,\log}}, \tag{2.7.63}$$

where

$$k_2 = \left(\log\frac{b}{a}\right)^{\Re(\alpha-\gamma)} \frac{\Gamma[\Re(\alpha)]|\Gamma(1-\Re(\gamma)|}{|\Gamma(\alpha)|\Gamma[1+\Re(\alpha-\gamma)]}. \tag{2.7.64}$$

In particular, $\mathcal{J}^{\alpha}_{a+}$ and $\mathcal{J}^{\alpha}_{b-}$ are bounded in $C_{\gamma,\log}[a,b]$.

(b) *If $\Re(\alpha) \geqq 0$, $n = [\Re(\alpha)] + 1$ and $y(x) \in C^{n}_{\delta,\gamma}[a,b]$, then the fractional derivatives $\mathcal{D}^{\alpha}_{a+}y$ and $\mathcal{D}^{\alpha}_{b-}y$ exist on $(a,b]$ and can be represented by (2.7.35) and (2.7.36), respectively. In particular, when $0 \leqq \Re(\alpha) < 1$ $(\alpha \neq 0)$ and $y(x) \in C_{\gamma,\log}[a,b]$, then $\mathcal{D}^{\alpha}_{a+}y$ and $\mathcal{D}^{\alpha}_{b-}y$ are given by (2.7.37) and (2.7.38), respectively.*

Proof. Estimates (2.7.61) and (2.7.63) are proved by using the definition (1.1.27) of the space $C_{\gamma,\log}[a,b]$. Statement (b) of Lemma 2.36 is proved by using Lemma 1.5.

Assertions (a)-(c) of Lemma 2.37 below for fractional calculus operators (2.7.1), (2.7.2) and (2.7.7), (2.7.8), being analogous to those in Properties 2.26 and 2.28 and Theorem 2.3, are proved directly.

Lemma 2.37 *Let $\Re(\alpha) > 0$, $\Re(\beta) > 0$, and $0 \leqq \Re(\gamma) < 1$. The following assertions are then true:*

(a) *If $f(x) \in C_{\gamma,\log}[a,b]$, then the first and the second relations in (2.7.32) hold at any point $x \in (a,b]$ and $x \in [a,b)$, respectively. When $f(x) \in C[a,b]$, these relations are valid at any point $x \in [a,b]$.*

(b) *If $f(x) \in C_{\gamma,\log}[a,b]$, then the first and the second equalities in (2.7.33) hold at any point $x \in (a,b]$ and $x \in [a,b)$, respectively. When $f(x) \in C[a,b]$, then these equalities are valid at any point $x \in [a,b]$.*

(c) *Let $\Re(\alpha) > \Re(\beta) > 0$. If $f(x) \in C_{\gamma,\log}[a,b]$, then the first and the second relations in (2.7.39) hold at any point $x \in (a,b]$ and $x \in [a,b)$, respectively. When $f(x) \in C[a,b]$, then these relations are valid at any point $x \in [a,b]$. In particular, when $\beta = k \in \mathbb{N}$ and $\Re(\alpha) > k$, the relations in (2.7.40) are valid in their respective cases.*

(d) *Let $n = [\Re(\alpha)] + 1$. Also let $f_{n-\alpha}(x) = (\mathcal{J}_{a+}^{n-\alpha} f)(x)$ and $g_{n-\alpha}(x) = (\mathcal{J}_{b-}^{n-\alpha} g)(x)$ be the fractional integrals (2.7.1) and (2.7.2) of order $n-\alpha$.*

If $f(x) \in C_{\gamma,\log}[a,b]$ and $f_{n-\alpha}(x) \in C_{\delta,\gamma}^n[a,b]$, then the relation (2.7.48) holds at any point $x \in (a,b]$. In particular, when $0 < \Re(\alpha) < 1$ and $f_{1-\alpha}(x) \in C_{\delta,\gamma}^1[a,b]$, then

$$\left(\mathcal{J}_{a+}^{\alpha} \mathcal{D}_{a+}^{\alpha} f\right)(x) = (x) - \frac{f_{1-\alpha}(a)}{\Gamma(\alpha)} \left(\log \frac{x}{a}\right)^{\alpha-1}. \tag{2.7.65}$$

If $f(x) \in C[a,b]$ and $f_{n-\alpha}(x) \in C_{\delta}^n[a,b]$, then (2.7.48) holds at any point $x \in [a,b]$. In particular, if $f(x) \in C_{\delta}^n[a,b]$, the relation (2.7.49) is valid at any point $x \in [a,b]$.

To conclude this section, we give relations for the Mellin transform (1.4.23) of the Hadamard type fractional integrals $(\mathcal{J}_{0,\mu}^{\alpha} f)(x)$, $(\mathcal{J}_{-,\mu}^{\alpha} f)(x)$ defined in (2.7.5), (2.7.6) and for the corresponding Hadamard type fractional derivatives $(\mathcal{D}_{0+,\mu}^{\alpha} f)(x)$, $(\mathcal{D}_{-,\mu}^{\alpha} f)(x)$ given in (2.7.11), (2.7.12). First we consider the Mellin transforms of the Hadamard type fractional integrals.

Lemma 2.38 *Let $\Re(\alpha) > 0$ and $\mu \in \mathbb{C}$. Also let a function $f(x)$ be such that its Mellin transform $(\mathcal{M}f)(s)$ exists for $s \in \mathbb{C}$.*

(a) *If $\Re(\mu - s) > 0$, then*

$$(\mathcal{M}\mathcal{J}_{0+,\mu}^{\alpha} f)(s) = (\mu - s)^{-\alpha} (\mathcal{M}f)(s). \tag{2.7.66}$$

In particular, when $\Re(s) < 0$, then

$$(\mathcal{M}\mathcal{J}_{0+}^{\alpha} f)(s) = (-s)^{-\alpha} (\mathcal{M}f)(s). \tag{2.7.67}$$

(b) *If $\Re(\mu + s) > 0$, then*

$$(\mathcal{M}\mathcal{J}_{-,\mu}^{\alpha} f)(s) = (\mu + s)^{-\alpha} (\mathcal{M}f)(s). \tag{2.7.68}$$

In particular, when $\Re(s) > 0$, then

$$(\mathcal{M}\mathcal{J}_{-}^{\alpha} f)(s) = s^{-\alpha} (\mathcal{M}f)(s). \tag{2.7.69}$$

Proof. Let $\alpha > 0$ and $s \in \mathbb{R}$ be such that $\mu - s > 0$. Using (1.4.23) and (2.7.5), changing the order of integration, and making the changes of variable $\log\left(\frac{x}{t}\right) = u$ and $u(\mu - s) = \tau$, we have

$$(\mathcal{M}\mathcal{J}_{0+,\mu}^{\alpha} f)(s) = \frac{1}{\Gamma(\alpha)} \int_0^{\infty} x^{s-1} dx \int_0^x \left(\frac{t}{x}\right)^{\mu} \left(\log \frac{x}{t}\right)^{\alpha-1} \frac{f(t)dt}{t}$$

$$= \frac{1}{\Gamma(\alpha)} \int_0^{\infty} \frac{f(t)dt}{t} \int_t^{\infty} x^{s-1} \left(\frac{t}{x}\right)^{\mu} \left(\log \frac{x}{t}\right)^{\alpha-1} dx$$

$$= \frac{1}{\Gamma(\alpha)} \int_0^{\infty} t^{s-1} f(t)dt \int_0^{\infty} e^{-u(\mu-s)} u^{\alpha-1} du$$

$$= \frac{1}{\Gamma(\alpha)}(\mu - s)^{-\alpha} \int_0^\infty \tau^{\alpha-1} e^{-\tau} d\tau (\mathcal{M}f)(s), \tag{2.7.70}$$

which, in accordance with (1.5.1), yields (2.7.66). This formula remains true for complex $\alpha \in \mathbb{C}$ and $\mu, s \in \mathbb{C}$ by analytic continuation. Relation (2.7.68) is proved similarly.

Remark 2.18 For a function $f(x)$ belonging to the space $X_c^p(\mathbb{R})$, defined in (1.1.3)-(1.1.4), with $1 \leqq p \leqq 2$ and $c \in \mathbb{R}$, the relations (2.7.66), (2.7.67) and (2.7.68), (2.7.69) were proved in Butzer et al. ([114], Theorems 1 and 3).

The Mellin transforms of the Hadamard type fractional derivatives are given by the following result.

Lemma 2.39 *Let $\Re(\alpha) > 0$ and $\mu \in \mathbb{C}$. Also let a function $y(x)$ be such that its Mellin transform $(\mathcal{M}y)(s)$ exists for $s \in \mathbb{C}$.*

(a) *If $\Re(\mu - s) > 0$ and $(\mathcal{M}\mathcal{D}_{0+,\mu}^\alpha y)(s)$ exists, then*

$$(\mathcal{M}\mathcal{D}_{0+,\mu}^\alpha y)(s) = (\mu - s)^\alpha (\mathcal{M}y)(s). \tag{2.7.71}$$

In particular, when $\Re(s) < 0$, then

$$(\mathcal{M}\mathcal{D}_{0+}^\alpha y)(s) = (-s)^\alpha (\mathcal{M}y)(s). \tag{2.7.72}$$

(b) *If $\Re(\mu + s) > 0$ and $(\mathcal{M}\mathcal{D}_{-,\mu}^\alpha y)(s)$ exists, then*

$$(\mathcal{M}\mathcal{D}_{-,\mu}^\alpha y)(s) = (\mu + s)^\alpha (\mathcal{M}y)(s). \tag{2.7.73}$$

In particular, when $\Re(s) > 0$, then

$$(\mathcal{M}\mathcal{D}_{-}^\alpha y)(s) = s^\alpha (\mathcal{M}y)(s). \tag{2.7.74}$$

Proof. Let $n = [\Re(\alpha)] + 1$. By (2.7.11) and (1.4.26), we have

$$\left(\mathcal{M}\mathcal{D}_{0+,\mu}^\alpha y\right)(s) = \left(\mathcal{M}M_{-\mu}\delta^n M_\mu(\mathcal{J}_{0+,\mu}^{n-\alpha} y\right)(s). \tag{2.7.75}$$

Applying (1.4.28), (1.4.35) , (1.4.28), and (2.7.66), with α replaced by $n - \alpha$, we find that

$$\left(\mathcal{M}\mathcal{D}_{0+,\mu}^\alpha y\right)(s) = \left(\mathcal{M}\delta^n M_\mu(\mathcal{J}_{0+,\mu}^{n-\alpha} y\right)(s - \mu)$$

$$= (\mu - s)^n \left(\mathcal{M}M_\mu(\mathcal{J}_{0+,\mu}^{n-\alpha} y\right)(s - \mu) = (\mu - s)^n \left(\mathcal{M}(\mathcal{J}_{0+,\mu}^{n-\alpha} y\right)(s) = (\mu - s)^\alpha (\mathcal{M}y)(s), \tag{2.7.76}$$

which yields (2.7.71). The formula in (2.7.73) is proved similarly.

2.8 Grünwald-Letnikov Fractional Derivatives

In this section we give the definition of the Grünwald-Letnikov fractional derivatives and give some of their properties as presented in the book by Samko et al. ([729], Section 20). The above operation of fractional differentiation is based on a generalization of the usual differentiation of a function $y(x)$ of order $n \in \mathbb{N}$ of the form

$$y^{(n)}(x) = \lim_{h \to 0} \frac{(\Delta_h^n y)(x)}{h^n}. \tag{2.8.1}$$

Here $(\Delta_h^n y)(x)$ is a finite difference of order $n \in \mathbb{N}_0$ of a function $y(x)$ with a step $h \in \mathbb{R}$ and centered at the point $x \in \mathbb{R}$ defined in (2.8.5).

Property (2.8.1) is used to define a fractional derivative by directly replacing $n \in \mathbb{N}$ in (2.8.1) by $\alpha > 0$. For this, h^n is replaced by h^α, while the finite difference $(\Delta_h^n y)(x)$ is replaced by the *difference* $(\Delta_h^\alpha y)(x)$ *of a fractional order* $\alpha \in \mathbb{R}$ defined by the following infinite series:

$$(\Delta_h^\alpha y)(x) := \sum_{k=0}^{\infty} (-1)^k \binom{\alpha}{k} y(x - kh) \quad (x, h \in \mathbb{R};\ \alpha > 0), \tag{2.8.2}$$

where $\binom{\alpha}{k}$ are the binomial coefficients (1.5.22). When $h > 0$, the difference (2.8.2) is called *left-sided difference*, while for $h < 0$ it is called a *right-sided difference* .

The series in (2.8.2) converges absolutely and uniformly for each $\alpha > 0$ and for every bounded function $y(x)$. The fractional difference $(\Delta_h^\alpha y)(x)$ has the following properties.

Property 2.29 *If $\alpha > 0$ and $\beta > 0$, then the semigroup property*

$$(\Delta_h^\alpha \Delta_h^\beta y)(x) = (\Delta_h^{\alpha+\beta} y)(x) \tag{2.8.3}$$

is valid for any bounded function $f(x)$.

Property 2.30 *If $\alpha > 0$ and $y(x) \in L_1(\mathbb{R})$, then the Fourier transform* (1.3.1) *of Δ_h^α is given by*

$$(\mathcal{F} \Delta_h^\alpha y)(x) = \left(1 - e^{ixh}\right)^\alpha (\mathcal{F} y)(x). \tag{2.8.4}$$

In particular, when $\alpha = n \in \mathbb{N}$, then, in accordance with (1.5.23) and (1.5.24), (2.8.2) coincides with (2.8.5):

$$(\Delta_h^n y)(x) = \sum_{k=0}^{n} (-1)^k \binom{n}{k} y(x - kh) \quad (x, h \in \mathbb{R};\ n \in \mathbb{N}). \tag{2.8.5}$$

By (1.5.22) and (1.5.23), we can define the difference $(\Delta_h^\alpha y)(x)$ in (2.8.2) for $\alpha = 0$ by

$$(\Delta_h^0 f)(x) = f(x). \tag{2.8.6}$$

Following (2.8.1), the *left-* and *right-sided Grünwald-Letnikov derivatives* $y_+^{\alpha)}(x)$ and $y_-^{\alpha)}(x)$ are defined by

$$y_+^{\alpha)}(x) := \lim_{h\to+0} \frac{(\Delta_h^\alpha y)(x)}{h^\alpha} \quad (\alpha > 0) \tag{2.8.7}$$

and

$$y_-^{\alpha)}(x) := \lim_{h\to+0} \frac{(\Delta_{-h}^\alpha y)(x)}{h^\alpha} \quad (\alpha > 0), \tag{2.8.8}$$

respectively. These constructions coincide with the Marchaud fractional derivatives for $y(x) \in L_p(\mathbb{R})$ $(1 \leqq p < \infty)$ [see Samko et al. ([729], Theorem 20.4)].

The definition (2.8.2) of the fractional difference $(\Delta_h^\alpha y)(x)$ assumes that the function $y(x)$ is given at least on the half-axis. For the function $y(x)$ given on a finite interval $[a,b]$, such a difference can be defined as follows by a continuation of $y(x)$ as a vanishing function beyond $[a,b]$:

$$(\Delta_h^\alpha y)(x) = (\Delta_h^\alpha y^*)(x) := \sum_{k=0}^{\infty}(-1)^k \binom{\alpha}{k} y^*(x-kh) \quad (x,h \in \mathbb{R};\ \alpha > 0), \tag{2.8.9}$$

where

$$y^*(x) = \begin{cases} y(x), & x \in [a,b], \\ 0, & x \notin [a,b]. \end{cases} \tag{2.8.10}$$

It is acceptable to rewrite the fractional difference (2.8.9) in terms of the function $y(x)$ itself, avoiding its continuation as a vanishing function, in the forms

$$(\Delta_{h,a+}^\alpha y)(x) := \sum_{k=0}^{\left[\frac{x-a}{h}\right]} (-1)^k \binom{\alpha}{k} y(x-kh) \quad (x \in \mathbb{R};\ h > 0;\ \alpha > 0) \tag{2.8.11}$$

and

$$(\Delta_{h,b-}^\alpha y)(x) := \sum_{k=0}^{\left[\frac{b-x}{h}\right]} (-1)^k \binom{\alpha}{k} y(x+kh) \quad (x \in \mathbb{R};\ h > 0;\ \alpha > 0). \tag{2.8.12}$$

Then, by analogy with (2.8.7) and (2.8.8), the *left-* and *right-sided Grünwald-Letnikov fractional derivatives of order* $\alpha > 0$ *on a finite interval* $[a,b]$ are defined by

$$y_{a+}^{\alpha)}(x) := \lim_{h\to+0} \frac{(\Delta_{h,a+}^\alpha y)(x)}{h^\alpha} \tag{2.8.13}$$

and

$$y_{b-}^{\alpha)}(x) := \lim_{h\to+0} \frac{(\Delta_{h,b-}^\alpha y)(x)}{h^\alpha}, \tag{2.8.14}$$

respectively. Such Grünwald-Letnikov fractional derivatives coincide with the Marchaud fractional derivatives ([729], Theorem 20.6)] and can be represented in the following form:

$$y_{a+}^{\alpha)}(x) = \frac{y(x)}{\Gamma(1-\alpha)(x-a)^{\alpha}} + \frac{\alpha}{\Gamma(1-\alpha)} \int_a^x \frac{y(x)-y(t)}{(x-t)^{1+\alpha}} dt \quad (0 < \alpha < 1) \tag{2.8.15}$$

and

$$y_{b-}^{\alpha)}(x) = \frac{y(x)}{\Gamma(1-\alpha)(b-x)^{\alpha}} + \frac{\alpha}{\Gamma(1-\alpha)} \int_x^b \frac{y(x)-y(t)}{(t-x)^{1+\alpha}} dt \quad (0 < \alpha < 1). \tag{2.8.16}$$

Here the equalities are understood in the sense of a special convergence in the norm of $L_p(a, b)$.

From (2.8.15) and (2.8.16) we obtain the following formulas of the form (2.1.20):

$$1_{a+}^{\alpha)}(x) = \lim_{h \to +0} \frac{1}{h^{\alpha}} \sum_{k=0}^{\left[\frac{x-a}{h}\right]} (-1)^k \binom{\alpha}{k} = \frac{1}{\Gamma(1-\alpha)} (x-a)^{-\alpha} \quad (0 < \alpha < 1), \tag{2.8.17}$$

$$1_{b-}^{\alpha)}(x) = \lim_{h \to +0} \frac{1}{h^{\alpha}} \sum_{k=0}^{\left[\frac{b-x}{h}\right]} (-1)^k \binom{\alpha}{k} = \frac{1}{\Gamma(1-\alpha)} (b-x)^{-\alpha} \quad (0 < \alpha < 1). \tag{2.8.18}$$

2.9 Partial and Mixed Fractional Integrals and Fractional Derivatives

In this section we give the definitions and some properties of multidimensional partial and mixed fractional integrals and fractional derivatives presented in the book by Samko et al. ([729], Sections 24.1). Such operations of fractional integration and fractional differentiation in the n-dimensional Euclidean space $\mathbb{R}^n$ ($n \in \mathbb{N}$) are natural generalizations of the corresponding one-dimensional fractional integrals and fractional derivatives, being taken with respect to one or several variables.

For $\mathbf{x} = (x_1, \cdots, x_n) \in \mathbb{R}^n$ ($n \in \mathbb{N} \backslash \{1\}$) and $\alpha = (\alpha_1, \cdots, \alpha_n) \in \mathbb{C}^n$ ($n \in \mathbb{N} \backslash \{1\}$), we use the following notations:

$$\mathbf{x}^{\alpha} := x_1^{\alpha_1} \cdots x_n^{\alpha_n}; \quad \Gamma(\alpha) := (\Gamma(\alpha_1) \cdots \Gamma(\alpha_n)); \quad \frac{\partial}{\partial \mathbf{x}} := \frac{\partial}{\partial x_1} \cdots \frac{\partial}{\partial x_n}; \tag{2.9.1}$$

$$[\mathbf{a}, \mathbf{b}] = [a_1, b_1] \times \cdots \times [a_n, b_n]; \quad \mathbf{a} = (a_1, \cdots, a_n) \in \mathbb{R}^n; \ \mathbf{b} = (b_1, \cdots, b_n) \in \mathbb{R}^n, \tag{2.9.2}$$

and by $\mathbf{x} > \mathbf{a}$ we mean $x_1 > a_1, \cdots, x_n > a_n$.

On the basis of (2.1.1) and (2.1.2), the *partial Riemann-Liouville fractional integrals of order* $\alpha_k \in \mathbb{C}$ $(\Re(\alpha_k) > 0)$ *with respect to the* kth *variable* x_k are defined by

$$(I_{a_k+}^{\alpha_k} f)(\mathbf{x}) := \frac{1}{\Gamma(\alpha_k)} \int_{a_k}^{x_k} \frac{f(x_1, \cdots, x_{k-1}, t_k, x_{k+1}, \cdots, x_n)}{(x_k - t_k)^{1-\alpha_k}} dt_k \quad (x_k > a_k) \tag{2.9.3}$$

and

$$(I_{b_k-}^{\alpha_k}f)(\mathbf{x}) := \frac{1}{\Gamma(\alpha_k)}\int_{x_k}^{b_k}\frac{f(x_1,\cdots,x_{k-1},t_k,x_{k+1},\cdots,x_n)}{(t_k-x_k)^{1-\alpha_k}}dt_k \quad (x_k<b_k), \tag{2.9.4}$$

respectively. These definitions are valid for functions $f(\mathbf{x}) = f(x_1, \cdots, x_n)$ defined for $x_k > a_k$ and $x_k < b_k$, respectively. By analogy with the one-dimensional case, the fractional integrals (2.9.3) and (2.9.4) are called *left- and right-sided partial Riemann-Liouville fractional integrals.*

Next we define the *mixed Riemann-Liouville fractional integrals of order* $\alpha \in \mathbb{C}^n$ $(\Re(\alpha) > \mathbf{0})$ as

$$(I_{\mathbf{a}+}^{\alpha}f)(\mathbf{x}) = (I_{a_1+}^{\alpha}\cdots I_{a_n+}^{\alpha}f)(\mathbf{x}) = \frac{1}{\Gamma(\alpha)}\int_{a_1}^{x_1}\cdots\int_{a_n}^{x_n}\frac{f(\mathbf{t})d\mathbf{t}}{(\mathbf{x}-\mathbf{t})^{1-\alpha}} \quad (\mathbf{x}>\mathbf{a}) \tag{2.9.5}$$

and

$$(I_{\mathbf{b}-}^{\alpha}f)(\mathbf{x}) = (I_{b_1-}^{\alpha}\cdots I_{b_n-}^{\alpha}f)(\mathbf{x}) = \frac{1}{\Gamma(\alpha)}\int_{x_1}^{b_1}\cdots\int_{x_n}^{b_n}\frac{f(\mathbf{t})d\mathbf{t}}{(\mathbf{t}-\mathbf{x})^{1-\alpha}} \quad (\mathbf{x}<\mathbf{b}). \tag{2.9.6}$$

The mixed fractional integrals can be applied with respect to a part of the variables, i.e., $\alpha_k = 0$ for some $k = 1, \cdots, n$. In this case we set $\alpha_k = 0$ for $k = m+1, \cdots, n$ and $\alpha_k \in \mathbb{C}$ $(\Re(\alpha_k) > 0)$ for $k = 1, \cdots, m$. Using the following notations,

$$\mathbf{x}' = (x_1,\cdots,x_m);\ \mathbf{x}'' = (x_{m+1},\cdots,x_n);\ \alpha' = (\alpha_1,\cdots,\alpha_m);\ (\mathbf{x}')^{\alpha'} = x_1^{\alpha_1}\cdots x_m^{\alpha_m}$$

such *mixed Riemann-Liouville fractional integrals of complex order* α' are defined by

$$(I_{\mathbf{a}+}^{\alpha'}f)(\mathbf{x}) := \frac{1}{\Gamma(\alpha_1)\cdots\Gamma(\alpha_m)}\int_{a_1}^{x_1}\cdots\int_{a_m}^{x_m}\frac{f(\mathbf{t}',\mathbf{x}'')d\mathbf{t}'}{(\mathbf{x}'-\mathbf{t}')^{1-\alpha'}} \quad (\mathbf{x}>\mathbf{a}) \tag{2.9.7}$$

and

$$(I_{\mathbf{b}+}^{\alpha'}f)(\mathbf{x}) := \frac{1}{\Gamma(\alpha_1)\cdots\Gamma(\alpha_m)}\int_{x_1}^{b_1}\cdots\int_{x_m}^{b_m}\frac{f(\mathbf{t}',\mathbf{x}'')d\mathbf{t}'}{(\mathbf{t}'-\mathbf{x}')^{1-\alpha'}} \quad (\mathbf{x}<\mathbf{b}). \tag{2.9.8}$$

Analogous to (2.9.3) and (2.9.4), the fractional integrals (2.9.5), (2.9.7) and (2.9.6), (2.9.8) are called *left-* and *right-sided mixed Riemann-Liouville fractional integrals.*

The *left-* and *right-sided partial Riemann-Liouville fractional derivatives of order* $\alpha_k \in \mathbb{C}$ $(\Re(\alpha_k) \geqq 0)$ with respect to the kth variable x_k are defined by

$$(D_{a_k+}^{\alpha_k}y)(\mathbf{x}) = \left(\frac{\partial}{\partial x_k}\right)^{L_k}(I_{a_k+}^{L_k-\alpha_k}y)(\mathbf{x})$$

$$:= \frac{1}{\Gamma(L_k-\alpha_k)}\left(\frac{\partial}{\partial x_k}\right)^{L_k}\int_{a_k}^{x_k}\frac{y(x_1,\cdots,x_{k-1},t_k,x_{k+1},\cdots,x_n)dt_k}{(x_k-t_k)^{\alpha_k-L_k+1}} \quad (x_k>a_k) \tag{2.9.9}$$

and

$$(D_{b_k-}^{\alpha_k}y)(\mathbf{x}) = \left(-\frac{\partial}{\partial x_k}\right)^{L_k}(I_{b_k-}^{L_k-\alpha_k}y)(\mathbf{x})$$

$$:= \frac{1}{\Gamma(L_k-\alpha_k)}\left(-\frac{\partial}{\partial x_k}\right)^{L_k}\int_{x_k}^{b_k}\frac{y(x_1,\cdots,\ x_{k-1},t_k,x_{k+1},\cdots,\ x_n)dt_k}{(t_k-x_k)^{\alpha_k-L_k+1}} \quad (x_k<b_k) \tag{2.9.10}$$

where $L_k = [\Re(\alpha_k)]+1$.

In particular, when $0<\alpha_k<1$, the relations (2.1.9) and (2.1.10) take the following forms:

$$(D_{a_k+}^{\alpha_k}y)(\mathbf{x}) = \frac{\partial}{\partial x_k}(I_{a_k+}^{1-\alpha_k}y)(\mathbf{x})$$

$$:= \frac{1}{\Gamma(1-\alpha_k)}\frac{\partial}{\partial x_k}\int_{a_k}^{x_k}\frac{y(x_1,\cdots,\ x_{k-1},t_k,x_{k+1},\cdots,\ x_n)dt_k}{(x_k-t_k)^{\alpha_k}} \quad (x_k>a_k), \tag{2.9.11}$$

$$(D_{b_k-}^{\alpha_k}y)(\mathbf{x}) = -\frac{\partial}{\partial x_k}(I_{b_k-}^{1-\alpha_k}y)(\mathbf{x})$$

$$:= -\frac{1}{\Gamma(1-\alpha_k)}\frac{\partial}{\partial x_k}\int_{a_k}^{x_k}\frac{y(x_1,\cdots,\ x_{k-1},t_k,x_{k+1},\cdots,\ x_n)dt_k}{(t_k-x_k)^{\alpha_k}} \quad (x_k<b_k). \tag{2.9.12}$$

If $\alpha_k = L_k \in \mathbb{N}_0$, (2.1.9) and (2.1.10) yield the usual partial derivatives as follows:

$$(D_{a_k+}^{L_k}y)(\mathbf{x}) = \left(\frac{\partial}{\partial x_k}\right)^{L_k}y(\mathbf{x}) \ \textit{and}\ (D_{b_k-}^{L_k}y)(\mathbf{x}) = (-1)^{L_k}\left(\frac{\partial}{\partial x_k}\right)^{L_k}y(\mathbf{x}). \tag{2.9.13}$$

The *left-* and *right-sided mixed Riemann-Liouville fractional derivatives of order* $\alpha_k \in \mathbb{C}$ $(\Re(\alpha_k) \geqq 0)$, corresponding to the mixed fractional integrals (2.9.5) and (2.9.6), are defined as

$$(D_{\mathbf{a}+}^{\alpha}y)(\mathbf{x}) = (D_{a_1+}^{\alpha}\cdots D_{a_n+}^{\alpha}y)(\mathbf{x})$$

$$:= \left(\frac{\partial}{\partial x_1}\right)^{L_1}\cdots\left(\frac{\partial}{\partial x_n}\right)^{L_n}\frac{1}{\Gamma(\mathbf{L}-\alpha)}\int_{a_1}^{x_1}\cdots\int_{a_n}^{x_n}\frac{y(\mathbf{t})dt}{(\mathbf{x}-\mathbf{t})^{\alpha-\mathbf{L}+\mathbf{1}}} \quad (\mathbf{x}>\mathbf{a}), \tag{2.9.14}$$

and

$$(D_{\mathbf{b}-}^{\alpha}y)(\mathbf{x}) = (D_{b_1-}^{\alpha}\cdots D_{b_n-}^{\alpha}y)(\mathbf{x})$$

$$:= \left(-\frac{\partial}{\partial x_1}\right)^{L_1}\cdots\left(-\frac{\partial}{\partial x_n}\right)^{L_n}\frac{1}{\Gamma(\mathbf{L}-\alpha)}\int_{x_1}^{b_1}\cdots\int_{x_n}^{b_n}\frac{y(\mathbf{t})dt}{(\mathbf{t}-\mathbf{x})^{\alpha-\mathbf{L}+\mathbf{1}}} \quad (\mathbf{x}<\mathbf{b}), \tag{2.9.15}$$

where $\mathbf{L} = (L_1,\cdots,\ L_n)$ and $L_k = [\Re(\alpha_k)]+1$ $(k=1,\cdots,\ n)$.

In particular, when $0<\alpha<1$, the relations (2.9.14) and (2.9.15) take the following forms:

$$(D_{\mathbf{a}+}^{\alpha}y)(\mathbf{x}) = (D_{a_1+}^{\alpha}\cdots D_{a_n+}^{\alpha}y)(\mathbf{x})$$

$$:= \frac{\partial}{\partial \mathbf{x}} \frac{1}{\Gamma(\mathbf{1}-\alpha)} \int_{a_1}^{x_1} \cdots \int_{a_n}^{x_n} \frac{y(\mathbf{t})d\mathbf{t}}{(\mathbf{x}-\mathbf{t})^{\alpha}} \quad (\mathbf{x} > \mathbf{a}) \tag{2.9.16}$$

and

$$(D_{\mathbf{b}-}^{\alpha} y)(\mathbf{x}) = (D_{b_1-}^{\alpha} \cdots D_{b_n-}^{\alpha} y)(\mathbf{x})$$

$$:= (-1)^n \frac{\partial}{\partial \mathbf{x}} \frac{1}{\Gamma(1-\alpha)} \int_{x_1}^{b_1} \cdots \int_{x_n}^{b_n} \frac{y(\mathbf{t})d\mathbf{t}}{(\mathbf{t}-\mathbf{x})^{\alpha}} \quad (\mathbf{x} < \mathbf{b}). \tag{2.9.17}$$

If $\alpha = \mathbf{L} \in \mathbb{N}_0^n := \mathbb{N}_0 \times \cdots \times \mathbb{N}_0$ (n times), then (2.9.14) and (2.9.15) are reduced to

$$(D_{\mathbf{a}+}^{\mathbf{L}} y)(\mathbf{x}) = \left(\frac{\partial}{\partial \mathbf{x}}\right)^{\mathbf{L}} y(\mathbf{x}) \text{ and } (D_{\mathbf{b}}^{\mathbf{L}} y)(\mathbf{x}) = (-1)^{|\mathbf{m}|} \left(\frac{\partial}{\partial \mathbf{x}}\right)^{\mathbf{L}} y(\mathbf{x}). \tag{2.9.18}$$

The *left-* and *right-sided mixed Riemann-Liouville fractional derivatives of order* $\alpha_k \in \mathbb{C}$ ($\Re(\alpha_k) \geqq 0$), corresponding to the mixed fractional integrals (2.9.7) and (2.9.8), are given by

$$(D_{\mathbf{a}+}^{\alpha'} y)(\mathbf{x}) = \left(\frac{\partial}{\partial x_1}\right)^{L_1} \cdots \left(\frac{\partial}{\partial x_m}\right)^{L_m} (I_{a_1+}^{L_1-\alpha_1} \cdots I_{a_m+}^{L_m-\alpha_m} y)(\mathbf{x})$$

$$:= \left(\frac{\partial}{\partial x_1}\right)^{L_1} \cdots \left(\frac{\partial}{\partial x_m}\right)^{L_m} \frac{1}{\prod_{k=1}^{m} \Gamma(L_k-\alpha_k)} \int_{a_1}^{x_1} \cdots \int_{a_m}^{x_m} \frac{f(\mathbf{t}',\mathbf{x}'')d\mathbf{t}'}{(\mathbf{x}'-\mathbf{t}')^{\alpha'-\mathbf{L}'+1}} \tag{2.9.19}$$

with $(\mathbf{x}' > \mathbf{a}')$ and

$$(D_{\mathbf{b}-}^{\alpha'} y)(\mathbf{x}) = \left(-\frac{\partial}{\partial x_1}\right)^{L_1} \cdots \left(-\frac{\partial}{\partial x_m}\right)^{L_m} (I_{b_1-}^{L_1-\alpha_1} \cdots I_{b_m-}^{L_m-\alpha_m} y)(\mathbf{x})$$

$$:= \left(-\frac{\partial}{\partial x_1}\right)^{L_1} \cdots \left(-\frac{\partial}{\partial x_m}\right)^{L_m} \frac{1}{\prod_{k=1}^{m} \Gamma(L_k-\alpha_k)} \int_{x_1}^{b_1} \cdots \int_{x_m}^{b_m} \frac{f(\mathbf{t}',\mathbf{x}'')d\mathbf{t}'}{(\mathbf{t}'-\mathbf{x}')^{\alpha'-\mathbf{L}'+1}} \tag{2.9.20}$$

where $(\mathbf{x}' < \mathbf{b}')$, $\mathbf{L}' = (L_1, \cdots, L_m)$ and $L_k = [\Re(\alpha_k)] + 1$ $(k = 1, \cdots, m)$.

In particular, when $0 < \alpha' < 1$, the formulas (2.9.19) and (2.9.20) are reduced to the following forms:

$$(D_{\mathbf{a}+}^{\alpha'} y)(\mathbf{x}) = \frac{\partial}{\partial x_1} \cdots \frac{\partial}{\partial x_n} I_{a_1+}^{1-\alpha_1} \cdots I_{a_m+}^{1-\alpha_m} y)(\mathbf{x})$$

$$= \frac{\partial}{\partial x_1} \cdots \frac{\partial}{\partial x_m} \frac{1}{\Gamma(1-\alpha_1)\cdots\Gamma(1-\alpha_m)} \int_{a_1}^{x_1} \cdots \int_{a_m}^{x_m} \frac{y(\mathbf{t}',\mathbf{x}'')d\mathbf{t}'}{(\mathbf{x}'-\mathbf{t}')^{\alpha'}} \quad (\mathbf{x}' > \mathbf{a}') \tag{2.9.21}$$

and

$$(D_{\mathbf{b}-}^{\alpha'} y)(\mathbf{x}) = (-1)^n \frac{\partial}{\partial x_1} \cdots \frac{\partial}{\partial x_n} I_{b_1-}^{1-\alpha_1} \cdots I_{b_m-}^{1-\alpha_m} y)(\mathbf{x})$$

$$= \frac{\partial}{\partial x_1} \cdots \frac{\partial}{\partial x_m} \frac{1}{\Gamma(1-\alpha_1)\cdots\Gamma(1-\alpha_m)} \int_{x_1}^{b_1} \cdots \int_{x_m}^{b_m} \frac{y(\mathbf{t}',\mathbf{x}'')d\mathbf{t}'}{(\mathbf{t}'-\mathbf{x}')^{\alpha'}} \quad (\mathbf{x}' < \mathbf{b}'). \tag{2.9.22}$$

If $\alpha' = \mathbf{L}' \in \mathbb{N}_0^m$, then (2.9.19) and (2.9.20) are reduced to

$$(D_{\mathbf{a}+}^{\mathbf{L}'} y)(\mathbf{x}) = \left(\frac{\partial}{\partial x_1}\right)^{L_1} \cdots \left(\frac{\partial}{\partial x_m}\right)^{L_m} y(\mathbf{x}) \tag{2.9.23}$$

and

$$(D_{\mathbf{b}-}^{\mathbf{L}'} y)(\mathbf{x}) = (-1)^{L_1+\cdots+L_m} \left(\frac{\partial}{\partial x_1}\right)^{L_1} \cdots \left(\frac{\partial}{\partial x_m}\right)^{L_m} y(\mathbf{x}), \tag{2.9.24}$$

respectively. We define above the partial and mixed Riemann-Liouville fractional integrals and fractional derivatives on the finite domain

$$[\mathbf{a},\mathbf{b}] = [a_1, b_1] \times \cdots \times [a_n, b_n].$$

The *partial and mixed Liouville fractional derivatives of order* $\alpha \in \mathbb{C}$ *on the octant* $\mathbb{R}^n_{+\cdots+} = \{\mathbf{t} = (t_1, \cdots, t_n) \in \mathbb{R}^n : t_1 \geqq 0, \cdots, t_n \geqq 0\}$ are similarly defined. For this we must let $a_k = 0$ and $b_k = \infty$ in the formulas (2.9.3)-(2.9.8), (2.9.9)-(2.9.12), (2.9.14)-(2.9.17), and (2.9.19)-(2.9.22). By having $a_k = -\infty$ and $b_k = \infty$ in these relations, we define *partial and mixed Liouville fractional derivatives of order* $\alpha \in \mathbb{C}$ *on the whole Euclidean space* $\mathbb{R}^n$. Most of the properties of the one-dimensional Riemann-Liouville and Liouville fractional integrals and fractional derivatives, presented in Sections 2.1-2.3, can be extended to these multidimensional cases. [See, in this regard, Samko et al. ([729], Sections 24.1-24.2)].

2.10 Riesz Fractional Integro-Differentiation

In this section we give the definitions and some properties of multidimensional Riesz fractional integrals and fractional derivatives presented in the book by Samko et al. ([729], Sections 25 and 26). Such operations of fractional integration and fractional differentiation in the n-dimensional Euclidean space $\mathbb{R}^n$ ($n \in \mathbb{N}$) are fractional powers $(-\Delta)^{\alpha/2}$ of the Laplace operator (1.3.31). For $\alpha \in \mathbb{C}\backslash\{0\}$ and "sufficiently good" functions $f(\mathbf{x}) = f(x_1, \cdots, x_n)$, such fractional operations are defined in terms of the Fourier transform (1.3.22) by

$$(-\Delta)^{-\alpha/2} f = \mathcal{F}^{-1}|\mathbf{x}|^{-\alpha}\mathcal{F}f = \begin{cases} I^\alpha f, & \Re(\alpha) > 0, \\ \mathbf{D}^{-\alpha} f, & \Re(\alpha) < 0. \end{cases} \tag{2.10.1}$$

The operations I^α and $\mathbf{D}^\alpha$ defined in (2.10.1) for $\Re(\alpha) > 0$ are called the *Riesz fractional integration* and the *Riesz fractional differentiation*, respectively.

The Riesz fractional integration I^α is realized in the form of the *Riesz potential* defined as the Fourier convolution (1.3.33) of the form

$$(I^\alpha f)(\mathbf{x}) = \int_{\mathbb{R}^n} k_\alpha(\mathbf{x}-\mathbf{t}) f(\mathbf{t}) d\mathbf{t} \quad (\Re(\alpha) > 0). \tag{2.10.2}$$

Here a function $k_\alpha(\mathbf{x})$, called the *Riesz kernel*, is given by

$$k_\alpha(\mathbf{x}) := \frac{1}{\gamma_n(\alpha)} \begin{cases} |\mathbf{x}|^{\alpha-n}, & \alpha - n \neq 0, 2, 4, \cdots, \\ |\mathbf{x}|^{\alpha-n} \log\left(\frac{1}{|\mathbf{x}|}\right), & \alpha - n = 0, 2, 4, \cdots, \end{cases} \tag{2.10.3}$$

and the constant $\gamma_n(\alpha)$ has the form

$$\gamma_n(\alpha) := \begin{cases} 2^\alpha \pi^{n/2} \Gamma[\alpha/2][\Gamma((n-\alpha)/2)]^{-1}, & \alpha - n \neq 0, 2, 4, \cdots, \\ (-1)^{(n-\alpha)/2} 2^{\alpha-1} \pi^{n/2} \Gamma[1 + (\alpha - n)/2] \Gamma[\alpha/2], & \alpha - n = 0, 2, 4, \cdots. \end{cases} \tag{2.10.4}$$

By the Fourier convolution theorem (see Theorem 1.4 in Section 1.3), we have the following property.

Property 2.31 *If $\Re(\alpha) > 0$, then the Fourier transform of the Riesz potential* (2.10.2) *is given by*

$$(\mathcal{F} I^\alpha f)(\mathbf{x}) = \frac{1}{|x|^\alpha} (\mathcal{F} f)(\mathbf{x}). \tag{2.10.5}$$

This formula is true for a function f belonging to Lizorkin's space Φ defined in Section 1.2.

Corollary 2.5 *If $\Re(\alpha) > 2$, then the following formula holds:*

$$\Delta I^\alpha f = -I^{\alpha-2} f \quad (\Re(\alpha) > 2;\ f \in \Phi). \tag{2.10.6}$$

For functions f from the Lizorkin space Φ, presented in (1.2.17), the *semigroup property* is also valid [see Samko et al. ([729], Theorem 25.1)].

Property 2.32 *The Lizorkin space Φ is invariant with respect to the Riesz potential I^α. Moreover, $I^\alpha(\Phi) = \Phi$, and*

$$I^\alpha I^\beta f = I^{\alpha+\beta} f \quad (\Re(\alpha) > 0;\ \Re(\beta) > 0;\ f \in \Phi). \tag{2.10.7}$$

When $\alpha - n \neq 0, 2, 4, \cdots$, the Riesz potential (2.10.2) takes the following form:

$$(I^\alpha f)(\mathbf{x}) = \frac{1}{\gamma_n(\alpha)} \int_{\mathbb{R}^n} \frac{f(\mathbf{t})dt}{|\mathbf{x} - \mathbf{t}|^{n-\alpha}} \quad (\Re(\alpha) > 0;\ \alpha - n \neq 0, 2, 4, \cdots). \tag{2.10.8}$$

When $\alpha > 0$, then the following assertion holds.

Property 2.33 *If $0 < \alpha < n$ and $1 < p < n/\alpha$, then the Riesz potential $(I^\alpha f)(\mathbf{x})$ is defined for $f \in L_p(\mathbb{R}^n)$.*

The next statement, known as the *Sobolev theorem*, is a generalization of the Hardy-Littlewood theorem given in Lemmas 2.10 and 2.15 [see Samko et al. ([729], Theorem 25.2)].

Theorem 2.4 *Let $\alpha > 0$, $1 \leqq p \leqq \infty$ and $1 \leqq q \leqq \infty$. The operator I^α is bounded from $L_p(\mathbb{R}^n)$ into $L_q(\mathbb{R}^n)$ if, and only if,*

$$0 < \alpha < n, \quad 1 < p < \frac{n}{\alpha}, \quad \textit{and} \ \frac{1}{q} = \frac{1}{p} - \frac{\alpha}{n}. \tag{2.10.9}$$

The next assertion is a generalization of Theorem 2.4 to a *weighted* L_p-space with power weight depending on $|\mathbf{x}|$:

$$L_p(\mathbb{R}^n, |\mathbf{x}|^\gamma) := \left\{ f: \ ||f||_{p;\gamma} = \left(\int_{\mathbb{R}^n} |\mathbf{x}|^\gamma |f(\mathbf{x})|^p d\mathbf{x} \right)^{1/p} \quad (1 \leqq p < \infty) \right\}. \tag{2.10.10}$$

Theorem 2.5 *Let $\alpha > 0$, $1 < p < \infty$, $1 < r < \infty$, and $\gamma, \mu \in \mathbb{R}$ be such that*

$$\alpha p - n < \gamma < n(p-1), \quad \frac{1}{p} - \frac{\alpha}{n} \leqq \frac{1}{r} \leqq \frac{1}{p} \ \textit{and} \ \frac{\mu + n}{r} = \frac{\gamma + n}{p} - \alpha. \tag{2.10.11}$$

Then the operator I^α is bounded from $L_p(\mathbb{R}^n, |\mathbf{x}|^\gamma)$ into $L_q(\mathbb{R}^n, |\mathbf{x}|^\mu)$ as follows:

$$\left(\int_{\mathbb{R}^n} |\mathbf{x}|^\gamma |I^\alpha f(\mathbf{x})|^r d\mathbf{x} \right)^{1/r} \leqq c \left(\int_{\mathbb{R}^n} |\mathbf{x}|^\gamma |f(\mathbf{x})|^p d\mathbf{x} \right)^{1/p}, \tag{2.10.12}$$

where the constant $c > 0$ depends on α, n, p, r, γ, and μ.

Let $\dot{\mathbb{R}}^n$ be a compactification of $\mathbb{R}^n$ by an infinite point, that is, $\dot{\mathbb{R}}^n = \mathbb{R}^n \bigcup \{\infty\}$, and let $\rho(\mathbf{x})$ be a nonnegative function on $\dot{\mathbb{R}}^n$. If $0 < \alpha < 1$, then the Riesz potential (2.10.8) is defined for $f(\mathbf{x})$ belonging to a weighted space of Hölderian functions $H^\lambda\left(\dot{\mathbb{R}}^n, \rho\right)$:

$$H^\lambda(\mathbb{R}^n, \rho) := \left\{ f(\mathbf{x}): \ \rho(\mathbf{x}) f(\mathbf{x}) \in H^\lambda\left(\dot{\mathbb{R}}^n\right) \quad (0 < \lambda < 1) \right\}, \tag{2.10.13}$$

where

$$H^\lambda\left(\dot{\mathbb{R}}^n\right)$$

$$:= \left\{ f(\mathbf{x}): \ f(\mathbf{x}) \in C\left(\dot{\mathbb{R}}^n\right), |f(\mathbf{x}+\mathbf{h}) - f(\mathbf{x})| \leqq \frac{c|\mathbf{h}|^\lambda}{(1+|\mathbf{x}|)^\lambda (1+|\mathbf{x}+\mathbf{h}|)^\lambda} \right\} \tag{2.10.14}$$

with $c > 0$. The following assertion is valid [see Samko et al. ([729], Theorem 25.5)].

Theorem 2.6 *Let $0 < \alpha < 1$, $0 < \lambda < 1$ and $\alpha + \lambda < 1$. Then the operator I^α is isomorphic from $H^\lambda\left(\dot{\mathbb{R}}^n, (1+|\mathbf{x}|^2)^{(n+\alpha)/2}\right)$ onto $H^{\lambda+\alpha}\left(\dot{\mathbb{R}}^n, (1+|\mathbf{x}|^2)^{(n-\alpha)/2}\right)$.*

When $0 < \alpha < n$ and $1 < p < n/\alpha$, the Riesz potential (2.10.8) is connected with the *Poisson* and *Gauss-Weierstrass transforms* defined by

$$(P_t f)(\mathbf{x}) := \int_{\mathbb{R}^n} P(\mathbf{y}, t) f(\mathbf{x} - \mathbf{y}) d\mathbf{y} \quad (t > 0) \tag{2.10.15}$$

and

$$(W_t f)(\mathbf{x}) := \int_{\mathbb{R}^n} W(\mathbf{y}, t) f(\mathbf{x} - \mathbf{y}) d\mathbf{y} \quad (t > 0), \tag{2.10.16}$$

respectively. Here $P(\mathbf{x}, t)$ is the *Poisson kernel*:

$$P(\mathbf{y}, t) := \frac{d_n t}{(|\mathbf{x}|^2 + t^2)^{(n+1)/2}} \quad \text{and} \quad d_n = \pi^{-(n+1)/2} \Gamma\left(\frac{n+1}{2}\right), \tag{2.10.17}$$

while $W(\mathbf{x}, t)$ is the *Gauss-Weierstrass kernel*:

$$W(\mathbf{x}, t) := (4\pi t)^{-n/2} e^{-|\mathbf{x}|^2/4t}. \tag{2.10.18}$$

The following result holds [see Samko et al. ([729], Theorem 25.6)].

Theorem 2.7 *If $0 < \alpha < n$ and $1 < p < n/\alpha$, then the Riesz potential $(I^\alpha f)(\mathbf{x})$ with $f(\mathbf{x}) \in L_p(\mathbb{R}^n)$ has the following representations:*

$$(I^\alpha f)(\mathbf{x}) = \frac{1}{\Gamma(\alpha)} \int_0^\infty t^{\alpha-1} (P_t f)(\mathbf{x}) dt \tag{2.10.19}$$

and

$$(I^\alpha f)(\mathbf{x}) = \frac{1}{\Gamma(\alpha/2)} \int_0^\infty t^{(\alpha/2)-1} (W_t f)(\mathbf{x}) dt. \tag{2.10.20}$$

Moreover, the compositional relations of the Poisson and Gauss-Weierstrass transforms with the Riesz potential are expressed in terms of the Liouville fractional integration operator (2.3.2) *as follows:*

$$(P_t I^\alpha f)(\mathbf{x}) = \left(I_-^\alpha (P_\tau f)(\mathbf{x})\right)(t) \tag{2.10.21}$$

and

$$(W_t I^\alpha f)(\mathbf{x}) = \left(I_-^{\alpha/2} (W_\tau f)(\mathbf{x})\right)(t), \tag{2.10.22}$$

where the operators I_-^α and $I_-^{\alpha/2}$ are applied with respect to $\tau > 0$.

For $\alpha > 0$, the Riesz fractional derivative $\mathbf{D}^\alpha$ in (2.10.1) is realized in the form of the *hypersingular integral* defined by

$$(\mathbf{D}^\alpha y)(\mathbf{x}) := \frac{1}{d_n(l, \alpha)} \int_{\mathbf{R}^n} \frac{(\Delta_{\mathbf{t}}^l y)(\mathbf{x})}{|\mathbf{t}|^{n+\alpha}} d\mathbf{t} \quad (l > \alpha). \tag{2.10.23}$$

Here $(\Delta_{\mathbf{t}}^l y)(\mathbf{x})$ is a *finite difference of order l of a function $y(\mathbf{x})$ with a vector step $\mathbf{t} \in \mathbb{R}^n$ and centered at the point $\mathbf{x} \in \mathbb{R}^n$*:

$$(\Delta_{\mathbf{t}}^l y)(\mathbf{x}) = (E - \tau_{\mathbf{t}})^l f(\mathbf{x}) := \sum_{k=0}^{l} (-1)^k \binom{l}{k} y(\mathbf{x} - k\mathbf{t}), \tag{2.10.24}$$

coinciding with that in (2.8.5) for $x, t \in \mathbb{R}$, while $d_n(l,\alpha)$ is a constant defined by

$$d_n(l,\alpha) = \frac{2^{-\alpha}\pi^{1+n/2}}{\Gamma\left(1+\frac{\alpha}{2}\right)\Gamma\left(\frac{n+\alpha}{2}\right)} \frac{A_l(\alpha)}{\sin\left(\frac{\alpha\pi}{2}\right)}, \tag{2.10.25}$$

where

$$A_l(\alpha) = \sum_{k=0}^{l} (-1)^{k-1} \binom{l}{k} k^{\alpha}. \tag{2.10.26}$$

Note that the hypersingular integral $(\mathbf{D}^{\alpha}y)(\mathbf{x})$ does not depend on the choice of l $(l > \alpha)$. Such a construction is also called the *Riesz fractional derivative* of order $\alpha > 0$ in the sense that

$$(\mathcal{F}\mathbf{D}^{\alpha}y)(\mathbf{x}) = |x|^{\alpha}(\mathcal{F}y)(\mathbf{x}) \quad (\alpha > 0) \tag{2.10.27}$$

for "sufficiently good" functions $y(\mathbf{x})$. In particular, this relation is valid for f belonging to the Lizorkin space Φ. Equality (2.10.27) is also valid for differentiable functions $y(\mathbf{x})$. The following property holds [see Samko et al. ([729], Lemma 25.3)].

Property 2.34 *If $y(\mathbf{x})$ belongs to the space $C_0^{\infty}(\mathbb{R}^n)$ of infinitely differentiable functions $y(x)$ on $\mathbb{R}^n$ with a compact support , then the Fourier transform of $(\mathbf{D}^{\alpha}y)(\mathbf{x})$ is given by*

$$(\mathcal{F}\mathbf{D}^{\alpha}y)(\mathbf{x}) = |x|^{\alpha}(\mathcal{F}y)(\mathbf{x}) \quad (\alpha > 0). \tag{2.10.28}$$

The conditions for the existence of $\mathbf{D}^{\alpha}f)(\mathbf{x})$ are given by the following assertion [see Samko et al. ([729], Section 26.2)].

Lemma 2.40 *Let $\alpha > 0$ and $[\alpha]$ be the integer part of α. Also let a function $y(\mathbf{x})$ be bounded together with its derivatives $(\mathbf{D}^{\mathbf{k}}y)(\mathbf{x})$, $(|\mathbf{k}| = [\alpha]+1)$. Then the hypersingular integral $(\mathbf{D}^{\alpha}f)(\mathbf{x})$ in (2.10.23) is absolutely convergent. If $l > 2[\alpha/2]$, then this integral is only conditionally convergent.*

The Riesz fractional derivative (2.10.23) yields an operator inverse to the Riesz potential (2.10.2).

Property 2.35 *The formula*

$$\mathbf{D}^{\alpha}I^{\alpha}f = f \quad (\alpha > 0) \tag{2.10.29}$$

holds for "sufficiently good" functions f; in particular, for f belonging to the Lizorkin space Φ.

Moreover, the inversion (2.10.29) is also valid for the Riesz potential (2.10.2) in the frame of L_p-spaces: $f(\mathbf{x}) \in L_p(\mathbb{R}^n)$ for $1 \leqq p < n/\alpha$. Here the Riesz fractional derivative $\mathbf{D}^{\alpha}$ is understood to be conditionally convergent in the sense that

$$\mathbf{D}^{\alpha}y = \lim_{\varepsilon \to 0+} \mathbf{D}^{\alpha}_{\varepsilon}y, \tag{2.10.30}$$

where $\mathbf{D}_\varepsilon^\alpha y$ is the *truncated hypersingular integral* defined by

$$(\mathbf{D}_\varepsilon^\alpha y)(\mathbf{x}) := \frac{1}{d_n(l,\alpha)} \int_{|\mathbf{t}|>\varepsilon} \frac{(\Delta_{\mathbf{t}}^l y)(\mathbf{x})}{|\mathbf{t}|^{n+\alpha}} d\mathbf{t} \quad (l > \alpha;\ \alpha > 0;\ \varepsilon > 0), \tag{2.10.31}$$

and the limit is taken in the norm of the space $L_p(\mathbb{R}^n)$.

The following result holds [see Samko et al. ([729], Theorem 26.3)].

Theorem 2.8 *Let* $0 < \alpha < n$ *and let* $y = I^\alpha f$ *with* $f(\mathbf{x}) \in L_p(\mathbb{R}^n)$ $(1 \leqq p < n/\alpha)$. *Then*

$$f(\mathbf{x}) = (\mathbf{D}^\alpha y)(\mathbf{x}), \tag{2.10.32}$$

where $(\mathbf{D}^\alpha y)(\mathbf{x})$ *is understood in the sense of* (2.10.30), *with the limit being taken in the norm of the space* $L_p(\mathbb{R}^n)$.

Corollary 2.6 *If* $0 < \alpha < n$ *and* $f(\mathbf{x}) \in L_p(\mathbb{R}^n)$ $(1 \leqq p < n/\alpha)$, *then*

$$f(\mathbf{x}) = (\mathbf{D}^\alpha I^\alpha f)(\mathbf{x}), \tag{2.10.33}$$

where $(\mathbf{D}^\alpha I^\alpha f)(\mathbf{x})$ *is understood in the sense of* (2.10.30), *and*

$$(\mathbf{D}^\alpha I^\alpha f)(\mathbf{x}) = \lim_{\varepsilon \to 0+} (\mathbf{D}_\varepsilon^\alpha I^\alpha f)(\mathbf{x}), \tag{2.10.34}$$

with the limit being taken in the norm of the space $L_p(\mathbb{R}^n)$.

2.11 Comments and Observations

A. In the preceding sections of this chapter, we have not dealt with the fractional calculus operators which are based essentially upon the Cauchy-Goursat Integral Formula. In recent years, these fractional calculus operators have been used rather extensively in obtaining particular solutions to numerous families of homogeneous (as well as nonhomogeneous) linear ordinary and partial differential equations which are associated, among others, with many of the *celebrated* equations of mathematical physics such as (for example) the Gauss hypergeometric equation (1.6.13) and the Bessel equation (1.7.9). In the cases of (ordinary as well as partial) differential equations of *higher* orders, which have stemmed naturally from the Gauss hypergeometric equation (1.6.13), the Bessel equation (1.7.9), and their many relatives and extensions, there have been several seemingly independent attempts to present a remarkably large number of widely-scattered results in a *unified* manner (see, for details, the work of Tu et al.[829] and the references cited by them; see also numerous other recent works on this subject, which are cited in the Bibliography). item [**B.**] From few years ago many interesting applications of the so called Fractional Fourier Transform (FFT) have been published. Under the name FFT are known several different definitions of generalization of the ordinary Fourier transform. This fractional Transform have been widely applied mainly in Optics and Signal Processing Theory. The excellent book by Ozaktas et al. [658] is a good introduction in this matter (see, also, [657], [664], [119], [784], [904], [904], [110], and [111])

C. In many recent investigations (*especially* those using fractional calculus in specific problems of applied mathematics and mathematical physics), use is frequently made of such analytical tools as the Laplace and Fourier transforms (see Sections 1.3 and 1.4), and also of such generalized functions as (for example) the Dirac delta function $\delta(t)$, which occur naturally in the differential equations considered. In situations like these, it is important from the viewpoint of mathematical analysis that we choose the proper *distributional* test-function spaces which can conveniently validate not only the use of the Laplace, Fourier, and other integral transforms as analytical tools, but also the presence of the generalized functions in the differential equation modeling the physical problem (see also Section 1.2).

D. A number of such not-so-common functions as the Weierstrass function are known to emerge as solutions of fractional differential equations. Here we have chosen not to develop a detailed investigation of these functions.

E. The recent monograph by Martínez and Sanz [558] provides an excellent source-book for the theory and applications of *fractional* (especially negative and imaginary) powers of many different families of non-negative linear operators as well as such differential operators as (for example) those of Riemann-Liouville and Weyl considered in this chapter, (see, also, [60], [437], [253], [719], [876]).

F. Control theory has been discipline where many mathematical ideas and methods have melt to produce a rich crossing point of Engineering and Mathematics. Control theory is certainly one of the most interdisciplinary areas of research and this field of Mathematics arises in most modern applications, for example, in the control of Autonomous and Intelligent Robotic Vehicles (see, [258], [691], and [814]). In particular, during the last years the Fractional control models have produced very good solutions to so many interesting control problems with applications in many branch of the Engineering.

Here we do not include explicit examples of such applications of the fractional models, but we will included some references where the reader can find many interesting theoretical and applied results (see, for example, [653], [815], [204],[678], [565], [562], [612], [667], [683], [668], [816], [685], [685], [11], [357], and [598])

Chapter 3

ORDINARY FRACTIONAL DIFFERENTIAL EQUATIONS. EXISTENCE AND UNIQUENESS THEOREMS

This chapter is devoted to proving the existence and uniqueness of solutions to Cauchy type problems for ordinary differential equations of fractional order on a finite interval of the real axis in spaces of summable functions and continuous functions. Nonlinear and linear fractional differential equations in one-dimensional and vectorial cases are considered. The corresponding results for the Cauchy problem for ordinary differential equations are presented.

3.1 Introduction and a Brief Overview of Results

In this section we give a brief overview of the results of existence and uniqueness theorems for differential equations of fractional order on a finite interval of the real axis.

Most of the investigations in this field involve the existence and uniqueness of solutions to fractional differential equations with the Riemann-Liouville fractional derivative $(D_{a+}^{\alpha}y)(x)$ defined for $(\Re(\alpha) > 0)$ by (2.1.5). The "model" nonlinear differential equation of fractional order α $(\Re(\alpha) > 0)$ on a finite interval $[a, b]$ of the real axis $\mathbb{R} = (-\infty, \infty)$ has the form

$$(D_{a+}^{\alpha}y)(x) = f[x, y(x)] \quad (\Re(\alpha) > 0;\ x > a), \tag{3.1.1}$$

with initial conditions

$$(D_{a+}^{\alpha-k}y)(a+) = b_k,\ b_k \in \mathbb{C}\ (k = 1, \cdots, n), \tag{3.1.2}$$

where $n = \Re(\alpha) + 1$ for $\alpha \notin \mathbb{N}$ and $\alpha = n$ for $\alpha \in \mathbb{N}$. The notation $(D_{a+}^{\alpha-k}y)(a+)$ means that the limit is taken at almost all points of the right-sided neighborhood $(a, a+\varepsilon)$ $(\varepsilon > 0)$ of a as follows:

$$(D_{a+}^{\alpha-k}y)(a+) = \lim_{x\to a+}(D_{a+}^{\alpha-k}y)(x)\ \ (1 \leqq k \leqq n-1), \tag{3.1.3}$$

$$(D_{a+}^{\alpha-n}y)(a+) = \lim_{x\to a+}(I_{a+}^{n-\alpha}y)(x)\ (\alpha \neq n);\ (D_{a+}^{0}y)(a+) = y(a)\ (\alpha = n), \tag{3.1.4}$$

where I_{a+}^{α} is the Riemann-Liouville fractional integration operator of order $\alpha \in \mathbb{C}$ defined by (2.1.1).

In particular, if $\alpha = n \in \mathbb{N}$, then, in accordance with (2.1.7) and (3.1.4), the problem in (3.1.1)-(3.1.2) is reduced to the usual *Cauchy problem* for the ordinary differential equation of order $n \in \mathbb{N}$:

$$y^{(n)}(x) = f[x, y(x)],\ y^{(n-k)}(a) = b_k,\ b_k \in \mathbb{C}\ (k = 1, \cdots, n). \tag{3.1.5}$$

The problem (3.1.1)-(3.1.2) is therefore called, by analogy, a *Cauchy type problem* [see Samko et al.([729], Section 42)].

When $0 < \Re(\alpha) < 1$, the problem (3.1.1)-(3.1.2) takes the form

$$(D_{a+}^{\alpha}y)(x) = f[x, y(x)],\ \ (I_{a+}^{1-\alpha}y)(a+) = b\ \ \ (b \in \mathbb{C}) \tag{3.1.6}$$

and this problem can be rewritten as the weighted Cauchy type problem

$$(D_{a+}^{\alpha}y)(x) = f[x, y(x)],\ \ \lim_{x\to a+}(x-a)^{1-\alpha}y(x) = c\ \ (c \in \mathbb{C}). \tag{3.1.7}$$

We discuss investigations of the above problems following their historical chronology. Essentially, they were based on reducing problem (3.1.1)-(3.1.2) to the following nonlinear Volterra integral equation of the second kind:

$$y(x) = \sum_{j=1}^{n}\frac{b_j}{\Gamma(\alpha-j+1)}(x-a)^{\alpha-j} + \frac{1}{\Gamma(\alpha)}\int_a^x\frac{f[t, y(t)]dt}{(x-t)^{1-\alpha}}\ \ (x > a). \tag{3.1.8}$$

Pitcher and Sewell [675] in 1938, first considered the nonlinear fractional differential equation (3.1.1) with $0 < \alpha < 1$, provided that $f(x, y)$ is bounded in a special region G lying in $\mathbb{R} \times \mathbb{R}$ and satisfies the Lipschitz condition with respect to y:

$$|f(x, y_1) - f(x, y_2)| \leqq A|y_1 - y_2|, \tag{3.1.9}$$

where the constant $A > 0$ does not depend on x. They proved the existence of the continuous solution $y(x)$ for the corresponding nonlinear integral equation of the form (3.1.8) with $0 < \alpha < 1$, $n = 1$, and $b_1 = 0$. But the main result given in Pitcher and Sewell ([675], Theorem 4.2) for the fractional differential equation

(3.1.1) with $0 < \alpha < 1$ was not correct because they used the compositional relation $I_{a+}^{\alpha} D_{a+}^{\alpha} y = y$ instead of the correct one given in (2.1.40) to reduce the differential equation to an integral equation. However, the work of Pitcher and Sewell [675] did contain the idea of reducing the solution of the fractional differential equation (3.1.1) to that of a Volterra integral equation (3.1.8).

Barrett [68] in 1954 first considered the Cauchy type problem for the linear differential equation

$$(D_{a+}^{\alpha} y)(x) - \lambda y(x) = f(x) \ (n - 1 \leqq \Re(\alpha) < n, \ \alpha \neq n - 1), \tag{3.1.10}$$

where $n \in \mathbb{N}$ is the smallest integer such that $n > \Re(\alpha) > 0$, with initial conditions (3.1.2). He proved in Barret ([68], Theorem 2.1) that, if $f(x)$ belongs to $L(a, b)$, then such a problem has the unique solution $y(x)$ in some subspace of $L(a, b)$ given by

$$y(x) = \sum_{k=1}^{n} b_k x^{\alpha-k} E_{\alpha,\alpha-k+1} \left(\lambda(x-a)^{\alpha}\right) + \int_a^x (x-t)^{\alpha-1} E_{\alpha,\alpha} \left[\lambda(x-t)^{\alpha}\right] f(t)dt, \tag{3.1.11}$$

where $E_{\alpha,\alpha-k+1}(z)$ $(k = 1, \cdots, n)$ are the Mittag-Leffler functions defined by (1.8.17). Barrett's arguments were based on the formula (2.1.39) for the product $I_{a+}^{\alpha} D_{a+}^{\alpha} f$. By applying this formula, he implicitly used the method of reducing the Cauchy type problem (3.1.10),(3.1.2) to a linear Volterra integral equation of the second kind of the form (3.1.8) with $f[x, y(t)] = \lambda y(t) + f(t)$ and then applied the method of successive approximations to derive the solution (3.1.11).

Al-Bassam [17] first considered the Cauchy type problem (3.1.6) for a real $0 < \alpha \leqq 1$

$$(D_{a+}^{\alpha} y)(x) = f[x, y(x)] \quad (0 < \alpha \leqq 1), \tag{3.1.12}$$

$$(I_{a+}^{1-\alpha} y)(a+) = b, \quad b \in \mathbb{R}, \tag{3.1.13}$$

in the space of continuous functions $C[a, b]$, provided that $f(x, y)$ is a real-valued continuous function in a domain $G \subset \mathbb{R} \times \mathbb{R}$ such that $\sup_{(x,y)\in G} |f(x, y)| \leqq \infty$ and that it satisfies the Lipschitz condition (3.1.9). Applying the operator I_{a+}^{α} to both sides of (3.1.10), and using the relation (2.1.39) and the initial conditions (3.1.11), he reduced the problem (3.1.12)-(3.1.13) to the Volterra nonlinear integral equation (3.1.8) with $n = 1$:

$$y(x) = \frac{b_1 (x-a)^{\alpha-1}}{\Gamma(\alpha)} + \frac{1}{\Gamma(\alpha)} \int_a^x \frac{f[t, y(t)]dt}{(x-t)^{1-\alpha}} \ (x > a; \ 0 < \alpha \leqq 1). \tag{3.1.14}$$

Using the method of successive approximations, Al-Bassam [17] established the existence of the continuous solution $y(x)$ to the equation (3.1.14). Moreover, he was probably the first to indicate that the method of contractive mapping could be applied to prove the uniqueness of the solution $y(x)$ to (3.1.14), and formally presented such a proof. Al-Bassam [17] also indicated - *but did not prove* - the equivalence of the Cauchy type problem (3.1.12)-(3.1.13) and the integral equation (3.1.14), and therefore his results on the existence and uniqueness of the continuous solution $y(x)$, formulated in Al-Bassam ([17], Theorem 1), could be true only for the integral equation (3.1.14). We also note that the conditions suggested by Al-Bassam [17] are not suitable for solving the Cauchy type problem (3.1.12)-(3.1.13) in some simple cases; for example, for $f[x, y(x)] = y(x)$.

The same remarks apply to the existence and uniqueness results formulated without proof in Al-Bassam ([17], Theorems 2, 4, 5, 6) for more general Cauchy type problems of the form (3.1.1)-(3.1.2) for a real $\alpha > 0$:

$$(D_{a+}^{\alpha} y)(x) = f[x, y(x)] \ (n-1 < \alpha \leqq n; \ n \in \mathbb{N}), \tag{3.1.15}$$

$$(D_{a+}^{\alpha-k} y)(a+) = b_k, \ b_k \in \mathbb{R} \ (k = 1, 2, \cdots, n), \tag{3.1.16}$$

where the corresponding Volterra equation is given by (3.1.8) for the system of equations (3.1.12), and for more general nonlinear and linear fractional differential equations than (3.1.12). The above and some other results were presented in Al-Bassam ([18]-[21]) [see also Samko et al. ([729], Section 42.1)].

Dzhrbashyan and Nersesyan [217] studied the linear differential equation of fractional order

$$(D^{\sigma} y)(x) := (D^{\sigma_n} y)(x) + \sum_{k=0}^{n-1} a_k(x)(D^{\sigma_{n-k-1}} y)(x) + a_n(x) y(x) = f(x), \tag{3.1.17}$$

with sequential fractional derivatives $(D^{\sigma} y)(x)$ and $(D^{\sigma_{n-k-1}} y)(x)$ $(k = 0, 1, \cdots, n-1)$ defined in terms of the Riemann-Liouville fractional derivatives (2.1.5) by

$$D^{\sigma_k} = D_{0+}^{\alpha_k - 1} D_{0+}^{\alpha_{k-1}} \cdots D_{0+}^{\alpha_0} \ (k = 1, \cdots, n), \ D^{\sigma_0} = D_{0+}^{\alpha_0 - 1}, \tag{3.1.18}$$

where

$$\sigma_k = \sum_{j=0}^{k} \alpha_j - 1 \ (k = 0, 1, \cdots, n); \ 0 < \alpha_j \leqq 1 \ (j = 0, 1, \cdots, n) \tag{3.1.19}$$

$$(\alpha_k = \sigma_k - \sigma_{k-1} \ (k = 1, \cdots, n); \ \alpha_0 = \sigma_0 + 1). \tag{3.1.20}$$

Dzhrbashyan and Nersesyan [217] proved that, for $\alpha_0 > 1 - \alpha_n$, the Cauchy type problem

$$(D^{\sigma} y)(x) = f(x), \ (D^{\sigma_k} y)(x)|_{x=0} = b_k \ (k = 0, 1, \cdots, n-1) \tag{3.1.21}$$

has a unique continuous solution $y(x)$ on an interval $[0, d]$, provided that the functions $p_k(x)$ $(0 \leq k \leq n-1)$ and $f(x)$ satisfy some additional conditions. In particular, when $p_k(x) = 0$ $(k = 0, 1, \cdots, n)$, they obtained the explicit solution

$$y(x) = \sum_{k=0}^{n-1} \frac{b_k x^{\sigma_k}}{\Gamma(1+\sigma_k)} + \frac{1}{\Gamma(\sigma_n)} \int_a^x (x-t)^{\sigma_n - 1} f(t) dt \tag{3.1.22}$$

to the Cauchy type problem

$$(D^{\sigma_n} y)(x) = f(x), \ (D^{\sigma_k} y)(x)|_{x=0} = b_k \ (k = 0, 1, \cdots, n-1). \tag{3.1.23}$$

The Cauchy type problems (3.1.1)-(3.1.2), (3.1.12)-(3.1.13), and (3.1.15)-(3.1.16) were studied by Al-Abedeen [12], Al-Abedeen and Arora [13], Arora and Alshamani [37], Tazali [811], Tazali and Karim [812], Hadid and Alshamani [321], Leskovskij [477], Semenchuk [757], Luszczki and Rzepecki [511], El-Sayed [222]-[224] and [226]), El-Sayed and Ibrahim [232], El-Sayed and Gafar [231]-[230], Hadid [320], and others. But the above investigations were not complete, however. Most researchers have obtained results not for the initial value problems, but for the corresponding Volterra integral equations. Some authors considered only particular cases. Moreover, some of the results obtained contained errors in the proof of the equivalence of initial value problems and the Volterra integral equations and in the proof of the uniqueness theorem. In this regard, see the survey paper by Kilbas and Trujillo ([407], Sections 4 and 5).

Delbosco and Rodino [164] considered the nonlinear Cauchy problem

$$(D_{0+}^{\alpha}y)(x) = f(x, y(x)); \qquad y^{(k)}(0) = y_k \in \mathbb{R} \ (k = 0, 1, \cdots, [\alpha]), \tag{3.1.24}$$

with $0 \leqq x \leqq T$, $\alpha > 0$, and $f(x, y)$ a continuous function on $[0, 1] \times \mathbb{R}$. They proved the equivalence of this problem to the corresponding Volterra equation and used Schauder's fixed point theorem to prove that the equation considered has at least a continuous solution $y(x)$ defined on $[0, \delta]$ for a suitable $0 < \delta \leqq 1$, provided that $x^{\sigma} f(x, y)$ is continuous on $[0, 1] \times \mathbb{R}$ for some σ $(0 \leqq \sigma < \alpha < 1)$. Applying the contractive mapping method, Delbosco and Rodino ([164], Theorem 3.4) showed that if, additionally, $f[x, y(x)]$ satisfies the following Lipschitz type condition,

$$|f[x, y(x)] - f[x, Y(x)]| \leqq \frac{M}{x^{\sigma}}|y(x) - Y(x)|, \tag{3.1.25}$$

then the equation in (3.1.24) has a unique solution $y(x) \in C[0, 1]$. They also proved that if $f[x, y(x)] = f[y(x)]$ is such that $f(0) = 0$ and the Lipschitz condition (3.1.9) holds, then the Cauchy problem

$$(D_{0+}^{\alpha}y)(x) = f[y(x)], \ y(a) = b \ \ (0 < \alpha < 1; \ a > 0; \ b \in \mathbb{R}) \tag{3.1.26}$$

and the weighted Cauchy type problem

$$(D_{0+}^{\alpha}y)(x) = f(y(x)), \ \lim_{x \to 0} x^{1-\alpha}y(x) = c \ \ (0 < \alpha < 1; \ c \in \mathbb{R}) \tag{3.1.27}$$

have a unique solution $y(x)$ such that $x^{1-\alpha}y(x) \in C[0, h]$ for any $h > 0$.

Hayek et al. [335] investigated the following Cauchy type problem for a system of linear differential equations:

$$(D_{0+}^{\alpha}y)(x) = f(x, y(x)), \ y(a) = b \ \ (0 < \alpha \leqq 1; \ a > 0; \ b \in \mathbb{R}^n) \tag{3.1.28}$$

with a real-valued vector function $y(x)$, provided that $f(x, y)$ is continuous and Lipschitzian with respect to y. Applying the method of contractive mapping defined on a complete metric space, they proved the existence and uniqueness of a continuous solution $y(x)$ to this problem. In particular, they obtained such a result for a system of linear differential equations:

$$(D_{0+}^{\alpha}y)(x) = A(x)y(x) + B(x), \ y(a) = b \ \ (0 < \alpha \leqq 1; \ a > 0; \ b \in \mathbb{R}^n) \tag{3.1.29}$$

with continuous matrices $A(x)$ and $B(x)$.

Using the approach of Al-Bassam [17] and Dzhrbashyan and Nersesyan [217], Podlubny ([682], Section 3.2) considered the problem of the existence and uniqueness for the nonlinear Cauchy type problem of the form

$$(D^{\sigma}y)(x) = f(x, y(x)), \ (D^{\sigma_k}y)(x)|_{x=0} = b_k \in \mathbb{C} \ (k = 0, 1, \cdots, \ n-1) \qquad (3.1.30)$$

with the sequential fractional derivatives defined in (3.1.17)-(3.1.18) in the space of continuous functions. Podlubny [682] tried to prove the equivalence of the problem (3.1.30) and the corresponding Volterra nonlinear integral equation, but he did not give sufficient conditions for such an equivalence.

The Cauchy type problem (3.1.1)-(3.1.2) with complex $\alpha \in \mathbb{C}$ ($\Re(\alpha) > 0$) on a finite interval $[a, b]$ of the real axis $\mathbb{R}$ was studied by Kilbas et al. ([376]-[374]) in the space of summable functions $L(a, b)$. The equivalence of this problem and the nonlinear Volterra integral equation (3.1.8) was established. The existence and uniqueness of the solution $y(x)$ to such a problem was proved by using the method of successive expansions. In particular, the corresponding results were proved for the Cauchy problem (3.1.5) and for the Cauchy type problems (3.1.6) and (3.1.7). The results obtained were extended to the system of such problems by Bonilla et al. [95]. The conditions for a unique solution $y(x)$ to the Cauchy type problem (3.1.1)-(3.1.2), in particular to the Cauchy problem (3.1.5) and to Cauchy type problems (3.1.6) and (3.1.7), in the weighted space of continuous functions $C_{n-\alpha}[a, b]$ and $C^{\alpha}_{n-\alpha}[a, b]$, defined in (1.1.22) and, [see, (3.3.1)], were established by Kilbas et al. [375] and by Kilbas et al. [389]. These results were extended by Kilbas and Marzan [378] to the Cauchy type problem for nonlinear fractional differential equations of order $\alpha \in \mathbb{C}$ ($\Re(\alpha) > 0$), more general than those in (3.1.1) and (3.1.17):

$$(D^{\alpha}_{a+}y)(x) = f\left[x, y(x), (D^{\alpha_1}_{a+}y)(x), (D^{\alpha_2}_{a+}y)(x), \cdots, \ (D^{\alpha_{m-1}}_{a+}y)(x)\right] \qquad (3.1.31)$$

where $0 < \Re(\alpha_1) < \Re(\alpha_2) < \cdots < \Re(\alpha_{m-1}) < \Re(\alpha)$ and $m \geqq 2$.

Diethelm ([173], Theorem 4.1) proved the uniqueness and existence of a local continuous solution $C(0, h]$ to the Cauchy type problem (3.1.15)-(3.1.16), provided that f is a continuous and Lipschitzian function. He used the above approach, based on the equivalence of the problem (3.1.15)-(3.1.16) to the Volterra integral equation (3.1.8), and applied a variant of Banach's fixed point theorem presented in Section 1.13, Theorem 1.10.

The above investigations were devoted to fractional differential equations with the Riemann-Liouville fractional derivative $D^{\alpha}_{a+}y$ on a finite interval $[a, b]$ of the real axis $\mathbb{R}$. Such equations with the Caputo fractional derivative ${}^{C}D^{\alpha}_{a+}y$, defined in Section 2.4, are not studied extensively. Gorenflo and Mainardi [303] applied the Laplace transform to solve the fractional differential equation

$$({}^{C}D^{\alpha}_{0+}y)(x) - \lambda y(x) = f(x) \ \ (x > 0; \ \alpha > 0; \ \lambda > 0), \qquad (3.1.32)$$

with the Caputo fractional derivative (2.4.1) of order $\alpha > 0$ and with the initial conditions

$$y^{(k)}(0) = b_k \ (k = 0, 1, \cdots n-1; \ n-1 < \alpha \leqq n; \ n \in \mathbb{N}). \qquad (3.1.33)$$

They discussed the key role of the Mittag-Leffler function (1.8.17) for the cases $1 < \alpha < 2$ and $2 < \alpha < 3$. In this regard, see also the papers by Gorenflo and Mainardi [304], Gorenflo and Rutman [311], and Gorenflo et al. [310]. Luchko and Gorenflo [507] used the operational method to prove that the Cauchy problem (3.1.32)-(3.1.33) has the unique solution

$$y(x) = \sum_{k=0}^{n-1} b_k x^k E_{\alpha,k+1}(\lambda x^\alpha) + \int_0^x (x-t)^{\alpha-1} E_{\alpha,\alpha}[\lambda(x-t)^\alpha] f(t)dt, \quad (3.1.34)$$

in terms of the Mittag-Lefler functions (1.8.17) in a special space of functions on the half-axis $\mathbb{R}_+$. They also obtained the explicit solution to the Cauchy problem for the more general fractional differential equation

$$({}^C D_{0+}^\alpha y)(x) - \sum_{k=1}^{m} c_k ({}^C D_{0+}^{\alpha_k} y)(x) = f(x) \ (\alpha > \alpha_1 > \cdots > \alpha_m \geqq 0) \quad (3.1.35)$$

via certain multivariate Mittag-Leffler functions.

Seredynska and Hanyga [758] considered the equation

$$y''(x) + k\left({}^C D_{0+}^\alpha\right)(x) + F(u) \quad (0 \leqq t \leqq T;\ 0 < \alpha \leqq 2) \quad (3.1.36)$$

with constant k, and proved the existence and the uniqueness of a solution $y(x,\alpha) \in C^2[0,T]$ for the initial Cauchy conditions $y(0) = y_0$, $y'(0) = v_0$. In this regard see also Seredynska and Hanyga [759].

Diethelm and Ford [177] investigated the Cauchy problem for the nonlinear differential equation of order $\alpha > 0$

$$({}^C D_{0+}^\alpha y)(x) = f[x, y(x)] \quad (0 \leqq x \leqq b < \infty), \quad (3.1.37)$$

with initial conditions (3.1.33). They proved the uniqueness and existence of a local continuous solution $y(x) \in C[0,h]$ to this problem for continuous and Lipschitzian f; in this regard, see also Diethelm ([173], Theorems 5.4-5.5). The dependence of this solution $y(x)$ on the order α, on the initial data (3.1.33), and on the function f was investigated. Applications were given to present numerical schemes for the solution $y(x)$ to the simplest linear problem with $0 < \alpha < 1$:

$$({}^C D_{0+}^\alpha y)(x) = \lambda y(x) + f(x) \quad (0 \leqq x \leqq b;\ \lambda < 0), \quad y(0) = b \in \mathbb{R}. \quad (3.1.38)$$

Kilbas and Marzan ([380] and [382]) investigated the Cauchy problem of the form (3.1.37)-(3.1.33) with the Caputo derivative (2.4.1) of complex order $\alpha \in \mathbb{C}$ ($\Re(\alpha) > 0$):

$$({}^C D_{0+}^\alpha y)(x) = f[x, y(x)] \quad (a \leqq x \leqq b), \quad (3.1.39)$$

$$y^{(k)}(0) = b_k \in \mathbb{C} \quad (k = 0, 1, \cdots, n-1) \quad (3.1.40)$$

where $n = [\Re(\alpha)] + 1$ for $\alpha \notin \mathbb{N}$ and $n = \alpha$ for $\alpha \in \mathbb{N}$, on a finite interval $[a,b]$ of $\mathbb{R}$. They proved the equivalence of (3.1.38)-(3.1.37) and the Volterra integral equation of second kind of the form (3.1.8)

$$y(x) = \sum_{k=0}^{n-1} \frac{b_k}{\Gamma(k+1)}(x-a)^k + \frac{1}{\Gamma(\alpha)} \int_a^x \frac{f[t, y(t)]dt}{(x-t)^{1-\alpha}} \quad (x > a) \quad (3.1.41)$$

in the space $C^{n-1}[a,b]$ and applied this result to establish conditions for a unique solution $y(x) \in C^{n-1}[a,b]$ to the Cauchy problem (3.1.39)-(3.1.40).

Daftardar-Gejji and Babakhani [153] have studied the existence, uniqueness, and stability of solutions for the following system of fractional differential equations:

$$\left({}^C D_{0+}^{\alpha} Y\right)(x) = \mathbf{A}Y(x), \quad Y(0) = Y_0, \tag{3.1.42}$$

where D_{0+}^{α} is the Caputo derivative of order $0 < \alpha < 1$, $Y(x) = [y_1(x), \cdots, y_n(x)]^T$ and $\mathbf{A}$ is $n \times n$ real matrix, and investigated the dependence of solutions on the initial data for the corresponding nonlinear system of the form (3.1.42), in which $\mathbf{A}Y(x)$ is replaced by $f[x, Y(x)]$. In particular, they obtained the unique solution of (3.1.42) in the form $Y(x) = E_{\alpha}(\mathbf{A}x^{\alpha})$, where $E_{\alpha}(\mathbf{A})$ is a a Mittag-Leffler function (1.8.1) with matrix arguments:

$$E_{\alpha}(\mathbf{A}) = \sum_{k=0}^{\infty} \frac{\mathbf{A}^k}{\Gamma(\alpha k + 1)},$$

(see, also [91], [147], [174], [513], [148], [179], [190], [184], and [187])

Using the Adomian decomposition method, well-known for ordinary differential equations [see, for example, the following papers [3],[4] and [318]] Daftardar-Gejji and Jafari [154] derived analytical solutions of a more general system of differential equations with Caputo derivatives and illustrated results obtained for equations of the form (4.1.65) and (5.3.55).

Kilbas et al. [383] investigated the Cauchy problem for the nonlinear fractional differential equation of the form (3.1.1) with the Hadamard fractional derivative (2.7.7) of complex order $\alpha \in \mathbb{C}$ ($\Re(\alpha) > 0$):

$$(\mathcal{D}_{a+}^{\alpha} y)(x) = f[x, y(x)] \quad (a \leqq x \leqq b), \tag{3.1.43}$$

$$(\mathcal{D}_{a+}^{\alpha-k} y)(a) = b_k \in \mathbb{C} \quad (k = 1, \cdots, n) \tag{3.1.44}$$

where $n = [\Re(\alpha)] + 1$ for $\alpha \notin \mathbb{N}$ and $n = \alpha$ for $\alpha \in \mathbb{N}$, on a finite interval $[a,b]$ of $\mathbb{R}$. They proved the equivalence of (3.1.43)-(3.1.44) and a Volterra integral equation of the second kind of the form (3.1.8)

$$y(x) = \sum_{j=1}^{n} \frac{b_j}{\Gamma(n-j+1)} \left(\log \frac{x}{a}\right)^{\alpha-j} + \frac{1}{\Gamma(\alpha)} \int_a^x \left(\log \frac{x}{t}\right)^{\alpha-1} f[t, y(t)] \frac{dt}{t} \quad (x > a) \tag{3.1.45}$$

in the space $X_0^1[a,b]$, defined by (1.1.3), and applied this result to establish conditions for a unique solution $y(x) \in X_0^1[a,b]$ to the Cauchy problem (3.1.43)-(3.1.44).

Fukunaga ([272] and [273]) considered the homogeneous linear differential equation

$$\left(D_{c-}^{\alpha_m} y\right)(x) + \sum_{k=1}^{m-1} c_k \left(D_{c-}^{\alpha_k} y\right)(x) = 0 \quad (x > 0;\ \alpha_m > \alpha_{m-1} > \cdots > \alpha_1 > 0), \tag{3.1.46}$$

containing the right-sided Riemann-Liouville fractional derivatives $D_{c-}^{\alpha_m} y$ $(k = 1, \cdots, m)$ given by (2.1.6). He gave the required conditions for special cases of initial value problems associated with equation (3.1.46), with $c < 0$, to have a unique solution $y(x)$ in a closed interval $[0, T]$ $(T > 0)$ of the real axis $\mathbb{R}$ and in the half axis $\mathbb{R}_+$.

A series of papers was devoted to the study of positive solutions of fractional differential equations. Zhang [923] considered the Cauchy problem for a nonlinear equation (3.1.12) with $a = 0$ and $0 < \alpha < 1$:

$$\left(D_{0+}^{\alpha} y\right)(x) = f[x, y(x)]; \quad y(0) = 0, \quad (0 < x < 1), \tag{3.1.47}$$

with a continuous function f. Using the equivalence of this problem and the corresponding Volterra integral equation, proved by Delbosco and Rodino [164], he applied the so-called point theorem for the cone to prove the uniqueness of a positive solution $y(x) > 0$ of (3.1.47). Such a method was applied in Zhang [922] to derive conditions for one or several positive solutions to the Cauchy problem

$$\left(D_{\beta}^{\gamma,\delta} y\right)(x) = x^{\nu} f[y(x)] \quad (0 < x < 1; \ 0 < \alpha < 1), \quad y(0) = 0, \tag{3.1.48}$$

with the generalized fractional derivative $D_{\gamma}^{\gamma,\delta} y$, inverse to the Erdéélyi-Kober type integral $I_{0+;\beta,\gamma}^{\delta}$ defined by (2.6.1). Babakhani and Daftardar-Gejji [42], [43], studied the problem of the form (3.1.47), where D_{0+}^{α} is replaced by a linear combination of such Riemann-Liouville derivatives, and used a fixed point theorem to give conditions under which the problem has a unique positive solution and a unique, but not necessarily positive, solution. Daftardar-Gejji [152] studied the existence of positive solutions for the system of fractional differential equations (3.1.47). We note that Atanackovic and Stankovic [38] have analyzed lateral motion of an elastic column fixed at one end and loaded at the other in terms of a system of fractional differential equations.

Note that Grin'ko [315] studied the nonlinear equation of the form (3.1.47), where the Riemann-Liouville derivative $D_{0+}^{\alpha} y$ is replaced by a more general fractional differentiation operator involving the Gauss hypergeometric function (1.6.1). He proved the existence and uniqueness theorem for the solution of the considered equation and constructed its approximate solution in the respective spaces of Holder and continuous functions [see [729], Section 43.2]

In conclusion, we indicate the papers by Dzhrbashyan [206], Nakshushev ([613] and [616]), Aleroev ([29] and [30]), Veber [853], and Delbosco [163], which are devoted to the investigation of Dirichet-type problems for ordinary fractional differential equations and papers by Veber ([847], [848], [849], [850], [852] and [851]), Didenko ([168], [167] and [169]), Kochura and Didenko [430], and Malakhovskya and Shikhmanter [541], where ordinary fractional differential equations were studied in spaces of generalized functions. In this regard, see Sections 2 and 3 of the survey paper by Kilbas and Trujillo [408]. Some authors constructed formal partial solutions to ordinary differential equations with other fractional derivatives as presented in Chapter 2. Nishimoto [[629], Volume II, Chapter 6], Nishimoto et al [631], Srivatsava et al [792], [793] and Campos [118] constructed explicit solutions of some particular fractional differential equations with the so-called fractional derivatives of complex order (see, for example, Samko et al [729], Section 22.1). A series of papers by Wiener [879]-[888] were devoted to the investigation of ordinary linear fractional differential equations and systems of such equations involving the fractional derivatives defined in the sense of a finite part of Hadamard (see, for example, Kilbas and Trujillo [408], Section 4).

We also note that many authors have applied methods of fractional integro-differentiation to constructing solutions of ordinary and partial differential equations, to investigating integro-differential equations, and to obtaining a unified theory of special functions. The methods and results in these fields are presented in Samko et al ([729], Chapter 8) and in Kiryakova [417]. We mention here the papers by Al-Saqabi [25], by Al-Saqabi and Vu Kim Tuan [27] and by Kiryakova and Al-Saqabi [418], [419] and [26], where solutions in closed form were constructed for certain integro-differential equations with the Riemann-Liouville and Erdelyi-Kober-type fractional integrals and derivatives given in (2.2.1), (2.2.3) and (2.6.1), (2.6.29), respectively.

We also indicate the paper by Kilbas et al [399], where the explicit solution of the Cauchy type problem for the equation

$$\left(D_{a+}^{\alpha}y\right)(x) = \lambda\int_a^x (x-t)^{\alpha-1}E_{\rho,\alpha}^{\gamma}\left[\nu(x-t)^{\rho}\right] + f(x) \quad (a < x \leq b) \tag{3.1.49}$$

with $\alpha \in \mathbb{C}$, $\mathrm{Re}(\alpha) > 0$, and $\lambda, \gamma, \rho, \omega \in \mathbb{C}$, and the initial conditions (3.1.2), was established. That solution involves the generalized Mittag-Leffler function (1.9.1). The homogeneous equation corresponding to (3.1.49) ($f(x) = 0$) is a generalization of the equation which describes the unsaturated behavior of the free electron laser (see [156], [157], [106], [104], [102], [103], [105] and [28]).

3.2 Equations with the Riemann-Liouville Fractional Derivative in the Space of Summable Functions

In this section we give conditions for a unique global solution to the Cauchy type problem (3.1.1)-(3.1.2) in the space $\mathbf{L}^{\alpha}(a,b)$ defined for $\alpha \in \mathbb{C}$ ($\Re(\alpha) > 0$) by

$$\mathbf{L}^{\alpha}(a,b) := \left\{y \in L(a,b) : \; D_{a+}^{\alpha}y \in L(a,b)\right\}. \tag{3.2.1}$$

Here $L(a,b) := L_1(a,b)$ is the space of summable functions in a finite interval $[a,b]$ of the real axis $\mathbb{R}$ defined by (1.1.1) with $p = 1$. We also generalize the obtained result to the Cauchy type problem for fractional differential equations more general than (3.1.1) and to the system Cauchy type problem (3.1.1)-(3.1.2). Our investigations are based on reducing the problems considered to Volterra integral equations of the second kind and on using the Banach fixed point theorem.

3.2.1 Equivalence of the Cauchy Type Problem and the Volterra Integral Equation

In this subsection we prove that the Cauchy type problem (3.1.1)-(3.1.2) and the nonlinear Volterra integral equation (3.1.8) are equivalent in the sense that, if $y(x) \in L(a,b)$ satisfies one of these relations, then it also satisfies the other. We prove such a result by assuming that a function $f[x,y]$ belongs to $L(a,b)$ for any $y \in G \subset \mathbb{C}$. For this we need the auxiliary assertion following from Lemma 2.1(a).

Lemma 3.1 *The fractional integration operator I^{α}_{a+} with $\alpha \in \mathbb{C}$ $(\Re(\alpha) > 0)$ is bounded in $L(a,b)$:*

$$||I^{\alpha}_{a+}g||_1 \leqq \frac{(b-a)^{\Re(\alpha)}}{\Re(\alpha)|\Gamma(\alpha)|}||g||_1. \tag{3.2.2}$$

In particular, for $\alpha > 0$, the estimate (3.2.2) takes the form

$$||I^{\alpha}_{a+}g||_1 \leqq \frac{(b-a)^{\alpha}}{\Gamma(\alpha+1)}||g||_1. \tag{3.2.3}$$

Here, and elsewhere in this chapter, we shall understand all relations to hold almost everywhere (a.e.) on $[a,b]$.

First we consider the Cauchy type problem (3.1.1)-(3.1.2) with real $\alpha > 0$ given in (3.1.15)-(3.1.16):

$$(D^{\alpha}_{a+}y)(x) = f[x,y(x)] \quad (\alpha > 0), \tag{3.2.4}$$

$$(D^{\alpha-k}_{a+}y)(a+) = b_k, \quad b_k \in \mathbb{R} \ (k = 1, \cdots, n = -[-\alpha]). \tag{3.2.5}$$

Theorem 3.1 *Let $\alpha > 0$, $n = -[-\alpha]$. Let G be an open set in $\mathbb{R}$ and let $f : (a,b] \times G \to \mathbb{R}$ be a function such that $f[x,y] \in L(a,b)$ for any $y \in G$.*

If $y(x) \in L(a,b)$, then $y(x)$ satisfies a.e. the relations (3.2.4) and (3.2.5) if, and only if, $y(x)$ satisfies a.e. the integral equation (3.1.8).

Proof. First we prove the necessity. Let $y(x) \in L(a,b)$ satisfy a.e. the relations (3.2.4) and (3.2.5). Since $f[x,y] \in L(a,b)$, (3.2.4) means that there exists a.e. on $[a,b]$ the fractional derivative $(D^{\alpha}_{a+}y)(x) \in L(a,b)$. According to (2.1.10) and (2.1.7),

$$(D^{\alpha}_{a+}y)(x) = \left(\frac{d}{dx}\right)^n (I^{n-\alpha}_{a+}y)(x), \ n = -[-\alpha)], \ (I^{0}_{a+}y)(x) = y(x), \tag{3.2.6}$$

and hence, by Lemma 1.1, $(I^{n-\alpha}_{a+}y)(x) \in AC^n[a,b]$. Thus we can apply Lemma 2.5(b) (with $f(x)$ replaced by $y(x)$) and, in accordance with (2.1.39), we have

$$(I^{\alpha}_{a+}D^{\alpha}_{a+}y)(x) = y(x) - \sum_{j=1}^{n} \frac{y^{(n-j)}_{n-\alpha}(a)}{\Gamma(\alpha-j+1)}(x-a)^{\alpha-j}, \ y_{n-\alpha}(x) = (I^{n-\alpha}_{a+}y)(x). \tag{3.2.7}$$

By (2.1.5), we also have

$$y_{n-\alpha}^{(n-j)}(x) = \left(\frac{d}{dx}\right)^{n-j} (I_{a+}^{(n-j)-(\alpha-j)}y)(x) = (D_{a+}^{\alpha-j}y)(x). \tag{3.2.8}$$

Using this relation and (3.2.5), we rewrite (3.2.7) in the form

$$(I_{a+}^{\alpha} D_{a+}^{\alpha} y)(x) = y(x) - \sum_{j=1}^{n} \frac{(D_{a+}^{\alpha-j}y)(a+)}{\Gamma(\alpha-j+1)}(x-a)^{\alpha-j}$$

$$= y(x) - \sum_{j=1}^{n} \frac{b_j}{\Gamma(\alpha-j+1)}(x-a)^{\alpha-j}. \tag{3.2.9}$$

By Lemma 3.1, the integral $(I_{a+}^{\alpha} f[t, y(t)])(x) \in L(a, b)$ exists a.e. on $[a, b]$. Applying the operator I_{a+}^{α} to both sides of (3.2.4) and using (3.2.9) and (2.1.1), we obtain the equation (3.1.8), and hence the necessity is proved.

Now we prove the sufficiency. Let $y(x) \in L(a, b)$ satisfy a.e. the equation (3.1.8). Applying the operator D_{a+}^{α} to both sides of (3.1.8), we have

$$(D_{a+}^{\alpha} y)(x) = \sum_{j=1}^{n} \frac{b_j}{\Gamma(\alpha-j+1)} \left(D_{a+}^{\alpha}(t-a)^{\alpha-j}\right)(x) + \left(D_{a+}^{\alpha} I_{a+}^{\alpha} f[t, y(t)]\right)(x).$$

From here, in accordance with the formula (2.1.21) and Lemma 2.4 (with $f(x)$ replaced by $f[x, y(x)]$), we arrive at the equation (3.2.4).

Now we show that the relations in (3.2.5) also hold. For this we apply the operators $D_{a+}^{\alpha-k}$ $(k = 1, \cdots, n)$ to both sides of (3.1.8). If $1 \leqq k \leqq n-1$, then, in accordance with (2.1.17) and (2.1.21) and Property 2.2 (with $f(x)$ replaced by $f[x, y(x)]$), we have

$$(D_{a+}^{\alpha-k} y)(x) = \sum_{j=1}^{n} \frac{b_j}{\Gamma(\alpha-j+1)} \left(D_{a+}^{\alpha-k}(t-a)^{\alpha-j}\right)(x) + (D_{a+}^{\alpha-k} I_{a+}^{\alpha} f[t, y(t)])(x)$$

$$= \sum_{j=1}^{n} \frac{b_j}{\Gamma(k-j+1)}(x-a)^{k-j} + I_{a+}^{k} f[t, y(t)])(x).$$

Hence

$$(D_{a+}^{\alpha-k} y)(x) = \sum_{j=1}^{k} \frac{b_j}{(k-j)!}(x-a)^{k-j} + \frac{1}{(k-1)!}\int_a^x (x-t)^{k-1} f[t, y(t)]dt. \tag{3.2.10}$$

If $k = n$, then, in accordance with (3.1.4) and (2.1.16), similarly to (3.2.10), we obtain

$$(D_{a+}^{\alpha-n} y)(x) = \sum_{j=1}^{n} \frac{b_j}{(n-j)!}(x-a)^{n-j} + \frac{1}{(n-1)!}\int_a^x (x-t)^{n-1} f[t, y(t)]dt. \tag{3.2.11}$$

Taking in (3.2.10) and (3.2.11) a limit as $x \to a+$ a.e., we obtain the relations in (3.2.5). Thus the sufficiency is proved, which completes the proof of Theorem 3.1.

Corollary 3.1 *Let* $0 < \alpha < 1$, *let* G *be an open set in* $\mathbb{R}$, *and let* $f : (a, b] \times G \to \mathbb{R}$ *be a function such that* $f[x, y] \in L(a, b)$ *for any* $y \in G$.

If $y(x) \in L(a, b)$, *then* $y(x)$ *satisfies a.e. the relations* (3.1.12) *and* (3.1.13) *if, and only if,* $y(x)$ *satisfies a.e. the integral equation* (3.1.14).

Corollary 3.2 *Let* $n \in \mathbb{N}$, *let* G *be an open set in* $\mathbb{R}$, *and let* $f : [a, b] \times G \to \mathbb{R}$ *be a function such that* $f[x, y] \in L(a, b)$ *for any* $y \in G$.

If $y(x) \in L(a, b)$, *then* $y(x)$ *satisfies a.e. the relations in* (3.1.5) *with* $b_k \in \mathbb{R}$ $(k = 1, \cdots, n)$ *if, and only if,* $y(x)$ *satisfies a.e. the integral equation*

$$y(x) = \sum_{j=1}^{n} \frac{b_j}{(n-j)!}(x-a)^{n-j} + \frac{1}{(n-1)!}\int_a^x (x-t)^{n-1} f[t, y(t)]dt. \tag{3.2.12}$$

Theorem 3.1 is clearly extended from real $\alpha > 0$ to a complex $\alpha \in \mathbb{C}$ $(\Re(\alpha) > 0)$ as follows:

Theorem 3.2 *Let* $\alpha \in \mathbb{C}$ *and* $n - 1 < \Re(\alpha) < n$ $(n \in \mathbb{N})$. *Let* G *be an open set in* $\mathbb{C}$ *and let* $f : (a, b] \times G \to \mathbb{C}$ *be a function such that* $f[x, y] \in L(a, b)$ *for any* $y \in G$.

If $y(x) \in L(a, b)$, *then* $y(x)$ *satisfies a.e. the relations* (3.1.1) *and* (3.1.2) *if, and only if,* $y(x)$ *satisfies a.e. the equation* (3.1.8).

In particular, if $0 < \Re(\alpha) < 1$, *then* $y(x)$ *satisfies a.e. the relations in* (3.1.6) *if, and only if,* $y(x)$ *satisfies a.e. the equation* (3.1.14).

Remark 3.1 The results of Theorems 3.1-3.2 and Corollaries 3.1-3.2 were obtained by Kilbas, Bonilla, and Trujillo ([376], Theorem 1 and Corollaries 1-2) and ([374], Theorems 1-2 and Corollaries 1-2).

Remark 3.2 Theorem 3.2 yields conditions for the equivalence of the Cauchy type problem (3.1.1)-(3.1.2) and the Volterra integral equation (3.1.8) in the space $L(a, b)$ for $\alpha \in \mathbb{C}$ and $n-1 < \Re(\alpha) < n$ $(n \in \mathbb{N})$. The problem of the equivalence of the Cauchy type problem (3.1.1)-(3.1.2) and the Volterra integral equation (3.1.8) for the case of complex order $\alpha = m + i\theta$ $(m \in \mathbb{N};\ \theta \in \mathbb{R};\ \theta \neq 0)$ is still open. We only indicate that in this case the Cauchy type problem (3.1.1)-(3.1.2) takes the form

$$(D_{a+}^{m+i\theta} y)(x) = f[x, y(x)], \quad (D_{a+}^{m+i\theta-k} y)(a+) = b_k \in \mathbb{C} \ (k = 1, 2, \cdots, m+1), \tag{3.2.13}$$

and it is clear that, for suitable functions $y(x)$, this problem can be reduced, by using (2.1.39), to the Volterra integral equation of the form (3.1.8) with $n = m+1$ and $\alpha = m + i\theta$:

$$y(x) = \sum_{j=1}^{m+1} \frac{b_j}{\Gamma(\alpha - j + 1)}(x-a)^{m-j+i\theta} + \frac{1}{\Gamma(m+i\theta)}\int_a^x \frac{f[t, y(t)]dt}{(x-t)^{1-m-i\theta}} \quad (x > a). \tag{3.2.14}$$

3.2.2 Existence and Uniqueness of the Solution to the Cauchy Type Problem

In this subsection we establish the existence of a unique solution to the Cauchy type problem (3.1.1)-(3.1.2) in the space $\mathbf{L}^\alpha(a,b)$ defined in (3.2.1) under the conditions of Theorems 3.1 and 3.2, and an additional Lipschitzian-type condition on $f[x,y]$ with respect to the second variable: for all $x \in (a,b]$ and for all $y_1, y_2 \in G \subset \mathbb{C}$,

$$|f[x,y_1] - f[x,y_2]| \leqq A|y_1 - y_2| \quad (A > 0), \tag{3.2.15}$$

where $A > 0$ does not depend on $x \in [a,b]$. First we derive a unique solution to the Cauchy-problem (3.2.4)-(3.2.5) with a real $\alpha > 0$.

Theorem 3.3 *Let $\alpha > 0$, $n = -[-\alpha]$. Let G be an open set in $\mathbb{R}$ and let $f : (a,b] \times G \to \mathbb{R}$ be a function such that $f[x,y] \in L(a,b)$ for any $y \in G$ and the condition* (3.2.15) *is satisfied.*

Then there exists a unique solution $y(x)$ to the Cauchy type problem (3.2.4)-(3.2.5) *in the space $\mathbf{L}^\alpha(a,b)$.*

Proof. First we prove the existence of a unique solution $y(x) \in L(a,b)$. According to Theorem 3.1, it is sufficient to prove the existence of a unique solution $y(x) \in L(a,b)$ to the nonlinear Volterra integral equation (3.1.8). For this we apply the known method, for nonlinear Volterra integral equations, of proving first the result on a part of the interval $[a,b]$; [for example, see Kolmogorov and Fomin [434]].

Equation (3.1.8) makes sense in any interval $[a,x_1] \subset [a,b]$ $(a < x_1 < b)$. Choose x_1 such that the inequality

$$A\frac{(x_1 - a)^\alpha}{\Gamma(\alpha+1)} < 1 \tag{3.2.16}$$

holds, and then prove the existence of a unique solution $y(x) \in L(a,x_1)$ to the equation (3.1.8) on the interval $[a,x_1]$. For this we use the Banach fixed point theorem given as Theorem 1.9 in Section 1.13 for the space $L(a,x_1)$, which is clearly a complete metric space with the distance

$$d(y_1,y_2) = ||y_1 - y_2||_1 := \int_a^{x_1} |y_1(x) - y_2(x)|dx. \tag{3.2.17}$$

We rewrite the integral equation (3.1.8) in the form $y(x) = (Ty)(x)$, where

$$(Ty)(x) = y_0(x) + \frac{1}{\Gamma(\alpha)}\int_a^x \frac{f[t,y(t)]dt}{(x-t)^{1-\alpha}} \tag{3.2.18}$$

with

$$y_0(x) = \sum_{j=1}^n \frac{b_j}{\Gamma(\alpha-j+1)}(x-a)^{\alpha-j}. \tag{3.2.19}$$

To apply Theorem 1.9, we have to prove the following: (1) if $y(x) \in L(a,x_1)$, then $(Ty)(x) \in L(a,x_1)$; (2) for any $y_1, y_2 \in L(a,x_1)$, the following estimate holds:

$$\|Ty_1 - Ty_2\|_1 \leqq \omega \|y_1 - y_2\|_1, \quad \omega = A\frac{(x_1-a)^\alpha}{\Gamma(\alpha+1)}. \tag{3.2.20}$$

It follows from (3.2.19) that $y_0(x) \in L(a,x_1)$. Since $f[x,y] \in L(a,b)$, by Lemma 3.1 (with $b = x_1$ and $g(t) = f[t,y(t)]$), the integral in the right-hand side of (3.2.18) also belongs to $L(a,x_1)$, and hence $(Ty)(x) \in L(a,x_1)$. Now we prove the estimate (3.2.20). By (3.2.18)-(3.2.19) and (3.2.1), using the Lipschitzian condition (3.2.15) and applying the relation (3.2.3) (with $b = x_1$ and $g(x) = f[x,y_1(x)] - f[x,y_2(x)]$), we have

$$\|Ty_1 - Ty_2\|_{L(a,x_1)} \leqq \left\| I^\alpha_{a+}[|f[t,y_1(t)] - f[t,y_2)(t)]|] \right\|_{L(a,x_1)} \leqq$$

$$\leqq A \left\| I^\alpha_{a+}[|y_1(t) - y_2(t)|] \right\|_{L(a,x_1)} \leqq A\frac{(x_1-a)^\alpha}{\Gamma(\alpha+1)} \|y_1(x) - y_2(x)\|_{L(a,x_1)},$$

which yields (3.2.20). In accordance with (3.2.16), $0 < \omega < 1$, and hence (by Theorem 1.9) there exists a unique solution $y^*(x) \in L(a,x_1)$ to the equation (3.1.8) on the interval $[a,x_1]$.

By Theorem 1.9, the solution y^* is obtained as a limit of a convergent sequence $(T^m y_0^*)(x)$:

$$\lim_{m\to\infty} \|T^m y_0^* - y^*\|_{L(a,x_1)} = 0, \tag{3.2.21}$$

where $y_0^*(x)$ is any function in $L(a,b)$. If at least one $b_k \neq 0$ in the initial condition (3.2.5), we can take $y_0^*(x) = y_0(x)$ with $y_0(x)$ defined by (3.2.19).

By (3.2.18), the sequence $(T^m y_0^*)(x)$ is defined by the recursion formulas

$$(T^m y_0^*)(x) = y_0(x) + \frac{1}{\Gamma(\alpha)} \int_a^x \frac{f[t,(T^{m-1}y_0^*)(t)]dt}{(x-t)^{1-\alpha}} \quad (m = 1,2,\cdots).$$

If we denote $y_m(x) = (T^m y_0^*)(x)$, then the last relation takes the form

$$y_m(x) = y_0(x) + \frac{1}{\Gamma(\alpha)} \int_a^x \frac{f[t,y_{m-1}(t)]dt}{(x-t)^{1-\alpha}} \quad (m \in \mathbb{N}), \tag{3.2.22}$$

and (3.2.21) can be rewritten as follows:

$$\lim_{m\to\infty} \|y_m - y^*\|_{L(a,x_1)} = 0. \tag{3.2.23}$$

This means that we actually applied the method of successive approximations to find a unique solution $y^*(x)$ to the integral equation (3.1.8) on $[a,x_1]$.

Next we consider the interval $[x_1,x_2]$, where $x_2 = x_1 + h_1$ and $h_1 > 0$ are such that $x_2 < b$. Rewrite the equation (3.1.8) in the form

$$y(x) = \frac{1}{\Gamma(\alpha)} \int_{x_1}^x \frac{f[t,y(t)]dt}{(x-t)^{1-\alpha}} + \sum_{j=1}^n \frac{b_j}{\Gamma(\alpha-j+1)}(x-a)^{\alpha-j}$$

$$+\frac{1}{\Gamma(\alpha)}\int_a^{x_1}\frac{f[t,y(t)]dt}{(x-t)^{1-\alpha}}. \tag{3.2.24}$$

Since the function $y(t)$ is uniquely defined on the interval $[a, x_1]$, the last integral can be considered as the known function, and we rewrite the last equation as

$$y(x) = y_{01}(x) + \frac{1}{\Gamma(\alpha)}\int_{x_1}^{x}\frac{f[t,y(t)]dt}{(x-t)^{1-\alpha}}, \tag{3.2.25}$$

where $y_{01}(x)$ as defined by

$$y_{01}(x) = \sum_{j=1}^{n}\frac{b_j}{\Gamma(\alpha-j+1)}(x-a)^{\alpha-j} + \frac{1}{\Gamma(\alpha)}\int_a^{x_1}\frac{f[t,y(t)]dt}{(x-t)^{1-\alpha}} \tag{3.2.26}$$

is the known function. Using the same arguments as above, we derive that there exists a unique solution $y^*(x) \in L(x_1, x_2)$ to the equation (3.1.8) on the interval $[x_1, x_2]$. Taking the next interval $[x_2, x_3]$, where $x_3 = x_2 + h_2$ and $h_2 > 0$ are such that $x_3 < b$, and repeating this process, we conclude that there exists a unique solution $y^*(x) \in L(a, b)$ for the equation (3.1.8) on the interval $[a, b]$.

Thus, there exists a unique solution $y(x) = y^*(x) \in L(a, b)$ to the Volterra integral equation (3.1.8) and hence to the Cauchy type problem (3.2.4)-(3.2.5).

To complete the proof of Theorem 3.3, we must show that such a unique solution $y(x) \in L(a, b)$ belongs to the space $\mathbf{L}^\alpha(a, b)$. In accordance with (3.2.1), it is sufficient to prove that $(D^\alpha_{a+}y)(x) \in L(a, b)$. By the above proof, the solution $y(x) \in L(a, b)$ is a limit of the sequence $y_m(x) \in L(a, b)$:

$$\lim_{m\to\infty} ||y_m - y||_1 = 0, \tag{3.2.27}$$

with the choice of certain y_m on each $[a, x_1], \cdots, [x_{L-1}, b]$. By (3.2.4) and (3.2.15) we have

$$||D^\alpha_{a+}y_m - D^\alpha_{a+}y||_1 = ||f[x, y_m] - f[x, y]||_1 \leqq A||y_m - y||_1. \tag{3.2.28}$$

Thus, by (3.2.27), we get

$$\lim_{m\to\infty} ||D^\alpha_{a+}y_m - D^\alpha_{a+}y||_1 = 0, \tag{3.2.29}$$

and hence $(D^\alpha_{a+}y)(x) \in L(a, b)$. This completes the proof of Theorem 3.3.

Corollary 3.3 *Let $n \in \mathbb{N}$, let G be an open set in $\mathbb{C}$, and let $f : [a, b] \times G \to \mathbb{C}$ be a function such that $f[x, y] \in L(a, b)$ for any $y \in G$ and* (3.2.15) *holds.*

Then there exists a unique solution $y(x)$ to the Cauchy problem (3.1.5) *in the space $\mathbf{L}^n(a, b)$:*

$$\mathbf{L}^n(a, b) = \left\{y \in L(a, b) : \ y^{(n)} \in L(a, b)\right\}. \tag{3.2.30}$$

Theorem 3.3 is extended from a real $\alpha > 0$ to a complex $\alpha \in \mathbb{C}$ ($\Re(\alpha) > 0$).

Theorem 3.4 *Let $\alpha \in \mathbb{C}$, $n-1 < \Re(\alpha) < n$ $(n \in \mathbb{N})$. Let G be an open set in $\mathbb{C}$ and let $f : (a,b] \times G \to \mathbb{C}$ be a function such that $f[x,y] \in L(a,b)$ for any $y \in G$ and the condition* (3.2.15) *holds.*

Then there exists a unique solution $y(x)$ to the Cauchy type problem (3.1.1)-(3.1.2) *in the space $\mathbf{L}^\alpha(a,b)$ defined in* (3.2.1).

In particular, if $0 < \Re(\alpha) < 1$, then there exists a unique solution $y(x)$ to the Cauchy type problem (3.1.6)

$$(D^\alpha_{a+}y)(x) = f[x,y(x)] \ (0 < \Re(\alpha) < 1) \ \ (I^{1-\alpha}_{a+}y)(a+) = b \in \mathbb{C} \tag{3.2.31}$$

in the space $\mathbf{L}^\alpha(a,b)$.

Proof. The proof of Theorem 3.4 is similar to the proof of Theorem 3.3, if we use the inequality

$$A\frac{(x_1-a)^{\Re(\alpha)}}{\Re(\alpha)|\Gamma(\alpha)|} < 1 \tag{3.2.32}$$

instead of the one in (3.2.16).

3.2.3 The Weighted Cauchy Type Problem

When $0 < \Re(\alpha) < 1$, the result of Theorem 3.4 remains true for the weighted Cauchy type problem (3.1.7), with $c \in \mathbb{C}$

$$(D^\alpha_{a+}y)(x) = f[x,y(x)] \ (0 < \Re(\alpha) < 1), \quad \lim_{x\to a+}\left[(x-a)^{1-\alpha}y(x)\right] = c. \tag{3.2.33}$$

Its proof is based on the following preliminary assertion.

Lemma 3.2 *Let $\alpha \in \mathbb{C}$ $(0 < \Re(\alpha) < 1)$ and let $y(x)$ be a Lebesgue measurable function on $[a,b]$.*

(a) If there exists a.e. a limit

$$\lim_{x\to a+}\left[(x-a)^{1-\alpha}y(x)\right] = c \ \ (c \in \mathbb{C}), \tag{3.2.34}$$

then there also exists a.e. a limit

$$(I^{1-\alpha}_{a+}y)(a+) := \lim_{x\to a+}(I^{1-\alpha}_{a+}y)(x) = c\Gamma(\alpha). \tag{3.2.35}$$

(b) If there exists a.e. a limit

$$\lim_{x\to a+}(I^{1-\alpha}_{a+}y)(x) = b \ \ (b \in \mathbb{C}), \tag{3.2.36}$$

and if there exists the limit $\lim_{x\to a+}\left[(x-a)^{1-\alpha}y(x)\right]$, then

$$\lim_{x\to a+}\left[(x-a)^{1-\alpha}y(x)\right] = \frac{b}{\Gamma(\alpha)}. \tag{3.2.37}$$

Proof. Choose an arbitrary $\epsilon > 0$. By (3.2.34), there exists $\delta = \delta(\epsilon) > 0$ such that

$$|(t-a)^{1-\alpha}y(t) - c| < \epsilon \frac{|\Gamma(1-\alpha)|}{\Gamma[\Re(\alpha)]\Gamma[1-\Re(\alpha)]} \tag{3.2.38}$$

for $a < t < a+\delta$. According to (2.1.16),

$$\Gamma(\alpha) = \left(I_{a+}^{\alpha}(t-a)^{\alpha-1}\right)(x) \quad (0 < \Re(\alpha) < 1). \tag{3.2.39}$$

Using this equality and taking (2.1.1) into account, we have

$$\begin{aligned} \left|\left(I_{a+}^{1-\alpha}y\right)(x) - c\Gamma(\alpha)\right| &= \left|\left(I_{a+}^{1-\alpha}y\right)(x) - c\left(I_{a+}^{1-\alpha}(t-a)^{\alpha-1}\right)(x)\right| \\ &\leqq \frac{1}{|\Gamma(1-\alpha)|}\int_a^x (x-t)^{-\Re(\alpha)}|y(t) - c(t-a)^{\alpha-1}|dt \\ &\leqq \frac{1}{|\Gamma(1-\alpha)|}\int_a^x (x-t)^{-\Re(\alpha)}(t-a)^{\Re(\alpha)-1}|(t-a)^{1-\alpha}y(t) - c|dt. \end{aligned}$$

If we choose $a < x < a+\delta$, then $a < t < x < a+\delta$, and we can apply the estimate (3.2.26) and the formula (2.1.16) to obtain

$$\left|\left(I_{a+}^{1-\alpha}y\right)(x) - c\Gamma(\alpha)\right| \leqq \frac{\epsilon}{\Gamma[\Re(\alpha)]}\left(I_{a+}^{1-\Re(\alpha)}(t-a)^{\Re(\alpha)-1}\right)(x) = \epsilon,$$

which proves the assertion (a) of Lemma 3.2.

Suppose that the limit in (3.2.37) is equal to c:

$$\lim_{x\to a+}\left[(x-a)^{1-\alpha}y(x)\right] = c.$$

Then, by Lemma 3.2(a), we have

$$(I_{a+}^{1-\alpha}y)(a+) := \lim_{x\to a+}(I_{a+}^{1-\alpha}y)(x) = c\,\Gamma(\alpha),$$

and hence, in accordance with (3.2.36), $c = b/\Gamma(\alpha)$, which proves (3.2.37).

From Theorem 3.4 and Lemma 3.2, we obtain the existence and uniqueness result for the weighted Cauchy type problem (3.2.33).

Theorem 3.5 *Let $\alpha \in \mathbb{C}$ and $0 < \Re(\alpha) < 1$. Let G be an open set in $\mathbb{C}$ and let $f : (a,b] \times G \to \mathbb{C}$ be a function such that $f[x,y] \in L(a,b)$ for any $y \in G$ and* (3.2.15) *holds.*

Then there exists a unique solution $y(x)$ to the weighted Cauchy type problem (3.2.33) *in the space $\mathbf{L}^{\alpha}(a,b)$ defined in* (3.2.1).

Proof. If $y(x)$ satisfies the conditions (3.2.33), then, by Lemma 3.2(a), $y(x)$ also satisfies the conditions (3.2.31) with $b = c\Gamma(\alpha)$:

$$(D_{a+}^{\alpha}y)(x) = f[x,y(x)] \ (0 < \Re(\alpha) < 1), \quad (I_{a+}^{1-\alpha}y)(a+) = c\Gamma(\alpha) \in \mathbb{C}.$$

According to Theorem 3.4, there exists a unique solution $y(x) \in \mathbf{L}^\alpha(a,b)$ to this problem. By Lemma 3.2(b), $y(x)$ is also a solution to the weighted Cauchy problem (3.2.33). This $y(x)$ will be a unique solution to (3.2.33). Indeed, if we suppose that the weighted problem (3.2.33) has two different solutions in $\mathbf{L}^\alpha(a,b)$, then by Lemma 3.2(a), there will be also two different solutions to the Cauchy type problem (3.2.31) in $\mathbf{L}^\alpha(a,b)$, which contradicts the uniqueness of the solution.

Remark 3.3 The results in Theorems 3.3-3.5 and Corollaries 3.3-3.4 were established by Kilbas et al. in ([376], Theorems 2-6 and Corollaries 3-4) and ([374], Theorems 3-6 and Corollaries 3-6) under stricter conditions in special function spaces of $y(x)$.

Remark 3.4 In Theorem 3.4, we gave conditions for a unique solution to the Cauchy type problem (3.1.1)-(3.1.2) in the space $\mathbf{L}^\alpha(a,b)$ for $\alpha \in \mathbb{C}$ with $n-1 < \Re(\alpha) < n$ $(n \in \mathbb{N})$. The problem of a unique solution to the Cauchy type problem (3.1.1)-(3.1.2) in the case of complex order $\alpha = m + i\theta$ $(m \in \mathbb{N};\ \theta \in \mathbb{R};\ \theta \neq 0)$ is still open [see Remark 3.2 in this regard].

3.2.4 Generalized Cauchy Type Problems

The results obtained in Sections 3.2.1-3.2.3 extend to Cauchy type problems more general than (3.1.1)-(3.1.2) for the differential equation (3.1.31) with initial conditions (3.1.1):

$$(D_{a+}^\alpha y)(x) = f\left[x, y(x), (D_{a+}^{\alpha_1} y)(x), \cdots, (D_{a+}^{\alpha_l} y)(x)\right], \tag{3.2.40}$$

$$(D_{a+}^{\alpha-k} y)(a+) = b_k, \quad b_k \in \mathbb{C} \quad (k = 1, \cdots, n). \tag{3.2.41}$$

The following assertion generalizing Theorems 3.1 and 3.2 holds.

Theorem 3.6 *Let* $\alpha \in \mathbb{C}$ $(n-1 < \Re(\alpha) < n;\ \ n \in \mathbb{N})$ *, and let* $n = [\Re(\alpha)] + 1$ *for* $\alpha \notin \mathbb{N}$ *and* $n = \alpha$ *for* $\alpha \in \mathbb{N}$.

Let $l \in \mathbb{N}$ *and* $\alpha_j \in \mathbb{C}$ $(j = 1, \cdots, l)$ *be such that*

$$0 = \alpha_0 < \Re(\alpha_1) < \cdots < \Re(\alpha_l) < \Re(\alpha). \tag{3.2.42}$$

Let G *be an open set in* $\mathbb{C}^{l+1}$ *and let* $f : (a,b] \times G \to \mathbb{C}$ *be a function such that* $f[x, y, y_1, \cdots, y_l] \in L(a,b)$ *for any* $(y, y_1, \cdots, y_l) \in G$.

If $y(x) \in L(a,b)$, *then* $y(x)$ *satisfies a.e. the relations* (3.2.40) *and* (3.2.41) *if, and only if,* $y(x)$ *satisfies a.e. the integral equation*

$$y(x) = \sum_{j=1}^{n} \frac{b_j}{\Gamma(\alpha - j + 1)} (x-a)^{\alpha - j}$$

$$+ \frac{1}{\Gamma(\alpha)} \int_a^x \frac{f\left[t, y(t), (D_{a+}^{\alpha_1} y)(t), \cdots, (D_{a+}^{\alpha_l} y)(t)\right] dt}{(x-t)^{1-\alpha}} \quad (x > a). \tag{3.2.43}$$

In particular, if $0 < \Re(\alpha) < 1$, $x > a$ *and* $b_0 \in \mathbb{C}$, *then* $y(x)$ *satisfies a.e. the relations*

$$(D^{\alpha}_{a+}y)(x) = f\left[x, y(x), (D^{\alpha_1}_{a+}y)(x), \cdots, (D^{\alpha_l}_{a+}y)(x)\right]; \quad (I^{1-\alpha}_{a+}y)(a+) = b_0 \tag{3.2.44}$$

if, and only if, $y(x)$ *satisfies a.e. the integral equation*

$$y(x) = \frac{b_0(x-a)^{\alpha-1}}{\Gamma(\alpha+1)} + \frac{1}{\Gamma(\alpha)} \int_a^x \frac{f\left[t, y(t), (D^{\alpha_1}_{a+}y)(t), \cdots, (D^{\alpha_l}_{a+}y)(t)\right] dt}{(x-t)^{1-\alpha}}. \tag{3.2.45}$$

Proof. The proof of Theorem 3.6 is similar to that of Theorem 3.1, using the properties of the Riemann-Liouville fractional integrals and derivatives.

The next statement generalizes Theorems 3.3-3.5.

Theorem 3.7 *Let the conditions of Theorem* 3.6 *be valid, let* $f[x, y, y_1, \cdots, y_l]$ *satisfy the Lipschitzian condition:*

$$|f[x, y, y_1, \cdots, y_l)(x)] - f[x, Y, Y_1, \cdots, Y_l]| \leqq A_l \left[\sum_{j=0}^{l} |y_j - Y_j|\right], \tag{3.2.46}$$

for all $x \in (a, b]$ *and* $y, y_1, \cdots, y_l$; $Y, Y_1, \cdots, Y_l \in G$, *and where* $A_l > 0$ *does not depend on* $x \in [a, b]$, *and let* $(D^{\alpha_j - k_j}_{a+}y)(a+) = b_{k_j} \in \mathbb{C}$ $(j = 1, \cdots, n_j)$ *be fixed numbers, where* $n_j = [\Re(\alpha_j)] + 1$ *for* $\alpha_j \notin \mathbb{N}$ *and* $n_j = \alpha_j$ *for* $\alpha_j \in \mathbb{N}$.

Then there exists a unique solution $y(x)$ *to the Cauchy type problem* (3.2.40)-(3.2.41) *in the space* $\mathbf{L}^{\alpha}(a, b)$.

In particular, if $0 < \Re(\alpha) < 1$ *and* $(I^{1-\alpha_j}_{a+}y)(a+) = b_j \in \mathbb{C}$ $(j = 1, \cdots, l)$ *are fixed numbers, then there exists a unique solution* $y(x) \in \mathbf{L}^{\alpha}(a, b)$ *to the Cauchy type problem* (3.2.44) *and the weighted Cauchy type problem*

$$(D^{\alpha}_{a+}y)(x) = f\left[x, y(x), (D^{\alpha_1}_{a+}y)(x), \cdots, (D^{\alpha_l}_{a+}y)(x)\right], \tag{3.2.47}$$

$$\lim_{x \to a+} \left[(x-a)^{1-\alpha} y(x)\right] = c \in \mathbb{C}. \tag{3.2.48}$$

Proof. Theorem 3.7 is proved in a way similar to the proofs of Theorems 3.3-3.5. By Theorem 3.6, it is sufficient to establish the existence of a unique solution $y(x) \in L(a, b)$ to the integral equation (3.2.45). We choose $x_1 \in (a, b)$ such that the condition

$$A \sum_{j=0}^{l} \left[\frac{(x_1 - a)^{\Re(\alpha - \alpha_j)}}{\Re(\alpha - \alpha_j)|\Gamma(\alpha - \alpha_j)|}\right] < 1, \tag{3.2.49}$$

generalizing the one in (3.2.32), holds and apply the Banach fixed point theorem to prove the existence of a unique solution $y(x) = y^*(x) \in L(a, x_1)$. We use the space $L(a, b)$ and rewrite the equation (3.2.43) in the form $y(x) = (Ty)(x)$, where

$$(Ty)(x) = y_0(x) + \frac{1}{\Gamma(\alpha)} \int_a^x \frac{f\left[t, y(t), (D^{\alpha_1}_{a+}y)(t), \cdots, (D^{\alpha_l}_{a+}y)(t)\right]}{(x-t)^{1-\alpha}}. \tag{3.2.50}$$

By (3.2.46), (2.1.30) and (2.1.39) we have

$$\left|\left(I_{a+}^{\alpha}\left\{f[x,y_1,D_{a+}^{\alpha_1}y_1,\cdots,\ D_{a+}^{\alpha_l}y_1]-f[x,y_2,D_{a+}^{\alpha_1}y_2,\cdots,\ D_{a+}^{\alpha_l}y_2]\right\}\right)(x)\right|$$

$$\leqq\left(I_{a+}^{\Re(\alpha)}\left|f[x,y_1,D_{a+}^{\alpha_1}y_1,\cdots,\ D_{a+}^{\alpha_l}y_1]-f[x,y_2,D_{a+}^{\alpha_1}y_2,\cdots,\ D_{a+}^{\alpha_l}y_2]\right|\right)(x)$$

$$\leqq A\left(I_{a+}^{\Re(\alpha}\left|\sum_{j=0}^{l}D_{a+}^{\alpha_j}(y_1-y_2)\right|\right)(x)$$

$$=A\sum_{j=0}^{l}\left(I_{a+}^{\Re(\alpha-\alpha_j)}\left|I_{a+}^{\alpha_j}D_{a+}^{\alpha_j}(y_1-y_2)\right|\right)(x)$$

$$=A\sum_{j=0}^{l}\left(I_{a+}^{\Re(\alpha-\alpha_j)}\left|\left[(y_1-y_2)(t)\right.\right.\right.$$

$$\left.\left.\left.-\sum_{k_j=1}^{n_j}\frac{(D_{a+}^{\alpha_j-k_j}[y_1-y_2])(a+)}{\Gamma(\alpha_j-k_j+1)}(x-a)^{\alpha_j-k_j}\right]\right|\right)(x).$$

By the hypothesis of Theorem 3.7, $(D_{a+}^{\alpha_j-k_j}y_1)(a+)=(D_{a+}^{\alpha_j-k_j}y_2)(a+)$, and hence, for any $x\in[a.b]$,

$$\left|\left(I_{a+}^{\alpha}\left\{f[x,y_1,D_{a+}^{\alpha_1}y_1,\cdots,\ D_{a+}^{\alpha_l}y_1]-f[x,y_2,D_{a+}^{\alpha_1}y_2,\cdots,\ D_{a+}^{\alpha_l}y_2]\right\}\right)(x)\right|$$

$$\leqq A\sum_{j=0}^{l}\left(I_{a+}^{\Re(\alpha-\alpha_j)}|y_1-y_2|\right)(x). \tag{3.2.51}$$

Using this relation with $x=x_1$ and (3.2.2) with $b=x_1$, we derive the estimate

$$\|(Ty_1)(x)-(Ty_2)(x)\|_{L(a,x_1)}\leqq\omega\|y_1-y_2\|_1,\quad \omega=A\sum_{j=0}^{l}\left[\frac{(x_1-a)^{\Re(\alpha-\alpha_j)}}{\Re(\alpha-\alpha_j)|\Gamma(\alpha-\alpha_j)|}\right],$$

which, by (3.2.49) and Theorem 1.9, yields the existence of a unique solution $y^*(x)$ to the equation (3.2.43) in $L(a,x_1)$. This solution is obtained as a limit of the convergent sequence $T^m y_0(x)=y_m(x)$, for which the relations (3.2.21) and (3.2.23) hold. Next we show, using the same arguments as in the proof of Theorem 3.3, that there exists a unique solution $y(x)\in L(a,b)$ to the integral equation (3.2.43) and hence to the Cauchy type problem (3.2.40)-(3.2.41) such that $D_{a+}^{\alpha}y\in L(a,b)$, which completes the proof of Theorem 3.7 in accordance with (3.2.1).

In particular, if $\Re(\alpha)<1$, then there exists a unique solution $y(x)\in\mathbf{L}^{\alpha}(a,b)$ to the Cauchy type problem (3.2.44). Then the same result for the weighted Cauchy type problem (3.2.47) is proved as in Theorem 3.5.

The results in Theorems 3.1-3.2 and Theorems 3.3-3.5 can be also extended to the system Cauchy type problem (3.1.1)-(3.1.2):

$$(D_{a+}^{\alpha_r}y_r)(x)=f_r[x;y_1(x),\cdots,y_m(x)]\quad(r=1,\cdots,\ m), \tag{3.2.52}$$

$$(D_{a+}^{\alpha_k - j_k} y_k)(a+) = b_{j_k} \ , b_{j_k} \in \mathbb{C} \ \ (k = 1, \cdots, m; \ j_k = 1, \cdots, n_k), \tag{3.2.53}$$

in a product of the spaces $\mathbf{L}^{\alpha_1}(a, b)$, $\cdots$, $\mathbf{L}^{\alpha_m}(a, b)$ defined by (3.2.1). We shall use the following notation:

$$L^m(a, b) := L(a, b) \times \cdots \times L(a, b) \ (m \text{ times}), \tag{3.2.54}$$

$$\mathbf{L}^{|\alpha|}[(a, b)^m] = \mathbf{L}^{\alpha_1}(a, b) \times \cdots \times \mathbf{L}^{\alpha_m}(a, b). \tag{3.2.55}$$

The following statements generalizing Theorems 3.1-3.2 and Theorems 3.3-3.5, respectively, hold.

Theorem 3.8 *Let $k = 1, \cdots , m$, $\alpha_k \in \mathbb{C}$, $n_k - 1 < \Re(\alpha_k) < n_k$ $(n_k \in \mathbb{N})$ or $\alpha_k \in \mathbb{N}$. Let G be open set in $\mathbb{C}^m$ and let $f_k : (a, b] \times G \to \mathbb{C}$ be functions such that $f_k[x, y_1, \cdots , y_m] \in L(a, b)$ for any $(y_1, \cdots , y_m) \in G$.*

If $y_k(x) \in L(a, b)$, then $y_k(x)$ satisfy a.e. the system of relations (3.2.52) *and* (3.2.53) *if, and only if, $y_k(x)$ satisfy a.e. the system of integral equations*

$$y_k(x) = \sum_{j_k=1}^{n_k} \frac{b_{j_k}}{\Gamma(\alpha_k - j_k + 1)} (x - a)^{\alpha_i - j_k}$$

$$+ \frac{1}{\Gamma(\alpha_k)} \int_a^x \frac{f_k[t; y_1(t), \cdots , y_m(t)]dt}{(x - t)^{1-\alpha_k}} \ \ (k = 1, \cdots, m). \tag{3.2.56}$$

In particular, if $0 < \Re(\alpha_k) < 1$ and $k = 1, \cdots , m$, then $y_k(x)$ satisfy a.e. the relations

$$(D_{a+}^{\alpha_k} y_k)(x) = f_k[x; y_1(x), \cdots, y_m(x)] \tag{3.2.57}$$

$$(I_{a+}^{1-\alpha_k} y_k)(a+) = b_k, \ b_k \in \mathbb{C}, \tag{3.2.58}$$

if, and only if, $y_k(x)$ satisfy a.e. the system of integral equations

$$y_k(x) = \frac{b_k}{\Gamma(\alpha - 1)} (x - a)^{\alpha - 1} + \frac{1}{\Gamma(\alpha_k)} \int_a^x \frac{f_k[t; y_1(t), \cdots, y_m(t)]dt}{(x - t)^{1-\alpha_k}}, \tag{3.2.59}$$

for all $k = 1, \cdots , m$

Theorem 3.9 *Let the conditions of Theorem 3.8 be valid and let $f_k(x, y_1, \cdots , y_m)$ satisfy the Lipschitzian conditions*

$$|f_k(x, y_1, \cdots , y_m) - f_k(x, Y_1, \cdots , Y_m)| \leqq \sum_{k=1}^{m} A_k |y_k - Y_k| \ \ (A_k > 0; \ k = 1, \cdots , m), \tag{3.2.60}$$

for all $x \in (a, b]$ and for all $y_1, \cdots , y_m$; $Y_1, \cdots , Y_m \in G$, and where $A_k > 0$ do not depend on $x \in [a, b]$.

Then there exists a unique solution $(y_1, \cdots , y_m)$ to the system Cauchy type problem (3.2.52)-(3.2.53) *in the space $\mathbf{L}^{|\alpha|}[(a, b)^m]$.*

In particular, if $0 < \Re(\alpha_k) < 1$, *then there exists a unique solution* $(y_1, \cdots, y_m) \in \mathbf{L}^{|\alpha|}[(a,b)^m]$ *to the system Cauchy type problem* (3.2.57)-(3.2.58) *and to the weighted Cauchy type problem*

$$(D^{\alpha_k}_{a+} y_k)(x) = f_k[x; y_1(x), \cdots, y_m(x)] \quad (k = 1, \cdots, m), \tag{3.2.61}$$

$$\lim_{x \to a+} \left[(x-a)^{1-\alpha_k} y_k(x)\right] = b_k, \; b_k \in \mathbb{C} \quad (k = 1, \cdots, m). \tag{3.2.62}$$

Proof. The proof of Theorem 3.8 is similar to the proofs of Theorems 3.1 and 3.2, using the properties of the Riemann-Liouville fractional integrals and derivatives. The proof of Theorem 3.9 is similar to the proofs of Theorems 3.3-3.5; we set $x_{j1} \in [a, b]$ $(j = 1, \cdots, m)$ and use the inequality

$$\sum_{k=1}^{m} A_k \frac{(x_{k1} - a)^{\Re(\alpha_k)}}{\Re(\alpha_k)|\Gamma(\alpha_k)|} < 1 \tag{3.2.63}$$

instead of the one in (3.2.32).

Remark 3.5 The results in Theorem 3.8 were obtained by Bonilla et al. ([95], Theorems 2.1-2-2).

Remark 3.6 The results of Theorems 3.9 were established by Bonilla et al. ([95], Theorems 3.1-3.4).

Remark 3.7 The results presented in Theorems 3.6 and 3.7 can be extended to a system Cauchy type problem (3.2.40)-(3.2.41) and to a system of Volterra integral equations (3.2.43).

3.2.5 Cauchy Type Problems for Linear Equations

From Theorems 3.4-3.5 we derive the corresponding results for the Cauchy type problems for linear differential equations of fractional order $\alpha \in \mathbb{C}$ $(\Re(\alpha) > 0)$.

Corollary 3.4 *Let* $\alpha = n \in \mathbb{N}$ *or* $\alpha \in \mathbb{C}$ *be such that* $n - 1 < \Re(\alpha) < n$ $(n \in \mathbb{N})$ *and let* $g(x) \in L(a, b)$.

If $a(x) \in L_\infty(a, b)$ *or if* $a(x)$ *is bounded on* $[a, b]$, *then the Cauchy type problem for the following linear differential equation of order* α *and* $b_k \in \mathbb{C}$ $(k = 1, \cdots, n)$

$$(D^{\alpha}_{a+} y)(x) = a(x)y(x) + g(x); \quad (D^{\alpha-k}_{a+} y)(a+) = b_k \tag{3.2.64}$$

has a unique solution $y(x)$ *in the space* $\mathbf{L}^{\alpha}(a, b)$.

In particular, there exists a unique solution $y(x) \in \mathbf{L}^{\alpha}(a, b)$ *to the problem*

$$(D^{\alpha}_{a+} y)(x) = \lambda(x-a)^{\beta} y(x) + g(x); \quad (D^{\alpha-k}_{a+} y)(a+) = b_k, \tag{3.2.65}$$

with complex $\lambda, \beta \in \mathbb{C}$ $(\Re(\beta) \geqq 0)$.

Corollary 3.5 *Let $\alpha = 1$ or $\alpha \in \mathbb{C}$ $(0 < \Re(\alpha) < 1)$. Let $c, b_0 \in \mathbb{C}$ and let $g(x) \in L(a,b)$. If $a(x) \in L_\infty(a,b)$ or if $a(x)$ is bounded on $[a,b]$, then the Cauchy type problem*

$$(D_{a+}^{\alpha}y)(x) = a(x)y(x) + g(x), \quad (I_{a+}^{1-\alpha}y)(a+) = b_0 \tag{3.2.66}$$

and the weighted Cauchy type problem

$$(D_{a+}^{\alpha}y)(x) = a(x)y(x) + g(x), \quad \lim_{x \to a+} \left[(x-a)^{1-\alpha}y(x)\right] = c \tag{3.2.67}$$

have a unique solution $y(x)$ in the space $\mathbf{L}^{\alpha}(a,b)$.

In particular, either of the following Cauchy type problems,

$$(D_{a+}^{\alpha}y)(x) = \lambda(x-a)^{\beta}y(x) + g(x), \quad (I_{a+}^{1-\alpha}y)(a+) = b_0, \tag{3.2.68}$$

$$(D_{a+}^{\alpha}y)(x) = \lambda(x-a)^{\beta}y(x) + g(x), \quad \lim_{x \to a+} \left[(x-a)^{1-\alpha}y(x)\right] = c \tag{3.2.69}$$

with $\lambda, \beta \in \mathbb{C}$ $(\Re(\beta) \geqq 0)$ has a unique solution $y(x) \in \mathbf{L}^{\alpha}(a,b)$.

The results given in Corollaries 3.4 and 3.5 can be extended to more general fractional differential equations. Thus, from Theorem 3.7, we derive the following assertions where we use the notation $D^{\alpha_0}y = y$.

Corollary 3.6 *Let $\alpha \in \mathbb{C}$ $(n-1 < \Re(\alpha) < n;\ n \in \mathbb{N})$, and let $n = [\Re(\alpha)] + 1$ for $\alpha \notin \mathbb{N}$ and $n = \alpha$ for $\alpha \in \mathbb{N}$. Let $l \in \mathbb{N}$ and $\alpha_j \in \mathbb{C}$ $(j = 1, \cdots, l)$ be such that the conditions in (3.2.42) are satisfied, and let $g(x) \in L(a,b)$.*

If $a_j(x) \in L_\infty(a,b)$ or if $a_j(x)$ $(j = 0, 1, \cdots, l)$ are bounded on $[a,b]$, then the following Cauchy type problem for the linear differential equation of order α, with $b_k \in \mathbb{C}$ and $k = 1, \cdots, n$

$$(D_{a+}^{\alpha}y)(x) + \sum_{j=0}^{l} a_j(x)(D_{a+}^{\alpha_j}y)(x) = g(x); \quad (D_{a+}^{\alpha-k}y)(a+) = b_k \tag{3.2.70}$$

has a unique solution $y(x)$ in the space $\mathbf{L}^{\alpha}(a,b)$.

In particular, there exists a unique solution $y(x) \in \mathbf{L}^{\alpha}(a,b)$ to the Cauchy type problem for the equation

$$(D_{a+}^{\alpha}y)(x) + \sum_{j=0}^{l} \lambda_j(x-a)^{\beta_j}(D_{a+}^{\alpha_j}y)(x) = g(x), \ (D_{a+}^{\alpha-k}y)(a+) = b_k \tag{3.2.71}$$

with $\lambda_j, \beta_j \in \mathbb{C}$ $(\Re(\beta_j) \geqq 0)$ $(j = 0, \cdots, l)$.

Corollary 3.7 *Let $\alpha = 1$ or $\alpha \in \mathbb{C}$ be such that $0 < \Re(\alpha) < 1$. Let $l \in \mathbb{N}$ and $\alpha_j \in \mathbb{C}$ $(j = 1, \cdots, l)$ be such that the conditions in (3.2.42) are satisfied, and let $g(x) \in L(a,b)$.*

If $a_j(x) \in L_\infty(a,b)$ or if $a_j(x)$ $(j = 0,1,\cdots, l)$ are bounded on $[a,b]$, and $c, b_0 \in \mathbb{C}$, then the Cauchy type problem

$$(D_{a+}^{\alpha}y)(x) + \sum_{j=0}^{l} a_j(x)(D_{a+}^{\alpha_j}y)(x) = g(x), \quad (I_{a+}^{1-\alpha}y)(a+) = b_0 \tag{3.2.72}$$

and the weighted Cauchy type problem

$$(D_{a+}^{\alpha}y)(x) + \sum_{j=0}^{l} a_j(x)(D_{a+}^{\alpha_j}y)(x) = g(x), \quad \lim_{x\to a+}\left[(x-a)^{1-\alpha}y(x)\right] = c \tag{3.2.73}$$

have a unique solution $y(x)$ in the space $\mathbf{L}^{\alpha}(a,b)$.

In particular, cithcr of thc following Cauchy type problems,

$$(D_{a+}^{\alpha}y)(x) + \sum_{j=0}^{l} \lambda_j(x-a)^{\beta_j}(D_{a+}^{\alpha_j}y)(x) = g(x), \quad (I_{a+}^{1-\alpha}y)(a+) = b_0, \tag{3.2.74}$$

$$(D_{a+}^{\alpha}y)(x) + \sum_{j=0}^{l} \lambda_j(x-a)^{\beta_j}(D_{a+}^{\alpha_j}y)(x) = g(x), \quad \lim_{x\to a+}\left[(x-a)^{1-\alpha}y(x)\right] = c \tag{3.2.75}$$

with $\lambda_j, \beta_j \in \mathbb{C}$ $(\Re(\beta_j) \geqq 0)$ $(j = 0,1,\cdots, l)$ has a unique solution $y(x) \in \mathbf{L}^{\alpha}(a,b)$.

From Theorem 3.9 we derive the results for the systems of Cauchy type problems for linear differential equations of fractional order.

Corollary 3.8 *Let $j = 1,\cdots, m$. Also let $\alpha_j = n_j \in \mathbb{N}$ or $\alpha_j \in \mathbb{C}$ be such that $n_j - 1 < \Re(\alpha_j) < n_j$ $(n_j \in \mathbb{N})$, and let $g_j(x) \in L(a,b)$.*

If $a_j(x) \in L_\infty(a,b)$ or if $a_j(x)$ are bounded on $[a,b]$, then the system Cauchy type problem

$$(D_{a+}^{\alpha_j}y_j)(x) = a_j(x)y_j(x) + g_j(x) \tag{3.2.76}$$

$$(D_{a+}^{\alpha_j-k_j}y)(a+) = b_{k_j} \in \mathbb{C} \; (k_j = 1,\cdots, n_j;\; j = 1,\cdots, m) \tag{3.2.77}$$

has a unique solution $(y_1,\cdots, y_m)$ in the space $\mathbf{L}^{|\alpha|}[(a,b)^m]$.

In particular, the system Cauchy type problem

$$(D_{a+}^{\alpha_j}y)(x) = \lambda_j(x-a)^{\beta_j}y_j(x) + g_j(x) \tag{3.2.78}$$

where $\lambda_j, \beta_j \in \mathbb{C}$; $\Re(\beta_j) \geqq 0$; $j = 1,\cdots, m$, and with the initial conditions (3.2.77) has a unique solution $(y_1,\cdots, y_m) \in \mathbf{L}^{|\alpha|}[(a,b)^m]$.

Corollary 3.9 *Let $j = 1,\cdots, m$. Also let $\alpha_j = 1$ or $\alpha_j \in \mathbb{C}$ be such that $0 < \Re(\alpha_j) < 1$, and let $g_j(x) \in L(a,b)$.*

If $a_j(x) \in L_\infty(a,b)$ or if $a_j(x)$ are bounded on $[a,b]$, then the system of Cauchy type problems, with $b_j, c_j \in \mathbb{C}$ and $j = 1,\cdots, m$

$$(D_{a+}^{\alpha_j}y_j)(x) = a_j(x)y_j(x) + g_j(x), \quad (I_{a+}^{1-\alpha_j}y_j)(a+) = b_j \tag{3.2.79}$$

and the system of weighted Cauchy type problems

$$(D_{a+}^{\alpha_j}y_j)(x) = a_j(x)y_j(x) + g_j(x), \quad \lim_{x\to a}[(x-a)^{1-\alpha_j}y_j(x)] = c_j \tag{3.2.80}$$

have a unique solution $(y_1, \cdots, y_m)$ *in the space* $\mathbf{L}^{|\alpha|}[(a,b)^m]$.

In particular, either system of the following Cauchy type problems,

$$(D_{a+}^{\alpha_j}y_j)(x) = \lambda_j(x-a)^{\beta_j}y_j(x) + g_j(x), \quad (I_{a+}^{1-\alpha_j}y_j)(a+) = b_j, \tag{3.2.81}$$

$$(D_{a+}^{\alpha_j}y_j)(x) = \lambda_j(x-a)^{\beta_j}y_j(x) + g_j(x), \quad \lim_{x\to a}[(x-a)^{1-\alpha_j}y_j(x)] = c_j \tag{3.2.82}$$

with $j = 1, \cdots, m$, $\lambda_j, \beta_j \in \mathbb{C}$ $(\Re(\beta_j) \geqq 0)$ *has* $(y_1, \cdots, y_m) \in \mathbf{L}^{|\alpha|}[(a,b)^m]$ *as the unique solution.*

Remark 3.8 It follows from the proofs of the theorems in this section that the method of successive expansion can be applied to construct exact unique solutions for all Cauchy type problems considered in Section 3.2.5.

3.2.6 Miscellaneous Examples

In Sections 3.2.3-3.2.5 we gave sufficient conditions for the Cauchy type problems and systems of such problems to have a unique solution in some subspaces of the space of integrable functions. Actually these conditions show that the differential equations of fractional order and systems of such equations can have integrable solutions, provided that their right-hand sides are integrable. But these conditions are not sufficient. Below we present two examples and discuss these examples in connection with the results obtained in Sections 3.2.2 and 3.2.3.

Example 3.1 Consider the following differential equation of fractional order $\alpha > 0$:

$$(D_{a+}^{\alpha}y)(x) = \lambda(x-a)^{\beta}[y(x)]^2 \quad (x > a; \ \ \lambda, \beta \in \mathbb{R}; \ \lambda \neq 0). \tag{3.2.83}$$

It can be directly shown that if $\alpha+\beta < 1$, then this equation has the exact solution

$$y(x) = \frac{\Gamma(1-\alpha-\beta)}{\lambda\Gamma(1-2\alpha-\beta)}(x-a)^{-(\alpha+\beta)} \tag{3.2.84}$$

and this solution belongs to the space $L(a,b)$.

In this case the right-hand side of the equation (3.2.83) takes the form

$$f[x,y(x)] = \frac{1}{\lambda}\left[\frac{\Gamma(1-\alpha-\beta)}{\Gamma(1-2\alpha-\beta)}\right]^2 (x-a)^{-(2\alpha+\beta)}, \tag{3.2.85}$$

and this function, generally speaking, does not belong to the space $L(a,b)$.

If we suppose that $2\alpha+\beta < 1$, then the right-hand side of the equation (3.2.83) belongs to the space $L(a,b)$ and, since $\alpha+\beta < 2\alpha+\beta$, then $\alpha+\beta < 1$, and the equation (3.2.83) has the exact solution (3.2.84) which belongs to $L(a,b)$.

Example 3.2 Consider the following differential equation of fractional order $\alpha > 0$:

$$(D_{a+}^{\alpha} y)(x) = \lambda (x-a)^{\beta} [y(x)]^{1/2} \quad (x > a;\ \lambda, \beta \in \mathbb{R};\ \lambda \neq 0). \tag{3.2.86}$$

It can be directly shown that, if $2(\alpha + \beta) > -1$, then this equation has the exact solution

$$y(x) = \left[\frac{\lambda \Gamma(\alpha + 2\beta + 1)}{\Gamma(2\alpha + 2\beta + 1)}\right]^2 (x-a)^{2(\alpha+\beta)} \tag{3.2.87}$$

and this solution $y(x)$ belongs to the space $L(a, b)$.

The right-hand side of the equation (3.2.86) is given by

$$f[x, y(x)] = \lambda^2 \left[\frac{\Gamma(\alpha + 2\beta + 1)}{\Gamma(2\alpha + 2\beta + 1)}\right] (x-a)^{\alpha+2\beta}, \tag{3.2.88}$$

which, generally speaking, does not belong to the space $L(a, b)$.

If we suppose that $\alpha + 2\beta > -1$, then the right-hand side of the equation (3.2.86) belongs to the space $L(a, b)$ and, since $2(\alpha + \beta) > \alpha + 2\beta$, then $2(\alpha + \beta) > 1$, and the equation (3.2.86) has the exact solution (3.2.87), which belongs to $L(a, b)$.

The above examples show that even if the right-hand sides $f[x, y(x)]$ of fractional differential equations are not integrable on $L(a, b)$, the solutions to these equations can still belong to $L(a, b)$.

On the other hand, it is possible to give a more detailed characterization of the solution $y(x)$ to fractional differential equations in the case when their right-hand sides $f[x, y(x)]$ are integrable. Namely, there are three possible cases:

1. The right-hand side $f[x, y(x)]$ is integrable on $[a, b]$ together with the solution $y(x)$, and they both have integrable singularities at the point a.

2. The right-hand side $f[x, y(x)]$ is integrable on $[a, b]$ with an integrable singularity at point a and the solution $y(x)$ is bounded on $[a, b]$.

3. The right-hand side $f[x, y(x)]$ is bounded on $[a, b]$ together with the solution $y(x)$.

The first one case is presented in Example 3.1 with $0 < 2\alpha + \beta < 1$ and $0 < \alpha + \beta < 1$, and in Example 3.2 with $-1 < \alpha + 2\beta < 0$ and $-1/2 < \alpha + \beta < 0$.

The second case is presented in Example 3.1 with $0 < 2\alpha + \beta < 1$ and $\alpha + \beta \leqq 0$, and in Example 3.2 with $-1 < \alpha + 2\beta < 0$ and $\alpha + \beta \geqq 0$.

The third case is presented in Example 3.1 with $2\alpha + \beta \leqq 0$ (so that $\alpha + \beta \leqq 0$), and in Example 3.2 with $\alpha + 2\beta \geqq 0$ (so that $\alpha + \beta \geqq 0$).

The spaces for $f[x, y(x)]$ and $y(x)$ in Examples 3.1 and 3.2 can be characterized in more detail by using the space $C[a, b]$ of continuous functions and the weighted space $C_\gamma[a, b]$ of continuous functions defined in (1.1.21). Using these spaces, Cases 1-3 in Example 3.1 are characterized as follows:

1. If $0 < 2\alpha + \beta < 1$ and $0 < \alpha + \beta < 1$, then $f[x, y(x)] \in C_{2\alpha+\beta}[a, b]$ and $y(x) \in C_{\alpha+\beta}[a, b]$;

2. If $0 < 2\alpha + \beta < 1$ and $\alpha + \beta \leqq 0$, then $f[x, y(x)] \in C[a, b]C_{2\alpha+\beta}[a, b]$ and $y(x) \in C[a, b]$;

3. If $2\alpha + \beta \leqq 0$, then $f[x, y(x)] \in C[a, b]$ and $y(x) \in C[a, b]$.

Similarly, Cases 1-3 in Example 3.2 are characterized as follows:

1. If $-1 < \alpha + 2\beta < 0$ and $-1/2 < \alpha + \beta < 0$, then $f[x, y(x)] \in C_{-(\alpha+2\beta)}[a, b]$ and $y(x) \in C_{-(\alpha+\beta)}[a, b]$;

2. If $-1 < \alpha + 2\beta < 0$ and $\alpha + \beta \geqq 0$, then $f[x, y(x)] \in C_{-(\alpha+2\beta)}[a, b]$ and $y(x) \in C[a, b]$;

3. If $\alpha + 2\beta \geqq 0$, then $f[x, y(x)] \in C[a, b]$ and $y(x) \in C[a, b]$.

The above arguments show that the spaces of continuous functions are preferable when considering solutions to differential equations of fractional order.

Remark 3.9 In this section, we established conditions for the existence of unique summable solutions of Cauchy type problems for nonlinear differential equations on the basis of their equivalence to the corresponding Volterra integral equations and a Banach fixed point theorem. We began our investigation in Sections 3.2.1 and 3.2.2 from the case of equations of positive order $\alpha > 0$ with a real-valued function $f[x, y]$ with y in a domain G of $\mathbb{R}$ and real parameters λ and $b_k \in \mathbb{R}$ $(k = 1, \cdots, -[-\alpha])$, and then extended them to the case of equations of complex order $\alpha \in \mathbb{C}$ $(n - 1 < \Re(\alpha) < n)$ with a complex-valued function $f[x, y]$ with y in a domain G of $\mathbb{C}$ and complex parameters λ and $b_k \in \mathbb{C}$ $(k = 1, \cdots, n)$. As we have seen, the proofs of the corresponding results are similar in both cases.

Therefore, further in this Chapter, we shall present results for the Cauchy type problems for fractional differential equations in the first case. The corresponding results in the second case can be derived analogously.

3.3 Equations with the Riemann-Liouville Fractional Derivative in the Space of Continuous Functions. Global Solution

In this section we give conditions for a unique solution $y(x)$ to the Cauchy type problem (3.1.1)-(3.1.2) in the space $\mathbf{C}^{\alpha}_{n-\alpha}[a, b]$ defined for $n - 1 < \alpha \leqq n$ $(n \in \mathbb{N})$ by

$$\mathbf{C}^{\alpha}_{n-\alpha}[a, b] = \{y(x) \in C_{n-\alpha}[a, b] : \ (D^{\alpha}_{a+}y)(x) \in C_{n-\alpha}[a, b]\}. \tag{3.3.1}$$

Here $C_{n-\alpha}[a, b]$ is a weighted space of continuous functions of the form (1.1.21):

$$C_{n-\alpha}[a, b] = \{g(x) : \ (x - a)^{n-\alpha}g(x) \in C[a, b], \ \|g\|_{C_{n-\alpha}} = \|(x - a)^{n-\alpha}g(x)\|_C\}. \tag{3.3.2}$$

In particular, when $\alpha = n \in \mathbb{N}$, $\mathbf{C}^n_0[a, b]$ coincides with the space $C^n[a, b]$ of functions g continuously differentiable up to order n on $[a, b]$: $\mathbf{C}^n_0[a, b] := C^n[a, b]$. We give conditions for the existence of a global continuous solution $y(x)$ to this Cauchy problem on $[a, b]$. As in Section 3.2, our approach is based on reducing this problem to a Volterra integral equation of the second kind of the form (3.1.8) and on using the Banach fixed point theorem. The arguments here are basically the same as the ones used in the previous section, using the properties of the Riemann-Liouville fractional integrals and derivatives in weighted spaces $C_{n-\alpha}[a,b]$ instead of that in $L(a,b)$. Therefore, we shall give schematic proofs in the main, except for the proof of the uniqueness in Section 3.3.2 where the technique of working with functions in the spaces $C_{n-\alpha}[a,b]$ is demonstrated.

3.3.1 Equivalence of the Cauchy Type Problem and the Volterra Integral Equation

In this subsection we prove the equivalence of the Cauchy type problem (3.2.4)-(3.2.5) and the nonlinear Volterra integral equation (3.1.8) in the sense that, if $y(x) \subset C_{n-\alpha}[a,b]$ satisfies one of these relations, then it also satisfies the other one. To establish such a result, we assume that a function $f[x,y]$ belongs to $C_{n-\alpha}[a,b]$ for any $y \in G \subset \mathbb{R}$. For this we need the auxiliary assertion which follows from Lemma 2.8(a).

Lemma 3.3 *If $\gamma \in \mathbb{R}$ $(0 \leqq \gamma < 1)$, then the fractional integration operator I_{a+}^{α} with $\alpha \in \mathbb{C}$ $(\Re(\alpha) > 0)$ is bounded in $C_\gamma[a,b]$:*

$$||I_{a+}^{\alpha} g||_{C_\gamma} \leqq (b-a)^{\alpha} \frac{\Gamma(1-\gamma)}{\Gamma(1+\alpha-\gamma)} ||g||_{C_\gamma}. \tag{3.3.3}$$

First we establish the equivalence of the Cauchy type problem (3.2.4)-(3.2.5) and the integral equation (3.1.8) in the space (3.3.2).

Theorem 3.10 *Let $\alpha > 0$, $n = -[-\alpha]$. Let G be an open set in $\mathbb{R}$ and let $f : (a,b] \times G \to \mathbb{R}$ be a function such that $f[x,y] \in C_{n-\alpha}[a,b]$ for any $y \in G$.*

If $y(x) \in C_{n-\alpha}[a,b]$, then $y(x)$ satisfies the relations (3.2.4) *and* (3.2.5) *if, and only if, $y(x)$ satisfies the Volterra integral equation* (3.1.8).

Proof. First we prove the necessity. Let $y(x) \in C_{n-\alpha}[a,b]$ satisfy (3.2.4)-(3.2.5). Since $f[x,y] \in C_{n-\alpha}[a,b]$, (3.2.4) means that there exists on $[a,b]$ the fractional derivative $(D_{a+}^{\alpha} y)(x) \in C_{n-\alpha}[a,b]$. According to (2.1.10) and (2.1.7), the derivative $(D_{a+}^{\alpha} y)(x)$ has the form (3.2.6), and so by Lemma 3.3, $(I_{a+}^{n-\alpha} y)(x) \in C_{n-\alpha}^{n}[a,b]$. Hence we can apply Lemma 2.9(d) (with $f(x) = y(x)$ and $\gamma = n - \alpha$) and in accordance with (2.1.39) relation (3.2.7) holds. Using (3.2.5) and taking the same arguments as in the proof of Theorem 3.1, we represent (3.2.7) in the form (3.2.9):

$$(I_{a+}^{\alpha} D_{a+}^{\alpha} y)(x) = y(x) - \sum_{j=1}^{n} \frac{b_j}{\Gamma(\alpha-j+1)} (x-a)^{\alpha-j}. \tag{3.3.4}$$

By Lemma 3.3, the integral $(I_{a+}^{\alpha} f[t,y(t)])(x) \in C_{n-\alpha}[a,b]$. Applying the operator I_{a+}^{α} to both sides of (3.2.4) and using (3.3.4) and (2.1.1), we obtain the equation (3.1.8), and hence the necessity is proved.

Now we prove the sufficiency. Let $y(x) \in C_{n-\alpha}[a,b]$ satisfy the equation (3.1.8). Applying the operator D_{a+}^{α} to both sides of (3.1.8), we have

$$(D_{a+}^{\alpha}y)(x) = \sum_{j=1}^{n} \frac{b_j}{\Gamma(\alpha - j + 1)} \left(D_{a+}^{\alpha}(t-a)^{\alpha-j}\right)(x) + \left(D_{a+}^{\alpha} I_{a+}^{\alpha} f[t, y(t)]\right)(x). \tag{3.3.5}$$

From here, in accordance with the formula (2.1.21) and Lemma 2.9(b) (with $f(x)$ replaced by $f[x, y(x)]$), we arrive at the equation (3.2.4).

Applying the operators $D_{a+}^{\alpha-k}$ $(k = 1, \cdots, n)$ to both sides of (3.1.8) and using the same arguments as in the proof of Theorem 3.1, we obtain relations of the forms (3.2.10) and (3.2.11):

$$(D_{a+}^{\alpha-k}y)(x) = \sum_{j=1}^{k} \frac{b_j}{(k-j)!}(x-a)^{k-j} + \frac{1}{(k-1)!} \int_a^x (x-t)^{k-1} f[t, y(t)]dt \tag{3.3.6}$$

and

$$(D_{a+}^{\alpha-n}y)(x) = \sum_{j=1}^{n} \frac{b_j}{(n-j)!}(x-a)^{n-j} + \frac{1}{(n-1)!} \int_a^x (x-t)^{n-1} f[t, y(t)]dt. \tag{3.3.7}$$

Applying the limit as $x \to a+$ in (3.3.6) and (3.3.7), we obtain the relations in (3.2.5). Thus sufficiency is also proved, which completes the proof of Theorem 3.10.

Corollary 3.10 *Let $0 < \alpha < 1$, let G be an open set in $\mathbb{R}$ and let $f : (a,b] \times G \to \mathbb{R}$ be a function such that $f[x,y] \in C_{1-\alpha}[a,b]$ for any $y \in G$.*

If $y(x) \in C_{1-\alpha}[a,b]$, then $y(x)$ satisfies the relations (3.1.12) *and* (3.1.13) *if, and only if, $y(x)$ satisfies the integral equation* (3.1.14).

Corollary 3.11 *Let $n \in \mathbb{N}$, let G be an open set in $\mathbb{R}$ and let $f : [a,b] \times G \to \mathbb{R}$ be a function such that $f[x,y] \in C[a,b]$ for any $y \in G$.*

If $y(x) \in C[a,b]$, then $y(x)$ satisfies the relations in (3.1.5) *if, and only if, $y(x)$ satisfies the integral equation* (3.2.12).

Remark 3.10 Theorem 3.10 and Corollaries 3.10-3.11 were proved in Kilbas et al. ([375], Theorems 1 and 2).

3.3.2 Existence and Uniqueness of the Global Solution to the Cauchy Type Problem

In this subsection we establish the existence of a unique solution to the Cauchy type problem (3.2.4)-(3.2.5) in the space $\mathbf{C}_{n-\alpha}^{\alpha}[a,b]$ defined in (3.3.1) under the conditions of Theorem 3.10, and an additional Lipschitzian condition (3.2.15).

We need the following preliminary assertion.

Lemma 3.4 *Let $\gamma \in \mathbb{R}$, $a < c < b$, $g \in C_\gamma[a,c]$ and $g \in C_\gamma[c,b]$. Then $g \in C_\gamma[a,b]$ and*

$$\|g\|_{C_\gamma[a,b]} \leqq \max\left[\|g\|_{C_\gamma[a,c]}, \|g\|_{C_\gamma[c,b]}\right].$$

There holds the following result.

Theorem 3.11 *Let $\alpha > 0$ and $n = -[-\alpha]$. Let G be an open set in $\mathbb{R}$ and let $f : (a,b] \times G \to \mathbb{R}$ be a function such that $f[x,y] \in C_{n-\alpha}[a,b]$ for any $y \in G$ and the condition* (3.2.15) *holds.*

Then there exists a unique solution $y(x)$ to the Cauchy type problem (3.2.4)-(3.2.5) *in the space $\mathbf{C}^\alpha_{n-\alpha}[a,b]$.*

Proof. First we prove the existence of a unique solution $y(x) \in C_{n-\alpha}[a,b]$. According to Theorem 3.10, it is sufficient to prove the existence of a unique solution $y(x) \in C_{n-\alpha}[a,b]$ to the nonlinear Volterra integral equation (3.1.8). For this, as in the proof of Theorem 3.3, we apply the known method for nonlinear Volterra integral equations.

Equation (3.1.8) makes sense in any interval $[a,x_1] \in [a,b]$ $(a < x_1 < b)$. Choose x_1 such that the inequality

$$A(x_1 - a)^\alpha \frac{\Gamma(\alpha - n + 1)}{\Gamma(2\alpha - n + 1)} < 1 \tag{3.3.8}$$

holds, and then prove the existence of a unique solution $y(x) \in C_{n-\alpha}[a,x_1]$ to equation (3.1.8) on the interval $[a,x_1]$. For this we use the Banach fixed point theorem (Theorem 1.9 in Section 1.13) for the space $C_{n-\alpha}[a,b]$, which is the complete metric space with the distance given by

$$d(y_1,y_2) = \|y_1 - y_2\|_{C_{n-\alpha}[a,x_1]} := \max_{x\in[a,x_1]} |(x-a)^{n-\alpha}[y_1(x) - y_2(x)]|. \tag{3.3.9}$$

We rewrite the integral equation (3.1.8) in the form

$$y(x) = (Ty)(x), \tag{3.3.10}$$

where T is the operator defined in (3.2.18) with $y_0(x)$ given by (3.2.19). To apply Theorem 1.9, we have to prove the following: (1) if $y(x) \in C_{n-\alpha}[a,x_1]$, then $(Ty)(x) \in C_{n-\alpha}[a,x_1]$; and (2) for any $y_1, y_2 \in C_{n-\alpha}[a,x_1]$, the following estimate holds:

$$\|Ty_1 - Ty_2\|_{C_{n-\alpha}[a,x_1]} \leqq \omega\|y_1 - y_2\|_{C_{n-\alpha}[a,x_1]}, \quad \omega = A_{n-\alpha}(x_1 - a)^\alpha \frac{\Gamma(\alpha - n + 1)}{\Gamma(2\alpha - n + 1)}. \tag{3.3.11}$$

It follows from (3.2.19) that $y_0(x) \in C_{n-\alpha}[a,x_1]$. Since $f[x,y(x)] \in C_{n-\alpha}[a,x_1]$, then, by Lemma 3.3 (with $\gamma = n - \alpha$, $b = x_1$ and $g(t) = f[t,y(t)]$), the integral in the right-hand side of (3.2.18) also belongs to $C_{n-\alpha}[a,x_1]$, and hence $(Ty)(x) \in C_{n-\alpha}[a,x_1]$. Now we prove the estimate in (3.3.11). By (3.2.18)-(3.2.19) and (2.1.1), using the Lipschitzian condition (3.2.15) and applying the relation (3.3.3) (with $\gamma = n - \alpha$, $b = x_1$ and $g(x) = f[x, y_1(x)] - f[x, y_2(x)]$), we have

$$\|Ty_1 - Ty_2\|_{C_{n-\alpha}[a,x_1]} \leqq \left\| I_{a+}^{\alpha}[|f[t, y_1(t)] - f[t, y_2)(t)|]\right\|_{C_{n-\alpha}[a,x_1]}$$

$$\leqq A \left\| I_{a+}^{\alpha}[|y_1(t) - y_2(t)|]\right\|_{C_{n-\alpha}[a,x_1]} \leqq A(x_1 - a)^{\alpha} \frac{\Gamma(\alpha - n + 1)}{\Gamma(2\alpha - n + 1)} \|y_1 - y_2\|_{C_{n-\alpha}[a,x_1]},$$

which yields the estimate (3.3.11). In accordance with (3.3.8), $0 < \omega < 1$, and hence by Theorem 1.9, there exists a unique solution $y^*(x) = y^{*0}(x) \in C_{n-\alpha}[a, x_1]$ to the equation (3.1.8) on the interval $[a, x_1]$.

By Theorem 1.9, this solution $y^*(x)$ is a limit of a convergent sequence $(T^m y_0^*)(x)$:

$$\lim_{m\to\infty} \|T^m y_0^* - y^*\|_{C_{n-\alpha}[a,x_1]} = 0, \tag{3.3.12}$$

where $y_0^*(x)$ is any function in $C_{n-\alpha}[a, b]$. If at least one $b_k \neq 0$ in the initial condition (3.2.5), we can take $y_0^*(x) = y_0(x)$ with $y_0(x)$ defined by (3.2.19). The last relation can be rewritten in the form

$$\lim_{m\to\infty} \|y_m - y\|_{C_{n-\alpha}[a,x_1]} = 0, \tag{3.3.13}$$

where

$$y_m(x) := (T^m y_0^*)(x) = y_0(x) + \frac{1}{\Gamma(\alpha)} \int_a^x \frac{f[t, (T^{m-1}y_0^*)(t)]dt}{(x - t)^{1-\alpha}} \quad (m \in \mathbb{N}). \tag{3.3.14}$$

Next we consider the interval $[x_1, x_2]$, where $x_2 = x_1 + h_1$ and $h_1 > 0$ are such that $x_2 < b$. Rewrite the equation (3.1.8) in the form (3.2.25), where $y_{01}(x)$ defined by (3.2.26) is the known function. Using the same arguments as above, we derive that there exists a unique solution $y_1^*(x) \in C_{n-\alpha}[x_1, x_2]$ to the equation (3.1.8) on the interval $[x_1, x_2]$. Taking the next interval $[x_2, x_3]$, where $x_3 = x_2 + h_2$ and $h_2 > 0$ are such that $x_3 < b$, and repeating this process, we find that there exists a unique solution $y(x)$ to the equation (3.1.8) such that $y(x) = y^{*k}(x)$ and $y^{*k}(x) \in C_{n-\alpha}[x_{k-1}, x_k]$ $(k = 1, \cdots, L)$, where $a = x_0 < x_1 < \cdots < x_L = b$. Then, by Lemma 3.4, there exists a unique solution $y(x) \in C_{n-\alpha}[a, b]$ on the whole interval $[a, b]$.

Thus there exists a unique solution $y(x) = y^*(x) \in C_{n-\alpha}[a, b]$ to the Volterra integral equation (3.1.8), and hence to the Cauchy type problem (3.2.4)-(3.2.5).

To complete the proof of Theorem 3.11, we must show that such a unique solution $y(x) \in C_{n-\alpha}[a, b]$ belongs to the space $\mathbf{C}_{n-\alpha}^{\alpha}[a, b]$. In accordance with the definition (3.3.1), it is sufficient to prove that $(D_{a+}^{\alpha} y)(x) \in C_{n-\alpha}[a, b]$. By the above proof, the solution $y(x) \in C_{n-\alpha}[a, b]$ is a limit of the sequence $y_m(x)$, where $y_m(x) := (T^m y_0^*)(x) \in C_{n-\alpha}[a, b]$:

$$\lim_{n\to\infty} \|y_m - y\|_{C_{n-\alpha}[a,b]} = 0, \tag{3.3.15}$$

with the choice of certain y_m on each $[a, x_1], \cdots, [x_{L-1}, b]$. By (3.2.4) and (3.2.15), we have

$$\|D_{a+}^{\alpha} y_m - D_{a+}^{\alpha} y\|_{C_{n-\alpha}} = \|f[x, y_m] - f[x, y]\|_{C_{n-\alpha}} \leqq A_{n-\alpha} \|y_m - y\|_{C_{n-\alpha}}.$$

Thus, by (3.3.15),

$$\lim_{n\to\infty} \|D^{\alpha}_{a+}y_m - D^{\alpha}_{a+}y\|_{C_{n-\alpha}[a,b]} = 0,$$

and hence $(D^{\alpha}_{a+}y)(x) \in C_{n-\alpha}[a,b]$. This completes the proof of Theorem 3.11.

Corollary 3.12 *Let $0 < \alpha < 1$, let G be an open set in $\mathbb{R}$ and let $f : (a,b] \times G \to \mathbb{R}$ be a function such that $f[x,y] \in C_{1-\alpha}[a,b]$ for any $y \in G$ and* (3.2.15) *holds.*

Then there exists a unique solution $y(x)$ to the Cauchy type problem (3.1.12)-(3.1.13) *in the space $\mathbf{C}^{\alpha}_{1-\alpha}[a,b]$.*

Corollary 3.13 *Let $n \in \mathbb{N}$, let G be an open set in $\mathbb{R}$ and let $f : [a,b] \times G \to \mathbb{R}$ be a function such that $f[x,y] \in C[a,b]$ for any $y \in G$ and* (3.2.15) *is valid.*

Then there exists a unique solution $y(x)$ to the Cauchy problem (3.1.5) *in the space $C^n[a,b]$.*

3.3.3 The Weighted Cauchy Type Problem

When $0 < \alpha < 1$, the result of Corollary 3.12 remains true for the following weighted Cauchy type problem (3.1.7), with $c \in \mathbb{C}$:

$$(D^{\alpha}_{a+}y)(x) = f[x,y(x)]; \quad \lim_{x\to a+} \left[(x-a)^{1-\alpha}y(x)\right] = c \quad (0 < \Re(\alpha) < 1). \tag{3.3.16}$$

Its proof is based on the following preliminary assertion.

Lemma 3.5 *Let $0 < \alpha < 1)$ and let $y(x) \in C_{1-\alpha}[a,b]$.*

(a) *If*

$$\lim_{x\to a+} \left[(x-a)^{1-\alpha}y(x)\right] = c, \ c \in \mathbb{R}, \tag{3.3.17}$$

then

$$(I^{1-\alpha}_{a+}y)(a+) := \lim_{x\to a+} (I^{1-\alpha}_{a+}y)(x) = c\Gamma(\alpha). \tag{3.3.18}$$

(b) *If*

$$\lim_{x\to a+} (I^{1-\alpha}_{a+}y)(x) = b, \ b \in \mathbb{R}, \tag{3.3.19}$$

and if there exists the limit $\lim_{x\to a+}\left[(x-a)^{1-\alpha}y(x)\right]$, then

$$\lim_{x\to a+} \left[(x-a)^{1-\alpha}y(x)\right] = \frac{b}{\Gamma(\alpha)}. \tag{3.3.20}$$

Proof. The proof of this lemma is similar to that of Lemma 3.2.

From Corollary 3.12 and Lemma 3.5, we obtain the existence and uniqueness result for the weighted Cauchy type problem (3.3.16).

Theorem 3.12 *Let $0 < \alpha < 1$, let G be an open set in $\mathbb{R}$ and let $f : (a,b] \times G \to \mathbb{R}$ be a function such that $f[x,y] \in C_{1-\alpha}[a,b]$ for any $y \in G$ and the relation* (3.2.15) *holds.*

Then there exists a unique solution $y(x)$ to the weighted Cauchy type problem (3.3.16) *in the space $\mathbf{C}^{\alpha}_{1-\alpha}[a,b]$.*

Proof. By applying Theorem 3.11 and Lemma 3.5, Theorem 3.12 is proved just as Theorem 3.5.

Remark 3.11 The results in Theorems 3.11-3.12 and Corollaries 3.12-3.13 were established by Kilbas et al. ([375], Theorems 3-6) under stricter conditions in some special function spaces of $y(x)$.

Remark 3.12 It follows from the proof of Theorem 3.11 that the method of successive expansions can be applied to construct the unique solution $y(x) = y^*(x) \in C_{n-\alpha}[a,b]$ to the Cauchy type problems (3.2.4)-(3.2.5).

Remark 3.13 The particular case of the weighted Cauchy type problem (3.3.16) with $a = 0$, $0 < \alpha < 1$, $f[x, y(x)] = f[y(x)]$ in the form (3.1.27) was considered by Delbosco and Rodino [164], who proved its unique solution $y(x) \in C_{1-\alpha}[0,h]$ for $h > 0$ [see Section 3.1 in this regard].

3.3.4 Generalized Cauchy Type Problems

The results obtained in Sections 3.3.1-3.3.2 are extended to the more general Cauchy type problems (3.2.40)-(3.2.41) than (3.1.1)-(3.1.2).

The following assertion generalizing Theorem 3.10 holds.

Theorem 3.13 *Let $\alpha > 0$ be such that $n - 1 < \alpha \leqq n$ $(n \in \mathbb{N})$. Let $l \in \mathbb{N}\backslash\{1\}$ and $\alpha_j \in \mathbb{C}$ $(j = 1, \cdots, l)$ be such that*

$$0 = \alpha_0 < \alpha_1 < \cdots < \alpha_l < \alpha, \tag{3.3.21}$$

Let G be an open set in $\mathbb{R}^{l+1}$ and let $f : (a,b] \times G \to \mathbb{R}$ be a function such that $f[x, y, y_1, \cdots, y_l] \in C_{n-\alpha}[a,b]$ for any $(y, y_1, \cdots, y_l) \in G$.

If $y(x) \in C_{n-\alpha}[a,b]$, then $y(x)$ satisfies the relations (3.2.40) *and* (3.2.41) *if, and only if, $y(x)$ satisfies the integral equation* (3.2.43).

In particular, if $0 < \alpha < 1$, then $y(x) \in C_{1-\alpha}[a,b]$ satisfies the relations (3.2.44)-(3.2.45) *if, and only if, $y(x)$ satisfies the integral equation* (3.2.45).

Proof. The proof of Theorem 3.13 is similar to that of Theorem 3.10, using the properties of the Riemann-Liouville fractional integrals and derivatives.

The next statement generalizes Theorems 3.11-3.12.

Theorem 3.14 *Let the conditions of Theorem* 3.13 *be valid, let $f[x, y, y_1, \cdots, y_l]$ satisfy the Lipschitzian condition* (3.2.46), *and let $(D_{a+}^{\alpha_j - k_j} y)(a+) = b_{k_j} \in \mathbb{R}$ $(j = 1, \cdots, n_j)$ be fixed numbers, where $n_j = [\Re(\alpha_j)] + 1$ for $\alpha_j \notin \mathbb{N}$ and $n_j = \alpha_j$ for $\alpha_j \in \mathbb{N}$.*

Then there exists a unique solution $y(x)$ to the Cauchy type problem (3.2.40)-(3.2.41) *in the space $\mathbf{C}_{n-\alpha}^{\alpha}[a,b]$.*

In particular, if $0 < \alpha < 1$ and $(I_{a+}^{1-\alpha_j} y)(a+) = b_j \in \mathbb{R}$ $(j = 1, \cdots, l)$ are fixed numbers, then there exists a unique solution $y(x) \in \mathbf{C}_{1-\alpha}^{\alpha}[a,b]$ to the Cauchy type problem (3.2.44) *with $b_0 \in \mathbb{R}$ and the weighted Cauchy type problem* (3.2.47) *with $c \in \mathbb{R}$.*

Proof. The proof of Theorem 3.14 is similar to the proofs of Theorem 3.11-3.12 in the same manner as Theorem 3.7. By Theorem 3.13, it is sufficient to establish the existence of a unique solution $y(x) \in C_{n-\alpha}[a,b]$ to the integral equation (3.2.43). We choose $x_1 \in (a,b)$ such that the relation

$$\omega = A_l \sum_{j=0}^{l} (x_1 - a)^{\alpha-\alpha_j} \frac{\Gamma(\alpha-\alpha_j)\Gamma(\alpha)-n+1)}{\Gamma(\alpha_j)\Gamma(2\alpha-\alpha_j-n+1)} < 1$$

holds, and rewrite the equation (3.2.43) in the form $y(x) = (Ty)(x)$, where T is operator (3.2.50). Using the relation (3.2.46) and the estimate (3.3.3) (with $\gamma = n-\alpha$, $b = x_1$ and α being replaced by $\alpha - \alpha_j$), we have

$$||Ty_1 - Ty_2||_{C_{n-\alpha}[a,x_1]} \leqq A_l \sum_{j=0}^{l} ||I_{a+}^{\alpha-\alpha_j}[|y_1 - y_2|]||_{C_{n-\alpha}[a,x_1]}$$

$$\leqq A_l \sum_{j=0}^{l} (x_1 - a)^{\alpha-\alpha_j} \frac{\Gamma(\alpha-\alpha_j)\Gamma(\alpha)-n+1)}{\Gamma(\alpha_j)\Gamma(2\alpha-\alpha_j-n+1)} ||y_1 - y_2||_{C_{n-\alpha}[a,x_1]},$$

which yields

$$||Ty_1 - Ty_2||_{C_{n-\alpha}[a,x_1]} \leqq \omega ||y_1 - y_2||_{C_{n-\alpha}[a,x_1]} \quad (0 < \omega < 1).$$

Applying the Banach fixed point theorem (Theorem 1.9), we prove the existence of a unique solution $y(x) = y^*(x) \in C_{n-\alpha}[a,x_1]$ for the equation (3.2.43) on $[a,x_1]$. Further arguments are similar to those in Theorem 3.11.

The results of Theorem 3.10 and Theorems 3.11-3.12 can be also extended to the system Cauchy type problem (3.2.52)-(3.2.53) in a product of the spaces $\mathbf{C}_{n_1-\alpha_1}^{\alpha_1}[a,b], \cdots, \mathbf{C}_{n_m-\alpha_m}^{\alpha_m}[a,b]$ defined in (3.3.1). We shall use the following notation:

$$C_{n-\alpha}\left[[a,b]^m\right] = C_{n_1-\alpha_1}[a,b] \times \cdots \times C_{n_m-\alpha_m}[a,b]\ , \ C_0\left[[a,b]^m\right] = C\left[[a,b]^m\right],$$

$$\mathbf{C}_{\mathbf{n}-\alpha}^{|\alpha|}\left[[a,b]^m\right] = \mathbf{C}_{n_1-\alpha_1}^{\alpha_1}[a,b] \times \cdots \times \mathbf{C}_{n_m-\alpha_m}^{\alpha_m}[a,b]. \tag{3.3.22}$$

The following statements, generalizing Theorem 3.10 and Theorems 3.11-3.12, respectively, hold.

Theorem 3.15 *Let $j = 1, \cdots, m$, $\alpha_j \in \mathbb{R}$, $n_j - 1 < \alpha_j \leqq n_j$ $(n_j \in \mathbb{N})$. Let G be an open set in $\mathbb{R}^m$ and let $f_j : (a,b] \times G \to \mathbb{R}$ be functions such that $f_j[x, y_1, \cdots, y_m] \in C_{n_j-\alpha_j}[a,b]$ for any $(y_1, \cdots, y_m) \in G$.*

If $y_j(x) \in C_{n_j-\alpha_j}[a,b]$, then $y_j(x)$ satisfy the system of relations (3.2.52) *and* (3.2.53) *if, and only if, $y_j(x)$ satisfy the system of integral equations* (3.2.56).

In particular, if $0 < \alpha_j < 1$ and $y_j(x) \in C_{1-\alpha_j}[a,b]$, then $y_j(x)$ satisfy the system of relations (3.2.57) *and* (3.2.58) *if, and only if, $y_j(x)$ satisfy the system of integral equations* (3.2.59).

Theorem 3.16 *Let the conditions of Theorem* 3.15 *be valid, and let there hold the Lipschitzian conditions* (3.2.60).

Then there exists a unique solution $(y_1, \cdots, y_m)$ *to the system Cauchy type problem* (3.2.52)-(3.2.53) *in the space* $\mathbf{C}^{|\alpha|}_{\mathbf{n}-\alpha}[[a,b]^m]$.

In particular, if $0 < \alpha_j < 1$ $(j = 1, \cdots, m)$, *then there exists a unique solution* $(y_1, \cdots, y_m) \in \mathbf{C}^{|1|}_{1-\alpha}[[a,b]^m|]$ *to the system Cauchy type problem* (3.2.57)-(3.2.58) *and to the system of weighted Cauchy type problems* (3.2.61)-(3.2.62).

Remark 3.14 The results of Theorems 3.13 and 3.14 were proved by Kilbas and Marzan ([378], Theorems 1 and 3) under stricter conditions on f in some special function spaces of $y(x)$.

Remark 3.15 The results presented in Theorems 3.15 and 3.16 can be extended to the system Cauchy type problem (3.2.40)-(3.2.41) and to the system of the Volterra integral equations (3.2.43).

3.3.5 Cauchy Type Problems for Linear Equations

In this subsection we present the existence and uniqueness of solutions for linear fractional differential equations considered in Section 3.2.5. First from Theorem 3.11, Corollary 3.12 and Theorem 3.12, we derive the corresponding results for the Cauchy type problems for linear differential equations of order $\alpha > 0$.

Corollary 3.14 *Let* $\alpha > 0$ $(n - 1 < \alpha \leqq n,\ n \in \mathbb{N})$, *and let* $a(x) \in C[a,b]$ *and* $g(x) \in C_{n-\alpha}[a,b]$.

Then the Cauchy type problem (3.2.64) *for the linear differential equation of order* α *has a unique solution* $y(x)$ *in the space* $\mathbf{C}^{\alpha}_{n-\alpha}[a,b]$.

In particular, there exists a unique solution $y(x) \in \mathbf{C}^{\alpha}_{n-\alpha}[a,b]$ *to the Cauchy type problem* (3.2.65).

Corollary 3.15 *Let* $0 < \alpha \leqq 1$, *and let* $a(x) \in C[a,b]$ *and* $g(x) \in C_{1-\alpha}[a,b]$.

Then the Cauchy type problem (3.2.66) *and the weighted Cauchy type problem* (3.2.67) *have a unique solution* $y(x)$ *in the space* $\mathbf{C}^{\alpha}_{1-\alpha}[a,b]$.

In particular, either of the Cauchy type problems (3.2.68) *and* (3.2.69) *has a unique solution* $y(x) \in \mathbf{C}^{\alpha}_{1-\alpha}[a,b]$.

The results of Corollaries 3.14 and 3.15 can be extended to more general fractional differential equations. Thus, from Theorem 3.14, we derive the following assertions.

Corollary 3.16 *Let* $\alpha > 0$ $(n - 1 < \alpha \leqq n;\ n \in \mathbb{N})$. *Let* $l \in \mathbb{N}$ *and* $\alpha_j > 0$ $(j = 1, \cdots, l)$ *be such that the conditions in* (3.3.21) *are satisfied. Let* $a_j(x) \in C[a,b]$ $(j = 0, 1, \cdots, l)$ *and* $g(x) \in C_{n-\alpha}[a,b]$.

Then the Cauchy type problem (3.2.70) *for the linear differential equation of order* α *has a unique solution* $y(x)$ *in the space* $\mathbf{C}^{\alpha}_{n-\alpha}[a,b]$.

In particular, there exists a unique solution $y(x) \in \mathbf{C}^{\alpha}_{n-\alpha}[a,b]$ *to the Cauchy type problem* (3.2.71).

Corollary 3.17 *Let $0 < \alpha \leqq 1$. Let $l \in \mathbb{N}$ and $\alpha_j \in \mathbb{R}$ $(j = 1, \cdots, l)$ be such that the conditions in* (3.3.21) *are satisfied. Let $a_j(x) \in C[a,b]$ $(j = 0, 1, \cdots, l)$ and $g(x) \in C_{1-\alpha}[a,b]$.*

Then the Cauchy type problem (3.2.72) *and the weighted Cauchy type problem* (3.2.73) *have a unique solution $y(x)$ in the space $\mathbf{C}^{\alpha}_{1-\alpha}[a,b]$.*

In particular, either of the Cauchy type problems (3.2.74) *and* (3.2.75) *has a unique solution $y(x) \in \mathbf{C}^{\alpha}_{1-\alpha}[a,b]$.*

Theorem 3.16 yields the results for systems of Cauchy type problems of linear differential equations of fractional order.

Corollary 3.18 *Let $j = 1, \cdots, m$, and let $\alpha_j > 0$ be such $n_j - 1 < \alpha_j \leqq n_j$ $(n_j \in \mathbb{N})$, and let $a_j(x) \in C[a,b]$ and $g_j(x) \in C_{n_j-\alpha_j}[a,b]$.*

Then the system Cauchy type problem (3.2.76)-(3.2.77) *has a unique solution $(y_1, \cdots, y_m)$ in the space $\mathbf{C}^{|\alpha|}_{\mathbf{n}-\alpha}[[a,b]^m]$.*

In particular, the system Cauchy type problem (3.2.78)-(3.2.77) *has a unique solution $(y_1, \cdots, y_m) \in \mathbf{C}^{|\alpha|}_{\mathbf{n}-\alpha}[[a,b]^m]$.*

Corollary 3.19 *Let $j = 1, \cdots, m$, $\alpha_j > 0$ be such that $0 < \alpha_j \leqq 1$, and let $a_j(x) \in C[a,b]$ and $g_j(x) \in C_{1-\alpha_j}[a,b]$.*

Then the system of Cauchy type problems (3.2.79) *and the system of weighted Cauchy type problems* (3.2.80) *have a unique solution $(y_1, \cdots, y_m)$ in the space $\mathbf{C}^{|\alpha|}_{\mathbf{1}-\alpha}[[a,b]^m]$.*

In particular, either of the systems of Cauchy type problems (3.2.81) *and weighted Cauchy type problems* (3.2.82) *has a unique solution*

$$(y_1, \cdots, y_m) \in \mathbf{C}^{|\alpha|}_{\mathbf{1}-\alpha}[[a,b]^m]$$

Remark 3.16 It follows from the proofs of the theorems in this section that the method of successive expansions can be applied to construct exact unique solutions to all Cauchy type problems considered in Section 3.3.

3.3.6 More Exact Spaces

In Sections 3.3.2 and 3.3.3, we gave sufficient conditions for the Cauchy type problems (3.2.4)-(3.2.5) and for the weighted Cauchy type problem (3.3.16) to have a unique solution $y(x) \in C^{\alpha}_{n-\alpha}[a,b]$, provided that the right-hand side $f[x, y(x)]$ of the fractional differential equation (3.2.4) belongs to the same space $C_{n-\alpha}[a,b]$ of weighted continuous functions $y(x)$ on $[a,b]$ defined in (3.3.2). When $n-1 < \alpha < n$ $(n \in \mathbb{N})$, then $f[x, y(x]$ and $y(x)$ have a power singularity at the point $x = a$ and belong to the space $L(a,b)$. This case reflects one of three possible cases characterizing $f[x, y(x]$ and $y(x)$, indicated in Section 3.2.6 while considering Examples 3.1 and 3.2:

Case 1. The right-hand side $f[x, y(x)]$ is integrable on $[a,b]$ together with the solution $y(x)$, and they both have integrable singularities at the point a.

Two other possible cases reflected in Section 3.2.6 are the following:

Case 2. The right-hand side $f[x, y(x)]$ is integrable on $[a, b]$ with an integrable singularity at the point a and the solution $y(x)$ is bounded on $[a, b]$.

Case 3. The right-hand side $f[x, y(x)]$ is bounded on $[a, b]$ together with the solution $y(x)$.

From the above, we can derive that the spaces $C_{n-\alpha}[a, b]$ for $f[x, y(x)]$ and $y(x)$, considered in Section 3.3.2, can be characterized more precisely. This also follows from the fact that the Riemann-Liouville fractional derivative D^{α}_{a+}, as a usual derivative, can transform "good" functions at point a into "bad" functions at a. For example, by (2.1.17),

$$(D^{\alpha}_{a+}1)(x) = \frac{1}{\Gamma(1-\alpha)}(x-a)^{1-\alpha} \quad (\alpha > 0), \tag{3.3.23}$$

so that, if $0 < \alpha < 1$, $(x-a)^{1-\alpha}$ has an integrable singularity at point a.

In fact, Theorems 3.10, Corollary 3.12 and Theorems 3.11-3.12 remain true if the condition $f[\cdot, y] : C_{n-\alpha}[a, b] \to C_{n-\alpha}[a, b]$ is replaced by a more general condition. Indeed, there holds the following assertion.

Property 3.1 *Let $\alpha > 0$ be such that $n - 1 < \alpha \leqq n$ $(n \in \mathbb{N})$. Let G be an open set in $\mathbb{R}$ and let $f : (a, b] \times G \to \mathbb{R}$ be a function such that, for any $y \in G$, $f[x, y] \in C_{\gamma}[a, b]$ with $\gamma \in \mathbb{R}$ $(n - \alpha \leqq \gamma < 1)$.*

(a) *If $y(x) \in C_{n-\alpha}[a, b]$, then $y(x)$ satisfies the relations* (3.2.4) *and* (3.2.5) *if, and only if, $y(x)$ satisfies the integral equation* (3.1.8).

(b) *Let $f[x, y]$ satisfy the Lipschitz condition* (3.2.15).

Then there exists a unique solution $y(x)$ to the Cauchy type problem (3.2.4)-(3.2.5) *in the space $\mathbf{C}^{\alpha}_{n-\alpha,\gamma}[a, b]$:*

$$\mathbf{C}^{\alpha}_{n-\alpha,\gamma}[a, b] = \{y(x) \in C_{n-\alpha}[a, b] : \ (D^{\alpha}_{a+}y)(x) \in C_{\gamma}[a, b]\}. \tag{3.3.24}$$

In particular, if $0 < \alpha < 1$, then there exists a unique solution of the Cauchy type problems (3.1.12)-(3.1.13) *and* (3.3.16) *in $\mathbf{C}^{\alpha}_{1-\alpha,\gamma}[a, b]$.*

(c) *In particular,* (a) *and* (b) *hold for $\gamma = 0$ when $f[x, y] \in C[a, b]$ for any $y \in G$.*

Proof. The proof of (a) is similar to the proofs of Theorems 3.10 and 3.11 by using Lemma 2.9(a) instead of Lemma 3.3 for the proof of necessity.

To prove (b), we note that if $f[\cdot, y] : [a, b] \times \mathbb{R} \to C_{\gamma}[a, b]$ with $\gamma \in \mathbb{R}$ $(0 \leqq \gamma < 1)$, then the second term in (3.1.8) given by

$$\left(I^{\alpha}_{a+}f[t, y(t)]\right)(x) = \frac{1}{\Gamma(\alpha)}\int_a^x \frac{f[t, y(t)]dt}{(x-t)^{1-\alpha}},$$

in accordance with Lemma 2.9(a) and 2.9(b), belongs to $C_{\gamma-\alpha}[a, b]$ for $\gamma > \alpha$ and to $C[a, b]$ for $\gamma \leqq \alpha$. Thus it is clear that the first term $y_0(x)$ in the right-hand side of (3.1.8), defined by (3.2.19), belongs to $C_{n-\alpha}[a, b]$. Since $n \in \mathbb{N}$, then

$\gamma - \alpha < n - \alpha$ and $n - \alpha \geqq 0$. Then, by Property 1.1 in Section 1.1, the right-hand side of (3.1.8) belongs to $C_{n-\alpha}[a, b]$. Therefore, the operator T defined by (3.2.18) maps $C_{n-\alpha}[a, b]$ onto $C_{n-\alpha}[a, b]$. Now we choose $x_1 \in (a, b)$ such that the condition (3.3.8) is satisfied and repeat the proof of Theorem 3.11 to prove that there exists a unique solution $y(x)$ to the Cauchy type problem (3.2.4)-(3.2.5) in the space $C_{n-\alpha}[a, b]$ such that (3.3.15) holds.

Now we show that $(D^\alpha_{a+}y)(x) \in C_\gamma[a, b]$. Since $\gamma \geqq n - \alpha$, then applying (3.2.4), (3.2.15) and (1.1.37), we have

$$||D^\alpha_{a+}y_m - D^\alpha_{a+}y||_{C_\gamma[a,b]} = ||f[x, y_m] - f[x, y]||_{C_\gamma[a,b]}$$

$$\leqq A||y_m - y||_{C_\gamma[a,b]} \leqq A(b-a)^{\gamma-n+\alpha}||y_m - y||_{C_{n-\alpha}[a,b]}.$$

Thus, by the above two estimates and (3.3.15), we find that

$$\lim_{m\to\infty} ||D^\alpha_{a+}y_m - D^\alpha_{a+}y||_{C_\gamma[a,b]} = 0.$$

Hence a unique solution $y(x) \in C_{n-\alpha}[a, b]$ has the property $(D^\alpha_{a+}y)(x) \in C_\gamma[a, b]$. By (3.3.24), this means that $y(x) \in \mathbf{C}^\alpha_{n-\alpha,\gamma}[a, b]$. This completes the proof of (b).

From Property 3.1 we obtain the fourth possible case characterizing $f[x, y(x)]$ and $y(x)$ for the Cauchy type problems (3.2.4)-(3.2.5) and (3.3.16).

Case 4. The right-hand side $f[x, y(x)]$ is bounded on $[a, b]$ and the solution $y(x)$ is integrable on $[a, b]$ with an integrable singularity at point a.

A property, similar to Property 3.1, exists for the generalized Cauchy type problem (3.2.40)-(3.2.41), (3.2.47) and for systems of Cauchy type problems (3.2.52)-(3.2.53), (3.2.61)-(3.2.62).

Property 3.2 *Let $\alpha > 0$ be such that $n - 1 < \alpha \leqq n$ $(n \in \mathbb{N})$. Let $l \in \mathbb{N}\backslash\{1\}$ and $\alpha_j > 0$ $(j = 1, \cdots, l)$ be such that the condition* (3.3.21) *is satisfied. Let G be an open set in $\mathbb{R}^{l+1}$ and let $f : (a, b] \times G \to \mathbb{R}$ be a function such that for any $(y, y_1, \cdots, y_l) \in G$, $f[x, y, y_1, \cdots, y_l] \in C_\gamma[a, b]$ with $\gamma \in \mathbb{R}$ $(n - \alpha \leqq \gamma < 1)$.*

(a) *If $y(x) \in C_{n-\alpha}[a, b]$, then $y(x)$ satisfies the relations* (3.2.40) *and* (3.2.41) *if, and only if, $y(x)$ satisfies the integral equation* (3.2.43).

(b) *Let the Lipschitz condition* (3.2.46) *be satisfied, and for $j = 1, \cdots, n_j$, let $(D^{\alpha_j-k_j}_{a+}y)(a+) = b_{k_j} \in \mathbb{R}$ be fixed numbers, where $n_j = -[-\alpha_j]$.*

Then there exists a unique solution $y(x)$ to the Cauchy type problem (3.2.40)-(3.2.41) *in the space $\mathbf{C}^\alpha_{n-\alpha,\gamma}[a, b]$.*

(c) *In particular,* (a) *and* (b) *hold for $\gamma = 0$ when $f[x, y, y_1, \cdots, y_l] \in C[a, b]$ for any $(y, y_1, \cdots, y_l) \in G$.*

Property 3.3 *Let $j = 1, \cdots, m$, $\alpha_j > 0$ be such that $n_j - 1 < \alpha_j \leqq n_j$ $(n_j \in \mathbb{N})$. Let G be an open set in $\mathbb{R}^m$ and let $f_j : (a, b] \times G \to \mathbb{R}$ be functions such that for any $(y_1, \cdots y_m) \in G$, $f_j[x, y_1, \cdots, y_m] \in C_{\gamma_j}[a, b]$ with $\gamma_j \in \mathbb{R}$ $(n_j - \alpha_j \leqq \gamma_j < 1)$.*

(a) *If $y_j(x) \in C_{n-\alpha_j}[a, b]$, then $y_j(x)$ satisfy the system of relations* (3.2.52) *and* (3.2.53) *if, and only if, $y_j(x)$ satisfy the system of integral equations* (3.2.56).

(b) *Let $f_j(x, y_1, \cdots, y_m)$ satisfy the Lipschitz conditions* (3.2.60).

Then there exists a unique solution $(y_1, \cdots, y_m)$ to the system Cauchy type problem (3.2.52)-(3.2.53) *in the space $\mathbf{C}^{|\alpha|}_{\mathbf{n}-\alpha,\gamma}[[a,b]^m]$:*

$$\mathbf{C}^{|\alpha|}_{\mathbf{n}-\alpha,\gamma}[[a,b]^m] = \mathbf{C}^{\alpha_1}_{n_1-\alpha_1,\gamma_1}[a,b] \times \cdots \times \mathbf{C}^{\alpha_m}_{n_m-\alpha_m,\gamma_m}[a,b], \tag{3.3.25}$$

where $\mathbf{C}^{\alpha_j}_{n_j-\alpha_j,\gamma_j}[a,b]$ $(j = 1, \cdots, m)$ are defined in (3.3.24).

In particular, if $0 < \alpha_j < 1$, then there exists a unique solution $(y_1, \cdots, y_m) \in \mathbf{C}^{|1|}_{1-\alpha}[[a,b]^m]$ to the system Cauchy type problem (3.2.57)-(3.2.58) *and to the system of weighted Cauchy type problems* (3.2.61)-(3.2.62).

(c) *In particular,* (a) *and* (b) *hold for $\gamma_j = 0$ when $f_j[x, y_1, \cdots, y_m] \in C[a,b]$ for any $(y_1, \cdots, y_m) \in G$.*

Cases 2 and 3 above appear when we consider the Cauchy type problem (3.2.4)-(3.2.5) with $b_n = 0$. The results will be different for $0 < \alpha \leqq 1$ and for $\alpha > 1$. First we consider the Cauchy type problem (3.2.31) with $0 < \alpha \leqq 1$:

$$(D^{\alpha}_{a+}y)(x) = f[x, y(x)] \ (0 < \alpha \leqq 1), \quad (I^{1-\alpha}_{a+}y)(a+) = 0 \tag{3.3.26}$$

and the weighted Cauchy type problem (3.2.33):

$$(D^{\alpha}_{a+}y)(x) = f[x, y(x)] \ (0 < \alpha \leqq 1), \quad \lim_{x \to a}(x-a)^{1-\alpha}y(x) = 0. \tag{3.3.27}$$

The integral equation (3.1.14), corresponding to the problem (3.3.26), takes the form

$$y(x) = \frac{1}{\Gamma(\alpha)} \int_a^x \frac{f[t, y(t)]dt}{(x-t)^{1-\alpha}} \ (x > a; \ 0 < \alpha \leqq 1). \tag{3.3.28}$$

By Lemma 3.9(a), if $f[x, y(x)] \in C_\gamma[a,b]$ with any $\gamma \in \mathbb{R}$ $(0 \leqq \gamma < 1)$, then the right-hand side of (3.3.28) and hence the solution $y(x)$ belong to $C_{\gamma-\alpha}[a,b]$ for $\gamma > \alpha$ and to $C[a,b]$ for $\gamma \leqq \alpha$. From here we derive the result.

Property 3.4 *Let $0 < \alpha \leqq 1$ and $0 \leqq \gamma < 1$. Let G be an open set in $\mathbb{R}$ and let $f : (a,b] \times G \to \mathbb{R}$ be a function such that $f[x,y] \in C_\gamma[a,b]$ for any $y \in G$ and the Lipschitz condition* (3.2.15) *holds.*

(a) *If $\gamma > \alpha$, then the Cauchy type problems* (3.3.26) *and* (3.3.27) *have a unique solution $y(x)$ in the space $\mathbf{C}^{\alpha}_{\gamma-\alpha,\gamma}[a,b]$:*

$$\mathbf{C}^{\alpha}_{\gamma-\alpha,\gamma}[a,b] = \{g(x) \in C_{\gamma-\alpha}[a,b] : \ (D^{\alpha}_{a+}g)(x) \in C_\gamma[a,b]\}. \tag{3.3.29}$$

(b) *If $\gamma \leqq \alpha$, then the Cauchy type problems* (3.3.26) *and* (3.3.27) *have a unique solution in the space $\mathbf{C}^{\alpha}_{0,\gamma}[a,b]$:*

$$\mathbf{C}^{\alpha}_{0,\gamma}[a,b] = \{g(x) \in C[a,b] : \ (D^{\alpha}_{a+}g)(x) \in C_\gamma[a,b]\}. \tag{3.3.30}$$

(c) *In particular,* (b) *holds for $\gamma = 0$, provided that $f[x,y] \in C[a,b]$ for any $y \in G$.*

Proof. The derivation of Property 3.4 for the Cauchy type problem (3.3.26) is similar to that of Property 3.1 due to the equivalence of this problem and the integral equation (3.3.28) in the spaces $C_{\gamma-\alpha}[a,b]$ and $C[a,b]$ in Cases (a) and (b), respectively, in accordance with Property 1.1. The result for the weighted Cauchy type problem (3.3.27) follows from that for (3.3.26) on the basis of Lemma 3.5.

Next we consider the Cauchy type problem (3.3.1)-(3.3.2) with $\alpha > 1$.

$$(D^{\alpha}_{a+}y)(x) = f[x, y(x)] \quad (\alpha > 1), \tag{3.3.31}$$

with initial conditions

$$(D^{\alpha-k}_{a+}y)(a+) = b_k, \; b_k \in \mathbb{R} \; (k = 1, \cdots, n-1), \quad b_n = 0, \tag{3.3.32}$$

where $n = -[-\alpha]$. The Volterra integral equation (3.1.8) corresponding to this problem takes the form

$$y(x) = \sum_{j=1}^{n-1} \frac{b_j}{\Gamma(\alpha - j + 1)}(x-a)^{\alpha-j} + \frac{1}{\Gamma(\alpha)} \int_a^x \frac{f[t, y(t)]dt}{(x-t)^{1-\alpha}} \quad (x > a). \tag{3.3.33}$$

If $f[x, y(x)] \in C_\gamma[a,b]$ $(0 \leqq \gamma < 1)$, then, by Lemma 2.9(a), the integral in the right-hand side of (3.3.33) is a continuous function on $[a,b]$ (because $\gamma < 1 < \alpha$). The first term in the right-hand side of (3.3.33) is also a continuous function on $[a,b]$. Then we obtain the result.

Property 3.5 *Let $\alpha > 1$ and $n = -[-\alpha]$. Let G be an open set in $\mathbb{R}$ and let $f : (a,b] \times G \to \mathbb{R}$ be a function such that for any $y \in G$, $f[x,y] \in C_\gamma[a,b]$ with $\gamma \in \mathbb{R}$ $(0 \leq \gamma < 1)$ and the Lipschitz condition* (3.2.15) *is satisfied.*

Then the Cauchy type problems (3.3.31) *and* (3.3.32) *have a unique solution $y(x) \in \mathbf{C}^{\alpha}_{0,0}[a,b]$.*

In particular, this assertion holds for $\gamma = 0$ when $f[x,y] \in C[a,b]$ for any $y \in G$.

Proof. The proof of Property 3.5 is similar to that of Property 3.1 on the basis of the equivalence of the Cauchy type problem (3.3.31) and the integral equation (3.3.33) in the space $C[a,b]$, by using Property 1.1 and Lemma 3.5.

Remark 3.17 Properties 3.4 and 3.5 contain Cases 2 and 3 above.

Properties, similar to Properties 3.4 and 3.5, are valid for the generalized Cauchy type problems (3.2.44), (3.2.47) and (3.2.40)-(3.2.41), respectively.

Property 3.6 *Let $0 < \alpha \leqq 1$ and $0 \leqq \gamma < 1$. Let $l \in \mathbb{N}\backslash\{1\}$ and $\alpha_j > 0$ $(j = 1, \cdots, l)$ be such that the condition* (3.3.21) *is satisfied. Let G be an open set in $\mathbb{R}^{l+1}$ and let $f : (a,b] \times G \to \mathbb{R}$ be a function such that $f[x, y, y_1, \cdots, y_l] \in C_\gamma[a,b]$ for any $(y, y_1, \cdots, y_l) \in G$, and the Lipschitz condition* (3.2.46) *is satisfied. Also let $(I^{1-\alpha_j}_{a+}y)(a+) = b_j \in \mathbb{R}$ $(j = 1, \cdots, l)$ be fixed numbers.*

(a) *If* $\gamma > \alpha$, *then there exists a unique solution* $y(x)$ *to the Cauchy type problems* (3.2.44) *with* $b_0 = 0$ *and* (3.2.47) *with* $c = 0$ *in the space* $\mathbf{C}^{\alpha}_{\gamma-\alpha,\gamma}[a,b]$.

(b) *If* $\gamma \leqq \alpha$, *then there exists a unique solution* $y(x)$ *to the Cauchy type problems* (3.2.44) *with* $b_0 = 0$ *and* (3.2.47) *with* $c = 0$ *in the space* $\mathbf{C}^{\alpha}_{0,\gamma}[a,b]$.

(c) *In particular,* (b) *holds for* $\gamma = 0$, *provided that* $f[x, y, y_1, \cdots, y_l] \in C[a,b]$ *for any* $(y, y_1, \cdots, y_l) \in G$.

Property 3.7 *Let* $\alpha > 1$ *and* $n = -[-\alpha)]$. *Let* $l \in \mathbb{N}\backslash\{1\}$ *and* $\alpha_j > 0$ $(j = 1, \cdots, l)$ *be such that the condition* (3.3.21) *is satisfied. Let the fractional derivatives* $D^{\alpha_j}_{a+}y$ $(j = 1, \cdots, l)$ *exist on* $[a,b]$. *Let* G *be an open set in* $\mathbb{R}^{l+1}$ *and let* $f : (a,b] \times G \to \mathbb{R}$ *be a function such that* $f[x, y, y_1, \cdots, y_l] : C[a,b] \to C_\gamma[a,b]$ *for any* $(y, y_1, \cdots, y_l) \in G$ *and the Lipschitz condition* (3.2.46) *is satisfied. Also let* $(I^{1-\alpha_j}_{a+}y)(a+) = b_j \in \mathbb{R}$ $(j = 1, \cdots, l)$ *be fixed numbers.*

Then the Cauchy type problem (3.2.40)-(3.2.41) *with* $b_n = 0$ *has a unique solution in the space* $\mathbf{C}^{\alpha}_{0,0}[a,b]$.

In particular, this assertion holds for $\gamma = 0$, *for any* $(y, y_1, \cdots, y_l) \in G$, *provided that* $f[x, y, y_1, \cdots, y_l] \in C[a,b]$.

Properties, similar to Properties 3.4 and 3.5, are also true for the systems of Cauchy type problems (3.2.61), (3.2.62) and (3.2.52)-(3.2.53), respectively.

Property 3.8 *Let* $j = 1, \cdots, m$, $0 < \alpha_j \leqq 1$, *and* $0 \leqq \gamma_j < 1$. *Let* G *be an open set in* $\mathbb{R}^m$. *Also let the functions* $f_j : (a,b] \times G \to \mathbb{R}$ *be such that* $f_j[x, y_1, \cdots, y_m] \in C_{\gamma_j}[a,b]$ *for any* $(y_1, \cdots, y_m) \in G$ *and the Lipschitz conditions* (3.2.60) *are satisfied.*

(a) *If* $\gamma_j > \alpha_j$, *then the systems of Cauchy type problems* (3.2.57)-(3.2.58) *and* (3.2.61)-(3.2.62) *with* $b_j = 0$ $(j = 1, \cdots, m)$ *have a unique solution* $(y_1, \cdots, y_m)$ *in the space* $\mathbf{C}^{|\alpha|}_{\gamma-\alpha,\gamma}[[a,b]^m]$:

$$\mathbf{C}^{|\alpha|}_{\gamma-\alpha,\gamma}[[a,b]^m] = \mathbf{C}^{\alpha_1}_{\gamma_1-\alpha_1,\gamma_1}[a,b] \times \cdots \times \mathbf{C}^{\alpha_m}_{\gamma_m-\alpha_m,\gamma_m}[a,b], \tag{3.3.34}$$

where the spaces $\mathbf{C}^{\alpha_j}_{\gamma_j-\alpha_j,\gamma_j}[a,b]$ *are defined in* (3.3.29).

(b) *If* $\gamma_j \leqq \alpha_j$, *then the systems of Cauchy type problems* (3.2.57)-(3.2.58) *and* (3.2.61)-(3.2.62) *with* $b_j = 0$ $(j = 1, \cdots, m)$ *have a unique solution* $(y_1, \cdots, y_m)$ *in the space* $\mathbf{C}^{|\alpha|}_{\mathbf{0},\gamma}[[a,b]^m]$:

$$\mathbf{C}^{|\alpha|}_{\mathbf{0},\gamma}[[a,b]^m] = \mathbf{C}^{\alpha_1}_{0,\gamma_1}[a,b] \times \cdots \times \mathbf{C}^{\alpha_m}_{0,\gamma_m}[a,b], \tag{3.3.35}$$

where the spaces $\mathbf{C}^{\alpha_j}_{0,\gamma_j}[a,b]$ *are defined in* (3.3.30).

(c) *In particular,* (b) *holds for* $\gamma = 0$, *provided that* $f_j[x, y_1, \cdots, y_m] \in C[a,b]$ *for any* $(y_1, \cdots, y_m) \in G$.

Property 3.9 *Let* $j = 1, \cdots, m$, $\alpha_j > 1$ *and* $n_j = -[-\alpha_j]$, *and let* $0 \leqq \gamma_j < 1$. *Let* G *be an open set in* $\mathbb{R}^m$ *and let the functions* $f_j : (a,b] \times G \to \mathbb{R}$ *be such* *that for any* $(y_1, \cdots, y_m) \in G$, $f_j[x, y_1, \cdots, y_m) \in C_{\gamma_j}[a, b]$, *and the Lipschitz conditions* (3.2.60) *are satisfied.*

Then the system Cauchy type problem (3.2.52)-(3.2.53) *with* $b_{n_j} = 0$ $(j = 1, \cdots, m)$ *has a unique solution* $(y_1, \cdots, y_m) \in \mathbf{C}^{|\alpha|}_{\mathbf{0},\gamma}[[a, b]^m]$.

In particular, this assertion holds for $\gamma_j = 0$ *when the functions*

$$f_j : (a, b] \times G \to \mathbb{R}$$

are such that $f_j[x, y_1, \cdot, y_m] \in C[a, b]$ *for any* $(y_1, \cdots, y_m) \in G$.

Remark 3.18 The results in Properties 3.1-3.9 remain true for the Cauchy type problems of linear fractional differential equations and systems of such equations considered in Section 3.3.5. In particular, Properties 3.1, 3.4, and 3.5 remain true for the Cauchy type problems (3.2.64)-(3.2.69), Properties 3.2, 3.6, and 3.7 for the Cauchy type problems (3.2.70)-(3.2.75), and Properties 3.3, 3.8, and 3.9 for the systems of Cauchy type problems (3.2.76)-(3.2.82).

3.3.7 Further Examples

In Sections 3.2.3-3.2.5 we gave sufficient conditions for the Cauchy type problems and systems of such problems to have a unique solution in some subspaces of the weighted spaces of continuous functions. Indeed, these conditions show that the differential equations of fractional order and systems of such equations can have such solutions, provided that their right-hand sides are weighted continuous functions. But these conditions are not sufficient. Below we present two examples and discuss them in connection with the results obtained in Sections 3.3.2 and 3.3.3. The first example generalizes Examples 3.1 and 3.2 considered in Section 3.2.6.

Example 3.3 Consider the following differential equation of fractional order $\alpha > 0$:

$$(D^{\alpha}_{a+}y)(x) = \lambda(x-a)^{\beta}[y(x)]^m \quad (x > a;\ \ m > 0;\ m \neq 1) \tag{3.3.36}$$

with real $\lambda, \beta \in \mathbb{R}$ $(\lambda \neq 0)$. Using (2.1.17), it is directly verified that, if

$$\frac{\beta + \alpha}{1 - m} > -1, \tag{3.3.37}$$

then the equation (3.3.36) has the explicit solution

$$y(x) = \left[\frac{\Gamma(\frac{\beta+\alpha}{m-1} + 1)}{\lambda\Gamma(\frac{\beta+\alpha m}{m-1} + 1)}\right]^{1/(m-1)} (x-a)^{(\beta+\alpha)/(1-m)}. \tag{3.3.38}$$

Let

$$\gamma = \frac{\beta + m\alpha}{m-1}, \quad \gamma - \alpha = \frac{\beta+\alpha}{m-1}. \tag{3.3.39}$$

The above solution $y(x) \in C_{\gamma-\alpha}[a,b]$ for $0 < \gamma - \alpha < 1$, while $y(x) \in C[a,b]$ for $\gamma - \alpha \leqq 0$:

$$y(x) \in C_{\gamma-\alpha}[a,b], \quad \text{if} \quad 0 < \gamma - \alpha = \frac{\beta+\alpha}{m-1} < 1, \tag{3.3.40}$$

$$y(x) \in C[a,b], \quad \text{if} \quad \gamma - \alpha = \frac{\beta+\alpha}{m-1} \leqq 0. \tag{3.3.41}$$

In this case the right-hand side of the equation (3.3.36) takes the form

$$f[x,y(x)] = \lambda \left[\frac{\Gamma(\frac{\beta+\alpha}{m-1}+1)}{\lambda\Gamma(\frac{\beta+\alpha m}{m-1}+1)} \right]^{m/(m-1)} (x-a)^{(\beta+\alpha m)/(1-m)}. \tag{3.3.42}$$

The function $f[x,y(x)] \in C_\gamma[a,b]$ when $0 < \gamma < 1$, and $f[x,y(x)] \in C[a,b]$ when $\gamma \leqq 0$:

$$f[x,y(x)] \in C_\gamma[a,b], \quad \text{if} \quad 0 < \gamma = \frac{\beta+m\alpha}{m-1} < 1, \tag{3.3.43}$$

$$f[x,y(x)] \in C[a,b], \quad \text{if} \quad \gamma = \frac{\beta+m\alpha}{m-1} \leqq 0. \tag{3.3.44}$$

In accordance with (3.3.39), the following cases are possible for the spaces of the right-hand side (3.3.42) and of the solution (3.3.47):

(**1**) (3.3.43) and (3.3.40); (**2**) (3.3.43) and (3.3.41); (**3**) (3.3.4) and (3.3.41).

It is directly verified that the first case is possible when $0 < \alpha < 1$ and either of the following conditions holds:

$$m > 1, \; -\alpha < \beta < m - 1 - m\alpha; \tag{3.3.45}$$

$$0 < m < 1, \; m - 1 - m\alpha < \beta < -\alpha. \tag{3.3.46}$$

The second case is possible when $\alpha > 0$ and

$$m > 1, \; -m\alpha < \beta < m - 1 - m\alpha, \; \beta \leqq -\alpha;$$

or

$$0 < m < 1, \; m - 1 - m\alpha < \beta < -m\alpha, \; \beta \geqq -\alpha.$$

When $0 < \alpha < 1$, these conditions take the following forms:

$$m > 1, \; -m\alpha < \beta \leqq -\alpha; \tag{3.3.47}$$

$$0 < m < 1, \; -\alpha \leqq \beta < -m\alpha; \tag{3.3.48}$$

and if $\alpha \geqq 1$, then

$$m > 1, \; -m\alpha < \beta < m - 1 - m\alpha; \tag{3.3.49}$$

$$0 < m < 1, \; m - 1 - m\alpha < \beta < -m\alpha. \tag{3.3.50}$$

It follows from (3.3.41) and (3.3.44) that the third case in (3.3.7) is possible for $\alpha > 0$ and either of the following conditions holds:

$$m > 1, \ \beta \leqq -m\alpha; \tag{3.3.51}$$

$$0 < m < 1, \ \ \beta \geqq -m\alpha. \tag{3.3.52}$$

It is directly verified that (3.3.38) is the explicit solution to the following Cauchy type problem:

$$(D_{a+}^{\alpha}y)(x) = \lambda(x-a)^{\beta}[y(x)]^m, \ \ (I_{a+}^{1-\alpha}y)(a+) = 0 \ \ (0 < \alpha < 1) \tag{3.3.53}$$

and to the following weighted Cauchy type problem:

$$(D_{a+}^{\alpha}y)(x) = \lambda(x-a)^{\beta}[y(x)]^m, \ \lim_{t\to a+}\left[(x-a)^{1-\alpha}y(x)\right] = 0 \ (0 < \alpha < 1), \tag{3.3.54}$$

provided that any of the conditions (3.3.45),(3.3.46), (3.3.47), (3.3.48), (3.3.51) and (3.3.52) holds, and to the following Cauchy type problem

$$(D_{a+}^{\alpha}y)(x) = \lambda(x-a)^{\beta}[y(x)]^m, \ (D_{a+}^{\alpha-k}y)(a+) = 0 \ (\alpha > 0; \ k = 1, \cdots, n = -[-\alpha]), \tag{3.3.55}$$

provided that any of (3.3.49), (3.3.50), (3.3.51) and (3.3.52) is valid.

Now we establish the conditions for the uniqueness of the solution (3.3.38) to the above problems by using Properties 3.4 and 3.5. For this we have to choose the domain G and check when the Lipschitz condition (3.2.15) with

$$f(x,y) = \lambda(x-a)^{\beta}y^m \tag{3.3.56}$$

is valid. We choose the following domain:

$$D = \{(x,y) \subset \mathbb{C}: \ \ a < x \leqq b, \ 0 < y < K(x-a)^{\omega}; \ \ \omega \in \mathbb{R}; \ \ K > 0\}. \tag{3.3.57}$$

This means that the domain G for y is unbounded for $\omega < 0$:

$$G = \{y \in \mathbb{R}: \ \ 0 < y < \frac{K}{(x-a)^{-\omega}}; \ \ \omega < 0\}, \tag{3.3.58}$$

and bounded for $\omega \geqq 0$:

$$G = \{y \in \mathbb{R}: \ \ 0 < y < K(x-a)^{\omega}; \ \ \omega > 0\}. \tag{3.3.59}$$

In particular, when $\omega = 0$,

$$G = \{y \in \mathbb{R}: \ \ 0 < y < K\}. \tag{3.3.60}$$

To prove the Lipschitz condition (3.2.15) with $f[x,y]$ given by (3.3.56), we have, for any $(x,y_1),(x,y_2) \in D$,

$$|f[x,y_1] - f[x,y_2]| \leqq |\lambda|(x-a)^{\beta}|y_1^m - y_2^m|. \tag{3.3.61}$$

There hold the following relations:

$$|y_1^m - y_2^m| \leqq m\,[\max[|y_1|,|y_2|]]^{m-1}\,|y_1 - y_2| \quad (m > 1), \tag{3.3.62}$$

$$|y_1^m - y_2^m| \leqq m\,[\min[|y_1|,|y_2|]]^{m-1}\,|y_1 - y_2| \quad (0 < m < 1). \tag{3.3.63}$$

By (3.3.57), $|y| < K(x-a)^{\omega}$. Therefore

$$\max\,[|y_1|,|y_2|] < K(x-a)^{\omega}, \quad \min\,[|y_1|,|y_2|] < K(x-a)^{\omega},$$

and hence, for any $m > 0$ $(m \neq 0)$,

$$|y_1^m - y_2^m| \leqq m < mK(x-a)^{(m-1)\omega}|y_1^m - y_2^m|. \tag{3.3.64}$$

Substituting this estimate into (3.3.61), we obtain

$$|f[x,y_1] - f[x,y_2]| \leqq |\lambda| mK(x-a)^{\beta+(m-1)\omega}|y_1 - y_2|. \tag{3.3.65}$$

If $\beta + (m-1)\omega \geqq 0$, then the last relation yields the following estimate:

$$|f[x,y_1] - f[x,y_2]| \leqq A|y_1 - y_2|, \quad A = |\lambda| mK(b-a)^{\beta+(m-1)\omega}. \tag{3.3.66}$$

Thus the Lipschitz condition (3.2.15) is satisfied, provided that $\beta + (m-1)\omega \geqq 0$.

Using the above results and Properties 3.4 and 3.5, we derive the uniqueness result for the Cauchy type problems (3.3.53), (3.3.54) and (3.3.55).

Proposition 3.1 *Let* $0 < \alpha < 1$, $\lambda, \beta \in \mathbb{R}$ $(\lambda \neq 0)$, $m > 0$ $(m \neq 1)$ *and* $\gamma = (\beta + m\alpha)/(m-1)$. *Let* D *be the domain* (3.3.57), *where* $\omega \in \mathbb{R}$ *is such that* $\beta + (m-1)\omega \geqq 0$.

(a) *If either of the conditions* (3.3.45) *or* (3.3.46) *holds, then the Cauchy type problems* (3.3.53) *and* (3.3.54) *have a unique solution* $y(x) \in \mathbf{C}^{\alpha}_{\gamma-\alpha,\gamma}[a,b]$ *given by* (3.3.38).

(b) *If either of the conditions* (3.3.47) *or* (3.3.48) *is satisfied, then the Cauchy type problems* (3.3.53) *and* (3.3.54) *have a unique solution* $y(x) \in \mathbf{C}^{\alpha}_{0,\gamma}[a,b]$ *given by* (3.3.38).

(c) *If either of the conditions* (3.3.51) *or* (3.3.52) *is valid, then the Cauchy type problems* (3.3.53) *and* (3.3.54) *have a unique solution* $y(x) \in \mathbf{C}^{\alpha}_{0,0}[a,b]$ *given by* (3.3.38).

Proposition 3.2 *Let* $\alpha \geqq 1$, $n = -[-\alpha]$, $\lambda \in \mathbb{R}$ $(\lambda \neq 0)$, $m > 0$ $(m \neq 1)$ *and* $\beta \in \mathbb{R}$. *Let* D *be the domain* (3.3.57), *where* $\omega \in \mathbb{R}$ *is such that* $\beta + (m-1)\omega \geqq 0$.

If any of the conditions (3.3.49), (3.3.50), (3.3.51) *and* (3.3.52) *holds, then the Cauchy type problem* (3.3.55) *has a unique solution* $y(x) \in \mathbf{C}^{\alpha}_{0,0}[a,b]$ *given by* (3.3.38).

Remark 3.19 In particular, when $m = 2$ and $m = 1/2$, the conditions in (3.3.43) and (3.3.44) and the results in Propositions 3.1 and 3.2 yield more exact characterizations of weighted and usual spaces of continuous functions for the right-hand side (3.3.42) and for the solution (3.3.38) of the Cauchy type problems (3.3.53) and (3.3.54) than was possible in Section 3.2 in the space of integrable functions.

Example 3.4 Consider the following nonlinear inhomogeneous fractional differential equation of order $\alpha > 0$:

$$(D_{a+}^{\alpha}y)(x) = \lambda(x-a)^{\beta}[y(x)]^m + b(x-a)^{\nu} \quad (x > a;\ \ m > 0;\ m \neq 1) \tag{3.3.67}$$

with real a $(\lambda \neq 0)$, b, β and ν. We suppose that

$$\nu = \frac{\beta + \alpha m}{1-m}, \tag{3.3.68}$$

the condition (3.3.37) is satisfied and the transcendental equation

$$\Gamma\left(\frac{\alpha+\beta}{1-m} + 1 - \alpha\right)[a\xi^m + b] - \Gamma\left(\frac{\alpha+\beta}{1-m} + 1\right)\xi = 0 \tag{3.3.69}$$

is solvable and $\xi = \mu$ is its solution. Then it is easily verified that the equation (3.3.67) has the solution

$$y(x) = \mu(x-a)^{(\beta+\alpha)/(1-m)}. \tag{3.3.70}$$

In this case the right-hand side of (3.3.67) takes the form

$$f[x, y(x)] = (\lambda + b)(x-a)^{(\beta+\alpha m)/(1-m)}. \tag{3.3.71}$$

The above solution $y(x)$ and the function $f[x, y(x)]$ belong to the spaces (3.3.40), (3.3.41) and (3.3.43), (3.3.44), respectively. Then, by using the same arguments as in Example 3.3, from Properties 3.4 and 3.5 we obtain the uniqueness theorem for the Cauchy type problem

$$(D_{a+}^{\alpha}y)(x) = \lambda(x-a)^{\beta}[y(x)]^m + b(x-a)^{\nu}, \quad (I_{a+}^{1-\alpha}y)(a+) = 0 \quad (0 < \alpha < 1), \tag{3.3.72}$$

for the weighted Cauchy type problem

$$(D_{a+}^{\alpha}y)(x) = \lambda(x-a)^{\beta}[y(x)]^m + b(x-a)^{\nu}, \quad \lim_{t \to a+}\left[(x-a)^{1-\alpha}y(x)\right] = 0 \ (0 < \alpha < 1), \tag{3.3.73}$$

and for the Cauchy type problem

$$(D_{a+}^{\alpha}y)(x) = \lambda(x-a)^{\beta}[y(x)]^m + b(x-a)^{\nu}, \quad (D_{a+}^{\alpha-k}y)(a+) = 0, \tag{3.3.74}$$

with $\alpha > 0$ and $k = 1, \cdots, n$ $(n = -[-\alpha])$.

Proposition 3.3 *Let* $0 < \alpha < 1$, $\lambda, \beta \in \mathbb{R}$ $(\lambda \neq 0)$, $m > 0$ $(m \neq 1)$ *and* $\gamma = (\beta + m\alpha)/(m-1)$. *Let* D *be the domain* (3.3.57), *where* $\omega \in \mathbb{R}$ *is such that* $\beta + (m-1)\omega \geqq 0$. *Let* ν *be given by* (3.3.68) *and let the transcendental equation* (3.3.69) *have a unique solution* $\xi = \mu$.

(a) *If either of the conditions* (3.3.45) *or* (3.3.46) *holds, then the Cauchy type problems* (3.3.72) *and* (3.3.73) *have a unique solution* $y(x) \in \mathbf{C}_{\gamma-\alpha,\gamma}^{\alpha}[a, b]$ *given by* (3.3.70).

(b) *If either of the conditions* (3.3.47) *or* (3.3.48) *is satisfied, then the Cauchy type problems* (3.3.72) *and* (3.3.73) *have a unique solution* $y(x) \in \mathbf{C}_{0,\gamma}^{\alpha}[a,b]$ *given by* (3.3.70).

(c) *If either of the conditions* (3.3.51) *or* (3.3.52) *is valid then the Cauchy type problems* (3.3.72) *and* (3.3.73) *have a unique solution* $y(x) \in \mathbf{C}_{0,0}^{\alpha}[a,b]$ *given by* (3.3.70).

Proposition 3.4 *Let* $\alpha > 0$, $\lambda \in \mathbb{R}$ $(\lambda \neq 0)$, $m > 0$ $(m \neq 1)$ *and* $\beta \in \mathbb{R}$. *Let* ν *be given by* (3.3.68). *Let* D *be the domain* (3.3.57), *where* $\omega \in \mathbb{R}$ *is such that* $\beta \geqq (m-1)\omega$, *and let the transcendental equation* (3.3.69) *have a unique solution* $\xi = \mu$.

If any of the conditions (3.3.49), (3.3.50), (3.3.51) *and* (3.3.52) *holds, then the Cauchy type problem* (3.3.74) *has a unique solution* $y(x) \in \mathbf{C}_{0,0}^{\alpha}[a,b]$ *given by* (3.3.70).

Remark 3.20 It is directly verified that (3.3.38) and (3.3.70) are solutions of the Cauchy type problems (3.3.53)-(3.3.55) and (3.3.72)-(3.3.74) provided that the condition (3.3.37) holds, or the equivalent one

$$\beta < m-1-\alpha \text{ for } m > 1; \text{ or } m-1-\alpha < \beta \text{ for } 0 < m < 1. \tag{3.3.75}$$

In Propositions 3.1-3.4 we have proved the uniqueness of the above solutions in a domain D given in (3.3.57), provided that $\omega \in \mathbb{R}$ be such that $\beta + (m-1)\omega \geqq 0$. The problem of the uniqueness of these solutions when $\beta + (m-1)\omega < 0$ is open. It is also an open problem to prove the uniqueness theorems for other domains.

We also note that the problem of uniqueness of the solutions (3.3.38) and (3.3.70) to the equations (3.3.36) and (3.3.67) was discussed by Kilbas and Saigo [396]. The solvability of the nonlinear equation (3.3.67) depends on the solvability of the transcendental equation (3.3.69). The positive solutions of such a transcendental equation were investigated by Karapetyants et al. [362].

3.4 Equations with the Riemann-Liouville Fractional Derivative in the Space of Continuous Functions. Semi-Global and Local Solutions

In Sections 3.2 and 3.3 we established sufficient conditions for the Cauchy type problems (3.1.1)-(3.1.2), (3.1.6), and (3.1.7), the generalized Cauchy type problems (3.2.40)-(3.2.41), (3.2.44), and (3.2.47), and the systems of Cauchy type problems (3.2.52)-(3.2.53), (3.2.57)-(3.2.58), and (3.2.61)-(3.2.62) to have a unique global solution in spaces of integrable and continuous functions. This section is devoted to the derivation of the conditions for the existence of a unique local continuous solution to such problems when the initial conditions are applied at any point $x_0 \in [a,b]$. We shall use the same methods which were developed in Sections 3.2 and 3.3 based on reducing the problems in question to Volterra integral equations and using the Banach fixed point theorem. The results will be different in the cases $x_0 = a$ and $x_o > a$. We begin with the simpler case.

3.4.1 The Cauchy Type Problem with Initial Conditions at the Endpoint of the Interval. Semi-Global Solution

In this section we give conditions for the existence of a unique continuous solution $y(x)$ to problem (3.2.4)-(3.2.5) in the right neighborhood around point a. Such solutions given in part of the interval $(a, b]$ we will call *semi-global solutions*. From Property 3.1 we derive the first result.

Theorem 3.17 *Let $\alpha > 0$ be such that $n-1 < \alpha \leqq n$ $(n \in \mathbb{N})$, and let $b_k \in \mathbb{R}$ $(k = 1, \cdots, n)$. Let $h \in \mathbb{R}$ be a fixed number such that $0 < h < b - a$, and let $\gamma \in \mathbb{R}$ $(0 \leqq \gamma < 1)$. Let G be an open set in $\mathbb{R}$ and let $f : (a, a+h] \times G \to \mathbb{R}$ be a function such that, for any $y \in G$, $f[x, y] \in C_\gamma[a, b]$ and satisfies the Lipschitz condition* (3.2.15) *for all $x \in (a, a+h]$.*

If $\gamma \geqq n - \alpha$, then there exists a unique solution $y(x)$ to the Cauchy type problem (3.2.4)-(3.2.5) *in the space $\mathbf{C}^\alpha_{n-\alpha,\gamma}[a, a+h]$:*

$$\mathbf{C}^\alpha_{n-\alpha,\gamma}[a, a+h] = \left\{y \in C_{n-\alpha}[a, a+h],\ D^\alpha_{a+}y \in C_\gamma[a, a+h]\right\}. \tag{3.4.1}$$

The next result yields the semi-global solution to the Cauchy type problem (3.2.4)-(3.2.5) in the set $G \subset \mathbb{R}$ involving real numbers $b_k \in \mathbb{R}$ in the initial condition (3.2.5).

Theorem 3.18 *Let $\alpha > 0$ be such that $n - 1 < \alpha \leqq n$ $(n \in \mathbb{N})$, and let $K > 0$, $h_1 > 0$ $(0 < h_1 < b - a)$ and $b_k \in \mathbb{R}$ $(k = 1, \cdots, n)$. Let G be the set given by*

$$G = \left\{(x, y) \in \mathbb{R}^2 : \ a < x \leqq a + h_1;\ \left|(x-a)^{n-\alpha}y - \sum_{j=1}^{n} \frac{b_j(x-a)^{n-j}}{\Gamma(\alpha - j + 1)}\right| \leqq K\right\}. \tag{3.4.2}$$

Let $\gamma \in \mathbb{R}$ be such that $n - \alpha \leqq \gamma < 1$. Let $f[x, y] : G \to \mathbb{R}$ be a function such that $(x-a)^\gamma f[x, y] \in C(\bar{G})$ and the Lipschitz condition is satisfied: for any (x, y_1), $(x, y_2) \in \bar{G}$:

$$|f[x, y_1] - f[x, y_2]| \leqq A|y_1 - y_2| \quad (A > 0), \tag{3.4.3}$$

where the constant $A > 0$ does not depend on x.

Then there exists an h, $0 < h \leqq h_1$, such that the Cauchy type problem (3.2.4)-(3.2.5) *has a unique semi-global solution $y(x)$ in the space $\mathbf{C}^\alpha_{n-\alpha,\gamma}[a, a+h]$ given in* (3.4.1).

Proof. As earlier in Section 3.3, first we prove the existence of a unique solution $y(x) \in C_{n-\alpha}[a, a+h]$. By Property 3.1(b), it is sufficient to prove the existence of such a unique solution to the Volterra equation (3.1.8). We again use the Banach fixed point theorem (Theorem 1.9 in Section 1.13).

Let

$$M_\gamma = \max_{(x,y)\in\bar{G}} |(x-a)^\gamma f(x, y)|. \tag{3.4.4}$$

We choose h such that

$$h = \min\left[h_1, \left(\frac{\Gamma(1+\alpha-\gamma)K}{M_\gamma \Gamma(1-\gamma)}\right)^{1/(n-\gamma)}\right] \tag{3.4.5}$$

and

$$\frac{A\Gamma(\alpha-n+1)}{\Gamma(2\alpha-n+1)}h^\alpha < 1. \tag{3.4.6}$$

Let $U = C_{n-\alpha}[a, a+h]$ be the space (1.1.21):

$$C_{n-\alpha}[a, a+h] = \left\{y(x) \in C(a, a+h] : \; \|(x-a)^{n-\alpha}y(x)\|_{C_{n-\alpha}[a,a+h]} < \infty\right\}.$$

We introduce the subset $C^0_{n-\alpha}[a, a+h]$ of $C_{n-\alpha}[a, a+h]$ as follows:

$$C^0_{n-\alpha}[a, a+h]$$

$$= \left\{y(x) \in C(a, a+h] : \; \max_{x\in[a,a+h]}\left|(x-a)^{n-\alpha}y(x) - \sum_{j=1}^{n}\frac{b_j(x-a)^{n-j}}{\Gamma(\alpha-j+1)}\right| \leqq K\right\}. \tag{3.4.7}$$

It directly follows that $C_{n-\alpha}[a, a+h]$ is the complete metric space with the distance given by

$$d(y_1, y_2) = \|y_1 - y_2\|_{C_{n-\alpha}} = \max_{x\in[a,a+h]}\left|(x-a)^{n-\alpha}[y_1(x) - y_2(x)]\right| \tag{3.4.8}$$

and that $C^0_{n-\alpha}[a, a+h]$ is a complete subset of $C_{n-\alpha}[a, a+h]$.

We define in the space $C^0_{n-\alpha}[a, a+h]$ the operator T by (3.2.18):

$$(Ty)(x) = \sum_{j=1}^{n}\frac{b_j}{\Gamma(\alpha-j+1)}(x-a)^{\alpha-j} + \frac{1}{\Gamma(\alpha)}\int_a^x \frac{f[t, y(t)]dt}{(x-t)^{1-\alpha}}. \tag{3.4.9}$$

Using Theorem 1.9, we shall prove that there exists a unique function $y^*(x) \in C^0_{n-\alpha}[a, a+h]$ such that

$$(Ty^*)(x) = y^*(x). \tag{3.4.10}$$

For this it is sufficient to prove the following: (1) if $y(x) \in C^0_{n-\alpha}[a, a+h]$, then $(Ty)(x) \in C^0_{n-\alpha}[a, a+h]$; and (2) for any $y_1, y_2 \in C_{n-\alpha}[a, a+h]$, the following estimate holds:

$$\|Ty_1 - Ty_2\|_{C_{n-\alpha}} \leqq \omega\|y_1 - y_2\|_{C_{n-\alpha}}, \quad 0 \leqq \omega < 1. \tag{3.4.11}$$

Let $y(x \in C^0_{n-\alpha}[a, a+h]$. Using (3.4.9) and (3.4.4), we have, for any $x \in [a, a+h]$,

$$\left|(x-a)^{n-\alpha}(Ty)(x) - \sum_{j=1}^{n}\frac{b_j(x-a)^{n-j}}{\Gamma(\alpha-j+1)}\right| \leqq \frac{(x-a)^{n-\alpha}}{\Gamma(\alpha)}\int_a^x \frac{|f[t, y(t)]|dt}{(x-t)^{1-\alpha}}$$

$$\leqq M_\gamma \frac{(x-a)^{n-\alpha}\Gamma(1-\gamma)}{\Gamma(1+\alpha-\gamma)}(x-a)^{\alpha-\gamma} \leqq M_\gamma h^{n-\gamma}\frac{\Gamma(1-\gamma)}{\Gamma(+\alpha-\gamma)}.$$

In accordance with (3.4.5), we also have

$$M_\gamma h^{n-\gamma}\frac{\Gamma(1-\gamma)}{\Gamma(1+\alpha-\gamma)} \leqq K,$$

so that

$$\max_{x\in[a,a+h]}\left|(x-a)^{n-\alpha}(Ty)(x) - \sum_{j=1}^{n}\frac{b_j(x-a)^{\alpha-j}}{\Gamma(\alpha-j+1)}\right| \leqq K,$$

and hence $(Ty)(x) \in C^0_{n-\alpha}[a, a+h]$.

Now we prove the estimate (3.4.11). Using (3.4.9), applying the Lipschitzian condition (3.4.3), and taking (3.4.8) into account, we have

$$\left|(x-a)^{n-\alpha}[(Ty_1)(x) - (Ty_2)(x)]\right| \leqq A\left|\frac{(x-a)^{n-\alpha}}{\Gamma(\alpha)}\int_a^x \frac{[y_1(t)] - y_2(t]dt}{(x-t)^{1-\alpha}}\right|$$

$$\leqq A\left|\frac{(x-a)^{n-\alpha}}{\Gamma(\alpha)}\int_a^x \frac{(t-a)^{\alpha-n}dt}{(x-t)^{1-\alpha}}\right| \|y_1 - y_2\|_{C_{n-\alpha}}$$

$$\leqq A\frac{(x-a)^{n-\alpha}(x-a)^{2\alpha-n}\Gamma(\alpha-n+1)}{\Gamma(2\alpha-n+1)}\|y_1 - y_2\|_{C_{n-\alpha}}$$

$$\leqq \frac{A\Gamma(\alpha-n+1)}{\Gamma(2\alpha-n+1)}h^\alpha\|y_1 - y_2\|_{C_{n-\alpha}},$$

and hence

$$\|Ty_1 - Ty_2\|_{C_{n-\alpha}} \leqq \omega\|y_1 - y_2\|_{AC_{n-\alpha}}, \quad \omega = \frac{A\Gamma(\alpha-n+1)}{\Gamma(2\alpha-n+1)}h^\alpha. \tag{3.4.12}$$

According to (3.4.6), $\omega < 1$, and hence, by Theorem 1.9, there exists a unique solution $y^*(x) \in C_{n-\alpha}[a, a+h]$. By Theorem 1.9, this solution is a limit of the convergent sequence $y_m(x) = T^m y_0^*(x)$:

$$\lim_{m\to\infty}\|y_m - y^*\|_{C_{n-\alpha}} = 0, \tag{3.4.13}$$

where $y_0^*(x)$ is any function in $C_{n-\alpha}[a, a+h]$. If at least one $b_k \neq 0$ in the initial condition (3.2.5), we can take $y_0^*(x) = y_0(x)$ with $y_0(x)$ defined by (3.2.19).

Further, it is proved similarly as in Property 3.1, that, if $\gamma \geqq n-\alpha$, then $D^\alpha y \in C_\gamma[a, a+h]$, and hence $y(x) \in \mathbf{C}^\alpha_{n-\alpha,\gamma}[a, a+h]$.

Corollary 3.20 *Let* $\alpha \in \mathbb{R}$ $(0 < \alpha < 1)$, *and let* $K > 0$, $h_1 > 0$ $(0 < h_1 < b-a)$ *and* $b \in \mathbb{R}$. *Let* G *be the set given by*

$$G = \left\{(x,y) \in \mathbb{R}^2 :\ a < x \leqq a+h_1;\ \left|(x-a)^{1-\alpha}y - \frac{b}{\Gamma(\alpha)}\right| \leqq K\right\}. \tag{3.4.14}$$

Let $\gamma \in \mathbb{R}$ be such that $1-\alpha \leqq \gamma < 1$. Let $f[x,y]: G \to \mathbb{R}$ be a function such that $(x-a)^{\gamma} f[x,y] \in C(\bar{G})$ and the Lipschitz condition (3.4.3) *is satisfied.*

Then there exists an h, $0 < h \leqq h_1$, such that the Cauchy type problem (3.1.6) *and the weighted Cauchy type problem* (3.1.7) *have a unique semi-global solution $y(x) \in C^{\alpha}_{1-\alpha,\gamma}[a, a+h]$.*

Remark 3.21 When the conditions of Theorem 3.18 are satisfied with $\gamma = 0$, Diethelm ([173], Theorem 4.1) proved the uniqueness of a solution $y(x) \in C(0,h]$ to the Cauchy type problem (3.2.4)-(3.2.5) by applying Theorem 1.10 (see Section 1.13) to the operator T defined by (3.4.9). The result in Theorem 3.18 provides a more precise characterization of the solution to the problem considered.

Remark 3.22 The results of Theorems 3.17-3.18 and Corollary 3.20 can be extended to systems of Cauchy type problems (3.2.52)-(3.2.53), (3.2.57)-(3.2.58), and (3.2.61)-(3.2.62).

3.4.2 The Cauchy Problem with Initial Conditions at the Inner Point of the Interval. Preliminaries

In Subsection 3.4.1 we considered the Cauchy type problem for the differential equation (3.2.4) with the initial conditions (3.2.5) and with the weighted initial conditions given in (3.1.7). It does not seem to be possible to consider the usual Cauchy conditions at the point a, because a solution $y(x)$ to this problem, in general, has a singularity at a and therefore it is not bounded and continuous at the point a. Such usual conditions can be imposed at any inner point $x_0 \in (a,b)$, because in this case the solution $y(x)$ does not have a singularity at this point.

In this subsection we consider the nonlinear fractional differential equation (3.2.4) with the initial conditions given at the inner point x_0 of a finite interval $[a,b]$:

$$(D^{\alpha}_{a+}y)(x) = f[x, y(x)] \;\; (\alpha > 0;\; x > a) \tag{3.4.15}$$

$$y^{(k)}(x_0) = b_k \in \mathbb{R} \; (a < x_0 < b;\; k = 0, 1, \cdots, n-1), \tag{3.4.16}$$

In particular, for $0 < \alpha < 1$, the simplest such problem has the form

$$(D^{\alpha}_{a+}y)(x) = f[x, y(x)] \; (0 < \alpha < 1), \;\; y(x_0) = b_0 \in \mathbb{R}. \tag{3.4.17}$$

We give the conditions for the existence of a local continuous solution $y(x)$ to this Cauchy problem in a neighborhood around the point x_0.

When $n \in \mathbb{N}$, then, by (2.1.7), the problem (3.4.15)-(3.4.16) is reduced to the Cauchy problem for the ordinary differential equation of order $n \in \mathbb{N}$ on $[a,b]$:

$$y^{(n)}(x) = f[x, y(x)], \;\; y^{(k)}(x_0) = b_k \in \mathbb{R} \, (a < x_0 < b;\; k = 0, 1, \cdots, n-1). \tag{3.4.18}$$

The existence of a local continuous solution $y(x)$ to this problem in a neighborhood of x_0 is well known [see, for example, Erugin [251]]. The proof of this result is based on the equivalence of (3.4.18) to the Volterra integral equation of the second kind of the form (3.2.12):

$$y(x) = \frac{1}{(n-1)!}\int_{x_0}^{x}(x-t)^{n-1}f[t,y(t)]dt + \sum_{j=0}^{n-1}\frac{b_j}{j!}(x-x_0)^j \quad (a \leqq x \leqq b). \tag{3.4.19}$$

Therefore, we shall consider the Cauchy problem (3.4.15)-(3.4.16) for $\alpha \notin \mathbb{N}$ with $n = [\alpha]+1$. We show that this problem, for the case when the point x_0 does not coincide with the endpoint a, can also be reduced to a Volterra integral equation. Such a Volterra integral equation has the form (3.1.8):

$$y(x) = \left(I_{a+}^{\alpha} f[t,y(t)]\right)(x) + \sum_{j=0}^{n-1}\frac{d_j}{\Gamma(\alpha+j-n+1)}\left(\frac{x_0-a}{x-a}\right)^{n-\alpha-j}, \tag{3.4.20}$$

where I_{a+}^{α} is the Riemann-Liouville fractional integration operator (2.1.1), and constants d_j $(j = 0, 1, \cdots, n-1)$ are solutions to the following system of algebraic equations:

$$\sum_{j=0}^{n-1}\frac{d_j}{\Gamma(\alpha+j-n-k+1)} = (x_0-a)^k\left[b_k - \left(I_{a+}^{\alpha-k}f[t,y(t)]\right)(x_0)\right] \tag{3.4.21}$$

$$(k = 0, 1, \cdots, n-1).$$

The determinant Δ_n of this system is given by

$$\Delta_n = \begin{vmatrix} \frac{1}{\Gamma(\alpha-n+1)} & \frac{1}{\Gamma(\alpha-n+2)} & \cdots & \frac{1}{\Gamma(\alpha)} \\ \frac{1}{\Gamma(\alpha-n)} & \frac{1}{\Gamma(\alpha-n+1)} & \cdots & \frac{1}{\Gamma(\alpha-1)} \\ \cdots & \cdots & \cdots & \cdots \\ \frac{1}{\Gamma(\alpha-2n+2)} & \frac{1}{\Gamma(\alpha-2n+3)} & \cdots & \frac{1}{\Gamma(\alpha-n+1)} \end{vmatrix}. \tag{3.4.22}$$

Lemma 3.6 *The following formula holds for the determinant* Δ_n*:*

$$\Delta_n = \frac{0!1!2!3!\cdots(n-1)!}{\Gamma(\alpha)\Gamma(\alpha-1)\cdots\Gamma(\alpha-n+1)} = \prod_{j=0}^{n-1}\frac{j!}{\Gamma(\alpha-j)} \quad (n \in \mathbb{N}), \tag{3.4.23}$$

where $0! = 1$ *and* $j! = 1\cdot 2\cdot 3\cdots j$ $(j \in \mathbb{N})$.

In particular,

$$\Delta_1 = \frac{1}{\Gamma(\alpha)}, \quad \Delta_2 = \frac{1}{\Gamma(\alpha)\Gamma(\alpha-1)}, \quad \Delta_3 = \frac{2}{\Gamma(\alpha)\Gamma(\alpha-1)\Gamma(\alpha-2)}. \tag{3.4.24}$$

Proof. The formula (3.4.23) is proved fairly easily.

Using (3.4.21) and (3.4.23), the constants d_j $(j = 0, 1, \cdots, n-1)$ in (3.4.20) can be found by Cramer's rule.

Lemma 3.7 *The constants d_j $(j = 0, 1, \cdots, n-1)$ in (3.4.20) are given by the following relations:*

$$d_0 = \frac{1}{\Delta_n}$$

$$\cdot \begin{vmatrix} \left[b_0 - \left(I^{\alpha}_{a+} f[t, y(t)]\right)(x_0)\right] & \frac{1}{\Gamma(\alpha-n+2)} & \cdots & \frac{1}{\Gamma(\alpha)} \\ (x_0 - a)\left[b_1 - \left(I^{\alpha-1}_{a+} f[t, y(t)]\right)(x_0)\right] & \frac{1}{\Gamma(\alpha-n+1)} & \cdots & \frac{1}{\Gamma(\alpha-1)} \\ \cdots\cdots\cdots\cdots\cdots\cdots \\ (x_0 - a)^{n-1}\left[b_{n-1} - \left(I^{\alpha-n+1}_{a+} f[t, y(t)]\right)(x_0)\right] & \frac{1}{\Gamma(\alpha-2n+3)} & \cdots & \frac{1}{\Gamma(\alpha-n+1)} \end{vmatrix},$$

$$d_1 = \frac{1}{\Delta_n}$$

$$\cdot \begin{vmatrix} \frac{1}{\Gamma(\alpha-n+1)} & \left[b_0 - \left(I^{\alpha} a+ f[t, y(t)]\right)(x_0)\right] & \cdots & \frac{1}{\Gamma(\alpha)} \\ \frac{1}{\Gamma(\alpha-n)} & (x_0 - a)\left[b_1 - \left(I^{\alpha-1}_{a+} f[t, y(t)]\right)(x_0)\right] & \cdots & \frac{1}{\Gamma(\alpha-1)} \\ \cdots\cdots\cdots\cdots\cdots\cdots \\ \frac{1}{\Gamma(\alpha-2n+2)} & (x_0 - a)^{n-1}\left[b_{n-1} - \left(I^{\alpha-n+1}_{a+} f[t, y(t)]\right)(x_0)\right] & \cdots & \frac{1}{\Gamma(\alpha-n+1)} \end{vmatrix}, \ldots,$$

$$d_{n-1} = \frac{1}{\Delta_n}$$

$$\cdot \begin{vmatrix} \frac{1}{\Gamma(\alpha-n+1)} & \frac{1}{\Gamma(\alpha-n+2)} & \cdots & \left[b_0 - \left(I^{\alpha}_{a+} f[t, y(t)]\right)(x_0)\right] \\ \frac{1}{\Gamma(\alpha-n)} & \frac{1}{\Gamma(\alpha-n+1)} & \cdots & (x_0 - a)\left[b_1 - \left(I^{\alpha-1}_{a+} f[t, y(t)]\right)(x_0)\right] \\ \cdots\cdots\cdots\cdots\cdots\cdots \\ \frac{1}{\Gamma(\alpha-2n+2)} & \frac{1}{\Gamma(\alpha-2n+3)} & \cdots & (x_0 - a)^{n-1}\left[b_{n-1} - \left(I^{\alpha-n+1}_{a+} f[t, y(t)]\right)(x_0)\right] \end{vmatrix}. \tag{3.4.25}$$

In particular, when $0 < \alpha < 1$, then

$$d_0 = b_0 - \left(I^{\alpha}_{a+} f[t, y(t)]\right)(x_0), \tag{3.4.26}$$

and the equation (3.4.20) takes the following form:

$$y(x) = \left(I^{\alpha}_{a+} f[t, y(t)]\right)(x) + \left(\frac{x_0 - a}{x - a}\right)^{1-\alpha} \left[b_0 - \left(I^{\alpha}_{a+} f[t, y(t)]\right)(x_0)\right]. \tag{3.4.27}$$

If $1 < \alpha < 2$, then

$$d_0 = \frac{1}{\Gamma(\alpha - 1)} \left[b_0 - \left(I^{\alpha}_{a+} f[t, y(t)]\right)(x_0)\right] - \frac{x_0 - a}{\Gamma(\alpha)} \left[b_1 - \left(I^{\alpha-1}_{a+} f[t, y(t)]\right)(x_0)\right], \tag{3.4.28}$$

$$d_1 = \frac{x_0 - a}{\Gamma(\alpha - 1)} \left[b_1 - \left(I^{\alpha-1}_{a+} f[t, y(t)]\right)(x_0)\right] - \frac{1}{\Gamma(\alpha - 2)} \left[b_0 - \left(I^{\alpha}_{a+} f[t, y(t)]\right)(x_0)\right], \tag{3.4.29}$$

and (3.4.20) takes the following more complicated form:

$$y(x) = \left(I^{\alpha}_{a+} f[t, y(t)]\right)(x)$$

$$+ \frac{d_0}{\Gamma(\alpha - 1)} \left(\frac{x_0 - a}{x - a}\right)^{2-\alpha} + \frac{d_1}{\Gamma(\alpha)} \left(\frac{x_0 - a}{x - a}\right)^{1-\alpha}, \tag{3.4.30}$$

where d_0 and d_1 are given by (3.4.28) and (3.4.29), respectively.

To prove the equivalence of the Cauchy problem (3.4.15)-(3.4.16) and the Volterra integral equation (3.4.20), we need a preliminary assertion which follows directly from Lemmas 2.9(b) and 2.9(d).

Lemma 3.8 *Let $\alpha > 0$ and $n = -[-\alpha]$. Also let $(I_{a+}^{n-\alpha}g)(x)$ be the Riemann-Liouville fractional integral* (2.1.1).

(a) *If $g(x) \in C(a,b]$, then, at any point $x \in (a,b]$, the following formula holds:*

$$\left(D_{a+}^{\alpha} I_{a+}^{\alpha} g\right)(x) = g(x). \tag{3.4.31}$$

(b) *If $g(x) \in C(a,b]$ and $(I_{a+}^{n-\alpha}g)(x) \in C(a,b]$, then a relation of the form* (2.1.39)*:*

$$(I_{a+}^{\alpha} D_{a+}^{\alpha} g)(x) = g(x) - \sum_{j=0}^{n-1} \frac{(I_{a+}^{n-\alpha}g)^{(j)}(a+)}{\Gamma(\alpha+j-n+1)}(x-a)^{\alpha+j-n} \tag{3.4.32}$$

holds at any point $x \in (a,b]$.

In particular, if $0 < \alpha < 1$, then

$$(I_{a+}^{\alpha} D_{a+}^{\alpha} g)(x) = g(x) - \frac{(I_{a+}^{1-\alpha}g)(a+)}{\Gamma(\alpha)}(x-a)^{\alpha-1}. \tag{3.4.33}$$

3.4.3 Equivalence of the Cauchy Problem and the Volterra Integral Equation

The following statement yields the equivalence of the Cauchy problem (3.4.15)-(3.4.16) and the Volterra integral equation (3.4.20) in a neighborhood of a point $x_0 \in (a,b)$.

Theorem 3.19 *Let $\alpha > 0$ be such that $n-1 < \alpha < n$ $(n \in \mathbb{N})$, and let $b_k \in \mathbb{R}$ $(k = 0,1,\cdots, n-1)$ be given constants. Let $a < x_0 < b$ and $h > 0$ be such that $[x_0-h, x_0+h] \subset (a,b]$, and let U be an open and connected set in $\mathbb{R}$. Let $D = [x_0-h, x_0+h] \times \bar{U}$ and let $f(x,y)$ be a function continuous in D.*

If $y(x) \in C[x_0-h, x_0+h]$, then $y(x)$ satisfies the relations in (3.4.15) *and* (3.4.16) *if, and only if, $y(x)$ satisfies the Volterra integral equation* (3.4.20)*, in which the constants d_j $(j = 0,1,\cdots, n-1)$ are given by* (3.4.25).

Proof. Let $y(x)$ satisfy the relations (3.4.15) and (3.4.16). Since $a < x_0 - h$ and $x_0 + h \leqq b$ and $y(x) \subset C[x_0 - h, x_0 + h]$, we can extend a function $y(x)$ to the continuous function y^* on $(a,b]$:

$$y^*(x) \in C(a,b], \quad y^*(x) = y(x) \quad (x \in [x_0-h, x_0+h]). \tag{3.4.34}$$

Therefore, we can apply Lemma 3.8(b). Using (3.4.32) with $g(t) = y^*(t)$, we have

$$(I_{a+}^{\alpha} D_{a+}^{\alpha} y^*)(x) = y^*(x) - \sum_{j=0}^{n-1} \frac{(I_{a+}^{n-\alpha}y^*)^{(j)}(a+)}{\Gamma(\alpha+j-n+1)}(x-a)^{\alpha+j-n}$$

at any point $x \in (a, b]$. Thus, in accordance with (3.4.34), the relation

$$(I_{a+}^{\alpha} D_{a+}^{\alpha} y)(x) = y(x) - \sum_{j=0}^{n-1} \frac{(I_{a+}^{n-\alpha} y)^{(j)}(a+)}{\Gamma(\alpha + j - n + 1)} (x - a)^{\alpha+j-n} \tag{3.4.35}$$

holds at any point $x \in [x_0 - h, x_0 + h]$.

Applying the operator I_{a+}^{α} to both sides of (3.4.14) and taking (3.4.35) into account, we obtain, for $x \in [x_0 - h, x_0 + h]$,

$$y(x) = \left(I_{a+}^{\alpha} f[t, y(t)]\right)(x) + \sum_{j=0}^{n-1} \frac{(I_{a+}^{n-\alpha} y)^{(j)}(a+)}{\Gamma(\alpha + j - n + 1)} (x - a)^{\alpha+j-n} \quad (x_0 - \delta \leqq x \leqq x_0 + \delta). \tag{3.4.36}$$

Differentiation of this expression $n - 1$ times gives for $k = 0, 1, \cdots, n - 1$

$$y^{(k)}(x_0) = \left(I_{a+}^{\alpha-k} f[t, y(t)]\right)(x_0) + \sum_{j=0}^{n-1} \frac{(I_{a+}^{n-\alpha} y)^{(j)}(a+)}{\Gamma(\alpha + j - n - k + 1)} (x_0 - a)^{\alpha+j-n-k}. \tag{3.4.37}$$

Multiplying both sides of this relation by $(x_0 - a)^k$ and using (3.4.16), we have

$$\sum_{j=0}^{n-1} \frac{(I_{a+}^{n-\alpha} y)^{(j)}(a+)}{\Gamma(\alpha + j - n - k + 1)} (x_0 - a)^{\alpha+j-n}$$

$$= (x_0 - a)^k \left[y^{(k)}(x_0) - \left(I_{a+}^{\alpha-k} f[t, y(t)]\right)(x_0) \right]$$

$$= (x_0 - a)^k \left[b_k - \left(I_{a+}^{\alpha-k} f[t, y(t)]\right)(x_0) \right] \quad (k = 0, 1, \cdots, n - 1). \tag{3.4.38}$$

Setting

$$d_j = (I_{a+}^{n-\alpha} y)^{(j)}(a+)(x_0 - a)^{\alpha+j-n} \quad (j = 0, 1, \cdots, n - 1), \tag{3.4.39}$$

from (3.4.36) and (3.4.38) we obtain the relations (3.4.20) and (3.4.21), respectively. According to Lemma 3.8, the constants d_j $(j = 0, 1, \cdots, n - 1)$ in (3.4.20) are given by (3.4.25). Thus the necessity is proved.

Now let the relation (3.4.20) hold, in which the constants d_j $(j = 0, 1, \cdots, n-1)$ are given by (3.4.25). Then, in accordance with the proof of the necessity, these constants satisfy the system (3.4.21). By the above proof, (3.4.21) is equivalent to (3.4.38), and substitution of (3.4.38) into (3.4.37) yields the relations in (3.4.16). Now we prove (3.4.15). Since $y(x) \in C[x_0 - h, x_0 + h]$ and $f(x, y)$ is a continuous function in $D = [x_0 - h, x_0 + h] \times \bar{U}$, the function $f[x, y(x)]$ is also continuous on $[x_0 - h, x_0 + h]$. Then, similarly as (3.4.34), we can extend $f[x, y(x)]$ to the continuous function $f^*[x, y(x]$ on $(a, b]$:

$$f^*[x, y(x)] \in C(a, b], \quad f^*[x, y^*(x)] = f[x, y(x)] \quad (x \in [x_0 - h, x_0 + h]). \tag{3.4.40}$$

Thus the integral equation (3.4.20) can be extended from $[x_0 - h, x_0 + h]$ to $(a, b]$:

$$y^*(x) = \left(I^\alpha_{a+} f^*[t, y^*(t)]\right)(x) + \sum_{j=0}^{n-1} \frac{d_j}{\Gamma(\alpha + j - n + 1)} \left(\frac{x_0 - a}{x - a}\right)^{n-\alpha-j} \quad (a < x \leqq b). \tag{3.4.41}$$

We can apply Lemma 3.8(a) to the function $g(t) = f^*[t, y^*(t)]$ and obtain, according to (3.4.31),

$$\left(D^\alpha_{a+} I^\alpha_{a+} f^*[t, y^*(t)]\right)(x) = f^*[t, y^*(x)]. \tag{3.4.42}$$

By (2.1.21), we have

$$\left(D^\alpha_{a+} (t - a)^{\alpha+j-n} y\right)(x) = 0 \quad (j = 0, 1, \cdots, n - 1). \tag{3.4.43}$$

Applying the operator D^α_{a+} to both sides of (3.4.41) and using (3.4.42) and (3.4.43), we obtain

$$\left(D^\alpha_{a+} y^*\right)(x) = f^*[x, y^*(x)] \quad (a < x \leqq b).$$

From here, in accordance with (3.4.34) and (3.4.40), we derive (3.4.15), and the sufficiency is proved. This completes the proof of Theorem 3.19.

Corollary 3.21 *Let* $0 < \alpha < 1$ *and* $b_0 \in \mathbb{R}$. *Let* $a < x_0 < b$, *let* $h > 0$ *be such that* $[x_0 - h, x_0 + h] \subset (a, b]$, *and let* U *be an open and connected set in* $\mathbb{R}$. *Let* $D = [x_0 - h, x_0 + h] \times \bar{U}$ *and let* $f(x, y)$ *be a continuous function on* D.

If $y(x) \in C[x_0 - h, x_0 + h]$, *then* $y(x)$ *satisfies the relation in* (3.4.17) *at any point* $x \in [x_0 - h, x_0 + h]$ *if, and only if,* $y(x)$ *satisfies the Volterra integral equation* (3.4.27).

Remark 3.23 The results of Theorem 3.19 and Corollary 3.21 can be extended to systems of the Cauchy type problems (3.2.52)-(3.2.53), (3.2.57)-(3.2.58), and (3.2.61)-(3.2.62).

3.4.4 The Cauchy Problem with Initial Conditions at the Inner Point of the Interval. The Uniqueness of Semi-Global and Local Solutions

Now we establish the existence of a unique semi-global solution $y(x)$ to the Cauchy problem (3.4.15)-(3.4.16) in the neighborhood of an inner point x_0 of (a, b) and of a unique local solution at the point x_0. We begin from the conditions for a unique continuous solution $y(x) \in C[x_0 - h, x_0 + h]$ under the conditions of Theorem 3.20, and an additional Lipschitzian-type condition of $f(x, y)$ with respect to the second variable of the form (3.4.3).

First we derive a unique solution to the Cauchy type problem (3.4.17).

Theorem 3.20 *Let* $0 < \alpha \leqq 1$, *and let* $x_0 \in (a, b)$, $K > 0$, *and* $h_1 > 0$ *be such that* $h_1 < \min[x_0 - a, b - x_0]$, *and let* $b_0 \in \mathbb{R}$. *Let* D *be a set*

$$D = \left\{(x, y) \in \mathbb{R}^2 : \ |x - x_0| \leqq h_1, \ \left|(x - a)^{1-\alpha} y - (x_0 - a)^{1-\alpha} b_0\right| \leqq K\right\}. \tag{3.4.44}$$

Let $f(x,y) : D \to \mathbb{R}$ be a continuous function in D such that the Lipschitz condition (3.4.3) with constant $A > 0$ holds for any (x, y_1), $(x, y_2) \in D$.

Then there exists an h, $0 < h \leqq h_1$, such that the Cauchy type problem

$$(D_{a+}^{\alpha} y)(x) = f[x, y(x)] \ (x \in [x_0 - h, x_0 + h]), \quad y(x_0) = b_0 \in \mathbb{R} \tag{3.4.45}$$

has a unique semi-global solution $y(x) \in C[x_0 - h, x_0 + h]$.

Proof. Since $f(x,y) : D \to \mathbb{R}$ is a continuous function in D, it is bounded in D, and we have

$$M = \|f(x,y)\|_{C(D)} = \max_{(x,y)\in D} |f(x,y)|. \tag{3.4.46}$$

We choose $h > 0$ such that

$$h = \min \left[h_1, \frac{x_0 - 3}{3}, \left(\frac{\Gamma(\alpha+1)K}{4M(x_0 - a)} \right)^{1/\alpha} \right], \quad \frac{4Ah^{\alpha}}{\Gamma(\alpha+1)} < 1. \tag{3.4.47}$$

By Corollary 3.21, it is sufficient to prove the existence of a unique solution $y(x) \in C(J)$ to the corresponding Volterra integral equation of the form (3.4.27):

$$y(x) = \left(I_{a+}^{\alpha} g[t, y(t)]\right)(x) - \left(\frac{x_0 - a}{x - a}\right)^{1-\alpha} \left[y(x_0) - \left(I_{a+}^{\alpha} g[t, y(t)]\right)(x_0)\right]. \tag{3.4.48}$$

For this we use the Banach fixed point (Theorem 1.9 from Section 1.13). We denote $J_h = [x_0 - h, x_0 + h]$. We introduce the subspace $U = C_{h,\alpha}(J)$ of the space $C(J)$:

$$C_{h,\alpha}(J) = \left\{ y(x) \in C(J) : \ \|(x-a)^{1-\alpha} y(x) - (x_0 - a)^{1-\alpha} y(x_0)\|_{C(J)} \leqq K \right\}. \tag{3.4.49}$$

It directly follows that $C_{h,\alpha}(J)$ is a complete metric subspace of $C(J)$ with the distance given by

$$d(y_1, y_2) = \|y_1 - y_2\|_{C(J)} := \max_{x\in J} |y_1(x) - y_2(x)|. \tag{3.4.50}$$

Let T be the operator defined, for $y \in C_{h,\alpha}(J)$, by the left-hand side of (3.4.48):

$$(Ty)(x) = \left(I_{a+}^{\alpha} g[t, y(t)]\right)(x) + \left(\frac{x_0 - a}{x - a}\right)^{1-\alpha} \left[y(x_0) - \left(I_{a+}^{\alpha} g[t, y(t)]\right)(x_0)\right]. \tag{3.4.51}$$

Using Theorem 1.9, we shall prove that there exists a unique function $y^*(x) \in C_{h,\alpha}(J)$ such that

$$(Ty^*)(x) = y^*(x). \tag{3.4.52}$$

It is enough to prove that $(Ty)(x) \in C_{h,\alpha}(J)$ and that $T : C_{h,\alpha}(J) \to C_{h,\alpha}(J)$ satisfies the relation (1.13.2), which, in accordance with (3.4.50), takes the form

$$\|Ty_1 - Ty_2\|_{C(J)} \leqq \omega \|y_1 - y_2\|_{C(J)} \quad (0 \leqq \omega < 1). \tag{3.4.53}$$

First we show that $(Ty)(x) \in C_{h,\alpha}(J)$. Since $y(x) \in C(J)$, then $f[x, y(x)] \in C(J)$ since it is a composition of two continuous functions. Thus, by Lemma 3.3 (with $\gamma = 0$), $\left(I_{a+}^{\alpha} f[t, y(t)]\right)(x) \in C(J)$, and hence $(Ty)(x) \in C(J)$. Now we establish the following estimate:

$$\|(x-a)^{1-\alpha}(Ty)(x) - (x_0-a)^{1-\alpha}(Ty)(x_0)\|_{C(J)} \leqq K. \tag{3.4.54}$$

Using (3.4.51) and (3.4.46), we have, for any $x \in J = [x_0 - \delta, x_0 + \delta]$,

$$|x-a)^{1-\alpha}(Ty)(x) - (x_0-a)^{1-\alpha}(Ty)(x_0)|$$

$$\leqq \frac{M(x-a)^{1-\alpha}}{\Gamma(\alpha)} \int_{x_0-h}^{x} (x-t)^{\alpha-1} dt + \frac{M(x_0-a)^{1-\alpha}}{\Gamma(\alpha)} \int_{x_0-h}^{x_0} (x_0-t)^{\alpha-1} dt$$

$$= \frac{M}{\Gamma(\alpha+1)} \left[(x-a)^{1-\alpha}(x-x_0+h)^{\alpha} + (x_0-a)^{1-\alpha} h^{\alpha}\right].$$

Since $x \leqq x_0 + h$ and, by the first relation in (3.4.47), $h \leqq (x_0 - a)/3$, then

$$\|(x-a)^{1-\alpha}(Ty)(x) - (x_0-a)^{1-\alpha}(Ty)(x_0)\|_{C(J)}$$

$$\leqq \frac{M}{\Gamma(\alpha+1)} \left[(x_0-a+h)^{1-\alpha}(2h)^{\alpha} + (x_0-a)^{1-\alpha} h^{\alpha}\right]$$

$$\leqq \frac{3Mh^{\alpha}}{\Gamma(\alpha+1)}(x_0-a+h)^{1-\alpha} \leqq \frac{4Mh^{\alpha}}{\Gamma(\alpha+1)}(x_0-a)^{1-\alpha}$$

and

$$\|(x-a)^{1-\alpha}(Ty)(x) - (x_0-a)^{1-\alpha}(Ty)(x_0)\|_{C(J)} \leqq \frac{4Mh^{\alpha}(x_0-a)}{\Gamma(\alpha+1)}. \tag{3.4.55}$$

By the first relation in (3.4.47), $\frac{4Mh^{\alpha}(x_0-a)}{\Gamma(\alpha+1)} \leqq K$, and thus (3.4.55) yields (3.4.54), and hence $(Ty)(x) \in C_{h,\alpha}(J)$.

Next we establish the estimate (3.4.53) with

$$\omega = \frac{4Ah^{\alpha}}{\Gamma(\alpha+1)}, \tag{3.4.56}$$

where A is the Lipschitz constant in (3.4.3). Let $y_1(x), y_2(x) \in C_{h,\alpha}(J)$. Using (3.4.51) and the Lipschitz condition (3.4.3), we have, for any $x \in J$,

$$|(Ty_1)(x) - (Ty_2)(x)|$$

$$\leqq A\|y_1 - y_2\|_{C(J)} \left[\frac{1}{\Gamma(\alpha)} \int_{x_0-h}^{x} (x-t)^{\alpha-1} dt\right.$$

$$\left. + \left(\frac{x_0-a}{x-a}\right)^{1-\alpha} \frac{1}{\Gamma(\alpha)} \int_{x_0-h}^{x_0} (x-t)^{\alpha-1} dt\right]$$

$$= \frac{A\|y_1 - y_2\|_{C(J)}}{\Gamma(\alpha+1)} \left[(x - x_0 + h)^\alpha + \left(\frac{x_0 - a}{x - a}\right)^{1-\alpha} h^\alpha\right].$$

Since $x_0 - h \leqq x \leqq x_0 + h$ and $h \leqq (x_0 - a)/3$, then

$$\|(Ty_1)(x) - (Ty_2)(x)\|_{C(J)}$$

$$\leqq \frac{A\|y_1 - y_2\|_{C(J)}}{\Gamma(\alpha+1)} \left[(2h)^\alpha + \left(\frac{x_0 - a}{x_0 - a - h}\right)^{1-\alpha} h^\alpha\right]$$

$$\leqq \frac{Ah^\alpha\|y_1 - y_2\|_{C(J)}}{\Gamma(\alpha+1)} \left[2 + \left(\frac{x_0 - a}{x_0 - a - h}\right)^{1-\alpha}\right]$$

$$\leqq \frac{Ah^\alpha\|y_1 - y_2\|_{C(J)}}{\Gamma(\alpha+1)} \left[2 + \frac{3}{2}\right] \leqq \frac{4Ah^\alpha}{\Gamma(\alpha+1)}\|y_1 - y_2\|_{C(J)}.$$

This proves the relation (3.4.53) with ω given in (3.4.56).

According to the second formula in (3.4.47), $\omega < 1$, and hence, by Theorem 1.9, there exists a unique function $y^*(x) \in C_{h,\alpha}(J)$ such that the formula (3.4.52) holds. Thus $y(x) = y^*(x)$ is a unique solution to the Volterra integral equation (3.4.52), and hence, in accordance with (3.4.51), also to the integral equation (3.4.27).

From Theorem 3.20 we derive the existence of a unique continuous local solution to the Cauchy problem (3.4.18).

Theorem 3.21 *Let $0 < \alpha \leqq 1$. Let U be an open and connected set in $\mathbb{R}$ and $\Omega = (a,b) \times U$. Let $f(x,y) : (a,b) \times U \to \mathbb{R}$ be a continuous function satisfying the Lipschitz condition* (3.4.3).

Then, for any $(x_0, y_0) \in \Omega$, there exists a positive number $h > 0$ such that $[x_0 - h, x_0 + h] \subset (a,b)$ and there exists a unique function $y(x) : [x_0 - h, x_0 + h] \to U$ such that $y(x) \in C[x_0 - h, x_0 + h]$ and

$$y(x_0) = y_0, \tag{3.4.57}$$

$$(D^\alpha_{a+}y)(x) = f[x, y(x)] \quad for \ any \quad x \in [x_0 - h, x_0 + h]. \tag{3.4.58}$$

Proof. Let $P = (x_0, y_0) \in \Omega$ be any point on Ω. We fix this point P and choose $h_1 > 0$ and $K > 0$, depending on P and such that $h_1 < \min[x_0 - a, b - x_0]$, and denote by D the closed set (3.4.44). Since $f(x,y) : (a,b) \times U \to \mathbb{R}$ is a continuous function satisfying the Lipschitz condition (3.4.3) on Ω, this function is also continuous and satisfies the Lipschitz condition on D. Then, on the basis of Theorem 3.20, there exists a positive number $h > 0$, depending on the point P, and there exists a unique continuous function $y(x) \in C[x_0 - h, x_0 + h]$ such that the relations (3.4.57) and (3.4.58) hold. Since (x_0, y_0) is any point on Ω, at any such point there exists a local solution $y(x)$ to the Cauchy problem (3.4.57)-(3.4.58).

The next result, generalizing Theorem 3.20, gives conditions for a unique semi-global solution to the Cauchy problem (3.4.15)-(3.4.16).

Theorem 3.22 *Let $\alpha > 0$ be such that $n-1 < \alpha \leqq n$ $(n \in \mathbb{N})$. Let $x_0 \in (a, b)$, $K > 0$, and $h_1 > 0$ be such that $h_1 < \min[x_0 - a, b - x_0]$, and let $b_k \in \mathbb{R}$ $(k = 0, 1, \cdots, n-1)$. Let D be a set given by*

$$D = \left\{(x, y) \in \mathbb{R}^2 : \ |x - x_0| \leqq h_1, \ \left|(x-a)^{n-\alpha} y - (x_0 - a)^{n-\alpha} b_0\right| \leqq K\right\}. \tag{3.4.59}$$

Let $f(x, y) : D \to \mathbb{R}$ be a continuous function in D such that the Lipschitz condition (3.4.3) with the constant $A > 0$ holds for any (x, y_1), $(x, y_2) \in D$.

Then there exists an h, $0 < h \leqq h_1$, such that the Cauchy type problem

$$(D^\alpha_{a+}y)(x) = f[x, y(x)] \ (x \in [x_0 - h, x_0 + h]), \ \ y^{(k)}(x_0) = b_k \in \mathbb{R} \ (k = 0, 1, \cdots, n-1) \tag{3.4.60}$$

has a unique solution $y(x) \in C^r[x_0 - h, x_0 + h]$.

Proof. When $\alpha = n \in \mathbb{N}$, this theorem is well known. If $\alpha \notin \mathbb{N}$, Theorem 3.22 is proved similarly as Theorem 3.20. Indeed, on the basis of Theorem 3.19, it is sufficient to prove that there exists a unique solution $y(x) \in C^{n-1}[x_0 - h, x_0 + h]$ to the integral equation (3.4.20). For this we use the fixed point theorem by introducing the subspace $U = C^{n-1}_{h,\alpha}(J)$ $(J = [x_0 - h, x_0 + h])$ of the space $C^{n-1}(J)$:

$$C^{n-1}_{h,\alpha}(J) = \left\{y(x) \in C^{n-1}(J) : \ \sum_{k=0}^{n-1} \|(x-a)^{n+k-\alpha} y^{(k)}(x) \right.$$

$$\left. -(x_0 - a)^{n+k-\alpha} y^{(k)}(x_0)\|_{C(J)} \leqq K\right\} \tag{3.4.61}$$

and taking for T the operator in the right-hand side of (3.4.20):

$$(Ty)(x) = \left(I^\alpha_{a+} f[t, y(t)]\right)(x) + \sum_{j=0}^{n-1} \frac{d_j}{\Gamma(\alpha + j - n + 1)} \left(\frac{x_0 - a}{x - a}\right)^{n-\alpha-j}. \tag{3.4.62}$$

From Theorem 3.22 we derive the existence of a unique continuous local solution to the Cauchy problems (3.4.15)-(3.4.16).

Theorem 3.23 *Let $n-1 < \alpha \leqq n$ $(n \in \mathbb{N})$, and let $b_k \in \mathbb{R}$ $(k = 0, 1, \cdots, n-1)$ be given constants. Let U be an open and connected set in $\mathbb{R}$ and $\Omega = (a, b) \times U$. Let $f(x, y) : (a, b) \times U \to \mathbb{R}$ be a continuous function satisfying the Lipschitz condition (3.4.3).*

Let $r = n$ for $\alpha = n \in \mathbb{N}$ and $r = n - 1$ for $\alpha \notin \mathbb{N}$. Then, for any $(x_0, y_0) \in \Omega$, there exists a positive number $h > 0$ such that $[x_0 - h, x_0 + h] \subset (a, b)$, and there exists a unique function $y(x) : [x_0 - h, x_0 + h] \to U$ such that $y(x) \in C^r[x_0 - h, x_0 + h]$ and

$$y(x_0) = b_0, \ \ y'(x_0) = b_1, \ \ \cdots, \ \ y^{(n-1)}(x_0) = b_{n-1}, \tag{3.4.63}$$

$$(D^\alpha_{a+}y)(x) = f[x, y(x)] \ \ for \ \ any \ \ x \in [x_0 - h, x_0 + h]. \tag{3.4.64}$$

Proof. Theorem 3.23 is proved similarly as Theorem 3.21 by using Theorem 3.22.

Corollary 3.22 *Let $n-1 < \alpha \leqq n$ $(n \in \mathbb{N})$. Let $x_0 \in (a,b)$ and $b_k \in \mathbb{R}$ $(k = 0, 1, \cdots, n-1)$. Let $a(x), g(x) \in C(a,b)$ and $\beta \geqq 0$. Let U be an open and connected set in $\mathbb{R}$ and let $f(x,y) : (a,b) \times U \to \mathbb{R}$ be a continuous function.*

Let $r = n$ for $\alpha = n \in \mathbb{N}$ and $r = n-1$ for $\alpha \notin \mathbb{N}$. Then, for any $(x_0, y_0) \in \Omega$, there exists a positive number $h > 0$ such that $[x_0 - h, x_0 + h] \subset (a,b)$ and there exists a unique function $y(x) : [x_0 - h, x_0 + h] \to U$ such that

$$y(x) \in C^r[x_0 - h, x_0 + h]; \quad \text{and} \quad y^{(k)}(x_0) = b_k \in \mathbb{R} \; (k = 0, 1, \cdots, n-1), \tag{3.4.65}$$

$$(D^\alpha_{a+}y)(x) = (x-a)^\beta a(x) y(x) + g(x) \; (x \in [x_0 - h, x_0 + h]). \tag{3.4.66}$$

In particular, if $0 < \alpha < 1$, then there exists a unique continuous function $y(x) \in C[x_0 - h, x_0 + h]$ such that $y(x_0) = b \in \mathbb{R}$, and

$$(D^\alpha_{a+}y)(x) = (x-a)^\beta a(x) y(x) + g(x) \; (x \in [x_0 - h, x_0 + h]). \tag{3.4.67}$$

Remark 3.24 The space $C_{h,\alpha}(J)$ $(J = [x_0 - h, x_0 + h])$ with $0 < \alpha < 1$ in (3.4.49) characterizes the behavior of the unique solution $y(x)$ to the Cauchy problem (3.4.17) near the point a in the case when h is chosen in such a way that x_0 is near the point a. When $0 < \alpha < 1$ and $\alpha \to 1$, then (3.4.49) tends to the space $S_{h,1}(J) = S_h(J)$:

$$C_h(J) = \left\{ y(x) \in C(J) : \; ||y(x) - (x_0)||_{C(J)} \leqq K \right\}. \tag{3.4.68}$$

Such a space is usually applied in the proof of a unique solution to the Cauchy problem (3.4.18) for the ordinary differential equation of order $n \in \mathbb{N}$.

Similarly, the space $C^{n-1}_{h,\alpha}(J)$ in (3.4.61) characterizes the behavior of the unique solution $y(x)$ to the Cauchy problem (3.4.15)-(3.4.16) and of its derivatives $y^{(k)}(x)$ $(k = 1, \cdots, n-1)$ near the point a for the case when h is chosen in such a way that x_0 is near the point a.

Remark 3.25 The results of Theorems 3.20-3.23 can be extended to the systems of the Cauchy type problems (3.2.52)-(3.2.53), (3.2.57)-(3.2.58), and (3.2.61)-(3.2.62).

The results of Theorems 3.20-3.23 can also be extended to the vectorial case when a function $y(x)$ is replaced by a vector-function $y = (y_1, y_2, \cdots, y_m)$. Such an extension of Theorem 3.23 in the case $0 < \alpha < 1$ was established by Hayek et al. ([335], Theorem 3.1).

3.4.5 A Set of Examples

In Sections 3.4.1 and 3.4.4 we gave sufficient conditions for a unique semi-global continuous solution of Cauchy type problems with initial conditions at the end and inner points of a finite interval, respectively. Below we present two examples. The first one generalizes Examples 3.1 considered in Section 3.2.6.

Example 3.5 Consider the following Cauchy type problems for the differential equation of fractional order $0 < \alpha < 1$:

$$(D_{a+}^{\alpha}y)(x) = \lambda(x-a)^{\beta}[y(x)]^{m} + g(x) \ (x > a), \quad (I_{a+}^{1-\alpha}y)(a+) = b \in \mathbb{R}, \tag{3.4.69}$$

$$(D_{a+}^{\alpha}y)(x) = \lambda(x-a)^{\beta}[y(x)]^{m} + g(x) \ (x > a), \quad \lim_{x\to a+}\left[(x-a)^{1-\alpha}y(x)\right] = b \in \mathbb{R}, \tag{3.4.70}$$

with real $m > 0$ ($m \neq 1$), $\lambda \in \mathbb{R}$ ($\lambda \neq 0$), $\beta \in \mathbb{R}$ such that $\beta \geqq m(1-\alpha)$. (3.4.69) and (3.4.70) are Cauchy type problem (3.1.6) and (3.1.7), respectively, with

$$f(x,y) = \lambda(x-a)^{\beta}y^{m} + g(x). \tag{3.4.71}$$

Let $K > 0$, $h_1 > 0$ ($0 < h_1 < b-a$) and G be the set (3.4.14). Also let $g(x) \in C_{\gamma}[a, h_1]$ with $1-\alpha \leqq \gamma < 1$. According to (3.4.14), for $(x,y) \in G$, we have

$$|y| \leqq A(x-a)^{\alpha-1}, \quad A = \frac{|b|}{\Gamma(\alpha)} + K \quad ((x,y) \in G). \tag{3.4.72}$$

By (3.4.71) and (3.4.72), $|f(x,y)| \leqq |\lambda|A^{m}(x-a)^{\beta+m(\alpha-1)} + |g(x)|$ for any $(x,y) \in G$, and hence, since $\beta \geqq m(1-\alpha)$, $f(x,y)$ is bounded on $\bar{G}$,

$$|f(x,y)| \leqq M, \quad M = |\lambda|A^{m}(b-a)^{\beta+m(\alpha-1)} + \|g\|_{C} \quad \left((x,y) \in \bar{G}\right). \tag{3.4.73}$$

We note that, according to (3.4.72), $|y| \leqq A(x-a)^{\alpha-1}$. Using this relation and the formulas (3.3.62) and (3.3.63) in respective cases $m > 1$ and $0 < m < 1$ it is directly verified that the Lipschitz condition (3.4.3) holds for the $f(x,y)$ given by (3.4.71). Thus, from Corollary 3.20, we derive the result.

Proposition 3.5 *Let $0 < \alpha < 1$, $m > 0$ ($m \neq 1$) and $\beta \in \mathbb{R}$ be such that $\beta \geq m(1-\alpha)$, and let $\lambda \in \mathbb{R}$ ($\lambda \neq 0$) and $b \in \mathbb{R}$. Let $K > 0$, $h_1 > 0$ ($0 < h_1 < b-a$), and let G be the set* (3.4.14) *and $g(x) \in C_{\gamma}[a, a+h_1]$ with $1-\alpha \leqq \gamma < 1$.*

Then there exists an h, $0 < h \leqq h_1$, such that the Cauchy type problem (3.4.69) *and the weighted Cauchy type problem* (3.4.70) *have a unique semi-global solution $y(x)$ in the space $\mathbf{C}_{1-\alpha,\gamma}^{\alpha}[a, a+h]$.*

In particular, the Cauchy type problems (3.4.69) *and* (3.4.70) *with $g(x) = 0$:*

$$(D_{a+}^{\alpha}y)(x) = \lambda(x-a)^{\beta}[y(x)]^{m} \ (x > a), \quad (I_{a+}^{1-\alpha}y)(a+) = b \in \mathbb{R}, \tag{3.4.74}$$

$$(D_{a+}^{\alpha}y)(x) = \lambda(x-a)^{\beta}[y(x)]^{m} \ (x > a), \quad \lim_{x\to a+}\left[(x-a)^{1-\alpha}y(x)\right] = b \in \mathbb{R}, \tag{3.4.75}$$

have a unique semi-global solution $y(x) \in \mathbf{C}_{1-\alpha,\gamma}^{\alpha}[a, a+h]$.

Remark 3.26 It was proved in Example 3.3 that the homogeneous differential equation (3.3.36) has the explicit solution

$$y(x) = c(x-a)^{\omega}, \quad \omega = \frac{\beta+\alpha}{1-m}, \quad c = \left[\frac{\Gamma(\omega+1)}{\lambda\Gamma(\omega-\alpha)}\right]^{1/(m-1)}. \tag{3.4.76}$$

Applying the same arguments as in the proof of Proposition 3.1, we obtain the following result.

Proposition 3.6 *Let $0 < \alpha < 1$, $0 < m < 1$ and $\beta \in \mathbb{R}$ be such that $\beta \geqq m(1-\alpha)$, and let $\lambda \in \mathbb{R}$ $(\lambda \neq 0)$ and $b \in \mathbb{R}$. Let $K > 0$, $h_1 > 0$ $(0 < h_1 < b-a)$, let G be the set* (3.4.14), *and let $1-\alpha \leqq \gamma < 1$.*

Then there exists an h, $0 < h \leqq h_1$, such that the Cauchy type problem (3.4.74) *and the weighted Cauchy type problem* (3.4.75) *have a unique semi-global solution $y(x) \in \mathbf{C}^{\alpha}_{1-\alpha,\gamma}[a, a+h]$ and this solution is given by* (3.4.76).

Example 3.6 Consider the Cauchy problem for the following differential equation of fractional order $(0 < \alpha < 1)$:

$$(D^{\alpha}_{a+}y)(x) = \lambda(x-a)^{\beta}[y(x)]^m + g(x) \ (x > a), \quad y(x_0) = b_0 \in \mathbb{R} \ (a < x_0 < b), \tag{3.4.77}$$

where $m > 0$ $(m \neq 1)$, $\lambda \in \mathbb{R}$ $(\lambda \neq 0)$ and $\beta \in \mathbb{R}$ such that $\beta \geqq m(1-\alpha)$. (3.4.77) is Cauchy type problem (3.4.45) with $f(x,y) = \lambda(x-a)^{\beta}[y(x)]^m + g(x)$ given in (3.4.71). Let $K > 0$ and $h_1 > 0$ be such that $h_1 < \min[x_0 - a, b - x_0]$, let D be the set (3.4.44), and let $g(x) \in C[x_0 - h_1, x_0 + h_1]$. It is clear that $f(x,y) \in C(D)$. By (3.4.44), similarly as (3.4.73), it is shown that $f(x,y)$ is bounded on D: for any $(x,y) \in D$,

$$|f(x,y)| \leqq M, \quad M = |\lambda| A_1^m (b-a)^{\beta+m(\alpha-1)} + \|g\|_C, \ A_1 = |b_0|(x-x_0)^{\alpha-1} + K. \tag{3.4.78}$$

Applying (3.4.78) and using (3.3.62) and (3.3.63) in respective cases $m > 1$ and $0 < m < 1$, it is directly verified that the right-hand side of (3.4.77) satisfies the Lipschitz condition (3.4.3). Then Theorem 3.20 yields the following assertion.

Proposition 3.7 *Let $0 < \alpha < 1$, $m > 0$ $(m \neq 1)$ and $\beta \in \mathbb{R}$ be such that $\beta \geqq m(1-\alpha)$, and let $\lambda \in \mathbb{R}$ $(\lambda \neq 0)$ and $b_0 \in \mathbb{R}$. Let $K > 0$ and $h_1 > 0$ be such that $h_1 < \min[x_0 - a, b - x_0]$, and let D be the set* (3.4.44) *and $g(x) \in C[x_0 - h_1, x_0 + h_1]$.*

Then there exists an h, $0 < h \leqq h_1$, such that the Cauchy problem (3.4.77) *has a unique semi-global solution $y(x) \in C[x_0 - h, x_0 + h]$.*

In particular, the Cauchy problem

$$(D^{\alpha}_{a+}y)(x) = \lambda(x-a)^{\beta}[y(x)]^m \ (x > a), \quad y(x_0) = b_0 \in \mathbb{R} \ (a < x_0 < b) \tag{3.4.79}$$

has a unique semi-global solution $y(x) \in C[x_0 - h, x_0 + h]$.

3.5 Equations with the Caputo Derivative in the Space of Continuously Differentiable Functions

In Sections 3.3 and 3.4 we gave the conditions for the existence of unique global and semi-global and local continuous solutions for the Cauchy type problems (3.2.4)-(3.2.5), (3.1.6), (3.1.7) and (3.4.15)-(3.4.16), and (3.4.17) when the initial conditions are given at the Endpoint a of a finite interval $[a, b]$ and at any point

$x_0 \in (a,b)$, respectively. Here we consider similar problems for the nonlinear differential equations with the Caputo derivative $({}^C D_{a+}^{\alpha} y)(x)$ defined in (2.4.1). We shall use the same methods which were developed in Sections 3.4 based on reducing the problems considered to Volterra integral equations and using the Banach fixed point theorem. The results will be different for the cases $x_0 = a$ and $x_o > a$. We begin with the simpler case.

3.5.1 The Cauchy Problem with Initial Conditions at the Endpoint of the Interval. Global Solution

In this section we consider the nonlinear differential equation of order $\alpha > 0$:

$$({}^C D_{a+}^{\alpha} y)(x) = f[x, y(x)] \quad (\alpha > 0;\ a \leqq x \leqq b), \tag{3.5.1}$$

involving the Caputo fractional derivative $({}^C D_{a+}^{\alpha} y)(x)$, defined in (2.4.1), on a finite interval $[a,b]$ of the real axis $\mathbb{R}$, with the initial conditions

$$y^{(k)}(a) = b_k,\ b_k \in \mathbb{R}\ (k = 0, 1, \cdots, n-1;\ n = -[-\alpha]). \tag{3.5.2}$$

We give the conditions for a unique solution $y(x)$ to this problem in the space $\mathbf{C}_{\gamma}^{r}[a,b]$ defined for $\alpha > 0$, $r \in \mathbb{N}$ and $\gamma \in \mathbb{R}$ $(0 \leqq \gamma < 1)$ by

$$\mathbf{C}_{\gamma}^{\alpha,r}[a,b] = \left\{ y(x) \in C^r[a,b] :\ {}^C D_{a+}^{\alpha} y \in C_{\gamma}[a,b] \right\}, \quad \mathbf{C}_{\gamma}^{r,r}[a,b] = C_{\gamma}^{r}[a,b]. \tag{3.5.3}$$

As in Sections 3.2-3.4, our investigations are based on reducing the problem considered to the Volterra integral equation

$$y(x) = \sum_{j-0}^{n-1} \frac{b_j}{j!}(x-a)^j + \frac{1}{\Gamma(\alpha)} \int_a^x \frac{f[t, y(t)]dt}{(x-t)^{1-\alpha}} \quad (a \leqq x \leqq b). \tag{3.5.4}$$

When $\alpha \in \mathbb{N}$, then, in accordance with (2.4.14), $({}^C D_{a+}^{n} y)(x) = y^{(n)}(x)$ for a suitable function $y(x)$, and hence (3.5.1)-(3.5.2) is the Cauchy problem for the ordinary differential equation of order $n \in \mathbb{N}$:

$$y^{(n)}(x) = f[x, y(x)]\ (a \leqq x \leqq b), \quad y^{(k)}(a) = b_k \in \mathbb{R}\ (k = 0, 1, \cdots, n-1). \tag{3.5.5}$$

The corresponding integral equation (3.5.3) takes the form

$$y(x) = \sum_{j=0}^{n-1} \frac{b_j}{j!}(x-a)^j + \frac{1}{(n-1)!} \int_a^x (x-t)^{n-1} f[t, y(t)]dt \quad (a \leqq x \leqq b). \tag{3.5.6}$$

First we establish an equivalence between problem (3.5.1)-(3.5.2) and the integral equation (3.5.4) in the space $C^r[a,b]$ of continuously differentiable functions.

Theorem 3.24 *Let $\alpha > 0$ and $n = -[-\alpha]$. Let G be an open set in $\mathbb{C}$ and let $f : (a,b] \times G \to \mathbb{C}$ be a function such that, for any $y \in G$, $f[x,y] \in C_{\gamma}[a,b]$ with $0 \leqq \gamma < 1$ and $\gamma \leqq \alpha$. Let $r = n$ for $\alpha \in \mathbb{N}$ and $r = n-1$ for $\alpha \notin \mathbb{N}$.*

If $y(x) \in C^r[a,b]$, then $y(x)$ satisfies the relations (3.5.1) *and* (3.5.2) *if, and only if, $y(x)$ satisfies the Volterra integral equation* (3.5.4).

Proof. First we prove the necessity. Let $\alpha = n \in \mathbb{N}$ and $y(x) \in C^n[a,b]$ be the solution to the Cauchy problem (3.5.5). Applying the operator I^n_{a+} to the first relation in (3.5.5) and taking into account (2.1.41) and the second relation in (3.5.5), we arrive at the integral equation (3.5.6). Inversely, if $y(x) \in C^n[a,b]$ satisfies (3.5.6), then, by term-by-term differentiation of (3.5.6), we have

$$y^{(k)}(x) = \sum_{j=k}^{n-1} \frac{b_j}{(k-j)!}(x-a)^{k-j} + \frac{1}{(n-k)!}\int_a^x (x-t)^{n-k-1} f[t,y(t)]dt \quad (3.5.7)$$

for $k = 1, \cdots, n-1$. Taking the limit as $x \to a$, and taking into account the continuity of the integrands in (3.5.6) and (3.5.7), we obtain $y^{(k)}(a) = b_k$ for $k = 0, 1, \cdots, n-1$. Differentiating the equation (3.5.6) n times, we obtain $y^{(n)}(x) = f[x, y(x)]$. Thus Theorem 3.24 is proved for $\alpha \in \mathbb{N}$.

Let now $n-1 < \alpha < n$ and $y(x) \in C^{n-1}[a,b]$. According to (2.4.1) and (2.1.5),

$$({}^C D^\alpha_{a+} y)(x) = \left(\frac{d}{dx}\right)^n \left(I^{n-\alpha}_{a+}\left[y(t) - \sum_{j=0}^{n-1}\frac{y^{(j)}(a)}{j!}(t-a)^j\right]\right)(x).$$

By the hypotheses of Theorem 3.24, $f[x,y] \in C_\gamma[a,b]$, and it follows from (3.5.1) that $({}^C D^\alpha_{a+} y)(x) \in C_\gamma[a,b]$, and hence, by Lemma 1.3,

$$\left(I^{n-\alpha}_{a+}\left[y(t) - \sum_{j=0}^{n-1}\frac{y^{(j)}(a)}{j!}(t-a)^j\right]\right)(x) \in C^n_\gamma[a,b].$$

Applying Lemma 2.9(d) to $g(t) = y(t) - \sum_{j=0}^{n-1}\frac{y^{(j)}(a)}{j!}(t-a)^j$, we obtain

$$\left(I^\alpha_{a+}\, {}^C D^\alpha_{a+} y\right)(x) = \left(I^\alpha_{a+}\, {}^C D^\alpha_{a+}\left[y(t) - \sum_{j=0}^{n-1}\frac{y^{(j)}(a)}{j!}(t-a)^j\right]\right)(x)$$

$$= y(x) - \sum_{j=0}^{n-1}\frac{y^{(j)}(a)}{j!}(x-a)^j - \sum_{k=1}^{n}\frac{y^{(n-k)}_{n-\alpha}(a+)}{\Gamma(\alpha-k+1)}(x-a)^{\alpha-k}, \quad (3.5.8)$$

where

$$y_{n-\alpha}(x) = \left(I^{n-\alpha}_{a+}\left[y(t) - \sum_{j=0}^{n-1}\frac{y^{(j)}(a)}{j!}(t-a)^j\right]\right)(x). \quad (3.5.9)$$

Integrating (3.5.9) by parts, differentiating the obtained expression, and using the first formula in (2.1.33) with $k = 1$, we have

$$y'_{n-\alpha}(x) = \frac{d}{dx}\left(I^{n-\alpha}_{a+}\left[y(t) - \sum_{j=0}^{n-1}\frac{y^{(j)}(a)}{j!}(t-a)^{j-1}\right]\right)(x)$$

$$= \left(I_{a+}^{n-\alpha} \left[y'(t) - \sum_{j=1}^{n-1} \frac{y^{(j)}(a)}{j!}(t-a)^j \right] \right)(x). \qquad (3.5.10)$$

Repeating this process $n - k$ times ($k = 1, \cdots, n$), we arrive at the following relation:

$$y_{n-\alpha}^{(n-k)}(x) = \left(I_{a+}^{n-\alpha} \left[y^{(n-k)}(t) - \sum_{j=n-k}^{n-1} \frac{y^{(j)}(a)}{j!}(t-a)^{j-n+k} \right] \right)(x). \qquad (3.5.11)$$

Making the change of variable $t = a + s(x - a)$, we obtain, for $k = 1, \cdots, n$,

$$y_{n-\alpha}^{(n-k)}(x) = \frac{(x-a)^{n-\alpha}}{\Gamma(n-\alpha)} \int_0^1 (1-s)^{n-\alpha-1} \left[y^{(n-k)}[s + a((x-a)] \right.$$

$$\left. - \sum_{j=n-k}^{n-1} \frac{y^{(j)}(a)}{j!}[s(x-a)]^{j-n+k} \right]. \qquad (3.5.12)$$

Since $\alpha < n$ and $y^{(n-k)}(x) \in C[a, b]$ for $k = 1, \cdots, n$, then the last relations yield $y^{(n-k)}(a+) = 0$ ($k = 1, \cdots, n$), and hence (3.5.8) takes the form

$$\left(I_{a+}^{\alpha} \, {}^{C}D_{a+}^{\alpha} y \right)(x) = y(x) - \sum_{j=0}^{n-1} \frac{y^{(j)}(a)}{j!}(x-a)^j. \qquad (3.5.13)$$

Since $f[x, y] \in C_\gamma[a, b]$ and $\gamma \leqq \alpha$, by Lemma 2.8(b),$I_{a+}^{\alpha}[t, y(t)])(x) \in C[a, b]$. Applying the operator I_{a+}^{α} to both sides of (3.5.1) and using (3.5.13) and the initial conditions (3.5.2), we find that $y(x) \in C^{n-1}[a, b]$ is the solution to the integral equation (3.5.4), and thus the necessity is proved.

Now let $y(x) \in C^{n-1}[a, b]$ be the solution to the integral equation (3.5.4). Let us first show that $y(x)$ satisfies the initial conditions (3.5.2). Differentiating both sides of (3.5.4) and taking (2.1.33) into account, for all $k = 1, \cdots, n-1$, we have

$$y^{(k)}(x) = \sum_{j=k}^{n-1} \frac{b_j}{(j-k)!}(x-a)^{j-k} + \frac{1}{\Gamma(\alpha-k)} \int_a^x \frac{f[t, y(t)]dt}{(x-t)^{1-\alpha+k}}. \qquad (3.5.14)$$

Making the change of variable $t = a + s(x - a)$ in the integrals in (3.5.4) and (3.5.14), for all $k = 1, \cdots, n-1$, we find that

$$y^{(k)}(x) = \sum_{j=k}^{n-1} \frac{b_j}{(j-k)!}(x-a)^{j-k}$$

$$+ \frac{(x-a)^{\alpha-k}}{\Gamma(\alpha-k)} \int_0^1 \frac{f[a + s(x-a), y(a + s(x-a))]dt}{(x-t)^{1-\alpha+k}}.$$

Since $\alpha > n - 1$ and $f[x, y(x)]$ is continuous, the integrands in these relations are continuous, and by taking the limit as $x \to a+$, we obtain relation (3.5.2).

Now we show that $y(x)$ satisfies the equation (3.5.1). Applying the operator D^{α}_{a+} to (3.5.4), taking (3.5.2) and (2.4.2) into account and using the first relation in (2.1.31), we arrive at the equation (3.5.1). Thus Theorem 3.24 is proved for $\alpha \notin N$.

Corollary 3.23 *Let $n \in \mathbb{N}$, let G be an open set in $\mathbb{R}$ and let $f : (a,b] \times G \to \mathbb{R}$ be a function such that, for any $y \in G$, $f[x,y] \in C_{\gamma}[a,b]$ with $0 \leqq \gamma < 1$.*

If $y(x) \in C^n[a,b]$, then $y(x)$ satisfies the relation (3.5.5) if, and only if, $y(x)$ satisfies the Volterra integral equation (3.5.6).

Corollary 3.24 *Let $0 < \alpha < 1$ and $0 \leqq \gamma \leqq \alpha$. Let G be an open set in $\mathbb{R}$ and let $f : (a,b] \times G \to \mathbb{R}$ be a function such that, for any $y \in G$, $f[x,y] \in C_{\gamma}[a,b]$.*

If $y(x) \in C[a,b]$, then $y(x)$ satisfies the relations

$$({}^C D^{\alpha}_{a+} y)(x) = f[x, y(x)], \quad y(a) = b_0 \in \mathbb{C}, \tag{3.5.15}$$

if, and only if, $y(x)$ satisfies the Volterra integral equation

$$y(x) = b_0 + \frac{1}{\Gamma(\alpha)} \int_a^x \frac{f[t, y(t)]dt}{(x-t)^{1-\alpha}} \quad (a \leqq x \leqq b). \tag{3.5.16}$$

Now we establish the existence and uniqueness of the solution to the Cauchy problem (3.5.1)-(3.5.2) in the space $\mathbf{C}^{\alpha,n-1}_{\gamma}[a,b]$ under the conditions of Theorem 3.24 and an additional Lipschitzian condition (3.2.15). For this we need the preliminary assertion similar to Lemma 3.4.

Lemma 3.9 *Let $n \in \mathbb{N}$, $a < c < b$, $g \in C^n[a,c]$, and $g \in C^n[c,b]$. Then $g \in C^n[a,b]$ and*

$$||g||_{C^n[a,b]} \leqq \max \left[||g||_{C^n[a,c]}, ||g||_{C^n[c,b]} \right] . \tag{3.5.17}$$

Theorem 3.25 *Let $\alpha > 0$ and $n = -[-\alpha]$, and let $0 \leqq \gamma < 1$ and $\gamma \leqq \alpha$. Let G be an open set in $\mathbb{C}$ and let $f : (a,b] \times G \to \mathbb{C}$ be a function such that, for any $y \in G$, $f[x,y] \in C_{\gamma}[a,b]$ and the Lipschitz condition (3.2.15) holds.*

(i) *If $n-1 < \alpha < n$ $(n \in \mathbb{N})$, then there exists a unique solution $y(x)$ to the Cauchy problem (3.5.1)-(3.5.2) in the space $\mathbf{C}^{\alpha,n-1}_{\gamma}[a,b]$.*

(ii) *If $\alpha = n \in \mathbb{N}$, then there exists a unique solution $y(x)$ to the Cauchy problem (3.5.5) in the space $C^n_{\gamma}[a,b]$.*

(iii) *In particular, when $\gamma = 0$ and $f[x,y] \in C[a,b]$, there exist unique solutions to the Cauchy problem (3.5.1)-(3.5.2) in the space $\mathbf{C}^{\alpha,n-1}[a,b]$:*

$$\mathbf{C}^{\alpha,n-1}[a,b] := \mathbf{C}^{\alpha,n-1}_0[a,b] = \left\{ y(x) \in C^{n-1}[a,b] : \; {}^C D^{\alpha}_{a+} y \in C[a,b] \right\}, \tag{3.5.18}$$

and to the Cauchy problem (3.5.4) in the space $C^n[a,b]$.

Proof. Let $n-1 < \alpha < n$ $(n \in \mathbb{N})$. First we prove the existence of a unique solution $y(x) \in C^r[a,b]$. According to Theorem 3.24, it is sufficient to prove the existence of a unique solution $y(x) \in C^r[a,b]$ to the Volterra integral equation (3.5.4). For this we use a method similar to that given in Theorems 3.3 and 3.11. Equation (3.5.4) makes sense in any interval $[a, x_1] \in [a, b]$ $(a < x_1 < b)$. Choose x_1 such that the inequality

$$A \sum_{k=0}^{n-1} \frac{(x_1 - a)^{\alpha-k}}{\Gamma(\alpha - k + 1)} < 1 \tag{3.5.19}$$

holds, A being given in (3.2.15), and prove the existence of a the unique solution $y(x) \in C^{n-1}[a, x_1]$ to the equation (3.5.4) on the interval $[a, x_1]$. For this we use the Banach fixed point theorem (Theorem 1.9 in Section 1.13) for the space $C^{n-1}[a, b]$, which is the complete metric space with the distance given by

$$d(y_1, y_2) = ||y_1 - y_2||_{C^{n-1}[a,x_1]} := \sum_{k=0}^{n-1} ||[y_1^{(k)} - y_2^{(k)}]||_{C[a,x_1]}. \tag{3.5.20}$$

We rewrite the integral equation (3.5.4) in the form $y(x) = (Ty)(x)$, where

$$(Ty)(x) = y_0(x) + \frac{1}{\Gamma(\alpha)} \int_a^x \frac{f[t, y(t)]dt}{(x - t)^{1-\alpha}}, \quad y_0(x) = \sum_{j=0}^{n-1} \frac{b_j}{j!}(x - a)^j. \tag{3.5.21}$$

To apply Theorem 1.9, we have to prove the following: (1) if $y(x) \in C^{n-1}[a, x_1]$, then $(Ty)(x) \in C^{n-1}[a, x_1]$; and (2) for any $y_1, y_2 \in C^{n-1}[a, x_1]$,

$$||Ty_1 - Ty_2||_{C^{n-1}[a,x_1]} \leqq \omega ||y_1 - y_2||_{C^{n-1}[a,x_1]}, \tag{3.5.22}$$

with $\omega = A_0 \sum_{k=0}^{n-1} \frac{\Gamma(1-\gamma)(x_1-a)^{\alpha-k-\gamma}}{\Gamma(\alpha-k-\gamma+1)}$.

Let $y(x) \in C^{n-1}[a, x_1]$. Differentiating (3.5.21) k times $(k = 1, \cdots, n-1)$ and using the first relation in (2.1.33), which is true by Lemma 2.9(c) with $\gamma = 0$, we have, for $k = 0, 1, \cdots, n-1$,

$$(Ty)^{(k)}(x) = y_0^{(k)}(x) + \frac{1}{\Gamma(\alpha - k)} \int_a^x \frac{f[t, y(t)]dt}{(x - t)^{1-\alpha+k}}, \tag{3.5.23}$$

with $y_0^{(k)}(x) = \sum_{j=k}^{n} \frac{b_j}{(j-k)!}(x - a)^{j-k}$. For any $k = 0, 1, \cdots, n-1$, the first term in the right-hand side of (3.5.23) is clearly a continuous function on $[a, x_1]$, and by Lemma 3.3(a) (with $\gamma = 0$ and α replaced by $\alpha - k - 1$), the second term is also continuous on $[a, x_1]$ and the following estimates hold:

$$\left\| \frac{1}{\Gamma(\alpha - k)} \int_a^x \frac{f[t, y(t)]dt}{(x - t)^{1-\alpha+k}} \right\|_{C[a,x_1]} \leqq \frac{(x_1 - a)^{\alpha-k}}{\Gamma(\alpha - k + 1)} ||f[t, y(t)]||_{C[a,x_1]}, \tag{3.5.24}$$

for all $k = 0, 1, \cdots, n-1$. Hence $(Ty)(x) \in C^{n-1}[a, x_1]$. Now we prove (3.5.22). Using (3.5.20), (3.5.23), (3.5.24) and the Lipschitzian condition (3.2.15), we have

$$\|Ty_1 - Ty_2\|_{C^{n-1}[a,x_1]} = \sum_{k=0}^{n-1} \|(Ty_1)^{(k)} - (Ty_2)^{(k)}\|_{C[a,x_1]}$$

$$\leqq \sum_{k=0}^{n-1} \left\| \frac{1}{\Gamma(\alpha)} \int_a^x \frac{f[t, y_1(t)] - f[t, y_2)(t]}{(x-t)^{1-\alpha}} dt \right\|_{C[a,x_1]}$$

$$\leqq \sum_{k=0}^{n-1} \frac{(x_1 - a)^{\alpha-k}}{\Gamma(\alpha - k + 1)} \|f[t, y_1(t)] - f[t, y_2)(t]\|_{C[a,x_1]}$$

$$\leqq A_0 \sum_{k=0}^{n-1} \frac{\Gamma(1-\gamma)(x_1 - a)^{\alpha-k-\gamma}}{\Gamma(\alpha - k - \gamma + 1)} \|y_1 - y_2\|_{C[a,x_1]},$$

and hence, in accordance with (1.1.19),

$$\|Ty_1 - Ty_2\|_{C^{n-1}[a,x_1]} \leqq A_0 \sum_{k=0}^{n-1} \frac{(x_1 - a)^{\alpha-k}}{\Gamma(\alpha - k + 1)} \|y_1 - y_2\|_{C^{n-1}[a,x_1]},$$

which yields the estimate (3.5.22). By (3.5.19), $0 < \omega < 1$, and hence, by Theorem 1.9, there exists a unique solution $y^*(x) = y^{*0}(x) \in C^{n-1}[a, x_1]$ to the equations in (3.5.4) on the interval $[a, x_1]$.

By Theorem 1.9, this solution $y^*(x)$ is a limit of the convergent sequence $y_m(x) = (T^m y^*)(x)$:

$$\lim_{m\to\infty} \|y_m - y^*\|_{C^{n-1}[a,x_1]} = 0,$$

where $y_0^*(x)$ is any function in $C_{n-\alpha}[a, b]$. If at least one $b_k \neq 0$ in the initial condition (3.5.2), we can take $y_0^*(x) = y_0(x)$ with $y_0(x)$ defined in (3.5.21).

Applying the same arguments as in the proofs of Theorems 3.3 and 3.11, and taking Lemma 3.9 into account, we prove that there exists a unique solution $y(x) \in C^{n-1}[a, b]$ to the Cauchy problem (3.5.1)-(3.5.2), and this solution $y(x)$ is a limit of the sequence $y_m(x) = (T^m y^*)(x)$:

$$\lim_{m\to\infty} \|y_m - y\|_{C^{n-1}[a,b]} = 0. \tag{3.5.25}$$

Using this formula and taking (3.5.1) and (3.2.15) into account, we have

$$\|{}^C D_{a+}^\alpha y_m - {}^C D_{a+}^\alpha y\|_{C_\gamma[a,b]} \leqq \|f[t, y_m(t)] - f[t, y(t)]\|_{C_\gamma[a,b]}$$

$$\leqq A\|y_m - y\|_{C_\gamma[a,b]} \leqq A(b-a)^\gamma \|y_m - y\|_{C[a,b]} \leqq A\|y_m - y\|_{C^{n-1}[a,b]}.$$

From this and (3.5.25) we obtain

$$\lim_{m\to\infty} \| {}^C D_{a+}^\alpha y_m - {}^C D_{a+}^\alpha y\|_{C_\gamma[a,b]} = 0.$$

Thus ${}^C D_{a+}^\alpha y \in C_\gamma[a, b]$, and hence, in accordance with (3.5.3), $y \in \mathbf{C}_\gamma^{\alpha,n-1}[a, b]$.

This completes the proof of the assertion (i) of Theorem 3.25. The assertion (ii) is proved similarly by using Corollary 3.23 and the Banach fixed point theorem.

Corollary 3.25 *Let $n \in \mathbb{N}$. Let G be an open set in $\mathbb{R}$ and let $f : (a, b] \times G \to \mathbb{R}$ be a function such that, for any $y \in G$, $f[x, y] \in C[a, b]$ and the condition* (3.2.15) *is valid. Then there exists a unique solution $y(x) \in C^n[a, b]$ to the Cauchy problem* (3.5.4).

Corollary 3.26 *Let $0 < \alpha < 1$ and $0 \leqq \gamma \leqq \alpha$. Let G be an open set in $\mathbb{R}$ and let $f : (a, b] \times G \to \mathbb{C}$ be a function such that, for any $y \in G$, $f[x, y] \in C_\gamma[a, b]$ and the relation* (3.2.15) *holds.*

Then there exists a unique solution $y(x)$ to the Cauchy problem (3.5.15) *in the space $\mathbf{C}^\alpha_\gamma[a, b]$:*

$$\mathbf{C}^\alpha_\gamma[a, b] := \mathbf{C}^{\alpha,0}_\gamma[a, b] = \left\{y(x) \in C[a, b] : \ {}^C D^\alpha_{a+} y \in C_\gamma[a, b]\right\}. \tag{3.5.26}$$

Corollary 3.27 *Let $n - 1 < \alpha \geqq n$ $(n \in \mathbb{N})$ and $0 \leqq \gamma < 1$ be such that $\gamma \leqq \alpha$, and let $g(x) \in C_\gamma[a, b]$.*

If $a(x) \in C[a, b]$, then the Cauchy problem

$$({}^C D^\alpha_{a+} y)(x) = a(x)y(x) + g(x), \ \ y^{(k)}(a) = b_k \tag{3.5.27}$$

with $k = 0, 1, \cdots, n - 1$ and $b_k \in \mathbb{R}$, has a unique solution $y(x)$ in the space $\mathbf{C}^{\alpha,n-1}[a, b]$ when $\alpha \notin \mathbb{N}$, and in the space $C^n_\gamma[a, b]$ when $\alpha = n \in \mathbb{N}$.

In particular, the Cauchy problem

$$({}^C D^\alpha_{a+} y)(x) = \lambda(x - a)^\beta y(x) + g(x); \quad y^{(k)}(a) = b_k, \tag{3.5.28}$$

with $k = 0, 1, \cdots, n - 1$, $b_k \in \mathbb{R}$, $\lambda \in \mathbb{C}$ and $\beta \geqq 0$, has a unique solution $y(x)$ in $\mathbf{C}^{\alpha,n-1}[a, b]$ when $\alpha \notin \mathbb{N}$, while in $C^n_\gamma[a, b]$ when $\alpha = n \in \mathbb{N}$.

Remark 3.27 The results presented in Section 3.5.1 were proved by Kilbas and Marzan ([380] and [382]).

Remark 3.28 The results of Theorems 3.24-3.25 and Corollaries 3.23-3.27 can be extended to the systems of Cauchy problems (3.5.1)-(3.5.2), (3.5.5), (3.5.15), and (3.5.27)-(3.5.28).

3.5.2 The Cauchy Problems with Initial Conditions at the End and Inner Points of the Interval. Semi-Global and Local Solutions

In this subsection we give the conditions for the existence of a unique semi-global and local solution to Cauchy problems for differential equations with a Caputo fractional derivative for the case where the initial conditions are given at any point x_0 of the interval $[a, b]$. The first result gives a unique continuous solution $y(x)$ to the Cauchy problem (3.5.1)-(3.5.2) in the right neighborhood of a.

Theorem 3.26 *Let $\alpha > 0$, $h_1 > 0$, $K > 0$ and $b_k \in \mathbb{R}$ $(k = 0, 1, \cdots, n-1)$. Let G be a set given by*

$$G = \left\{ (x, y) \in \mathbb{R}^2 : \ a \leqq x \leqq a + h_1, \ \ |y - \sum_{j=0}^{n-1} \frac{b_j}{j!}(x-a)^j| \leqq K \right\}. \tag{3.5.29}$$

Let $0 \leqq \gamma < 1$ be such that $\gamma \leqq \alpha$, and let $f : G \to \mathbb{R}$ be a function such that, for any $y \in G$, $f[x, y] \in C_\gamma[a, b]$ and the Lipschitz condition (3.2.15) holds.

Then there exists h, $0 < h \leqq h_1$, such that the following assertions hold:

(i) *If $\alpha \notin \mathbb{N}$ and $n = [\alpha] + 1$, then there exists a unique solution $y(x)$ to the Cauchy problem (3.5.1)-(3.5.2) in the space $\mathbf{C}_\gamma^{\alpha, n-1}[a, a+h]$.*

(ii) *If $\alpha = n \in \mathbb{N}$, then there exists a unique solution $y(x)$ to the Cauchy problem (3.5.5) in the space $C_\gamma^n[a, a+h]$.*

(iii) *In particular, when $\gamma = 0$ and $f[x, y] \in C[a, b]$, there exist unique solutions to the Cauchy problem (3.5.1)-(3.5.2) in the space $\mathbf{C}^{\alpha, n-1}[a, a+h]$ and Cauchy problem (3.5.4) in the space $C^n[a, a+h]$.*

Proof. The proof is similar to that of Theorem 3.18. It is based on the equivalence of the Cauchy problem (3.5.1)-(3.5.2) and the Volterra integral equation (3.5.4) in the space $C^{n-1}[a, a+h]$ for $\alpha \notin \mathbb{N}$, and of the Cauchy problem (3.5.5) and the integral equation (3.5.6) for $\alpha \in \mathbb{N}$. For this we take $M = \max_{(x,y) \in G} |f(x, y)|$, choose h as

$$h = \min\left[h_1, \left(\frac{\Gamma(\alpha+1)K}{M} \right)^{1/\alpha} \right], \tag{3.5.30}$$

and apply the Banach fixed point theorem to the operator T given by (3.5.21) to prove the uniqueness of the solution $y(x) \in C^r[a, a+h]$ to the equation (3.5.4).

Corollary 3.28 *Let $0 < \alpha < 1$, and let $h_1 > 0$, $K > 0$, and $b_0 \in \mathbb{R}$. Let G be the set given by*

$$G = \left\{ (x, y) \in \mathbb{R}^2 : \ a \leqq x \leqq a + h_1, \ \ |y - b| \leqq K \right\}. \tag{3.5.31}$$

Let $0 \leqq \gamma < 1$ be such that $\gamma \leqq \alpha$, and let $f : G \to \mathbb{R}$ be a function such that, for any $y \in G$, $f[x, y] \in C_\gamma[a, b]$ and the Lipschitz condition (3.2.15) holds. Then there exists an h, $0 < h \leqq h_1$, such that the Cauchy problem (3.5.15) has a unique solution $y(x) \in \mathbf{C}_\gamma^\alpha[a, a+h]$.

Remark 3.29 When the conditions of Theorem 3.26 with $\gamma = 0$ are satisfied, Diethelm and Ford ([177], Theorems 2.1 and 2.2) proved the existence a local continuous solution $y(x) \in C[a, a+h]$ [see also Diethelm ([173], Theorems 5.4 and 5.5)]. The result in Theorem 3.26 characterizes more precisely the solution of the problem considered.

The following result yields the conditions for a unique local continuous solution to the Cauchy problem for the differential equation (3.5.1) with the initial conditions at any inner point $x_0 \in (a, b)$:

$$({}^C D_{a+}^\alpha y)(x) = f[x, y(x)] \ \ (\alpha > 0; \ a \leqq x \leqq b), \tag{3.5.32}$$

$$y^{(k)}(x_0) = b_k, \ b_k \in \mathbb{R} \ (k = 0, 1, \cdots, n-1). \tag{3.5.33}$$

Theorem 3.27 *Let $n-1 < \alpha \leqq n$ $(n \in \mathbb{N})$, and let $b_k \in \mathbb{R}$ $(k = 0, 1, \cdots, n-1)$ be given constants. Let U be an open and connected set in $\mathbb{R}$ and $\Omega = (a, b) \times U$. Let $f(x, y) : (a, b) \times U \to \mathbb{R}$ be a continuous function satisfying the Lipschitz condition* (3.2.15).

Then, for any $(x_0, y_0) \in \Omega$, there exists $h > 0$ such that $[x_0 - h, x_0 + h] \subset (a, b)$ and there exists a unique function $y(x) : [x_0 - h, x_0 + h] \to U$ such that $y(x) \in C^r[x_0 - h, x_0 + h]$, where $r = n$ for $\alpha = n \in \mathbb{N}$ and $r = n - 1$ for $\alpha \notin \mathbb{N}$, and

$$y(x_0) = b_0, \ \ y'(x_0) = b_1, \ \ \cdots, \ \ y^{(n-1)}(x_0) = b_{n-1}, \tag{3.5.34}$$

$$({}^C D^{\alpha}_{a+} y)(x) = f[x, y(x)] \ \ \text{for any} \ \ x \in [x_0 - h, x_0 + h]. \tag{3.5.35}$$

Proof. When $\alpha \in \mathbb{N}$, then, by the first relation in (2.4.14), $({}^C D^{\alpha}_{a+} y)(x) = y^{(n)}(x)$ and the result of Theorem 3.27 follows from that of Theorem 3.23.

Let now $\alpha \notin \mathbb{N}$. First of all, we note that the constants $y^{(k)}(a)$ $(k = 0, 1, \cdots, n-1)$ can be expressed uniquely via the constants $y^{(k)}(x_0) = b_k$ $(k = 0, 1, \cdots, n-1)$. Indeed, by Theorem 3.24, the Cauchy problem

$$({}^C D^{\alpha}_{a+} y)(x) = f[x, y(x)] \ (a \leqq x \leqq b), \ y^{(k)}(a) \in \mathbb{C} \ (k = 0, 1, \cdots, n-1) \tag{3.5.36}$$

is equivalent to the following integral equation:

$$y(x) = \sum_{j=0}^{n-1} \frac{y^{(j)}(a)}{j!}(x-a)^j + \left(I^{\alpha}_{a+} f[t, y(t)]\right)(x) \ \ (a \leqq x \leqq b). \tag{3.5.37}$$

Differentiating this relation k times $(k = 1, \cdots, n-1)$ and using the first formula in (2.1.33), asserted by Lemma 2.9(c) (with $\gamma = 0$), we have

$$y^{(k)}(x) = \sum_{j=k}^{n-1} \frac{y^{(j)}(a)}{(j-k)!}(x-a)^{j-k} + \left(I^{\alpha-k}_{a+} f[t, y(t)]\right)(x) \ \ (k = 0, 1, \cdots, n-1). \tag{3.5.38}$$

Taking the limit as $x \to x_0$, we obtain

$$y^{(k)}(x_0) = \sum_{j=k}^{n-1} \frac{y^{(j)}(a)}{(j-k)!}(x_0-a)^{j-k} + \left(I^{\alpha-k}_{a+} f[t, y(t)]\right)(x_0) \ \ (k = 0, 1, \cdots, n-1),$$

or, in accordance with (3.5.34),

$$b_k = \sum_{j=k}^{n-1} \frac{y^{(j)}(a)}{(j-k)!}(x_0-a)^{j-k} + \left(I^{\alpha-k}_{a+} f[t, y(t)]\right)(x_0) \ \ (k = 0, 1, \cdots, n-1). \tag{3.5.39}$$

This formula yields the following recurrence relations for finding $y^{(k)}(a)$:

$$b_{n-1} = y^{(n-1)}(a) + \left(I^{\alpha-n+1}_{a+} f[t, y(t)]\right)(x_0),$$

$$b_{n-2} = \sum_{j=n-2}^{n-1} \frac{y^{(j)}(a)}{(j-n+1)!}(x_0-a)^{j-n+1} + \left(I_{a+}^{\alpha-n+2} f[t,y(t)]\right)(x_0), \cdots,$$

$$b_0 = \sum_{j=0}^{n-1} \frac{y^{(j)}(a)}{j!}(x_0-a)^j + \left(I_{a+}^{\alpha} f[t,y(t)]\right)(x_0) \quad (k=0,1,\cdots,\ n-1). \quad (3.5.40)$$

Since $x_0 \in (a,b)$, then, for any $x \in (a,b)$, the existence of the Caputo fractional derivative $({}^C D_{a+}^{\alpha} y)(x)$ is equivalent to the existence of the Riemann-Liouville fractional derivative $(D_{a+}^{\alpha} y)(x)$. Then, by (2.4.6),

$$({}^C D_{a+}^{\alpha} y)(x) = (D_{a+}^{\alpha} y)(x) - \sum_{k=0}^{n-1} \frac{y^{(k)}(a)}{\Gamma(k-\alpha+1)}(x-a)^{k-\alpha} \ (\alpha \notin \mathbb{N}_0;\ n=[\alpha]+1), \quad (3.5.41)$$

with $\alpha \notin \mathbb{N}_0$ and $n = [\alpha]+1$. Now we rewrite (3.5.35) in the form

$$(D_{a+}^{\alpha} y)(x) = g[x, y(x)], \quad (3.5.42)$$

for any $x \in [x_0-h, x_0+h]$, and where

$$g[x,y(x)] = f[x,y(x)] + \sum_{k=0}^{n-1} \frac{y^{(k)}(a)}{\Gamma(k-\alpha+1)}(x-a)^{k-\alpha}. \quad (3.5.43)$$

Therefore, the problem (3.5.34)-(3.5.35) is equivalent to the following Cauchy problem:

$$y(x_0) = b_0, \quad y'(x_0) = b_1, \quad \cdots, \quad y^{(n-1)}(x_0) = b_{n-1}, \quad (3.5.44)$$

$$(D_{a+}^{\alpha} y)(x) = g[x,y(x)] \ \text{ for any } \ x \in [x_0-h, x_0+h]. \quad (3.5.45)$$

By the hypotheses of Theorem 3.27, $f[x,y(x)]$ is a continuous function on (a,b), the second term in (3.5.43) is also continuous on (a,b), and thus the function $g[x,y(x)] \in C(a,b)$. Therefore, by Theorem 3.23, there exists a positive number $h > 0$ such that $[x_0-h, x_0+h] \subset (a,b)$ and there exists a unique function $y(x) : [x_0-h, x_0+h] \to U$ such that $y(x) \in C^r[x_0-h, x_0+h]$, and the relations (3.5.44) and (3.5.45) are valid. This completes the proof of Theorem 3.27.

Corollary 3.29 *Let $0 < \alpha < 1$, and let $b_0 \in \mathbb{R}$ be a given constant. Let U be an open and connected set in $\mathbb{R}$ and $\Omega = (a,b) \times U$. Let $f(x,y) : (a,b) \times U \to \mathbb{R}$ be a continuous function satisfying the Lipschitz condition* (3.2.15).

Then, for any $(x_0, y_0) \in \Omega$, there exists $h > 0$ such that the real interval $[x_0-h, x_0+h] \subset (a,b)$ and there exists a unique function $y(x) : [x_0-h, x_0+h] \to U$ such that $y(x) \in C[x_0-h, x_0+h]$, and for any $x \in [x_0-h, x_0+h]$

$$({}^C D_{a+}^{\alpha} y)(x) = f[x, y(x)]; \quad y(x_0) = b_0. \quad (3.5.46)$$

Remark 3.30 The results of Theorems 3.26-3.27 and Corollaries 3.28-3.29 can be extended to the systems of the corresponding Cauchy problems.

3.5.3 Illustrative Examples

In Sections 3.5.1 and 3.5.2 we gave the conditions when the Cauchy problems for differential equations with Caputo derivatives have a unique global and local solution in the space of continuously differentiable functions and a semi-global solution in the space of continuous functions. Here we present several illustrative examples.

Example 3.7 Consider the following differential equation of fractional order α $(\alpha > 0)$:

$$({}^C D^\alpha_{a+} y)(x) = \lambda (x-a)^\beta [y(x)]^m \quad (x > a;\ \ m > 0;\ \ m \neq 1) \tag{3.5.47}$$

with $\lambda, \beta \in \mathbb{R}$ $(\lambda \neq 0)$. Using (2.4.28), it is directly verified that this equation has the following explicit solution:

$$y(x) = \left[\frac{\Gamma(\gamma - \alpha + 1)}{\lambda \Gamma(\gamma + 1)}\right]^{1/(m-1)} (x-a)^{\alpha - \gamma}, \quad \gamma = \frac{\beta + m\alpha}{m-1}. \tag{3.5.48}$$

By the definition (2.4.1) of the Caputo derivative ${}^C D^\alpha_{a+} y$, this solution must be continuous and hence bounded in a right neighborhood $[a, a+\epsilon)$ $(\epsilon > 0)$ of a, which yields the condition $\alpha - \gamma \geqq 0$, or $(\beta + \alpha)/(m-1) \geqq 0$, and hence $y(x) \in C[a,b]$. The right-hand side of the equation (3.5.47),

$$f[x, y(x)] = \lambda \left[\frac{\Gamma(\gamma - \alpha + 1)}{\lambda \Gamma(\gamma + 1)}\right]^{m/(m-1)} (x-a)^{-\gamma}, \tag{3.5.49}$$

belongs to $C_\gamma[a,b]$ and to $C[a,b]$ in the respective cases $0 < \gamma < 1$ and $\gamma \leqq 0$. This leads to Cases (2) and (3), given in (3.3.7), which means that any of the conditions (3.3.47), (3.3.48), (3.3.49) and (3.3.50) is satisfied in the first case, while either of (3.3.51) and (3.3.52) in the second. The validity of the Lipschitz condition (3.2.15) was proved in Example 3.3 for the domain D defined by (3.3.57). Then Theorem 3.25 yields the following results.

Proposition 3.8 *Let $0 < \alpha < 1$, $\lambda, \beta \in \mathbb{R}$ $(\lambda \neq 0)$ and $m > 0$ $(m \neq 1)$. Let D be the domain (3.3.57), where $\omega \in \mathbb{R}$ is such that $\beta + (m-1)\omega \geqq 0$.*

(i) *If either of the conditions (3.3.47) or (3.3.48) holds, then the Cauchy problem*

$$({}^C D^\alpha_{a+} y)(x) = \lambda (x-a)^\beta [y(x)]^m \quad (x \geqq a), \quad y(a) = 0 \tag{3.5.50}$$

has a unique solution $y(x)$ in the space $\mathbf{C}^\alpha_\gamma[a,b]$, $\gamma = (\beta + m\alpha)/(m-1)$, and this solution is given by (3.5.48).

(ii) *If either of the conditions (3.3.51) or (3.3.52) is valid, then the Cauchy problem (3.5.50) has a unique solution $y(x) \in \mathbf{C}^\alpha_0[a,b]$ given by (3.5.48).*

Proposition 3.9 *Let $n-1 < \alpha < n$ $(n \in \mathbb{N}\backslash\{1\})$, $\lambda, \beta \in \mathbb{R}$ $(\lambda \neq 0)$ and $m > 0$ $(m \neq 1)$. Let D be the domain (3.3.57), where $\omega \in \mathbb{R}$ is such that $\beta + (m-1)\omega \geqq 0$.*

(i) *If either of the conditions* (3.3.49) *or* (3.3.50) *holds, then the Cauchy problem*

$$(D_{a+}^{\alpha}y)(x) = \lambda(x-a)^{\beta}[y(x)]^{m} \quad (x \geqq a), \quad y^{(k)}(a) = 0 \quad (k = 0, 1, \cdots, n-1) \tag{3.5.51}$$

has a unique solution $y(x) \in \mathbf{C}_{\gamma}^{\alpha,n-1}[a,b]$, $\gamma = (\beta + m\alpha)/(m-1)$, *and this solution is given by* (3.5.48).

(ii) *If either of the conditions* (3.3.51) *or* (3.3.52) *is valid, then the Cauchy problem* (3.5.51) *has a unique solution* $y(x) \in \mathbf{C}_{0}^{\alpha,n-1}[a,b]$ *given by* (3.5.48).

Example 3.8 Consider the following nonlinear inhomogeneous fractional differential equation of order $\alpha > 0$:

$$({}^{C}D_{a+}^{\alpha}y)(x) = \lambda(x-a)^{\beta}y^{m}(x) + b(x-a)^{\nu} \quad (x \geqq a; \;\; m \in \mathbb{R}; \;\; m \neq 1) \tag{3.5.52}$$

with λ, $b \in \mathbb{R}$ $(\lambda \neq 0)$ and β, $\nu \in \mathbb{R}$. Applying the same arguments as in the proof of Propositions 3.3-3.4, from Theorem 3.25 we obtain the uniqueness theorem for the Cauchy problem for the equation (3.5.51).

Proposition 3.10 *Let* $0 < \alpha < 1$, $\lambda, b \in \mathbb{R}$ $(\lambda \neq 0)$, $m > 0$ $(m \neq 1)$ *and* $\beta \in \mathbb{R}$. *Let* D *be the domain* (3.3.57), *where* $\omega \in \mathbb{R}$ *is such that* $\beta + (m-1)\omega \geqq 0$. *Let* $\nu = (\beta + \alpha m)/(1-m)$ *and let the transcendental equation* (3.3.69) *have a unique solution* $\xi = \mu$.

(i) *If either of the conditions* (3.3.47) *or* (3.3.48) *holds, then the Cauchy problem*

$$({}^{C}D_{a+}^{\alpha}y)(x) = \lambda(x-a)^{\beta}[y(x)]^{m} + b(x-a)^{\nu}, \quad y(a) = 0 \tag{3.5.53}$$

has a unique solution $y(x)$ *in the space* $\mathbf{C}_{\gamma}^{\alpha}[a,b]$, $\gamma = (\beta + m\alpha)/(m-1)$, *and this solution is given by*

$$y(x) = \mu(x-a)^{\alpha-\gamma} \in C[a,b], \tag{3.5.54}$$

(ii) *If either of the conditions* (3.3.51) *or* (3.3.52) *is valid, then the Cauchy problem* (3.5.53) *has a unique solution* $y(x) \in \mathbf{C}^{\alpha}[a,b]$ *given by* (3.5.54).

Proposition 3.11 *Let* $\alpha \geqq 1$, $\lambda, b \in \mathbb{R}$ $(\lambda \neq 0)$, $m > 0$ $(m \neq 1)$ *and* $\beta \in \mathbb{R}$. *Let* D *be the domain* (3.3.57), *where* $\omega \in \mathbb{R}$ *is such that* $\beta + (m-1)\omega \geqq 0$. *Let* $\nu = (\beta + \alpha m)/(1-m)$ *and let the transcendental equation* (3.3.69) *have a unique solution* $\xi = \mu$.

(i) *If either of the conditions* (3.3.49) *or* (3.3.50) *holds, then the Cauchy problem*

$$(D_{a+}^{\alpha}y)(x) = \lambda(x-a)^{\beta}[y(x)]^{m} + b(x-a)^{\nu}; \quad y^{(k)}(a) = 0 \quad (x \geqq a), \tag{3.5.55}$$

for all $k = 0, 1, \cdots, \;\; n-1$, *has a unique solution* $y(x) \in \mathbf{C}_{\gamma}^{\alpha,n-1}[a,b]$, *where* $\gamma = (\beta + m\alpha)/(m-1)$, *and this solution is given by* (3.5.54).

(ii) *If either of the conditions* (3.3.51) *or* (3.3.52) *is valid, then the Cauchy problem* (3.5.55) *has a unique solution* $y(x) \in \mathbf{C}_{0}^{\alpha,n-1}[a,b]$ *given by* (3.5.54).

Remark 3.31 By Propositions 3.1-3.4 and 3.8-3.11, the Cauchy type problems (3.3.53)-(3.3.55) and (3.3.72)-(3.3.74) for the differential equations (3.3.36) and (3.3.67) with the Riemann-Liouville derivatives D_{a+}^{α} and the Cauchy problems (3.5.50)-(3.5.51) and (3.5.53), (3.5.55) for the differential equations (3.5.47) and (3.5.52) with the Caputo derivatives ${}^{C}D_{a+}^{\alpha}$ have the same solutions as in (3.3.38), (3.5.48) and as in (3.3.70), (3.5.54). But the solutions given in (3.3.38) and (3.3.70) can be continuous on $[a,b]$ or continuous on $(a,b]$ with an integrable singularity at the point a. The solutions in (3.5.48) and (3.5.54) can be only continuous on $[a,b]$.

Example 3.9 Consider the following Cauchy problems for the differential equation of fractional order $\alpha > 0$ $(n-1 < \alpha \leqq n, \ \ n \in \mathbb{N})$:

$$({}^{C}D_{a+}^{\alpha}y)(x) = \lambda(x-a)^{\beta}[y(x)]^{m} + g(x), \quad (a \leqq x \leqq b) \tag{3.5.56}$$

$$y^{(k)}(a) = b_k, \tag{3.5.57}$$

with $b_k \in \mathbb{R}$, $k = 0, 1, \cdots, n-1$, $m > 0$ $(m \neq 1)$, $\lambda \in \mathbb{R}$ $(\lambda \neq 0)$ and $\beta \in \mathbb{R}$. Let $K > 0$, $h_1 > 0$ $(0 < h_1 < b-a)$ and G be the set (3.5.29). We suppose that $g(x) \in C[a, a+h_1]$ and the constants α, β and m are such that $\beta \geqq m(1-\alpha)$. (3.5.56)-(3.5.57) is the Cauchy problem (3.5.1)-(3.5.2). Analogous to Example 3.5, it is proved that the right-hand side of (3.5.56): $f(x,y) = \lambda(x-a)^{\beta}y^{m} + g(x)$ satisfies the Lipschitz condition (3.2.15) on G. Then, from Theorem 3.26, we derive the uniqueness of a semi-global solution $y(x)$ of the Cauchy problem (3.5.56)-(3.5.57).

Proposition 3.12 *Let* $\alpha > 0$ $(n-1 < \alpha < n;\ n \in \mathbb{N})$, $m > 0$ $(m \neq 1)$, *and* $\beta \in \mathbb{R}$ *be such that* $\beta \geqq m(1-\alpha)$, *and let* $\lambda \in \mathbb{R}$ $(\lambda \neq 0)$ *and* $b, b_k \in \mathbb{R}$ $(k = 0, 1, \cdots, n-1)$. *Let* $K > 0$, $h_1 > 0$ $(0 < h_1 < b-a)$, *let* G *be the set* (3.5.29) *and let* $g(x) \in C_{\gamma}[a, a+h_1]$, *where* $0 \leqq \gamma < 1$ *is such that* $\gamma \leqq \alpha$.

Then there exists an h, $0 < h \leqq h_1$, *such that the Cauchy problem* (3.5.56)-(3.5.57) *has a unique semi-global solution* $y(x) \in \mathbf{C}_{\gamma}^{\alpha,n-1}[a, a+h]$.

In particular, if $0 < \alpha < 1$, *the Cauchy problem*

$$({}^{C}D_{a+}^{\alpha}y)(x) = \lambda(x-a)^{\beta}[y(x)]^{m} + g(x) \quad (a \leqq x \leqq b;\ 0 < \alpha < 1), \quad y(a) = b_0 \in \mathbb{R}, \tag{3.5.58}$$

has a unique semi-global solution $y(x) \in \mathbf{C}_{1-\alpha,\gamma}^{\alpha}[a, a+h]$.

Example 3.10 Consider the Cauchy problems for the differential equation (3.5.56) of order $\alpha > 0$ with the following initial conditions:

$$y^{(k)}(x_0) = b_k \in \mathbb{R} \quad (k = 0, 1, \cdots, n-1). \tag{3.5.59}$$

Here $f(x,y) = \lambda(x-a)^{\beta}y^{m} + g$ is a continuous function on $(a,b) \times \Omega$ and satisfies the Lipschitz condition (3.2.15). Then, from Theorem 3.27, we obtain the uniqueness of a local solution of the Cauchy problem (3.5.56), (3.5.59).

Proposition 3.13 *Let $\alpha > 0$ $(n-1 < \alpha \leqq n;\ n \in \mathbb{N})$, $m > 0$ $(m \neq 1)$ and $\beta \in \mathbb{R}$ be such that $\beta \geqq m(1-\alpha)$, let $\lambda \in \mathbb{R}$ $(\lambda \neq 0)$ and let $b_k \in \mathbb{R}$ $(k = 0, 1, \cdots, n-1)$ be given constants. Let U be an open and connected set in $\mathbb{R}$ and $g(x) \in C(U)$.*

Then, for any $(x_0, y_0) \in (a,b) \times \Omega$, there exists an $h > 0$ such that the real interval $[x_0 - h, x_0 + h] \subset (a,b)$ and there exists a unique real function $y(x) : [x_0 - h, x_0 + h] \to U$ such that $y(x) \in C^r[x_0 - h, x_0 + h]$, with $r = n$ for $\alpha = n \in \mathbb{N}$ and $r = n-1$ for $\alpha \notin \mathbb{N}$, and the relations (3.5.56) *and* (3.5.59) *are valid.*

In particular, if $0 < \alpha < 1$, $b_0 \in \mathbb{R}$ and $a \leqq x \leqq b$, then $y(x) \in C[x_0 - h, x_0 + h]$ and it satisfies the following equation:

$$({}^C D^{\alpha}_{a+} y)(x) = \lambda (x-a)^{\beta} [y(x)]^m + g(x); \quad y(x_0) = b_0. \tag{3.5.60}$$

3.6 Equations with the Hadamard Fractional Derivative in the Space of Continuous Functions

In this section we give the conditions for the existence of a unique global continuous solution to the Cauchy type problem (3.1.43)-(3.1.44) in the space $\mathbf{C}^{\alpha}_{\delta;n-\alpha,\gamma}[a,b]$ defined, for $n-1 < \alpha \leqq n$ $(n \in \mathbb{N})$ and $0 \leqq \gamma < 1$, by

$$\mathbf{C}^{\alpha}_{\delta;n-\alpha,\gamma}[a,b] = \{y(x) \in C_{n-\alpha,\log}[a,b] : (\mathcal{D}^{\alpha}_{a+} y)(x) \in C_{\gamma,\log}[a,b]\}. \tag{3.6.1}$$

Here $\mathcal{D}^{\alpha}_{a+} y$ is the Hadamard fractional derivative (2.7.7), and $C_{\gamma,\log}[a,b]$ is the weighted space of continuous functions (1.1.27):

$$C_{\gamma,\log}[a,b] = \left\{ g(x) : \left(\log \frac{x}{a}\right)^{\gamma} g(x) \in C[a,b],\ \|y\|_{C_{\gamma,\log}} = \left\| \left(\log \frac{x}{a}\right)^{\gamma} g(x) \right\|_C \right\}. \tag{3.6.2}$$

We consider the following Cauchy type problem (3.1.43)-(3.1.44):

$$(\mathcal{D}^{\alpha}_{a+} y)(x) = f[x, y(x)] \quad (x > a;\ \alpha > 0), \tag{3.6.3}$$

$$(\mathcal{D}^{\alpha-k}_{a+} y)(a+) = b_k,\ b_k \in \mathbb{R}\ (k = 1, \cdots, n;\ n = -[-\alpha]). \tag{3.6.4}$$

The notation $(\mathcal{D}^{\alpha-k}_{a+} y)(a+)$ means that the limit is taken at all points of the right-sided neighborhood $(a, a+\epsilon)$ $(\epsilon > 0)$ of a:

$$(\mathcal{D}^{\alpha-k}_{a+} y)(a+) = \lim_{x \to a+} (\mathcal{D}^{\alpha-k}_{a+} y)(x) \quad (k = 1, \cdots, n-1), \tag{3.6.5}$$

$$(\mathcal{D}^{\alpha-n}_{a+} y)(a+) = \lim_{x \to a+} (\mathcal{J}^{n-\alpha}_{a+} y)(x)\ (\alpha \neq n);\ (\mathcal{D}^{0}_{a+} y)(a+) = y(a)\ (\alpha = n), \tag{3.6.6}$$

where $\mathcal{J}^{n-\alpha}_{a+}$ is the Hadamard fractional operator of order $n-\alpha$ defined in (3.1.45).

Using the properties of the Hadamard integration and differentiation operators given in 2.35 and 2.36 and applying the same arguments as in the proofs of Theorems 3.1 and 3.10, we prove the equivalence of the problem (3.6.3)-(3.6.4) and the Volterra integral equation (3.1.45).

Theorem 3.28 *Let $\alpha > 0$, $n = -[-\alpha]$ and $0 \leqq \gamma < 1$. Let G be an open set in $\mathbb{R}$ and let $f : (a, b] \times G \to \mathbb{R}$ be a function such that $f[x, y] \in C_{\gamma,\log}[a, b]$ for any $y \in G$.*

If $y(x) \in C_{n-\alpha,\log}[a, b]$, then $y(x)$ satisfies the relations (3.6.3) and (3.6.4) if, and only if, $y(x)$ satisfies the Volterra integral equation (3.1.58).

In particular, if $0 < \alpha \leqq 1$ and $y(x) \in C_{1-\alpha,\log}[a, b]$, then $y(x)$ satisfies the relations

$$(\mathcal{D}^{\alpha}_{a+}y)(x) = f[x, y(x)], \ \left(\mathcal{J}^{1-\alpha}_{a+}y\right)(a+) = b \in \mathbb{R} \tag{3.6.7}$$

if, and only if, $y(x)$ satisfies the following integral equation:

$$y(x) = \frac{b}{\Gamma(\alpha)}\left(\log\frac{x}{a}\right)^{\alpha-1} + \frac{1}{\Gamma(\alpha)}\int_a^x \left(\log\frac{x}{t}\right)^{\alpha-1} f[t, y(t)]\frac{dt}{t} \quad (x > a), \tag{3.6.8}$$

To prove a uniqueness result for (3.6.3)-(3.6.4), we need a preliminary assertion.

Lemma 3.10 *Let $\gamma \in \mathbb{C}$, $a < c < b$, $g \in C_{\gamma,\log}[a, c]$ and $g \in C_{\gamma,\log}[c, b]$. Then $g \in C_{\gamma}[a, b]$ and*

$$||g||_{C_{\gamma,\log}[a,b]} \leqq \max\left[||g||_{C_{\gamma,\log}[a,c]}, ||g||_{C_{\gamma,\log}[c,b]}\right].$$

Theorem 3.29 *Let $\alpha > 0$, $n = -[-\alpha]$ and $0 \leqq \gamma < 1$ be such that $\gamma \geqq n - \alpha$. Let G be an open set in $\mathbb{C}$ and let $f : (a, b] \times G \to \mathbb{C}$ be a function such that, for any $y \in G$, $f[x, y] \in C_{\gamma,\log}[a, b]$ and the Lipschitz condition (3.2.15) holds.*

Then there exists a unique solution $y(x)$ to the Cauchy type problem (3.6.3)-(3.6.4) in the space $\mathbf{C}^{\alpha}_{\delta;n-\alpha,\gamma}[a, b]$.

In particular, if $0 < \alpha < 1$ and $\gamma \geqq 1 - \alpha$, then there exists a unique solution $y(x)$ to the Cauchy type problem (3.6.7) in the space $\mathbf{C}^{\alpha}_{\delta;1-\alpha,\gamma}[a, b]$.

Proof. The proof is similar to the proofs of Theorems 3.3 and 3.11 and is based on Theorem 3.28 and the Banach fixed point theorem (Theorem 1.9 in Section 1.13) for the space $C_{n-\alpha,\log}[a, x_1]$, which is the complete metric space with the distance given by

$$d(y_1, y_2) = ||y_1 - y_2||_{C_{n-\alpha,\log}[a,x_1]} := \max_{x\in[a,x_1]}\left|\left(\log\frac{x}{a}\right)^{n-\alpha}[y_1(x) - y_2(x)]\right|. \tag{3.6.9}$$

Choosing x_1 ($a < x_1 < b$) such that there hold the following inequalities:

$$\omega_1 := A\left(\log\frac{x_1}{a}\right)^{n-\gamma+\alpha}\frac{\Gamma(1-\gamma)}{\Gamma(1+\alpha-\gamma)} < 1 \ \text{ for } \gamma > \alpha, \tag{3.6.10}$$

$$\omega_2 := A\left(\log\frac{x_1}{a}\right)^{n-\gamma}\frac{\Gamma(1-\gamma)}{\Gamma(1+\alpha-\gamma)} < 1 \ \text{ for } \gamma \leqq \alpha, \tag{3.6.11}$$

rewriting the equation (3.1.45) in the form $y(x) = (Ty)(x)$, where

$$(Ty)(x) = y_0(x) + \frac{1}{\Gamma(\alpha)}\int_a^x \left(\log\frac{x}{t}\right)^{\alpha-1} f[t, y(t)]\frac{dt}{t}, \tag{3.6.12}$$

$$y_0(x) = \sum_{j=1}^{n} \frac{b_j}{\Gamma(\alpha - j + 1)} \left(\log \frac{x}{a}\right)^{\alpha - j},$$

and applying Theorem 1.9, we prove that, if $y(x) \in C_{n-\alpha,\log}[a, x_1]$, then $(Ty)(x) \in C_{n-\alpha,\log}[a, x_1)$, and, for any $y_1, y_2 \in C_{n-\alpha,\log}[a, x_1]$, there hold the following estimates:

$$\|Ty_1 - Ty_2\|_{C_{n-\alpha,\log}[a,x_1]} \leqq \omega_j \|y_1 - y_2\|_{C_{n-\alpha,\log}[a,x_1]}, \quad 0 < \omega_j < 1 \ (j = 1, 2). \tag{3.6.13}$$

This means that there exists a unique solution $y^*(x) \in C_{n-\alpha,\log}[a, x_1]$ to the equation (3.1.45) such that

$$\lim_{m \to \infty} \|y_m(x) - y(x)\|_{C_{n-\alpha,\log}[a,x_1]} = 0, \tag{3.6.14}$$

$$y_m(x) := y_0(x) + \frac{1}{\Gamma(\alpha)} \int_a^x \left(\log \frac{x}{t}\right)^{\alpha-1} f[t, (T^{m-1}y)(t)] \frac{dt}{t} \quad (m \in \mathbb{N}). \tag{3.6.15}$$

Considering the interval $[x_1, x_2]$, where $x_2 = x_1 + h_1$ and $h_1 > 0$ are such that $x_2 < b$, and applying the same arguments as in proofs of Theorems 3.3 and 3.11, we prove, using Lemma 3.10, that there exists a unique solution $y(x) = y^*(x) \in C_{n-\alpha,\log}[a, b]$ to the integral equation (3.1.45) and hence to the Cauchy type problem (3.6.3)-(3.6.4). To complete the proof of Theorem 3.29, in accordance with the definition (3.3.1), it is sufficient to prove that $(\mathcal{D}_{a+}^{\alpha} y)(x) \in C_\gamma[a, b]$. By the above proof, the solution $y(x) \in C_{n-\alpha,\log}[a, b]$ is a limit of the sequence $y_m(x) \in C_{n-\alpha,\log}[a, b]$, and it is directly proved that

$$\lim_{n \to \infty} \|\mathcal{D}_{a+}^{\alpha} y_m - \mathcal{D}_{a+}^{\alpha} y\|_{C_{\gamma,\log}[a,b]} = 0,$$

and hence $(\mathcal{D}_{a+}^{\alpha} y)(x) \in C_{\gamma,\log}[a, b]$. This completes the proof of Theorem 3.29.

Corollary 3.30 *Let $n \in \mathbb{N}$ and $0 \leqq \gamma < 1$. Let G be an open set in $\mathbb{R}$ and let $f : (a, b] \times G \to \mathbb{R}$ be a function such that, for any $y \in G$, $f[x, y] \in C_{\gamma,\log}[a, b]$ and the relation* (3.2.15) *holds.*

Then there exists a unique solution $y(x)$ to the Cauchy problem

$$(\delta^n y)(x) = f[x, y(x)], \quad (\delta^{n-k} y)(a) = b_k \in \mathbb{R} \ (k = 1, \cdots, n) \tag{3.6.16}$$

in the space $C_\delta^n[a, b] = \{y \in C[a, b] : \delta^n y \in C[a, b]\}$.

The spaces of the solution $y(x)$ of the Cauchy type problem (3.6.3)-(3.6.4) are characterized more precisely when $b_n = 0$ and the results will be different for $0 < \alpha \leqq 1$ and for $\alpha > 1$. First we consider the first case:

$$(\mathcal{D}_{a+}^{\alpha} y)(x) = f[x, y(x)] \ (0 < \alpha \leqq 1), \quad (\mathcal{J}_{a+}^{1-\alpha} y)(a+) = 0. \tag{3.6.17}$$

The integral equation (3.6.8), corresponding to the problem (3.6.17), takes the following form:

$$y(x) = \frac{1}{\Gamma(\alpha)} \int_a^x \left(\log \frac{x}{t}\right)^{\alpha-1} f[t, y(t)] \frac{dt}{t} \quad (x > a). \tag{3.6.18}$$

By Lemma 2.36(a), if $f[x, y(x)] \in C_{\gamma,\log}[a, b]$ $(0 \leqq \gamma < 1)$, then the right-hand side of (3.3.28) and hence the solution $y(x)$ belong to $C_{\gamma-\alpha}[a, b]$ for $\gamma > \alpha$ and to $C[a, b]$ for $\gamma \leqq \alpha$. From here we derive the result.

Theorem 3.30 *Let $0 < \alpha \leqq 1$ and $0 \leqq \gamma < 1$. Let G be an open set in $\mathbb{C}$ and let $f : (a, b] \times G \to \mathbb{R}$ be a function such that, for any $y \in G$, $f[x, y] \in C_{\gamma,\log}[a, b]$ and the Lipschitz condition* (3.2.15) *is satisfied.*

(a) *If $\gamma > \alpha$, then the Cauchy type problem* (3.6.17) *has a unique solution $y(x) \in \mathbf{C}^{\alpha}_{\delta;\gamma-\alpha,\gamma}[a, b]$.*

(b) *If $\gamma \leqq \alpha$, then the Cauchy type problem* (3.6.17) *has a unique solution $y(x) \in \mathbf{C}^{\alpha}_{\delta;0,\gamma}[a, b]$.*

Similarly, we can prove the result for the problem (3.6.3)-(3.6.4) with $\alpha > 1$:

$$(\mathcal{D}^{\alpha}_{a+}y)(x) = f[x, y(x)] \quad (\alpha > 1), \tag{3.6.19}$$

$$(\mathcal{D}^{\alpha-k}_{a+}y)(a+) = b_k \in \mathbb{R} \ (k = 1, \cdots, n-1;\ n = -[-\alpha]), \quad b_n = 0. \tag{3.6.20}$$

Theorem 3.31 *Let $\alpha > 1$, $n = -[-\alpha]$. Let G be an open set in $\mathbb{R}$ and let $f : (a, b] \times G \to \mathbb{R}$ be a function such that, for any $y \in G$, $f[x, y] \in C_{\gamma,\log}[a, b]$ and the Lipschitz condition* (3.2.15) *is satisfied.*

Then the Cauchy type problem (3.6.19)-(3.6.20) *has a unique solution $y(x) \in \mathbf{C}^{\alpha}_{\delta}[a, b]$.*

In particular, if $f[x, y] \in C[a, b]$ for any $y \in G$, then the Cauchy type problem (3.6.19)-(3.6.20) *has a unique solution $y(x) \in \mathbf{C}^{\alpha}_{\delta}[a, b]$.*

When $0 < \alpha < 1$, the results of Theorem 3.29 remain true for the following problem:

$$(\mathcal{D}^{\alpha}_{a+}y)(x) = f[x, y(x)] \ (0 < \alpha < 1), \quad \lim_{x \to a} \left[\left(\log \frac{x}{a} \right)^{1-\alpha} y(x) \right] = c \in \mathbb{R}. \tag{3.6.21}$$

Theorem 3.32 *Let $0 < \alpha < 1$ and $0 \leqq \gamma < 1$ be such that $\gamma \geqq 1 - \alpha$. Let G be an open set in $\mathbb{R}$ and let $f : (a, b] \times G \to \mathbb{R}$ be a function such that, for any $y \in G$, $f[x, y] \in C_{\gamma,\log}[a, b]$ and the Lipschitz condition* (3.2.15) *is satisfied.*

Then there exists a unique solution $y(x)$ to the weighted Cauchy type problem (3.6.21) *in the space $\mathbf{C}^{\alpha}_{\delta;1-\alpha,\gamma}[a, b]$.*

Theorem 3.30 yields the result for the weighted problem (3.6.21) with $c = 0$:

$$(\mathcal{D}^{\alpha}_{a+}y)(x) = f[x, y(x)] \ (0 < \alpha < 1), \quad \lim_{x \to a} \left[\left(\log \frac{x}{a} \right)^{1-\alpha} y(x) \right] = 0. \tag{3.6.22}$$

Theorem 3.33 *Let $0 < \alpha < 1$ and $0 \leqq \gamma < 1$. Let G be an open set in $\mathbb{R}$ and let $f : (a, b] \times G \to \mathbb{R}$ be a function such that, for any $y \in G$, $f[x, y] \in C_{\gamma,\log}[a, b]$ and the Lipschitz condition* (3.2.15) *is satisfied.*

(a) *If $\gamma > \alpha$, then the weighted Cauchy type problem* (3.6.22) *has a unique solution $y(x)$ in the space $\mathbf{C}^{\alpha}_{\delta;\gamma-\alpha,\gamma}[a,b]$.*

(b) *If $\gamma \leqq \alpha$, then the weighted Cauchy type problem* (3.6.22) *has a unique solution $y(x)$ in the space $\mathbf{C}^{\alpha}_{\delta;0,\gamma}[a,b]$.*

(c) *In particular, if for any $y \in G$, $f[x,y] \in C[a,b]$, then the weighted Cauchy type problem* (3.6.22) *has a unique solution $y(x)$ in the space $\mathbf{C}^{\alpha}_{\delta}[a,b]$.*

The above results can be extended to the following equation, which is more general than (3.6.3):

$$(\mathcal{D}^{\alpha}_{a+}y)(x) = f\left[x, y(x), (\mathcal{D}^{\alpha_1}_{a+}y)(x), \cdots, (\mathcal{D}^{\alpha_l}_{a+}y)(x)\right]. \tag{3.6.23}$$

Theorem 3.34 *Let $\alpha > 0$, $n = -[-\alpha]$ and $0 \leqq \gamma < 1$ be such that $\gamma \geqq n - \alpha$. Let $l \in \mathbb{N}\backslash\{1\}$ and $\alpha_j > 0$ $(j = 1, \cdots, l)$ be such that the conditions in* (3.3.21) *are satisfied. Let G be an open set in $\mathbb{R}^{l+1}$ and let $f : (a,b] \times G \to \mathbb{R}$ be a function such that $f[x,y,y_1,\cdots,y_l] \in C_{\gamma,\log}[a,b]$ for any $(y,y_1,\cdots,y_l) \in G$, and the Lipschitz condition* (3.2.46) *is satisfied.*

Then there exists a unique solution $y(x)$ to the Cauchy type problem (3.6.23)-(3.6.4) *in the space $\mathbf{C}^{\alpha}_{\delta;n-\alpha,\gamma}[a,b]$.*

In particular, if $0 < \alpha < 1$ and $\gamma \geqq 1-\alpha$, then there exists a unique solution $y(x) \in \mathbf{C}^{\alpha}_{\delta;1-\alpha,\gamma}[a,b]$ to the Cauchy type problem for the equation (3.6.23) *with the initial conditions $\left(\mathcal{J}^{1-\alpha}_{a+}y\right)(a+) = b \in \mathbb{R}$ and $\lim_{x\to a}\left[\left(\log\frac{x}{a}\right)^{1-\alpha} y(x)\right] = c \in \mathbb{R}$.*

Theorem 3.35 *Let $\alpha > 0$ and $0 \leqq \gamma < 1$. Let $l \in \mathbb{N}\backslash\{1\}$ and $\alpha_j > 0$ $(j = 1, \cdots, l)$ be such that the condition* (3.3.21) *is satisfied. Let G be an open set in $\mathbb{R}^{l+1}$ and let $f : (a,b] \times G \to \mathbb{R}$ be a function such that $f[x,y,y_1,\cdots,y_l] \in C_{\gamma,\log}[a,b]$ for any $(y,y_1,\cdots,y_l) \in G$, and the Lipschitz condition* (3.2.46) *is satisfied.*

(a) *If $0 < \alpha \leqq 1$ and $\gamma > \alpha$, then there exists a unique solution $y(x) \in \mathbf{C}^{\alpha}_{\delta;\gamma-\alpha,\gamma}[a,b]$ to the Cauchy type problem for the equation* (3.6.23) *with the initial conditions $\left(\mathcal{J}^{1-\alpha}_{a+}y\right)(a+) = 0$ and $\lim_{x\to a}\left[\left(\log\frac{x}{a}\right)^{1-\alpha} y(x)\right] = 0$.*

(b) *If $0 < \alpha \leqq 1$ and $\gamma \leqq \alpha$, then there exists a unique solution $y(x) \in \mathbf{C}^{\alpha}_{\delta;0,\gamma}[a,b]$ to the Cauchy type problem for the equation* (3.6.23) *with the initial conditions $\left(\mathcal{J}^{1-\alpha}_{a+}y\right)(a+) = 0$ and $\lim_{x\to a}\left[\left(\log\frac{x}{a}\right)^{1-\alpha} y(x)\right] = 0$.*

(c) *If $\alpha > 1$ and $n = -[-\alpha]$, then the Cauchy type problem for the equation* (3.6.23) *with the initial conditions*

$$(\mathcal{D}^{\alpha-k}_{a+}y)(a+) = b_k, \quad b_k \in \mathbb{C} \quad (k = 1, 2, \cdots, n-1), \quad b_n = 0. \tag{3.6.24}$$

has a unique solution $y(x) \in \mathbf{C}^{\alpha}_{\delta}[a,b]$.

(d) *In particular, if for any $(y,y_1,\cdots,y_l) \in G$, $f[x,y,y_1,\cdots,y_l] \in C_{0,\log}[a,b]$, then the Cauchy type problems in* (a)-(c) *have a unique solution $y(x) \in \mathbf{C}^{\alpha}_{\delta}[a,b]$.*

Corollary 3.31 *Let $n, l \in \mathbb{N}$ and $\alpha_j > 0$ ($\alpha_j < n$; $j = 1, \cdots, l$) be such that the conditions in* (3.3.21) *with $\alpha = n$ are satisfied. Let G be an open set in $\mathbb{R}^{l+1}$ and let $f : (a, b] \times G \to \mathbb{R}$ be a function such that $f[x, y, y_1, \cdots, y_l] \in C_{0,\log}[a, b]$ for any $(y, y_1, \cdots, y_l) \in G$, and the Lipschitz condition* (3.2.46) *is satisfied.*

Then there exists a unique solution $y(x) \in \mathbf{C}^n_{\delta;0,\gamma}[a, b]$ for the Cauchy problem

$$(\delta^n y)(x) = f\left[x, y(x), (\mathcal{D}^{\alpha_1}_{a+}y)(x), \cdots, (\mathcal{D}^{\alpha_l}_{a+}y)(x)\right] \quad \left(x > a;\ \delta = x\frac{d}{dx}\right), \tag{3.6.25}$$

$$(\delta^{n-k}y)(a) = b_k \in \mathbb{R} \quad (k = 1, \cdots, n). \tag{3.6.26}$$

Remark 3.32 The results in Theorems 3.28-3.35 can be extended to a system of corresponding Cauchy type problems. In this regard, we note that Annaby and Butzer [36] studied a first-order system of equations of the form

$$(\delta + c)y(x) = f_j\left[x, y_1(x), \cdots, y_l(x)\right] \quad (c \in \mathbb{R};\ j = 1, \cdots, l). \tag{3.6.27}$$

They developed the theory for (3.6.27): existence and uniqueness of solutions, linear systems, Sturm-Liouville problems, asymptotic of eigenvalues, etc., and proposed an application to sampling theory. Indeed it will be an interesting problem to develop such a theory for the generalized fractional equation (3.6.23).

From Theorems 3.29-3.35 we can derive the corresponding results for the Cauchy type problems for linear fractional differential equations. For example, from Theorem 3.34 we derive the following assertion.

Corollary 3.32 *Let $\alpha > 0$, $n = -[-\alpha]$ and $0 \leqq \gamma < 1$ be such that $\gamma \geqq n - \alpha$. Let $l \in \mathbb{N}\backslash\{1\}$ and $\alpha_j > 0$ ($j = 1, \cdots, l$) be such that the conditions in* (3.3.21) *are satisfied, and let $a_j(x) \in C[a, b]$ ($j = 1, \cdots, l$) and $g(x) \in C_{\gamma,\log}[a, b]$.*

Then the Cauchy type problem for the following linear differential equation of order α,

$$(\mathcal{D}^{\alpha}_{a+}y)(x) + \sum_{j=1}^{l} a_j(x)(\mathcal{D}^{\alpha_j}_{a+}y)(x) + a_0(x)y(x) = g(x) \quad (x > a) \tag{3.6.28}$$

with the initial conditions (3.6.4), *has a unique solution $y(x)$ in the space $\mathbf{C}^{\alpha}_{\delta;n-\alpha,\gamma}[a, b]$.*

In particular, there exists a unique solution $y(x) \in \mathbf{C}^{\alpha}_{\delta;n-\alpha,\gamma}[a, b]$ to the Cauchy type problem for the equation with $\lambda_j \in \mathbb{R}$ and $\beta_j \geqq 0$ ($j = 0, 1, \cdots, l$):

$$(\mathcal{D}^{\alpha}_{a+}y)(x) + \sum_{j=1}^{l} \lambda_j (x-a)^{\beta_j}(\mathcal{D}^{\alpha_j}_{a+}y)(x) + \lambda_j(x-a)^{\beta_j}y(x) = g(x) \quad (x > a). \tag{3.6.29}$$

Example 3.11 Consider the following differential equation of fractional order $\alpha > 0$:

$$(D^{\alpha}_{a+}y)(x) = \lambda\left(\log\frac{x}{a}\right)^{\beta}[y(x)]^m \quad (x > a > 0;\ m > 0;\ m \neq 1) \tag{3.6.30}$$

with $\lambda, \beta \in \mathbb{R}$ $(\lambda \neq 0)$. Using (2.7.16), it is easily verified that, if the condition $[(\beta+\alpha)/(1-m)] > -1$ holds, then this equation has the explicit solution

$$y(x) = \left[\frac{\Gamma(\frac{\beta+\alpha}{m-1}+1)}{\lambda\Gamma(\frac{\beta+\alpha m}{m-1}+1)}\right]^{1/(m-1)} \left(\log\frac{x}{a}\right)^{(\beta+\alpha)/(1-m)}. \tag{3.6.31}$$

It is clear that (3.6.31) is also the explicit solution to the following Cauchy type problems:

$$(\mathcal{D}_{a+}^{\alpha}y)(x) = \lambda\left(\log\frac{x}{a}\right)^{\beta}[y(x)]^m, \quad (\mathcal{J}_{a+}^{1-\alpha}y)(a+) = 0 \quad (0 < \alpha < 1), \tag{3.6.32}$$

$$(\mathcal{D}_{a+}^{\alpha}y)(x) = \lambda\left(\log\frac{x}{a}\right)^{\beta}[y(x)]^m, \quad \lim_{t\to a+}\left[(x-a)^{1-\alpha}y(x)\right] = 0 \quad (0 < \alpha < 1), \tag{3.6.33}$$

provided that any of the conditions (3.3.45), (3.3.46), (3.3.47), (3.3.48), (3.3.51) and (3.3.52) holds, and to the following Cauchy type problem of order $\alpha > 0$:

$$(\mathcal{D}_{a+}^{\alpha}y)(x) = \lambda\left(\log\frac{x}{a}\right)^{\beta}[y(x)]^m, \quad (\mathcal{D}_{a+}^{\alpha-k}y)(a+) = 0 \quad (k = 1, \cdots, n = -[-\alpha]), \tag{3.6.34}$$

provided that any of the conditions (3.3.49), (3.3.50), (3.3.51) and (3.3.52) is valid.

We choose the following domain:

$$D = \left\{(x,y) \in \mathbb{C}: \ a < x \leqq b, \ 0 < y < K\left(\log\frac{x}{a}\right)^{\omega}; \ \ \omega \in \mathbb{R}, \ \ K > 0\right\}. \tag{3.6.35}$$

It is directly verified that the Lipschitz condition (3.2.15) is satisfied, provided that $\beta + (m-1)\omega \geqq 0$. Using the same arguments as in the proofs of Propositions 3.1-3.2, from Theorems 3.29-3.31 we derive the uniqueness result.

Proposition 3.14 *Let $0 < \alpha < 1$, $\lambda \in \mathbb{R}$ $(\lambda \neq 0)$, $m > 0$ $(m \neq 1)$, $\beta \in \mathbb{R}$ and $0 \leqq \gamma = [(\beta + m\alpha)/(m-1)] < 1$. Let D be the domain* (3.6.35), *where $\omega \in \mathbb{R}$ is such that $\beta + (m-1)\omega \geqq 0$.*

(a) *If either of the conditions* (3.3.45) *or* (3.3.46) *holds, then the problems* (3.6.32) *and* (3.6.33) *have a unique solution $y(x) \in \mathbf{C}_{\delta;\gamma-\alpha,\gamma}^{\alpha}[a,b]$ given by* (3.6.31).

(b) *If either of the conditions* (3.3.47) *or* (3.3.48) *is satisfied, then the problems* (3.6.32) *and* (3.6.33) *have a unique solution $y(x) \in \mathbf{C}_{\delta;0,\gamma}^{\alpha}[a,b]$ given by* (3.6.31).

(c) *If either of the conditions* (3.3.51) *or* (3.3.52) *is valid, then the problems* (3.6.32) *and* (3.6.33) *have a unique solution $y(x) \in \mathbf{C}_{\delta}^{\alpha}[a,b]$ given by* (3.6.31).

Proposition 3.15 *Let $\alpha \geqq 1$, $n = -[-\alpha]$, $\lambda \in \mathbb{R}$ $(\lambda \neq 0)$, $m > 0$ $(m \neq 1)$, $\beta \in \mathbb{R}$ and $0 \leqq \gamma < 1$. Let D be the domain* (3.6.35), *where $\omega \in \mathbb{R}$ is such that $\beta + (m-1)\omega \geqq 0$.*

If any of the conditions (3.3.49), (3.3.50), (3.3.51) *and* (3.3.52) *holds, then the problem* (3.6.34) *has a unique solution $y(x) \in \mathbf{C}_{0,0}^{\alpha}[a,b]$ given by* (3.6.31).

Example 3.12 Consider the following nonlinear inhomogeneous fractional differential equation of order $\alpha > 0$:

$$(\mathcal{D}_{a+}^{\alpha}y)(x) = \lambda \left(\log \frac{x}{a}\right)^{\beta} y^{m}(x) + b\left(\log \frac{x}{a}\right)^{\nu} \quad (x > a; \quad m > 0; \ m \neq 1) \quad (3.6.36)$$

with $a, \lambda \in \mathbb{R}$ $(\lambda \neq 0)$, b, β and $\nu \in \mathbb{R}$. Applying the same arguments as in the proof of Propositions 3.3-3.4, from Theorems 3.29-3.31 we obtain the following results.

Proposition 3.16 *Let* $0 < \alpha \leqq 1$, $\lambda, \beta, b \in \mathbb{R}$ $(\lambda \neq 0)$, $m > 0$ $(m \neq 1)$ *and* $0 \leqq \gamma = [(\beta + m\alpha)/(m-1)] < 1$. *Let D be the domain* (3.6.35), *where* $\omega \in \mathbb{R}$ *is such that* $\beta + (m-1)\omega \geqq 0$. *Let* $\nu = (\beta + m\alpha)/(1-m)$ *and let the transcendental equation* (3.3.69) *have a unique solution* $\xi = \mu$.

(a) *If either of the conditions* (3.3.45) *or* (3.3.46) *holds, then the Cauchy type problems for the equation* (3.6.36) *with the initial conditions* $(\mathcal{J}_{a+}^{1-\alpha}y)(a+) = 0$ *and* $\lim_{t\to a+}\left[(x-a)^{1-\alpha}y(x)\right] = 0$ *have a unique solution* $y(x) \in \mathbf{C}_{\delta;\gamma-\alpha,\gamma}^{\alpha}[a,b]$.

(b) *If either of the conditions* (3.3.47) *or* (3.3.48) *is satisfied, then the Cauchy type problems in* (a) *have a unique solution* $y(x) \in \mathbf{C}_{\delta;0,\gamma}^{\alpha}[a,b]$.

(c) *If either of the conditions* (3.3.51) *or* (3.3.52) *is valid, then the Cauchy type problems in* (a) *have a unique solution* $y(x) \in \mathbf{C}_{\delta}^{\alpha}[a,b]$.

In all cases (a)-(c), *the unique solution* $y(x)$ *is given by* (3.6.31).

Proposition 3.17 *Let* $\alpha > 1$, $\lambda, \beta, b \in \mathbb{R}$ $(\lambda \neq 0)$, $m > 0$ $(m \neq 1)$ *and* $0 \leqq \gamma = [(\beta + m\alpha)/(m-1)] < 1$. *Let D be the domain* (3.3.57), *where* $\omega \in \mathbb{R}$ *is such that* $\beta + (m-1)\omega \geqq 0$. *Let* $\nu = (\beta + m\alpha)/(1-m)$ *and let the transcendental equation* (3.3.69) *have a unique solution* $\xi = \mu$.

If any of the conditions (3.3.49), (3.3.50), (3.3.51) *and* (3.3.52) *holds, then the Cauchy type problem for the equation* (3.6.36) *with the initial conditions* (3.6.4) *has a unique solution* $y(x) \in \mathbf{C}_{\delta}^{\alpha}[a,b]$ *given by* (3.6.31).

Chapter 4

METHODS FOR EXPLICITLY SOLVING FRACTIONAL DIFFERENTIAL EQUATIONS

The present chapter is devoted to explicit and numerical solutions to fractional differential equations and boundary problems associated with them. The approaches based on the reduction to Volterra integral equations, on compositional relations, and on operational calculus are presented to give explicit solutions to linear differential equations. For simplicity, we give results involving fractional differential equations of real order $\alpha > 0$ and given real numbers. They also remain true for fractional differential equations of complex order α $(n - 1 < \Re(\alpha) < n;\ n \in \mathbb{N})$ and complex numbers. Numerical treatment of fractional differential equations is also discussed.

4.1 Method of Reduction to Volterra Integral Equations

In this section we present the method for finding closed-form solutions to boundary value problems for linear ordinary differential equations based on the reduction to Volterra integral equations. We derive explicit solutions to the Cauchy type and Cauchy problems for linear ordinary fractional differential equations involving the Riemann-Liouville, Caputo, and Hadamard fractional derivatives on a finite interval $[a, b]$ of the real axis $\mathbb{R}$. The problems considered have a unique solution in accordance with the results obtained in Sections 3.3, 3.5, and 3.6.

4.1.1 The Cauchy Type Problems for Differential Equations with the Riemann-Liouville Fractional Derivatives

In this subsection we construct explicit solutions to linear fractional differential equations with the Riemann-Liouville fractional derivative $\left(D_{a+}^{\alpha}y\right)(x)$ of order $\alpha > 0$ given by (2.1.10) in the space $C_{n-\alpha,\gamma}^{\alpha}[a,b]$ $(n = -[-\alpha],\ 0 \leqq \gamma < 1)$, defined in (3.3.24).

First we consider the Cauchy type problem for the fractional differential equation (3.1.10) of order $\alpha > 0$ with the initial conditions (3.1.2):

$$(D_{a+}^{\alpha}y)(x) - \lambda y(x) = f(x) \quad (a < x \leqq b;\ \alpha > 0;\ \lambda \in \mathbb{R}), \tag{4.1.1}$$

$$(D_{a+}^{\alpha-k}y)(a+) = b_k, \quad (b_k \in \mathbb{R};\ k = 1, \cdots, n = -[-\alpha]). \tag{4.1.2}$$

We suppose that $f(x) \in C_{\gamma}[a,b]$ $(0 \leqq \gamma < 1)$. Then, by Property 3.1(a), (4.1.1)-(4.1.2) is equivalent in the space $C_{n-\alpha}[a,b]$ to the following Volterra integral equation:

$$y(x) = \sum_{j=1}^{n} \frac{b_j}{\Gamma(\alpha - j + 1)}(x-a)^{\alpha-j} + \frac{\lambda}{\Gamma(\alpha)}\int_a^x \frac{y(t)dt}{(x-t)^{1-\alpha}} + \frac{1}{\Gamma(\alpha)}\int_a^x \frac{f(t)dt}{(x-t)^{1-\alpha}} \tag{4.1.3}$$

We apply the method of successive approximations to solve this integral equation. According to this method, we set

$$y_0(x) = \sum_{j=1}^{n} \frac{b_j}{\Gamma(\alpha - j + 1)}(x-a)^{\alpha-j}, \tag{4.1.4}$$

$$y_m(x) = y_0(x) + \frac{\lambda}{\Gamma(\alpha)}\int_a^x \frac{y_{m-1}(t)dt}{(x-t)^{1-\alpha}} + \frac{1}{\Gamma(\alpha)}\int_a^x \frac{f(t)dt}{(x-t)^{1-\alpha}} \quad (m \in \mathbb{N}). \tag{4.1.5}$$

Using (2.1.1), (4.1.4) and taking (2.1.16) into account, we find for $y_1(x)$ that

$$y_1(x) = y_0(x) + \lambda\left(I_{a+}^{\alpha}y_0\right)(x) + \left(I_{a+}^{\alpha}f\right)(x)$$

that is,

$$y_1(x) = \sum_{j=1}^{n} \frac{b_j}{\Gamma(\alpha - j + 1)}(x-a)^{\alpha-j} + \lambda\sum_{j=1}^{n} \frac{b_j}{\Gamma(2\alpha - j + 1)}(x-a)^{2\alpha-j} + \left(I_{a+}^{\alpha}f\right)(x)$$

$$= \sum_{j=1}^{n} b_j \sum_{k=1}^{2} \frac{\lambda^{k-1}(x-a)^{\alpha k-j}}{\Gamma(\alpha k - j + 1)} + \frac{1}{\Gamma(\alpha)}\int_a^x (x-t)^{\alpha-1}f(t)dt \tag{4.1.6}$$

Similarly, using (4.1.4)-(4.1.6), we find for $y_2(x)$ that

$$y_2(x) = y_0(x) + \lambda\left(I_{a+}^{\alpha}y_1\right)(x) + \left(I_{a+}^{\alpha}f\right)(x)$$

that is,

$$y_2(x) = \sum_{j=1}^{n} \frac{b_j}{\Gamma(\alpha - j + 1)}(x - a)^{\alpha - j}$$

$$+\lambda \sum_{j=1}^{n} b_j \sum_{k=1}^{2} \frac{\lambda^{k-1}}{\Gamma(\alpha k - j + 1)} \left(I_{a+}^{\alpha}(t - a)^{\alpha k - j}\right)(x) + \left(I_{a+}^{\alpha} f\right)(x) + \left(I_{a+}^{\alpha} I_{a+}^{\alpha} f\right)(x)$$

Taking into account the first relation in (2.1.30), we get

$$y_2(x) = \sum_{j=1}^{n} b_j \sum_{k=1}^{3} \frac{\lambda^{k-1}(x - a)^{\alpha k - j}}{\Gamma(\alpha k - j + 1)} + \int_a^x \left[\sum_{k=1}^{2} \frac{\lambda^{k-1}}{\Gamma(\alpha k)}(x - t)^{\alpha k - 1}\right] f(t)dt. \quad (4.1.7)$$

Continuing this process, we derive the following relation for $y_m(x)$ $(m \in \mathbb{N})$:

$$y_m(x) = \sum_{j=1}^{n} b_j \sum_{k=1}^{m+1} \frac{\lambda^{k-1}(x - a)^{\alpha k - j}}{\Gamma(\alpha k - j + 1)} + \int_a^x \left[\sum_{k=1}^{m} \frac{\lambda^{k-1}}{\Gamma(\alpha k)}(x - t)^{\alpha k - 1}\right] f(t)dt. \quad (4.1.8)$$

Taking the limit as $m \to \infty$, we obtain the following explicit solution $y(x)$ to the integral equation (4.1.3):

$$y(x) = \sum_{j=1}^{n} b_j \sum_{k=1}^{\infty} \frac{\lambda^{k-1}(x - a)^{\alpha k - j}}{\Gamma(\alpha k - j + 1)} + \int_a^x \left[\sum_{k=1}^{\infty} \frac{\lambda^{k-1}}{\Gamma(\alpha k)}(x - t)^{\alpha k - 1}\right] f(t)dt,$$

or, by replacing the index of summation k by $k - 1$,

$$y(x) = \sum_{j=1}^{n} b_j \sum_{k=0}^{\infty} \frac{\lambda^{k}(x - a)^{\alpha k + \alpha - j}}{\Gamma(\alpha k + \alpha - j + 1)} + \int_a^x \left[\sum_{k=0}^{\infty} \frac{\lambda^{k}}{\Gamma(\alpha k + \alpha)}(x - t)^{\alpha k + \alpha - 1}\right] f(t)dt. \quad (4.1.9)$$

Taking into account the relation (1.8.17), we rewrite this solution in terms of the Mittag-Leffler function $E_{\alpha,\beta}(z)$:

$$y(x) = \sum_{j=1}^{n} b_j (x - a)^{\alpha - j} E_{\alpha, \alpha - j + 1}\left[\lambda(x - a)^{\alpha}\right] + \int_a^x (x - t)^{\alpha - 1} E_{\alpha,\alpha}\left[\lambda(x - t)^{\alpha}\right] f(t)dt. \quad (4.1.10)$$

This yields an explicit solution to the Volterra integral equation (4.1.3) and hence to the Cauchy type problem (4.1.1)-(4.1.2).

It is clear that $f[x, y] = \lambda y + f(x)$ satisfies the Lipschitz condition for any $x_1, x_2 \in (a, b]$ and any $y \in G$, where G is any open set of $\mathbb{C}$. If $\gamma \geqq n - \alpha$, then, by Property 3.1(b) and Remark 3.18, there exists a unique solution to the Cauchy type problem (4.1.1)-(4.1.2) in the space $\mathbf{C}_{n-\alpha,\gamma}^{\alpha}[a, b]$. Equation (4.1.10) yields this solution. Thus we derive the following result.

Theorem 4.1 *Let $\alpha > 0$, $n = -[-\alpha]$ and γ $(0 \leqq \gamma < 1)$ be such that $\gamma \geqq n - \alpha$. Also let $\lambda \in \mathbb{R}$. If $f \in C_\gamma[a,b]$, then the Cauchy type problem (4.1.1)-(4.1.2) has a unique solution $y(x) \in \mathbf{C}^\alpha_{n-\alpha,\gamma}[a,b]$ and this solution is given by (4.1.10).*

In particular, if $f(x) = 0$, then the Cauchy type problem involving the homogeneous differential equation (4.1.1)

$$(D^\alpha_{a+}y)(x) - \lambda y(x) = 0 \quad (a < x \leqq b;\ \alpha > 0;\ \lambda \in \mathbb{R}), \tag{4.1.11}$$

with the initial conditions (4.1.2), has a unique solution $y(x) \in C^\alpha_{n-\alpha}[a,b]$ $(C^\alpha_{n-\alpha}[a,b] := C^\alpha_{n-\alpha,0}[a,b])$ of the form

$$y(x) = \sum_{j=1}^{n} b_j (x-a)^{\alpha-j} E_{\alpha,\alpha-j+1}\left[\lambda(x-a)^\alpha\right]. \tag{4.1.12}$$

Remark 4.1 Barrett [68] first constructed the explicit solutions (4.1.10) and (4.1.12) to the Cauchy type problems (4.1.1), (4.1.2) and (4.1.11), (4.1.2) in a certain subspace of $L(a,b)$.

Example 4.1 The solution to the Cauchy type problem

$$(D^\alpha_{a+}y)(x) - \lambda y(x) = f(x), \quad (D^{\alpha-1}_{a+}y)(a+) = b \quad (b \in \mathbb{R}) \tag{4.1.13}$$

with $0 < \alpha < 1$ and $\lambda \in \mathbb{R}$ has the following form:

$$y(x) = b(x-a)^{\alpha-1} E_{\alpha,\alpha}\left[\lambda(x-a)^\alpha\right] + \int_a^x (x-t)^{\alpha-1} E_{\alpha,\alpha}\left[\lambda(x-t)^\alpha\right] f(t)dt, \tag{4.1.14}$$

while the solution to the problem

$$(D^\alpha_{a+}y)(x) - \lambda y(x) = 0, \quad (D^{\alpha-1}_{a+}y)(a+) = b \quad (b \in \mathbb{R}) \tag{4.1.15}$$

is given by

$$y(x) = b(x-a)^{\alpha-1} E_{\alpha,\alpha}\left[\lambda(x-a)^\alpha\right]. \tag{4.1.16}$$

In particular, the Cauchy type problem

$$(D^{1/2}_{a+}y)(x) - \lambda y(x) = f(x), \quad (I^{1/2}_{a+}y)(a+) = b \quad (b \in \mathbb{R}) \tag{4.1.17}$$

has the solution given by

$$y(x) = \frac{b}{(x-a)^{1/2}} E_{1/2,1/2}\left[\lambda(x-a)^{1/2}\right] + \int_a^x E_{1/2,1/2}\left[\lambda(x-t)^{1/2}\right] \frac{f(t)dt}{(x-t)^{1/2}}, \tag{4.1.18}$$

and the solution to the problem

$$(D^{1/2}_{a+}y)(x) - \lambda y(x) = 0, \quad (I^{1/2}_{a+}y)(a+) = b \quad (b \in \mathbb{R}) \tag{4.1.19}$$

is given by

$$y(x) = b(x-a)^{-1/2} E_{1/2,1/2}\left[\lambda(x-a)^{1/2}\right]. \tag{4.1.20}$$

Example 4.2 The solution to the Cauchy type problem

$$(D_{a+}^{\alpha}y)(x) - \lambda y(x) = f(x), \ \ (D_{a+}^{\alpha-1}y)(a+) = b, \ \ (D_{a+}^{\alpha-2}y)(a+) = d, \tag{4.1.21}$$

with $1 < \alpha < 2$ and $\lambda, b, d \in \mathbb{R}$ has the form

$$y(x) = b(x-a)^{\alpha-1}E_{\alpha,\alpha}\left[\lambda(x-a)^{\alpha}\right] + d(x-a)^{\alpha-2}E_{\alpha,\alpha-1}\left[\lambda(x-a)^{\alpha}\right]$$

$$+ \int_a^x (x-t)^{\alpha-1}E_{\alpha,\alpha}\left[\lambda(x-t)^{\alpha}\right]f(t)dt. \tag{4.1.22}$$

In particular, the solution to the problem

$$(D_{a+}^{\alpha}y)(x) - \lambda y(x) = 0, \ \ (D_{a+}^{\alpha-1}y)(a+) = b, \ \ (D_{a+}^{\alpha-2}y)(a+) = d \tag{4.1.23}$$

is given by

$$y(x) = b(x-a)^{\alpha-1}E_{\alpha,\alpha}\left[\lambda(x-a)^{\alpha}\right] + d(x-a)^{\alpha-2}E_{\alpha,\alpha-1}\left[\lambda(x-a)^{\alpha}\right]. \tag{4.1.24}$$

Next we consider the Cauchy type problem for the following more general homogeneous fractional differential equation than (4.1.11):

$$(D_{a+}^{\alpha}y)(x) - \lambda(x-a)^{\beta}y(x) = 0 \ \ (a < x \leqq b; \ \alpha > 0; \ \lambda \in \mathbb{R}), \tag{4.1.25}$$

$$(D_{a+}^{\alpha-k}y)(a+) = b_k \ (b_k \in \mathbb{R}; \ k = 1, \cdots, n; \ n = -[-\alpha]), \tag{4.1.26}$$

with $\beta > -\{\alpha\}$. By Property 3.1 and Remark 3.18, the problem (4.1.25)-(4.1.26) is equivalent in the space $C_{n-\alpha}[a,b]$ to the Volterra integral equation of the second kind

$$y(x) = \sum_{j=1}^{n} \frac{b_j}{\Gamma(\alpha-j+1)}(x-a)^{\alpha-j} + \frac{\lambda}{\Gamma(\alpha)}\int_a^x \frac{(t-a)^{\beta}y(t)dt}{(x-t)^{1-\alpha}}. \tag{4.1.27}$$

We again apply the method of successive approximations to solve this integral equation. We use the notation $y_0(x)$ in (4.1.4) and set

$$y_m(x) = y_0(x) + \frac{\lambda}{\Gamma(\alpha)}\int_a^x \frac{(t-a)^{\beta}y_{m-1}(t)dt}{(x-t)^{1-\alpha}} \ \ (m \in \mathbb{N}). \tag{4.1.28}$$

Using the same arguments as above, we find for $y_1(x)$ that

$$y_1(x) = y_0(x) + \lambda\left(I_{a+}^{\alpha}y_0\right)(x)$$

$$= \sum_{j=1}^{n} \frac{b_j}{\Gamma(\alpha-j+1)}(x-a)^{\alpha-j} + \lambda\sum_{j=1}^{n} \frac{b_j}{\Gamma(\alpha-j+1)}\left(I_{a+}^{\alpha}(t-a)^{\alpha+\beta-j}\right)(x).$$

By the condition $\beta > -\{\alpha\}$, all integrals $\left(I_{a+}^{\alpha}(t-a)^{\alpha+\beta-j}\right)(x)$ $(j = 1, \cdots, n)$ are convergent, and applying the first relation in (2.2.16) yields the following result:

$$y_1(x) = \sum_{j=1}^{n} \frac{b_j}{\Gamma(\alpha - j + 1)}(x-a)^{\alpha-j}$$

$$+\lambda \sum_{j=1}^{n} \frac{b_j}{\Gamma(\alpha - j + 1)} \frac{\Gamma(\alpha+\beta-j+1)}{\Gamma(2\alpha+\beta-j+1)}(x-a)^{2\alpha+\beta-j}. \tag{4.1.29}$$

Similarly, using (4.1.28) with $m = 2$ and taking (4.1.29) into account, we derive

$$y_2(x) = y_0(x)+\lambda \left(I_{a+}^{\alpha} y_1\right)(x) = y_0(x)+\lambda \left(I_{a+}^{\alpha} y_0\right)(x)$$

$$+\lambda^2 \sum_{j=1}^{n} \frac{b_j}{\Gamma(\alpha - j + 1)} \frac{\Gamma(\alpha+\beta-j+1)}{\Gamma(2\alpha+\beta-j+1)} \left(I_{a+}^{\alpha}(t-a)^{2\alpha+\beta-j}\right)(x)$$

$$= \sum_{j=1}^{n} \frac{b_j}{\Gamma(\alpha - j + 1)}(x-a)^{\alpha-j} \left[1 + c_1\left(\lambda(x-a)^{\alpha+\beta-j}\right) + c_2\left(\lambda(x-a)^{\alpha+\beta}\right)^2\right],$$

where

$$c_1 = \frac{\Gamma(\alpha+\beta-j+1)}{\Gamma(2\alpha+\beta-j+1)}, \quad c_2 = \frac{\Gamma(\alpha+\beta-j+1)}{\Gamma(2\alpha+\beta-j+1)} \frac{\Gamma(2\alpha+2\beta-j+1)}{\Gamma(3\alpha+2\beta-j+1)}.$$

Continuing this process, we derive the following relation for $y_m(x)$ ($m \in \mathbb{N}$):

$$y_m(x) = \sum_{j=1}^{n} \frac{b_j}{\Gamma(\alpha - j + 1)}(x-a)^{\alpha-j} \left[1 + \sum_{k=1}^{m} c_k \left(\lambda(x-a)^{\alpha+\beta}\right)^k\right], \tag{4.1.30}$$

where

$$c_k = \prod_{r=1}^{k} \frac{\Gamma[r(\alpha+\beta)-j+1]}{\Gamma[r(\alpha+\beta)+\alpha-j+1]} \quad (k \in \mathbb{N}). \tag{4.1.31}$$

Taking the limit as $m \to \infty$, we obtain the following explicit solution $y(x)$ to the integral equation (4.1.27) and hence to the Cauchy type problem (4.1.25)-(4.1.26):

$$y(x) = \sum_{j=1}^{n} \frac{b_j}{\Gamma(\alpha - j + 1)}(x-a)^{\alpha-j} \left[1 + \sum_{k=1}^{\infty} c_k \left(\lambda(x-a)^{\alpha+\beta}\right)^k\right]. \tag{4.1.32}$$

According to the relations (1.9.19) and (1.9.20), we rewrite this solution in terms of the generalized Mittag-Leffler function $E_{\alpha,m,l}(z)$:

$$y(x) = \sum_{j=1}^{n} \frac{b_j}{\Gamma(\alpha - j + 1)}(x-a)^{\alpha-j} E_{\alpha,1+\beta/\alpha,1+(\beta-j)/\alpha}\left[\lambda(x-a)^{\alpha+\beta}\right]. \tag{4.1.33}$$

If $\beta \geqq 0$, then $f[x,y] = \lambda(x-a)^{\beta}$ satisfies the Lipschitz condition for any $x_1, x_2 \in (a,b]$ and any $y \in G$, where G is any open set of $\mathbb{C}$. If $\gamma \geqq n-\alpha$, then, by Property 3.1(b) and Remark 3.18, there exists a unique solution to the Cauchy type problem (4.1.25)-(4.1.26) in the space $\mathbf{C}_{n-\alpha}^{\alpha}[a,b]$, and thus this solution has the form (4.1.33). This leads to the following result.

Theorem 4.2 *Let $\alpha > 0$, $n = -[-\alpha]$, $\lambda \in \mathbb{R}$ and $\beta \geqq 0$. Then the Cauchy type problem (4.1.25)-(4.1.26) has a unique solution $y(x)$ in the space $C^{\alpha}_{n-\alpha}[a,b] := C^{\alpha}_{n-\alpha,0}[a,b]$ and this solution is given by (4.1.33).*

Remark 4.2 For $a = 0$, the solution (4.1.33) to the Cauchy type problem (4.1.25)-(4.1.26) in the space $I_{loc}[0,d]$ of locally integrable functions on $[0,d]$ $(0 < d < \infty)$ was first given by Kilbas and Saigo ([397], Equation 53).

Remark 4.3 If $\alpha > 1$ and $b_n = 0$, then, in accordance with Property 3.5, (4.1.10), and (4.1.12) and (4.1.33) are unique solutions to the respective Cauchy type problems (4.1.1), (4.1.2), and (4.1.11), (4.1.2), and (4.1.25), (4.1.26) in the space $\mathbf{C}^{\alpha}_{0,0}[a,b]$.

Remark 4.4 If $f(x) \in L(a,b)$, then, by Theorem 3.3, $y(x)$ in (4.1.10) and (4.1.33) would yield unique solutions to the Cauchy type problem (4.1.1),(4.1.2) and , by (4.1.25), (4.1.26) in the space $\mathbf{L}^{\alpha}(a,b)$ defined by (3.2.1). In this case one may formulate statements similar to those of Theorems 4.1 and 4.2.

Remark 4.5 Equation (4.1.33) yields the explicit solution to the Cauchy type problem (4.1.25), (4.1.26) for any $\beta > -\{\alpha\}$, but, according to Theorem 4.2 and Remark 4.2, this solution is unique in spaces $\mathbf{C}^{\alpha}_{n-\alpha,\gamma}[a,b]$ and $\mathbf{L}^{\alpha}(a,b)$ when $\beta \geqq 0$. The problem of the uniqueness of this solution in the case $-\{\alpha\} < \beta < 0$ remains open.

Example 4.3 The solution to the Cauchy type problem

$$(D^{\alpha}_{a+}y)(x) - \lambda(x-a)^{\beta}y(x) = 0; \quad (D^{\alpha-1}_{a+}y)(a+) = b \quad (b \in \mathbb{R}) \tag{4.1.34}$$

with $0 < \alpha < 1$, $\beta \in \mathbb{R}$ $(\beta > -\{\alpha\})$ and $\lambda \in \mathbb{R}$ is given by

$$y(x) = \frac{b}{\Gamma(\alpha)}(x-a)^{\alpha-1}E_{\alpha,1+\beta/\alpha,1+(\beta-1)/\alpha}\left[\lambda(x-a)^{\alpha+\beta}\right]. \tag{4.1.35}$$

In particular, the Cauchy type problem

$$(D^{1/2}_{a+}y)(x) - \lambda(x-a)^{\beta}y(x) = 0; \quad (D^{-1/2}_{a+}y)(a+) = b \quad (b \in \mathbb{R}) \tag{4.1.36}$$

has a unique solution given by

$$y(x) = \frac{b}{\sqrt{\pi}}(x-a)^{-1/2}E_{1/2,1+2\beta,2\beta-1}\left[\lambda(x-a)^{\beta+1/2}\right]. \tag{4.1.37}$$

Example 4.4 The solution to the Cauchy type problem

$$(D^{\alpha}_{a+}y)(x) - \lambda(x-a)^{\beta}y(x) = 0; \quad (D^{\alpha-1}_{a+}y)(a+) = b, \quad (D^{\alpha-2}_{a+}y)(a+) = d \tag{4.1.38}$$

with $b, d \in \mathbb{R}$, $1 < \alpha < 2$, $\beta \in \mathbb{R}$ $(\beta > -\{\alpha\})$ and $\lambda \in \mathbb{R}$ has the form

$$y(x) = \frac{b}{\Gamma(\alpha)}(x-a)^{\alpha-1}E_{\alpha,1+\beta/\alpha,1+(\beta-1)/\alpha}\left[\lambda(x-a)^{\alpha+\beta}\right]$$

$$+\frac{d}{\Gamma(\alpha-1)}(x-a)^{\alpha-2}E_{\alpha,1+\beta/\alpha,1+(\beta-2)/\alpha}\left[\lambda(x-a)^{\alpha+\beta}\right]. \tag{4.1.39}$$

Remark 4.6 When $\beta = 0$, then, $E_{\alpha,1,1-j/\alpha}(z) = \Gamma(\alpha - j + 1)\ E_{\alpha,\alpha-j+1}(z)$, by (1.9.23). Hence the solution (4.1.33) to the Cauchy type problem (4.1.26)-(4.1.27) yields the solution (4.1.10) to the Cauchy type problem (4.1.1)-(4.1.2). In particular, solutions (4.1.35), (4.1.37) and (4.1.39) to the problems (4.1.34), (4.1.36) and (4.1.38) yield the solutions (4.1.16), (4.1.20) and (4.1.24) to the problems (4.1.15), (4.1.19) and (4.1.23), respectively.

4.1.2 The Cauchy Problems for Ordinary Differential Equations

In this section we use the results of Section 4.1.1 with $\alpha = n \in \mathbb{N}$ to derive the explicit solutions to the Cauchy problems for ordinary differential equations of order n on a finite interval $[a, b]$. First we consider the Cauchy problem

$$y^{(n)}(x) - \lambda y(x) = f(x); \quad y^{(n-k)}(a) = b_k \quad (\lambda, b_k \in \mathbb{R}, \quad n, k \in \mathbb{N}). \tag{4.1.40}$$

By (2.1.7), this problem is a particular case of the problem (4.1.1)-(4.1.2) with $\alpha = n \in \mathbb{N}$. Therefore, from (4.1.10) we derive the solution to (4.1.40) in the following form:

$$y(x) = \sum_{j=1}^{n} b_j (x-a)^{n-j} E_{n,n-j+1}\left[\lambda(x-a)^n\right] + \int_a^x (x-t)^{n-1} E_{n,n}\left[\lambda(x-t)^n\right] f(t)dt. \tag{4.1.41}$$

This is the unique explicit solution to the Cauchy problem (4.1.40) in the space $C^n[a, b]$. In particular, when $f(x) = 0$, the solution to the problem

$$y^{(n)}(x) - \lambda y(x) = 0; \quad y^{(n-k)}(a) = b_k \quad (\lambda, b_k \in \mathbb{R}, \quad n, k \in \mathbb{N}) \tag{4.1.42}$$

is given by

$$y(x) = \sum_{j=1}^{n} b_j (x-a)^{n-j} E_{n,n-j+1}\left[\lambda(x-a)^n\right]. \tag{4.1.43}$$

Example 4.5 The solution to the Cauchy problem

$$y'(x) - \lambda y(x) = f(x); \quad y(a) = b \ (b \in \mathbb{R}) \tag{4.1.44}$$

is given by

$$y(x) = b\ E_{1,1}[\lambda(x-a)] + \int_a^x E_{1,1}[\lambda(x-t)] f(t)dt,$$

which, in accordance with (1.8.18) and (1.8.2), takes the well-known form

$$y(x) = b\ e^{\lambda(x-a)} + \int_a^x e^{\lambda(x-t)} f(t)dt. \tag{4.1.45}$$

Example 4.6 The solution to the Cauchy problem

$$y''(x) - \lambda y(x) = f(x); \quad y(a) = b, \ y'(x) = d \ (b, d \in \mathbb{R}) \tag{4.1.46}$$

is given by

$$y(x) = b(x-a)E_{2,2}[\lambda(x-a)^2] + dE_{2,1}[\lambda(x-a)^2] + \int_a^x (x-t)E_{2,2}[\lambda(x-t)^2]f(t)dt. \tag{4.1.47}$$

In particular, for $f(x) = 0$ the solution to the problem

$$y''(x) - \lambda y(x) = 0; \quad y(a) = b, \ y'(x) = d \ (b, d \in \mathbb{R}) \tag{4.1.48}$$

has the form

$$y(x) = b(x-a)E_{2,2}[\lambda(x-a)^2] + dE_{2,1}[\lambda(x-a)^2]. \tag{4.1.49}$$

If $\lambda > 0$, then, according to (1.8.19), we can rewrite (1.8.18) and (1.8.2), (4.1.47) and (4.1.49) in the respective forms

$$y(x) = \frac{b}{\sqrt{\lambda}}(x-a)^{1/2}\sinh[\sqrt{\lambda}(x-a)] + d\cosh[\sqrt{\lambda}(x-a)]$$

$$+\frac{1}{\sqrt{\lambda}}\int_a^x (x-t)^{1/2}\sinh[\sqrt{\lambda}(x-t)]f(t)dt \tag{4.1.50}$$

and

$$y(x) = \frac{b}{\sqrt{\lambda}}(x-a)^{1/2}\sinh[\sqrt{\lambda}(x-a)] + d\cosh[\sqrt{\lambda}(x-a)]. \tag{4.1.51}$$

Next we consider the following Cauchy problem:

$$y^{(n)}(x) - \lambda(x-a)^{\beta}y(x) = f(x); \quad y^{(n-k)}(a) = b_k, \quad (\lambda, b_k \in \mathbb{R}, \ \ n, k \in \mathbb{N}) \tag{4.1.52}$$

with real $\lambda, \beta \geqq 0$. By (2.1.7), this problem is a particular case of the problem (4.1.25)-(4.1.26) with $\alpha = n$, and (4.1.33) yields the solution to (4.1.52):

$$y(x) = \sum_{j=1}^{n} \frac{b_j}{\Gamma(n-j+1)}(x-a)^{n-j}E_{n,1+\beta/n,1+(\beta-j)/n}\left[\lambda(x-a)^{n+\beta}\right]. \tag{4.1.53}$$

It is directly verified, on the basis of the formula (1.9.25), that (4.1.53) is the solution to the problem (4.1.52) for any $\beta > -n$. By Property 3.1(b) (with $\alpha = n$ and $\gamma = 0$) and Remark 3.18, the solution (4.1.53) is unique in the space $C^n[a,b] := C^n_{0,0}[a,b]$.

Remark 4.7 The problem of the uniqueness of the solution (4.1.53) in the case $-n < \beta < 0$ remains open.

Example 4.7 The solution to the Cauchy problem

$$y'(x) - \lambda(x-a)^{\beta}y(x) = f(x); \quad y(a) = b (b \in \mathbb{R}) \tag{4.1.54}$$

with $\beta > -1$ has the form

$$y(x) = bE_{1,1+\beta,\beta}\left[\lambda(x-a)^{1+\beta}\right]. \tag{4.1.55}$$

Example 4.8 The solution to the Cauchy problem

$$y''(x) - \lambda(x-a)^{\beta}y(x) = f(x), \quad y(a) = b, \; y'(a) = d(b, d \in \mathbb{R}) \tag{4.1.56}$$

with $\beta > -2$ is given by

$$y(x) = b(x-a)E_{2,1+\beta/2,(\beta+1)/2}\left[\lambda(x-a)^{2+\beta}\right] + dE_{2,1+\beta/2,\beta/2}\left[\lambda(x-a)^{2+\beta}\right]. \tag{4.1.57}$$

4.1.3 The Cauchy Problems for Differential Equations with the Caputo Fractional Derivatives

In this section we construct the explicit solutions to linear fractional differential equations with the Caputo fractional derivative $\left({}^{C}D_{a+}^{\alpha}y\right)(x)$ of order $\alpha > 0$ ($\alpha \notin \mathbb{N}$) given by (2.4.1) in the space $C_{\gamma}^{\alpha,n-1}[a,b]$ ($n = [\alpha]+1$), defined in (3.5.3).

First we consider the Cauchy problem for the fractional differential equation (3.1.38) of order $\alpha > 0$ with the initial conditions (3.1.37):

$$({}^{C}D_{a+}^{\alpha}y)(x) - \lambda y(x) = f(x) \quad (a \leqq x \leqq b; \; n-1 < \alpha < n; \; n \in \mathbb{N}; \; \lambda \in \mathbb{R}), \tag{4.1.58}$$

$$y^{(k)}(a) = b_k \quad (b_k \in \mathbb{R}; \; k = 0, \cdots, n-1). \tag{4.1.59}$$

We suppose that $f(x) \in C_{\gamma}[a,b]$ with $0 \leqq \gamma < 1$ and $\gamma \leqq \alpha$. Then, by Theorem 3.24, the problem (4.1.58)-(4.1.59) is equivalent in the space $C^{n-1}[a,b]$ to the Volterra integral equation of the second kind of the form (3.5.4):

$$y(x) = \sum_{j=0}^{n-1}\frac{b_j}{j!}(x-a)^j + \frac{\lambda}{\Gamma(\alpha)}\int_a^x \frac{y(t)dt}{(x-t)^{1-\alpha}} + \frac{1}{\Gamma(\alpha)}\int_a^x \frac{f(t)dt}{(x-t)^{1-\alpha}} \tag{4.1.60}$$

To solve (4.1.60), we apply the method of successive approximations by setting

$$y_0(x) = \sum_{j=0}^{n-1}\frac{b_j}{j!}(x-a)^j \tag{4.1.61}$$

and

$$y_m(x) = y_0(x) + \frac{\lambda}{\Gamma(\alpha)}\int_a^x \frac{y_{m-1}(t)dt}{(x-t)^{1-\alpha}} + \frac{1}{\Gamma(\alpha)}\int_a^x \frac{f(t)dt}{(x-t)^{1-\alpha}} \quad (m \in \mathbb{N}).$$

Arguments similar to those in Section 4.1.1 yield the following expression for $y_m(x)$:

$$y_m(x) = \sum_{j=0}^{n-1} b_j \sum_{k=0}^{m}\frac{\lambda^k(x-a)^{\alpha k+j}}{\Gamma(\alpha k+j+1)} + \int_a^x\left[\sum_{k=1}^{m}\frac{\lambda^{k-1}}{\Gamma(\alpha k)}(x-t)^{\alpha k-1}\right]f(t)dt.$$

Taking the limit as $m \to \infty$, and taking into account (1.8.17), we obtain

$$y(x) = \sum_{j=0}^{n-1} b_j(x-a)^j E_{\alpha,j+1}\left[\lambda(x-a)^{\alpha}\right] + \int_a^x (x-t)^{\alpha-1}E_{\alpha,\alpha}\left[\lambda(x-t)^{\alpha}\right]f(t)dt. \tag{4.1.62}$$

This yields an explicit solution $y(x)$ of the Volterra integral equation (4.1.60) and hence of the Cauchy problem (4.1.58)-(4.1.59).

Since $f[x, y] = \lambda y + f(x)$ satisfies the Lipschitz condition, by Theorem 3.25 there exists a unique solution $y(x)$ to the Cauchy problem (4.1.58)-(4.1.59) in the space $\mathbf{C}_\gamma^{\alpha,n-1}[a, b]$ defined by (3.5.3). From here we obtain the following result.

Theorem 4.3 *Let $n-1 < \alpha < n$ $(n \in \mathbb{N})$ and let $0 \leqq \gamma < 1$ be such that $\gamma \leqq \alpha$. Also let $\lambda \in \mathbb{R}$. If $f(x) \in C_\gamma[a, b]$, the Cauchy problem* (4.1.58)-(4.1.59) *has a unique solution $y(x) \in \mathbf{C}_\gamma^{\alpha,n-1}[a, b]$ and this solution is given by* (4.1.62).

In particular, if $\gamma = 0$ and $f(x) \in C[a, b]$, then the solution $y(x)$ in (4.1.62) *belongs to the space $\mathbf{C}^{\alpha,n-1}[a, b]$ defined in* (3.5.18).

The Cauchy problem (4.1.59) *involving the homogeneous differential equation* (4.1.58)

$$({}^C D_{a+}^{\alpha} y)(x) - \lambda y(x) = 0 \quad (a \leqq x \leqq b;\ n-1 < \alpha < n;\ n \in \mathbb{N};\ \lambda \in \mathbb{R}) \qquad (4.1.63)$$

has a unique solution $y(x) \in \mathbf{C}_\gamma^{\alpha,n-1}[a, b]$ of the form

$$y(x) = \sum_{j=0}^{n-1} b_j (x-a)^j E_{\alpha,j+1}\left[\lambda(x-a)^\alpha\right]. \qquad (4.1.64)$$

Example 4.9 The solution to the Cauchy problem

$$({}^C D_{a+}^{\alpha} y)(x) - \lambda y(x) = f(x), \quad y(a) = b \ (b \in \mathbb{R}) \qquad (4.1.65)$$

with $0 < \alpha < 1$ and $\lambda \in \mathbb{R}$ has the form

$$y(x) = bE_\alpha\left[\lambda(x-a)^\alpha\right] + \int_a^x (x-t)^{\alpha-1} E_{\alpha,\alpha}\left[\lambda(x-t)^\alpha\right] f(t)dt, \qquad (4.1.66)$$

while the solution to the problem

$$({}^C D_{a+}^{\alpha} y)(x) - \lambda y(x) = 0, \quad y(a) = b \ (b \in \mathbb{R}) \qquad (4.1.67)$$

is given by

$$y(x) = bE_\alpha\left[\lambda(x-a)^\alpha\right]. \qquad (4.1.68)$$

In particular, the Cauchy problem

$$({}^C D_{a+}^{1/2} y)(x) - \lambda y(x) = f(x), \quad y(a) = b \ (b \in \mathbb{R}) \qquad (4.1.69)$$

has the solution given by

$$y(x) = bE_{1/2}\left[\lambda(x-a)^{1/2}\right] + \int_a^x (x-t)^{-1/2} E_{1/2,1/2}\left[\lambda(x-t)^{1/2}\right] f(t)dt, \quad (4.1.70)$$

and the solution to the problem

$$({}^C D_{a+}^{1/2} y)(x) - \lambda y(x) = 0, \quad y(a) = b \ (b \in \mathbb{R}) \qquad (4.1.71)$$

is given by

$$y(x) = bE_{1/2}\left[\lambda(x-a)^{1/2}\right]. \qquad (4.1.72)$$

Example 4.10 The solution to the Cauchy problem

$$({}^C D_{a+}^{\alpha} y)(x) - \lambda y(x) = f(x), \quad y(a) = b, \quad y'(a) = d \ (b, d \in \mathbb{R}) \tag{4.1.73}$$

with $1 < \alpha < 2$ and $\lambda \in \mathbb{R}$ has the form

$$y(x) = bE_{\alpha}\left[\lambda(x-a)^{\alpha}\right] + d(x-a)E_{\alpha,2}\left[\lambda(x-a)^{\alpha}\right] + \int_a^x (x-t)^{\alpha-1}E_{\alpha,\alpha}\left[\lambda(x-t)^{\alpha}\right] f(t)dt. \tag{4.1.74}$$

In particular, the solution to the problem

$$({}^C D_{a+}^{\alpha} y)(x) - \lambda y(x) = 0 \ (1 < \alpha < 2), \quad y(a) = b, \quad y'(a) = d \ (b, d \in \mathbb{R}) \tag{4.1.75}$$

is given by

$$y(x) = bE_{\alpha}\left[\lambda(x-a)^{\alpha}\right] + d(x-a)E_{\alpha,2}\left[\lambda(x-a)^{\alpha}\right]. \tag{4.1.76}$$

Now we consider the Cauchy problem for the following more general homogeneous fractional differential equation than (4.1.63):

$$({}^C D_{a+}^{\alpha} y)(x) - \lambda(x-a)^{\beta} y(x) = 0 \quad (a \leqq x \leqq b; \ n-1 < \alpha < n; \ n \in \mathbb{N}; \ \lambda \in \mathbb{R}), \tag{4.1.77}$$

$$y^{(k)}(0) = b_k \ (b_k \in \mathbb{R}; \ k = 0, \cdots, n-1), \tag{4.1.78}$$

with $\beta > -\alpha$. Note again that, in accordance with Theorem 3.24, the problem (4.1.77)-(4.1.78) is equivalent in the space $C^{n-1}[a,b]$ to the following Volterra integral equation of the second kind:

$$y(x) = \sum_{j=0}^{n-1} \frac{b_j}{j!}(x-a)^j + \frac{\lambda}{\Gamma(\alpha)} \int_a^x \frac{(t-a)^{\beta} y(t)dt}{(x-t)^{1-\alpha}}. \tag{4.1.79}$$

We again apply the method of successive approximations to solve this integral equation. We use the notation $y_0(x)$ in (4.1.61) and set

$$y_m(x) = y_0(x) + \frac{\lambda}{\Gamma(\alpha)} \int_a^x \frac{(t-a)^{\beta} y_{m-1}(t)dt}{(x-t)^{1-\alpha}} \quad (m \in \mathbb{N}).$$

The same arguments as in Section 4.1.1 lead to the following expression for $y_m(x)$:

$$y_m(x) = \sum_{j=0}^{n-1} \frac{b_j}{j!}(x-a)^j \left[1 + \sum_{k=1}^{m} d_k \left(\lambda(x-a)^{\alpha+\beta}\right)^k\right], \tag{4.1.80}$$

where

$$d_k = \prod_{r=1}^{k} \frac{\Gamma(r\alpha + r\beta + \beta + j - \alpha + 1)}{\Gamma(r\alpha + r\beta\alpha + \beta + 1)} \quad (k \in \mathbb{N}). \tag{4.1.81}$$

Taking the limit as $m \to \infty$, and taking (1.9.19) and (1.9.20) into account, we obtain the explicit solution $y(x)$ of the integral equation (4.1.79) and hence of the Cauchy problem (4.1.77)-(4.1.78) in terms of the generalized Mittag-Leffler function (1.9.19):

$$y(x) = \sum_{j=0}^{n-1} \frac{b_j}{j!}(x-a)^j E_{\alpha,1+\beta/\alpha,(\beta+j)/\alpha}\left[\lambda(x-a)^{\alpha+\beta}\right]. \tag{4.1.82}$$

If $\beta \geqq 0$, then $f[x,y] = \lambda(x-a)^\beta y$ satisfies the Lipschitz condition. Hence, by Theorem 3.25, (4.1.28) is a unique explicit solution to the Cauchy problem (4.1.77)-(4.1.78) in the space $C_\gamma^{\alpha,n-1}[a,b]$.

Theorem 4.4 *Let* $n-1<\alpha<n$ $(n \in \mathbb{N})$ *and let* $0 \leqq \gamma < 1$ *be such that* $\gamma \leqq \alpha$. *Also let* $\lambda \in \mathbb{R}$ *and* $\beta \geqq 0$. *If* $f \in C_\gamma[a,b]$, *then the Cauchy problem* (4.1.77)-(4.1.78) *has a unique solution* $y(x) \in \mathbf{C}_\gamma^{\alpha,n-1}[a,b]$ *and this solution is given by* (4.1.82).

Remark 4.8 Equation (4.1.82) yields the explicit solution to the Cauchy problem (4.1.77)-(4.1.78) for any $\beta > -\{\alpha\}$. But, in accordance with Theorem 4.4, this solution is unique in spaces $\mathbf{C}_\gamma^{\alpha,n-1}[a,b]$ when $\beta \geqq 0$. The problem of the uniqueness of this solution in the case $-\{\alpha\} < \beta < 0$ remains open.

Example 4.11 The solution to the Cauchy problem

$$({}^C D_{a+}^\alpha y)(x) - \lambda(x-a)^\beta y(x) = 0, \quad y(a) = b \ (b \in \mathbb{R}) \tag{4.1.83}$$

with $0<\alpha<1$, $\beta > -\alpha$ and $\lambda \in \mathbb{R}$ has the form:

$$y(x) = bE_{\alpha,1+\beta/\alpha,\beta/\alpha}\left[\lambda(x-a)^{\alpha+\beta}\right]. \tag{4.1.84}$$

In particular, the solution to the Cauchy problem

$$({}^C D_{a+}^{1/2} y)(x) - \lambda(x-a)^\beta y(x) = 0, \quad y(a) = b \ (b \in \mathbb{R}) \tag{4.1.85}$$

with $\beta > -1/2$ is given by

$$y(x) = bE_{1/2,2\beta+1,2\beta}\left[\lambda(x-a)^{\beta+1/2}\right]. \tag{4.1.86}$$

Example 4.12 The solution to the Cauchy problem

$$({}^C D_{a+}^\alpha y)(x) - \lambda(x-a)^\beta y(x) = 0, \quad y(a) = b, \quad y'(a) = d \ (b,d \in \mathbb{R}) \tag{4.1.87}$$

with $1<\alpha<2$, $\beta > -\alpha$ and $\lambda \in \mathbb{R}$ has the form

$$y(x) = bE_{\alpha,1+\beta/\alpha,\beta/\alpha}\left[\lambda(x-a)^{\alpha+\beta}\right] + d(x-a)E_{\alpha,1+\beta/\alpha,(\beta+1)/\alpha}\left[\lambda(x-a)^{\alpha+\beta}\right]. \tag{4.1.88}$$

Remark 4.9 When $\beta = 0$, in accordance with (1.9.23) we have

$$E_{\alpha,1,j/\alpha}(z) = \Gamma(j+1)E_{\alpha,j+1}(z)$$

Thus the solution (4.1.82) to the Cauchy problem (4.1.77)-(4.1.78) yields the solution (4.1.64) to the Cauchy problem (4.1.58)-(4.1.59). In particular, solutions (4.1.84), (4.1.86) and (4.1.88) to problems (4.1.83), (4.1.83) and (4.1.87) yield solutions (4.1.68), (4.1.72) and (4.1.76) to problems (4.1.67), (4.1.71) and (4.1.75), respectively.

4.1.4 The Cauchy Type Problems for Differential Equations with Hadamard Fractional Derivatives

In this section we construct the explicit solutions to linear fractional differential equations with the Riemann-Liouville fractional derivative $\left(\mathcal{D}_{a+}^{\alpha}y\right)(x)$ of order $\alpha > 0$ given by (2.7.7) in the space $C_{\delta;n-\alpha,\gamma}^{\alpha}[a,b]$ $(n = -[-\alpha];\ 0 \leqq \gamma < 1)$, defined in (3.6.1).

First we consider the following Cauchy type problem:

$$(\mathcal{D}_{a+}^{\alpha}y)(x) - \lambda y(x) = f(x) \quad (a < x \leqq b;\ \alpha > 0;\ \lambda \in \mathbb{R}), \tag{4.1.89}$$

$$(\mathcal{D}_{a+}^{\alpha-k}y)(a+) = b_k \quad (b_k \in \mathbb{R};\ k = 1, \cdots, n;\ n = -[-\alpha]), \tag{4.1.90}$$

with $f(x) \in C_{\gamma,\log}[a,b]$ $(0 \leqq \gamma < 1)$ [see (1.1.27)]. Then, by Theorem 3.28, (4.1.89)-(4.1.90) is equivalent in the space $C_{n-\alpha,\log}[a,b]$ to the following Volterra integral equation of the second kind:

$$y(x) = \sum_{j=1}^{n} \frac{b_j}{\Gamma(\alpha-j+1)} \left(\log\frac{x}{a}\right)^{\alpha-j}$$

$$+\frac{\lambda}{\Gamma(\alpha)} \int_a^x \left(\log\frac{x}{t}\right)^{\alpha-1} y(t)\frac{dt}{t} + \frac{1}{\Gamma(\alpha)} \int_a^x \left(\log\frac{x}{t}\right)^{\alpha-1} f(t)\frac{dt}{t}. \tag{4.1.91}$$

Applying the method of successive approximations, we set

$$y_0(x) = \sum_{j=1}^{n} \frac{b_j}{\Gamma(\alpha-j+1)} \left(\log\frac{x}{a}\right)^{\alpha-j}, \tag{4.1.92}$$

$$y_m(x) = y_0(x) + \frac{\lambda}{\Gamma(\alpha)} \int_a^x \left(\log\frac{x}{t}\right)^{\alpha-1} y(t)\frac{dt}{t} + \frac{1}{\Gamma(\alpha)} \int_a^x \left(\log\frac{x}{t}\right)^{\alpha-1} f(t)\frac{dt}{t}, \tag{4.1.93}$$

with $m \in \mathbb{N}$; and we derive the relation for $y_m(x)$ $(m \in \mathbb{N})$:

$$y_m(x) = \sum_{j=1}^{n} b_j \sum_{l=1}^{m+1} \frac{\lambda^{l-1}}{\Gamma(\alpha l-j+1)} \left(\log\frac{x}{a}\right)^{\alpha l-j}$$

$$+\int_a^x \left[\sum_{l=1}^{m} \frac{\lambda^{l-1}}{\Gamma(\alpha l)} \left(\log\frac{x}{a}\right)^{\alpha l-1}\right] f(t)\frac{dt}{t}. \tag{4.1.94}$$

Taking the limit as $m \to \infty$, we obtain the solution to the integral equation (4.1.91) and hence to the problem (4.1.89)-(4.1.90) in terms of the function (1.8.17):

$$y(x) = \sum_{j=1}^{n} b_j (x-a)^{\alpha-j} E_{\alpha,\alpha-j+1}\left[\lambda\left(\log\frac{x}{a}\right)^{\alpha}\right]$$

$$+\int_a^x \left(\log\frac{x}{t}\right)^{\alpha-1} E_{\alpha,\alpha}\left[\lambda(\left(\log\frac{x}{t}\right)^{\alpha}\right] f(t)\frac{dt}{t}. \tag{4.1.95}$$

It is clear that $f[x,y] = \lambda y + f(x)$ satisfies the Lipschitz condition. If $\gamma \geqq n-\alpha$, then, by Theorem 3.29, there exists a unique solution to the problem (4.1.89)-(4.1.90) in the space $\mathbf{C}^{\alpha}_{\delta;,n-\alpha,\gamma}[a,b]$. Equation (4.1.95) yields this solution. Thus we derive the following result.

Theorem 4.5 *Let* $\alpha > 0$, $n = -[-\alpha]$ *and let* γ $(0 \leqq \gamma < 1)$ *be such that* $\gamma \geqq n-\alpha$. *Also let* $\lambda \in \mathbb{R}$. *If* $f \in C_{\gamma,\log}[a,b]$, *then the Cauchy type problem* (4.1.89)-(4.1.90) *has a unique solution* $y(x) \in \mathbf{C}^{\alpha}_{\delta;n-\alpha,\gamma}[a,b]$ *and this solution is given by* (4.1.95).

In particular, the Cauchy type problem involving the homogeneous differential equation (4.1.89)

$$(\mathcal{D}^{\alpha}_{a+}y)(x) - \lambda y(x) = 0 \quad (a < x \leqq b;\ \alpha > 0;\ \lambda \in \mathbb{R}), \tag{4.1.96}$$

with the initial conditions (4.1.90), *has a unique solution* $y(x)$ *in the space* $C^{\alpha}_{\delta;n-\alpha}[a,b] := C^{\alpha}_{\delta;n-\alpha,0}[a,b]$ *of the form*

$$y(x) = \sum_{j=1}^{n} b_j \left(\log\frac{x}{a}\right)^{\alpha-j} E_{\alpha,\alpha-j+1}\left[\lambda\left(\log\frac{x}{a}\right)^{\alpha}\right]. \tag{4.1.97}$$

Remark 4.10 The explicit solution (4.1.95) to the Cauchy type problem (4.1.89)-(4.1.90) was given by Kilbas et al. [383].

Example 4.13 The solution to the Cauchy type problem

$$(\mathcal{D}^{\alpha}_{a+}y)(x) - \lambda y(x) = f(x), \quad (\mathcal{D}^{\alpha-1}_{a+}y)(a+) = b \quad (b \in \mathbb{R}) \tag{4.1.98}$$

with $0 < \alpha < 1$ and $\lambda \in \mathbb{R}$ has the form

$$y(x) = b\left(\log\frac{x}{a}\right)^{\alpha-1} E_{\alpha,\alpha}\left[\lambda\left(\log\frac{x}{a}\right)^{\alpha}\right] + \int_a^x \left(\log\frac{x}{t}\right)^{\alpha-1} E_{\alpha,\alpha}\left[\lambda\left(\log\frac{x}{t}\right)^{\alpha}\right] f(t)\frac{dt}{t}, \tag{4.1.99}$$

while the solution to the problem

$$(\mathcal{D}^{\alpha}_{a+}y)(x) - \lambda y(x) = 0; \quad (\mathcal{D}^{\alpha-1}_{a+}y)(a+) = b \quad (b \in \mathbb{R}) \tag{4.1.100}$$

is given by

$$y(x) = b\left(\log\frac{x}{a}\right)^{\alpha-1} E_{\alpha,\alpha}\left[\lambda\left(\log\frac{x}{a}\right)^{\alpha}\right]. \tag{4.1.101}$$

In particular, the Cauchy type problem

$$(\mathcal{D}^{1/2}_{a+}y)(x) - \lambda y(x) = f(x); \quad (\mathcal{D}^{-1/2}_{a+}y)(a+) = b \quad (b \in \mathbb{R}) \tag{4.1.102}$$

has the solution given by

$$y(x) = b\left(\log\frac{x}{a}\right)^{-1/2} E_{1/2,1/2}\left[\lambda\left(\log\frac{x}{a}\right)^{1/2}\right]$$

$$+\int_a^x \left(\log\frac{x}{t}\right)^{-1/2} E_{1/2,1/2}\left[\lambda\left(\log\frac{x}{t}\right)^{1/2}\right] f(t)\frac{dt}{t}, \tag{4.1.103}$$

and the solution to the problem

$$(\mathcal{D}_{a+}^{1/2}y)(x) - \lambda y(x) = 0; \quad (\mathcal{D}_{a+}^{-1/2}y)(a+) = b \quad (b \in \mathbb{R}) \tag{4.1.104}$$

is given by

$$y(x) = b\left(\log\frac{x}{a}\right)^{-1/2} E_{1/2,1/2}\left[\lambda\left(\log\frac{x}{a}\right)^{1/2}\right]. \tag{4.1.105}$$

Example 4.14 Let $b, d \in \mathbb{R}$. The solution to the Cauchy type problem

$$(\mathcal{D}_{a+}^{\alpha}y)(x) - \lambda y(x) = f(x); \quad (\mathcal{D}_{a+}^{\alpha-1}y)(a+) = b, \quad (\mathcal{D}_{a+}^{\alpha-2}y)(a+) = d \tag{4.1.106}$$

with $1 < \alpha < 2$ and $\lambda \in \mathbb{R}$ has the form

$$y(x) = b\left(\log\frac{x}{a}\right)^{\alpha-1} E_{\alpha,\alpha}\left[\lambda(x-a)^{\alpha}\right] + d(x-a)^{\alpha-2} E_{\alpha,\alpha-1}\left[\lambda\left(\log\frac{x}{a}\right)^{\alpha}\right]$$

$$+\int_a^x \left(\log\frac{x}{t}\right)^{\alpha-1} E_{\alpha,\alpha}\left[\lambda\left(\log\frac{x}{t}\right)^{\alpha}\right] f(t)dt. \tag{4.1.107}$$

In particular, the solution to the problem

$$(\mathcal{D}_{a+}^{\alpha}y)(x) - \lambda y(x) = 0; \quad (\mathcal{D}_{a+}^{\alpha-1}y)(a+) = b, \quad (\mathcal{D}_{a+}^{\alpha-2}y)(a+) = d \tag{4.1.108}$$

with $1 < \alpha < 2$ and $b, d \in \mathbb{R}$, is given by

$$y(x) = b\left(\log\frac{x}{a}\right)^{\alpha-1} E_{\alpha,\alpha}\left[\lambda\left(\log\frac{x}{a}\right)^{\alpha}\right] + d(x-a)^{\alpha-2} E_{\alpha,\alpha-1}\left[\lambda\left(\log\frac{x}{a}\right)^{\alpha}\right]. \tag{4.1.109}$$

Next we consider the Cauchy type problem for the following more general homogeneous fractional differential equation than (4.1.89):

$$(\mathcal{D}_{a+}^{\alpha}y)(x) - \lambda(x-a)^{\beta}y(x) = 0 \quad (a < x \leqq b; \ \alpha > 0; \ \lambda \in \mathbb{R})), \tag{4.1.110}$$

$$(\mathcal{D}_{a+}^{\alpha-k}y)(a+) = b_k \ (b_k \in \mathbb{R}; \ k = 1, \cdots, n; \ n = -[-\alpha]), \tag{4.1.111}$$

with $\beta > -\{\alpha\}$. Again, by Theorem 3.28, the problem (4.1.110)-(4.1.11) is equivalent in the space $C_{n-\alpha,\log}[a,b]$ to the following Volterra integral equation:

$$y(x) = \sum_{j=1}^{n} \frac{b_j}{\Gamma(\alpha-j+1)}\left(\log\frac{x}{a}\right)^{\alpha-j} + \frac{\lambda}{\Gamma(\alpha)}\int_a^x \left(\log\frac{x}{t}\right)^{\beta} y(t)\frac{dt}{t}. \tag{4.1.112}$$

We again apply the method of successive approximations to solve this integral equation. We use the notation $y_0(x)$ in (4.1.92) and set

$$y_m(x) = y_0(x) + \frac{\lambda}{\Gamma(\alpha)} \int_a^x \left(\log \frac{x}{t}\right)^{\alpha-1} (t-a)^\beta y_{m-1}(t) \frac{dt}{t} \quad (m \in \mathbb{N}). \tag{4.1.113}$$

Using the same arguments as above, we derive the following relation for $y_m(x)$ $(m \in \mathbb{N})$:

$$y_m(x) = \sum_{j=1}^{n} \frac{b_j}{\Gamma(\alpha - j + 1)} \left(\log \frac{x}{a}\right)^{\alpha-j} \left[1 + \sum_{k=1}^{m} c_k \left(\lambda \left(\log \frac{x}{a}\right)^{\alpha+\beta}\right)^k\right], \tag{4.1.114}$$

where c_k are given by (4.1.31). Taking the limit as $m \to \infty$, we obtain the solution $y(x)$ to the integral equation (4.1.112) and hence to the Cauchy type problem 4.1.110)-(4.1.111) in terms of the generalized Mittag-Lefffler function (1.9.19)-(1.9.20):

$$y(x) = \sum_{j=1}^{n} \frac{b_j}{\Gamma(\alpha - j + 1)} \left(\log \frac{x}{a}\right)^{\alpha-j} E_{\alpha,1+\beta/\alpha,1+(\beta-j)/\alpha} \left[\lambda \left(\log \frac{x}{a}\right)^{\alpha+\beta}\right]. \tag{4.1.115}$$

If $\beta \geqq 0$, then $f[x,y] = \lambda(x-a)^\beta$ satisfies the Lipschitz condition. If $\gamma \geqq n-\alpha$, then, by Theorem 3.29, there exists a unique solution to the Cauchy type problem (4.1.110)-(4.1.111) in the space $\mathbf{C}^\alpha_{\delta;n-\alpha,\gamma}[a,b]$, and thus this solution has the form (4.1.115). This leads to the following result.

Theorem 4.6 *Let $\alpha > 0$, $n = -[-\alpha]$, $\lambda \in \mathbb{R}$ and $\beta \geqq 0$. Then the Cauchy type problem (4.1.110)-(4.1.111) has a unique solution $y(x)$ in the space $C^\alpha_{\delta;n-\alpha}[a,b] := C^\alpha_{\delta;n-\alpha,0}[a,b]$ and this solution is given by (4.1.115).*

Remark 4.11 If $\alpha > 1$ and $b_n = 0$, then, in accordance with Theorem 3.31, (4.1.95), and (4.1.97) and (4.1.115) are unique solutions to the respective Cauchy type problems (4.1.89), (4.1.90), and (4.1.96), (4.1.90), and (4.1.110), (4.1.111) in the space $\mathbf{C}^\alpha_\delta[a,b] := \mathbf{C}^\alpha_{\delta;0,0}[a,b$.

Remark 4.12 Equation (4.1.115) yields the explicit solution to the Cauchy type problem (4.1.110), (4.1.111) for any $\beta > -\{\alpha\}$. But, in accordance with Theorem 3.29, this solution is unique in the space $\mathbf{C}^\alpha_{\delta;n-\alpha,\gamma}[a,b]$ when $\beta \geqq 0$. The problem of the uniqueness of this solution in the case $-\{\alpha\} < \beta < 0$ remains open.

Example 4.15 The solution to the Cauchy type problem

$$(\mathcal{D}^\alpha_{a+}y)(x) - \lambda \left(\log \frac{x}{a}\right)^\beta y(x) = f(x); \quad (\mathcal{D}^{\alpha-1}_{a+}y)(a+) = b \tag{4.1.116}$$

with $0 < \alpha < 1$, b,$\beta \in \mathbb{R}$ $(\beta > -\{\alpha\})$ and $\lambda \in \mathbb{R}$ is given by

$$y(x) = \frac{b}{\Gamma(\alpha)} \left(\log \frac{x}{a}\right)^{\alpha-1} E_{\alpha,1+\beta/\alpha,1+(\beta-1)/\alpha} \left[\lambda \left(\log \frac{x}{a}\right)^{\alpha+\beta}\right]. \tag{4.1.117}$$

In particular, the Cauchy type problem

$$(\mathcal{D}_{a+}^{1/2}y)(x) - \lambda(x-a)^\beta y(x) = f(x); \quad (\mathcal{D}_{a+}^{-1/2}y)(a+) = b \tag{4.1.118}$$

with $b \in \mathbb{R}$, has a unique solution given by

$$y(x) = \frac{b}{\sqrt{\pi}}\left(\log\frac{x}{a}\right)^{-1/2} E_{1/2,1+2\beta,2\beta-1}\left[\lambda\left(\log\frac{x}{a}\right)^{\beta+1/2}\right]. \tag{4.1.119}$$

Example 4.16 The solution to the Cauchy type problem

$$(\mathcal{D}_{a+}^{\alpha}y)(x) - \lambda(x-a)^\beta y(x) = f(x); \quad (\mathcal{D}_{a+}^{\alpha-1}y)(a+) = b, \quad (D_{a+}^{\alpha-2}y)(a+) = d \tag{4.1.120}$$

with $1 < \alpha < 2$, $b, d, \beta \in \mathbb{R}$ $(\beta > -\{\alpha\})$ and $\lambda \in \mathbb{R}$, has the form

$$y(x) = \frac{b}{\Gamma(\alpha)}\left(\log\frac{x}{a}\right)^{\alpha-1} E_{\alpha,1+\beta/\alpha,1+(\beta-1)/\alpha}\left[\lambda\left(\log\frac{x}{a}\right)^{\alpha+\beta}\right]$$

$$+\frac{d}{\Gamma(\alpha-1)}\left(\log\frac{x}{a}\right)^{\alpha-2} E_{\alpha,1+\beta/\alpha,1+(\beta-2)/\alpha}\left[\lambda\left(\log\frac{x}{a}\right)^{\alpha+\beta}\right]. \tag{4.1.121}$$

Remark 4.13 When $\beta = 0$, in accordance with (1.9.23), the solutions (4.1.115), (4.1.117), (4.1.119) and (4.1.121) to the problems (4.1.110)-(4.1.111), (4.1.116), (4.1.118) and (4.1.120) yield the solutions (4.1.97), (4.1.101), (4.1.105) and (4.1.107) to the problems (4.1.96), (4.1.100), (4.1.104) and (4.1.106), respectively.

4.2 Compositional Method

In this section we present the method for solving, in closed form, some linear fractional differential equations and boundary value problems for these equations based on compositions of fractional integrals and derivatives with special functions of the Mittag-Leffler and Bessel type. For simplicity, we present our results for equations with fractional derivatives of order $\alpha > 0$. The corresponding results for equations with fractional derivatives of complex order are derived similarly.

4.2.1 Preliminaries

We illustrate the idea of the compositional method with the following formula for the Riemann-Liouville fractional derivative (2.1.6) given in (2.1.17):

$$(D_{a+}^{\alpha}(t-a)^{\beta-1})(x) = \frac{\Gamma(\beta)}{\Gamma(\beta-\alpha)}(x-a)^{\beta-\alpha-1} \quad (\beta > \alpha > 0). \tag{4.2.1}$$

From this equality we immediately derive that the simplest differential equation

$$(D_{a+}^{\alpha}y)(x) = (x-a)^{\beta-\alpha-1} \tag{4.2.2}$$

has the explicit solution

$$y(x) = \frac{\Gamma(\beta-\alpha)}{\Gamma(\beta)}(x-a)^{\beta-1} \quad (\beta > \alpha > 0). \tag{4.2.3}$$

Moreover, (4.2.1) means that the composition of the Riemann-Liouville fractional derivative D_{a+}^{α} with the power function $(x-a)^{\beta-1}$ leads to the same function except for a certain factor. It follows that

$$y(x) = (x-a)^{\beta-1} \tag{4.2.4}$$

is a solution to the homogeneous differential equation

$$(D_{a+}^{\alpha}y)(x) = \frac{\Gamma(\beta)(x-a)^{-\alpha}}{\Gamma(\beta-\alpha)}y(x). \tag{4.2.5}$$

These arguments lead us to the conjecture that compositions of fractional derivatives with elementary functions can give solutions to fractional differential equations. Moreover, from here we derive another assumption about the possibility of such results for compositions of fractional calculus operators with special functions of the type Mittag-Leffler and Bessel. Further in this section we develop such an approach for solving, in closed form, certain classes of linear fractional differential equations with the Riemann-Liouville and Liouville fractional derivatives.

4.2.2 Compositional Relations

In this section we present compositions of the Riemann-Liouville and Liouville fractional derivatives D_{a+}^{α} $(a \in \mathbb{R})$ and D_{-}^{α} with the generalized Mittag-Leffler function $E_{\alpha,m,l}(z)$ defined in (1.9.19)-(1.9.20), and also compositions of D_{-}^{α} with the Bessel type functions $Z_{\rho}^{\nu}(z)$ and $\lambda^{(\beta)}\nu,\sigma(z)$ defined in (1.7.42) and (1.7.51), respectively.

Compositions of D_{a+}^{α}, D_{-}^{α} with $E_{\alpha,m,l}(z)$ are given by the following statements.

Proposition 4.1 *Let* $\alpha > 0$ *and* $m > 0$ *such that*

$$l > m - 1 - \frac{1}{\alpha}, \quad \alpha(jm+l) \notin \mathbb{Z}^{-} \ (j \in \mathbb{N}_0). \tag{4.2.6}$$

Then there holds the formula

$$\left(\mathcal{D}_{a+}^{\alpha}\left[(t-a)^{\alpha(l-m+1)}E_{\alpha,m,l}\left(\lambda(t-a)^{\alpha m}\right)\right]\right)(x)$$

$$= \frac{\Gamma[\alpha(l-m+1)+1]}{\Gamma[\alpha(l-m)+1]}(x-a)^{\alpha(l-m)} + \lambda(x-a)^{\alpha l}E_{\alpha,m,l}\left(\lambda(x-a)^{\alpha m}\right). \tag{4.2.7}$$

In particular, if $\alpha(l-m) = -j$ *for some* $j = 1, \cdots, -[-\alpha]$, *then*

$$\left(\mathcal{D}_{a+}^{\alpha}\left[(t-a)^{\alpha(l-m+1)}E_{\alpha,m,l}\left(\lambda(t-a)^{\alpha m}\right)\right]\right)(x) = \lambda(x-a)^{\alpha l}E_{\alpha,m,l}\left(\lambda(x-a)^{\alpha m}\right). \tag{4.2.8}$$

Corollary 4.1 *If $\alpha > 0$, $\beta > 0$ and $\lambda \in \mathbb{R}$, then*

$$\left(\mathcal{D}_{a+}^{\alpha}\left[(t-a)^{\beta-1}E_{\alpha,\beta}\left(\lambda(t-a)^{\alpha}\right)\right]\right)(x)$$

$$= \frac{1}{\Gamma(\beta-\alpha)}(x-a)^{\beta-\alpha-1} + \lambda(x-a)^{\beta-1}E_{\alpha,\beta}\left(\lambda(x-a)^{\alpha}\right). \tag{4.2.9}$$

In particular, if $\beta-\alpha \in \mathbb{Z}_0^-$, then

$$\left(\mathcal{D}_{a+}^{\alpha}\left[(t-a)^{\beta-1}E_{\alpha,\beta}\left(\lambda(t-a)^{\alpha}\right)\right]\right)(x) = \lambda(x-a)^{\beta-1}E_{\alpha,\beta}\left(\lambda(x-a)^{\alpha}\right). \tag{4.2.10}$$

Corollary 4.2 *If $\alpha > 0$ and $\lambda \in \mathbb{R}$, then*

$$\left(\mathcal{D}_{a+}^{\alpha}\left[E_{\alpha}\left(\lambda(t-a)^{\alpha}\right)\right]\right)(x) = \frac{1}{\Gamma(1-\alpha)}(x-a)^{-\alpha} + \lambda E_{\alpha}\left(\lambda(x-a)^{\alpha}\right). \tag{4.2.11}$$

In particular, if $0 < \alpha < 1$, then

$$\left(\mathcal{D}_{a+}^{\alpha}\left[E_{\alpha}\left(\lambda(t-a)^{\alpha}\right)\right]\right)(x) = (x-a)^{-\alpha}E_{\alpha,1-\alpha}\left(\lambda(x-a)^{\alpha}\right). \tag{4.2.12}$$

Proposition 4.2 *Let $\lambda \in \mathbb{R}$ and let $\alpha > 0$ and $m > 0$ such that $l > m - \{\alpha\}/\alpha$, where $\{\alpha\}$ is the fractional part of α.*

Then there holds the formula

$$\left(\mathcal{D}_{-}^{\alpha}\left[t^{\alpha(m-l)}E_{\alpha,m,l}\left(\lambda t^{-\alpha m}\right)\right]\right)(x)$$

$$= \frac{\Gamma[\alpha(l-m+1)+1]}{\Gamma[\alpha(l-m)+1]}x^{\alpha(m-l-1)-1} + \lambda x^{-\alpha(l+1)-1}E_{\alpha,m,l}\left(\lambda x^{-\alpha m}\right). \tag{4.2.13}$$

Corollary 4.3 *If $\alpha > 0$ and $\beta > [\alpha] + 1$, then*

$$\left(\mathcal{D}_{-}^{\alpha}\left[t^{\alpha-\beta}E_{\alpha,\beta}\left(\lambda t^{-\alpha}\right)\right]\right)(x) = \frac{1}{\Gamma(\beta-\alpha)}x^{-\beta} + \lambda x^{-\alpha-\beta}E_{\alpha,\beta}\left(\lambda x^{-\alpha}\right). \tag{4.2.14}$$

By (2.1.7), when $\alpha = n \in \mathbb{N}$, $(D_{a+}^{n}y)(x) = y^{(n)}(x)$. Therefore, from Proposition 4.1 and Corollaries 4.1 and 4.2, we derive the following assertions.

Proposition 4.3 *Let $\lambda, a \in \mathbb{R}$. Also let $n \in \mathbb{N}$, $m > 0$ and $l \in \mathbb{R}$ be such that*

$$n(jm+l) \neq -1, -2, \cdots, -n \quad (j \in \mathbb{N}_0). \tag{4.2.15}$$

Suppose that $D = d/dx$.

Then there holds the formula

$$D^{n}\left[(x-a)^{n(l-m+1)}E_{n,m,l}\left(\lambda(x-a)^{nm}\right)\right]$$

$$= \prod_{j=1}^{n}[n(l-m)+j](x-a)^{n(l-m)} + \lambda(x-a)^{nl}E_{n,m,l}\left(\lambda(x-a)^{nm}\right). \tag{4.2.16}$$

In particular, if $n(l-m) = -j$ for $j = 1, \cdots, n$, then

$$D^{n}\left[(x-a)^{n(l-m+1)}E_{n,m,l}\left(\lambda(x-a)^{nm}\right)\right] = \lambda(x-a)^{nl}E_{\alpha,m,l}\left(\lambda(x-a)^{nm}\right). \tag{4.2.17}$$

Corollary 4.4 *If $n \in \mathbb{N}$, $\beta > 0$ and $\lambda \in \mathbb{R}$, then*

$$D^n \left[(x-a)^{\beta-1} E_{n,\beta}\left(\lambda(x-a)^n\right)\right]$$

$$= \frac{1}{\Gamma(\beta-n)}(x-a)^{\beta-n-1} + \lambda(x-a)^{\beta-1} E_{n,\beta}\left(\lambda(x-a)^n\right). \tag{4.2.18}$$

In particular, if $\beta = 1, \cdots, n$, then

$$D^n \left[(x-a)^{\beta-1} E_{n,\beta}\left(\lambda(x-a)^n\right)\right] = \lambda(x-a)^{\beta-1} E_{n,\beta}\left(\lambda(x-a)^n\right), \tag{4.2.19}$$

$$D^n \left[E_n\left(\lambda(x-a)^n\right)\right] = \lambda E_n\left(\lambda(x-a)^n\right). \tag{4.2.20}$$

Remark 4.14 The condition $l > m - 1 - 1/\alpha$ in Proposition 4.1, needed for the existence of the left-hand side of (4.2.7), is omitted in Proposition 4.3.

The Proposition 4.3 can be proved directly by using the definition (1.9.24) for $E_{n,m,l}(z)$. This definition is directly used to establish the next assertion.

Proposition 4.4 *Let $\lambda, l \in \mathbb{R}$ and let $n \in \mathbb{N}$ and $m > 0$ such that the condition* (4.2.14) *is satisfied. Also let $D = d/dx$.*

Then there holds the formula

$$D^n \left[x^{n(m-l)-1} E_{n,m,l}\left(\lambda x^{-nm}\right)\right]$$

$$= \prod_{j=1}^{n} [n(m-l)-j] x^{n(m-l-1)-1} + (-1)^n \lambda x^{-n(l+1)-1} E_{n,m,l}\left(\lambda x^{-nm}\right). \tag{4.2.21}$$

In particular, if $n(m-l) = j$ for some $j = 1, \cdots, n$, then

$$D^n \left[x^{n(m-l)-1} E_{n,m,l}\left(\lambda x^{-nm}\right)\right] = (-1)^n \lambda x^{-n(l+1)-1} E_{n,m,l}\left(\lambda x^{-nm}\right). \tag{4.2.22}$$

Corollary 4.5 *If $n \in \mathbb{N}$, $\beta > 0$ and $\lambda \in \mathbb{R}$, then*

$$D^n \left[x^{n-\beta} E_{n,\beta}\left(\lambda x^{-n}\right)\right] = \prod_{j=1}^{n} (j-\beta) \frac{x^{-\beta}}{\Gamma(\beta)} + (-1)^n \lambda x^{-n-\beta} E_{n,\beta}\left(\lambda x^{-n}\right). \tag{4.2.23}$$

In particular, if $\beta = 1, \cdots, n$, then

$$D^n \left[x^{n-\beta} E_{n,\beta}\left(\lambda x^{-n}\right)\right] = (-1)^n \lambda x^{-n-\beta} E_{n,\beta}\left(\lambda x^{-n}\right), \tag{4.2.24}$$

$$D^n \left[x^{n-\beta} E_n\left(\lambda x^{-n}\right)\right] = (-1)^n x^{-n-1} E_n\left(\lambda x^{-n}\right). \tag{4.2.25}$$

Remark 4.15 By (2.2.5), $(D_-^n y)(x) = (-1)^n y^{(n)}(x)$, and thus (4.2.13) with $\alpha = n$ yields (4.2.21). The condition $l > m - \{\alpha\}/\alpha$ in Proposition 4.2, needed for the existence of the left-hand side of (4.2.13), is omitted in Proposition 4.4.

Remark 4.16 Results presented in Propositions 4.1 and 4.4 and Corollaries 4.1 and 4.5 were established by Kilbas and Saigo [397] and Saigo and Kilbas [725] (see also [391] and [724]).

Finally, we give the compositions of the operator of Liouville fractional differentiation D_-^α with the Bessel type functions $Z_\rho^\nu(z)$ and $\lambda_{\nu,\sigma}^{(\beta)}(z)$.

Proposition 4.5 *If $\alpha > 0$, $\nu \in \mathbb{R}$ and $\rho > 0$, then there holds the formula*

$$\left(D_-^\alpha Z_\rho^\nu\right)(x) = Z_\rho^{\nu-\alpha}(x). \tag{4.2.26}$$

In particular, if $\alpha = m \in \mathbb{N}$, then

$$D^m Z_\rho^\nu(x) = (-1)^m Z_\rho^{\nu-m}(x). \tag{4.2.27}$$

Proposition 4.6 *Let $\alpha > 0$ and $\nu \in \mathbb{R}$. Also let $\beta > 0$ and $\sigma \in \mathbb{R}$ be such that $\sigma > (1/\beta) - 1$.*

Then there holds the formula

$$\left(D_-^\alpha \lambda_{\nu,\sigma}^{(\beta)}\right)(x) = \lambda_{\nu,\sigma+\alpha}^{(\beta)}(x). \tag{4.2.28}$$

In particular, if $\alpha = m \in \mathbb{N}$, then

$$D^m \lambda_{\nu,\sigma}^{(\beta)}(x) = (-1)^m \lambda_{\nu,\sigma+m}^{\beta)}(x). \tag{4.2.29}$$

Remark 4.17 Results presented in Propositions 4.5 and 4.6 were proved by [373] and by Glaeske et al. [283], respectively.

4.2.3 Homogeneous Differential Equations of Fractional Order with Riemann-Liouville Fractional Derivatives

We consider the homogeneous differential equation (4.1.25) of order $\alpha > 0$:

$$(D_{a+}^\alpha y)(x) = \lambda(x-a)^\beta y(x) \quad (a < x < b \leqq \infty). \tag{4.2.30}$$

In what follows $I_{\text{loc}}(\Omega)$ and $B_{\text{loc}}(\Omega)$ denote spaces of locally integrable functions on a subset Ω of $\mathbb{R}$.

There holds the following result.

Theorem 4.7 *Let $n-1 < \alpha < n$ $(n \in \mathbb{N})$, $\beta > -\alpha$ and $\lambda \in \mathbb{R}$.*

(i) *If $0 < \alpha < 1$, then equation* (4.2.29) *has the solution given by*

$$y(x) = (x-a)^{\alpha-1} E_{\alpha,1+\beta/\alpha,1+((\beta-1)/\alpha}\left(\lambda(x-a)^{\alpha+\beta}\right) \in I_{loc}(a,b). \tag{4.2.31}$$

(ii) *If $\alpha > 1$ and*

$$(\alpha+\beta)(k+1) \neq 1, \cdots, [\alpha] \quad (k \in \mathbb{N}_0), \tag{4.2.32}$$

then the equation (4.2.29) *has n solutions given by*

$$y_j(x) = (x-a)^{\alpha-j} E_{\alpha,1+\beta/\alpha,1+(\beta-j)/\alpha}\left(\lambda(x-a)^{\alpha+\beta}\right) \quad (j = 1, \cdots, n). \tag{4.2.33}$$

Moreover, $y_j(x) \in B_{\text{loc}}(a,b)$ $(j = 1, \cdots, n-1)$ and $y_n(x) \in I_{\text{loc}}(a,b)$.

Proof. We apply Proposition 4.1 with $m = 1 + \beta/\alpha$ and $l = l_j = 1 + (\beta - j)/\alpha$ $(j = 1, \cdots, n)$. Condition (4.2.6) acquires the form

$$1 + \frac{\beta}{\alpha} > 0, \quad (\alpha + \beta)(k + 1) - j \notin \mathbb{Z}^- \quad (k \in \mathbb{N}_0). \tag{4.2.34}$$

If $0 < \alpha < 1$, then $j = 1$, and condition (4.2.34) is satisfied. If $\alpha > 1$ $(\alpha \notin \mathbb{N}\backslash\{1\})$, then (4.2.34) is satisfied for $(\alpha+\beta)(k+1) \leqq j-1$; hence we obtain (4.2.32). Since

$$l = l_j = 1 + \frac{\beta}{\alpha} - \frac{j}{\alpha} \geqq 1 + \frac{\beta}{\alpha} - \frac{[\alpha] + 1]}{\alpha} > \frac{\beta}{\alpha} - \frac{1}{\alpha} = m - 1 - \frac{1}{\alpha}$$

for any $j = 1, \cdots, n$, then the hypotheses of Proposition 4.1 are satisfied.

Substituting $y(x) = y_j(x)$ given by (4.2.33) into $(D_{a+}^{\alpha} y)(x)$, and using (4.2.8) with $m = 1 + \beta/\alpha$ and $l = l_j = 1 + (\beta - j)/\alpha$ $(l_j = m - j/\alpha)$, we have

$$(D_{a+}^{\alpha} y_j)(x) = \left(D_{a+}^{\alpha} \left[(t-a)^{\alpha(l_j - m+1)} E_{\alpha,m,l_j} \left(\lambda(t-a)^{\alpha+\beta}\right)\right]\right)(x)$$

$$= \lambda(x-a)^{\alpha-j+\beta} E_{\alpha,m,l_j} \left(\lambda(x-a)^{\alpha+\beta}\right) = \lambda(x-a)^{\beta} y_j(x)$$

for $j = 1, \cdots, [\alpha] + 1$. This means that $y_j(x)$ in (4.2.33) are solutions to (4.2.30).

Now we prove that the solutions $y_j(x)$ in (4.2.33) are linearly independent for $\beta > -\{\alpha\}$. For this we introduce the function $W_\alpha(x)$ defined by

$$W_\alpha(x) = \det\left(\left(D_{a+}^{\alpha-k} y_j\right)(x)\right)_{k,j=1}^{n} \quad (n = [\alpha] + 1; \ \ a \leqq x \leqq b). \tag{4.2.35}$$

This function is an analog of the Wronskian suitable to ordinary linear n-order differential equations [see, for example, Erugin ([251], p. 225)]. We have the following statement, whose proof is similar to that of the corresponding theorem for ordinary differential equations [see, for example, Erugin ([251], p. 226)].

Lemma 4.1 *The solutions $y_1(x), y_2(x), \cdots, y_n(x)$ in (4.2.33) are linearly independent if, and only if, $W_\alpha(x_0) \neq 0$ at some point $x_0 \in [a, b]$.*

Now we prove the following uniqueness result.

Theorem 4.8 *Let $n - 1 < \alpha < n$ $(n \in \mathbb{N})$ and $\lambda \in \mathbb{R}$. If $\beta > -\alpha$, then the solutions $y_1(x), \cdots, y_n(x)$ in (4.2.33) are linearly independent.*

Proof. Let $k, j = 1, \cdots, n$. By (4.2.8) and (1.9.19), we find that

$$\left(D_{a+}^{\alpha-k} y_j\right)(x) = \left(D_{a+}^{\alpha-k} \left[(t-a)^{\alpha-j} E_{\alpha,1+\beta/\alpha,1+(\beta-j)/\alpha} \left(\lambda(t-a)^{\alpha+\beta}\right)\right]\right)(x)$$

$$= \sum_{p=0}^{\infty} c_p \lambda^p \left(D_{a+}^{\alpha-k} (t-a)^{\alpha-j+(\alpha+\beta)p}\right)(x),$$

where $c_0 = 1$ and c_p $(p \in \mathbb{N})$ are given in (1.9.20) with $m = 1 + \beta/\alpha$ and $l = l + (\beta - j)/\alpha$. Using (2.1.17) for $1 \leqq p < n$ and (2.1.16) for $p = n$, we have

$$\left(D_{a+}^{\alpha-k} y_j\right)(x) = \sum_{p=0}^{\infty} d_p(k, j) \lambda^p (x-a)^{(\alpha+\beta)p+k-j} \quad (j, k = 1, \cdots, n), \tag{4.2.36}$$

where

$$d_0(k,j) = \frac{\Gamma(1+\alpha-j)}{\Gamma(1-j+k)}, \quad d_p(k,j) = c_p \frac{\Gamma[1+\alpha-j+(\alpha+\beta)p]}{\Gamma[1-j+k+(\alpha+\beta)p]} \quad (p \in \mathbb{N}). \quad (4.2.37)$$

In particular, when $k = j = 1, \cdots, n$, we get

$$d_0(k,k) = \Gamma(1+\alpha-k), \quad d_p(k,k) = c_p \frac{\Gamma[1+\alpha-k+(\alpha+\beta)p]}{\Gamma[1+(\alpha+\beta)p]} \quad (p \in \mathbb{N}). \quad (4.2.38)$$

If $k < j$, then $d_0(k,j) = 0$ and (4.2.36) takes the form

$$\left(D_{a+}^{\alpha-k} y_j\right)(x) = \sum_{p=1}^{\infty} d_p(k,j) \lambda^p (x-a)^{(\alpha+\beta)p+k-j} \quad (k,j = 1, \cdots, n; \; k < j). \quad (4.2.39)$$

If $k \geqq j$, then, taking the limit as $x \to a+$ in (4.2.36), we have

$$\left(D_{a+}^{\alpha-k} y_j\right)(a) := \lim_{x \to a+} \left(D_{a+}^{\alpha-k} y_j\right)(x) = 0 \quad (k,j = 1, \cdots, n; \; k > j), \quad (4.2.40)$$

$$\left(D_{a+}^{\alpha-k} y_k\right)(a) := \lim_{x \to a+} \left(D_{a+}^{\alpha-k} y_k\right)(x) = \Gamma(1+\alpha-k) \quad (k = 1, \cdots, n). \quad (4.2.41)$$

If $k < j$, then $(\alpha+\beta)p + k - j \geqq \{\alpha\} + \beta > 0$ for any $p \in \mathbb{N}$ and $j, k = 1, \cdots, n$, in accordance with the condition $\beta > -\{\alpha\}$. Then, taking the limit as $x \to a+$ in (4.2.39), we derive

$$\left(D_{a+}^{\alpha-k} y_j\right)(a) := \lim_{x \to a+} \left(D_{a+}^{\alpha-k} y_j\right)(x) = 0 \quad (k,j = 1, \cdots, n; \; k < j). \quad (4.2.42)$$

It follows from (4.2.40)-(4.2.42) that

$$W_\alpha(a) = \prod_{k=1}^{n} \Gamma(1+\alpha-k) \neq 0$$

for an analog of the Wronskian (4.2.35). Therefore, by Lemma 4.1, the solutions $y_1(x), \cdots, y_n(x)$ in (4.2.33) are linearly independent.

The next assertion follows from Theorems 4.7 and 4.8 when $\beta = 0$.

Corollary 4.6 *If $n - 1 < \alpha < n$ $(n \in \mathbb{N})$, then the equation*

$$(D_{a+}^{\alpha} y)(x) = \lambda y(x) \quad (a < x \leqq b < \infty) \quad (4.2.43)$$

has $n = [\alpha] + 1$ linearly independent solutions given by

$$y_j(x) = (x-a)^{\alpha-j} E_{\alpha,\alpha+1-j}\left(\lambda (x-a)^{\alpha}\right) \quad (j = 1, \cdots, n). \quad (4.2.44)$$

Moreover, $y_j(x) \in B_{\rm loc}(a,b)$ $(j = 1, \cdots, n-1]$ and $y_n(x) \in I_{\rm loc}(a,b)$.

In particular, for $0 < \alpha < 1$, the unique solution has the form

$$y(x) = (x-a)^{\alpha-1} E_{\alpha,\alpha}\left(\lambda (x-a)^{\alpha}\right) \in I_{\rm loc}(a,b). \quad (4.2.45)$$

Remark 4.18 If we consider the equation (4.2.30) on a finite interval (a, b) $(b < \infty)$, the spaces of the solutions (4.2.31), (4.2.45) and (4.2.33), (4.2.44) can be characterized more precisely. Thus the solutions $y(x)$ in (4.2.31) and (4.2.45) belong to the weighted space $C_{1-\alpha}[a, b]$, while in (4.2.33) and (4.2.44) the solutions $y_j(x)$ $(j = 1, \cdots, n-1) \in C[a, b]$ are continuous and $y_n(x) \in C_{n-\alpha}[a, b]$ with $n = -[-\alpha]$.

4.2.4 Nonhomogeneous Differential Equations of Fractional Order with Riemann-Liouville and Liouville Fractional Derivatives with a Quasi-Polynomial Free Term

We consider the nonhomogeneous differential equation of fractional order $\alpha > 0$ with a quasi-polynomial free term $f(x) = \sum_{r=1}^{k} f_r(x-a)^{\mu_r}$:

$$(D_{a+}^{\alpha} y)(x) = \lambda(x-a)^{\beta} y(x) + \sum_{r=1}^{k} f_r(x-a)^{\mu_r} \ (a < x < b \leqq \infty), \tag{4.2.46}$$

where $\lambda, \beta \in \mathbb{R}$ and f_r, $\mu_r \in \mathbb{R}$ $(r = 1, \cdots, k)$ are given real constants.

Theorem 4.9 *Let $n-1 < \alpha < n$ $(n \in \mathbb{N})$, $\beta > -\alpha$, $\mu_r \in \mathbb{R}$ $(r = 1, \cdots, k)$ be such that*

$$\mu_r > -1-\alpha, \quad \mu_r \neq -j \quad (j = 1, \cdots, [\alpha]+1; \quad r = 1, \cdots, k), \tag{4.2.47}$$

$$(j+1)(\alpha+\beta) + \mu_r \notin \mathbb{Z}^- \ (r = 1, \cdots, k; \ j \in \mathbb{N}_0). \tag{4.2.48}$$

(i) *Equation* (4.2.46) *is solvable in the space $I_{loc}(a, b)$ if there exists a $j \in \{1, \cdots, k\}$ such that $\mu_j < -\alpha$ and in the space $B_{loc}(a, b)$ if $\mu_r \geqq -\alpha$ for all $r \in \{1, \cdots, k\}$. Furthermore, there is a particular solution of the form*

$$y_0(x) = \sum_{r=1}^{k} \frac{\Gamma(\mu_r+1) f_r}{\Gamma(\mu_r+\alpha+1)} (x-a)^{\alpha+\mu_r} E_{\alpha, 1+\beta/\alpha, 1+((\beta+\mu_r)/\alpha} \left(\lambda(x-a)^{\alpha+\beta}\right). \tag{4.2.49}$$

(ii) *If $\beta > -\{\alpha\}$, then the general solution to the equation* (4.2.46) *is given by*

$$y(x) = \sum_{j=1}^{n} c_j (x-a)^{\alpha-j} E_{\alpha, 1+\beta/\alpha, 1+(\beta-j)/\alpha} \left(\lambda(x-a)^{\alpha+\beta}\right) + y_0(x), \tag{4.2.50}$$

where c_j $(j = 1, \cdots, n)$ are arbitrary real constants.

Proof. We use Proposition 4.1 with $m = 1 + \beta/\alpha$ and $l = l_r = 1 + (\beta + \mu_r)/\alpha$ $(r = 1, \cdots, k)$. Condition (4.2.6) acquires the form (4.2.47)-(4.2.48). Substituting $y(x)$ given by (4.2.49) into $(D_{a+}^{\alpha} y)(x)$ and using the formula (4.2.7), we have

$$(D_{a+}^{\alpha} y)(x) = \sum_{r=1}^{k} \frac{\Gamma(\mu_r+1) f_r}{\Gamma(\mu_r+\alpha+1)} \left(D_{a+}^{\alpha} \left[(t-a)^{\alpha(l_r-m+1)} E_{\alpha, m, l_r} \left(\lambda(t-a)^{\alpha+\beta}\right)\right]\right)(x)$$

$$= \sum_{r=1}^{k} f_r (x-a)^{\mu_r} + \lambda \sum_{r=1}^{k} \frac{\Gamma(\mu_r+1) f_r}{\Gamma(\mu_r+\alpha+1)} (x-a)^{\alpha+\mu_r+\beta} E_{\alpha,m,l_r}\left(\lambda (x-a)^{\alpha+\beta}\right)$$

$$= \sum_{r=1}^{k} f_r (x-a)^{\mu_r} + \lambda (x-a)^{\beta} y(x).$$

Thus $y(x)$ in (4.2.49) is a particular solution to the equation (4.2.46), and (i) is proved. The second assertion (ii) of Theorem 4.9 follows from Theorem 4.7.

The next assertion contains sufficient conditions for the validity of (4.2.47) and (4.2.48).

Corollary 4.7 *Let $\alpha > 0$ $(\alpha \notin \mathbb{N})$, $\beta > -\alpha$, and $\mu_r \in \mathbb{R}$ $(r = 1, \cdots, k)$. Also let either of the following conditions be satisfied:*

(i) $\mu_r > -1-\alpha, \quad \mu_r > -1-\alpha-\beta, \quad \mu_r \neq -j \ (r = 1, \cdots, k; \ j = 1, \cdots, n = -[-\alpha]).$ (4.2.51)

(ii) $\mu_r > -1 \ (r = 1, \cdots, n).$

Then the first assertion of Theorem 4.9 is valid. If, in addition, $\beta > -\{\alpha\}$, then the second assertion of Theorem 4.9 holds.

Setting $\beta = 0$, from Theorem 4.9 we derive the corresponding result for the differential equation (4.2.46) with $\beta = 0$:

$$(D_{a+}^{\alpha} y)(x) = \lambda y(x) + \sum_{r=1}^{k} f_r (x-a)^{\mu_r} \quad (a < x < b \leqq \infty). \tag{4.2.52}$$

Corollary 4.8 *Let $n-1 < \alpha < n$ $(n \in \mathbb{N})$ and $\mu_r \in \mathbb{R}$ $(r = 1, \cdots, k)$ be such that the conditions (4.2.47)-(4.2.48) are satisfied. Then the differential equation (4.2.52) is solvable in the space $I_{loc}(a,b)$ and its general solution has the form*

$$y(x) = \sum_{j=1}^{n} c_j (x-a)^{\alpha-j} E_{\alpha-j+1}\left(\lambda (x-a)^{\alpha}\right)$$

$$+ \sum_{r=1}^{k} \Gamma(\mu_r+1) f_r (x-a)^{\alpha+\mu_r} E_{\alpha,\alpha+\mu_r+1}\left(\lambda (x-a)^{\alpha}\right), \tag{4.2.53}$$

where c_j $(j = 1, \cdots, n)$ are arbitrary real constants.

Example 4.17 Let $n-1 < \alpha < n$ $(n \in \mathbb{N})$, $\beta \in \mathbb{R}$ and $\mu > -1-\alpha$ be such that

$$\mu > -1-\alpha, \quad \mu \neq -j \quad (j = 1, \cdots, n), \quad (r+1)(\alpha+\beta) + \mu \notin \mathbb{Z}^{-} (r \in \mathbb{N}_0).$$

If $\beta > -\alpha$, then the equation

$$(D_{a+}^{\alpha} y)(x) = \lambda (x-a)^{\beta} y(x) + f (x-a)^{\mu} \quad (a < x < b \leqq \infty; \ \lambda, f \in \mathbb{R}) \tag{4.2.54}$$

has a particular solution given by

$$y_0(x) = \frac{\Gamma(\mu+1)f}{\Gamma(\mu+\alpha+1)}(x-a)^{\alpha+\mu}E_{\alpha,1+\beta/\alpha,1+((\beta+\mu)/\alpha}\left(\lambda(x-a)^{\alpha+\beta}\right). \quad (4.2.55)$$

If $\beta > -\{\alpha\}$, then the general solution to equation (4.2.54) is given by

$$y(x) = \sum_{j=1}^{n} c_j(x-a)^{\alpha-j}E_{\alpha,1+\beta/\alpha,1+(\beta-j)/\alpha}\left(\lambda(x-a)^{\alpha+\beta}\right) + y_0(x), \quad (4.2.56)$$

where c_j $(j=1,\cdots,n)$ are arbitrary real constants.

Example 4.18 If $n-1 < \alpha < n$ $(n \in \mathbb{N})$, $\mu > -1-\alpha$ and $\mu \neq -j$ $(j = 1,\cdots,n)$, then the differential equation (4.2.54) with $\beta = 0$

$$(D^{\alpha}_{a+}y)(x) = \lambda y(x) + f(x-a)^{\mu} \quad (a < x < b \leqq \infty;\ \lambda, f \in \mathbb{R}) \quad (4.2.57)$$

has the general solution to the form

$$y(x) = \sum_{j=1}^{n} c_j(x-a)^{\alpha-j}E_{\alpha,\alpha-j+1}\left(\lambda(x-a)^{\alpha}\right)$$

$$+\Gamma(\mu+1)f(x-a)^{\alpha+\mu}E_{\alpha,\alpha+\mu+1}\left(\lambda(x-a)^{\alpha}\right), \quad (4.2.58)$$

where c_j $(j=1,\cdots,n)$ are arbitrary real constants.

In particular, for $0 < \alpha < 1$ and $\mu > -1-\alpha$ $(\mu \neq -1)$, the equation (4.2.57) has the explicit solution given by

$$y(x) = c(x-a)^{\alpha-1}E_{\alpha,\alpha}\left(\lambda(x-a)^{\alpha}\right) + +\Gamma(\mu+1)f(x-a)^{\alpha+\mu}E_{\alpha,\alpha+\mu+1}\left(\lambda(x-a)^{\alpha}\right), \quad (4.2.59)$$

where c is an arbitrary real constant.

Now we consider the following differential equation with the Liouville fractional derivative $(D^{\alpha}_{-}y)(x)$ defined in (2.2.4):

$$(D^{\alpha}_{-}y)(x) = \lambda x^{\beta}y(x) + \sum_{r=1}^{l} f_r x^{-\mu_r} \quad (0 \leqq d < x < \infty), \quad (4.2.60)$$

where $\lambda, \beta \in \mathbb{R}$ and $f_r,\ \mu_r \in \mathbb{R}$ $(r = 1,\cdots,k)$ are given real constants.

Theorem 4.10 *Let* $n-1 < \alpha < n$ $(n \in \mathbb{N})$, $\beta < -\alpha$ *and* $\mu_r > n$ *for* $r = 1,\cdots,k$. *Equation* (4.2.60) *is solvable in the space* $I_{loc}(d,\infty)$ *and has a particular solution*

$$y(x) = \sum_{r=1}^{k} \frac{\Gamma(\mu_r-\alpha)f_r}{\Gamma(\mu_r)}x^{\alpha-\mu_r}E_{\alpha,-1-\beta/\alpha,-2+(\mu_r-\beta-1)/\alpha}\left(\lambda x^{\alpha+\beta}\right). \quad (4.2.61)$$

Proof. We use Proposition 4.2 with $m = -1 - \beta/\alpha$ and, for $r = 1, \cdots, k$, $l = l_r = -2 + (\mu_r - \beta - 1)/\alpha$. The condition $l > m - \alpha/\{\alpha\}$ acquires the form $\mu_r > n$. Substituting $y(x)$ in (4.2.61) into $(D_-^\alpha y)(x)$ and using the formula (4.2.13), we have

$$(D_-^\alpha y)(x) = \sum_{r=1}^{k} \frac{\Gamma(\mu_r - \alpha) f_r}{\Gamma(\mu_r)} \left(D_-^\alpha \left[t^{\alpha(m-l_r)-1} E_{\alpha,m,l_r}\left(\lambda t^{-\alpha m}\right)\right]\right)(x)$$

$$= \sum_{r=1}^{k} f_r x^{-\mu_r} + \lambda \sum_{r=1}^{k} \frac{\Gamma(\mu_r - \alpha) f_r}{\Gamma(\mu_r)} x^{\alpha-\mu_r+\beta} E_{\alpha,m,l_r}\left(\lambda x^{-\alpha m}\right) = \sum_{r=1}^{k} f_r x^{\mu_r} + \lambda x^\beta y(x).$$

Thus $y(x)$ in (4.2.61) is a particular solution to the equation (4.2.60).

Corollary 4.9 *Let $\alpha > 0$ $(\alpha \notin \mathbb{N})$ and $\mu_r > [\alpha] + 1$ $(r = 1, \cdots, k)$. Then the equation*

$$(D_-^\alpha y)(x) = \lambda x^{-2\alpha} y(x) + \sum_{r=1}^{k} f_r x^{-\mu_r} \quad (0 \leqq d < x < \infty;\ \lambda, f_r \in \mathbb{R}), \tag{4.2.62}$$

is solvable in the space $I_{\text{loc}}(d, \infty)$ and has a particular solution given by

$$y(x) = \sum_{r=1}^{k} \Gamma(\mu_r - \alpha) f_r x^{\alpha-\mu_r} E_{\alpha,\mu_r}\left(\lambda x^{-\alpha}\right). \tag{4.2.63}$$

Example 4.19 Let $\alpha > 0$ $(\alpha \notin \mathbb{N})$, $\beta < -\alpha$ and $\mu > [\alpha] + 1$. Then the equation

$$(D_-^\alpha y)(x) = \lambda (x-a)^\beta y(x) + f(x-a)^{-\mu} \quad (0 \leqq d < x < \infty) \tag{4.2.64}$$

with $\lambda, f \in \mathbb{R}$ has a particular solution given by

$$y(x) = \frac{\Gamma(\mu - \alpha) f_k}{\Gamma(\mu)} x^{\alpha-\mu} E_{\alpha,-1-\beta/\alpha,-2+(\mu-\beta-1)/\alpha}\left(\lambda x^{\alpha+\beta}\right). \tag{4.2.65}$$

4.2.5 Differential Equations of Order 1/2

In this section we present solutions to the differential equations of order 1/2. First we consider the differential equations (4.2.30) and (4.2.46) with $a = 0$:

$$(D_{0+}^{1/2} y)(x) = \lambda x^\beta y(x) \quad (0 < x < b \leqq \infty), \tag{4.2.66}$$

$$(D_{0+}^{1/2} y)(x) = \lambda x^\beta y(x) + \sum_{r=1}^{l} f_r x^{\mu_r} \quad (0 < x < b \leqq \infty), \tag{4.2.67}$$

with given $\lambda,\ f_r, \mu_r \in \mathbb{R}$ $(r = 1, \cdots, l)$. Theorems 4.7 and 4.9 with $a = 0$ imply the following result.

Theorem 4.11 *There hold the following assertions:*

(i) *If $\beta > -1/2$, then the equation (4.2.66) has the solution given by*

$$y(x) = x^{-1/2}E_{1/2,2\beta+1,2\beta-1}\left(\lambda x^{\beta+1/2}\right) \in I_{loc}(a,b). \tag{4.2.68}$$

(ii) *If $\beta > -1/2$ and $\mu_r > -3/2$, $\mu_r \neq -1$ $(r = 1, \cdots, l)$, then the equation (4.2.67) is solvable in the space $I_{\rm loc}(a,b)$. Its particular solution is given by*

$$y_0(x) = \sum_{r=1}^{l} \frac{\Gamma(\mu_r+1)f_r}{\Gamma(\mu_r+\alpha+1)}(x-a)^{\mu_r+1/2}E_{1/2,2\beta+1,2((\beta+\mu_r)+1}\left(\lambda(x-a)^{\beta+1/2}\right), \tag{4.2.69}$$

and its general solution has the form

$$y(x) = cx^{-1/2}E_{1/2,2\beta+1,2\beta-1}\left(\lambda x^{\beta+1/2}\right) + y_0(x), \tag{4.2.70}$$

where c is an arbitrary real constant.

Corollary 4.10 *If $\mu_r > -3/2$ $(\mu_r \neq -1)$ $(r = 1, \cdots, l)$, then the equation*

$$(D_{0+}^{1/2}y)(x) = \lambda y(x) + \sum_{r=1}^{l} f_r x^{\mu_r} \quad (0 < x < b \leqq \infty), \tag{4.2.71}$$

has the general solution given by

$$y(x) = cx^{-1/2}E_{1/2,1/2}\left(\lambda x^{1/2}\right) + \sum_{r=1}^{l}\Gamma(\mu_r+1)f_r x^{\mu_r+1/2}E_{1/2,\mu_r+3/2}\left(\lambda x^{1/2}\right), \tag{4.2.72}$$

where c is an arbitrary real constant.

Example 4.20 If $\beta > -1/2$ and $\mu > -3/2$ $(\mu \neq -1)$, then the equation

$$(D_{0+}^{1/2}y)(x) = \lambda x^{\beta}y(x) + fx^{\mu} \quad (0 < x < b \leqq \infty;\ \lambda, f \in \mathbb{R}) \tag{4.2.73}$$

has the general solution given by

$$y(x) = cx^{-1/2}E_{1/2,2\beta+1,2\beta-1}\left(\lambda x^{\beta+1/2}\right)$$
$$+\frac{\Gamma(\mu+1)f}{\Gamma(\mu+\alpha+1)}x^{\mu+1/2}E_{1/2,2\beta+1,2(\beta+\mu)+1}\left(\lambda x^{\beta+1/2}\right), \tag{4.2.74}$$

where c is an arbitrary real constant.

In particular, the equation (4.2.73) with $\beta = 0$

$$(D_{0+}^{1/2}y)(x) = \lambda y(x) + fx^{\mu} \quad (0 < x < b \leqq \infty) \tag{4.2.75}$$

has the general solution given by

$$y(x) = cx^{-1/2}E_{1/2,1/2}\left(\lambda x^{1/2}\right) + \Gamma(\mu+1)fx^{\mu+1/2}E_{1/2,\mu+3/2}\left(\lambda x^{1/2}\right), \tag{4.2.76}$$

where c is an arbitrary real constant.

Equations of the form (4.2.73) arise in applications. For example, the voltmeter equation in electrochemistry has the form

$$x^{1/2}(D_{0+}^{1/2}y)(x) + x^{w}y(x) = 1 \quad (0 < x \leqq b < \infty;\ 0 < w \leqq \frac{1}{2}) \tag{4.2.77}$$

[see Oldham and Spanier ([643], p. 159 and [644])] and the equation of the polarography theory is given by

$$(D_{0+}^{1/2}y)(x) = \lambda x^{\beta}y(x) + x^{-1/2} \quad \left(x > 0;\ -\frac{1}{2} < \beta \leqq 0;\ \lambda \in \mathbb{R}\right), \tag{4.2.78}$$

[see Wiener [881]]. From the above considerations, we derive the solutions to these two equations as follows.

Corollary 4.11 *A particular solution to the equation* (4.2.77) *has the form*

$$y_0(x) = \sqrt{\pi}E_{1/2,2w,2w-1}\left(-x^{w}\right) \tag{4.2.79}$$

and the general solution to such a voltmeter equation is given by

$$y(x) = cx^{-1/2}E_{1/2,2w,2w-2}\left(-x^{w}\right) + \sqrt{\pi}E_{1/2,2w,2w-1}\left(-x^{w}\right), \tag{4.2.80}$$

where c is an arbitrary real constant.

Corollary 4.12 *A particular solution to the equation* (4.2.78) *has the form*

$$y_0(x) = \sqrt{\pi}E_{1/2,2\beta+1,2\beta}\left(\lambda x^{\beta+1/2}\right) \tag{4.2.81}$$

and its general solution is given by

$$y(x) = cx^{-1/2}E_{1/2,2\beta+1,2\beta-1}\left(\lambda x^{\beta+1/2}\right) + \sqrt{\pi}E_{1/2,2\beta+1,2\beta}\left(\lambda x^{\beta+1/2}\right), \tag{4.2.82}$$

where c is an arbitrary real constant.

Remark 4.19 Using the definition (1.9.19)-(1.9.20) of the generalized Mittag-Leffler function $E_{\alpha,m,l}(z)$ and the relation $\Gamma(1/2) = \sqrt{\pi}$ (see the second formula in (1.5.8)), we can rewrite the particular solution $y_0(x)$ given in (4.2.79) in the form

$$y_0(x) = \sum_{k=0}^{\infty} d_k\left(-x^{w}\right)^{k}, \quad d_k = \prod_{r=0}^{k}\frac{\Gamma(wr+1/2)}{\Gamma(wr+1)} \quad (k \in \mathbb{N}_0). \tag{4.2.83}$$

Such a form for a particular solution to the voltmeter equation (4.2.77) was obtained by Oldham and Spanier ([643], p. 160) by using the formal representation of $y_0(x)$ in the form of a quasi-power series.

From Theorem 4.10 we obtain the following solution to the equation (4.2.60) with $\alpha = 1/2$:

$$(D_{-}^{1/2}y)(x) = \lambda x^{\beta}y(x) + \sum_{r=1}^{l} f_r x^{-\mu_r} \quad (0 \leqq d < x < \infty), \tag{4.2.84}$$

where $\lambda, \beta \in \mathbb{R}$ and $f_r,\ \mu_r \in \mathbb{R}$ $(r = 1, \cdots, l)$ are given real constants.

Theorem 4.12 *If $\beta < -1/2$ and $\mu_r > 1$ $(r = 1, \cdots, l)$, then the equation* (4.2.84) *is solvable in the space $I_{loc}(d, \infty)$ and its particular solution has the form*

$$y(x) = \sum_{r=1}^{l} \frac{\Gamma(\mu_r - 1/2) f_r}{\Gamma(\mu_r)} x^{-\mu_r + 1/2} E_{1/2, -2\beta-1, 2(\mu_r - \beta)-4} \left(\lambda x^{\beta+1/2}\right), \quad (4.2.85)$$

Corollary 4.13 *If $\mu_r > 1$ $(r = 1, \cdots, l)$, the equation* (4.2.84) *with $\beta = -1$*

$$(D_-^{1/2} y)(x) = \lambda \frac{y(x)}{x} + \sum_{r=1}^{l} f_r x^{-\mu_r} \ (0 \leqq d < x < \infty) \quad (4.2.86)$$

has a particular solution given by

$$y(x) = \sum_{r=1}^{l} \Gamma(\mu_k - 1/2) f_r x^{-\mu_r + 1/2} E_{1/2, \mu_r} \left(\lambda x^{-1/2}\right). \quad (4.2.87)$$

Example 4.21 If $\beta < -1/2$ and $\mu > 1$, then the equation

$$(D_-^{1/2} y)(x) = \lambda x^{\beta} y(x) + f x^{-\mu} \ (0 \leqq d < x < \infty). \quad (4.2.88)$$

has a particular solution given by

$$y(x) = \frac{\Gamma(\mu - 1/2) f}{\Gamma(\mu)} x^{-\mu + 1/2} E_{1/2, -2\beta-1, 2(\mu - \beta)-4} \left(\lambda x^{\beta+1/2}\right), \quad (4.2.89)$$

Moreover, the equation (4.2.88) with $\beta = -1$

$$(D_-^{1/2} y)(x) = \lambda \frac{y(x)}{x} + f x^{-\mu} \ (0 \leqq d < x < \infty) \quad (4.2.90)$$

has a particular solution given by

$$y(x) = \Gamma(\mu - 1/2) f x^{-\mu + 1/2} E_{1/2, \mu} \left(\lambda x^{-1/2}\right). \quad (4.2.91)$$

4.2.6 Cauchy Type Problem for Nonhomogeneous Differential Equations with Riemann-Liouville Fractional Derivatives and with a Quasi-Polynomial Free Term

We consider the Cauchy type problem for the differential equation (4.2.46):

$$(D_{a+}^{\alpha} y)(x) = \lambda (x-a)^{\beta} y(x) + \sum_{r=1}^{l} f_r (x-a)^{\mu_r} \ (a < x < b \leqq \infty; \ \lambda \in \mathbb{R}), \quad (4.2.92)$$

$$(D_{a+}^{\alpha-k} y)(a+) = b_k \ \ (b_k \in \mathbb{R}; \ k = 1, \cdots, n; \ n = -[-\alpha]), \quad (4.2.93)$$

where $\lambda, \beta \in \mathbb{R}$ and f_r, $\mu_r \in \mathbb{R}$ $(r = 1, \cdots, l)$, and $b_k \in \mathbb{R}$ $(k = 1, \cdots, n)$.

Theorem 4.13 *Let $n-1<\alpha<n$ $(n\in\mathbb{N})$, $\beta>-\{\alpha\}$ and $\mu_r>-1$ $(r=1,\cdots,l)$.*

Then the solution to the Cauchy type problem (4.2.92)-(4.2.93) *exists in* $I_{\rm loc}(a,b)$*, is unique and given by the formula*

$$y(x)=\sum_{j=1}^{n}\frac{b_j}{\Gamma(1+\alpha-j)}(x-a)^{\alpha-j}E_{\alpha,1+\beta/\alpha,1+(\beta-j)/\alpha}\left(\lambda(x-a)^{\alpha+\beta}\right)+y_0(x), \tag{4.2.94}$$

where

$$y_0(x)=\sum_{r=1}^{l}\frac{\Gamma(\mu_r+1)f_r}{\Gamma(\mu_r+\alpha+1)}(x-a)^{\alpha+\mu_r}E_{\alpha,1+\beta/\alpha,1+(\beta+\mu_r)/\alpha}\left(\lambda(x-a)^{\alpha+\beta}\right). \tag{4.2.95}$$

Proof. By Corollary 4.7, (4.2.94) is a solution to equation (4.2.92). By Theorem 4.2 (in Section 4.1.1), the first term in (4.2.94) is the solution to the Cauchy type problem (4.2.30), (4.2.93) for the corresponding homogeneous equation, and thus

$$\left(D_{a+}^{\alpha-k}\left[\sum_{j=1}^{n}\frac{b_j(t-a)^{\alpha-j}}{\Gamma(1+\alpha-j)}E_{\alpha,1+\beta/\alpha,1+(\beta-j)/\alpha}\left(\lambda(t-a)^{\alpha+\beta}\right)\right]\right)(a+)=b_k \tag{4.2.96}$$

for $k=1,\cdots,n$. By (1.9.19) and (2.1.17), we have

$$\left(D_{a+}^{\alpha-k}\left[t-a)^{\alpha+\mu_r}E_{\alpha,1+\beta/\alpha,1+((\beta+\mu_r)/\alpha}\left(\lambda(t-a)^{\alpha+\beta}\right)\right]\right)(x)$$

$$=\sum_{n=0}^{\infty}c_n\lambda^n\frac{\Gamma[\alpha+\mu_r+(\alpha+\beta)n+1]}{\Gamma[k+\mu_r+(\alpha+\beta)n+1]}(x-a)^{\mu_r+(\alpha+\beta)n+k}$$

for $r=1,\cdots,l$ and $k=1,\cdots,n$. By the condition $\mu_r>-1$, we have $\mu_r+(\alpha+\beta)m+k>0$ for any $r=1,\cdots,l$; $k=1,\cdots,n$ and $m\in\mathbb{N}_0$. Thus

$$\left(D_{a+}^{\alpha-k}\left[t-a)^{\alpha+\mu_r}E_{\alpha,1+\beta/\alpha,1+((\beta+\mu_r)/\alpha}\left(\lambda(t-a)^{\alpha+\beta}\right)\right]\right)(a+)=0$$

and hence $\left(D_{a+}^{\alpha-k}y_0\right)(a+)=0$ for any $k=1,\cdots,n$. Therefore, in accordance with (4.2.96), the solution (4.2.94) satisfies the relations (4.2.93). The uniqueness of the solution follows from (4.2.94)-(4.2.95). This completes the proof of Theorem 4.13

When (a,b) is a finite interval and $\mu_r>-1$ $(r=1,\cdots,l)$, the space $I_{\rm loc}(a,b)$ of the solution $y(x)$ in (4.2.94) can be characterized more precisely in terms of the space $\mathbf{C}_{n-\alpha,\gamma}^{\alpha}[a,b]$ defined in (3.3.24).

Theorem 4.14 *Let $\alpha>0$ $(\alpha\notin\mathbb{N})$, $n=[\alpha]+1$, $\beta>-\{\alpha\}$, $\mu_r>-1$ $(r=1,\cdots,l)$, and let (a,b) be a finite interval.*

If $b_n\neq 0$, then the unique solution (4.2.94) *of the Cauchy type problem* (4.2.92)-(4.2.93) *belongs to one of the spaces $\mathbf{C}_{n-\alpha,\gamma}^{\alpha}[a,b]$ as follows:*

(i) *Let $\mu_r \geqq 0$ $(r = 1, \cdots, l)$. Then $y(x) \in \mathbf{C}^{\alpha}_{n-\alpha,0}[a,b]$ for $\beta \geqq 0$, while $y(x) \in \mathbf{C}^{\alpha}_{n-\alpha,-\beta}[a,b]$ for $-\{\alpha\} < \beta < 0$.*

(ii) *Let there exist $r_1, \cdots, r_k \in \{1, \cdots, l\}$ $(1 \leqq k \leqq l)$ such that $\mu_{r_j} < 0$ $(j = 1, \cdots, k)$ and let μ be a minimum of such μ_{r_j}:*

$$\mu = \min_{\mu_{r_j}<0;\ r_1,\cdots,\ r_k\in\{1,\cdots,\ l\};\ 1\leqq k\leqq l} \mu_{r_j}. \tag{4.2.97}$$

Then $y(x) \in \mathbf{C}^{\alpha}_{n-\alpha,-\mu}[a,b]$ for $\beta \geqq 0$, while $y(x) \in \mathbf{C}^{\alpha}_{n-\alpha,-\min[\mu,\beta]}[a,b]$ for $-\{\alpha\} < \beta < 0$.

Proof. By Theorem 4.13, the unique solution $y(x)$ given by (4.2.94) has the form

$$y(x) = y_1(x) + y_0(x),$$

where

$$y_1(x) = \sum_{j=1}^{n} \frac{b_j}{\Gamma(1+\alpha-j)} (x-a)^{\alpha-j} E_{\alpha,1+\beta/\alpha,1+(\beta-j)/\alpha}\left(\lambda(x-a)^{\alpha+\beta}\right),$$

and $y_0(x)$ is given by (4.2.95). Since $b_n \neq 0$, $y_1(x) \in C_{n-\alpha}[a,b]$. Since $\mu_r > -1$, $\mu_r + \alpha > -1$ $(r = 1, \cdots, l)$. If $\mu_r + \alpha \geqq 0$, then $y_1(x) \in C[a,b]$, and hence $y(x) \in C_{n-\alpha}[a,b]$. If there exist $r_0, r_1, \cdots, r_k \in \{1, \cdots, l\}$ $(0 \leqq k \leqq l)$ such that $\mu_{r_j} + \alpha < 0$ $(j = 0, 1, \cdots, k)$ and μ^* is a minimum of such μ_{r_j}, then $y_1(x) \in C_{-(\mu^*+\alpha)}[a,b]$. Hence $y(x) \in C_\gamma[a,b]$, where $\gamma = \max[n-\alpha, -\mu^* - \alpha]$. Since $n = [\alpha] + 1 \geqq 1 > -\mu^*$, $\gamma = n - \alpha$, and hence again $y(x) \in C_{n-\alpha}[a,b]$.

Further, since for any $y \in G \subset \mathbb{R}$ the first term $\lambda(x-a)^\beta y$ in the right-hand side of (4.2.92) belongs to the space $C[a,b]$ for $\beta \geqq 0$, and to the space $C_{-\beta}[a,b]$ when $-\{\alpha\} < \beta < 0$. If $\mu_r \leqq 0$ $(r = 1, \cdots, l)$, then the second term in the right-hand side of (4.2.92) belongs to $C[a,b]$, but if there exist $r_0, r_1, \cdots, r_k \in \{1, \cdots, l\}$ $(0 \leqq k \leqq l)$ such that $\mu_{r_j} < 0$ $(j = 0, 1, \cdots, k)$ and μ is a minimum of such μ_{r_j}, then the second term of the right-hand side of (4.2.92) belongs to $C_{-\mu}[a,b]$.

Thus, if $\mu_r \geqq 0$ $(r = 0, 1, \cdots, l)$, then the right-hand side of (4.2.92) and hence $D^\alpha_{a+}y$ belong to $C[a,b]$ for $\beta \geqq 0$, and to $C_{-\beta}[a,b]$ when $-\{\alpha\} < \beta < 0$. If there exist $r_0, r_1, \cdots, r_k \in \{0, 1, \cdots, l\}$ $(0 \leqq k \leqq l)$ such that $\mu_{r_j} < 0$ $(j = 0, 1, \cdots, k)$ and μ is a minimum of such μ_{r_j}, then the right-hand side of (4.2.92) and hence $D^\alpha_{a+}y$ belong to $C_{-\mu}[a,b]$ for $\beta \geqq 0$, and to $C_{\max[-\mu,-\beta}[a,b] = C_{-\min[\mu,\beta]}$ when $-\{\alpha\} < \beta < 0$.

By the definition (3.3.4) of the space $\mathbf{C}^{\alpha}_{n-\alpha,\gamma}[a,b]$, the above arguments prove the assertions (i) and (ii), and Theorem 4.14 is proved.

Corollary 4.14 *Let $n - 1 < \alpha < n$ $(n \in \mathbb{N})$ and $\mu \in \mathbb{R}$ be such that either $\mu > -1 - \alpha$, $\mu \neq -j$ $(j = 1, \cdots, n)$ or $\mu > -1$.*

Then the solution to the Cauchy type problem

$$(D^\alpha_{a+}y)(x) = \lambda y(x) + \sum_{r=1}^{l} f_r (x-a)^{\mu_r} \quad (a < x < b \leqq \infty), \tag{4.2.98}$$

$$(D_{a+}^{\alpha-k}y)(a+) = b_k \ (b_k \in \mathbb{R};\ k = 1, \cdots, n) \tag{4.2.99}$$

has the form

$$y(x) = \sum_{j=1}^{n} b_j (x-a)^{\alpha-j} E_{\alpha,\alpha-j+1}\left(\lambda(x-a)^\alpha\right)$$

$$+ \sum_{r=1}^{l} \Gamma(\mu+1) f_r (x-a)^{\alpha+\mu} E_{\alpha,\alpha+\mu_r+1}\left(\lambda(x-a)^\alpha\right). \tag{4.2.100}$$

In particular, if $0 < \alpha < 1$ *and either* $\mu > -1-\alpha$ *or* $\mu > -1$*, then the solution to the Cauchy type problem*

$$(D_{a+}^{\alpha}y)(x) = \lambda y(x) + f(x-a)^\mu; \quad (D_{a+}^{\alpha-1}y)(a+) = d \tag{4.2.101}$$

with $a < x < b \leqq \infty$ *and* $d \in \mathbb{R}$*, is given by*

$$y(x) = bd(x-a)^{\alpha-1} E_{\alpha,\alpha}\left(\lambda(x-a)^\alpha\right) + \Gamma(\mu+1) f (x-a)^{\alpha+\mu} E_{\alpha,\alpha+\mu+1}\left(\lambda(x-a)^\alpha\right). \tag{4.2.102}$$

Example 4.22 Let $n-1 < \alpha < n$ $(n \in \mathbb{N})$, $\beta > -\{\alpha\}$ and $\mu \in \mathbb{R}$ be such that either

$$\mu > -1-\alpha, \quad \mu > -1-\alpha-\beta, \quad \mu \neq -j \ (j = 1, \cdots, n) \tag{4.2.103}$$

or $\mu > -1$. Then the solution to the following Cauchy type problem for the equation (4.2.54) with the initial conditions (4.2.93)

$$(D_{a+}^{\alpha}y)(x) = \lambda(x-a)^\beta y(x) + f(x-a)^\mu \ (a < x < b \leqq \infty;\ \lambda, f \in \mathbb{R}), \tag{4.2.104}$$

$$(D_{a+}^{\alpha-k}y)(a+) = b_k \ (b_k \in \mathbb{R};\ k = 1, \cdots, n). \tag{4.2.105}$$

has the form

$$y(x) = \sum_{j=1}^{n} \frac{b_j}{\Gamma(1+\alpha-j)} (x-a)^{\alpha-j} E_{\alpha,1+\beta/\alpha,1+(\beta-j)/\alpha}\left(\lambda(x-a)^{\alpha+\beta}\right)$$

$$+ \frac{\Gamma(\mu+1) f}{\Gamma(\mu+\alpha+1)} (x-a)^{\alpha+\mu} E_{\alpha,1+\beta/\alpha,1+((\beta+\mu)/\alpha}\left(\lambda(x-a)^{\alpha+\beta}\right). \tag{4.2.106}$$

In particular, if $0 < \alpha < 1$, $\beta > -\alpha$ and $\mu \in \mathbb{R}$ be such that either of the conditions (4.2.103) or $\mu > -1$ is satisfied, then the solution to the Cauchy type problem

$$(D_{a+}^{\alpha}y)(x) = \lambda(x-a)^\beta y(x) + f(x-a)^\mu; \quad (D_{a+}^{\alpha-1}y)(a+) = d \tag{4.2.107}$$

with $a < x < b \leqq \infty$ and $d \in \mathbb{R}$, is given by

$$y(x) = \frac{d}{\Gamma(\alpha)} (x-a)^{\alpha-1} E_{\alpha,1+\beta/\alpha,1+(\beta-1)/\alpha}\left(\lambda(x-a)^{\alpha+\beta}\right)$$

$$+ \frac{\Gamma(\mu+1) f}{\Gamma(\mu+\alpha+1)} (x-a)^{\alpha+\mu} E_{\alpha,1+\beta/\alpha,1+((\beta+\mu)/\alpha}\left(\lambda(x-a)^{\alpha+\beta}\right). \tag{4.2.108}$$

Setting $\alpha = 1/2$ in Theorems 4.13 and 4.14, we derive the following assertions.

Theorem 4.15 *Let $\beta > -1/2$ and $\mu_r \in \mathbb{R}$ $(r = 1, \cdots, l)$ be such that either of the following conditions is satisfied:*

$$\textbf{(i)}\ \mu_r > -\frac{3}{2},\quad \mu_r > -\frac{3}{2} - \beta,\quad \mu_r \neq -1\ (r = 1, \cdots, l), \tag{4.2.109}$$

(ii) $\mu_r > -1$ $(r = 1, \cdots, l)$.

Then the solution to the Cauchy type problem (4.2.92)-(4.2.93) with $\alpha = 1/2$

$$(D_{a+}^{1/2} y)(x) = \lambda(x-a)^{\beta} y(x) + \sum_{r=1}^{l} f_r (x-a)^{\mu_r};\quad (D_{a+}^{-1/2} y)(a+) = d, \tag{4.2.110}$$

with $a < x < b \leqq \infty$ and $d \in \mathbb{R}$, exists in $I_{\mathrm{loc}}(a, b)$, is unique and given by

$$y(x) = \frac{d}{\sqrt{\pi}} (x-a)^{-1/2} E_{1/2, 2\beta+1, 2\beta-1}\left(\lambda(x-a)^{\beta+1/2}\right)$$

$$+ \sum_{r=1}^{l} \frac{\Gamma(\mu_r + 1) f_r}{\Gamma(\mu_r + \alpha + 1)} (x-a)^{\mu_r + 1/2} E_{1/2, 2\beta+1, 2\beta}\left(\lambda(x-a)^{\beta+1/(2)}\right). \tag{4.2.111}$$

Theorem 4.16 *Let $\beta > -1/2$, $\mu_r > -1$ $(r = 1, \cdots, l)$, and let (a, b) $(-\infty < a < b < \infty)$ be a finite interval.*

If $b \neq 0$, then the unique solution (4.2.111) of the Cauchy type problem (4.2.110) belongs to one of the spaces $\mathbf{C}^{\alpha}_{1-\alpha,\gamma}[a, b]$ as follows:

(i) *Let $\mu_r \geqq 0$ $(r = 1, \cdots, l)$.*

Then $y(x) \in \mathbf{C}^{\alpha}_{1-\alpha,0}[a, b]$ for $\beta \geqq 0$, while $y(x) \in \mathbf{C}^{\alpha}_{1-\alpha,-\beta}[a, b]$ for $-1/2 < \beta < 0$.

(ii) *Let there exist $r_1, \cdots, r_k \in \{1, \cdots, l\}$ $(1 \leqq k \leqq l)$ such that $\mu_{r_j} < 0$ $(j = 0, 1, \cdots, k)$ and let μ be a minimum of such μ_{r_j} given in (4.2.97).*

Then $y(x) \in \mathbf{C}^{\alpha}_{1-\alpha,-\mu}[a, b]$ for $\beta \geqq 0$, while $y(x) \in \mathbf{C}^{\alpha}_{1-\alpha,-\min[\mu,\beta]}[a, b]$ for $-1/2 < \beta < 0$.

Example 4.23 If $\beta > -1/2$ and $\mu > -3/2$ $(\mu \neq -1)$, then the Cauchy type problem

$$(D_{0+}^{1/2} y)(x) = \lambda x^{\beta} y(x) + f x^{\mu}\ \ (0 < x < b \leqq \infty)\ \ (D_{a+}^{-1/2} y)(a+) = d \in \mathbb{R} \tag{4.2.112}$$

has the solution given by

$$y(x) = \frac{d x^{-1/2}}{\sqrt{\pi}} E_{1/2, \beta+1, 2\beta-1}\left(\lambda x^{\beta+1/2}\right)$$

$$+ \frac{\Gamma(\mu+1) f x^{\mu+1/2}}{\Gamma(\mu+\alpha+1)} E_{1/2, 2\beta+1, 2(\beta+\mu)+1}\left(\lambda x^{\beta+1/2}\right). \tag{4.2.113}$$

In particular, the Cauchy type problem for equation (4.2.112) with $\beta = 0$

$$(D_{0+}^{1/2}y)(x) = \lambda y(x) + f x^{\mu}; \quad (D_{a+}^{-1/2}y)(a+) = d \tag{4.2.114}$$

with $d \in \mathbb{R}$ and $0 < x < b \leqq \infty$, has the solution given by

$$y(x) = \frac{d}{\sqrt{\pi}} x^{-1/2} E_{1/2,1/2}\left(\lambda x^{1/2}\right) + \Gamma(\mu+1) f x^{\mu+1/2} E_{1/2,\mu+3/2}\left(\lambda x^{1/2}\right). \tag{4.2.115}$$

Corollary 4.15 *The following Cauchy type problem for the voltmeter equation* (4.2.77) *in electrochemistry*

$$x^{1/2}(D_{0+}^{1/2}y)(x) + x^{w}y(x) = 1; \quad (D_{a+}^{-1/2}y)(a+) = d \tag{4.2.116}$$

with $0 < x \leqq b < \infty$, $0 < w \leqq 1/2$ *and* $d \in \mathbb{R}$, *has the solution given by*

$$y(x) = \frac{d}{\sqrt{\pi}} x^{-1/2} E_{1/2,2w,2w-2}\left(-x^{w}\right) + \sqrt{\pi} E_{1/2,2w,2w-1}\left(-x^{w}\right), \tag{4.2.117}$$

Corollary 4.16 *The following Cauchy type problem for the polarography equation* (4.2.78)

$$(D_{0+}^{1/2}y)(x) = \lambda x^{\beta} y(x) + x^{-1/2}; \quad (D_{a+}^{-1/2}y)(a+) = d, \tag{4.2.118}$$

with $x > 0$, $-\frac{1}{2} < \beta \leqq 0$, *and* $\lambda, d \in \mathbb{R}$, *has the solution given by*

$$y(x) = \frac{d}{\sqrt{\pi}} x^{-1/2} E_{1/2,2\beta+1,2\beta-1}\left(\lambda x^{\beta+1/2}\right) + \sqrt{\pi} E_{1/2,2\beta+1,2\beta}\left(\lambda x^{\beta+1/2}\right), \tag{4.2.119}$$

Remark 4.20 Results presented in Sections 4.2.3-4.2.6 for differential equations involving the fractional derivatives $D_{0+}^{\alpha}y$ and $D_{-}^{\alpha}y$ in spaces of integrable functions $I_{\mathrm{loc}}(0,d)$ $(0 < d \leqq \infty)$ and $I_{\mathrm{loc}}(d,\infty)$ $(0 \leqq d < \infty)$, respectively, were established by Kilbas and Saigo [397] [see also Kilbas and Saigo [393], [394] and Saigo and Kilbas [724])].

Remark 4.21 Statements of Propositions 4.3 and 4.4 and Corollaries 4.4 and 4.5 can be used to find solutions to homogeneous and nonhomogeneous ordinary nth-order differential equations of the form

$$y^{(n)}(x) = \lambda (x-a)^{\beta} y(x), \tag{4.2.120}$$

$$y^{(n)}(x) = \lambda (x-a)^{\beta} y(x) + \sum_{r=0}^{l} f_r (x-a)^{\mu_r} \tag{4.2.121}$$

and Cauchy problems for them. Such results were given in Saigo and Kilbas [725].

4.2.7 Solutions to Homogeneous Fractional Differential Equations with Liouville Fractional Derivatives in Terms of Bessel-Type Functions

In this section we apply the relations (4.2.26) and (4.2.28) to find explicit solutions to certain homogeneous differential equations of fractional order in terms of the Bessel-type functions $Z_\rho^\nu(z)$ and $\lambda_{\nu,\sigma}^{(\beta)}(z)$ defined in (1.7.42) and (1.7.51). Using the usual notation $D = d/dx$, first we consider the fractional differential operators:

$$L_\rho^\nu y := x^{\nu-\rho+1} D x^{\rho-\nu} D_-^\rho y = -x D_-^{\rho+1} y + (\rho - \nu) D_-^\rho y, \tag{4.2.122}$$

$$M_\rho^\nu y := x^{\nu-\rho+1} D x^{1-\rho} D x^{2\rho-\nu} D_-^\rho y$$
$$= x^2 D_-^{2\rho+2} y + (2\nu - 3\rho - 1) x D_-^{2\rho+1} y + (\nu - \rho(\nu - 2\rho) D_-^{2\rho} y. \tag{4.2.123}$$

When $\rho = 1$, the operator in (4.2.122) is a differential operator of second order:

$$L_1^\nu y := -x^\nu \frac{d}{dx} x^{1-\nu} \frac{dy}{dx} = -x \frac{d^2 y}{dx^2} - (1 - \nu) \frac{dy}{dx} \tag{4.2.124}$$

The Bessel-Clifford operator [see Kiryakova [417]]

$$B_\nu y := \frac{d}{dx} x^{1-\nu} \frac{d}{dx} x^\nu y = x \frac{d^2 y}{dx^2} + (1 + \nu) \frac{dy}{dx}, \tag{4.2.125}$$

and the operator of axisymmetric potential theory [see Gilbert [281]]

$$L_\nu y := x^{-2\nu-1} \frac{d}{dx} x^{2\nu+1} \frac{dy}{dx} = \frac{d^2 y}{dx^2} + \frac{2\nu + 1}{x} \frac{dy}{dx}, \tag{4.2.126}$$

are expressed in terms of the operator (4.2.124) as follows:

$$B_\nu y = -x^{-\nu} L_1^\nu x^\nu y, \quad L_\nu y = -\frac{1}{x} L^{-2\nu} y. \tag{4.2.127}$$

A function $Z_\rho^\nu(x)$ is invariant with respect to the operators $L_\rho^\nu(z)$ and M_ρ^ν, apart from a constant multiplier factor.

Lemma 4.2 *If* $\nu \in \mathbb{R}$ *and* $\rho > 0$, *then*

$$\left(L_\rho^\nu Z_\rho^\nu\right)(x) = -\rho Z_\rho^\nu(x), \tag{4.2.128}$$

$$\left(M_\rho^\nu Z_\rho^\nu\right)(x) = -\rho^2 Z_\rho^\nu(x). \tag{4.2.129}$$

In particular, if $\rho = 1$, *then*

$$\left(L_1^\nu Z_1^\nu\right)(x) = \left(M_1^\nu Z_1^\nu\right)(x) = -Z_1^\nu(x), \tag{4.2.130}$$

$$x^\nu \left(B_\nu x^{-\nu} Z_1^\nu\right)(x) = Z_1^\nu \tag{4.2.131}$$

and

$$x \left(L_\nu Z_1^{-2\nu}\right)(x) = Z_1^{-2\nu}. \tag{4.2.132}$$

Remark 4.22 Formula (4.2.128) provides a more precise version of the result in Rodríguez et al. [713]. Also Lemma 4.2 provides the explicit solutions to two fractional differential equations.

Theorem 4.17 *If $\nu \in \mathbb{R}$ and $\rho > 0$, then the function $y(x) = Z_\rho^\nu(x)$ is a solution to the following differential equation of order $\rho + 1$:*

$$x\left(D_-^{\rho+1}y\right)(x) + (\nu - \rho)\left(D_-^{\rho}y\right)(x) - \rho y(x) = 0 \tag{4.2.133}$$

and to the following differential equation of order $2\rho + 2$:

$$x^2\left(D_-^{2\rho+2}y\right)(x) + (2\nu - 3\rho - 1)x\left(D_-^{2\rho+1}y\right)(x)$$
$$+(\nu - \rho)(\nu - 2\rho)\left(D_-^{2\rho}y\right)(x) + \rho^2 y(x) = 0. \tag{4.2.134}$$

In particular, $y(x) = Z_1^\nu(x)$ is the solution to the following differential equations of the second and fourth orders:

$$xy''(x) + (1 - \nu)y'(x) - y(x) = 0, \tag{4.2.135}$$

$$x^2y^{(4)}(x) - 2(\nu - 2)xy^{(3)}(x) + (\nu - 1)\nu - 2)y''(x) + y(x) = 0. \tag{4.2.136}$$

Corollary 4.17 *If $K_\nu(x)$ ($\nu \in \mathbb{R}$) is the Macdonald function (1.7.25), then the function $y(x) = x^\nu K_\nu(x)$ is a solution to the differential equations*

$$xy''(x) + (1 - 2\nu)y'(x) - xy(x) = 0, \tag{4.2.137}$$

$$x^3y^{(4)}(x) - 2(2\nu - 1)y^{(3)}(x) + (4\nu^2 - 1)xy''(x) - (4\nu^2 - 1)y'(x) + x^3y(x) = 0. \tag{4.2.138}$$

Corollary 4.18 *The functions $y(x) = x^{-\nu}Z_1^\nu(x)$ and $y(x) = Z_1^{-2\nu}(x)$ with $\nu \in \mathbb{R}$ are solutions to the Bessel-Clifford equation*

$$xy''(x) + (\nu + 1)y'(x) - y(x) = 0, \tag{4.2.139}$$

and to the following equation of axisymmetric potential theory:

$$xy''(x) + (2\nu + 1)y'(x) - y(x) = 0. \tag{4.2.140}$$

Now we consider the fractional differentiation operator $L_\sigma^{\alpha,\beta}$ defined by

$$L_\sigma^{\alpha,\beta} := x\left(D_-^{\alpha+\beta+1} - D_-^{\alpha+1}\right) + (\sigma + \alpha + 1 - \beta)D_-^{\alpha}, \tag{4.2.141}$$

with $\alpha \geqq 0$, $\beta > 0$ and $\gamma, \sigma \in \mathbb{R}$. There holds the following assertion.

Lemma 4.3 *Let $\alpha \geqq 0$, $\beta > 0$ and $\nu, \sigma \in \mathbb{R}$ be such that $\nu > (1/\beta) - 1$. Then, for $x > 0$, the relation*

$$\left(L_\sigma^{\alpha,\beta}\lambda_{\nu,\sigma-\beta}^{(\beta)}\right)(x) = (\nu\beta + \sigma + \alpha)\left(D_-^{\alpha+\beta}\lambda_{\nu,\sigma-\beta}^{(\beta)}\right)(x) \tag{4.2.142}$$

holds for the Bessel-type function $\lambda_{\nu,\sigma-\beta}^{(\beta)}(x)$ defined in (1.7.51).

From Lemma 4.3 we derive solutions to two fractional differential equations.

Theorem 4.18 *If $\alpha \geqq 0$, $\beta > 0$ and $\sigma, \nu \in \mathbb{R}$ are such that $\nu > (1/\beta) - 1$, then $y(x) = \lambda^{(\beta)}_{\nu,\sigma-\beta}(x)$ is a solution to the following differential equation of fractional order:*

$$x\left[\left(D_-^{\alpha+\beta+1}y\right)(x) - \left(D_-^{\alpha+1}y\right)(x)\right]$$
$$-(\nu\beta+\sigma+\alpha)\left(D_-^{\alpha+\beta}y\right)(x) + (\sigma+\alpha+1-\beta)\left(D_-^{\alpha}y\right)(x) \qquad (4.2.143)$$

and, in particular, to the following differential equation:

$$x\left(D_-^{\beta+1}y\right)(x) - (\nu\beta+\sigma)\left(D_-^{\beta}y\right)(x) - xy'(x) + (\sigma+1-\beta)y(x). \qquad (4.2.144)$$

It is known that, if $\beta = 1$, $\nu \in \mathbb{C}$ ($\Re(\nu) > 0$), $\sigma \in \mathbb{R}$ and $z \in \mathbb{C}$ ($\Re(z) > 0$), then

$$\lambda^{(1)}_{\nu,\sigma}(z) = e^{-z}\Psi(\nu, \nu+\sigma+1; z), \qquad (4.2.145)$$

where $\Psi(\nu, \nu+\sigma+1; z)$ is the Tricomi confluent hypergeometric function defined in (1.6.25). In particular

$$\lambda^{(1)}_{\nu,m}(z) = z^{-\nu}e^{-z}\sum_{k=0}^{m}\frac{(-m)_k(\nu)_k}{k!}\frac{(-1)^k}{z^k} \quad (m \in \mathbb{N}_0). \qquad (4.2.146)$$

(See, in this regard, Glaeske et al. [283], Kilbas et al. [406], Bonilla et al. [93] and Kilbas et al. [390]). Then, from Theorem 4.18, we obtain the following assertions.

Corollary 4.19 *If $\alpha \geqq 0$, $\nu > 0$ and $\sigma \in \mathbb{R}$, then the following differential equation of fractional order*

$$x\left(D_-^{\alpha+2}y\right)(x) - (\nu+\sigma+\alpha+x)\left(D_-^{\alpha+1}y\right)(x) + (\sigma+\alpha)\left(D_-^{\alpha}y\right)(x) = 0 \quad (4.2.147)$$

has the solution $y(x) = e^{-x}\Psi(\nu, \nu+\sigma; x)$.

In particular, if $\sigma = m+1$ ($m \in \mathbb{N}_0$), then

$$y(x) = x^{-\nu}e^{-x}\sum_{k=0}^{m}\frac{(-m)_k(\nu)_k}{k!}\frac{(-1)^k}{x^k} = \lambda^{(1)}_{\nu,m}(x) \qquad (4.2.148)$$

is the solution to the following fractional differential equation:

$$x\left(D_-^{\alpha+2}y\right)(x) - (\nu+m+1+\alpha+x)\left(D_-^{\alpha+1}y\right)(x) + (m+1+\alpha)\left(D_-^{\alpha}y\right)(x) = 0. \qquad (4.2.149)$$

Corollary 4.20 *If $\alpha \geqq 0$, $\nu > -1/2$ and $m \in \mathbb{N}_0$, then the following differential equation*

$$x\left(D_-^{\alpha+3}y\right)(x) - (2\nu+m+\alpha+1)\left(D_-^{\alpha+2}y\right)(x) - x\left(D_-^{\alpha+1}y\right)(x) + (m+\alpha)\left(D_-^{\alpha}y\right)(x) = 0 \qquad (4.2.150)$$

has the solution $y(x) = \lambda^{(2)}_{\nu,m}(x)$.

Example 4.24 Taking $m = 0$ in Corollary 4.19, we see that $y(x) = x^{-\nu}e^{-x}$ is the solution to the fractional differential equation of fractional order

$$x\left(D_{-}^{\alpha+2}y\right)(x) - (\nu+1+\alpha+x)\left(D_{-}^{\alpha+1}y\right)(x) + (1+\alpha)\left(D_{-}^{\alpha}y\right)(x) = 0. \quad (4.2.151)$$

Example 4.25 It is known (see, for example, Bonilla et al. [93]) that $\lambda_{\nu,0}^{(2)}(z)$ and $\lambda_{\nu,1}^{(2)}(z)$ can be expressed in terms of the Macdonald function (1.7.25):

$$\lambda_{\nu,0}^{(2)}(z) = 2^{\nu+1}\pi^{-1/2}z^{-\nu}K_{-\nu}(z), \quad \lambda_{\nu,1}^{(2)}(z) = 2^{\nu+1}\pi^{-1/2}z^{-\nu}K_{-(\nu+1)}(z), \quad (4.2.152)$$

provided that $z, \nu \in \mathbb{C}$ $(\Re(z) > 0;\ \Re(\nu) > -1/2)$. Therefore, taking $m = 0$ and $m = 1$ in Corollary 4.20, we derive that the following differential equation of fractional order

$$x\left(D_{-}^{\alpha+3}y\right)(x) - (2\nu+\alpha+1)\left(D_{-}^{\alpha+2}y\right)(x) - x\left(D_{-}^{\alpha+1}y\right)(x) + \alpha\left(D_{-}^{\alpha}y\right)(x) = 0 \quad (4.2.153)$$

has the solution $y(x) = x^{-\nu}K_{-\nu}(x)$, while $y(x) = z^{-\nu}K_{-(\nu+1)}(x)$ is the solution to the following differential equation of fractional order:

$$x\left(D_{-}^{\alpha+3}y\right)(x) - (2\nu+2+\alpha)\left(D_{-}^{\alpha+2}y\right)(x) - x\left(D_{-}^{\alpha+1}y\right)(x) + (1+\alpha)\left(D_{-}^{\alpha}y\right)(x) = 0. \quad (4.2.154)$$

Remark 4.23 The results in Theorem 4.18, Corollaries 4.19-4.20 and Examples 4.24-4.25 give solutions to ordinary differential equations, obtained from (4.2.143), (4.2.144), (4.2.147), (4.2.149)-(4.2.151) and (4.2.153)-(4.2.154) for $\alpha = m \in \mathbb{N}_0$ and $\beta = l \in \mathbb{N}$. For example, $y(x) = x^{-\nu}e^{-x}$ is a solution to the equation

$$xy''(x) + (\nu+1+x)y'(x) + y(x) = 0, \quad (4.2.155)$$

being the equation of the form (4.2.151) with $\alpha = 0$.

Remark 4.24 Lemma 4.2, Theorem 4.17 and Corollaries 4.17-4.18 were established by Kilbas et al. [373] for $\alpha \in \mathbb{C}$ $(\Re(\alpha) > 0)$, $\beta > 0$ and $\nu \in \mathbb{C}$, while Lemma 4.3, Theorem 4.18 and Corollaries 4.19-4.20 were proved by Bonilla et al. [93] for $\beta > 0$ $\sigma \in \mathbb{R}$ and $\nu \in \mathbb{C}$ $(\Re(\nu) > (1/\beta) - 1)$.

4.3 Operational Method

The usefulness of operational calculus in solving ordinary differential equations is well known (see, for example, Ditkin and Prudnikov [195],[196],[197]). The basis of such an operational calculus for the operators of differentiation was developed by Mikusiński [599]. It is based on the interpretation of the Laplace convolution (1.4.10)

$$(f * g)(x) = \int_0^x f(x-t)g(t)dt \quad (4.3.1)$$

as a multiplication of elements f and g in the ring of functions continuous on the half-axis $\mathbb{R}_+$. Mikusiński [599] applied his operational calculus to solve ordinary differential equations with constant coefficients. We also mention that Mikusiński's scheme was used by Ditkin [193], Ditkin and Prudnikov [194], Meller [572] and Rodríguez [712], to develop the operational calculus for Bessel-type differential operators with nonconstant coefficients. The transform approach to the development of operational calculus was considered by Dimovski [191],[192]. Rodríguez et al. [713] were probably the first to apply operational calculus for a Kratzel transform defined in [451] to solve the partial differential equation of fractional order in the form (6.2.4).

A series of papers was devoted to developing operational calculus for the fractional calculus operators so as to solve differential equations of fractional order. In Sections 4.3.1-4.3.2 we shall present an operational approach based on the construction of the operational calculus for the Riemann-Liouville fractional derivative $D_{0+}^{\alpha}y$ in a special space C_{-1} of functions $y(x)$ such as $x^{-p}(D_{0+}^{\alpha})^k y(x) \in C[0,\infty)$ $(k = 1, \cdots, m)$ for some $p > -1$.

4.3.1 Liouville Fractional Integration and Differentiation Operators in Special Function Spaces on the Half-Axis

For $\mu \in \mathbb{R}$ we denote by C_{μ} the space of functions $f(x)$ given on the half-axis $\mathbb{R}_+$ and represented in the form $f(x) = x^p f_1(x)$ for some $p > \mu$, where the function $f_1(x)$ is continuous on $[0, \infty)$:

$$C_{\mu} = \{f(x) : \ f(x) = x^p f_1(x), \ f_1(x) \in C[0,\infty), \text{ for some } p > \mu \in \mathbb{R}\}. \tag{4.3.2}$$

Let $I_{0+}^{\alpha} f$ and $D_{0+}^{\alpha} y$ be the Liouville fractional integral and fractional derivative of order $\alpha > 0$ defined on $\mathbb{R}_+$ by (2.2.1) and (2.2.3), respectively. Using (4.3.2), we can directly prove the following properties of these operators.

Property 4.1 *If $\alpha > 0$, then the operator I_{0+}^{α} is a linear map of the space C_{-1} into the space $C_{\alpha-1}$, that is, $I_{0+}^{\alpha} : C_{-1} \to C_{\alpha-1}$. In particular, I_{0+}^{α} is a linear map of C_{-1} into itself: $I_{0+}^{\alpha} : C_{-1} \to C_{-1}$.*

Property 4.2 *If $\alpha > 0$ and $\beta > 0$, then the semigroup property* (2.2.25) *holds:*

$$(I_{0+}^{\alpha} I_{0+}^{\beta} f)(x) = (I_{0+}^{\alpha+\beta} f)(x) \quad (f(x) \in C_{-1}). \tag{4.3.3}$$

Property 4.3 *For $\alpha > 0$ and $f(x) \in C_{-1}$, then relation* (2.2.26) *holds:*

$$(D_{0+}^{\alpha} I_{0+}^{\alpha} f)(x) = f(x). \tag{4.3.4}$$

For $k \in \mathbb{N}$ we denote by I_{α}^{k} and D_{α}^{k} the compositions of k operators I_{0+}^{α} and D_{0+}^{α}:

$$\left(I_{\alpha}^{k} f\right)(x) = \left(I_{0+}^{\alpha} \cdots I_{0+}^{\alpha} f\right)(x), \quad \left(D_{\alpha}^{k} y\right)(x) = \left(D_{0+}^{\alpha} \cdots D_{0+}^{\alpha} y\right)(x). \tag{4.3.5}$$

For $\alpha > 0$ and $m \in \mathbb{N}$ we introduce the space $\Omega_{\alpha}^{m}(C_{-1})$ of functions $f(x) \in C_{-1}$ such that $\left(D_{\alpha}^{k} f\right)(x) \in C_{-1}$ for $k = 1, \cdots, m$:

$$\Omega_{\alpha}^{m}(C_{-1}) = \left\{f(x) \in C_{-1} : \ \left(D_{\alpha}^{k} f\right)(x) \in C_{-1} \ \ (k = 1, \cdots, m)\right\}. \tag{4.3.6}$$

The following assertion follows from Property 4.3.

Property 4.4 *The space* $\Omega_\alpha^m(C_{-1})$ $(\alpha > 0;\ m \in \mathbb{N})$ *contains the functions* $y(x) \in C_{-1}$ *which are representable in the form*

$$y(x) = (I_\alpha^m f)(x) \equiv (I_{0+}^{m\alpha} f)(x) \quad (y(x) \in C_{-1}). \tag{4.3.7}$$

The next assertion gives sufficient conditions for the operator D_{0+}^α to be a right inverse of the operator I_{0+}^α.

Lemma 4.4 *Let* $\alpha > 0$, $m \in \mathbb{N}$, $f(x) \in C_{-1}$ *and* $y(x) = (I_{0+}^{m\alpha} f)(x)$. *Then*

$$(D_\alpha^m y)(x) = (D_{0+}^{m\alpha} y)(x) \tag{4.3.8}$$

and

$$(I_{0+}^\alpha D_{0+}^\alpha y)(x) = y(x). \tag{4.3.9}$$

Remark 4.25 It follows from (4.3.9) that the operator D_{0+}^α is a right inverse of the operator I_{0+}^α, but only in the subspace of $\Omega_\alpha^1(C_{-1})$ which consists of functions $y(x) \in \Omega_\alpha^1(C_{-1})$ that are representable in the form

$$y(x) = (I_{0+}^\alpha f)(x) \quad (f(x) \in C_{-1}).$$

In the case of the whole space $\Omega_\alpha^1(C_{-1})$, there holds the following statement for the composition $(I_{0+}^\alpha D_{0+}^\alpha y)(x)$.

Lemma 4.5 *Let* $\alpha > 0$ *be such that* $m - 1 < \alpha \leqq m$ $(m \in \mathbb{N})$.
If $y(x) \in \Omega_\alpha^1(C_{-1})$, *then*

$$(I_{0+}^\alpha D_{0+}^\alpha y)(x) = y(x) - \sum_{k=1}^{m} \frac{(D_{0+}^{\alpha-k} y)(0+)}{\Gamma(\alpha - k + 1)} x^{k-\alpha}. \tag{4.3.10}$$

Remark 4.26 If we denote by E an identity operator in the space $\Omega_\alpha^1(C_{-1})$, then (4.3.10) can be rewritten in the form

$$(E - I_{0+}^\alpha D_{0+}^\alpha) y(x) = \sum_{k=1}^{m} \frac{(D_{0+}^{\alpha-k} y)(0+)}{\Gamma(\alpha - k + 1)} x^{k-\alpha}. \tag{4.3.11}$$

The operator $E - I_{0+}^\alpha D_{0+}^\alpha$ is called a *projector* of the operator I_{0+}^α.

Remark 4.27 Lemma 4.5 yields the formula for the composition $(I_{0+}^\alpha D_{0+}^\alpha y)(x)$ in the space $\Omega_\alpha^1(C_{-1})$ of functions $y(x)$ defined on the half-axis $\mathbb{R}_+$. It is analogous to the formula (2.1.39) in Lemmas 2.5(b) and 2.9(d) giving the composition $(I_{a+}^\alpha D_{a+}^\alpha y)(x)$ of the Riemann-Liouville fractional integrals and derivatives (2.1.1) and (2.1.5) in special spaces of functions $y(x)$ defined on a finite interval $[a, b]$.

4.3.2 Operational Calculus for the Liouville Fractional Calculus Operators

To construct the operational calculus for the Liouville fractional integral and derivative operators I_α and D_α, defined by (2.2.1) and (2.2.3), we construct an analog of the Laplace convolution (4.3.1). For this we use the general definition of the convolution of a linear operator A in a linear space X given by Dimovski [192]. Thus, a bilinear, commutative, and associative operation $o: \ X \times X \to X$ is said to be a convolution of the linear operator A if

$$A(f \ o \ g) = (Af \ o \ g) \text{ for any } f, g \in X.$$

Lemma 4.6 *For $\lambda \geqq 1$ the operation o_λ given by*

$$(f \ o_\lambda \ g)(x) = \left(I_{0+}^{\lambda-1} f * g\right)(x) := \int_0^x (I_{0+}^{\lambda-1} f)(x-t) g(t) dt \tag{4.3.12}$$

is the convolution (without divisors of zero) of the linear operator I_{0+}^α $(\alpha > 0)$ in the space C_{-1}.

The following statement plays an important role in the further development of the operational calculus.

Lemma 4.7 *If $\alpha > 0$ and $1 \leqq \lambda < \alpha + 1$, then the Liouville fractional integral operator I_{0+}^α has the following convolutional representation:*

$$(I_{0+}^\alpha f)(x) = (h \ o_\lambda \ f)(x), \quad h(x) = \frac{x^{\alpha-\lambda}}{\Gamma(\alpha - \lambda + 1)}. \tag{4.3.13}$$

Corollary 4.21 *Let $\alpha > 0$, $n \in \mathbb{N}$, and $1 \leqq \lambda < n\alpha + 1$. Also let*

$$h^n(x) = \frac{x^{n\alpha-\lambda}}{\Gamma(1 + n\alpha - \lambda)}. \tag{4.3.14}$$

Then

$$\left(I_\alpha^n f\right)(x) = (h \ o_\lambda \ \cdots o_\lambda \ h \ o_\lambda \ f)(x) = \left(h^n \ o_\lambda \ f\right)(x). \tag{4.3.15}$$

It is easily verified that the operations o_λ and $+$ possess the property of distributivity in the space C_{-1}, that is, that

$$\left(f \ o_\lambda \ (g+h)\right)(x) = \left(f \ o_\lambda \ g)\right)(x) + \left(f \ o_\lambda \ h\right)(x), \quad \text{for any } f, g, h \in C_{-1}. \tag{4.3.16}$$

By the property (4.3.16) and Lemma 4.6, the space C_{-1} with the operations o_λ and $+$ becomes a commutative ring without divisors of zero. This ring can be extended to the quotient field $\mathcal{P}$, along the lines of Mikusiński [599]:

$$\mathcal{P} = C_{-1} \times (C_{-1} \backslash \{0\}) / \sim, \tag{4.3.17}$$

where the equivalence relation ($\sim$) is defined, as usual, such that $(f, g) \sim (f_1, g_1)$ if, and only if, $(f\ o_\lambda\ g_1)(x) = (g\ o_\lambda\ f_1)(x)$. Thus, we can consider the elements of the field $\mathcal{P}$ as convolution quotients f/g and define the operations in $\mathcal{P}$ as follows:

$$\frac{f}{g} + \frac{f_1}{g_1} = \frac{(f\ o_\lambda\ g_1) + (g\ o_\lambda\ f_1)}{(g\ o_\lambda\ g_1)}, \quad \left(\frac{f}{g}\right)\left(\frac{f_1}{g_1}\right) = \frac{(f\ o_\lambda\ f_1)}{(g\ o_\lambda\ g_1)}, \tag{4.3.18}$$

It is clearly seen that the ring C_{-1} can be embedded in the field $\mathcal{P}$ by the map: $f(x) \to \frac{(h\ o_\lambda\ f)(x)}{h(x)}$, where $h(x)$ is given in (4.3.13). Moreover, the field $\mathbb{C}$ of complex numbers can also be embedded in $\mathcal{P}$ by the map: $\alpha \to \frac{\alpha h(x)}{h(x)}$.

On the basis of these facts we define the algebraic inverse of I_α as an element S of the field $\mathcal{P}$, which is reciprocal to the element $h(x)$ in the field $\mathcal{P}$; that is,

$$S = \frac{I}{h} := \frac{h}{(h\ o_\lambda\ h)} := \frac{h}{h^2}, \tag{4.3.19}$$

where $I = h/h$ denotes the identity element of the field $\mathcal{P}$ with respect to the operation of multiplication.

The relation between the Liouville fractional derivative operator D_α and the aforementioned algebraic inverse of I_α is contained in the following assertion.

Theorem 4.19 *Let $\alpha > 0$, $m \in \mathbb{N}$ and $f(x) \in \Omega_\alpha^m(C_{-1})$. Then there holds the following relation in the field $\mathcal{P}$ of convolution quotients:*

$$\left(D_\alpha^m f\right)(x) = S^m f - \sum_{k=0}^{m-1} S^{m-k} F D_\alpha^k f, \tag{4.3.20}$$

where $F = E - I_\alpha D_\alpha$ is the projector of the operator I_α given by (4.3.11). This result means that the Liouville fractional derivative operator D_α is reduced to the operator of multiplication in the field $\mathcal{P}$.

Proof. We represent the identity operator E in $\Omega_\alpha^m(C_{-1})$ in the following fashion:

$$E = \sum_{k=0}^{m-1} I_\alpha^k F D_\alpha^k + I_\alpha^m D_\alpha^m.$$

Then, for $f(x) \in \Omega_\alpha^m(C_{-1})$, we have

$$f(x) = \sum_{k=0}^{m-1} \left(I_\alpha^k F D_\alpha^k f\right)(x) + \left(I_\alpha^m D_\alpha^m f\right)(x). \tag{4.3.21}$$

Multiplying both sides of (4.3.21) by S^m and applying the relations (4.3.15) and (4.3.19), we obtain (4.3.20). This completes the proof of Theorem 4.19.

For applications it is important to know the functions of S in $\mathcal{P}$ which can be represented by means of the elements of the ring C_{-1}. One useful class of such functions is given in the following assertion.

Property 4.5 *If the power series*

$$\sum_{k=0}^{\infty} c_k z^k \quad (z, c_k \in \mathbb{C})$$

is convergent at a point $z_0 \neq 0$*, then the power series*

$$\sum_{k=0}^{\infty} c_k S^{-k} = \sum_{k=0}^{\infty} c_k h^k(x),$$

where $h^k(x)$ $(k \in \mathbb{N}_0)$ *are given in* (4.3.14)*, defines an element of the ring* $\mathcal{C}_{-1}$.

Remark 4.28 Applying Property 4.5, we may obtain various functions of S in the field $\mathcal{P}$, which can be represented by using the elements of the ring $\mathcal{C}_{-1}$. In particular, we have the following corollaries.

Corollary 4.22 *If* $\alpha > 0$, $1 \leqq \lambda < \alpha + 1$ *and* $\omega \in \mathbb{C}$*, then there holds the following operational relation:*

$$\frac{I}{S-\omega} = \frac{h}{E - \omega h} = h(E + \omega h + \omega^2 h^2 + \cdots) = x^{\alpha-\lambda} E_{\alpha,\alpha-\lambda+1}\left(\omega x^{\alpha}\right), \tag{4.3.22}$$

where $E_{\alpha,\alpha-\lambda+1}(z)$ *is the Mittag-Leffler function* (1.8.17).

Furthermore, if $m \in \mathbb{N}$*, then*

$$\frac{I}{(S-\omega)^m} = x^{m\alpha-\lambda} E^m_{\alpha,m\alpha-\lambda+1}\left(\omega x^{\alpha}\right), \tag{4.3.23}$$

where $E^m_{\alpha,m\alpha-\lambda+1}(z)$ *is the generalized Mittag-Leffler function* (1.9.1).

Corollary 4.23 *If* $\alpha > 0$, $1 \leqq \lambda < \alpha + 1$ *and* $\omega \in \mathbb{C}$*, then there hold the following operational relations:*

$$\frac{I}{S^2 + \omega^2} = \frac{x^{\alpha-\lambda}}{\omega} \sin_{\lambda,\alpha}\left(\omega x^{\alpha}\right) \quad (\omega \neq 0) \tag{4.3.24}$$

and

$$\frac{S}{S^2 + \omega^2} = x^{\alpha-\lambda} \cos_{\lambda,\alpha}\left(\omega x^{\alpha}\right), \tag{4.3.25}$$

where

$$\sin_{\lambda,\alpha}(z) = \sum_{k=0}^{\infty} \frac{(-1)^k z^{2k+1}}{\Gamma(2\alpha k + 2\alpha - \lambda + 1)} = z E_{2\alpha,2\alpha-\lambda+1}\left(-z^2\right) \tag{4.3.26}$$

and

$$\cos_{\lambda,\alpha}(z) = \sum_{k=0}^{\infty} \frac{(-1)^k z^{2k}}{\Gamma(2\alpha k + 2\alpha - \lambda + 1)} = E_{2\alpha,2\alpha-\lambda+1}\left(-z^2\right). \tag{4.3.27}$$

Remark 4.29 Other operational relations of the types contained in Corollaries 4.22 and 4.23 can be derived similarly from Property 4.5, especially for those functions of S which can be represented as a finite sum of partial fractions.

4.3.3 Solutions to Cauchy Type Problems for Fractional Differential Equations with Liouville Fractional Derivatives

In this section we apply the results of Section 4.3.2 to derive the explicit solution to the Cauchy type problem for differential equations with the Liouville fractional derivative D_{0+}^{α}.

Let $P_m(z)$ be a polynomial of degree $m \in \mathbb{N}$ in $z \in \mathbb{C}$ with complex coefficients:

$$P_m(z) = \sum_{r=0}^{m} c_r z^r, \quad (c_r \in \mathbb{C};\ c_0 \neq 0).$$

We consider the following boundary-value problem involving the Liouville fractional derivative operator D_{0+}^{α} of order $\alpha > 0$ $(n-1 < \alpha \leqq n;\ n \in \mathbb{N})$:

$$\left(P_m(D_{0+}^{\alpha})y\right)(x) = f(x), \quad (FD_{\alpha}^{r}y)(x) = \gamma_r(x) \tag{4.3.28}$$

$$(r = 0, 1, \cdots, m-1;\quad \gamma_r(x) = \ \ker D_{\alpha}),$$

where $F = E - I_{0+}^{\alpha} D_{0+}^{\alpha}$ is the projector of the operator I_{0+}^{α}, the function $f(x) \in \mathcal{C}_{-1}$ is given, and the unknown function $y(x)$ is to be determined in the space $\Omega_{\alpha}^{m}(\mathcal{C}_{-1})$.

Making use of (4.3.11), we rewrite the Cauchy type problem (4.3.28) in the following more familiar form:

$$\left(P_m(D_{0+}^{\alpha})y\right)(x) = f(x), \quad \lim_{x\to 0}\left((D_{\alpha-k}D_{\alpha}^{r})y\right)(x) = b_{rk} \tag{4.3.29}$$

$$(r = 0, 1, \cdots m-1;\ k = 1, \cdots, n;\ d_{rk} \in \mathbb{C})$$

Since $y(x) \in \Omega_{\alpha}^{m}(\mathcal{C}_{-1})$, by using the formula (4.3.20), we reduce the Cauchy type problem (4.3.28) to the following algebraic equation in the field $\mathcal{P}$:

$$P_m(S)y = f + \sum_{q=0}^{m-1} P_q(S)\gamma_q, \tag{4.3.30}$$

where

$$P_q(S) = \sum_{j=1}^{m-r} \gamma_{q+j} S^j,\ \gamma_q = \sum_{r=1}^{n} \frac{d_{qr}}{\Gamma(\alpha - r + 1)} x^{\alpha - r} \quad (q = 0, 1, \cdots, m-1). \tag{4.3.31}$$

Equation (4.3.30) has a unique solution in the field $\mathcal{P}$ given by

$$y = \frac{I}{P_m(S)} + \sum_{q=0}^{m-1} \frac{P_q(S)}{P_m(S)} \gamma_q. \tag{4.3.32}$$

This leads us to the following result.

Theorem 4.20 *If $\alpha > 0$ $(n-1 < \alpha \leqq n;\ n \in \mathbb{N})$ and $m \in \mathbb{N}$, then the Cauchy type problem (4.3.29) has a unique solution $y(x)$ in the space $\Omega_\alpha^m(C_{-1})$ and this solution is represented in the form*

$$y(x) = \sum_{j=1}^{l}\sum_{u=1}^{m_j} c_{ju} \int_0^x E_{\alpha,u\alpha}^u\left(z_j(x-t)^\alpha\right) f(t)dt + \sum_{r=1}^{n} \frac{d_{0r}}{\Gamma(\alpha-r+1)} x^{\alpha-r}$$

$$+\sum_{r=1}^{n}\sum_{q=0}^{m-1}\sum_{j=1}^{l}\sum_{u=1}^{m_j} A_{juq}\frac{d_{qr}}{\Gamma(\alpha-r+1)} x^{u\alpha+\alpha-r} E_{\alpha,u\alpha+\alpha-r+1}^u\left(z_j x^\alpha\right), \tag{4.3.33}$$

where the constants z_j, c_{ju} and A_{juq} are given by

$$\frac{1}{P(z)} = \frac{1}{\lambda_m \prod_{j=1}^{l}(z-z_j)^{m_j}} = \sum_{j=1}^{l}\sum_{u=1}^{m_j}\frac{c_{ju}}{(z-z_j)^u} \quad (m = m_1 + \cdots + m_l), \tag{4.3.34}$$

$$\frac{P_0(z)}{P(z)} = 1 + \sum_{j=1}^{l}\sum_{u=1}^{m_j}\frac{A_{ju0}}{(z-z_j)^u}, \quad \frac{P_q(z)}{P(z)} = \sum_{j=1}^{l}\sum_{u=1}^{m_j}\frac{A_{juq}}{(z-z_j)^u} \quad (q = 1, \cdots, m-l). \tag{4.3.35}$$

Proof. Problem (4.3.29) is equivalent to (4.3.28). Above we have reduced (4.3.28) to the algebraic equation (4.3.30) in the field $\mathcal{P}$, which has a unique solution (4.3.32). Using (4.3.22) and (4.3.23), we express the rational functions $1/P(z)$ and $P_q/P(z)$ $(q = 0, 1, \cdots, m-1)$ as the sums of partial fractions, and then from (4.3.32) we derive, taking (4.3.12) (with $1 \leqq \lambda < \alpha + 1$) into account, the solution to (4.3.28) in the form

$$y(x) = \sum_{j=1}^{l}\sum_{u=1}^{m_j} c_{ju}\left(f(t)\ o_\lambda\ t^{u\alpha-\lambda}E_{\alpha,u\alpha-\lambda+1}^u\left(z_j t^\alpha\right)\right)(x) + \sum_{r=1}^{n}\frac{d_{0r}}{\Gamma(\alpha-r+1)}x^{\alpha-r}$$

$$+\sum_{r=1}^{n}\sum_{q=0}^{m-1}\sum_{j=1}^{l}\sum_{u=1}^{m_j} A_{juq}\frac{d_{qr}}{\Gamma(\alpha-r+1)} x^{u\alpha+\alpha-r} E_{\alpha,u\alpha+\alpha-r+1}^u\left(z_j x^\alpha\right), \tag{4.3.36}$$

where $1 \leqq \lambda < \alpha + 1$. According to (4.3.12), we have

$$\left(f(t)\ o_\lambda\ t^{u\alpha-\lambda}E_{\alpha,u\alpha-\lambda+1}^u\left(z_j t^\alpha\right)\right)(x) = \left(t^{u\alpha-\lambda}E_{\alpha,u\alpha-\lambda+1}^u\left(z_j t^\alpha\right)\ o_\lambda f(t)\right)(x)$$

$$= \int_0^x \left(I_{0+}^{\lambda-1}\left[t^{u\alpha-\lambda}E_{\alpha,u\alpha-\lambda+1}^u\left(z_j t^\alpha\right)\right]\right)(x-t)f(t)dt.$$

From (2.2.1) and (1.9.1), by term-by-term fractional integration using (2.2.10), we get

$$\left(I_{0+}^{\lambda-1}\left[t^{u\alpha-\lambda}E_{\alpha,u\alpha-\lambda+1}^u\left(z_j t^\alpha\right)\right]\right)(x)$$

$$= \left(I_{0+}^{\lambda-1}\left[t^{u\alpha-\lambda}E_{\alpha,u\alpha-\lambda+1}^u\left(z_j t^\alpha\right)\right]\right)(x) = x^{\alpha-1}E_{\alpha,u\alpha}^u\left(z_j x^\alpha\right)$$

and hence

$$\left(f(t)\ o_\lambda\ t^{u\alpha-\lambda}E^u_{\alpha,u\alpha-\lambda+1}\left(z_j t^\alpha\right)\right)(x) = \int_0^x E^u_{\alpha,u\alpha}\left(z_j(x-t)^\alpha\right) f(t)dt.$$

Substituting this relation into (4.3.36), we obtain (4.3.33).

Setting $\gamma_r(x) = 0$ $(r = 0, 1, \cdots, m-1)$ in (4.3.28) and applying Lemma 4.4, we obtain the following corollary.

Corollary 4.24 *If $\alpha > 0$ $(n-1 < \alpha \leqq n;\ n \in \mathbb{N})$ and $m \in \mathbb{N}$, then the unique solution to the Cauchy type problem*

$$\sum_{r=0}^{m} \lambda_r (D^{r\alpha}_{0+}y)(x) = f(x), \quad \lim_{x\to 0}(D^{r\alpha+\alpha-k}_{0+}y)(x) = 0 \tag{4.3.37}$$

$$(r = 0, 1, \cdots, m-1;\ k = 1, \cdots, n)$$

in the space $\Omega^m_\alpha(C_{-1})$ may be represented in the form

$$y(x) = (I^m_\alpha g)(x \quad (g(x) \in C_1). \tag{4.3.38}$$

Example 4.26 We consider the Cauchy type problem

$$(D^\alpha_{0+}y)(x) - \omega y(x) = f(x), \quad \lim_{x\to 0}(D^{\alpha-k}_{0+}y)(x) = b_k \in \mathbb{C} \quad (k = 1, \cdots, n), \tag{4.3.39}$$

where $\omega \in \mathbb{C}$ and $(D^\alpha_{0+}y)(x)$ is the Liouville fractional derivative (2.2.3) of order $\alpha > 0$ such that $n-1 < \alpha \leqq n$ $(n \in \mathbb{N})$. Using the relations (4.3.20) and (4.3.11), we reduce the problem (4.3.39) to the following algebraic equation:

$$Sy - \omega y = f + S\gamma_0, \quad \gamma_0 = \sum_{j=1}^{n} \frac{b_j}{\Gamma(\alpha-j+1)} x^{\alpha-j}, \tag{4.3.40}$$

whose solution in the field $\mathcal{P}$ of convolution quotients has the form

$$y = \frac{I}{S-\omega} f + \frac{S}{S-\omega}\gamma_0. \tag{4.3.41}$$

In order to evaluate the first term on the right-hand side of (4.3.41) by means of (4.3.22) with $\lambda = 1$, we observe that

$$\frac{I}{S-\omega} = x^{\alpha-1}E_{\alpha,\alpha}\left(\omega x^\alpha\right). \tag{4.3.42}$$

On the other hand, using (4.3.40) and taking (4.3.22) with $\lambda = 1$ and (1.9.1) into account, we have, in accordance with (1.8.17),

$$\frac{S}{S-\omega}\gamma_0 = \sum_{r=1}^{n} b_r x^{\alpha-r} E_{\alpha,\alpha-r+1}\left(\omega x^\alpha\right) \tag{4.3.43}$$

Substituting the expressions (4.3.42) and (4.3.43) into the right-hand side of (4.3.41) and taking into account the convolution (4.3.12), we obtain the solution to the Cauchy type problem (4.3.39) in the form

$$y(x)=\int_0^x (x-t)^{\alpha-1}E_{\alpha,\alpha}\left(\omega(x-t)^\alpha\right)f(tdt+\sum_{r=1}^n b_r x^{\alpha-r}E_{\alpha,\alpha-r+1}\left(\omega x^\alpha\right). \quad (4.3.44)$$

By Theorem 4.21, this solution is unique in the space $\Omega_\alpha^1(C_{-1})$, provided that $f(x)\in C_{-1}$.

Remark 4.30 By Theorem 4.20, the relation (4.3.44) yields the explicit solution $y(x)$ of the Cauchy type problem (4.3.39) with $f(x)\in C_{-1}$ in the space $\Omega_\alpha^1(C_{-1})$:

$$\Omega_\alpha^1(C_{-1})=\{y(x)\in C_{-1}:\ \ (D_{0+}^\alpha y)(x)\in C_{-1}\}. \quad (4.3.45)$$

By (4.3.38), (4.3.39) is the Cauchy type problem (4.1.1)-(4.1.2) with $a=0$ and $\lambda=\omega$, and its solution (4.3.43) coincides with (4.1.10) (when $a=0$ and $\lambda=\omega$). Thus, by Theorem (4.1), (4.3.44) gives the explicit solution to the problem (4.3.38) with $f(x)\in C_\gamma[0,b]$ $(0\leqq\gamma<1)$ in the space $\mathbf{C}_{n-\alpha,\gamma}^\alpha[0,b]$ of functions defined on any finite interval $[0,b]\subset\mathbb{R}$ and having, in accordance with (3.3.24), the form

$$\mathbf{C}_{n-\alpha,\gamma}^\alpha[0,b]=\{y(x)\in C_{n-\alpha}[0,b]:\ \ (D_{0+}^\alpha y)(x)\in C_\gamma[0,b]\}. \quad (4.3.46)$$

Example 4.27 We consider the following Cauchy type problem:

$$(D_{0+}^{3/2}y)(x)-\omega y'(x)+\lambda^2(D_{0+}^{1/2}y)(x)-\omega\lambda y(x)=f(x), \quad (4.3.47)$$

$$\lim_{x\to0}(D_{0+}^{1/2}y)(x)=\lim_{x\to0}y(x)=\lim_{x\to0}(I_{0+}^{1/2}y)(x)=0. \quad (4.3.48)$$

This problem in the field $\mathcal{P}$ of convolution quotients (with $\alpha=1/2$) is reduced to the algebraic equation

$$S^3y-\omega S^2+\lambda^2Sy-\omega\lambda^2y=f, \quad (4.3.49)$$

so that

$$y(x)=\frac{1}{\omega^2+\lambda^2}\left(-\frac{S}{S^2+\lambda^2}-\frac{\omega}{S^2+\lambda^2}+\frac{I}{S-\omega}\right)f. \quad (4.3.50)$$

Now, using (4.3.22), (4.3.24) and (4.3.25) with $\alpha=1/2$ and $\omega=1$, we obtain

$$K(x):=\frac{1}{\omega^2+\lambda^2}\left(-\frac{S}{S^2+\lambda^2}-\frac{\omega}{S^2+\lambda^2}+\frac{I}{S-\omega}\right)$$

$$=\frac{1}{\omega^2+\lambda^2}\left[x^{\ 1/2}E_{1/2,1/2}\left(\omega x^{1/2}\right)-x^{\ 1/2}E_{1,1/2}\left(-\lambda^2x\right)-\omega E_{1,1}\left(-\lambda^2x\right)\right]. \quad (4.3.51)$$

Then, from (4.3.12), we obtain the solution to (4.3.47)-(4.3.47)

$$y(x)=\int_0^x K(x-t)f(t)dt, \quad (4.3.52)$$

where $K(x)$ is given by (4.3.50). By Theorem 4.20, this solution is unique in the space $\Omega_{3/2}^3(C_{-1})$, provided that $f(x)\in C_{-1}$.

Remark 4.31 Results presented in Sections 4.1-4.4 were established by Luchko and Srivastava [508].

Remark 4.32 Theorem 4.20 yields conditions for the unique solution $y(x)$ of the Cauchy type problem for the differential equation in (4.3.28), involving integer powers of the Liouville fractional derivative operator $D_\alpha y = D_{0+}^\alpha y$. Such constructions are known as sequential (see Section 3.1) and therefore, such an equation can be considered as an equation with sequential fractional derivatives.

Remark 4.33 Elizarrazaz and Verde-Star [237] obtained the explicit general solution to the differential equation in (4.3.29) and the explicit solution to the Cauchy type problem (4.3.29) by using linear algebraic construction and classical methods of operational calculus. Their approach was based on introducing divided differences of fractional order, coinciding with the Riemann-Liouville fractional derivative operator in a certain space of functions, and some generalized exponential polynomials, which are related to Mittag-Leffler type functions.

4.3.4 Other Results

In Section 4.3 we presented the operational method for solving, in closed form, differential equations of fractional order of the form (4.3.52) involving integer powers of the Liouville fractional derivative $D_\alpha y := D_{0+}^\alpha$ on the positive half-axis $\mathbb{R}^+$. Such an approach was also applied to solving Cauchy type and Cauchy problems involving other differential equations of fractional other.

Luchko and Yakubovich [509], [510] and Al-Basssam and Luchko [24] developed operational calculus for the Erdélyi-Kober type fractional derivative operator $D_{0+;\sigma,\eta}^\alpha$ defined in (2.6.29) (with $a = 0$) on the half-axis $\mathbb{R}^+$ and applied the results obtained to solve, in closed form and in a special function space, the Cauchy type problems for fractional differential equations of the form (4.3.52) in which the Liouville fractional derivative $D_{0+}^\alpha y$ is replaced by the Erdélyi-Kober type fractional derivative $D_{0+;\sigma,\eta}^\alpha y$. The explicit solutions to the problems considered were expressed via the generalized Mittag-Leffler function $E_\varrho\left((\alpha_j,\beta_j)_m; z\right)$ defined in (1.9.14).

Hadid and Luchko [323] have applied the operational calculus of Section 4.2 to solve the Cauchy type problem for the linear fractional differential equation

$$\left(D_{0+}^\alpha y\right)(x) - \sum_{r=1}^{m} \lambda_r \left(D_{0+}^{\alpha_r} y\right)(x) + \lambda_0 y(x) = f(x) \quad (x > 0) \tag{4.3.53}$$

$$\left(D_{a+}^{\alpha-k} y\right)(a+) = b_k \in \mathbb{R} \quad (k = 1, \cdots, n;\ n = -[-\alpha]). \tag{4.3.54}$$

involving the Liouville fractional derivatives $D_{0+}^\alpha y$, $D_{0+}^{\alpha_r} y$ $(r = 1, \cdots, m)$ $(0 < \alpha_1 < \cdots < \alpha_m < \alpha)$ and constant coefficients $\lambda_r \in \mathbb{R}$; $r = 0, 1, \cdots, m$. They suggested that $f(x) \in C_{-1}$ and established a unique solution $y(x)$ of the Cauchy type (4.3.53), (4.3.54) in the space $\Omega_\alpha^n(C_{-1})$ (see (4.3.6)), in terms of the multivariable Mittag-Leffler function $E_{(a_1,\cdots,\ a_n),b}(z_1, \cdots, z_n)$ defined by (1.9.27).

Luchko and Gorenflo [507] developed the operational calculus for the Caputo derivative $({}^C D_0^\alpha y)(x)$ defined by (2.4.1) with $a = 0$ in the space C_{-1} and gave an application to solve the following Cauchy problem:

$$({}^C D_{0+}^\alpha y)(x) - \lambda y(x) = f(x) \quad (x \geqq 0;\ \ n-1 < \alpha \leqq n;\ n \in \mathbb{N};\ \lambda \in \mathbb{R}), \tag{4.3.55}$$

$$y^{(k)}(0) = b_k \quad (b_k \in \mathbb{R};\ k = 0, 1, \cdots n-1). \tag{4.3.56}$$

They introduced the space C_α^m ($m \in \mathbb{N}_0$; $\alpha > 0$) of functions in the following space:

$$C_\alpha^n = \{y(x) :\ y^{(m)}(x) \in C_{-1},\quad m \in \mathbb{N}_0;\ \alpha > 0\} \tag{4.3.57}$$

and established a unique solution $y(x) \in C_{-1}^n$ of (4.3.55)-(4.3.56) in the form

$$y(x) = \sum_{k=0}^{n-1} b_k x^k E_{\alpha,k+1}(\lambda x^\alpha) + \int_0^x t^{\alpha-1} E_{\alpha,\alpha}(\lambda t^\alpha) f(x-t)dt, \tag{4.3.58}$$

provided that $f(x) \in C_{-1}^1$ for $\alpha \notin \mathbb{N}$ and $f(x) \in C_{-1}$ for $\alpha \in \mathbb{N}$.

Remark 4.34 When $\alpha \notin \mathbb{N}$, (4.3.58) yields the explicit solution $y(x)$ of the Cauchy problem (4.3.55)-(4.3.56) with $f(x) \in C_{-1}^1$ in the space C_{-1}^n defined by (4.3.57). (4.3.55)-(4.3.56) is the Cauchy problem (4.1.58)-(4.1.59) with $a = 0$, and its solution (4.3.58) coincides with the solution (4.1.62) (when $a = 0$). Thus, by Theorem (4.3), (4.3.58) gives the explicit solution to the problem (4.3.55)-(4.3.56) with $f(x) \in C_\gamma[0, b]$ ($0 \leqq \gamma < 1$) in the space $\mathbf{C}_\gamma^{\alpha,n-1}[0, b]$ of functions defined on any finite interval $[0, b] \subset \mathbb{R}$ and having, in accordance with (3.5.3), the form

$$\mathbf{C}_\gamma^{\alpha,n-1}[0, b] = \{y(x) \in C_{n-1}[0, b] :\ \ ({}^C D_{0+}^\alpha y)(x) \in C_\gamma[0, b]\}. \tag{4.3.59}$$

Luchko and Gorenflo [507] also investigated the Cauchy problem for the following more general fractional differential equation of the form (4.3.53) than (4.3.55):

$$({}^C D_*^\alpha y)(x) - \sum_{r=1}^m \lambda_r ({}^C D_*^{\alpha_r} y)(x) = f(x) \quad (x \geqq 0) \tag{4.3.60}$$

with the initial conditions (4.3.56). Making the same assumptions for $f(x)$ as in the above Cauchy problem (4.3.55)-(4.3.56), they proved that a unique solution $y(x)$ of the Cauchy problem (4.3.60), (4.3.56) in the space C_{-1}^n has the form

$$y(x) = y_0(x) + \sum_{k=0}^{n-1} b_k y_k(x),\ y_0(x) = \int_0^x t^{\alpha-1} E_{(\cdot),\alpha}(x-t) f(t)dt. \tag{4.3.61}$$

$$y_k(x) = \frac{x^k}{k!} + \sum_{r=l_k+1}^m \lambda_r x^{k+\alpha-\alpha_r} E_{(\cdot),k+1+\alpha-\alpha_r}(x) \quad (k = 0, 1, \cdots, n-1). \tag{4.3.62}$$

Here $y_0(x)$ is a solution to the problem (4.3.60), (4.3.56) with zero initial conditions $b_k = 0$ ($k = 0, 1, \cdots, n-1$), and the function

$$E_{(\cdot),\beta}(x) = E_{(\alpha-\alpha_1,\cdots,\ \alpha-\alpha_m),\beta}\left(\lambda_l t^{\alpha-\alpha_1}, \cdots,\ \alpha - \alpha_m\right) \tag{4.3.63}$$

is a particular case of the multivariable Mittag-Leffler function (1.9.27), in which the natural numbers l_k $(k = 0, 1, \cdots, n-1)$ are determined from the condition

$$m_{l_k} \geqq k+1, \quad m_{l_k+1} \leqq k. \tag{4.3.64}$$

In the case when $m_r \leqq k$ $(r = 0, 1, \cdots, n-1)$ we set $l_k = 0$, and if $m_r \geqq k+1$ $(r = 0, 1, \cdots, n-1)$, then $l_k = m$.

The above results of Luchko and Gorenflo [507] were also presented in a survey paper by Luchko [502] where historical remarks and an extensive bibliography may be found.

4.4 Numerical Treatment

In this section we discuss the numerical treatment of fractional differential equations. We note that many authors have discussed numerical methods for Abel-Volterra type integral equations of the first and second kind, called fractional integral equations, which contain the Riemann-Liouville fractional integral (2.1.1) [see the book by Gorenflo and Vessella [312] and the paper by Gorenflo [291]]. However, such investigations into fractional differential equations began only recently. Here we discuss the numerical treatment of fractional differential equations based mainly on the available good approximations of the fractional derivatives, analogous to the approximations of fractional integrals for fractional integral equations.

Shkhanukov [762] first applied the difference methods to study the Dirichlet problem for the differential equation

$$Ly \equiv \frac{d}{dx}\left[k(x)\frac{d}{dx}\right] - r(x)(D_{0+}^{\alpha}y)(x) - q(x)y(x) = -f(x) \ (0 < x < 1; \ 0 < \alpha < 1), \tag{4.4.1}$$

$$y(0) = y(1) = 0; \ k(x) \geqq c_0 > 0, \ r(x) \geqq 0, \ q(x) \geqq 0 \tag{4.4.2}$$

with the Riemann-Liouville fractional derivative (2.1.5). His approach was based on the following approximation of $(D_{0+}^{\alpha}y)(x)$ by its discrete analog:

$$(D_{0+}^{\alpha}y)(x_i) = \Delta_{0x_i}^{\alpha}y + O(h), \ \Delta_{0x_i}^{\alpha}y = \frac{1}{\Gamma(2-\alpha)}\sum_{k=1}^{i}\left(x_{i-k+1}^{1-\alpha} - x_{i-k}^{1-\alpha}\right)y_{\bar{x}k} \tag{4.4.3}$$

in a uniform grid $\{x_j = jh : \ j = 0, 1, \cdots, N-1\}$ of the interval $(0,1)$ with step $h = 1/N$. Using (4.4.3), Shkhanukov obtained a one-parameter family of difference schemes for the problem (4.4.1)-(4.4.2) and proved the stability and convergence of these difference schemes in the uniform metric. Using the approximation (4.4.3), Shkhanukov [762] also constructed the difference scheme for the first initial-boundary problem for the partial differential equation

$$(D_{0+;t}^{\alpha}u)(x,t) = \frac{\partial^2 u(x,t)}{\partial x^2} + f(x,t) \ \ (0 < x < 1; \ 0 < t < T; \ 0 < \alpha < 1), \tag{4.4.4}$$

$$u(0,t) = u(1,t) = 0 \ (0 \leqq t \leqq T); \ u(x,0) = 0,$$
$$(D^{\alpha}_{0+;t}u)(x,t)|_{t=0} = 0 \quad (0 \leqq x \leqq 1), \tag{4.4.5}$$

with the partial Riemann-Liouville fractional derivative (2.9.9), and proved the stability and convergence of this difference scheme in the uniform metric.

Blank [91] applied the collocation method with polynomial splines in the numerical treatment of the ordinary fractional differential equation

$$\left(D^{\alpha}_{0+}\left[y(t) - \sum_{k=0}^{N} \frac{y^{(k)}(0)}{k!}t^k\right]\right)(x) = -\lambda^{\alpha}y(x) + f(x) \ (0 < x < 1), \tag{4.4.6}$$

with the Caputo derivative (2.4.1) of order $\alpha > 0$ ($N < \alpha \leqq N+1$; $N \in \mathbb{N}_0$), where $\lambda > 0$, $y^{(k}(0)$ ($k = 0, 1, \cdots, N$) are given initial data and $f(x)$ is a given function on $[0,1]$. Blank's method is based on replacing (4.4.6) by

$$\left(D^{\alpha}_{0+}\left[y(t) - \sum_{s=0}^{N} \frac{y^{s}(0)}{s!}t^s\right]\right)(x_{n,i}) = -\lambda^{\alpha}y(x_{n,i}) + f(x_{n,i}) \tag{4.4.7}$$

in the collocation points $x_{n,j} = nh + c_jh$ ($j = 1, \cdots, k = m-(r+1)$; $n \in \mathbb{N}$) and on evaluating the Riemann-Liouville fractional derivative $(D^{\alpha}_{0+}y)(x_{n,i})$ at these collocation points. Here $c_1, \cdots, c_k$ ($0 < c_1 < \cdots < c_k \leqq 1$) are collocation parameters, that are chosen for the r times continuously differentiable numerical solution $y(x)$ and for the spline y defined on any interval $[x_j, x_{j+1}]$ ($j \in \mathbb{N}_0$) by the piecewise polynomial

$$y(x_j + vh) = \sum_{l=0}^{m-1} a^j_l v^l \ (v \in [0,1]) \tag{4.4.8}$$

with constants a^j_l ($l = 0, 1, \cdots, m-1$; $j \in \mathbb{N}_0$). In this way the systems of equations characterizing the numerical solution to (4.4.6) were determined. Such a system was simplified for the case of an N-times differentiable spline (i.e. $r = N$), and numerical results were given for $N = 1$ and $N = 2$.

Diethelm [170] suggested another approach for the numerical treatment of the fractional differential equation (4.4.6) with $0 < \alpha < 1$:

$$(D^{\alpha}_{0+}[y(t) - y(0)])(x) = \beta y(x) + f(x) \ (0 < x < 1; \ \beta \leqq 0). \tag{4.4.9}$$

For $\alpha = 1/2$ such an equation describes the behavior of a damping model in mechanics [see Gaul et al. [275]]. The algorithm suggested by Diethelm is based on the interpretation of the Riemann-Liouville fractional derivative (2.1.8) as a Hadamard finite part integral

$$(D^{\alpha}_{0+}y)(x) \equiv y^{\alpha)}(x) = \frac{1}{\Gamma(-\alpha)} \ p.f. \int_0^x \frac{y(t)}{(x-t)^{1+\alpha}}dt, \tag{4.4.10}$$

in the sense of a finite part of Hadamard [see Samko et al. ([729], Section 5.1)], and on the algorithm for the quadrature of finite-part integrals developed by Diethelm in [171]. Using (4.4.10) the relation (4.4.9) is discretized at the equispaced grid $x_j = j/N$ $(j = 0, 1, \cdots, N-1)$ with a given N, and then the obtained integral

$$\frac{1}{\Gamma(-\alpha)} \, p.f. \int_0^{x_j} \frac{[y(t) - y(0)]}{(x_j - t)^{1+\alpha}} dt = \frac{x_j^{-\alpha}}{\Gamma(-\alpha)} \, p.f. \int_0^1 \frac{y(x_j - x_j\tau) - y(0)}{\tau^{1+\alpha}} d\tau, \quad (4.4.11)$$

is replaced by a first degree compound quadrature formula with the equispaced nodes $0, 1/j, \cdots, 1$, which yields the resulting equation for the approximate values y_j of $y(x_j)$ $(j = 1, \cdots, n)$ in the form

$$\left[a_{j0} - \left(\frac{j}{n}\right)^{\alpha} \Gamma(-\alpha)\beta\right] y_j = \left[\left(\frac{j}{n}\right)^{\alpha} \Gamma(-\alpha) f(x_j) - \sum_{k=1}^{j} a_{kj} y_{j-k} - \frac{y_0}{\alpha}\right]. \quad (4.4.12)$$

The error of this approximation method was found for sufficiently smooth involved functions, and two numerical examples for $\beta = -1$ and $\beta = -2$, with the choice of $f(x)$ such that the exact solutions to (4.4.9) have the form $y(x) = x^2$ and $y(x) = \cos(\pi x)$, respectively, confirmed the obtained error bounds for $\alpha = 0.5$, $\alpha = 0.75$ and $\alpha = 0.25$.

The relation (4.4.12) was used by Diethelm and Walz [189] to obtain the asymptotic expansion for the sequence of approximate values $\{y_n\}$ in the form

$$y_n = y(x_n) + \sum_{l=2}^{M_1} a_i n^{l-\alpha} + \sum_{j=1}^{M_2} b_j n^{-2j} + o\left(x^{-\lambda_M}\right) \; (n \to \infty), \quad (4.4.13)$$

where the natural numbers M_1 and M_2 are defined by the smoothness of $f(x)$ (and therefore of $y(x)$), a_k $(k = 2, \cdots, M_1)$ and b_j $(j = 1, \cdots, M_2)$ are certain constants depending on $k - \alpha$ and $2j$ and $M = \min[\alpha - M_1, 2M_2]$. They applied the asymptotic estimate in (4.4.13) to state an extrapolation algorithm for the numerical solution to the problem (4.4.9) and illustrated the results by numerical examples.

Diethelm and Freed [186] interpreted the nonlinear differential equation

$$\left(D_{0+}^{\alpha}[y(t) - y(0)]\right)(x) = f[x, y(x)] \; (0 < x < 1; \; 0 < \alpha < 1) \quad (4.4.14)$$

as the Volterra integral equation of the second kind

$$y(x) = y(0) + \frac{1}{\Gamma(\alpha)} \int_0^x \frac{f[t, y(t)]}{(x - t)^{1-\alpha}} dt \quad (4.4.15)$$

and implemented a product integration method for the latter by using a product trapezoidal quadrature formula with nodes t_j $(j = 0, 1, \cdots, n+1)$ taken with respect to the weight function $(t_{n+1} - t)^{\alpha-1}$.

For the numerical solution to fractional differential equations, Podlubny ([681] and [682], Chapter 8) has used the approach based on the approximation of the Riemann-Liouville fractional derivative $(D_{0+}^{\alpha} y)(x)$ of order $\alpha > 0$ by the relation

$$h^{-\alpha}(\Delta_h^{\alpha} y)(x), \quad (4.4.16)$$

where

$$(\Delta_h^\alpha y)(x) = \sum_{j=1}^{[x/h]} (-1)^j \frac{\alpha(\alpha-1)\cdots(\alpha-j+1)}{j!} f(x-jh) \tag{4.4.17}$$

is the "truncated" difference of fractional order $\alpha > 0$ of $y(x)$ with a step $h > 0$ and with a center at the point x [see Section 2.10 and Samko et al. ([729], Sections 5.6 and 20.1)]. Such a construction in (4.4.16) for $x = nh$ approximates $(D_{0+}^\alpha y)(x)$ with an accuracy $O(h)$. It should be noted that Zheludev [926] first applied such a method to numerically solve the Abel integral equation $(I_{0+}^\alpha y)(x) = f(x)$, and also a more general convolution integral equation.

Assuming that the step h and the number $n \in \mathbb{N}$ of nodes are related by $x = nh$, Podlubny ([681] and [682], Chapter 8) presented a numerical solution with order of approximation $O(h)$ for the following Cauchy type-problems:

$$(D_{0+}^\alpha y)(x) + Ay(x) = f(x) \ (x > 0), \ y^{(k)}(0) = 0 \ (k = 0, 1, \cdots, n-1), \tag{4.4.18}$$

with the Riemann-Liouville fractional derivative (1.1) of order α ($n - 1 < \alpha \leqq n$; $n \in \mathbb{N}$); the problem

$$ay''(x) + b(D_{0+}^{3/2} y)(x) + cy(x) = f(x) \ (x > 0; \ a > 0), \ y(0) = y'(0) = 0, \tag{4.4.19}$$

with $b, c \in \mathbb{R}$, first considered by Torvik and Bagley [820] while studying the behavior of certain viscoelastic materials; also for the mathematical model which describes the dynamics of certain gases dissolved in a fluid,

$$\frac{d}{dx}(f(x)[y(x) + 1]) + \lambda(D_{0+}^{1/2} y)(x) = 0 \ (0 < x < 1), \ y(0) = 0, \tag{4.4.20}$$

considered by Babenko [44]; and finally for the nonlinear problem of the form

$$(D_{0+}^{1/2} y)(x) - \alpha[u_0 - y(x)]^4 = 0 \ (x > 0; \ \alpha \in \mathbb{R}; \ u_0 \in \mathbb{R}), \ y(0) = 0, \tag{4.4.21}$$

which models the dynamics of the cooling process of a semi-infinite body by radiation. It should be noted that Podlubny claims that the order of approximation of his algorithm is $O(h)$, but he does not give a proof.

Numerical treatment of the equation (4.4.19) was studied by Diethelm and Ford [178] on the basis of the reformulation of this equation as a system of fractional differential equations of order 1/2. This allowed them to propose numerical methods for its solution which are consistent and stable and have an arbitrary high order. In this way they especially looked at fractional linear multistep methods and Adams type predictor-corrector methods. An Adams type predictor-corrector method for the numerical solution of fractional differential equations, used for both linear and nonlinear problems, was discussed by Diethelm et al. [182].

Edwards et al. [218] showed how the numerical approximation of the solution to a linear multi-term fractional differential equation

$$\sum_{k=1}^{m} c_k \left(D_{0+}^{\alpha_k} y\right)(x) + c_0 y(x) = f(x) \quad (x > 0; \ \alpha_m < \alpha_{m-1} > \cdots > \alpha_1 > 0) \tag{4.4.22}$$

can be obtained by reducing the problem to a system of ordinary and fractional differential equations each of order unity, at most. They solved the problem (4.4.19) as an example, and showed how the method can be applied to a general linear equation (4.4.22). Note that numerical solutions to multi-order fractional differential equations were also discussed by Diethelm and Ford [179].

Ford and Simpson [261] used the fixed memory principle described by Podlubny ([682],Chapter 8) to numerically solve the Cauchy problem of the form (3.5.1)-(3.5.2):

$$\left({}^{C}D_{0+}^{\alpha}y\right)(x) = f[x,y(x)]\ (x>0);\quad y^{(k)}(0)=b_k\ (k=0,1,\cdots,\ n-1) \qquad (4.4.23)$$

with the Caputo derivative ${}^{C}D_{0+}^{\alpha}y$ of order $n-1<\alpha<n$ $(n\in\mathbb{N})$ given by (2.4.15). Diethelm et al. [183] suggested an algorithm for the numerical solution to (4.4.23), which is a generalization of the classical one-step Adams-Bashforth-Moulton scheme for first-order equations.

We also mention Boyjadziev et al. [106] who have applied the tau-method of approximation to obtain the numerical solution of the fractional integro-differential equation with the Liouville fractional derivative (2.2.4) of order $\alpha>0$

$$\left(D_{-}^{\alpha}y\right)(x) = \int_0^x y(x-t)te^{i\nu t}dt\quad (\sigma\in\mathbb{C};\nu\in\mathbb{R}), \qquad (4.4.24)$$

with the initial conditions (3.1.2), and Boyadziev and Dobner [102] who computed numerical values of the solution $y(x)$ of this problem.

In concluding this section, we indicate that from our point of view the approach suggested by Lubich ([499]-[501]) for the numerical treatment of the Abel-Volterra integral equations and more general weakly singular Volterra integral equations, can be also used to obtain numerical solutions for both ordinary and partial fractional differential equations. This approach is based on the numerical approximation of the Riemann-Liouville fractional integral $(I_{0+}^{\alpha}f)(x)$ $(0<\alpha<1)$, defined by (2.1.1), for $x=nh$ $(n\in\mathbb{N};\ 0\leqq nh\leqq T<\infty;\ h>0)$ by a discrete convolution quadrature:

$$h^{\alpha}\sum_{l=0}^{n}\omega_{n-l}^{\alpha}y(lh). \qquad (4.4.25)$$

Here the weights ω_l^{α} are the coefficients of ζ^l in the expansions of $\omega^{\alpha}(\zeta)$, where $\omega(\zeta)=\sigma(1/\zeta)/\rho(1/\zeta)$ with polynomials $\sigma(z)$ and $\rho(z)$. In particular, when $\sigma(z)=1$ and $\rho(z)=1-z$, $\omega(z)=1/(1-z)$ and $\omega(\zeta)^{\alpha}=(1-z)^{-\alpha}$ and (4.4.25) takes the form (4.4.16) with α replaced by $-\alpha$. Such an approach was applied by Sanz-Serna [732] to numerically solve the homogeneous partial integro-differential equation

$$\frac{\partial u(x,t)}{\partial t} = \int_0^t (t-s)^{-1/2}u_{xx}(x,t)dt\ (0\leqq x\leqq 1;\ x>0) \qquad (4.4.26)$$

with the boundary conditions

$$u(0,t)=u(1,t)=0\ (t\leqq 0),\ u(x,0)=f(x)\ (0\leqq x\leqq 1). \qquad (4.4.27)$$

We note that the above approach by Lubich as well as other known numerical methods for solving Abel-Volterra integral equations will be useful for the numerical treatment of initial value problems for fractional differential equations which are equivalent to these integral equations. The results presented in Chapter 3, which established an equivalence for Cauchy type problems for differential equations with Riemann-Liouiville, Caputo, and Hadamard fractional derivatives, and the corresponding Volterra integral equation, permit us to apply the above mentioned numerical methods. For example, we could use successful the corresponding routines of the known numerical mathematical libraries called NAG and IMSL. In this regard we refer the reader to Kilbas and Marzan [379] and Marzan [559], where approximate solutions were obtained to the Cauchy type problem (3.1.1)-(3.1.2) and to the Cauchy problem (4.4.31), respectively. In the bibliography of this book, the reader can find more references on numerical methods to solve fractional differential equation.

Chapter 5

INTEGRAL TRANSFORM METHOD FOR EXPLICIT SOLUTIONS TO FRACTIONAL DIFFERENTIAL EQUATIONS

The present chapter is devoted to the application of Laplace, Mellin, and Fourier integral transforms to construct explicit solutions to linear differential equations involving Liouville, Caputo, and Riesz fractional derivatives.

5.1 Introduction and a Brief Survey of Results

First we present a scheme for solving the one-dimensional fractional nonhomogeneous differential equation with constant coefficients of the form

$$\sum_{k=1}^{m} A_k (D_{0+}^{\alpha_k} y)(x) + A_0 y(x) = f(x) \quad (x > 0) \tag{5.1.1}$$

with $m \in \mathbb{N}$; $0 < \Re(\alpha_1) < \cdots < \Re(\alpha_m)$; $A_0, A_1, \cdots, A_m \in \mathbb{R}$, and involving the Liouville fractional derivatives $D_{0+}^{\alpha_k} y$ $(k = 1, \cdots, m)$, given by (2.2.3). By (2.2.38), for suitable functions y, the Laplace transform (1.4.1) of $D_{0+}^{\alpha} y$ is given by

$$(\mathcal{L} D_{0+}^{\alpha} y)(s) = s^{\alpha} (\mathcal{L} y)(s). \tag{5.1.2}$$

Applying the Laplace transform to (5.1.1) and taking (5.1.2) into account, we have

$$\left[A_0 + \sum_{k=1}^{m} A_k s^{\alpha_k}\right] (\mathcal{L}y)(s) = (\mathcal{L}f)(s). \tag{5.1.3}$$

Using the inverse Laplace transform $\mathcal{L}^{-1}$, given by (1.4.2), from here we obtain a particular solution to the equation (5.1.1) in the form

$$y(x) = \left(\mathcal{L}^{-1}\left[\frac{(\mathcal{L}f)(s)}{A_0 + \sum_{k=1}^{m} A_k s^{\alpha_k}}\right]\right)(x). \tag{5.1.4}$$

We note that the Laplace transform method was applied earlier by Hille and Tamarkin [349] to solve the integral equation

$$\varphi(x) - \frac{\lambda}{\Gamma(\alpha)} \int_0^x \frac{\varphi(t)dt}{(x-t)^{1-\alpha}} = f(x) \quad (x > 0), \tag{5.1.5}$$

with $\alpha > 0$ and $\lambda \in \mathbb{R}$, in terms of the
Mittag-Leffler function (1.8.1):

$$\varphi(x) = \frac{d}{dx} \int_0^x E_\alpha\left[\lambda(x-t)^\alpha\right] f(t)dt. \tag{5.1.6}$$

Maravall [549], [550],[553], suggested a formal approach based on the Laplace transform for obtaining the explicit solution to a particular case of the fractional differential equation (5.1.1). Probably he was first to present applied fractional models, connecting with anomalous oscillatory phenomena (see, also, [551],[552], [554], and [555]).

The Laplace transform was used by Oldham and Spanier ([643], Section 8.5) to solve the homogeneous case of the following equation ($f(x) = 0$):

$$y'(x) + b(D_{0+}^\alpha y)(x) + cy(x) = f(x) \quad (0 < \alpha < 1;\ b, c \in \mathbb{R}), \tag{5.1.7}$$

with $\alpha = 1/2$, $b = 1$ and $c = -2$, by Dorta [200] to obtain solutions to some simple fractional differential equations (4.1.11) with rational $\alpha > 0$, and by Seitkazieva [756] to construct the solution to the equation (5.1.7).

Miller and Ross ([603], Sections V.5-V.9) applied the Laplace transform method to give explicit representations of solutions for the particular case of the equation (5.1.1) with derivatives of orders $\alpha_k = k\alpha$ ($\alpha = \frac{1}{n}; n \in \mathbb{N}$) for the homogeneous equation

$$\sum_{k=1}^{m} A_k (D_{0+}^{k\alpha} y)(x) + A_0 y(x) = 0 \quad \left(\frac{1}{\alpha} \in \mathbb{N}\right), \tag{5.1.8}$$

and for the nonhomogeneous equation

$$\sum_{k=1}^{m} A_k (D_{0+}^{k\alpha} y)(x) + A_0 y(x) = f(x) \quad \left(\frac{1}{\alpha} \in \mathbb{N}\right). \tag{5.1.9}$$

It should be noted that Miller and Ross ([603], Section V.6) practically found linear independent solutions to the homogeneous equation (5.1.8). This fact and the results of Section 4.2.3 mean that the homogeneous linear differential equation of the form (5.1.1)

$$\sum_{k=1}^{m} A_k(D_{a+}^{\alpha_k}y)(x) + A_0 y(x) = 0 \quad (0 < \Re(\alpha_1) < \cdots < \Re(\alpha_m)), \tag{5.1.10}$$

can have nontrivial linearly independent solutions by analogy with the ordinary differential equation of order m obtained from (5.1.10) for $\alpha_k = k$ $(k = 1, \cdots, m)$. We note that Leskovskij [477] constructed linearly independent solutions to the equation (5.1.10) with specific distinct real exponents $0 < \alpha_k < 1$ $(k = 1, \cdots, m)$ in terms of the Mittag-Leffler function (1.8.17).

Miller [see Miller and Ross ([603], Sections V.7-V.9)] introduced a fractional analog of the Green function $G_\alpha(x)$ defined, via the inverse Laplace transform (1.4.2), by

$$G_\alpha(x) = \left(\mathcal{L}^{-1}\left[\frac{1}{P(s^\alpha)}\right]\right)(x), \quad P(s) = \sum_{k=1}^{m} A_k s^k + A_0, \tag{5.1.11}$$

represented a particular solution to the nonhomogeneous equation (5.1.9) in the form of the convolution of $G_\alpha(x)$ and $f(x)$:

$$y(x) = \int_0^x G_\alpha(x-t)f(t)dt, \tag{5.1.12}$$

and proved that this formula yields a unique solution $y(x)$ to the Cauchy problem for the equation (5.1.9) with the following initial conditions:

$$y(0) = y'(0) = \cdots = y^{(m-1)}(0) = 0. \tag{5.1.13}$$

Miller and Ross ([603], Section VI.3) indicated that the Laplace transform can be applied to solve homogeneous differential equations of the form (5.1.8):

$$\sum_{k=1}^{m} A_k(x)(D_{0+}^{k\alpha}y)(x) + A_0 y(x) = 0 \quad \left(\frac{1}{\alpha} = 1, 2, \cdots\right), \tag{5.1.14}$$

with polynomial coefficients $A_k(x)$, and illustrated such an approach by solving the following equation:

$$(D_{0+}^{1/2}y)(x) = \frac{y}{x} \quad (x > 0). \tag{5.1.15}$$

Miller [600] obtained linearly independent solutions to the equation (5.1.14) in terms of an analog of the Green function (5.1.11).

It should be noted that the relation (5.1.15) was the first known differential equation of fractional order discussed by O'Shaughnessay [650] and Post [686] [see in this regard Samko et al. ([729], Section 42.1) and Kilbas and Trujllo ([407], Section 3)].

Miller and Ross [603] also solved the so-called sequential fractional differential equations of the form (5.1.9):

$$\left(P_m(D_{0+}^{\alpha})y\right)(x) \equiv \sum_{k=1}^{m} A_k\left((D_{0+}^{\alpha})^k y\right)(x) + c_0 y(x) = f(x) \quad \left(\frac{1}{\alpha} \in \mathbb{N}\right), \tag{5.1.16}$$

by using the roots $\lambda = \lambda_j$ of the characteristic polynomial $P_m(\lambda) = \sum_{k=0}^{m} A_k \lambda^k$. In the case $k = 2$ the solution to such an equation was given by Miller [600].

The above investigations of Miller and Ross [603] were developed by Podlubny [679], [681] [see also ([682], Chapter 5)], who defined a fractional analog of the Green function $G_\alpha(x,t)$ for the equation (3.1.17) and showed that a particular solution $y(x)$ of the Cauchy type problem (3.1.22) with $b_k = 0$ $(k = 0, \cdots, n-1)$

$$\left(D^{\sigma}y\right)(x) = f(x), \quad \left(D^{\sigma_k}y\right)(x)|_{x=0} = 0 \quad (k = 0, \cdots, n-1) \tag{5.1.17}$$

can be expressed in terms of $G_\alpha(x,t)$ as follows:

$$y(x) = \int_0^x G_\alpha(x,t)f(t)dt. \tag{5.1.18}$$

When $a_k(x) = a_k \in \mathbb{C}$ are constants, $G_\alpha(x,t) = G_\alpha(x-t)$, and the explicit solution to the Cauchy type problem (3.1.22) has the form

$$y(x) = \sum_{k=0}^{n-1} b_k y_k(x) + \int_0^x G_\alpha(x-t)f(t)dt, \quad y_k(x) = \left(D_{0+}^{\alpha_n} D_{0+}^{\alpha_{n-1}} \cdots D_{0+}^{\alpha_k} G_\alpha\right)(x). \tag{5.1.19}$$

In particular, for the equation (5.1.1), Podlubny ([682], Section 5.6) constructed the explicit formula for $G_\alpha(x)$ as a multiple series containing the Mittag-Leffler functions (1.8.17).

Examples of linear fractional differential equations of the forms (5.1.8), (5.1.9), (5.1.10), (5.1.14) and (5.1.16), solved by using the Laplace transform method and the analogs of the Green function, were given by Miller and Ross ([603], Chapters V and VI) and Podlubny ([682], Sections 4.1.1 and 4.2.1).

Gorenflo and Mainardi [304] applied the Laplace transform to solve the Cauchy problem for the fractional differential equation with the Caputo derivative (2.4.52)

$$({}^{C}D_{0+}^{\alpha}y)(x) + y(x) = f(x) \quad (x > 0;\ m-1 < \alpha \leqq m;\ m \in \mathbb{N}), \tag{5.1.20}$$

with the initial conditions (5.1.13), and the Cauchy problems

$$y'(x) + a({}^{C}D_{0+}^{\alpha}y)(x) + y(x) = f(x), \quad y(0) = c_0 \in \mathbb{R} \quad (x > 0;\ 0 < \alpha < 1;\ a > 0) \tag{5.1.21}$$

$$y''(x) + a({}^{C}D_{0+}^{\alpha}y)(x) + y(x) = f(x), \quad y(0) = c_0 \in R,\ y'(0) = c_1 \in \mathbb{R} \tag{5.1.22}$$

$$(x > 0;\ 1 < \alpha < 2).$$

See also Gorenflo and Rutman [311], Gorenflo and Mainardi [303], Gorenflo et al. [310] and Debnath [162].

Podlubny ([682], Section 6.1) indicated that the one-dimensional Mellin transform (1.4.23) can be applied to solve the Cauchy problem for the fractional differential equation

$$x^{\alpha+1}\left(D^{\alpha+1}y\right)(x)+x^{\alpha}\left(D^{\alpha}y\right)(x)=f(x)\quad(x>0)\tag{5.1.23}$$

with the Liouville $D^{\alpha+k}y=D_{0+}^{\alpha+k}y$ or the Caputo $D^{\alpha+k}y={}^{C}D_{0+}^{\alpha+k}y$ $(k=0,1)$ fractional derivatives (2.2.6) and (2.4.52) of order $0<\alpha<1$. Stanković [800] studied the Cauchy problem

$$y''(x)+a(D_{0+}^{\alpha}y)(x)+g(x)y(x)=f(x)\quad(x>0;\ \ 0<\alpha<1)\tag{5.1.24}$$

$$y(0)=b_0,\quad y'(0)=b_1\quad(b_1,b_2\in\mathbb{R}),\tag{5.1.25}$$

where $a>0$, $f(x)\in L_{loc}(\mathbb{R}_+)$ and $g(x)$ is a real-analytic function on an open set containing $\mathbb{R}_+$. He reduced the above problem to a differential equation of the second order in a special space of hyperfunctions with support on $\mathbb{R}_+$ and used the Laplace transform to prove the existence of a unique solution for the Cauchy problem (5.1.24)-(5.1.25) in $L_{loc}(\mathbb{R}_+)$. Stanković showed that, in the case $y(0)=0$, this solution $y(x)$ belongs to $C^2(\mathbb{R}_+)$ and is a classical solution, while for $y(0)\neq 0$, $y(x)\in\mathbb{C}^1(\mathbb{R}_+)$ and is a solution in the sense of distributions. He also proved that if $f(x)$ is a continuous function such that $|f(x)|\leqq Ke^{\lambda x}$ for $x\geqq 0$ with constants K and λ, then the solution $y(x)$ will have the same estimate.

5.2 Laplace Transform Method for Solving Ordinary Differential Equations with Liouville Fractional Derivatives

Our discussion involving the Laplace transform method for the Liouville fractional derivatives will be presented under several subsections as follows.

5.2.1 Homogeneous Equations with Constant Coefficients

In this section we apply the Laplace transform method to derive explicit solutions to homogeneous equations of the form (5.1.10)

$$\sum_{k=1}^{m}A_k\left(D_{0+}^{\alpha_k}y\right)(x)+A_0y(x)=0\quad(x>0;\ m\in\mathbb{N};\ 0<\alpha_1<\cdots<\alpha_m),\tag{5.2.1}$$

with the Liouville fractional derivatives $D_{0+}^{\alpha_k}y$ $(k=1,\cdots,\ m)$, given by (2.2.3). Here $A_k\in\mathbb{R}$ $(k=0,\cdots,\ m)$ are real constants, and, generally speaking, we can take $A_m=1$. We give the conditions when the solutions $y_1(x),\cdots,\ y_l(x)$ of the equation (5.2.1) with $l-1<\alpha:=\alpha_m\leqq l$ $(l\in\mathbb{N})$ will be linearly independent, and when these linearly independent solutions form the fundamental system of solutions, which (by analogy with the ordinary case) is defined by

$$\left(D_{0+}^{\alpha-k}y_j\right)(0+)=0\quad(k,j=1,\cdots,\ l;\ k\neq j),$$

$$\left(D_{0+}^{\alpha-k}y_k\right)(0+) = 1 \quad (k = 1, \cdots, l). \tag{5.2.2}$$

The Laplace transform method is based on the relation (2.2.37) which, in accordance with (2.2.3), is equivalent to the following one:

$$\left(\mathcal{L}D_{0+}^{\alpha}y\right)(s) = s^{\alpha}\left(\mathcal{L}y\right)(s) - \sum_{j=1}^{l} d_j s^{j-1} \quad (l-1 < \alpha \leqq l; \; l \in \mathbb{N}), \tag{5.2.3}$$

$$d_j = \left(D_{0+}^{\alpha-j}y\right)(0+) \quad (j = 1, \cdots, l). \tag{5.2.4}$$

First we derive explicit solutions to the equation (5.2.1) with $m = 1$ in the form (4.1.11)

$$\left(D_{0+}^{\alpha}y\right)(x) - \lambda y(x) = 0 \quad (x > 0; \; l-1 < \alpha \leqq l; \; l \in \mathbb{N}; \; \lambda \in \mathbb{R}), \tag{5.2.5}$$

in terms of the Mittag-Lefler functions (1.8.17). There holds the following statement.

Theorem 5.1 *Let $l-1 < \alpha \leqq l$ $(l \in \mathbb{N})$ and $\lambda \in \mathbb{R}$. Then the functions*

$$y_j(x) = x^{\alpha-j}E_{\alpha,\alpha+1-j}\left(\lambda x^{\alpha}\right) \quad (j = 1, \cdots, l) \tag{5.2.6}$$

yield the fundamental system of solutions to equation (5.2.5).

Proof. Applying the Laplace transform (1.4.1) to (5.2.5) and taking (5.2.3) into account, we have

$$\left(\mathcal{L}y\right)(s) = \sum_{j=1}^{l} d_j \frac{s^{j-1}}{s^{\alpha} - \lambda}, \tag{5.2.7}$$

where d_j $(j = 1, \cdots, l)$ are given by (5.2.4). Formula (1.10.9) with $\beta = \alpha + 1 - j$ yields

$$\mathcal{L}\left[t^{\alpha-j}E_{\alpha,\alpha+1-j}\left(\lambda t^{\alpha}\right)\right](s) = \frac{s^{j-1}}{s^{\alpha} - \lambda} \quad \left(|s^{-\alpha}\lambda| < 1\right). \tag{5.2.8}$$

Thus, from (5.2.7), we derive the following solution to the equation (5.2.5):

$$y(x) = \sum_{j=1}^{l} d_j y_j(x), \quad y_j(x) = x^{\alpha-j}E_{\alpha,\alpha+1-j}\left(\lambda x^{\alpha}\right). \tag{5.2.9}$$

It is easily verified that the functions $y_j(x)$ are solutions to the equation (5.2.5)

$$\left(D_{0+}^{\alpha}\left[t^{\alpha-j}E_{\alpha,\alpha+1-j}\left(\lambda t^{\alpha}\right)\right]\right)(x) = \lambda x^{\alpha-j}E_{\alpha,\alpha+1-j}\left(\lambda x^{\alpha}\right) \quad (j = 1, \cdots, l) \tag{5.2.10}$$

and, moreover,

$$\left(D_{0+}^{\alpha-k}y_j\right)(x) = \sum_{n=0}^{\infty} \frac{\lambda^n}{\Gamma(\alpha n + k + 1 - j)} x^{\alpha n + k - j}. \tag{5.2.11}$$

It follows from (5.2.11) that

$$\left(D_{0+}^{\alpha-k}y_j\right)(0+)=0\ \ (k,j=1,\cdots,\ l;\ k>j),\ \ \left(D_{0+}^{\alpha-k}y_k\right)(0+)=1\ (k=1,\cdots,\ l). \tag{5.2.12}$$

If $k<j$, then

$$\left(D_{0+}^{\alpha-k}y_j\right)(x)=\sum_{n=1}^{\infty}\frac{\lambda^n}{\Gamma(\alpha n+k+1-j)}x^{\alpha n+k-j}$$

$$=\sum_{n=0}^{\infty}\frac{\lambda^{n+1}}{\Gamma(\alpha n+\alpha+k+1-j)}x^{\alpha n+\alpha+k-j} \tag{5.2.13}$$

and, since $\alpha+k-j\geqq\alpha+1-l>0$ for any $k,j=1,\cdots,l$, the following relations hold:

$$\left(D_{0+}^{\alpha-k}y_j\right)(0+)=0\ \ (k,j=1,\cdots,\ l;\ k<j). \tag{5.2.14}$$

By (5.2.12) and (5.2.14), $W_\alpha(0)=1$ for the analog of the Wronskian $W_\alpha(x)$ defined by (4.2.34). Then the result of the Theorem 5.1 follows from Lemma 4.1 and (5.2.2).

Corollary 5.1 *The equation*

$$\left(D_{0+}^{\alpha}y\right)(x)-\lambda y(x)=0\ \ (x>0;\ \ 0<\alpha\leqq 1;\ \lambda\in\mathbb{R}) \tag{5.2.15}$$

has its solution given by

$$y(x)=x^{\alpha-1}E_{\alpha,\alpha}\left(\lambda x^{\alpha}\right), \tag{5.2.16}$$

while the equation

$$\left(D_{0+}^{\alpha}y\right)(x)-\lambda y(x)=0\ \ (x>0;\ \ 1<\alpha\leqq 2;\ \lambda\in\mathbb{R}) \tag{5.2.17}$$

has the fundamental system of solutions given by

$$y_1(x)=x^{\alpha-1}E_{\alpha,\alpha}\left(\lambda x^{\alpha}\right),\ \ y_2(x)=x^{\alpha-2}E_{\alpha,\alpha-1}\left(\lambda x^{\alpha}\right). \tag{5.2.18}$$

Example 5.1 The equation

$$\left(D_{0+}^{l-1/2}y\right)(x)-\lambda y(x)=0\ \ (x>0;\ \ l\in\mathbb{N},\ \lambda\in\mathbb{R}), \tag{5.2.19}$$

has the fundamental system of solutions given by

$$y_j(x)=x^{l-j-1/2}E_{l-1/2,l-j+1/2}\left(\lambda x^{l-1/2}\right)\ \ (j=1,\cdots,\ l). \tag{5.2.20}$$

Example 5.2 The following ordinary differential equation of order $l\in\mathbb{N}$

$$y^{(l)}(x)-\lambda y(x)=0\ \ (x>0;\ \ l\in\mathbb{N}) \tag{5.2.21}$$

has the fundamental system of solutions given by

$$y_j(x)=x^{l-j}E_{l,l+1-j}\left(\lambda x^{l}\right)\ \ (j=1,\cdots,\ l). \tag{5.2.22}$$

In particular, the equations

$$y'' - \lambda y(x) = 0 \quad (x > 0;\quad \lambda > 0) \tag{5.2.23}$$

and

$$y'' + \lambda y(x) = 0 \quad (x > 0;\quad \lambda > 0) \tag{5.2.24}$$

have their respective fundamental systems of solutions given by

$$y_1(x) = \frac{1}{\sqrt{\lambda}} \sinh\left(\sqrt{\lambda}x\right),\quad y_2(x) = \cosh\left(\sqrt{\lambda}x\right) \tag{5.2.25}$$

and

$$y_1(x) = \frac{1}{\sqrt{\lambda}} \sin\left(\sqrt{\lambda}x\right),\quad y_2(x) = \cos\left(\sqrt{\lambda}x\right). \tag{5.2.26}$$

Next we derive the explicit solutions to the equation (5.2.1) with $m = 2$ of the form

$$\left(D_{0+}^{\alpha}y\right)(x) - \lambda\left(D_{0+}^{\beta}y\right)(x) - \mu y(x) = 0 \quad (x > 0;\ l-1 < \alpha \leqq l;\ l \in \mathbb{N};\ \alpha > \beta > 0), \tag{5.2.27}$$

with $\lambda, \mu \in \mathbb{R}$, in terms of the generalized Wright function (1.11.14) with $p = q = 1$ of the form

$$ {}_1\Psi_1\left[\begin{array}{c|c} (n+1,1) & \\ & z \\ (\alpha n+\beta,\alpha) & \end{array}\right] := \sum_{j=0}^{\infty} \frac{\Gamma(n+j+1)}{\Gamma(\alpha n+\beta+\alpha j)} \frac{z^j}{j!} = \left(\frac{\partial}{\partial z}\right)^n E_{\alpha,\beta}(z). \tag{5.2.28}$$

The following assertion follows from Theorem 1.5

Lemma 5.1 *The generalized Wright function* (5.2.28) *is an entire function of* $z \in \mathbb{C}$ *for* $\alpha > 0$.

Theorem 5.2 *Let* $l - 1 < \alpha \leqq l$ $(l \in \mathbb{N})$, $0 < \beta < \alpha$ *and* $\lambda, \mu \in \mathbb{R}$. *Then the functions*

$$y_j(x) = \sum_{n=0}^{\infty} \frac{\mu^n}{n!} x^{\alpha n+\alpha-j}\ {}_1\Psi_1\left[\begin{array}{c|c} (n+1,1) & \\ & \lambda x^{\alpha-\beta} \\ (\alpha n+\alpha+1-j,\alpha-\beta) & \end{array}\right] \quad (j = 1,\cdots,\ l) \tag{5.2.29}$$

are solutions to the equation (5.2.27), *provided that the series in* (5.2.29) *is convergent.*

In particular, the equation

$$\left(D_{0+}^{\alpha}y\right)(x) - \lambda\left(D_{0+}^{\beta}y\right)(x) = 0 \quad (x > 0;\quad l-1 < \alpha \leqq l;\quad l \in \mathbb{N};\ \alpha > \beta > 0), \tag{5.2.30}$$

has its solutions given by

$$y_j(x) = x^{\alpha-j} E_{\alpha-\beta,\alpha+1-j}\left(\lambda x^{\alpha-\beta}\right) \quad (j = 1,\cdots,l). \tag{5.2.31}$$

If $\alpha - l + 1 \geqq \beta$, *then* $y_j(x)$ *in* (5.2.29) *and* (5.2.31) *are linearly independent solutions to equations* (5.2.27) *and* (5.2.30), *respectively. In particular, for* $\alpha - l + 1 > \beta$ *they yield the fundamental systems of solutions.*

Proof. Let $m-1<\beta \leqq m$ $(m \in \mathbb{N};\ m \leqq l)$. Applying the Laplace transform to (5.2.27) and using (5.2.3) as in (5.2.7), we obtain

$$(\mathcal{L}y)(s)=\sum_{j=1}^{l} d_j \frac{s^{j-1}}{s^{\alpha}-\lambda s^{\beta}-\mu}, \tag{5.2.32}$$

where, for all $j=m+1,\cdots,l$

$$d_j=\left(D_{0+}^{\alpha-j}y\right)(0+)+\left(D_{0+}^{\beta-j}y\right)(0+)\ (j=1,\cdots,m),\quad d_j=\left(D_{0+}^{\alpha-j}y\right)(0+).$$

For $s \in \mathbb{C}$ and $\left|\frac{\mu s^{-\beta}}{s^{\alpha-\beta}-\lambda}\right|<1$, we have

$$\frac{1}{s^{\alpha}-\lambda s^{\beta}-\mu}=\frac{s^{-\beta}}{s^{\alpha-\beta}-\lambda}\frac{1}{\left(1-\frac{\mu s^{-\beta}}{s^{\alpha-\beta}-\lambda}\right)}=\sum_{n=0}^{\infty}\frac{\mu^n s^{-\beta-\beta n}}{(s^{\alpha-\beta}-\lambda)^{n+1}}, \tag{5.2.33}$$

and hence (5.2.32) has the following representation:

$$(\mathcal{L}y)(s)=\sum_{j=1}^{l} d_j \sum_{n=0}^{\infty}\mu^n \frac{s^{j-1-\beta-\beta n}}{(s^{\alpha-\beta}-\lambda)^{n+1}}. \tag{5.2.34}$$

Using (1.10.10), with α replaced by $\alpha-\beta$ and β by $\alpha+\beta n+1-j$, and taking (5.2.28) into account, for $s \in \mathbb{C}$ and $|\lambda s^{\beta-\alpha}|<1$, we have

$$\frac{s^{j-1-\beta-\beta n}}{(s^{\alpha-\beta}-\lambda)^{n+1}}=\frac{s^{(\alpha-\beta)-(\alpha+\beta n+1-j)}}{(s^{\alpha-\beta}-\lambda)^{n+1}}$$

$$=\frac{1}{n!}\left(\mathcal{L}\left[t^{\alpha n+\alpha-j}\left(\frac{\partial}{\partial \lambda}\right)^n E_{\alpha-\beta,\alpha+\beta n+1-j}\left(\lambda t^{\alpha-\beta}\right)\right]\right)(s)$$

$$=\frac{1}{n!}\left(\mathcal{L}\left[t^{\alpha n+\alpha-j}\,{}_1\Psi_1\left[\begin{matrix}(n+1,1)\\ (\alpha n+\alpha+1-j,\alpha-\beta)\end{matrix}\ \middle|\ \lambda t^{\alpha-\beta}\right]\right]\right)(s). \tag{5.2.35}$$

From (5.2.34) and (5.2.35) we derive the solution to the equation (5.2.27)

$$y(x)=\sum_{j=1}^{l} d_j y_j(x), \tag{5.2.36}$$

where $y_j(x)$ $(j=1,\cdots,l)$ are given by (5.2.29). It is readily verified that these functions are solutions to the equation (5.2.27), which proves the first assertion of Theorem 5.2.

For $j,k=1,\cdots,l$, the direct evaluation yields

$$\left(D_{0+}^{\alpha-k}y_j\right)(x)=\sum_{n=0}^{\infty}\frac{\mu^n}{n!}x^{\alpha n+k-j}\,{}_1\Psi_1\left[\begin{matrix}(n+1,1)\\ (\alpha n+k+1-j,\alpha-\beta)\end{matrix}\ \middle|\ \lambda x^{\alpha-\beta}\right]. \tag{5.2.37}$$

It follows from (5.2.37) that the relations in (5.2.12) hold for $k \geqq j$. If $k < j$, then, in accordance with (5.2.28), we rewrite (5.2.37) as follows:

$$\left(D_{0+}^{\alpha-k}y_j\right)(x) = \sum_{q=0}^{\infty} \frac{\lambda^{q+1}}{\Gamma[(\alpha-\beta)(q+1)+k+1-j]} x^{(\alpha-\beta)q+\alpha-\beta+k-j}$$

$$+\sum_{n=1}^{\infty} \frac{\mu^n}{n!} x^{\alpha n+k-j} \, {}_1\Psi_1 \left[\begin{array}{c} (n+1,1) \\ \\ (\alpha n+k+1-j, \alpha-\beta) \end{array} \middle| \lambda x^{\alpha-\beta} \right] = I_1(x) + I_2(x). \tag{5.2.38}$$

If $\alpha - l + 1 \geqq \beta$, then $(\alpha-\beta)q + \alpha - \beta + k - j \geqq \alpha - \beta + 1 - l \geqq 0$ for any $j, k = 1, \cdots, l$ and $q \in \mathbb{N}_0$. Thus $\lim_{x\to 0+} I_1(x) = 0 \quad (k, j = 1, \cdots, l;\ k < j)$ except for the case $\alpha - l + 1 = \beta$ with $k = 1$ and $j = l$, for which $\lim_{x\to 0+} I_1(x) = \lambda$. Moreover, since $\alpha n + k - j \geqq \alpha + 1 - l > 0$ for any $j, k = 1, \cdots, l$ and $n \in \mathbb{N}$, then $\lim_{x\to 0+} I_2(x) = 0 \quad (k, j = 1, \cdots, l;\ k < j)$. Thus (5.2.38) yields the relation (5.2.14) for any solution $y_j(x)$ in (5.2.29), except for the case $\alpha - l + 1 = \beta$ with $k = 1$ and $j = l$, for which

$$\left(D_{0+}^{\alpha-1}y_l\right)(0+) = \lambda. \tag{5.2.39}$$

According to (5.2.12), (5.2.14) and (5.2.39) for the analog of the Wronskian $W_\alpha(x)$, given by (4.2.34), we have $W_\alpha(0) = 1$. Thus it follows from Lemma 4.1 that the functions $y_j(x)$ in (5.2.29) are linearly independent solutions to the equation (5.2.27). If $\alpha - l + 1 > \beta$, then the relations (5.2.2) are valid, and hence $y_j(x)$ in (5.2.29) yield the fundamental system of solutions to the equation (5.2.27). This completes the proof of the Theorem 5.2 for the equation (5.2.27).

Setting $\mu = 0$ and taking into account the formula

$${}_1\Psi_1 \left[\begin{array}{c} (1,1) \\ \\ (b,\beta) \end{array} \middle| z \right] = E_{\beta,b}(z) \quad (b \in \mathbb{C};\ \beta \in \mathbb{R}), \tag{5.2.40}$$

we obtain the assertion of the Theorem 5.2 for the equation (5.2.30).

Corollary 5.2 *The equation*

$$\left(D_{0+}^{\alpha}y\right)(x) - \lambda\left(D_{0+}^{\beta}y\right)(x) - \mu y(x) = 0 \quad (x > 0;\ \ 0 < \beta < \alpha \leqq 1;\ \lambda, \mu \in \mathbb{R}) \tag{5.2.41}$$

has its solution given by

$$y_1(x) = \sum_{n=0}^{\infty} \frac{\mu^n}{n!} x^{\alpha n+\alpha-1} \, {}_1\Psi_1 \left[\begin{array}{c} (n+1,1) \\ \\ (\alpha n+\alpha, \alpha-\beta) \end{array} \middle| \lambda x^{\alpha-\beta} \right]. \tag{5.2.42}$$

In particular,

$$y_1(x) = x^{\alpha-1} E_{\alpha-\beta,\alpha}\left(\lambda x^{\alpha-\beta}\right) \tag{5.2.43}$$

is the solution to the following equation:

$$\left(D_{0+}^{\alpha}y\right)(x) - \lambda\left(D_{0+}^{\beta}y\right)(x) = 0 \quad (x > 0;\ \ 0 < \beta < \alpha \leqq 1;\ \lambda \in \mathbb{R}). \tag{5.2.44}$$

Corollary 5.3 *The equation*

$$\left(D_{0+}^{\alpha}y\right)(x)-\lambda\left(D_{0+}^{\beta}y\right)(x)-\mu y(x)=0\ \ (x>0;\ 1<\alpha\leqq 2,\ 0<\beta<\alpha;\ \lambda,\mu\in\mathbb{R}) \tag{5.2.45}$$

has two solutions $y_1(x)$, *given by* (5.2.42), *and*

$$y_2(x)=\sum_{n=0}^{\infty}\frac{\mu^n}{n!}x^{\alpha n+\alpha-2}\,{}_1\Psi_1\left[\begin{array}{c}(n+1,1)\\ (\alpha n+\alpha-1,\alpha-\beta)\end{array}\middle|\,\lambda x^{\alpha-\beta}\right]. \tag{5.2.46}$$

In particular, the equation

$$\left(D_{0+}^{\alpha}y\right)(x)-\lambda\left(D_{0+}^{\beta}y\right)(x)=0\ \ (x>0;\ 1<\alpha\leqq 2,\ 0<\beta<\alpha;\ \lambda\in\mathbb{R}) \tag{5.2.47}$$

has two solutions $y_1(x)$, *given by* (5.2.43), *and*

$$y_2(x)=x^{\alpha-2}E_{\alpha-\beta,\alpha-1}\left(\lambda x^{\alpha-\beta}\right). \tag{5.2.48}$$

If $\alpha\geqq\beta+1$, *then the above functions* $y_1(x)$ *and* $y_2(x)$ *are linearly independent solutions of the equations* (5.2.45) *and* (5.2.47), *respectively. In particular, for* $\alpha>\beta+1$ *these functions provide the fundamental system of solutions.*

Example 5.3 The equation

$$y'(x)-\lambda\left(D_{0+}^{\beta}y\right)(x)-\mu y(x)=0\ \ (x>0;\ \ 0<\beta<1;\ \lambda,\mu\in\mathbb{R}) \tag{5.2.49}$$

has its solution given by

$$y(x)=\sum_{n=0}^{\infty}\frac{\mu^n}{n!}x^{n}\,{}_1\Psi_1\left[\begin{array}{c}(n+1,1)\\ (n+1,1-\beta)\end{array}\middle|\,\lambda x^{1-\beta}\right]. \tag{5.2.50}$$

In particular,

$$y(x)=\sum_{n=0}^{\infty}\frac{(\mu x)^n}{n!}x^{n}\,{}_1\Psi_1\left[\begin{array}{c}(n+1,1)\\ (n+1,1/2)\end{array}\middle|\,\lambda x^{1/2}\right] \tag{5.2.51}$$

is the solution to the equation

$$y'(x)-\lambda\left(D_{0+}^{1/2}y\right)(x)-\mu y(x)=0\ \ (x>0;\ \ \lambda,\mu\in\mathbb{R}). \tag{5.2.52}$$

Remark 5.1 Oldham and Spanier ([643], Section 8.5) expressed the solution (5.2.51) of the equation (5.2.52), with $\lambda=-1$ and $\mu=2$, in terms of the complementary error function $\operatorname{erfc}(z)$ defined via the incomplete gamma function (1.5.27) by (Abramowitz and Stegun [1], formula 6.5.16)

$$\operatorname{erfc}(z):=\frac{1}{\pi}\gamma\left(\frac{1}{2},z^2\right)=\frac{2}{\pi}\int_0^z e^{-t^2}dt. \tag{5.2.53}$$

Example 5.4 The equation

$$y''(x) - \lambda \left(D_{0+}^{\beta} y \right)(x) - \mu y(x) = 0 \quad (x > 0; \ \ 0 < \beta < 2; \ \lambda, \mu \in \mathbb{R}) \tag{5.2.54}$$

has its two solutions given by

$$y_1(x) = \sum_{n=0}^{\infty} \frac{\mu^n}{n!} x^{2n+1} \, {}_1\Psi_1 \left[\begin{array}{c|c} (n+1, 1) & \\ & \lambda x^{2-\beta} \\ (2n+2, 2-\beta) & \end{array} \right], \tag{5.2.55}$$

$$y_2(x) = \sum_{n=0}^{\infty} \frac{\mu^n}{n!} x^{2n} \, {}_1\Psi_1 \left[\begin{array}{c|c} (n+1, 1) & \\ & \lambda x^{2-\beta} \\ (2n+1, 2-\beta) & \end{array} \right]. \tag{5.2.56}$$

These solutions are linearly independent when $\beta \leqq 1$ and form the fundamental system of solutions when $\beta < 1$.

In particular, the equation

$$y''(x) - \lambda \left(D_{0+}^{1/2} y \right)(x) - \mu y(x) = 0 \quad (x > 0; \ \ \lambda, \mu \in \mathbb{R}) \tag{5.2.57}$$

has the fundamental system of solutions given by

$$y_1(x) = \sum_{n=0}^{\infty} \frac{\mu^n}{n!} x^{2n+1} \, {}_1\Psi_1 \left[\begin{array}{c|c} (n+1, 1) & \\ & \lambda x^{3/2} \\ (2n+2, 3/2) & \end{array} \right], \tag{5.2.58}$$

$$y_2(x) = \sum_{n=0}^{\infty} \frac{\mu^n}{n!} x^{2n} \, {}_1\Psi_1 \left[\begin{array}{c|c} (n+1, 1) & \\ & \lambda x^{3/} \\ (2n+1, 3/2) & \end{array} \right]. \tag{5.2.59}$$

Example 5.5 The following ordinary differential equation of order $l \in \mathbb{N}$

$$y^{(l)}(x) - \lambda y^{(m)}(x) - \mu y(x) = 0 \quad (x > 0; \ \ l, m \in \mathbb{N}; \ m < l; \ \ \lambda, \mu \in \mathbb{R}) \tag{5.2.60}$$

has l solutions given by

$$y_j(x) = \sum_{n=0}^{\infty} \frac{\mu^n}{n!} x^{ln+l-j} \, {}_1\Psi_1 \left[\begin{array}{c|c} (n+1, 1) & \\ & \lambda x^{l-m} \\ (ln+l+1-j, l-m) & \end{array} \right] \quad (j = 1, \cdots, l). \tag{5.2.61}$$

When $m = 1$, these solutions are linearly independent.

In particular, the following ordinary second order differential equation

$$y''(x) - \lambda y'(x) - \mu y(x) = 0 \quad (x > 0; \ \ \lambda, \mu \in \mathbb{R}) \tag{5.2.62}$$

has two linearly independent solutions given by

$$y_1(x) = \sum_{n=0}^{\infty} \frac{\mu^n}{n!} x^{2n+1} \, {}_1\Psi_1 \left[\begin{array}{c|c} (n+1, 1) & \\ & \lambda x \\ (2n+2, 1) & \end{array} \right], \tag{5.2.63}$$

$$y_2(x) = \sum_{n=0}^{\infty} \frac{\mu^n}{n!} x^{2n} \, {}_1\Psi_1 \left[\begin{array}{c|c} (n+1,1) & \\ & \lambda x \\ (2n+1,1) & \end{array} \right]. \tag{5.2.64}$$

Finally, we find the explicit solutions to the equation (5.2.1) with any $m \in \mathbb{N}\backslash\{1,2\}$ in terms of the generalized Wright function (5.2.28). It is convenient to rewrite (5.2.1) in the form

$$(D_{0+}^{\alpha} y)(x) - \lambda (D_{0+}^{\beta} y)(x) - \sum_{k=0}^{m-2} A_k (D_{0+}^{\alpha_k} y)(x) = 0 \quad (x > 0) \tag{5.2.65}$$

$$(m \in \mathbb{N}\backslash\{1,2\}; \quad 0 = \alpha_0 < \alpha_1 < \cdots < \alpha_{m-2} < \beta < \alpha; \quad \lambda, A_0, \cdots, A_{m-2} \in \mathbb{R}).$$

Theorem 5.3 *Let $m \in \mathbb{N}\backslash\{1,2\}$, $l-1 < \alpha \leqq l$ $(l \in \mathbb{N})$, and let β and $\alpha_1, \cdots, \alpha_{m-2}$ be such that $\alpha > \beta > \alpha_{m-2} > \cdots > \alpha_1 > \alpha_0 = 0$, and let $\lambda, A_0, \cdots, A_{m-2} \in \mathbb{R}$. Then the functions*

$$y_j(x) = \sum_{n=0}^{\infty} \left(\sum_{k_0+\cdots+k_{m-2}=n} \right) \frac{1}{k_0! \cdots k_{m-2}!}$$

$$\cdot \left[\prod_{\nu=0}^{m-2} (A_\nu)^{k_\nu} \right] x^{(\alpha-\beta)n+\alpha-j+\sum_{\nu=0}^{m-2}(\beta-\alpha_\nu)k_\nu}$$

$$\cdot {}_1\Psi_1 \left[\begin{array}{c|c} (n+1,1) & \\ & \lambda x^{\alpha-\beta} \\ \left((\alpha-\beta)n+\alpha+1-j+\sum_{\nu=0}^{m-2}(\beta-\alpha_\nu)k_\nu, \alpha-\beta\right) & \end{array} \right] \tag{5.2.66}$$

with $j = 1, \cdots, l$, are solutions to the equation (5.2.65), provided that the series in (5.2.66) are convergent. The inner sum is taken over all $k_0, \cdots, k_{m-2} \in \mathbb{N}_0$ such that $k_0 + \cdots + k_{m-2} = n$.

If $\alpha - l + 1 \geqq \beta$, then $y_j(x)$ in (5.2.66) are linearly independent solutions to the equation (5.2.65). In particular, for $\alpha - l + 1 > \beta$, they provide the fundamental system of solutions.

Proof. . Let

$$l_{m-1}-1 < \beta \leqq l_{m-1}, \; l_k - 1 < \alpha_k \leqq l_k \; (k = 1, \cdots, m-2; \; 0 \leqq l_1 \leqq \cdots \leqq l_{m-1} \leqq l).$$

Applying the Laplace transform to (5.2.65) and using (5.2.3) as in (5.2.32), we obtain

$$(\mathcal{L}y)(s) = \sum_{j=1}^{l} d_j \frac{s^{j-1}}{s^\alpha - \lambda s^\beta - \sum_{k=0}^{m-2} A_k s^{\alpha_k}}, \tag{5.2.67}$$

where

$$d_j = \left(D_{0+}^{\alpha-j} y\right)(0+) - \lambda \left(D_{0+}^{\beta-j} y\right)(0+) - \sum_{k=1}^{m-2} A_k \left(D_{0+}^{\alpha_k - j} y\right)(0+) \quad (j = 1, \cdots, l_1),$$

$$d_j = \left(D_{0+}^{\alpha-j}y\right)(0+) - \lambda\left(D_{0+}^{\beta-j}y\right)(0+) - \sum_{k=2}^{m-2} A_k \left(D_{0+}^{\alpha_k-j}y\right)(0+) \ (j = l_1+1, \cdots, l_2),$$

$$\cdots, d_j = \left(D_{0+}^{\alpha-j}y\right)(0+) - \lambda\left(D_{0+}^{\beta-j}y\right)(0+) \quad (j = l_{m-2}+1, \cdots, l_{m-1}),$$

$$d_j = \left(D_{0+}^{\alpha-j}y\right)(0+) \quad (j = l_{m-1}+1, \cdots, l);$$

here $\sum_{k=m}^{n} A_k := 0$ $(m > n)$. For $s \in \mathbb{C}$ and $\left|\frac{\sum_{k=0}^{m-2} A_k s^{\alpha_k-\beta}}{s^{\alpha-\beta}-\lambda}\right| < 1$, just as in (5.2.33), we have

$$\frac{1}{s^\alpha - \lambda s^\beta - \sum_{k=0}^{m-2} A_k s^{\alpha_k}} = \frac{s^{-\beta}}{s^{\alpha-\beta}-\lambda}\frac{1}{\left(1 - \frac{\sum_{k=0}^{m-2} A_k s^{\alpha_k-\beta}}{s^{\alpha-\beta}-\lambda}\right)}$$

$$= \sum_{n=0}^{\infty} \frac{s^{-\beta}}{(s^{\alpha-\beta}-\lambda)^{n+1}} \left(\sum_{k=0}^{m-2} A_k s^{\alpha_k-\beta}\right)^n$$

$$= \sum_{n=0}^{\infty} \left(\sum_{k_0+\cdots+k_{m-2}=n}\right) \frac{n!}{k_0!\cdots k_{m-2}!} \left[\prod_{\nu=0}^{m-2} (A_\nu)^{k_\nu}\right] \frac{s^{-\beta-\sum_{\nu=0}^{m-2}(\beta-\alpha_\nu)k_\nu}}{(s^{\alpha-\beta}-\lambda)^{n+1}}, \quad (5.2.68)$$

if we also take into account the following relation:

$$(x_0 + \cdots + x_{m-2})^n = \sum_{n=0}^{\infty} \left(\sum_{k_0+\cdots+k_{m-2}=n}\right) \frac{n!}{k_0!\cdots k_{m-2}!} \prod_{\nu=0}^{m-2} x_j^{k_\nu}, \qquad (5.2.69)$$

where the summation is taken over all $k_0, \cdots, k_{m-2} \in \mathbb{N}_0$ such that $k_0 + \cdots + k_{m-2} = n$ [see, for example, Abramowitz and Stegun ([1], p. 823)].

According to (1.10.10) and (5.2.28), just as in (5.2.35), for $s \in \mathbb{C}$ and $|\lambda s^{\beta-\alpha}| < 1$, we have

$$\frac{s^{j-1-\beta-\sum_{\nu=0}^{m-2}(\beta-\alpha_\nu)k_\nu}}{(s^{\alpha-\beta}-\lambda)^{n+1}} = \frac{s^{(\alpha-\beta)-\left(\alpha+1-j+\sum_{\nu=0}^{m-2}(\beta-\alpha_\nu)k_\nu\right)}}{(s^{\alpha-\beta}-\lambda)^{n+1}}$$

$$= \frac{1}{n!}\left(\mathcal{L}\left[t^{(\alpha-\beta)n+\alpha-j+\sum_{\nu=0}^{m-2}(\beta-\alpha_\nu)k_\nu}\right.\right.$$

$$\left.\left.\cdot \left(\frac{\partial}{\partial\lambda}\right)^n E_{\alpha-\beta,\alpha+1-j+\sum_{\nu=0}^{m-2}(\beta-\alpha_\nu)k_\nu}\left(\lambda t^{\alpha-\beta}\right)\right]\right)(s)$$

$$= \frac{1}{n!}\left(\mathcal{L}\left[t^{(\alpha-\beta)n+\alpha-j+\sum_{\nu=0}^{m-2}(\beta-\alpha_\nu)k_\nu}\right.\right.$$

$$\left.\left.\cdot {}_1\Psi_1\left[\begin{matrix}(n+1,1)\\ \left((\alpha-\beta)n+\alpha+1-j+\sum_{\nu=0}^{m-2}(\beta-\alpha_\nu)k_\nu, \alpha-\beta\right)\end{matrix}\,\middle|\,\lambda t^{\alpha-\beta}\right]\right]\right). \qquad (5.2.70)$$

From (5.2.67), (5.2.68) and (5.2.70) we obtain the solution to the equation (5.2.65) in the form (5.2.36), where $y_j(x)$ $(j = 1, \cdots, l)$ are given by (5.2.66). It is easily verified that these functions are solutions to (5.2.65), and thus the first assertion of the Theorem 5.3 is proved.

For $j, k = 1, \cdots, l$, the direct evaluation leads to the following equations:

$$\left(D_{0+}^{\alpha-k} y_j\right)(x) = x^{k-j} \, {}_1\Psi_1 \left[\begin{matrix} (1,1) \\ (k+1-j, \alpha-\beta) \end{matrix} \middle| \lambda x^{\alpha-\beta} \right]$$

$$+ \sum_{n=1}^{\infty} \left(\sum_{k_0+\cdots+k_{m-2}=n} \right) \frac{1}{k_0! \cdots k_{m-2}!} \left[\prod_{\nu=0}^{m-2} (A_\nu)^{k_\nu} \right] x^{(\alpha-\beta)n+k-j+\sum_{\nu=0}^{m-2}(\beta-\alpha_\nu)k_\nu}$$

$$\cdot {}_1\Psi_1 \left[\begin{matrix} (n+1,1) \\ \left((\alpha-\beta)n+k+1-j+\sum_{\nu=0}^{m-2}(\beta-\alpha_\nu)k_\nu, \alpha-\beta\right) \end{matrix} \middle| \lambda x^{\alpha-\beta} \right]. \quad (5.2.71)$$

with $j = 1, \cdots, l$. If $k \geqq j$, then the last formula yields the relations in (5.2.12). If $k < j$, then, by (5.2.28), the first term in the right-hand side of (5.2.71) takes the form

$$x^{k-j} \, {}_1\Psi_1 \left[\begin{matrix} (1,1) \\ (k+1-j, \alpha-\beta) \end{matrix} \middle| \lambda x^{\alpha-\beta} \right]$$

$$= x^{k-j} \sum_{\mu=0}^{\infty} \frac{\lambda^\mu}{\Gamma[k+1-j+(\alpha-\beta)\mu]} x^{(\alpha-\beta)\mu}$$

$$= x^{k-j+\alpha-\beta} \sum_{\mu=0}^{\infty} \frac{\lambda^{\mu+1}}{\Gamma[k+1-j+(\alpha-\beta)(\mu+1)]} x^{(\alpha-\beta)\mu}. \quad (5.2.72)$$

If $\alpha - l + 1 \geqq \beta$, then, for any $k < j$, we have $k - j + \alpha - \beta \geqq \alpha - \beta + 1 - l \geqq 0$ and $(\alpha-\beta)n + k - j + \sum_{\nu=0}^{m-2}(\beta - \alpha_\nu)k_\nu \geqq \alpha - \beta + 1 - l > 0$ for any $n \in \mathbb{N}_0$. Then, from (5.2.71) and (5.2.72), we derive (5.2.14), except for the case $\alpha - l + 1 = \beta$ with $k = 1$ and $j = l$, for which the relation (5.2.39) holds. By (5.2.12), (5.2.14) and (5.2.39), we have $W_\alpha(0) = 1$ for the analog of the Wronskian in (4.2.34). Thus it follows from Lemma 4.1 that the functions $y_j(x)$ in (5.2.66) are linearly independent solutions to the equation (5.2.65). When $\alpha - l + 1 > \beta$, then the relations in (5.2.2) are valid, and thus $y_j(x)$ in (5.2.66) yield the fundamental system of solutions to the equation (5.2.65). This completes the proof of Theorem 5.3.

Corollary 5.4 *If $l \in \mathbb{N}\backslash\{1,2\}$ and $\lambda, A_0, \cdots, A_{l-2} \in \mathbb{R}$, then the following ordinary differential equation of order l*

$$y^{(l)}(x) - \lambda y^{(l-1)}(x) - \sum_{k=0}^{l-2} A_k y^{(k)}(x) = 0 \quad (x > 0) \quad (5.2.73)$$

has l solutions given by

$$y_j(x) = \sum_{n=0}^{\infty} \left(\sum_{k_0+\cdots+k_{l-2}=n} \right) \frac{1}{k_0! \cdots k_{l-2}!} \left[\prod_{\nu=0}^{l-2} (A_\nu)^{k_\nu} \right] x^{n+l-j+\sum_{\nu=0}^{l-2}(l-1-\nu)k_\nu}$$

$$\cdot {}_1\Psi_1 \left[\begin{matrix} (n+1,1) \\ (n+l+1-j+\sum_{\nu=0}^{l-2}(l-1-\nu)k_\nu, 1) \end{matrix} \middle| \lambda x \right] \quad (j = 1, \cdots, m). \qquad (5.2.74)$$

Example 5.6 The equation

$$\left(D_{0+}^{\alpha} y\right)(x) - \lambda \left(D_{0+}^{\beta} y\right)(x) - \delta \left(D_{0+}^{\gamma} y\right)(x) - \mu y(x) = 0 \quad (x > 0; \quad \lambda, \delta, \mu \in \mathbb{R}) \qquad (5.2.75)$$

with $l - 1 < \alpha \leqq l$ $(l \in \mathbb{N})$ and $0 < \gamma < \beta < \alpha$, has l solutions given by

$$y_j(x) = \sum_{n=0}^{\infty} \left(\sum_{q+\nu=n} \right) \frac{\mu^q \delta^\nu}{q! \nu!} x^{(\alpha-\beta)n+\alpha-j+\beta q+(\beta-\gamma)\nu}$$

$$\cdot {}_1\Psi_1 \left[\begin{matrix} (n+1,1) \\ ((\alpha-\beta)n+\alpha+1-j+\beta i+(\beta-\gamma)\nu, \alpha-\beta) \end{matrix} \middle| \lambda x^{\alpha-\beta} \right] \quad (j = 1, \cdots, l). \qquad (5.2.76)$$

If $\alpha - l + 1 \geqq \beta$, then the functions $y_j(x)$ in (5.2.76) are linearly independent solutions to the equation (5.2.75). In particular, for $\alpha - l + 1 > \beta$ these functions provide the fundamental system of solutions.

Example 5.7 The ordinary differential equation of order $l \in \mathbb{N}$

$$y^{(l)}(x) - \lambda y^{(m)}(x) - \delta y^{(k)}(x) - \mu y(x) = 0, \qquad (5.2.77)$$

where $x > 0$; $l, m, k \in \mathbb{N}$; $k < m < l$; $\lambda, \delta, \mu \in \mathbb{R}$, has l solutions given by

$$y_j(x) = \sum_{n=0}^{\infty} \left(\sum_{q+\nu=n} \right) \frac{\mu^q}{q!} \frac{\delta^\nu}{\nu!} x^{(l-m)n+l-j+mq+(m-k)\nu}$$

$$\cdot {}_1\Psi_1 \left[\begin{matrix} (n+1,1) \\ ((l-m)n+l+1-j+mq+(m-k)\nu, l-m) \end{matrix} \middle| \lambda x^{l-m} \right] \quad (j = 1, \cdots, l). \qquad (5.2.78)$$

Remark 5.2 (5.2.22) and (5.2.61) yield new systems of linearly independent solutions to the ordinary differential equation (5.2.21) and (5.2.60) (with $m = 1$) of order $l \in \mathbb{N}$.

Remark 5.3 It was proved in Theorems 5.2 and 5.3 that, if $\alpha - l + 1 \geqq \beta$, then $y_j(x)$ in (5.2.29) and (5.2.66) are linearly independent solutions to the equations (5.2.27) and (5.2.65), respectively. It would be interesting to see whether or not they have the same property in the case $\alpha - 1 + l < \beta$. In particular, it would also be interesting to see whether or not the functions $y_j(x)$ in (5.2.74) yield a system of linearly independent solutions to the ordinary differential equation (5.2.73) of order $m \in \mathbb{N}\backslash\{1, 2\}$.

5.2.2 Nonhomogeneous Equations with Constant Coefficients

In Section 5.2.1 we applied the Laplace transform method to derive explicit solutions to the homogeneous equations (5.2.1) with the Liouville fractional derivatives (2.1.10). Here we use this approach to find particular solutions to the corresponding nonhomogeneous equations

$$\sum_{k=1}^{m} A_k (D_{0+}^{\alpha_k} y)(x) + A_0 y(x) = f(x) \quad (x > 0;\ 0 < \alpha_1 \cdots < \alpha_m; m \in \mathbb{N}) \tag{5.2.79}$$

with real $A_k \in \mathbb{R}$ $(k = 0, \cdots, m)$ and a given function $f(x)$ on $\mathbb{R}_+$. Our arguments are based on a scheme for deducing a particular solution (5.1.4) to the equation (5.2.79), presented in Section 5.1. Using the Laplace convolution formula (1.4.12)

$$\left(\mathcal{L}\left(\int_0^x k(x-t)f(t)dt\right)\right)(s) = (\mathcal{L}k)(s)(\mathcal{L}f)(p), \tag{5.2.80}$$

just as in (5.1.11) we can introduce the *Laplace fractional analog* of the Green function as follows

$$G_{\alpha_1,\cdots,\ \alpha_m}(x) = \left(\mathcal{L}^{-1}\left[\frac{1}{P_{\alpha_1,\cdots,\ \alpha_m}(s)}\right]\right)(x), \quad P_{\alpha_1,\cdots,\ \alpha_m}(s) = A_0 + \sum_{k=1}^{m} A_k s^{\alpha_k}, \tag{5.2.81}$$

and express a particular solution (5.1.4) of the equation (5.2.79) in the form of the Laplace convolution of $G_{\alpha_1,\cdots,\ \alpha_m}(x)$ and $f(x)$

$$y(x) = \int_0^x G_{\alpha_1,\cdots,\ \alpha_m}(x-t)f(t)dt. \tag{5.2.82}$$

Generally speaking, we can consider the equation (5.2.79) with $A_m = 1$. First we derive a particular solution to the equation (5.2.79) with $m = 1$ in the form (4.1.1)

$$\left(D_{0+}^{\alpha} y\right)(x) - \lambda y(x) = f(x) \quad (x > 0;\ \ \alpha > 0) \tag{5.2.83}$$

in terms of the Mittag-Lefler function (1.8.17).

Theorem 5.4 *Let* $\alpha > 0$, $\lambda \in \mathbb{R}$ *and let* $f(x)$ *be a given function defined on* $\mathbb{R}_+$. *Then the equation* (5.2.83) *is solvable, and its particular solution has the form*

$$y(x) = \int_0^x (x-t)^{\alpha-1} E_{\alpha,\alpha}\left[\lambda(x-t)^{\alpha}\right] f(t)dt, \tag{5.2.84}$$

provided that the integral in the right-hand side of (5.2.84) *is convergent.*

Proof. (5.2.83) is the equation (5.2.79) with $m = 1$, $\alpha_1 = \alpha$, $A_1 = 1$, $A_0 = -\lambda$ and (5.2.81) takes the form

$$G_\alpha(x) = \left(\mathcal{L}^{-1}\left[\frac{1}{s^\alpha - \lambda}\right]\right)(x). \tag{5.2.85}$$

By (1.10.9) with $\beta = \alpha$, we have

$$\mathcal{L}\left[t^{\alpha-1}E_{\alpha,\alpha}\left(\lambda t^\alpha\right)\right] = \frac{1}{s^\alpha - \lambda} \quad \left(\Re(s) > 0;\ |\lambda s^{-\alpha}| < 1\right). \tag{5.2.86}$$

Hence

$$G_\alpha(x) = x^{\alpha-1}E_{\alpha,\alpha}\left(\lambda x^\alpha\right), \tag{5.2.87}$$

and thus (5.2.82), with $G_{\alpha_1,\cdots,\,\alpha_m}(x) = G_\alpha(x)$, yields (5.2.84). Theorem 5.4 is proved.

Example 5.8 The equation

$$\left(D_{0+}^{l-1/2}y\right)(x) - \lambda y(x) = f(x) \quad (x > 0;\ \ l \in \mathbb{N};\ \lambda \in \mathbb{R}), \tag{5.2.88}$$

has a particular solution given by

$$y(x) = \int_0^x (x-t)^{l-3/2}E_{l-1/2,l-1/2}\left[\lambda(x-t)^{l-1/2}\right]f(t)dt. \tag{5.2.89}$$

Example 5.9 The following ordinary differential equation of order $l \in \mathbb{N}$

$$y^{(l)}(x) - \lambda y(x) = f(x) \quad (x > 0;\ \ l \in \mathbb{N}) \tag{5.2.90}$$

has a particular solution given by

$$y(x) = \int_0^x (x-t)^{l-1}E_{l,l}\left[\lambda(x-t)^l\right]f(t)dt. \tag{5.2.91}$$

In particular, the following second order equation

$$y''(x) - \lambda y(x) = f(x) \quad (x > 0;\ \lambda \in \mathbb{R}) \tag{5.2.92}$$

has different particular solutions for $\lambda > 0$ and $\lambda < 0$:

$$y(x) = \int_0^x \frac{\sinh[\sqrt{\lambda}(x-t)]}{\sqrt{\lambda}}f(t)dt \quad (\lambda > 0), \tag{5.2.93}$$

$$y(x) = \int_0^x \frac{\sin[(-\lambda)^{1/2}(x-t)]}{(-\lambda)^{1/2}}f(t)dt \quad (\lambda < 0). \tag{5.2.94}$$

Next we derive a particular solution to the equation (5.2.79) with $m = 2$ of the form

$$\left(D_{0+}^{\alpha}y\right)(x) - \lambda\left(D_{0+}^{\beta}y\right)(x) - \mu y(x) = f(x) \quad (x > 0;\quad \alpha > \beta > 0) \tag{5.2.95}$$

in terms of the generalized Wright function (5.2.28).

Theorem 5.5 *Let $\alpha > \beta > 0$, $\lambda, \mu \in \mathbb{R}$ and let $f(x)$ be a given function defined on $\mathbb{R}_+$. Then the equation (5.2.95) is solvable, and its particular solution has the form*

$$y(x) = \int_0^x (x-t)^{\alpha-1} G_{\alpha,\beta;\lambda,\mu}(x-t) f(t)dt, \tag{5.2.96}$$

$$G_{\alpha,\beta;\lambda,\mu}(z) = \sum_{n=0}^{\infty} \frac{\mu^n}{n!} z^{\alpha n} {}_1\Psi_1\left[\begin{matrix}(n+1,1)\\ \\(\alpha n+\alpha, \alpha-\beta)\end{matrix}\,\middle|\, \lambda z^{\alpha-\beta}\right], \tag{5.2.97}$$

provided that the series in (5.2.97) and the integral in (5.2.96) are convergent.

In particular, the equation

$$\left(D_{0+}^{\alpha}y\right)(x) - \lambda\left(D_{0+}^{\beta}y\right)(x) = f(x) \quad (x > 0;\quad \alpha > \beta > 0) \tag{5.2.98}$$

has a particular solution given by

$$y(x) = \int_0^x (x-t)^{\alpha-1} E_{\alpha-\beta,\alpha}\left[\lambda(x-t)^{\alpha-\beta}\right] f(t)dt. \tag{5.2.99}$$

Proof. (5.2.95) is the same as the equation (5.2.79) with $m = 2$, $\alpha_2 = \alpha$, $\alpha_1 = \beta$, $A_2 = 1$, $A_1 = -\lambda$, $A_0 = -\mu$, and (5.2.81) is given by

$$G_{\alpha,\beta}(x) = \left(\mathcal{L}^{-1}\left[\frac{1}{s^\alpha - \lambda s^\beta - \mu}\right]\right)(x).$$

According to (5.2.33) for $s \in \mathbb{C}$ and $\left|\frac{\mu s^{-\beta}}{s^{\alpha-\beta}-\lambda}\right| < 1$, we have

$$G_{\alpha,\beta}(x) = \left(\mathcal{L}^{-1}\left[\sum_{n=0}^{\infty} \mu^n \frac{s^{-\beta-\beta n}}{(s^{\alpha-\beta}-\lambda)^{n+1}}\right]\right)(x). \tag{5.2.100}$$

By (1.10.10) and (5.2.28), for $s \in \mathbb{C}$ and $|\lambda s^{\beta-\alpha}| < 1$, we have

$$\frac{s^{-\beta-\beta n}}{(s^{\alpha-\beta}-\lambda)^{n+1}} = \frac{1}{n!}\left(\mathcal{L}\left[t^{(\alpha-\beta)n+(\alpha+\beta n)-1}\left(\frac{\partial}{\partial\lambda}\right)^n E_{\alpha-\beta,\alpha+\beta n}\left(\lambda t^{\alpha-\beta}\right)\right]\right)(s)$$

$$= \frac{1}{n!}\left(\mathcal{L}\left[t^{\alpha(n+1)-1} {}_1\Psi_1\left[\begin{matrix}(n+1,1)\\ \\(\alpha n+\alpha, \alpha-\beta)\end{matrix}\,\middle|\, \lambda t^{\alpha-\beta}\right]\right]\right)(s), \tag{5.2.101}$$

and hence (5.2.100) takes the following form:

$$G_{\alpha,\beta}(x) = \sum_{n=0}^{\infty} \mu^n x^{\alpha(n+1)-1} {}_1\Psi_1 \left[\begin{array}{c} (n+1,1) \\ (\alpha n+\alpha, \alpha-\beta) \end{array} \middle| \lambda x^{\alpha-\beta} \right]. \tag{5.2.102}$$

Thus the result in (5.2.96) follows from (5.2.82) with $G_{\alpha_1,\cdots,\ \alpha_m}(x) = G_{\alpha,\beta}(x)$. (5.2.96) with $\mu = 0$ yields (5.2.99) if we take (5.2.40) into account. Note that, in the limiting case $\beta \to 0$, the solution (5.2.99) of the equation (5.2.98) coincides with the solution (5.2.84) of the equation (5.2.83).

Example 5.10 The equation

$$y'(x) - \lambda \left(D_{0+}^{\beta} y\right)(x) - \mu y(x) = f(x) \quad (x > 0;\ 0 < \mathrm{Re}(\beta) < 1;\ \lambda, \mu \in \mathbb{R}) \tag{5.2.103}$$

has a particular solution given by

$$y(x) = \int_0^x G_{1,\beta;\lambda,\mu}(x-t) f(t) dt, \tag{5.2.104}$$

$$G_{1,\beta;\lambda,\mu}(z) = \sum_{n=0}^{\infty} \frac{(\mu z)^n}{n!} {}_1\Psi_1 \left[\begin{array}{c} (n+1,1) \\ (n+1, 1-\beta) \end{array} \middle| \lambda z^{1-\beta} \right]. \tag{5.2.105}$$

In particular, the equation

$$y'(x) - \lambda \left(D_{0+}^{1/2} y\right)(x) - \mu y(x) = f(x) \quad (x > 0;\ \ \lambda, \mu \in \mathbb{R}) \tag{5.2.106}$$

has a particular solution given by

$$y(x) = \int_0^x G_{1,1/2;\lambda,\mu}(x-t) f(t) dt, \tag{5.2.107}$$

$$G_{1,1/2;\lambda,\mu}(z) = \sum_{n=0}^{\infty} \frac{(\mu z)^n}{n!} {}_1\Psi_1 \left[\begin{array}{c} (n+1,1) \\ (n+1, 1/2) \end{array} \middle| \lambda z^{1/2} \right]. \tag{5.2.108}$$

Example 5.11 The equation

$$y''(x) - \lambda \left(D_{0+}^{\beta} y\right)(x) - \mu y(x) = f(x) \quad (x > 0;\ 0 < \beta < 2;\ \lambda, \mu \in \mathbb{R}) \tag{5.2.109}$$

has a particular solution given by

$$y(x) = \int_0^x (x-t) G_{2,\beta;\lambda,\mu}(x-t) f(t) dt, \tag{5.2.110}$$

$$G_{2,\beta;\lambda,\mu}(z) = \sum_{n=0}^{\infty} \frac{\mu^n}{n!} z^{2n} {}_1\Psi_1 \left[\begin{array}{c} (n+1,1) \\ (2n+2, 2-\beta) \end{array} \middle| \lambda z^{2-\beta} \right], \tag{5.2.111}$$

Example 5.12 The following ordinary differential equation of order $l \in \mathbb{N}$

$$y^{(l)}(x) - \lambda y^{(m)}(x) - \mu y(x) = f(x) \quad (x > 0;\ l, m \in \mathbb{N},\ m < l:\ \lambda, \mu \in \mathbb{R}) \tag{5.2.112}$$

has a particular solution given by

$$y(x) = \int_0^x (x-t)^{l-1} G_{l,m;\lambda,\mu}(x-t) f(t) dt, \tag{5.2.113}$$

$$G_{l,m;\lambda,\mu}(z) = \sum_{n=0}^{\infty} \frac{\mu^n}{n!} z^{ln} \, {}_1\Psi_1 \left[\begin{array}{c} (n+1,1) \\ \\ (ln+l, l-m) \end{array} \middle| \lambda z^{l-m} \right]. \tag{5.2.114}$$

In particular,

$$y(x) = \int_0^x (x-t) G_{2,1;\lambda,\mu}(x-t) f(t) dt, \tag{5.2.115}$$

$$G_{2,1;\lambda,\mu}(z) = \sum_{n=0}^{\infty} \frac{\mu^n}{n!} z^{2n} \, {}_1\Psi_1 \left[\begin{array}{c} (n+1,1) \\ \\ (2n+2,1) \end{array} \middle| \lambda z \right], \tag{5.2.116}$$

is a particular solution to the equation

$$y''(x) - \lambda y'(x) - \mu y(x) = f(x) \quad (x > 0). \tag{5.2.117}$$

Finally, we find a particular solution to the equation (5.2.79) with any $m \in \mathbb{N} \backslash \{1, 2\}$ in terms of the Wright function (5.2.28). It is convenient to rewrite (5.2.79), just as (5.2.65) in the form

$$(D_{0+}^{\alpha} y)(x) - \lambda (D_{0+}^{\beta} y)(x) - \sum_{k=1}^{m-2} A_k (D_{0+}^{\alpha_k} y)(x) - A_0 y(x) = f(x) \quad (x > 0), \tag{5.2.118}$$

with $m \in \mathbb{N} \backslash \{1, 2\}$, $0 < \alpha_1 < \cdots < \alpha_{m-2} < \beta < \alpha$, and $\lambda,\ A_0, \cdots,\ A_{m-2} \in \mathbb{R}$.

Theorem 5.6 *Let $m \in \mathbb{N} \backslash \{1, 2\}$, $\alpha > \beta > \alpha_{m-2} > \cdots > \alpha_1 > \alpha_0 = 0$, let $\lambda, A_0, \cdots,\ A_{m-2} \in \mathbb{R}$, and let $f(x)$ be a given function defined on $\mathbb{R}_+$. Then the equation* (5.2.118) *is solvable, and its particular solution has the form*

$$y(x) = \int_0^x (x-t)^{\alpha-1} G_{\alpha_1,\cdots,\alpha_{m-2},\beta,\alpha;\lambda}(x-t) f(t) dt, \tag{5.2.119}$$

$$G_{\alpha_1,\cdots,\ \alpha_{m-2},\beta,\alpha;\lambda}(z) = \sum_{n=0}^{\infty} \left(\sum_{k_0+\cdots+k_{m-2}=n} \right) \frac{1}{k_0! \cdots k_{m-2}!}$$

$$\cdot \left[\prod_{\nu=0}^{m-2} (A_\nu)^{k_\nu} \right] z^{(\alpha-\beta)n + \sum_{\nu=0}^{m-2} (\beta-\alpha_\nu) k_\nu}$$

$$\cdot {}_1\Psi_1 \left[\begin{array}{c} (n+1,1) \\ \\ \left((\alpha-\beta)n + \alpha + \sum_{\nu=0}^{m-2} (\beta-\alpha_\nu) k_\nu, \alpha-\beta \right) \end{array} \middle| \lambda z^{\alpha-\beta} \right], \tag{5.2.120}$$

provided that the series (5.2.120) *and the integral in* (5.2.119) *are convergent. The inner sum is taken over all $k_0, \cdots, k_{m-2} \in \mathbb{N}_0$ such that $k_0 + \cdots + k_{m-2} = n$.*

Proof. (5.2.118) is the same equation as the equation (5.2.79) with $\alpha_m = \alpha$, $\alpha_{m-1} = \beta$, $A_m = 1$, $A_{m-1} = -\lambda$, and with $-A_k$ instead of A_k for $k = 0, \cdots, m-2$. Since $\alpha_0 = 0$, (5.2.81) takes the form

$$G_{\alpha_1,\cdots,\alpha_m}(x) = \left(\mathcal{L}^{-1}\left[\frac{1}{s^\alpha - \lambda s^\beta - \sum_{k=0}^{m-2} A_k s^{\alpha_k}}\right]\right)(x).$$

For $s \in \mathbb{C}$ and $\left|\frac{\sum_{k=0}^{m-2} A_k s^{\alpha_k - \beta}}{s^{\alpha-\beta} - \lambda}\right| < 1$, in accordance with (5.2.68), we have

$$G_{\alpha_1,\cdots,\alpha_m}(x)$$
$$= \left(\mathcal{L}^{-1}\left[\sum_{n=0}^{\infty}\left(\sum_{k_0+\cdots+k_{m-2}=n}\right)\frac{n!}{k_0!\cdots k_{m-2}!}\right.\right.$$
$$\left.\left.\cdot\left[\prod_{\nu=0}^{m-2}(A_\nu)^{k_\nu}\right]\frac{s^{-\beta-\sum_{\nu=0}^{m-2}(\beta-\alpha_\nu)k_\nu}}{(s^{\alpha-\beta}-\lambda)^{n+1}}\right]\right)(x). \tag{5.2.121}$$

Using (5.2.28) and (1.10.10) just as in (5.2.70), for $s \in \mathbb{C}$ and $|\lambda s^{\beta-\alpha}| < 1$, we have

$$\frac{s^{-\beta-\sum_{\nu=0}^{m-2}(\beta-\alpha_\nu)k_\nu}}{(s^{\alpha-\beta}-\lambda)^{n+1}} = \frac{1}{n!}$$
$$\cdot\left(\mathcal{L}\left[t^{(\alpha-\beta)n+\alpha+\sum_{\nu=0}^{m-2}(\beta-\alpha_\nu)k_\nu-1}\left(\frac{\partial}{\partial\lambda}\right)^n E_{\alpha-\beta,\alpha+\sum_{\nu=0}^{m-2}(\beta-\alpha_\nu)k_\nu}\left(\lambda t^{\alpha-\beta}\right)\right]\right)(s)$$
$$= \frac{1}{n!}\left(\mathcal{L}\left[t^{(\alpha-\beta)n+\alpha+\sum_{\nu=0}^{m-2}(\beta-\alpha_\nu)k_\nu-1}\right.\right.$$
$$\left.\left.\cdot{}_1\Psi_1\left[\begin{matrix}(n+1,1)\\ \left((\alpha-\beta)n+\alpha+\sum_{\nu=0}^{m-2}(\beta-\alpha_\nu)k_\nu, \alpha-\beta\right)\end{matrix}\middle|\lambda t^{\alpha-\beta}\right]\right]\right)(s). \tag{5.2.122}$$

It follows from (5.2.122) that $G_{\alpha_1,\cdots,\alpha_m}(x)$ in (5.2.121) is given by

$$G_{\alpha_1,\cdots,\alpha_m}(x) = \sum_{n=0}^{\infty}\left(\sum_{k_0+\cdots+k_{m-2}=n}\right)\frac{n!}{k_0!\cdots k_{m-2}!}\left[\prod_{\nu=0}^{m-2}(A_\nu)^{k_\nu}\right]$$
$$\cdot x^{(\alpha-\beta)n+\alpha+\sum_{\nu=0}^{m-2}(\beta-\alpha_\nu)k_\nu-1}$$
$$\cdot{}_1\Psi_1\left[\begin{matrix}(n+1,1)\\ \left((\alpha-\beta)n+\alpha+\sum_{\nu=0}^{m-2}(\beta-\alpha_\nu)k_\nu, \alpha-\beta\right)\end{matrix}\middle|\lambda x^{\alpha-\beta}\right]. \tag{5.2.123}$$

Hence (5.2.82) yields the result in (5.2.119), which completes the proof of Theorem 5.6.

Corollary 5.5 *If $l \in \mathbb{N}\backslash\{1,2\}$ and $\lambda, A_0, \cdots, A_{l-2} \in \mathbb{R}$, then the ordinary differential equation of order l*

$$y^{(l)}(x) - \lambda y^{(l-1)}(x) - \sum_{k=0}^{l-2} A_k y^{(k)}(x) = f(x) \quad (x > 0) \tag{5.2.124}$$

is solvable, and its particular solution has the form

$$y(x) = \int_0^x (x-t)^{l-1} G_{\lambda,l}(x-t) f(t) dt, \tag{5.2.125}$$

$$G_{\lambda,l}(z) = \sum_{n=0}^{\infty} \left(\sum_{k_0+\cdots+k_{l-2}=n} \right) \frac{1}{k_0! \cdots k_{l-2}!} \left[\prod_{\nu=0}^{l-2} (A_\nu)^{k_\nu} \right] z^{ln - \sum_{\nu=1}^{l-2} \nu k_\nu}$$

$$\cdot {}_1\Psi_1 \left[\begin{array}{c|c} (n+1,1) & \\ & \lambda z \\ (ln + l - \sum_{\nu=1}^{l-2} \nu k_\nu, 1) & \end{array} \right], \tag{5.2.126}$$

provided that the series (5.2.126) *and the integral in* (5.2.125) *are convergent.*

Example 5.13 The equation

$$\left(D_{0+}^{\alpha} y\right)(x) - \lambda \left(D_{0+}^{\beta} y\right)(x) - \delta \left(D_{0+}^{\gamma} y\right)(x) - \mu y(x) = f(x) \quad (x > 0; \ \ \lambda, \delta, \mu \in \mathbb{R}) \tag{5.2.127}$$

with $l - 1 < \alpha \leqq l \ (l \in \mathbb{N})$, $0 < \gamma < \beta < \alpha$, has a particular solution given by

$$y(x) = \int_0^x (x-t)^{\alpha-1} G_{\gamma,\beta,\alpha;\lambda}(x-t) f(t) dt, \tag{5.2.128}$$

$$G_{\gamma,\beta,\alpha;\lambda}(z) = \sum_{n=0}^{\infty} \left(\sum_{i+\nu=n} \right) \frac{\mu^i \delta^\nu}{i! \nu!} z^{(\alpha-\beta)n + \beta i - (\beta-\gamma)\nu}$$

$$\cdot {}_1\Psi_1 \left[\begin{array}{c|c} (n+1,1) & \\ & \lambda z^{\alpha-\beta} \\ ((\alpha-\beta)n + \alpha + \beta i + (\beta-\gamma)\nu, \alpha-\beta) & \end{array} \right]. \tag{5.2.129}$$

Example 5.14 The following ordinary differential equation of order $l \in \mathbb{N}$

$$y^{(l)}(x) - \lambda y^{(m)}(x) - \delta y^{(k)}(x) - \mu y(x) = f(x) \quad (x > 0; \ \ l, m, k \in \mathbb{N}; \ k < m < l) \tag{5.2.130}$$

with $\lambda, \delta, \mu \in \mathbb{R}$, has a particular solution given by

$$y(x) = \int_0^x (x-t)^{l-1} G_{k,m,l;\lambda}(x-t) f(t) dt, \tag{5.2.131}$$

$$G_{k,m,l;\lambda}(z) = \sum_{n=0}^{\infty} \left(\sum_{q+\nu=n} \right) \frac{\mu^q}{q!} \frac{\delta^\nu}{\nu!} z^{(l-m)n + mq - (m-k)\nu}$$

$$\cdot {}_1\Psi_1 \left[\begin{array}{c|c} (n+1,1) & \\ & \lambda z^{l-m} \\ ((l-m)n + l + mq + (m-k)\nu, \alpha-\beta) & \end{array} \right]. \tag{5.2.132}$$

Remark 5.4 The equation of the form (5.2.103) was considered by Seitkazieva [756], of the form (5.2.109) with $0 < \beta < 1$ by Caputo [121], and of the form (5.2.109) with $\beta = 3/2$ by Bagley and Torvic [52].

Remark 5.5 The results obtained in Theorems 5.5 and 5.6 were presented in other forms by Podlubny ([682], Sections 5.4 and 5.6).

As in the case of ordinary differential equations, a general solution to the nonhomogeneous equation (5.2.79) is a sum of a particular solution to this equation and of the general solution to the corresponding homogeneous equation (5.2.1). Therefore, the results established in this section and in Section 5.2.1 can be used to derive general solutions to the nonhomogeneous equations (5.2.83), (5.2.95) and (5.2.118). The following statements can thus be derived from Theorems (5.1), 5.4, Theorems 5.2, 5.5 and Theorems 5.3, 5.6, respectively.

Theorem 5.7 *Let $l - 1 < \alpha \leqq l$ $(l \in \mathbb{N})$, $\lambda \in \mathbb{R}$ and let $f(x)$ be a given real function defined on $\mathbb{R}_+$. Then the equation (5.2.83) is solvable, and its general solution is given by*

$$y(x) = \int_0^x (x-t)^{\alpha-1} E_{\alpha,\alpha}\left[\lambda(x-t)^{\alpha}\right] f(t)dt + \sum_{j=1}^{l} c_j x^{\alpha-j} E_{\alpha,\alpha+1-j}\left(\lambda x^{\alpha}\right), \tag{5.2.133}$$

where c_j $(j = 1, \cdots, l)$ are arbitrary real constants.

Theorem 5.8 *Let $l-1 < \alpha \leqq l$ $(l \in \mathbb{N})$, $0 < \beta < \alpha$ be such that $\alpha - l + 1 \geqq \beta$, let $\lambda, \mu \in \mathbb{R}$ and let $f(x)$ be a given real function defined on $\mathbb{R}_+$. Then the equation (5.2.95) is solvable, and its general solution has the form*

$$y(x) = \int_0^x (x-t)^{\alpha-1} G_{\alpha,\beta;\lambda,\mu}(x-t) f(t)dt$$

$$+\sum_{j=1}^{l} c_j \sum_{n=0}^{\infty} \frac{\mu^n}{n!} x^{\alpha n+\alpha-j} \, {}_1\Psi_1\left[\begin{array}{c} (n+1,1) \\ \\ (\alpha n+\alpha+1-j, \alpha-\beta) \end{array} \middle| \lambda x^{\alpha-\beta}\right], \tag{5.2.134}$$

where $G_{\alpha,\beta;\lambda,\mu}(z)$ is given by (5.2.97) and c_j $(j = 1, \cdots, l)$ are arbitrary real constants.

In particular, the general solution to the equation (5.2.98) has the form

$$y(x) = \int_0^x (x-t)^{\alpha-1} E_{\alpha-\beta,\alpha}\left[\lambda(x-t)^{\alpha-\beta}\right] f(t)dt + \sum_{j=1}^{l} c_j x^{\alpha-j} E_{\alpha-\beta,\alpha+1-j}\left(\lambda x^{\alpha-\beta}\right). \tag{5.2.135}$$

Theorem 5.9 *Let $m \in \mathbb{N}\backslash\{1,2\}$, $l-1 < \alpha \leqq l$ $(l \in \mathbb{N})$, and let β and $\alpha_1, \cdots, \alpha_{m-2}$ be such that $\alpha > \beta > \alpha_{m-2} > \cdots > \alpha_1 > \alpha_0 = 0$ and $\alpha - l + 1 \geqq \beta$, and let*

$\lambda, A_0, \cdots, A_{m-2} \in \mathbb{R}$, and let $f(x)$ be a given real function on $\mathbb{R}_+$. Then the equation (5.2.118) is solvable, and its general solution is given by

$$y(x) = \int_0^x (x-t)^{\alpha-1} G_{\alpha_1,\cdots,\alpha_{m-2},\beta,\alpha;\lambda}(x-t) f(t)dt$$

$$+\sum_{j=1}^{l} c_j \sum_{n=0}^{\infty} \left(\sum_{k_0+\cdots+k_{m-2}=n} \right) \frac{1}{k_0! \cdots k_{m-2}!} \prod_{\nu=0}^{m-2} (A_\nu)^{k_\nu} x^{(\alpha-\beta)n+\alpha-j+\sum_{\nu=0}^{m-2}(\beta-\alpha_\nu)k_\nu}$$

$${}_1\Psi_1 \left[\begin{matrix} (n+1,1) \\ \left((\alpha-\beta)n+\alpha+1-j+\sum_{\nu=0}^{m-2}(\beta-\alpha_\nu)k_\nu, \alpha-\beta\right) \end{matrix} \middle| \lambda x^{\alpha-\beta} \right], \quad (5.2.136)$$

where $G_{\alpha_1,\cdots,\alpha_{m-2},\beta,\alpha;\lambda}(z)$ is given by (5.2.120) and c_j $(j = 1, \cdots, l)$ are arbitrary real constants.

Remark 5.6 Using the formulas obtained in Examples 5.8-5.14 and Corollary 5.5, the general solutions to the fractional and ordinary differential equations considered there can be found.

5.2.3 Equations with Nonconstant Coefficients

In Sections 5.2.1 and 5.2.2 we used the Laplace transform method to derive explicit solutions to fractional homogeneous and nonhomogeneous equations of the forms (5.2.1) and (5.2.79) with real constant coefficients A_k $(k = 0, \cdots, m)$. Such a method can be also applied to solve linear fractional differential equations with nonconstant coefficients. We illustrate such an approach to derive the explicit solution to the following fractional differential equation

$$\left(D_{0+}^{\alpha} y\right)(x) = \frac{\lambda y(x)}{x} \quad (x > 0; \ \lambda \in \mathbb{R}) \qquad (5.2.137)$$

of order $\alpha > 0$ $(l-1 < \alpha \leqq l, \ l \in \mathbb{N}; \ l \neq 1)$. First we consider the case $0 < \alpha < 1$ and give the solution in terms of the generalized Wright function (1.11.14) with $p = 0$ and $q = 1$ of the form

$${}_0\Psi_1 \left[\begin{matrix} \\ (b,\beta) \end{matrix} \middle| z \right] \quad (b \in \mathbb{C}; \ \beta \in \mathbb{R}). \qquad (5.2.138)$$

The following assertion is derivable from Theorem 1.5.

Lemma 5.2 (5.2.138) *is an entire function of $z \in \mathbb{C}$ for $\beta > -1$.*

The following assertion holds.

Theorem 5.10 *The differential equation* (5.2.137) *with* $0 < \alpha < 1$ *and* $\lambda \in \mathbb{R}$ *is solvable, and its solution is given by*

$$y(x) = cy_1(x) = cx^{\alpha-1} {}_0\Psi_1 \left[\begin{array}{c|c} \\ (\alpha, \alpha-1) \end{array} \; -\frac{\lambda}{1-\alpha} x^{\alpha-1} \right], \tag{5.2.139}$$

where c *is an arbitrary real constant.*

Proof. Rewrite the equation (5.2.137) in the form

$$x \left(D_{0+}^{\alpha} y \right)(x) = \lambda y(x) \quad (x > 0; \quad 0 < \alpha < 1, \quad \lambda \in \mathbb{R}). \tag{5.2.140}$$

Applying the Laplace transform to (5.2.140), and using the relation

$$D^k \left(\mathcal{L} y \right)(s) = (-1)^k \left(\mathcal{L} t^k y(t) \right)(s) \quad \left(D = \frac{\partial}{\partial s}; \; k \in \mathbb{N} \right) \tag{5.2.141}$$

with $k = 1$ and formula (5.2.3) with $l = 1$, we have

$$D \left(\mathcal{L} y \right)(s) = -\left(\alpha s^{-1} + \lambda s^{-\alpha} \right) \left(\mathcal{L} y \right)(s). \tag{5.2.142}$$

From (5.2.142) we derive the solution to this ordinary differential equation of the first order with respect to $(\mathcal{L}y)(s)$:

$$\left(\mathcal{L} y \right)(s) = cs^{-\alpha} \exp\left(-\frac{\lambda s^{1-\alpha}}{1-\alpha} \right), \tag{5.2.143}$$

where c is an arbitrary real constant. Expanding the exponential function in a series, we obtain

$$\left(\mathcal{L} y \right)(s) = \sum_{j=0}^{\infty} \frac{(-1)^j}{j!} \left(\frac{\lambda}{1-\alpha} \right)^j s^{-[\alpha+(\alpha-1)j]}. \tag{5.2.144}$$

By the formula

$$\left(\mathcal{L} x^{\beta-1} \right)(s) = \Gamma(\beta) s^{-\beta} \quad (\beta > -1), \tag{5.2.145}$$

we have

$$s^{-[\alpha+(\alpha-1)j]} = \left(\mathcal{L} \frac{x^{\alpha+(\alpha-1)j-1}}{\Gamma[\alpha+(\alpha-1)j]} \right)(s) \quad (j \in \mathbb{N}_0),$$

and from (5.2.143) we derive the following solution

$$y(x) = cx^{\alpha-1} \sum_{j=0}^{\infty} \frac{(-1)^j}{j!} \left(\frac{\lambda}{1-\alpha} \right)^j \frac{x^{(\alpha-1)j}}{\Gamma[\alpha+(\alpha-1)j]},$$

which, in accordance with (5.2.138), yields the result in (5.2.139). It is readily verified that the function $y_1(x)$ in (5.2.139) is the solution to the equation (5.2.137), thus proving Theorem 5.10.

Example 5.15 The equation

$$\left(D_{0+}^{1/2}y\right)(x) = \frac{\lambda y(x)}{x} \quad (x > 0;\ \ \lambda \in \mathbb{R}) \tag{5.2.146}$$

has its solution given by

$$y(x) = \frac{c}{\sqrt{x}}e^{-\lambda^2/x}, \tag{5.2.147}$$

where c is an arbitrary real constant. In particular, the equations

$$\left(D_{0+}^{1/2}y\right)(x) = \frac{y(x)}{x} \quad (x > 0;\ \ \lambda \in \mathbb{R}) \tag{5.2.148}$$

and

$$\left(D_{0+}^{1/2}y\right)(x) = -\frac{y(x)}{x} \quad (x > 0;\ \ \lambda \in \mathbb{R}) \tag{5.2.149}$$

have the same solution given by

$$y(x) = \frac{c}{\sqrt{x}}e^{-1/x}. \tag{5.2.150}$$

Note that (5.2.147) follows from (5.2.139), if we take into account the relation

$${}_0\Psi_1\left[\begin{array}{c|c} & \\ \left(\frac{1}{2},-\frac{1}{2}\right) & z\end{array}\right] = \frac{1}{\sqrt{\pi}}e^{-z^2/4}. \tag{5.2.151}$$

Remark 5.7 Solution (5.2.150) of the equation (5.2.148) was obtained by Miller and Ross ([603], Chapter VI, Section 3). As mentioned in Section 5.1, this is the first differential equation of fractional order 1/2 discussed by O'Shaughnessay [650] and Post [686].

Next we consider the equation (5.2.137) with $l-1 < \alpha \leqq l$ $(l \in \mathbb{N}\setminus\{1\})$ and express its solution in terms of the generalized Wright function (5.2.138) and the generalized Wright function (1.1.16) with $p = 1$ and $q = 2$:

$${}_1\Psi_2\left[\begin{array}{c|c} (a,\alpha_1) & \\ (b_1,\beta_1),(b_2,\beta_2) & z\end{array}\right] \quad (a, b_1, b_2 \in \mathbb{C};\ \ \alpha, \beta_1, \beta_2 \in \mathbb{R}). \tag{5.2.152}$$

The following assertion can be derived from Theorem 1.5.

Lemma 5.3 (5.2.152) *is an entire function of* $z \in \mathbb{C}$ *for* $\beta_1 + \beta_2 - \alpha_1 > -1$.

We also need the numbers $c_k(\alpha, m)$ defined, for $\alpha > 0$, $m = 2, \cdots, l$ $(\alpha \neq \frac{q+m-1}{q},\ q \in \mathbb{N})$ and $k \in \mathbb{N}_0$, by

$$c_k(\alpha, m) = \sum_{q,j=0,\cdots k;\ q+j=k} \frac{(-1)^q}{q!j![(1-\alpha)q + m - 1]}. \tag{5.2.153}$$

These numbers have the following property.

Lemma 5.4 *The numbers* $c_k(\alpha,m)$, *with* $\alpha > 0$, $m = 2,\cdots, l$ $(\alpha \neq \frac{q+m-1}{q}$, $q \in \mathbb{N})$ *and* $k \in \mathbb{N}_0$ *satisfy the following recurrence relations:*

$$\left(\frac{\alpha-m}{\alpha-1}+k\right)c_{k+1}(\alpha,m) = c_k(\alpha,m), \tag{5.2.154}$$

which yield the explicit expression for $c_k(\alpha,m)$ *in the form*

$$c_k(\alpha,m) = \frac{\Gamma\left(\frac{\alpha-m}{\alpha-1}\right)}{(m-1)\Gamma\left(\frac{\alpha-m}{\alpha-1}+k\right)} \quad (k \in \mathbb{N}_0). \tag{5.2.155}$$

Now we can derive the explicit solution to the equation (5.2.137).

Theorem 5.11 *The differential equation* (5.2.137) *with* $l-1<\alpha \leqq l$ $(l \in \mathbb{N}\backslash\{1\})$ *and* $\lambda \in \mathbb{R}$ *is solvable, and its solution is given by*

$$y(x) = c_1 x^{\alpha-1}\ {}_0\Psi_1\left[\begin{array}{c|c} & \\ (\alpha,\alpha-1) & \frac{\lambda}{\alpha-1}x^{\alpha-1}\end{array}\right]$$

$$+\sum_{m=2}^{l} c_m x^{\alpha-m}\ {}_1\Psi_2\left[\begin{array}{c|c} (1,1) & \\ (\alpha+1-m,\alpha-1),\left(\frac{\alpha-m}{\alpha-1},1\right) & \frac{\lambda}{\alpha-1}x^{\alpha-1}\end{array}\right], \tag{5.2.156}$$

where $c_1,\cdots, c_m$ *are arbitrary real constants.*

In particular, the solution to the equation (5.2.137) *with* $1<\alpha<2$ *is given by*

$$y(x) = c_1 x^{\alpha-1}\ {}_0\Psi_1\left[\begin{array}{c|c} & \\ (\alpha,\alpha-1) & \frac{\lambda}{\alpha-1}x^{\alpha-1}\end{array}\right]$$

$$+c_2 x^{\alpha-2}\ {}_1\Psi_2\left[\begin{array}{c|c} (1,1) & \\ (\alpha-1,\alpha-1),\left(\frac{\alpha-2}{\alpha-1},1\right) & \frac{\lambda}{\alpha-1}x^{\alpha-1}\end{array}\right], \tag{5.2.157}$$

where c_1 *and* c_2 *are arbitrary real constants.*

Proof. Applying the Laplace transform to (5.2.140), and using (5.2.141) with $k=1$ and (5.2.3), we have

$$D(\mathcal{L}y)(s) + \left(\alpha s^{-1}+\lambda s^{-\alpha}\right)(\mathcal{L}y)(s) = \sum_{m=2}^{l} d_m(m-1)s^{m-2-\alpha},$$

where $d_m = \left(D_{0+}^{\alpha-m}y\right)(0+)$ $(m=2,\cdots,l)$. From here we derive the solution to this ordinary differential equation of the first order with respect to $(\mathcal{L}y)(s)$:

$$(\mathcal{L}y)(s) = s^{-\alpha}\exp\left(-\frac{\lambda s^{1-\alpha}}{1-\alpha}\right)\left[c_1+\sum_{m=2}^{l} d_m(m-1)\int s^{m-2}\exp\left(\frac{\lambda s^{1-\alpha}}{1-\alpha}\right)ds\right],$$

where c_1 is an arbitrary real constant. Expanding the exponential function in the integrand in a series and using term-by-term integration, we obtain

$$(\mathcal{L}y)(s) = c(\mathcal{L}y_1)(s) + \sum_{m=2}^{l} d_m(m-1)(\mathcal{L}y_m^*)(s), \tag{5.2.158}$$

where

$$(\mathcal{L}y_1)(s) = s^{-\alpha}\exp\left(-\frac{\lambda s^{1-\alpha}}{1-\alpha}\right),$$

$$(\mathcal{L}y_m^*)(s) = s^{-\alpha}\exp\left(\frac{\lambda s^{1-\alpha}}{\alpha-1}\right)\sum_{j=0}^{\infty}\left(\frac{\lambda}{1-\alpha}\right)^j \frac{s^{(1-\alpha)j+m-1}}{[(1-\alpha)j+m-1]j!}.$$

By the proof of Theorem 5.10,

$$y_1(x) = x^{\alpha-1}\,{}_0\Psi_1\left[\begin{array}{c|c} & -\dfrac{\lambda}{1-\alpha}x^{\alpha-1} \\ (\alpha,\alpha-1) & \end{array}\right]$$

$$= x^{\alpha-1}\,{}_0\Psi_1\left[\begin{array}{c|c} & \dfrac{\lambda}{\alpha-1}x^{\alpha-1} \\ (\alpha,\alpha-1) & \end{array}\right]. \tag{5.2.159}$$

Expanding the exponential term $\exp\left(\frac{\lambda s^{\alpha-1}}{\alpha-1}\right)$ in power series, multiplying the resulting two series, and taking (5.2.153) into account, we have

$$(\mathcal{L}y_m^*)(s) = s^{m-\alpha-1}\left(\sum_{j=0}^{\infty}\left(\frac{\lambda}{\alpha-1}\right)^j \frac{s^{(1-\alpha)j}}{j!}\right)$$

$$\cdot\left(\sum_{q=0}^{\infty}\left(\frac{\lambda}{\alpha-1}\right)^q \frac{(-1)^q}{[(1-\alpha)q+m-1]}\frac{s^{(1-\alpha)q}}{q!}\right)$$

$$= \sum_{k=0}^{\infty} c_k(\alpha,m)\left(\frac{\lambda}{\alpha-1}\right)^k s^{(1-\alpha)k+m-\alpha-1} \quad (m=2,\cdots,l).$$

From here, by using the formula (5.2.145) with $\beta=(\alpha-1)k+\alpha+1-m$, we derive the following expression for $y_m^*(x)$:

$$y_m^*(x) = \sum_{k=0}^{\infty} c_k(\alpha,m)\left(\frac{\lambda}{\alpha-1}\right)^k \frac{\Gamma(k+1)}{\Gamma[\alpha+1-m+(\alpha-1)k]}\frac{x^{(\alpha-1)k+\alpha-m}}{k!}$$

or, in accordance with (5.2.155) and (5.2.152),

$$y_m^*(x) = \frac{1}{(m-1)}\Gamma\left(\frac{\alpha-m}{\alpha-1}\right)y_m(x), \tag{5.2.160}$$

where

$$y_m(x) = x^{\alpha-m}\ {}_1\Psi_2\left[\begin{array}{c|c} (1,1) & \\ & \dfrac{\lambda}{\alpha-1}x^{\alpha-1} \\ (\alpha+1-m,\alpha-1),\left(\frac{\alpha-m}{\alpha-1},1\right) & \end{array}\right]. \tag{5.2.161}$$

It follows from (5.2.158) that the solution to the equation (5.2.137) has the form

$$y(x) = c_1 y_1(x) + \sum_{m=2}^{l} d_m(m-1)y_m^*(x),$$

where $y_1(x)$ and $y_m^*(x)$ $(m = 2, \cdots, l)$ are given by (5.2.159) and (5.2.160), respectively.

It is easily verified that the functions $y_1(x)$ and $y_m(x)$ $(m = 2, \cdots, l)$ in (5.2.159) and (5.2.161) are solutions to the equation (5.2.137), and thus Theorem 5.11 is proved.

Corollary 5.6 *The following ordinary differential equation of order* $l \in \mathbb{N}\backslash\{1\}$

$$xy^{(l)}(x) = \lambda y(x)\ (\lambda \in \mathbb{R}) \tag{5.2.162}$$

is solvable, and its solution is given by

$$y(x) = c_1 x^{l-1}\ {}_0\Psi_1\left[\begin{array}{c|c} & \dfrac{\lambda}{l-1}x^{l-1} \\ (l,l-1) & \end{array}\right]$$

$$+\sum_{m=2}^{l} c_m x^{l-m}\ {}_1\Psi_2\left[\begin{array}{c|c} (1,1) & \\ & \dfrac{\lambda}{l-1}x^{l-1} \\ (l+1-m,l-1),\left(\frac{l-m}{l-1},1\right) & \end{array}\right], \tag{5.2.163}$$

where $c_1, \cdots, c_m$ *are arbitrary real constants.*

Example 5.16 The equation

$$\left(D_{0+}^{3/2}y\right)(x) = \frac{\lambda y(x)}{x} \quad (x > 0;\ \ \lambda \in \mathbb{R}) \tag{5.2.164}$$

has its solution given by

$$y(x) = c_1 x^{1/2}\ {}_0\Psi_1\left[\begin{array}{c|c} --- & 2\lambda x^{1/2} \\ \left(\frac{3}{2},\frac{1}{2}\right) & \end{array}\right] + c_2 x^{-1/2}\ {}_1\Psi_2\left[\begin{array}{c|c} (1,1) & 2\lambda x^{1/2} \\ \left(\frac{1}{2},\frac{1}{2}\right),(-1,1) & \end{array}\right], \tag{5.2.165}$$

where c_1 and c_2 are arbitrary real constants.

Remark 5.8 The generalized Wright functions ${}_1\Psi_2(z)$ in (5.2.156) coincide with special cases of the generalized Mittag-Leffler function $E_{\alpha,m,l}(z)$ defined in (1.9.19) except for the constant multiplier. Indeed, if $\lambda \in \mathbb{R}$, then, for any $m = 2, \cdots, l$, we have

$$\Gamma(\alpha - m)\Gamma\left(\frac{\alpha - m}{\alpha - 1} + 1\right) {}_1\Psi_2\left[\begin{array}{c|c} (1,1) & \\ (\alpha + 1 - m, \alpha - 1), \left(\frac{\alpha - m}{\alpha - 1}, 1\right) & \frac{\lambda}{\alpha - 1}x^{\alpha-1}\end{array}\right]$$

$$= E_{\alpha,1-1/\alpha,1-(1+m)/\alpha}\left(\lambda x^{\alpha-1}\right) \quad (m = 2, \cdots, l). \tag{5.2.166}$$

Therefore the explicit solutions $y_m(x)$ $(m = 1, \cdots, l$ $(l \in \mathbb{N}\backslash\{1\})$ in (5.2.156) can be expressed in terms of the generalized Mittag-Leffler functions presented in the right-hand sides of (5.2.16). Thus the solution (5.2.156) is equivalent to the following one:

$$y(x) = \sum_{m=1}^{l} c_m x^{\alpha-m} E_{\alpha,1-1/\alpha,1-(1+m)/\alpha}\left(\lambda x^{\alpha-1}\right). \tag{5.2.167}$$

Note that this solution can be derived from the solution (4.2.33) of the equation (4.2.30) with $\beta = -1$.

5.2.4 Cauchy Type Problems for Fractional Differential Equations

Explicit solutions to fractional differential equations, established in Sections 5.2.1-5.2.3, can be applied to obtain solutions to initial and boundary problems for such equations. Here we illustrate such an application to Cauchy type problems for fractional differential equations of order α $(l - 1 < \alpha \leqq l;\ l \in \mathbb{N})$ with initial conditions of the form (4.1.2)

$$\left(D_{0+}^{\alpha-k}y\right)(0+) = b_k \in \mathbb{R} \quad (k = 1, \cdots, l;\ \ l - 1 < \alpha \leqq l)). \tag{5.2.168}$$

The first results for the homogeneous equations (5.2.5), (5.2.27) and (5.2.65) follow from Theorems 5.1-5.3 if we take into account the relations (5.2.12), (5.2.14) and (5.2.39).

Proposition 5.1 *Let* $l - 1 < \alpha \leqq l$ $(l \in \mathbb{N})$ *and* $\lambda \in \mathbb{R}$. *Then the Cauchy type problem* (5.2.168) *for the equation* (5.2.5) *is solvable, and its solution is given by*

$$y(x) = \sum_{j=1}^{l} b_j x^{\alpha-j} E_{\alpha,\alpha+1-j}\left(\lambda x^{\alpha}\right). \tag{5.2.169}$$

Proposition 5.2 *Let* $l - 1 < \alpha \leqq l$ $(l \in \mathbb{N})$ *and* $0 < \beta < \alpha$ *be such that* $\alpha - l + 1 \geqq \beta$, *and let* $\lambda, \mu \in \mathbb{R}$. *Then the Cauchy type problem* (5.2.168) *for the equation* (5.2.27) *is solvable, and its solution has the form*

$$y(x) = \sum_{j=1}^{l} b_j y_j(x) \quad (\alpha - l + 1 > \beta) \tag{5.2.170}$$

or

$$y(x) = (b_1 - \lambda b_l)y_1(x) + \sum_{j=2}^{l} b_j y_j(x) \quad (\alpha - l + 1 = \beta), \tag{5.2.171}$$

where $y_j(x)$ $(j = 1, \cdots, l)$ *are given by* (5.2.29).

Proposition 5.3 *Let* $m \in \mathbb{N}\backslash\{1,2\}$, $l - 1 < \alpha \leqq l$ $(l \in \mathbb{N})$, *and let* β *and* $\alpha_1, \cdots, \alpha_{m-2}$ *be such that* $\alpha > \beta > \alpha_{m-2} > \cdots > \alpha_1 > \alpha_0 = 0$ *and* $\alpha - l + 1 \geqq \beta$, *and let* $\lambda, A_0, \cdots, A_{m-2} \in \mathbb{R}$. *Then the Cauchy type problem* (5.2.168) *for the equation* (5.2.65) *is solvable, and its solution has the forms* (5.2.170) *and* (5.2.171) *for the cases* $\alpha - l + 1 > \beta$ *and* $\alpha - l + 1 = \beta$ *respectively, where* $y_j(x)$ $(j = 1, \cdots, l)$ *are given by* (5.2.66).

Our next results for nonhomogeneous equations (5.2.83), (5.2.95) and (5.2.118) follow from Theorems 5.7-5.9.

Proposition 5.4 *Let* $l - 1 < \alpha \leqq l$ $(l \in \mathbb{N})$, $\lambda \in \mathbb{R}$ *and let* $f(x)$ *be a given real function defined on* $\mathbb{R}_+$. *Then the Cauchy type problem* (5.2.168) *for the equation* (5.2.83) *is solvable, and its solution is given by*

$$y(x) = \int_0^x (x-t)^{\alpha-1} E_{\alpha,\alpha}\left[\lambda(x-t)^\alpha\right] f(t)dt + \sum_{j=1}^{l} b_j x^{\alpha-j} E_{\alpha,\alpha+1-j}\left(\lambda x^\alpha\right). \tag{5.2.172}$$

Proposition 5.5 *Let* $l - 1 < \alpha \leqq l$ $(l \in \mathbb{N})$ *and* $0 < \beta < \alpha$ *be such that* $\alpha - l + 1 \geqq \beta$, *let* $\lambda, \mu \in \mathbb{R}$ *and let* $f(x)$ *be a given real function defined on* $\mathbb{R}_+$. *Then the Cauchy type problem* (5.2.168) *for the equation* (5.2.95) *is solvable, and its solution has the form*

$$y(x) = \int_0^x (x-t)^{\alpha-1} G_{\alpha,\beta;\lambda,\mu}(x-t) f(t)dt + \sum_{j=1}^{l} b_j y_j(x) \quad (\alpha - l + 1 > \beta), \tag{5.2.173}$$

or

$$y(x) = \int_0^x (x-t)^{\alpha-1} G_{\alpha,\beta;\lambda,\mu}(x-t) f(t)dt + (b_1 - \lambda b_l)y_1(x)$$

$$+ \sum_{j=2}^{l} b_j y_j(x) \quad (\alpha - l + 1 = \beta), \tag{5.2.174}$$

where $G_{\alpha,\beta;\lambda,\mu}(z)$ *and* $y_j(x)$ $(j = 1, \cdots, l)$ *are given by* (5.2.97) *and* (5.2.29), *respectively.*

Proposition 5.6 *Let* $m \in \mathbb{N}\backslash\{1,2\}$, $l - 1 < \alpha \leqq l$ $(l \in \mathbb{N})$, *and let* β *and* $\alpha_1, \cdots, \alpha_{m-2}$ *be such that* $\alpha > \beta > \alpha_{m-2} > \cdots > \alpha_1 > \alpha_0 = 0$ *and* $\alpha - l + 1 \geqq \beta$, *and let* $\lambda, A_0, \cdots, A_{m-2} \in \mathbb{R}$, *and let* $f(x)$ *be a given real function defined on* $\mathbb{R}_+$.

Then the Cauchy type problem (5.2.168) *for the equation* (5.2.118) *is solvable, and its solution has the form*

$$y(x) = \int_0^x (x-t)^{\alpha-1} G_{\alpha_1,\cdots,\alpha_{m-2},\beta,\alpha;\lambda}(x-t) f(t)dt + \sum_{j=1}^{l} b_j y_j(x) \quad (\alpha - l + 1 > \beta), \tag{5.2.175}$$

or

$$y(x) = \int_0^x (x-t)^{\alpha-1} G_{\alpha_1,\cdots,\alpha_{m-2},\beta,\alpha;\lambda}(x-t) f(t)dt$$

$$+(b_1 - \lambda b_l) y_1(x) + \sum_{j=2}^{l} b_j y_j(x) \quad (\alpha - l + 1 = \beta), \tag{5.2.176}$$

where $G_{\alpha_1,\cdots,\alpha_{m-2},\beta,\alpha;\lambda}(z)$ *and* $y_j(x)$ $(j = 1, \cdots, l)$ *are given by* (5.2.120) *and* (5.2.66), *respectively.*

Proof. The proofs of Propositions 5.4-5.6 follow from Theorems 5.7-5.9, on the basis of the respective formulas (5.2.133), (5.2.134) and (5.2.136) and the directly verified relations

$$\left(D_{0+}^{\alpha-k}\left[\int_0^t (t-\tau)^{\alpha-1} E_{\alpha,\alpha}\left[\lambda(t-\tau)^\alpha\right] f(\tau)d\tau\right]\right)(0+) = 0 \quad (k = 1, \cdots, l),$$

$$\left(D_{0+}^{\alpha-k}\left[\int_0^t (t-\tau)^{\alpha-1} G_{\alpha,\beta;\lambda,\mu}(t-\tau) f(\tau)d\tau\right]\right)(0+) = 0 \quad (k = 1, \cdots, l),$$

$$\left(D_{0+}^{\alpha-k}\left[\int_0^x (t-\tau)^{\alpha-1} G_{\alpha_1,\cdots,\alpha_{m-2},\beta,\alpha;\lambda}(t-\tau) f(\tau)d\tau\right]\right)(0+) = 0 \quad (k = 1, \cdots, l),$$

if we also take the relations (5.2.12), (5.2.14) and (5.2.39) into account.

Remark 5.9 Results presented in Sections 5.2.1-5.2.4 were established in Kilbas and Trujillo [411], [410]. Using the explicit formulas obtained in Examples 5.8-5.14 and Corollary 5.5, one may derive the solutions to the Cauchy type and Cauchy problems for the corresponding fractional and ordinary differential the equations.

Remark 5.10 In Sections 5.2.1 and 5.2.2, we applied the Laplace transform to derive explicit solutions to homogeneous and nonhomogeneous equations of the forms (5.2.1) and (5.2.79) considered on the the positive half-axis $\mathbb{R}_+$. It is easily verified that all obtained formulas remain true for these equations considered on a finite interval $[0, b]$ $(0 < b < \infty)$. Moreover, these formulas also apply on any finite interval $(a, b]$ $(0 \leqq a < b \leqq \infty)$, if the Liouville fractional derivative $D_{0+}^\alpha y$ is replaced by the Riemann-Liouville derivative $D_{a+}^\alpha y$, and (in the corresponding formulas for explicit solutions) x^α is replaced by $(x-a)^\alpha$, and the integral $\int_0^x$ by $\int_a^x$. The same applies to the ordinary differential equations and Cauchy type and Cauchy problems for fractional and ordinary differential equations. In particular, in this way explicit solutions can be obtained for homogeneous and nonhomogeneous fractional and ordinary differential equations with constant coefficients, and for the corresponding Cauchy type and Cauchy problems, established in Sections 4.1-4.3 by other methods. For example, from (5.2.6) we can derive the solution (4.2.44) to the equation (4.2.43).

5.3 Laplace Transform Method for Solving Ordinary Differential Equations with Caputo Fractional Derivatives

We present our development of the Laplace transform method for the Caputo fractional derivatives in the following subsections.

5.3.1 Homogeneous Equations with Constant Coefficients

In this section we apply the Laplace transform method to derive explicit solutions to homogeneous equations of the form (5.2.1)

$$\sum_{k=1}^{m} A_k({}^C D_{0+}^{\alpha_k} y)(x) + A_0 y(x) = 0 \quad (x > 0;\ m \in \mathbb{N};\ 0 < \alpha_1 < \cdots < \alpha_m), \qquad (5.3.1)$$

involving the Caputo fractional derivatives ${}^C D_{0+}^{\alpha_k} y$ $(k = 1, \cdots, m)$, given by (2.4.1), with real constants $A_k \in \mathbb{R}$ $(k = 0, \cdots, m-1)$ and $A_m = 1$. We give the conditions when the solutions $y_1(x), \cdots, y_l(x)$ of the equation (5.3.1) with $l-1 < \alpha := \alpha_m < l$ will be linearly independent, and when these solutions form the fundamental system of solutions

$$y_j^{(k)}(0) = 0 \quad (k, j = 0, \cdots, l-1;\ k \neq j), \quad y_k^{(k)}(0) = 1 \quad (k = 0, \cdots, l-1). \qquad (5.3.2)$$

The Laplace transform method is based on the formula (2.4.67):

$$\left(\mathcal{L}{}^C D_{0+}^{\alpha} y\right)(s) = s^{\alpha} \left(\mathcal{L}y\right)(s) - \sum_{j=0}^{l-1} d_j s^{\alpha-j-1} \quad (l-1 < \alpha \leqq l,\ \ l \in \mathbb{N}), \qquad (5.3.3)$$

$$d_j = y^{(j)}(0) \quad (j = 0, \cdots,\ l-1). \qquad (5.3.4)$$

First we derive the explicit solutions to the equation (5.3.1) with $m = 1$ in the form (5.2.5)

$$\left({}^C D_{0+}^{\alpha} y\right)(x) - \lambda y(x) = 0 \quad (x > 0;\ \ l-1 < \alpha \leqq l;\ \ l \in \mathbb{N};\ \lambda \in \mathbb{R}), \qquad (5.3.5)$$

in terms of the Mittag-Lefler functions (1.8.17). The following statement holds.

Theorem 5.12 *Let $l - 1 < \alpha \leqq l$ $(l \in \mathbb{N})$ and $\lambda \in \mathbb{R}$. Then the functions*

$$y_j(x) = x^j E_{\alpha,j+1}\left(\lambda x^{\alpha}\right) \quad (j = 0, \cdots,\ l-1) \qquad (5.3.6)$$

yield the fundamental system of solutions to the equation (5.3.5).

Proof. Applying the Laplace transform (1.4.1) to (5.3.5) and taking (5.3.3) into account just as in (5.2.7), we have

$$\left(\mathcal{L}y\right)(s) = \sum_{j=0}^{l-1} d_j \frac{s^{\alpha-j-1}}{s^{\alpha} - \lambda}, \qquad (5.3.7)$$

where d_j $(j = 0, \cdots, l-1)$ are given by (5.3.4). Formula (1.10.9) with $\beta = j+1$ yields

$$\mathcal{L}\left[t^j E_{\alpha,j+1}\left(\lambda t^\alpha\right)\right](s) = \frac{s^{\alpha-j-1}}{s^\alpha - \lambda} \quad \left(|s^{-\alpha}\lambda| < 1\right). \tag{5.3.8}$$

Thus, from (5.3.7), we derive the following solution to the equation (5.3.5):

$$y(x) = \sum_{j=0}^{l-1} d_j y_j(x), \quad y_j(x) = x^j E_{\alpha,j+1}\left(\lambda x^\alpha\right). \tag{5.3.9}$$

It is directly verified that the functions $y_j(x)$ are solutions to the equation (5.3.5):

$$\left(D_{0+}^\alpha \left[t^j E_{\alpha,j+1}\left(\lambda t^\alpha\right)\right]\right)(x) = \lambda x^j E_{\alpha,j+1}\left(\lambda x^\alpha\right) \quad (j = 0, \cdots, l-1), \tag{5.3.10}$$

and

$$y_j^{(k)}(x) = x^{j-k} E_{\alpha,j-k+1}\left(\lambda x^\alpha\right). \tag{5.3.11}$$

It follows from here that

$$y_j^{(k)}(0) = 0 \quad (k,j = 0, \cdots, l-1; \; j > k), \quad y_k^{(k)}(0) = 1 \; (k = 0, \cdots, l-1). \tag{5.3.12}$$

If $j < k$, then, on the basis of (1.8.17), the relation (5.3.11) takes the form

$$y_j^{(k)}(x) = \lambda x^{\alpha+j-k} E_{\alpha,\alpha+j-k+1}\left(\lambda x^\alpha\right), \tag{5.3.13}$$

and, since $\alpha + j - k > 0$ for any $k, j = 0, \cdots, l-1$,

$$y_j^{(k)}(0) = 0 \quad (k,j = 0, \cdots, \; l-1; \; j < k). \tag{5.3.14}$$

By (5.3.12) and (5.3.14), the Wronskian

$$W(x) = \det\left(y_j^{(k)}(x)\right)_{k,j=0}^{l-1} \tag{5.3.15}$$

at the point 0 is equal 1: $W(0) = 1$. Then the result of the Theorem 5.12 follows from the known theorem of classical analysis and (5.3.2).

Corollary 5.7 *The equation*

$$\left({}^C D_{0+}^\alpha y\right)(x) - \lambda y(x) = 0 \quad (x > 0; \;\; 0 < \alpha \leqq 1; \;\; \lambda \in \mathbb{R}) \tag{5.3.16}$$

has its solution given by

$$y(x) = E_\alpha\left(\lambda x^\alpha\right), \tag{5.3.17}$$

while the equation

$$\left({}^C D_{0+}^\alpha y\right)(x) - \lambda y(x) = 0 \quad (x > 0; \;\; 1 < \alpha \leqq 2; \;\; \lambda \in \mathbb{R}) \tag{5.3.18}$$

has its fundamental system of solutions given by

$$y_1(x) = E_\alpha\left(\lambda x^\alpha\right), \quad y_2(x) = x E_{\alpha,2}\left(\lambda x^\alpha\right). \tag{5.3.19}$$

Example 5.17 The equation

$$\left({}^{C}D_{0+}^{l-1/2}y\right)(x) - \lambda y(x) = 0 \quad (x > 0;\ \ l \in \mathbb{N};\ \ \lambda \in \mathbb{R}) \tag{5.3.20}$$

has its fundamental system of solutions given by

$$y_j(x) = x^j E_{l-1/2,j+1}\left(\lambda x^{l-1/2}\right) \quad (j = 0, \cdots, l-1) \tag{5.3.21}$$

Next we derive the explicit solutions to the equation (5.3.1) with $m = 2$ of the form (5.2.27):

$$\left({}^{C}D_{0+}^{\alpha}y\right)(x) - \lambda\left({}^{C}D_{0+}^{\beta}y\right)(x) - \mu y(x) = 0 \ (x > 0;\ l-1 < \alpha \leqq l;\ l \in \mathbb{N};\ \alpha > \beta > 0), \tag{5.3.22}$$

with complex $\lambda, \mu \in \mathbb{C}$, in terms of the generalized Wright function (5.2.28).

Theorem 5.13 *Let $\alpha > 0$ and $\beta > 0$ be such that*

$$0 < \beta < \alpha, \quad l-1 < \alpha \leqq l, \quad m-1 < \beta \leqq m \quad (l, m \in \mathbb{N};\ \ m \leqq l), \tag{5.3.23}$$

and let $\lambda, \mu \in \mathbb{R}$. Then the functions

$$y_j(x) = \sum_{n=0}^{\infty} \frac{\mu^n}{n!} x^{\alpha n + j}\ {}_1\Psi_1\left[\begin{array}{c} (n+1, 1) \\ (\alpha n + j + 1, \alpha - \beta) \end{array} \middle|\ \lambda x^{\alpha-\beta}\right] -$$

$$-\lambda \sum_{n=0}^{\infty} \frac{\mu^n}{n!} x^{\alpha n + j + \alpha - \beta}\ {}_1\Psi_1\left[\begin{array}{c} (n+1, 1) \\ (\alpha n + j + 1 + \alpha - \beta, \alpha - \beta) \end{array} \middle|\ \lambda x^{\alpha-\beta}\right] \tag{5.3.24}$$

$$(j = 0, \cdots, m-1),$$

$$y_j(x) = \sum_{n=0}^{\infty} \frac{\mu^n}{n!} x^{\alpha n + j}\ {}_1\Psi_1\left[\begin{array}{c} (n+1, 1) \\ (\alpha n + j + 1, \alpha - \beta) \end{array} \middle|\ \lambda x^{\alpha-\beta}\right] \quad (j = m, \cdots, l-1) \tag{5.3.25}$$

are solutions to the equation (5.3.22), provided that the series in (5.3.24) and (5.3.25) are convergent.

In particular, the equation

$$\left({}^{C}D_{0+}^{\alpha}y\right)(x) - \lambda\left({}^{C}D_{0+}^{\beta}y\right)(x) = 0 \quad (x > 0) \tag{5.3.26}$$

has its solutions given by

$$y_j(x) = x^j E_{\alpha-\beta,j+1}\left(\lambda x^{\alpha-\beta}\right) - \lambda x^{\alpha-\beta+j} E_{\alpha-\beta,\alpha-\beta+j+1}\left(\lambda x^{\alpha-\beta}\right) \tag{5.3.27}$$

$$(j = 0, \cdots, m-1),$$

$$y_j(x) = x^j E_{\alpha-\beta,j+1}\left(\lambda x^{\alpha-\beta}\right) \quad (j = m, \cdots, l-1). \tag{5.3.28}$$

If $\alpha - l + 1 \geqq \beta$, then the functions $y_j(x)$ in (5.3.24), (5.3.25) and (5.3.27), (5.3.28) are linearly independent solutions to the equations (5.3.22) and (5.3.26), respectively. In particular, when $\alpha - l + 1 > \beta$, these functions provide the fundamental system of solutions.

Proof. Applying the Laplace transform to (5.3.22) and using (5.3.3) as in (5.3.7), we obtain

$$(\mathcal{L}y)(s) = \sum_{j=0}^{l-1} d_j \frac{s^{\alpha-j-1}}{s^\alpha - \lambda s^\beta - \mu} - \lambda \sum_{j=0}^{m-1} d_j \frac{s^{\beta-j-1}}{s^\alpha - \lambda s^\beta - \mu},$$

where d_j $(j = 0, \cdots, l-1)$ are given by (5.3.4). For $s \in \mathbb{C}$ and $\left|\frac{\mu s^{-\beta}}{s^{\alpha-\beta}-\lambda}\right| < 1$, in accordance with (5.2.33), we have

$$(\mathcal{L}y)(s) = \sum_{j=0}^{l-1} d_j \sum_{n=0}^{\infty} \mu^n \frac{s^{(\alpha-\beta)-(\beta n+j+1)}}{(s^{\alpha-\beta} - \lambda)^{n+1}} - \lambda \sum_{j=0}^{m-1} d_j \sum_{n=0}^{\infty} \mu^n \frac{s^{(\alpha-\beta)-(\beta n+j+1+\alpha-\beta)}}{(s^{\alpha-\beta} - \lambda)^{n+1}}. \tag{5.3.29}$$

Using (1.10.10) and taking (5.2.28) into account, for $s \in \mathbb{C}$ and $|\lambda s^{\beta-\alpha}| < 1$, we have

$$\frac{s^{(\alpha-\beta)-(\beta n+j+1)}}{(s^{\alpha-\beta} - \lambda)^{n+1}} = \frac{1}{n!}\left(\mathcal{L}\left[t^{\alpha n+j}\left(\frac{\partial}{\partial\lambda}\right)^n E_{\alpha-\beta,\beta n+j+1}\left(\lambda t^{\alpha-\beta}\right)\right]\right)(s)$$

$$= \frac{1}{n!}\left(\mathcal{L}\left[t^{\alpha n+j}\, {}_1\Psi_1\left[\begin{matrix}(n+1,1)\\(\alpha n+j+1,\alpha-\beta)\end{matrix}\,\middle|\,\lambda t^{\alpha-\beta}\right]\right]\right)(s), \tag{5.3.30}$$

$$\frac{s^{(\alpha-\beta)-(\beta n+j+1+\alpha-\beta)}}{(s^{\alpha-\beta} - \lambda)^{n+1}}$$

$$= \frac{1}{n!}\left(\mathcal{L}\left[t^{\alpha n+j+\alpha-\beta}\left(\frac{\partial}{\partial\lambda}\right)^n E_{\alpha-\beta,\beta n+j+1+\alpha-\beta}\left(\lambda t^{\alpha-\beta}\right)\right]\right)(s)$$

$$= \frac{1}{n!}\left(\mathcal{L}\left[t^{\alpha n+j+\alpha-\beta}\, {}_1\Psi_1\left[\begin{matrix}(n+1,1)\\(\alpha n+j+1+\alpha-\beta,\alpha-\beta)\end{matrix}\,\middle|\,\lambda t^{\alpha-\beta}\right]\right]\right)(s). \tag{5.3.31}$$

From (5.3.29)-(5.3.31) we derive the following solution to the equation (5.3.22):

$$y(x) = \sum_{j=0}^{l-1} d_j y_j(x) - \lambda \sum_{j=0}^{l-1} d_j y_j(x), \tag{5.3.32}$$

where $y_j(x)$ $(j = 1, \cdots, l)$ are given by (5.3.24) for $j = 0, \cdots, m-1$ and by (5.3.25) for $j = m, \cdots, l-1$. It is directly verified that these functions are solutions to the equation (5.3.22) which proves the first assertion of the Theorem 5.13.

For $j,k=0,\cdots,l-1$, the direct evaluation yields

$$y_j^{(k)}(x)=\sum_{n=0}^{\infty}\frac{\mu^n}{n!}x^{\alpha n+j-k}\,{}_1\Psi_1\left[\begin{matrix}(n+1,1)\\(\alpha n+j+1-k,\alpha-\beta)\end{matrix}\,\middle|\,\lambda x^{\alpha-\beta}\right]-\lambda I(x) \tag{5.3.33}$$

$$(j=0,\cdots,m-1),$$

$$y_j^{(k)}(x)=\sum_{n=0}^{\infty}\frac{\mu^n}{n!}x^{\alpha n+j-k}\,{}_1\Psi_1\left[\begin{matrix}(n+1,1)\\(\alpha n+j+1-k,\alpha-\beta)\end{matrix}\,\middle|\,\lambda x^{\alpha-\beta}\right] \tag{5.3.34}$$

with $(j=m,\cdots,\ l-1)$, and where

$$I(x)=\sum_{n=0}^{\infty}\frac{\mu^n}{n!}x^{\alpha n+j-k+\alpha-\beta}\,{}_1\Psi_1\left[\begin{matrix}(n+1,1)\\(\alpha n+j+1-k+\alpha-\beta,\alpha-\beta)\end{matrix}\,\middle|\,\lambda x^{\alpha-\beta}\right], \tag{5.3.35}$$

It follows from (5.3.33)-(5.3.35) that the relations in (5.3.12) hold for $j\geqq k$. If $j<k$, then, in accordance with (5.2.28), we have

$$J(x)=\sum_{i=0}^{\infty}\frac{\lambda^{i+1}}{\Gamma[(\alpha-\beta)(i+1)+j+1-k]}x^{(\alpha-\beta)i+\alpha-\beta+j-k}$$

$$+\sum_{n=1}^{\infty}\frac{\mu^n}{n!}x^{\alpha n+j-k}\,{}_1\Psi_1\left[\begin{matrix}(n+1,1)\\(\alpha n+j+1-k,\alpha-\beta)\end{matrix}\,\middle|\,\lambda x^{\alpha-\beta}\right].$$

If $\alpha-l+1\geqq\beta$, then $(\alpha-\beta)q+\alpha-\beta+j-k\geqq\alpha-\beta+1-l\geqq 0$ for any $j,k=0,\cdots,\ l-1$ and $q\in\mathbb{N}_0$. Moreover, $\alpha n+j-\geqq\alpha+1-l>0$ for any $j,k=0,\cdots,l-1$ and $n\in\mathbb{N}$. Then $\lim_{x\to 0}J(x)=0$ except for the case $\alpha-l+1=\beta$ with $k=l-1$ and $j=0$, for which $\lim_{x\to 0}J(x)=\lambda$. If $\alpha-l+1>\beta$, such that, for any $n\in\mathbb{N}_0$ and $j,k=0,\cdots,l-1$, $\alpha n+j-k+\alpha-\beta>\alpha-\beta-1+l>0$, then $\lim_{x\to 0}I(x)=0$. Thus (5.3.33)-(5.3.35) yield the relation (5.3.14) except for the case $\alpha-l+1=\beta$ with $k=l-1$ and $j=0$, for which

$$y_0^{(l-1)}(0)=\lambda. \tag{5.3.36}$$

According to (5.3.12), (5.3.14) and (5.3.36) $W(0)=1$ for the Wronskian (5.3.15). Thus it follows from known theorems in mathematical analysis that the functions $y_j(x)$ in (5.3.24), (5.3.25) are linearly independent solutions to the equation (5.3.22). If $\alpha-l+1>\beta$, then the relations in (5.3.2) are valid, and hence $y_j(x)$ in (5.3.24), (5.3.25) yield the fundamental system of solutions. This completes the proof of Theorem 5.13 for the equation (5.3.22). Setting $\mu=0$ and taking the formula (5.2.40) into account, we arrive at the assertion of Theorem 5.13 for the equation (5.3.26).

Corollary 5.8 *The equation*

$$\left({}^C D_{0+}^{\alpha} y\right)(x) - \lambda \left({}^C D_{0+}^{\beta} y\right)(x) - \mu y(x) = 0 \quad (x > 0; \quad 0 < \beta < \alpha \leqq 1; \; \lambda, \mu \in \mathbb{R}) \tag{5.3.37}$$

has its solution given by

$$y_1(x) = \sum_{n=0}^{\infty} \frac{\mu^n}{n!} x^{\alpha n} \, {}_1\Psi_1 \left[\begin{matrix} (n+1, 1) \\ (\alpha n + 1, \alpha - \beta) \end{matrix} \middle| \lambda x^{\alpha-\beta} \right] -$$

$$-\lambda \sum_{n=0}^{\infty} \frac{\mu^n}{n!} x^{\alpha n + \alpha - \beta} \, {}_1\Psi_1 \left[\begin{matrix} (n+1, 1) \\ (\alpha n + 1 + \alpha - \beta, \alpha - \beta) \end{matrix} \middle| \lambda x^{\alpha-\beta} \right] \tag{5.3.38}$$

In particular,

$$y_1(x) = E_{\alpha-\beta}\left(\lambda x^{\alpha-\beta}\right) - \lambda x^{\alpha-\beta} E_{\alpha-\beta, \alpha-\beta+1}\left(\lambda x^{\alpha-\beta}\right) \tag{5.3.39}$$

is a solution to the equation

$$\left({}^C D_{0+}^{\alpha} y\right)(x) - \lambda \left({}^C D_{0+}^{\beta} y\right)(x) = 0 \quad (x > 0; \quad 0 < \beta < \alpha \leqq 1; \; \lambda \in \mathbb{R}). \tag{5.3.40}$$

Corollary 5.9 *The equation*

$$\left({}^C D_{0+}^{\alpha} y\right)(x) - \lambda \left({}^C D_{0+}^{\beta} y\right)(x) - \mu y(x) = 0 \tag{5.3.41}$$

where $x > 0$; $1 < \alpha \leqq 2$, $0 < \beta < \alpha$; $\lambda, \mu \in \mathbb{R}$, has one solution $y_1(x)$, given by (5.3.38), *and a second solution $y_2(x)$ given by*

$$y_2(x) = \sum_{n=0}^{\infty} \frac{\mu^n}{n!} x^{\alpha n + 1} \, {}_1\Psi_1 \left[\begin{matrix} (n+1, 1) \\ (\alpha n + 2, \alpha - \beta) \end{matrix} \middle| \lambda x^{\alpha-\beta} \right] -$$

$$-\lambda \sum_{n=0}^{\infty} \frac{\mu^n}{n!} x^{\alpha n + 1 + \alpha - \beta} \, {}_1\Psi_1 \left[\begin{matrix} (n+1, 1) \\ (\alpha n + 2 + \alpha - \beta, \alpha - \beta) \end{matrix} \middle| \lambda x^{\alpha-\beta} \right]; \tag{5.3.42}$$

for $j = 0, \cdots, m-1$ and $1 < \beta < \alpha$, while, for $0 < \beta \leqq 1$, by

$$y_2(x) = \sum_{n=0}^{\infty} \frac{\mu^n}{n!} x^{\alpha n + 1} \, {}_1\Psi_1 \left[\begin{matrix} (n+1, 1) \\ (\alpha n + 2, \alpha - \beta) \end{matrix} \middle| \lambda x^{\alpha-\beta} \right]. \tag{5.3.43}$$

In particular, the equation

$$\left({}^C D_{0+}^{\alpha} y\right)(x) - \lambda \left({}^C D_{0+}^{\beta} y\right)(x) = 0 \quad (x > 0; \; 1 < \alpha \leqq 2, \; 0 < \beta < \alpha; \; \lambda \in \mathbb{R}) \tag{5.3.44}$$

has one solution $y_1(x)$ given by (5.3.39), *and a second solution $y_2(x)$ given by*

$$y_2(x) = x E_{\alpha-\beta, 2}\left(\lambda x^{\alpha-\beta}\right) - \lambda x^{1+\alpha-\beta} E_{\alpha-\beta, 2+\alpha-\beta}\left(\lambda x^{\alpha-\beta}\right), \tag{5.3.45}$$

for $1 < \beta < \alpha$, *while, for* $0 < \beta \leqq 1$, *by*

$$y_2(x) = xE_{\alpha-\beta,2}\left(\lambda x^{\alpha-\beta}\right). \tag{5.3.46}$$

If $\alpha \geqq \beta+1$, *then* (5.3.38), (5.3.43) *and* (5.3.39), (5.3.46) *are linearly independent solutions to the equations* (5.3.41) *and* (5.3.44) *respectively. When* $\alpha > \beta+1$, *they provide the fundamental systems of solutions.*

Example 5.18 The equation

$$y'(x) - \lambda\left({}^C D_{0+}^{\beta} y\right)(x) - \mu y(x) = 0 \quad (x > 0;\ \ 0 < \beta < 1;\ \ \lambda, \mu \in \mathbb{R}). \tag{5.3.47}$$

has its solution given by

$$y_1(x) = \sum_{n=0}^{\infty} \frac{\mu^n}{n!} x^n \, {}_1\Psi_1\left[\begin{array}{c|c} (n+1,1) & \\ & \lambda x^{1-\beta} \\ (n+1, 1-\beta) & \end{array}\right] -$$

$$-\lambda \sum_{n=0}^{\infty} \frac{\mu^n}{n!} x^{n+1-\beta} \, {}_1\Psi_1\left[\begin{array}{c|c} (n+1,1) & \\ & \lambda x^{1-\beta} \\ (n+2-\beta, 1-\beta) & \end{array}\right]. \tag{5.3.48}$$

In particular,

$$y_1(x) = \sum_{n=0}^{\infty} \frac{\mu^n}{n!} x^n \, {}_1\Psi_1\left[\begin{array}{c|c} (n+1,1) & \\ & \lambda x^{1/2} \\ (n+1, 1/2) & \end{array}\right] -$$

$$-\lambda \sum_{n=0}^{\infty} \frac{\mu^n}{n!} x^{n+1/2} \, {}_1\Psi_1\left[\begin{array}{c|c} (n+1,1) & \\ & \lambda x^{1/2} \\ (n+3/2, 1/2) & \end{array}\right]. \tag{5.3.49}$$

is a solution to the equation

$$y'(x) - \lambda\left({}^C D_{0+}^{1/2} y\right)(x) - \mu y(x) = 0 \quad (x > 0;\ \ \lambda, \mu \in \mathbb{R}). \tag{5.3.50}$$

Example 5.19 The equation

$$y''(x) - \lambda\left({}^C D_{0+}^{\beta} y\right)(x) - \mu y(x) = 0 \quad (x > 0;\ \ 0 < \beta < 2;\ \ \lambda, \mu \in \mathbb{R}) \tag{5.3.51}$$

has two solutions given by

$$y_1(x) = \sum_{n=0}^{\infty} \frac{\mu^n}{n!} x^{2n} \, {}_1\Psi_1\left[\begin{array}{c|c} (n+1,1) & \\ & \lambda x^{2-\beta} \\ (2n+1, 2-\beta) & \end{array}\right] -$$

$$-\lambda \sum_{n=0}^{\infty} \frac{\mu^n}{n!} x^{2n+2-\beta} \, {}_1\Psi_1\left[\begin{array}{c|c} (n+1,1) & \\ & \lambda x^{2-\beta} \\ (2n+3-\beta, 2-\beta) & \end{array}\right] \tag{5.3.52}$$

and

$$y_2(x) = \sum_{n=0}^{\infty} \frac{\mu^n}{n!} x^{2n+1} \, {}_1\Psi_1 \left[\begin{array}{c} (n+1,1) \\ \\ (2n+2, 2-\beta) \end{array} \middle| \lambda x^{2-\beta} \right] -$$

$$-\lambda \sum_{n=0}^{\infty} \frac{\mu^n}{n!} x^{2n+3-\beta} \, {}_1\Psi_1 \left[\begin{array}{c} (n+1,1) \\ \\ (2n+4-\beta, 2-\beta) \end{array} \middle| \lambda x^{2-\beta} \right] \quad (j = 0, \cdots, m-1), \tag{5.3.53}$$

for $1 < \beta < 2$, while, for $0 < \beta \leqq 1$, the second solution $y_2(x)$ is given by

$$y_2(x) = \sum_{n=0}^{\infty} \frac{\mu^n}{n!} x^{2n+1} \, {}_1\Psi_1 \left[\begin{array}{c} (n+1,1) \\ \\ (2n+2, 2-\beta) \end{array} \middle| \lambda x^{2-\beta} \right] \tag{5.3.54}$$

The solutions (5.3.52) and (5.3.54) are linearly independent, and when $0 < \beta < 1$ they provide the fundamental system of solutions to the equation (5.3.51).

In particular, the equation

$$y''(x) - \lambda \left({}^C D_{0+}^{1/2} y \right)(x) - \mu y(x) = 0 \quad (x > 0; \quad \lambda, \mu \in \mathbb{R}) \tag{5.3.55}$$

has the fundamental system of solutions given by (5.3.52) and (5.3.54) with $\beta = 1/2$.

Finally, we find explicit solutions to the equation (5.3.1) with any $m \in \mathbb{N} \backslash \{1,2\}$ in terms of the generalized Wright function (5.2.28). It is convenient to rewrite (5.3.1) in the form (5.2.65)

$$({}^C D_{0+}^{\alpha} y)(x) - \lambda ({}^C D_{0+}^{\beta} y)(x) - \sum_{k=0}^{m-2} A_k ({}^C D_{0+}^{\alpha_k} y)(x) = 0 \quad (x > 0) \tag{5.3.56}$$

$$(m \in \mathbb{N} \backslash \{1,2\}; \quad 0 = \alpha_0 < \alpha_1 < \cdots < \alpha_{m-2} < \beta < \alpha; \quad \lambda, A_0, \cdots, \ A_{m-2} \in \mathbb{R}).$$

Theorem 5.14 *Let α, β, $\alpha_{m-2}, \cdots, \ \alpha_0$ and $l, l_{m-1}, \cdots, \ l_0 \in \mathbb{N}_0$ $(m \in \mathbb{N} \backslash \{1,2\})$ be such that*

$$0 = \alpha_0 < \alpha_1 \cdots < \alpha_{m-2} < \beta < \alpha, \quad 0 = l_0 \leqq l_1 \leqq \cdots l_{m-1} \leqq l,$$

$$l-1 < \alpha \leqq l, \ l_{m-1} - 1 < \beta \leqq l_{m-1}, \ l_k - 1 < \alpha_k \leqq l_k \ (k = 1, \cdots, \ m-2), \tag{5.3.57}$$

and let $\lambda, A_0, \cdots, \ A_{m-2} \in \mathbb{R}$. Then the solutions to the equation (5.3.56) are given by the formulas

$$y_j(x) = \sum_{n=0}^{\infty} \left(\sum_{k_0 + \cdots + k_{m-2} = n} \right) \frac{1}{k_0! \cdots k_{m-2}!} \left[\prod_{\nu=0}^{m-2} (A_\nu)^{k_\nu} \right] x^{(\alpha-\beta)n + j + \sum_{\nu=0}^{m-2} (\beta - \alpha_\nu) k_\nu}$$

$$\cdot \left\{ {}_1\Psi_1 \left[\begin{array}{c} (n+1,1) \\ \\ \left((\alpha-\beta)n + 1 + j + \sum_{\nu=0}^{m-2} (\beta - \alpha_\nu) k_\nu, \alpha - \beta \right) \end{array} \middle| \lambda x^{\alpha-\beta} \right] \right.$$

$$-\lambda x^{\alpha-\beta}\ {}_1\Psi_1\left[\begin{array}{c}(n+1,1)\\ \left((\alpha-\beta)(n+1)+1+j+\sum_{\nu=0}^{m-2}(\beta-\alpha_\nu)k_\nu,\alpha-\beta\right)\end{array}\Bigg|\ \lambda x^{\alpha-\beta}\right]$$

$$-\sum_{k=i}^{m-2}A_k x^{\alpha-\alpha_k}$$

$$\cdot{}_1\Psi_1\left[\begin{array}{c}(n+1,1)\\ \left((\alpha-\beta)n+\alpha-\alpha_k+1+j+\sum_{\nu=0}^{m-2}(\beta-\alpha_\nu)k_\nu,\alpha-\beta\right)\end{array}\Bigg|\ \lambda x^{\alpha-\beta}\right]\Bigg\}, \tag{5.3.58}$$

when $0 \leqq j \leqq l_{m-2}-1$ *with* $l_q \leqq j \leqq l_{q+1}-1$ *for* $q=0,\cdots,m-3$; *by*

$$y_j(x)=\sum_{n=0}^{\infty}\left(\sum_{k_0+\cdots+k_{m-2}=n}\right)\frac{1}{k_0!\cdots k_{m-2}!}\left[\prod_{\nu=0}^{m-2}(A_\nu)^{k_\nu}\right]x^{(\alpha-\beta)n+j+\sum_{\nu=0}^{m-2}(\beta-\alpha_\nu)k_\nu}$$

$$\cdot\Bigg\{{}_1\Psi_1\left[\begin{array}{c}(n+1,1)\\ \left((\alpha-\beta)n+1+j+\sum_{\nu=0}^{m-2}(\beta-\alpha_\nu)k_\nu,\alpha-\beta\right)\end{array}\Bigg|\ \lambda x^{\alpha-\beta}\right]-$$

$$-\lambda x^{\alpha-\beta}\ {}_1\Psi_1\left[\begin{array}{c}(n+1,1)\\ \left((\alpha-\beta)(n+1)+1+j+\sum_{\nu=0}^{m-2}(\beta-\alpha_\nu)k_\nu,\alpha-\beta\right)\end{array}\Bigg|\ \lambda x^{\alpha-\beta}\right]\Bigg\}, \tag{5.3.59}$$

for $l_{m-2} \leqq j \leqq l_{m-1}-1$, *and by*

$$y_j(x)=\sum_{n=0}^{\infty}\left(\sum_{k_0+\cdots+k_{m-2}=n}\right)\frac{1}{k_0!\cdots k_{m-2}!}\left[\prod_{\nu=0}^{m-2}(A_\nu)^{k_\nu}\right]x^{(\alpha-\beta)n+j+\sum_{\nu=0}^{m-2}(\beta-\alpha_\nu)k_\nu}$$

$$\cdot{}_1\Psi_1\left[\begin{array}{c}(n+1,1)\\ \left((\alpha-\beta)n+1+j+\sum_{\nu=0}^{m-2}(\beta-\alpha_\nu)k_\nu,\alpha-\beta\right)\end{array}\Bigg|\ \lambda x^{\alpha-\beta}\right], \tag{5.3.60}$$

for $l_{m-1} \leqq j \leqq l-1$, *provided that the series in* (5.3.58), (5.3.59) *and* (5.3.60) *are convergent. The inner sums are taken over all* $k_0,\cdots,k_{m-2}\in\mathbb{N}_0$ *such that* $k_0+\cdots+k_{m-2}=n$.

If $\alpha-l+1 \geqq \beta$, *then the functions* $y_j(x)$ *in* (5.3.58)-(5.3.60) *are linearly independent solutions to the equation* (5.3.56). *In particular, for* $\alpha-l+1>\beta$, *these functions provide the fundamental system of solutions.*

Proof. The proof of Theorem 5.14 is carried out along the lines of the proof of Theorems 5.12 and 5.13 using relations (5.3.3), (5.2.68), (1.10.10) and (5.2.28), which lead to the explicit solutions (5.3.58)-(5.3.60) of the equation (5.3.56). When

$\alpha - l + 1 \geqq \beta$, then it is directly verified that the relations in (5.3.12) hold in the case $j \geqq k$, while, for $j < k$, the formula (5.3.14) is valid for any $j, k = 1, \cdots, l$ except for the case $\alpha - l + 1 = \beta$, for which (5.3.36) is true. This means that the Wronskian $W(0) = 1$, which implies the linear independence of the solutions $y_j(x)$ in (5.3.58)-(5.3.60). If $\alpha - l + 1 > \beta$, then the relations in (5.3.2) hold, so that these functions form the fundamental system of solutions.

Corollary 5.10 *The equation*

$$\left({}^C D_{0+}^{\alpha} y\right)(x) - \lambda \left({}^C D_{0+}^{\beta} y\right)(x) - \delta \left({}^C D_{0+}^{\gamma} y\right)(x) - \mu y(x) = 0 \quad (x > 0;\quad \lambda, \delta, \mu \in \mathbb{R}) \tag{5.3.61}$$

with $0 < \gamma < \beta < \alpha$; $l_1 - 1 < \gamma \leqq l_1$, $l_2 - 1 < \beta \leqq l_2$, $l - 1 < \alpha \leqq l$ $(l_1, l_2, l \in \mathbb{N};\ l_1 \leqq l_2 \leqq l)$, *has* l *solutions given by*

$$y_j(x) = \sum_{n=0}^{\infty} \left(\sum_{i+\nu=n} \right) \frac{\mu^i \delta^\nu}{i!\nu!} x^{(\alpha-\beta)n+j+\beta i+(\beta-\gamma)\nu}$$

$$\cdot \left\{ {}_1\Psi_1 \left[\begin{matrix} (n+1,1) \\ ((\alpha-\beta)n+1+j+\beta i+(\beta-\gamma)\nu, \alpha-\beta) \end{matrix} \middle| \lambda x^{\alpha-\beta} \right] - \right.$$

$$-\lambda x^{\alpha-\beta}\ {}_1\Psi_1 \left[\begin{matrix} (n+1,1) \\ ((\alpha-\beta)(n+1)+1+j+\beta i+(\beta-\gamma)\nu, \alpha-\beta) \end{matrix} \middle| \lambda x^{\alpha-\beta} \right] -$$

$$-\delta x^{\alpha-\gamma}\ {}_1\Psi_1 \left[\begin{matrix} (n+1,1) \\ ((\alpha-\beta)n+\alpha-\gamma+1+j+\beta i+(\beta-\gamma)\nu, \alpha-\beta) \end{matrix} \middle| \lambda x^{\alpha-\beta} \right]$$

$$\left. -\mu x^{\alpha}\ {}_1\Psi_1 \left[\begin{matrix} (n+1,1) \\ ((\alpha-\beta)n+\alpha+1+j+\beta i+(\beta-\gamma)\nu, \alpha-\beta) \end{matrix} \middle| \lambda x^{\alpha-\beta} \right] \right\}, \tag{5.3.62}$$

when $0 \leqq j \leqq l_1 - 1$, *by*

$$y_j(x) = \sum_{n=0}^{\infty} \left(\sum_{i+\nu=n} \right) \frac{\mu^i \delta^\nu}{i!\nu!} x^{(\alpha-\beta)n+j+\beta i+(\beta-\gamma)\nu}$$

$$\cdot \left\{ {}_1\Psi_1 \left[\begin{matrix} (n+1,1) \\ ((\alpha-\beta)n+1+j+\beta i+(\beta-\gamma)\nu, \alpha-\beta) \end{matrix} \middle| \lambda x^{\alpha-\beta} \right] - \right.$$

$$\left. -\lambda x^{\alpha-\beta}\ {}_1\Psi_1 \left[\begin{matrix} (n+1,1) \\ ((\alpha-\beta)(n+1)+1+j+\beta i+(\beta-\gamma)\nu, \alpha-\beta) \end{matrix} \middle| \lambda x^{\alpha-\beta} \right] \right\}, \tag{5.3.63}$$

for $l_1 \leqq j \leqq l_2 - 1$, *and by*

$$y_j(x) = \sum_{n=0}^{\infty} \left(\sum_{i+\nu=n} \right) \frac{\mu^i \delta^\nu}{i!\nu!} x^{(\alpha-\beta)n+j+\beta i+(\beta-\gamma)\nu}$$

$$\cdot {}_1\Psi_1 \left[\begin{matrix} (n+1,1) \\ ((\alpha-\beta)n+1+j+\beta i+(\beta-\gamma)\nu, \alpha-\beta) \end{matrix} \middle| \lambda x^{\alpha-\beta} \right], \tag{5.3.64}$$

for $l_2 \leqq j \leqq l - 1$.

Example 5.20 The equation

$$\left({}^C D_{0+}^{5/2} y\right)(x) - \lambda \left({}^C D_{0+}^{3/2} y\right)(x) - \delta \left({}^C D_{0+}^{1/2} y\right)(x) - \mu y(x) = 0 \quad (x > 0;\ \ \lambda, \delta, \mu \in \mathbb{R}) \tag{5.3.65}$$

has three solutions $y_0(x)$, $y_1(x)$ and $y_2(x)$ given, respectively, by (5.3.62), (5.3.63) and (5.3.64) with $\alpha = 5/2$, $\beta = 3/2$ and $\delta = 1/2$.

Remark 5.11 In Sections 5.2.1 and 5.3.1 we constructed the explicit solutions to linear ordinary homogeneous differential equations with constant coefficients of the same forms (5.2.1) and (5.3.1) involving Liouville and Caputo fractional derivatives, respectively. Moreover, we gave conditions for when the obtained solutions are linearly independent and form fundamental systems of solutions. The results obtained show that these equations, generally speaking, have different solutions. This is due to the fact that Liouville and Caputo derivatives differ by a quasi-polynomial term (see (2.4.6)).

5.3.2 Nonhomogeneous Equations with Constant Coefficients

In Section 5.3.1 we applied the Laplace transform method to derive explicit solutions to the homogeneous equations (5.3.1) involving Caputo fractional derivatives (2.4.1). Here we use this approach to find general solutions to the corresponding nonhomogeneous equations

$$\sum_{k=1}^{m} A_k (D_{0+}^{\alpha_k} y)(x) + A_0 y(x) = f(x) \quad (x > 0;\ 0 < \alpha_1 \cdots < \alpha_m) \tag{5.3.66}$$

with real coefficients $A_k \in \mathbb{R}$ $(k = 0, \cdots, m)$ and a given function $f(x)$ defined on $\mathbb{R}_+$. As in the case of the equations with Liouville fractional derivatives, the general solution to the equation (5.3.66) is a sum of its particular solution and of the general solution to the corresponding homogeneous equation (5.3.1). It is sufficient to construct a particular solution to the equation (5.3.66). In fact, if $l_k - 1 < \alpha_k < l_k$ and $0 = l_0 \leqq l_1 \leqq \cdots \leqq l_m$, then applying the Laplace transform to (5.3.66), taking (5.3.3) into account and using the inverse Laplace transform, we derive the general solution to the equation (5.3.66) in the form

$$y(x) = \sum_{j=0}^{l_m - 1} c_j y_j(x) + \int_0^x G_{\alpha_1, \cdots, \alpha_m}(x-t) f(t) dt, \tag{5.3.67}$$

where $y_j(x)$ $(j = 1, \cdots, l_m)$ are solutions to the corresponding homogeneous equation (5.3.1)

$$y_j(x) = \sum_{k=q}^{l_m-1} A_k \left(\mathcal{L}^{-1} \left[\frac{s^{\alpha_k - j - 1}}{P_{\alpha_1, \cdots, \alpha_m}(s)} \right] \right)(x), \tag{5.3.68}$$

where $l_q \leqq j \leqq l_{q+1} - 1$ for $q = 0, \cdots, m-1$, $P_{\alpha_1, \cdots, \alpha_m}(s)$ is given in (5.2.81), and c_j are arbitrary real constants.

The latter means that a particular solution to the nonhomogeneous equation with Caputo fractional derivatives (5.3.66) coincides with a particular solution to the nonhomogeneous equation with Liouville fractional derivatives (5.2.79). Therefore from Theorems 5.12-5.14 and Theorems 5.4-5.6 we derive the following statements.

Theorem 5.15 *Let $l - 1 < \alpha \leqq l$ $(l \in \mathbb{N})$, $\lambda \in \mathbb{R}$ and let $f(x)$ be a given real function defined on $\mathbb{R}_+$. Then the equation*

$$\left({}^C D_{0+}^{\alpha} y\right)(x) - \lambda y(x) = f(x) \quad (x > 0) \tag{5.3.69}$$

is solvable, and its general solution is given by

$$y(x) = \int_0^x (x-t)^{\alpha-1} E_{\alpha,\alpha}\left[\lambda(x-t)^{\alpha}\right] f(t)dt + \sum_{j=0}^{l-1} c_j x^j E_{\alpha,j+1}\left(\lambda x^{\alpha}\right), \tag{5.3.70}$$

where c_j $(j = 0, \cdots, l-1)$ are arbitrary real constants.

In particular, the general solutions to the equation (5.3.69) with $0 < \alpha \leqq 1$ and $1 < \alpha \leqq 2$ have the forms

$$y(x) = \int_0^x (x-t)^{\alpha-1} E_{\alpha,\alpha}\left[\lambda(x-t)^{\alpha}\right] f(t)dt + c_1 E_{\alpha}\left(\lambda x^{\alpha}\right) \tag{5.3.71}$$

and

$$y(x) = \int_0^x (x-t)^{\alpha-1} E_{\alpha,\alpha}\left[\lambda(x-t)^{\alpha}\right] f(t)dt + c_1 E_{\alpha}\left(\lambda x^{\alpha}\right) + c_2 x E_{\alpha,2}\left(\lambda x^{\alpha}\right), \tag{5.3.72}$$

respectively, where c_1 and c_2 are arbitrary real constants.

Remark 5.12 Gorenflo and Mainardi ([304], Section 3) obtained the explicit solution to the equation (5.3.69) with $\lambda = -1$ in a form different from (5.3.70). Their arguments were based on applying the Riemann-Liouville fractional integration operator I_{0+}^{α} to both sides of (5.3.69) and then using the Laplace transform. They also discussed the solutions obtained in the cases $0 < \alpha < 1$ and $1 < \alpha < 2$ (see also Gorenflo et al. [310]).

Theorem 5.16 *Let $l-1 < \alpha \leqq l$ $(l \in \mathbb{N})$, $0 < \beta < \alpha$ be such that $\alpha - l + 1 \geqq \beta$, let $\lambda, \mu \in \mathbb{R}$, and let $f(x)$ be a given real function defined on $\mathbb{R}_+$. Then the equation*

$$\left({}^C D_{0+}^{\alpha} y\right)(x) - \lambda \left({}^C D_{0+}^{\beta} y\right)(x) - \mu y(x) = f(x) \quad (x > 0; \;\; \alpha > 0) \tag{5.3.73}$$

is solvable, and its general solution has the form

$$y(x) = \int_0^x (x-t)^{\alpha-1} G_{\alpha,\beta;\lambda,\mu}(x-t) f(t)dt + \sum_{j=0}^{l-1} c_j y_j(x), \tag{5.3.74}$$

where $G_{\alpha,\beta;\lambda,\mu}(z)$ *and* $y_j(x)$ $(j = 0, \cdots, l-1)$ *are given by* (5.2.97) *and* (5.3.24)-(5.3.25), *respectively, and* c_j $(j = 0, \cdots, l-1)$ *are arbitrary real constants.*

In particular, the general solution to the equation

$$\left({}^C D_{0+}^{\alpha} y\right)(x) - \lambda \left({}^C D_{0+}^{\beta} y\right)(x) = f(x) \quad (x > 0; \quad \alpha > 0) \tag{5.3.75}$$

is given by

$$y(x) = \int_0^x (x-t)^{\alpha-1} E_{\alpha-\beta,\alpha}\left[\lambda (x-t)^{\alpha-\beta}\right] f(t)dt + \sum_{j=0}^{l-1} c_j y_j(x), \tag{5.3.76}$$

where $y_j(x)$ $(j = 0, \cdots, l-1)$ *are given by* (5.3.27)-(5.3.28).

Theorem 5.17 *Let* $m \in \mathbb{N}\setminus\{1,2\}$, $l-1 < \alpha \leqq l$ $(l \in \mathbb{N})$, *and let* β *and* $\alpha_1, \cdots, \alpha_{m-2}$ *be such that* $\alpha > \beta > \alpha_{m-2} > \cdots > \alpha_1 > \alpha_0 = 0$ *with* $\alpha - l + 1 \geqq \beta$, *and let* $\lambda, A_0, \cdots, A_{m-2} \in \mathbb{R}$, *and let* $f(x)$ *be a given real function defined on* $\mathbb{R}_+$. *Then the equation*

$$({}^C D_{0+}^{\alpha} y)(x) - \lambda ({}^C D_{0+}^{\beta} y)(x) - \sum_{k=0}^{m-2} A_k ({}^C D_{0+}^{\alpha_k} y)(x) = f(x) \quad (x > 0) \tag{5.3.77}$$

$$(m \in \mathbb{N}\setminus\{1,2\}; \quad 0 = \alpha_0 < \alpha_1 < \cdots < \alpha_{m-2} < \beta < \alpha; \quad \lambda, A_0, \cdots, A_{m-2} \in \mathbb{R}).$$

is solvable, and its general solution is given by

$$y(x) = \int_0^x (x-t)^{\alpha-1} G_{\alpha_1,\cdots,\alpha_{m-2},\beta,\alpha;\lambda}(x-t) f(t)dt + \sum_{j=0}^{l-1} c_j y_j(x), \tag{5.3.78}$$

where $G_{\alpha_1,\cdots,\alpha_{m-2},\beta,\alpha;\lambda}(z)$ *is given by* (5.2.120), $y_j(x)$ $(j = 0, \cdots, l-1)$ *by* (5.3.58)-(5.3.60), *and* c_j $(j = 0, \cdots, l-1)$ *are arbitrary real constants.*

Example 5.21 The equation

$$y'(x) - \lambda \left({}^C D_{0+}^{\beta} y\right)(x) - \mu y(x) = f(x) \quad (x > 0; \quad 0 < \beta < 1; \; \lambda, \mu \in \mathbb{R}) \tag{5.3.79}$$

has its general solution given by

$$y(x) = \int_0^x G_{1,\beta;\lambda,\mu}(x-t) f(t)dt + c\left\{\sum_{n=0}^{\infty} \frac{\mu^n}{n!} x^n \, {}_1\Psi_1\left[\begin{array}{c} (n+1,1) \\ \\ (n+1,1-\beta) \end{array} \middle| \lambda x^{1-\beta}\right] - \right.$$

$$-\lambda \sum_{n=0}^{\infty} \frac{\mu^n}{n!} x^{n+1-\beta} \, {}_1\Psi_1 \left[\begin{array}{c|c} (n+1,1) & \\ & \lambda x^{1-\beta} \\ (n+2-\beta, 1-\beta) & \end{array} \right] \Bigg\}, \tag{5.3.80}$$

where $G_{1,\beta;\lambda,\mu}(z)$ is given by (5.2.105) and c is an arbitrary real constant.

In particular,

$$y(x) = \int_0^x G_{1,1/2;\lambda,\mu}(x-t) f(t) dt + c \left[E_{1-\beta}\left(\lambda x^{1-\beta}\right) - \lambda x^{1-\beta} E_{1-\beta,2-\beta}\left(\lambda x^{1-\beta}\right) \right] \tag{5.3.81}$$

is the general solution to the following equation:

$$y'(x) - \lambda \left({}^C D_{0+}^{\beta} y \right)(x) = f(x) \quad (x > 0; \;\; 0 < \beta < 1; \;\; \lambda \in \mathbb{R}), \tag{5.3.82}$$

while

$$y(x) = \int_0^x G_{1,1/2;\lambda,\mu}(x-t) f(t) dt + c \left\{ \sum_{n=0}^{\infty} \frac{\mu^n}{n!} x^n \, {}_1\Psi_1 \left[\begin{array}{c|c} (n+1,1) & \\ & \lambda x^{1/2} \\ (n+1, 1/2) & \end{array} \right] - \right.$$

$$\left. -\lambda \sum_{n=0}^{\infty} \frac{\mu^n}{n!} x^{n+1/2} \, {}_1\Psi_1 \left[\begin{array}{c|c} (n+1,1) & \\ & \lambda x^{1/2} \\ (n+3/2, 1/2) & \end{array} \right] \right\} \tag{5.3.83}$$

is the general solution to the equation

$$y'(x) - \lambda \left({}^C D_{0+}^{1/2} y \right)(x) - \mu y(x) = f(x) \quad (x > 0; \;\; \lambda, \mu \in \mathbb{R}). \tag{5.3.84}$$

Example 5.22 The equation

$$y''(x) - \lambda \left({}^C D_{0+}^{\beta} y \right)(x) - \mu y(x) = f(x) \quad (x > 0; \;\; 0 < \beta < 2; \;\; \lambda, \mu \in \mathbb{R}) \tag{5.3.85}$$

has its general solution given by

$$y(x) = \int_0^x (x-t) G_{2,\beta;\lambda,\mu}(x-t) f(t) dt + c_1 y_1(x) + c_2 y_2(x), \tag{5.3.86}$$

where $G_{2,\beta;\lambda,\mu}(z)$ is given by (5.2.111), $y_1(x)$ by (5.3.52), $y_2(x)$ by (5.3.53) and (5.3.54) for the cases $1 < \beta < 2$ and $0 < \beta \leqq 1$ respectively, and c_1 and c_2 are arbitrary real constants.

In particular, the equation

$$y''(x) - \lambda \left({}^C D_{0+}^{1/2} y \right)(x) - \mu y(x) = f(x) \quad (x > 0; \;\; \lambda, \mu \in \mathbb{R}) \tag{5.3.87}$$

has its general solution given by

$$y(x) = \int_0^x (x-t) G_{2,1/2;\lambda,\mu}(x-t) f(t) dt + c_1 y_1(x) + c_2 y_2(x), \tag{5.3.88}$$

where $G_{2,1/2;\lambda,\mu}(z)$, $y_1(x)$ and $y_2(x)$ are given by (5.2.111), (5.3.52) and (5.3.54) with $\beta = 1/2$, and c_1 and c_2 are arbitrary real constants.

Example 5.23 The equation

$$\left({}^C D_{0+}^{\alpha} y\right)(x) - \lambda \left({}^C D_{0+}^{\beta} y\right)(x) - \delta \left({}^C D_{0+}^{\gamma} y\right)(x) - \mu y(x) = f(x) \quad (x > 0;\ \lambda, \delta, \mu \in \mathbb{R}), \tag{5.3.89}$$

with $l - 1 < \alpha \leqq l$ ($l \in \mathbb{N}$) and $0 < \gamma < \beta < \alpha$, has its general solution given by

$$y(x) = \int_0^x (x-t)^{\alpha-1} G_{\gamma,\beta,\alpha;\lambda}(x-t) f(t)dt + \sum_{j=0}^{l-1} c_j y_j(x), \tag{5.3.90}$$

where $G_{\gamma,\beta,\alpha;\lambda}(x)$ and $y_j(x)$ ($j = 0, \cdots, l-1$) are given by (5.2.129) and (5.3.62)-(5.3.64), respectively, and c_j ($j = 0, \cdots,\ l-1$) are arbitrary real constants.

5.3.3 Cauchy Problems for Fractional Differential Equations

The results giving the explicit solutions to fractional differential equations, established in Sections 5.3.1-5.3.2, can be applied to obtain the explicit solutions to initial and boundary problems for such equations. Here we illustrate such an application to Cauchy problems for fractional differential equations of order α ($l - 1 < \alpha \leqq l$; $l \in \mathbb{N}$) with the initial conditions of the form (4.1.59)

$$y^{(k)}(0) = b_k \in \mathbb{R} \quad (k = 0, \cdots, l-1; \quad l - 1 < \alpha \leqq l)). \tag{5.3.91}$$

The first results for homogeneous equations (5.3.5), (5.3.22) and (5.3.56) follow from Theorems 5.12-5.14 if we take relations (5.3.12), (5.3.14) and (5.3.36) into account.

Proposition 5.7 *Let $l - 1 < \alpha \leqq l$ ($l \in \mathbb{N}$) and $\lambda \in \mathbb{R}$. Then the Cauchy problem* (5.3.91) *for the equation* (5.3.5) *is solvable, and its solution is given by*

$$y(x) = \sum_{j=0}^{l-1} b_j x^j E_{\alpha,j+1}\left(\lambda x^{\alpha}\right). \tag{5.3.92}$$

Proposition 5.8 *Let $\alpha > 0$ and $\beta > 0$ be such that $l - 1 < \alpha \leqq l$ ($l \in \mathbb{N}$), $0 < \beta < \alpha$ and $\alpha - l + 1 \geqq \beta$, and let $\lambda, \mu \in \mathbb{R}$. Then the Cauchy problem* (5.3.91) *for the equation* (5.3.22) *is solvable, and its solution has the form*

$$y(x) = \sum_{j=0}^{l-1} b_j y_j(x) \quad (\alpha - l - 1 > \beta) \tag{5.3.93}$$

or

$$y(x) = \sum_{j=0}^{l-2} b_j y_j(x) + (b_{l-1} - \lambda b_0) y_{l-1}(x) \quad (\alpha - l - 1 = \beta), \tag{5.3.94}$$

where $y_j(x)$ ($j = 0, \cdots, l-1$) are given by (5.3.24)-(5.3.25).

Proposition 5.9 *Let* $m \in \mathbb{N}\backslash\{1,2\}$, $l-1<\alpha \leqq l$ $(l \in \mathbb{N})$, *and let* β *and* $\alpha_1,\cdots,\alpha_{m-2}$ *be such that* $\alpha>\beta>\alpha_{m-2}>\cdots>\alpha_1>\alpha_0=0$ *and* $\alpha-l+1 \geqq \beta$, *and let* $\lambda, A_0,\cdots,A_{m-2} \in \mathbb{R}$. *Then the Cauchy type problem* (5.3.91) *for the equation* (5.3.56) *is solvable, and its solution has the forms* (5.3.93) *and* (5.3.94) *in the respective cases* $\alpha-l-1>\beta$ *and* $\alpha-l-1=\beta$, *where* $y_j(x)$ $(j=0,\cdots,l-1)$ *are given by* (5.3.58)-(5.3.60).

Our next results for nonhomogeneous equations (5.3.69), (5.3.73) and (5.3.77) follow from Theorems 5.15-5.17.

Proposition 5.10 *Let* $l-1<\alpha \leqq l$ $(l \in \mathbb{N})$, $\lambda \in \mathbb{R}$ *and let* $f(x)$ *be a given real function defined on* $\mathbb{R}_+$. *Then the Cauchy problem* (5.3.91) *for the equation* (5.3.69) *is solvable, and its solution is given by*

$$y(x)=\int_0^x (x-t)^{\alpha-1}E_{\alpha,\alpha}\left[\lambda(x-t)^{\alpha}\right]f(t)dt+\sum_{j=0}^{l-1} b_j x^j E_{\alpha,j+1}\left(\lambda x^{\alpha}\right). \qquad (5.3.95)$$

Proposition 5.11 *Let* $l-1<\alpha \leqq l$ $(l \in \mathbb{N})$ *and* $0<\beta<\alpha$ *be such that* $\alpha-l+1 \geqq \beta$, *let* $\lambda,\mu \in \mathbb{R}$, *and let* $f(x)$ *be a given real function defined on* $\mathbb{R}_+$. *Then the Cauchy problem* (5.3.91) *for the equation* (5.3.73) *is solvable, and its solution has the form*

$$y(x)=\int_0^x (x-t)^{\alpha-1}G_{\alpha,\beta;\lambda,\mu}(x-t)f(t)dt+\sum_{j=0}^{l-1} b_j y_j(x) \quad (\alpha-l-1>\beta), \quad (5.3.96)$$

or

$$y(x)=\int_0^x (x-t)^{\alpha-1}G_{\alpha,\beta;\lambda,\mu}(x-t)f(t)dt+\sum_{j=0}^{l-2} b_j y_j(x)+(b_{l-1}-\lambda b_0)y_{l-1}(x), \qquad (5.3.97)$$

with $\alpha-l-1=\beta$, *where* $G_{\alpha,\beta;\lambda,\mu}(z)$ *and* $y_j(x)$ $(j=0,\cdots,l-1)$ *are given by* (5.2.97) *and* (5.3.24)-(5.3.25), *respectively.*

Proposition 5.12 *Let* $m \in \mathbb{N}\backslash\{1,2\}$, $l-1<\alpha \leqq l$ $(l \in \mathbb{N})$, *and let* β *and* $\alpha_1,\cdots,\alpha_{m-2}$ *be such that* $\alpha>\beta>\alpha_{m-2}>\cdots>\alpha_1>\alpha_0=0$ *and* $\alpha-l+1 \geqq \beta$, *let* $\lambda, A_0,\cdots,A_{m-2} \in \mathbb{R}$, *and let* $f(x)$ *be a given real function defined on* $\mathbb{R}_+$. *Then the Cauchy problem* (5.3.91) *for the equation* (5.3.77) *is solvable, and its solution has the form*

$$y(x)=\int_0^x (x-t)^{\alpha-1}G_{\alpha_1,\cdots,\alpha_{m-2},\beta,\alpha;\lambda}(x-t)f(t)dt+\sum_{j=0}^{l-1} b_j y_j(x) \quad (\alpha-l-1>\beta), \qquad (5.3.98)$$

or

$$y(x)=\int_0^x (x-t)^{\alpha-1}G_{\alpha_1,\cdots,\alpha_{m-2},\beta,\alpha;\lambda}(x-t)f(t)dt$$

$$+\sum_{j=0}^{l-2} b_j y_j(x) + (b_{l-1} - \lambda b_0) y_{l-1}(x) \quad (\alpha - l - 1 = \beta), \tag{5.3.99}$$

where $G_{\alpha_1,\cdots,\alpha_{m-2},\beta,\alpha;\lambda}(z)$ *and* $y_j(x)$ $(j = 0, \cdots, l-1)$ *are given by* (5.2.120) *and* (5.3.58)-(5.3.60), *respectively.*

Proof. Propositions 5.10-5.12 follow from Theorems 5.15-5.17 on the basis of the respective formulas (5.3.70), (5.3.74) and (5.3.78) and the directly verified relations

$$\left(\frac{d}{dx}\right)^k \left[\int_0^t (t-\tau)^{\alpha-1} E_{\alpha,\alpha}\left[\lambda(t-\tau)^\alpha\right] f(\tau) d\tau\right](0) = 0 \quad (k = 0, \cdots, l-1),$$

$$\left(\frac{d}{dx}\right)^k \left[\int_0^t (t-\tau)^{\alpha-1} G_{\alpha,\beta;\lambda,\mu}(t-\tau) f(\tau) d\tau\right](0) = 0 \quad (k = 0, \cdots, l-1),$$

$$\left(\frac{d}{dx}\right)^k \left[\int_0^x (t-\tau)^{\alpha-1} G_{\alpha_1,\cdots,\alpha_{m-2},\beta,\alpha;\lambda}(t-\tau) f(\tau) d\tau\right](0) = 0 \quad (k = 0, \cdots, l-1),$$

if we also take the relations (5.3.12), (5.3.14) and (5.3.36) into account.

Example 5.24 The Cauchy problem

$$y'(x) - \lambda \left({}^C D_{0+}^\beta y\right)(x) - \mu y(x) = f(x), \quad y(0) = b \in \mathbb{R} \quad (x > 0; \; 0 < \beta < 1; \; \lambda, \mu \in \mathbb{R}) \tag{5.3.100}$$

has its solution given by

$$y(x) = \int_0^x G_{1,\beta;\lambda,\mu}(x-t) f(t) dt + b \left\{ \sum_{n=0}^\infty \frac{\mu^n}{n!} x^n \; {}_1\Psi_1 \left[\begin{matrix} (n+1, 1) \\ (n+1, 1-\beta) \end{matrix} \middle| \lambda x^{1-\beta} \right] - \right.$$

$$\left. -\lambda \sum_{n=0}^\infty \frac{\mu^n}{n!} x^{n+1-\beta} \; {}_1\Psi_1 \left[\begin{matrix} (n+1, 1) \\ (n+2-\beta, 1-\beta) \end{matrix} \middle| \lambda x^{1-\beta} \right] \right\}, \tag{5.3.101}$$

where $G_{1,\beta;\lambda,\mu}(z)$ is given by (5.2.105).

Example 5.25 The Cauchy problem

$$y''(x) - \lambda \left({}^C D_{0+}^\beta y\right)(x) - \mu y(x) = f(x) \quad (x > 0; \; 0 < \beta < 1; \; \lambda, \mu \in \mathbb{R}) \tag{5.3.102}$$

$$y(0) = b_0, \quad y'(0) = b_1 \quad (b_0, b_1 \in \mathbb{R}) \tag{5.3.103}$$

has its solution given by

$$y(x) = \int_0^x (x-t) G_{2,\beta;\lambda,\mu}(x-t) f(t) dt$$

$$+b_0 \left\{ \sum_{n=0}^\infty \frac{\mu^n}{n!} x^{2n} \; {}_1\Psi_1 \left[\begin{matrix} (n+1, 1) \\ (2n+1, 2-\beta) \end{matrix} \middle| \lambda x^{2-\beta} \right] \right.$$

$$-\lambda \sum_{n=0}^{\infty} \frac{\mu^n}{n!} x^{2n+2-\beta} \, {}_1\Psi_1 \left[\begin{array}{c} (n+1,1) \\ \\ (2n+3-\beta, 2-\beta) \end{array} \middle| \lambda x^{2-\beta} \right] \Bigg\}$$

$$+b_1 \sum_{n=0}^{\infty} \frac{\mu^n}{n!} x^{2n+1} \, {}_1\Psi_1 \left[\begin{array}{c} (n+1,1) \\ \\ (2n+2, 2-\beta) \end{array} \middle| \lambda x^{2-\beta} \right], \tag{5.3.104}$$

where $G_{2,\beta;\lambda,\mu}(z)$ is given by (5.2.111)

Remark 5.13 Gorenflo and Mainardi ([304], Section 4) used the Laplace transform to derive the explicit solutions to the Cauchy problems (5.3.100) and (5.3.102)-(5.3.103) with $\lambda < 0$ in terms of the inverse Laplace transform $\mathcal{L}^{-1}$ in (1.4.2), and discussed the solutions obtained by them. They also indicated that the explicit solutions to Cauchy problem (5.3.102)-(5.3.103) will be different in the cases $0 < \beta < 1$ and $1 < \beta < 2$ [see also Gorenflo et al. [310]].

Remark 5.14 Using Propositions 5.10 and 5.11, in Examples 5.24 and 5.25 we have presented the explicit solutions to two Cauchy problems (5.3.100) and (5.3.102)-(5.3.103). Using Propositions 5.7-5.9 and 5.10-5.12, we can also derive the explicit solutions to Cauchy problems for particular cases of the homogeneous equations (5.3.5), (5.3.22), (5.3.56) and of the nonhomogeneous equations (5.3.69), (5.3.73), (5.3.77), considered in Sections 5.3.1 and 5.3.2.

5.4 Mellin Transform Method for Solving Nonhomogeneous Fractional Differential Equations with Liouville Derivatives

In the following subsections, we present our investigations of the Mellin transform method involving the Liouville fractional derivatives.

5.4.1 General Approach to the Problems

In this section we apply the one-dimensional direct and inverse Mellin integral transforms (1.4.23) and (1.4.24) to derive particular solutions to such linear nonhomogeneous differential equations as

$$\sum_{k=0}^{m} A_k x^{\alpha+k} \left(D_{0+}^{\alpha+k} y\right)(x) = f(x) \quad (x > 0; \quad \alpha > 0) \tag{5.4.1}$$

and

$$\sum_{k=0}^{m} B_k x^{\alpha+k} \left(D_{-}^{\alpha+k} y\right)(x) = f(x) \quad (x > 0; \quad \alpha > 0), \tag{5.4.2}$$

with constants $A_k,\ B_k \in \mathbb{R}$ $(k = 0, \cdots, m)$ and the left- and right- sided Liouville fractional derivatives $\left(D_{0+}^{\alpha+k} y\right)(x)$ and $\left(D_{-}^{\alpha+k} y\right)(x)$ given by (2.2.3) and (2.2.4), respectively.

First we present a scheme for solving such equations by using the direct and inverse Mellin transforms $\mathcal{M}$ and $\mathcal{M}^{-1}$ given by (1.4.23) and (1.4.24), respectively. The Mellin transform method for solving the equations (5.4.1) and (5.4.2) is based on the following relations:

$$\left(\mathcal{M}x^{\alpha+k}D_{0+}^{\alpha+k}y\right)(s) = \frac{\Gamma(1-s)}{\Gamma(1-s-\alpha-k)}\left(\mathcal{M}y\right)(s), \tag{5.4.3}$$

$$\left(\mathcal{M}x^{\alpha+k}D_{-}^{\alpha+k}y\right)(s) = \frac{\Gamma(s+\alpha+k)}{\Gamma(s)}\left(\mathcal{M}y\right)(s). \tag{5.4.4}$$

These formulas, being valid for suitable functions $y(x)$, follow from (2.2.43) and (2.2.46) if we take the property (1.4.28) into account.

Applying the Mellin transform to (5.4.1) and (5.4.2) and using (5.4.3) and (5.4.4), we obtain

$$\left[\sum_{k=0}^{m} A_k \frac{\Gamma(1-s)}{\Gamma(1-s-\alpha-k)}\right]\left(\mathcal{M}y\right)(s) = \left(\mathcal{M}f\right)(s) \tag{5.4.5}$$

and

$$\left[\sum_{k=0}^{m} B_k \frac{\Gamma(s+\alpha+k)}{\Gamma(s)}\right]\left(\mathcal{M}y\right)(s) = \left(\mathcal{M}f\right)(s), \tag{5.4.6}$$

respectively. Using the inverse Mellin transform in (5.4.5) and (5.4.6), we derive the following solutions to the equations (5.4.1) and (5.4.2) in respective forms:

$$y(x) = \left(\mathcal{M}^{-1}\left[\frac{1}{P_\alpha^1(1-s)}\left(\mathcal{M}f\right)(s)\right]\right)(x), \quad P_\alpha^1(s) = \sum_{k=0}^{m} A_k \frac{\Gamma(s)}{\Gamma(s-\alpha-k)} \tag{5.4.7}$$

and

$$y(x) = \left(\mathcal{M}^{-1}\left[\frac{1}{P_\alpha^2(s)}\left(\mathcal{M}f\right)(s)\right]\right)(x), \quad P_\alpha^2(s) = \sum_{k=0}^{m} B_k \frac{\Gamma(s+\alpha+k)}{\Gamma(s)}. \tag{5.4.8}$$

By analogy with (5.2.81), we now introduce the *Mellin fractional analog* of the Green function:

$$G_\alpha^1(x) = \left(\mathcal{M}^{-1}\left[\frac{1}{P_\alpha^1(1-s)}\right]\right)(x), \quad P_\alpha^1(s) = \sum_{k=0}^{m} A_k \frac{\Gamma(s)}{\Gamma(s-\alpha-k)}, \tag{5.4.9}$$

$$G_\alpha^2(x) = \left(\mathcal{M}^{-1}\left[\frac{1}{P_\alpha^2(s)}\right]\right)(x), \quad P_\alpha^2(s) = \sum_{k=0}^{m} B_k \frac{\Gamma(s+\alpha+k)}{\Gamma(s)}. \tag{5.4.10}$$

Applying the property (1.4.41) for the Mellin convolution (1.4.39):

$$\left(\mathcal{M}\left(\int_0^\infty k\left(\frac{x}{t}\right) f(t)\frac{dt}{t}\right)\right)(s) = (\mathcal{M}k)(s)(\mathcal{M}f)(s), \tag{5.4.11}$$

we present the solution (5.4.7) in the form

$$y(x) = \int_0^\infty G_\alpha^1(t) f(xt)dt, \tag{5.4.12}$$

and the solution (5.4.8) in the form of the Mellin convolution of $G_\alpha^2(x)$ and $f(x)$

$$y(x) = \int_0^\infty G_\alpha^2\left(\frac{x}{t}\right) f(t)\frac{dt}{t}. \tag{5.4.13}$$

5.4.2 Equations with Left-Sided Fractional Derivatives

We consider the equation (5.4.1) with $m = 1$:

$$x^{\alpha+1}\left(D_{0+}^{\alpha+1}y\right)(x) + \lambda x^\alpha \left(D_{0+}^\alpha y\right)(x) = f(x) \quad (x > 0;\ \ \alpha > 0;\ \ \lambda \in \mathbb{R}). \tag{5.4.14}$$

Particular solutions to this equation are different in the cases when $\lambda \neq n+1$ and $\lambda = n+1$ with $n \in \mathbb{N}_0$. In the first case a particular solution to (5.4.14) is expressed in terms of the generalized Wright function (1.11.14) with $p = 1$ and $q = 2$ of the form

$${}_1\Psi_2\left[\begin{array}{c|c} (a,1) & \\ & z \\ (b,-1),(c,1) & \end{array}\right] = \frac{\gamma(a)}{\gamma(b)\gamma(c)}\,{}_2\mathcal{F}_1(a, 1-b; c; -z) \quad (a,b,c,z \in \mathbb{C}). \tag{5.4.15}$$

The following assertion can be derived from Theorem 1.5.

Lemma 5.5 *The generalized Wright function* (5.4.15) *is defined for complex* $z \in \mathbb{C}$ *such that* $|z| < 1$ *and when* $|z| = 1$, *provided that* $\Re(b + c - a) > 1$.

Theorem 5.18 *If* $\alpha > 0$ *and* $\lambda \in \mathbb{R}$ $(\lambda \neq n+1;\ n \in \mathbb{N}_0)$, *then the fractional differential equation* (5.4.14) *is solvable, and its particular solution is given by*

$$y(x) = \int_0^1 G_{\alpha,\lambda}^1(t) f(xt)dt, \tag{5.4.16}$$

$$G_{\alpha,\lambda}^1(x) = x^{-\alpha}\left\{\frac{\Gamma(1-\lambda)}{\Gamma(\alpha+1-\lambda)}x^{\lambda-1} - {}_1\Psi_2\left[\begin{array}{c|c} (1-\lambda,1) & \\ & -x \\ (\alpha,-1),(2-\lambda,1) & \end{array}\right]\right\}. \tag{5.4.17}$$

Proof. (5.4.14) is the same as the equation (5.4.1) with $m = 1$, $A_1 = 1$ and $A_0 = \lambda$. And (5.4.5) takes the form

$$\frac{(\lambda - s - \alpha)\Gamma(1-s)}{\Gamma(1-s-\alpha)}(\mathcal{M}y)(s) = (\mathcal{M}f)(s). \tag{5.4.18}$$

Hence the Mellin fractional analog of the Green function in (5.4.9) is given by

$$G_\alpha^1(x) = \left(\mathcal{M}^{-1}\left[\frac{\Gamma(s-\alpha)}{(s-\alpha-1+\lambda)\Gamma(s)}\right]\right)(x) =: G_{\alpha,\lambda}^1(x), \tag{5.4.19}$$

and, in accordance with (5.4.12), a particular solution to (5.4.14) has the form

$$y(x) = \int_0^\infty G^1_{\alpha,\lambda}(t) f(xt)dt, \tag{5.4.20}$$

First we show that $G^1_{\alpha,\lambda}(x) = 0$ for $x > 1$. By (5.4.19), we have

$$\left(\mathcal{M}G^1_{\alpha,\lambda}\right)(s) = \left(\mathcal{M}G_1\right)(s)\left(\mathcal{M}G_2\right)(s), \tag{5.4.21}$$

where

$$\left(\mathcal{M}G_1\right)(s) = \frac{\Gamma(s-\alpha)}{\Gamma(s)}, \quad \left(\mathcal{M}G_2\right)(s) = \frac{1}{s-\alpha-1+\lambda}. \tag{5.4.22}$$

The direct application of the Mellin transform leads to the following relations:

$$G_1(x) = \begin{cases} \frac{x^{-\alpha}(1-x)^{\alpha-1}}{\Gamma(\alpha)}, & 0 < x < 1, \\ 0, & x > 1; \end{cases} \qquad G_2(x) = \begin{cases} x^{-\alpha-1+\lambda}, & 0 < x < 1, \\ 0, & x > 1. \end{cases} \tag{5.4.23}$$

Thus, according to (5.4.11), we have

$$G^1_{\alpha,\lambda}(x) = \int_0^\infty G_1\left(\frac{x}{t}\right) G_2(t)\frac{dt}{t}, \tag{5.4.24}$$

and it follows from (5.4.23) that $G^1_{\alpha,\lambda}(x) = 0$ for $x > 1$. Therefore, (5.4.20) yields (5.4.16).

Now we show that $G^1_{\alpha,\lambda}(x)$ is given by (5.4.17). By (5.4.19) and (1.4.24), we have

$$G^1_{\alpha,\lambda}(x) = \frac{1}{2\pi i}\int_{\gamma-i\infty}^{\gamma+i\infty} \frac{\Gamma(s-\alpha)}{(s-\alpha-1+\lambda)\Gamma(s)} x^{-s}ds. \tag{5.4.25}$$

Since $\lambda \neq n+1$ ($n \in \mathbb{N}_0$), then the pole $s = \alpha+1-\lambda$ of the integrand in (5.4.25) does not coincide with any pole $s_k = \alpha - k$ ($k \in \mathbb{N}_0$) of $\Gamma(s-\alpha)$. If we choose $\gamma > \max[\alpha+1-\Re(\lambda), \alpha]$, then evaluation of the residues at the above poles yields

$$G^1_{\alpha,\lambda}(x) = \mathrm{Res}_{s=\alpha+1-\lambda}\left[\frac{\Gamma(s-\alpha)}{\Gamma(s)(s-\alpha-1+\lambda)}x^{-s}\right]$$

$$+\sum_{k=0}^\infty \mathrm{Res}_{s=\alpha-k}\left[\frac{\Gamma(s-\alpha)}{(s-\alpha-1+\lambda)\Gamma(s)}x^{-s}\right]$$

$$= \frac{\Gamma(1-\lambda)}{\Gamma(\alpha+1-\lambda)}x^{-\alpha+\lambda-1} + \sum_{k=0}^\infty \frac{(-1)^k}{k!}\frac{x^{-\alpha+k}}{(\lambda-k-1)\Gamma(\alpha-k)}$$

$$= \frac{\Gamma(1-\lambda)}{\Gamma(\alpha+1-\lambda)}x^{-\alpha+\lambda-1} - \sum_{k=0}^\infty \frac{(-1)^k}{k!}\frac{\Gamma(k+1-\lambda)}{\Gamma(k+2-\lambda)\Gamma(\alpha-k)}x^{-\alpha+k}.$$

Hence, by (5.4.15) and (1.11.14), $G^1_{\alpha,\lambda}(x)$ is given by (5.4.17), and Theorem 5.18 is completely proved.

Example 5.26 Equation (5.4.14) with $\alpha = 1/2$ and $\lambda \neq n+1$ $(n \in \mathbb{N}_0)$:

$$x^{3/2}\left(D_{0+}^{3/2}y\right)(x) + \lambda x^{1/2}\left(D_{0+}^{1/2}y\right)(x) = f(x) \quad (x > 0; \quad \alpha > 0; \quad \lambda \in \mathbb{R}) \tag{5.4.26}$$

has a particular solution given by

$$y(x) = \int_0^1 G^1_{1/2,\lambda}(t) f(xt)dt, \tag{5.4.27}$$

where $G^1_{1/2,\lambda}(x)$ is given by (5.4.17) with $\alpha = 1/2$, while

$$y(x) = \frac{1}{1-\lambda}\int_0^1 \left(t^{\lambda-2} - \frac{1}{t}\right) f(xt)dt \tag{5.4.28}$$

is a particular solution to the equation

$$x^2 y''(x) + \lambda x y'(x) = f(x) \quad (x > 0; \quad \lambda \neq 1; \; n \in \mathbb{N}_0). \tag{5.4.29}$$

Note that, for $\lambda \neq n+1$ $(n \in \mathbb{N}_0)$, (5.4.16) with $\alpha = 1$ yields the solution (5.4.28), but it is directly verified that (5.4.28) satisfies the equation (5.4.29) with any $\lambda \neq 1$.

Now we give a particular solution to the equation (5.4.14) with $\lambda = n+1$ $(n \in \mathbb{N}_0)$:

$$x^{\alpha+1}\left(D_{0+}^{\alpha+1}y\right)(x) + (n+1)x^{\alpha}\left(D_{0+}^{\alpha}y\right)(x) = f(x) \quad (x > 0; \quad \alpha > 0; \quad n \in \mathbb{N}_0), \tag{5.4.30}$$

in terms of the psi function $\psi(z)$ defined in (1.5.17) and of the Euler constant γ [see Abramowitz and Stegun ([1], formula 6.3.2)]:

$$\gamma = -\psi(1) = -\Gamma'(1). \tag{5.4.31}$$

The results will be different in the cases $n \in \mathbb{N}$ and $n = 0$.

Theorem 5.19 (a) *If $\alpha > 0$ and $n \in \mathbb{N}$ are such that $\alpha \neq 1, \cdots, n$, then the fractional differential equation* (5.4.30) *is solvable, and its particular solution has the form*

$$y(x) = \int_0^1 G^1_{\alpha,n+1}(t) f(xt)dt, \tag{5.4.32}$$

where

$$G^1_{\alpha,n+1}(x) = \frac{(-1)^{n-1}x^{n-\alpha}}{n!\Gamma(\alpha-n)}\left[\log(x) + \sum_{j=0}^{n-1}\frac{1}{j-n} + \psi(\alpha-n) + \gamma\right] + \sum_{k=0\ (k\neq n)}^{\infty}\frac{(-1)^k x^{k-\alpha}}{k!(n-k)\Gamma(\alpha-k)}. \tag{5.4.33}$$

(b) *If $n = 0$, then the fractional differential equation*

$$x^{\alpha+1}\left(D_{0+}^{\alpha+1}y\right)(x)+x^{\alpha}\left(D_{0+}^{\alpha}y\right)(x)=f(x)\quad (x>0;\quad \alpha>0) \tag{5.4.34}$$

is solvable, and its particular solution is given by

$$y(x)=\int_0^1 G_{\alpha,1}^1(t)f(xt)dt, \tag{5.4.35}$$

$$G_{\alpha,1}^1(x)=-\frac{x^{-\alpha}}{\Gamma(\alpha)}\left[\log(x)+\psi(\alpha)+\gamma\right]-\sum_{k=1}^{\infty}\frac{(-1)^k x^{k-\alpha}}{k!k\Gamma(\alpha-k)}. \tag{5.4.36}$$

Proof. By the proof of Theorem 5.18, the solution to the equation (5.4.30) has the form (5.4.16) with $\lambda=n+1$, which yields (5.4.32) and (5.4.35). To prove the explicit representations for $G_{\alpha,n+1}^1(x)$, we use the relation (5.4.25) with $\lambda=n+1$:

$$G_{\alpha,n+1}^1(x)=\frac{1}{2\pi i}\int_{\gamma-i\infty}^{\gamma+i\infty}\frac{\Gamma(s-\alpha)}{(s-\alpha+n)\Gamma(s)}x^{-s}ds. \tag{5.4.37}$$

The integrand in (5.4.37) has a pole of the second order at $s_n=\alpha-n$ and simple poles at $s_k=\alpha-k$ for $k\in\mathbb{N}_0$, $(k\neq n)$. If we choose $\gamma>\alpha$, then evaluation of the residues at the above poles, using the relation (1.5.4), yields

$$G_{\alpha,n+1}^1(x)=\operatorname{Res}_{s=\alpha-n}\left[\frac{\Gamma(s-\alpha)}{\Gamma(s)(s-\alpha+n)}x^{-s}\right]$$

$$+\sum_{k=0,\ (k\neq n)}^{\infty}\operatorname{Res}_{s=\alpha-k}\left[\frac{\Gamma(s-\alpha)}{(s-\alpha+n)\Gamma(s)}x^{-s}\right]$$

$$=\lim_{s\to\alpha-n}\left[\frac{\Gamma(s-\alpha+n+1)x^{-s}}{(s-\alpha)\cdots(s-\alpha+n-1)\Gamma(s)}\right]'+\sum_{k=0,\ (k\neq n)}^{\infty}\frac{(-1)^k x^{k-\alpha}}{k!(n-k)\Gamma(\alpha-k)}$$

$$=x^{n-\alpha}\frac{\Gamma'(1)-\Gamma(1)\log(x)-\Gamma(1)\left[\sum_{j=0}^{n-1}\frac{1}{j-n}+\frac{\Gamma'(\alpha-n)}{\Gamma(\alpha-n)}\right]}{(-n)\cdots(-1)\Gamma(\alpha-n)}$$

$$+\sum_{k=0,\ (k\neq n)}^{\infty}\frac{(-1)^k x^{k-\alpha}}{k!(n-k)\Gamma(\alpha-k)}.$$

From here, in accordance with (1.5.6), (1.5.17) and (5.4.31), $G_{\alpha,n+1}^1(x)$ is given by (5.4.33), which completes the proof of the assertion (a) of Theorem 5.19. The assertion (b) of Theorem 5.19 is proved similarly.

Example 5.27 Equations (5.4.30) and (5.4.34) with $\alpha=1/2$:

$$x^{3/2}\left(D_{0+}^{3/2}y\right)(x)+(n+1)x^{1/2}\left(D_{0+}^{1/2}y\right)(x)=f(x)\quad (x>0;\quad \alpha>0;\quad n\in\mathbb{N}) \tag{5.4.38}$$

and

$$x^{3/2}\left(D_{0+}^{3/2}y\right)(x)+x^{1/2}\left(D_{0+}^{1/2}y\right)(x)=f(x)\quad(x>0) \tag{5.4.39}$$

have their respective particular solutions given by

$$y(x)=\int_0^1 G_{1/2,n+1}^1(t)f(xt)dt \tag{5.4.40}$$

and

$$y(x)=\int_0^1 G_{1/2,1}^1(t)f(xt)dt, \tag{5.4.41}$$

where $G_{1/2,n+1}^1(x)$ and $G_{1/2,1}^1(x)$ are given by (5.4.33) and (5.4.36) with $\alpha=1/2$.

Example 5.28 The following ordinary differential equation of order $n+m+1$:

$$x^{n+m+1}y^{(n+m+1)}(x)+(n+1)x^{n+m}y^{(n+m)}(x)=f(x)\quad(x>0;\quad \alpha>0;\quad n,m\in\mathbb{N}) \tag{5.4.42}$$

has a particular solution given by

$$y(x)=\int_0^1 G_{n+m,n+1}^1(t)f(xt)dt, \tag{5.4.43}$$

where

$$G_{n+m,n+1}^1(x)=\frac{(-1)^{n-1}x^{-m}}{n!(m-1)!}\left[\log(x)+\sum_{j=0}^{n-1}\frac{1}{j-n}+\psi(m)+\gamma\right]$$
$$+\sum_{k=0,\ (k\neq n)}^{n+m-1}\frac{(-1)^k x^{k-n-m}}{k!(n-k)(n+m-k-1)!}. \tag{5.4.44}$$

Example 5.29 The following ordinary differential equation of order $m+1$:

$$x^{m+1}y^{(m+1)}(x)+x^m y^{(m)}(x)=f(x)\quad(x>0;\quad m\in\mathbb{N}) \tag{5.4.45}$$

has a particular solution of the form

$$y(x)=\int_0^1 G_{m,1}^1(t)f(xt)dt, \tag{5.4.46}$$

where

$$G_{m,1}^1(x)=-\frac{x^{-m}}{(m-1)!}\left[\log(x)+\psi(m)+\gamma\right]-\sum_{k=1}^{m-1}\frac{(-1)^k x^{k-m}}{k!k(m-k-1)!}. \tag{5.4.47}$$

In particular,

$$y(x)=-\int_0^1\frac{\log(t)}{t}f(xt)dt \tag{5.4.48}$$

is a particular solution to the equation:

$$x^2y''(x)+xy'(x)=f(x)\quad(x>0). \tag{5.4.49}$$

Remark 5.15 In Theorem 5.19 we obtained a particular solution to the equation (5.4.30) for all $\alpha > 0$ and $n \in \mathbb{N}_0$, except for the case when $n \in \mathbb{N}$ and $\alpha \neq 1, \cdots, n$. This is due to the fact that the function $G^1_{\alpha,n+1}(x)$, given by (5.4.33), is not defined for such $\alpha = k$ ($k = 1, \cdots, n$). This means that the Laplace transform method can not be used to obtain a particular solution to the following ordinary differential equation of order k ($k = 1, \cdots n$):

$$x^{k+1}y^{(k+1)}(x)+(n+1)x^k y^{(k)}(x) = f(x) \quad (x > 0; \;\; n \in \mathbb{N}; \;\; k = 1, \cdots, n). \tag{5.4.50}$$

Remark 5.16 The solution to the equation (5.4.34) in the form (5.4.35) was obtained by Podlubny ([682], Example 6.1), but his formula (6.10) for $G^1_{\alpha,1}(x)$ is in error. In fact, $\log(x) + \psi(\alpha)$ should be replaced by $-\log(x) - \psi(\alpha)$, as it is given in the right-hand side of (5.4.36).

5.4.3 Equations with Right-Sided Fractional Derivatives

We consider the equation (5.4.2) with $m = 1$:

$$x^{\alpha+1}\left(D^{\alpha+1}_{-}y\right)(x) + \lambda x^{\alpha}\left(D^{\alpha}_{-}y\right)(x) = f(x) \quad (x > 0; \;\; \alpha > 0; \;\; \lambda \in \mathbb{R}). \tag{5.4.51}$$

As in the case of the equation (5.4.14) with the left-sided Liouville derivatives, its particular solutions will be different for $\alpha + \lambda \neq n$ and $\alpha + \lambda = n$ ($n \in \mathbb{N}_0$). In the first case, a particular solution is expressed in terms of the generalized Wright function (5.4.15).

Theorem 5.20 *If $\alpha > 0$ and $\lambda \in \mathbb{C}$ are such that $\alpha + \lambda \neq n$ ($n \in \mathbb{N}_0$), then the fractional differential equation* (5.4.51) *is solvable, and its particular solution has the form*

$$y(x) = \int_x^\infty G^2_{\alpha,\lambda}\left(\frac{x}{t}\right) f(t)\frac{dt}{t}, \tag{5.4.52}$$

where

$$G^2_{\alpha,\lambda}(x) = \frac{\Gamma(-\lambda-\alpha)}{\Gamma(-\lambda)}x^{\alpha+\lambda} - {}_1\Psi_2\left[\begin{array}{c|c} (-\alpha-\lambda,1) & \\ & -x \\ (\alpha,-1),(1-\alpha-\lambda,1) & \end{array}\right]. \tag{5.4.53}$$

Proof. (5.4.51) is the same as the equation (5.4.2) with $m = 1$, $A_1 = 1$, $A_0 = \lambda$. The relation (5.4.6) takes the form

$$\frac{(s+\alpha+\lambda)\Gamma(s+\alpha)}{\Gamma(s)}(\mathcal{M}y)(s) = (\mathcal{M}f)(s). \tag{5.4.54}$$

Hence the Mellin fractional analog of the Green function in (5.4.10) is given by

$$G^2_\alpha(x) = \left(\mathcal{M}^{-1}\left[\frac{\Gamma(s)}{(s+\alpha+\lambda)\Gamma(s+\alpha)}\right]\right)(x) =: G^2_{\alpha,\lambda}(x), \tag{5.4.55}$$

and, in accordance with (5.4.13), a particular solution to the equation (5.4.51) is given by

$$y(x) = \int_0^\infty G^2_\alpha\left(\frac{x}{t}\right) f(t)\frac{dt}{t}. \tag{5.4.56}$$

We show that $G^2_{\alpha,\lambda}(x) = 0$ for $x < 1$. By (5.4.55),

$$\left(\mathcal{M}G^2_{\alpha,\lambda}\right)(s) = (\mathcal{M}G_3)(s)\,(\mathcal{M}G_4)(s), \tag{5.4.57}$$

where

$$(\mathcal{M}G_s)(s) = \frac{\Gamma(s)}{\Gamma(s+\alpha)}, \quad (\mathcal{M}G_4)(s) = \frac{1}{s+\alpha+\lambda}. \tag{5.4.58}$$

The direct application of the Mellin transform leads to the following relations:

$$G_3(x) = \begin{cases} \frac{x^{\alpha-1}}{\Gamma(\alpha)}, & 0 < x < 1, \\ 0, & x > 1; \end{cases} \qquad G_4(x) = \begin{cases} x^{\alpha+\lambda}, & 0 < x < 1, \\ 0, & x > 1. \end{cases} \tag{5.4.59}$$

Thus, according to (5.4.11), we have

$$G^2_{\alpha,\lambda}(x) = \int_0^\infty G_3\left(\frac{x}{t}\right) G_4(t)\frac{dt}{t}, \tag{5.4.60}$$

and it follows from (5.4.59) that $G^2_{\alpha,\lambda}(x) = 0$ for $x > 1$. Therefore, (5.4.56) yields (5.4.52).

Now we show that $G^2_{\alpha,\lambda}(x)$ is given by (5.4.53). By (5.4.55) and (1.4.24), we get

$$G^2_\alpha(x) = \frac{1}{2\pi i}\int_{\gamma-i\infty}^{\gamma+i\infty} \frac{\Gamma(s)}{(s+\alpha+\lambda)\Gamma(s+\alpha)} x^{-s}ds =: G^2_{\alpha,\lambda}(x). \tag{5.4.61}$$

Since $\alpha + \lambda \neq n$ ($n \in \mathbb{N}_0$), then the pole $s = -\lambda - \alpha$ of the integrand of (5.4.61) does not coincide with any pole $s_k = -k$ ($k \in \mathbb{N}_0$) of $\Gamma(s)$. If we choose $\gamma > \max[-\alpha - \Re(\lambda), 0]$, then evaluation of the residues at the above poles yields

$$G^2_{\alpha,\lambda}(x) = \text{Res}_{s=-\lambda-\alpha}\left[\frac{\Gamma(s)x^{-s}}{(s+\alpha+\lambda)\Gamma(s+\alpha)}\right] + \sum_{k=0}^{\infty}\text{Res}_{s=-k}\left[\frac{\Gamma(s)x^{-s}}{(s+\alpha+\lambda)\Gamma(s+\alpha)}\right]$$

$$= \frac{\Gamma(-\alpha-\lambda)}{\Gamma(-\lambda)}x^{\alpha+\lambda} - \sum_{k=0}^{\infty}\frac{(-1)^k}{k!}\frac{\Gamma(-\lambda-\alpha+k)}{\Gamma(1-\lambda-\alpha+k)\Gamma(\alpha-k)}x^k.$$

Hence, by (5.4.15) and (1.11.14), $G^2_{\alpha,\lambda}(x)$ is given by (5.4.53), and thus Theorem 5.20 is completely proved.

Example 5.30 The equation (5.4.51) with $\alpha = 1/2$ and $\lambda \neq n - 1/2$ ($n \in \mathbb{N}_0$):

$$x^{3/2}\left(D_-^{3/2}y\right)(x) + \lambda x^{1/2}\left(D_-^{1/2}y\right)(x) = f(x) \quad (x > 0;\; \alpha > 0;\; \lambda \in \mathbb{R}) \tag{5.4.62}$$

has a particular solution of the form

$$y(x) = \int_x^\infty G^2_{1/2,\lambda}\left(\frac{x}{t}\right) f(t)\frac{dt}{t}, \tag{5.4.63}$$

where $G^2_{1/2,\lambda}(x)$ is given by (5.4.53) with $\alpha = 1/2$, while

$$y(x) = \frac{1}{1+\lambda}\int_x^\infty \left[1 - \left(\frac{x}{t}\right)^{\lambda+1}\right] f(t)dt \tag{5.4.64}$$

is a particular solution to the equation

$$x^2 y''(x) - \lambda x y'(x) = f(x) \quad (x > 0;\ \ \lambda \neq -1). \tag{5.4.65}$$

Note that, for $\lambda \neq n-1$ ($n \in \mathbb{N}_0$), (5.4.53) with $\alpha = 1$ yields the solution (5.4.64), but it is directly verified that (5.4.64) satisfies the equation (5.4.65) with any $\lambda \neq -1$.

Now we give a particular solution to the equation (5.4.51) with $\lambda = n - \alpha$ ($n \in \mathbb{N}_0$):

$$x^{\alpha+1}\left(D_-^{\alpha+1}y\right)(x) + (n-\alpha)x^\alpha\left(D_-^\alpha y\right)(x) = f(x) \quad (x > 0;\ \ \alpha > 0;\ \ n \in \mathbb{N}_0). \tag{5.4.66}$$

As in the case of the equation (5.4.30) with the Liouville left-sided derivatives, the solution will be expressed in terms of the psi function $\psi(z)$ and the Euler constant γ, and the results will be different in the cases $n \in \mathbb{N}$ and $n = 0$.

Theorem 5.21 (a) *If $\alpha > 0$ and $n \in \mathbb{N}$ are such that $\alpha \neq 1, \cdots, n$, then the fractional differential equation* (5.4.66) *is solvable, and its particular solution has the form*

$$y(x) = \int_x^\infty G^2_{\alpha,n-\alpha}\left(\frac{x}{t}\right) f(t)\frac{dt}{t}, \tag{5.4.67}$$

where

$$G^2_{\alpha,n-\alpha}(x) = \frac{(-1)^{n-1}x^n}{n!\Gamma(\alpha-n)}\left[\log(x) + \sum_{j=0}^{n-1}\frac{1}{j-n} + \psi(\alpha-n) + \gamma\right]$$

$$+ \sum_{k=0,\ k\neq n}^{\infty} \frac{(-1)^k x^k}{k!(n-k)\Gamma(\alpha-k)}. \tag{5.4.68}$$

(b) *If $n = 0$, then the fractional differential equation*

$$x^{\alpha+1}\left(D_-^{\alpha+1}y\right)(x) - \alpha x^\alpha\left(D_-^\alpha y\right)(x) = f(x) \quad (x > 0;\ \ \alpha > 0) \tag{5.4.69}$$

is solvable, and its particular solution is given by

$$y(x) = \int_x^\infty G^2_{\alpha,-\alpha}\left(\frac{x}{t}\right) f(t)\frac{dt}{t}, \tag{5.4.70}$$

where

$$G^2_{\alpha,-\alpha}(x) = -\frac{1}{\Gamma(\alpha)}\left[\log(x) + \psi(\alpha) + \gamma\right] - \sum_{k=1}^{\infty}\frac{(-1)^k x^k}{k!k\Gamma(\alpha-k)}. \tag{5.4.71}$$

Proof. By the proof of Theorem 5.20, the solution to the equation (5.4.66) has the form (5.4.52) with $\lambda = n - \alpha$, which yields (5.4.67) and (5.4.70). To prove the explicit representations for $G^2_{\alpha,n-\alpha}(x)$, we use the relation (5.4.61) with $\lambda = n-\alpha$:

$$G^1_{\alpha,n-\alpha}(x) = \frac{1}{2\pi i}\int_{\gamma-i\infty}^{\gamma+i\infty} \frac{\Gamma(s)}{(s+n)\Gamma(s+\alpha)} x^{-s}ds. \tag{5.4.72}$$

The integrand in (5.4.72) has a pole of the second order at $s_n = -n$ and simple poles at $s_k = -k$ for $k \in \mathbb{N}_0$ $(k \neq n)$. If we choose $\gamma > 0$, then evaluation of the residues at the above poles, using the relation (1.5.4), yields

$$G^2_{\alpha,n-\alpha}(x) = \mathrm{Res}_{s=-n}\left[\frac{\Gamma(s)}{(s+n)\Gamma(s-\alpha)}x^{-s}\right]$$

$$+\sum_{k=0,\ (k\neq n)}^{\infty} \mathrm{Res}_{s=-k}\left[\frac{\Gamma(s)}{(s+n)\Gamma(s-\alpha)}x^{-s}\right]$$

$$= \lim_{s\to -n}\left[\frac{\Gamma(s+n+1)x^{-s}}{s\cdots(s+n-1)\Gamma(s-\alpha)}\right]' + \sum_{k=0,\ (k\neq n)}^{\infty} \frac{(-1)^k x^k}{k!(n-k)\Gamma(\alpha-k)}$$

$$= x^n \frac{\Gamma'(1) - \Gamma(1)\log(x) - \Gamma(1)\left[\sum_{j=0}^{n-1}\frac{1}{j-n} + \frac{\Gamma'(\alpha-n)}{\Gamma(\alpha-n)}\right]}{(-n)\cdots(-1)\Gamma(\alpha-n)}$$

$$+ \sum_{k=0,\ (k\neq n)}^{\infty} \frac{(-1)^k x^{k-\alpha}}{k!(n-k)\Gamma(\alpha-k)}.$$

From here, in accordance with (1.5.6), (1.5.17) and (5.4.31), $G_{\alpha,n-\alpha}(x)$ is given by (5.4.68), which completes the proof of the assertion (a) of Theorem 5.21. The assertion (b) of Theorem 5.21 is proved similarly.

Example 5.31 Equations (5.4.66) and (5.4.69) with $\alpha = 1/2$:

$$x^{3/2}\left(D_-^{3/2}y\right)(x) + \left(n-\frac{1}{2}\right)x^{1/2}\left(D_-^{1/2}y\right)(x) = f(x) \quad (x>0;\ \ \alpha>0;\ \ n\in\mathbb{N}) \tag{5.4.73}$$

and

$$x^{3/2}\left(D_-^{3/2}y\right)(x) - \frac{1}{2}x^{1/2}\left(D_-^{1/2}y\right)(x) = f(x) \quad (x>0) \tag{5.4.74}$$

have their respective particular solutions in the forms

$$y(x) = \int_x^\infty G^2_{1/2,n-1/2}\left(\frac{x}{t}\right) f(t)\frac{dt}{t} \tag{5.4.75}$$

and

$$y(x) = \int_x^\infty G^2_{1/2,-1/2}\left(\frac{x}{t}\right) f(t)\frac{dt}{t}, \tag{5.4.76}$$

where $G^2_{1/2,n-1/2}(x)$ and $G^2_{1/2,-1/2}(x)$ are given by (5.4.68) and (5.4.71) with $\alpha = 1/2$.

Example 5.32 The following ordinary differential equation of order $n+m+1$:

$$x^{n+m+1}y^{(n+m+1)}(x) + mx^{n+m}y^{(n+m)}(x) = f(x) \quad (x > 0; \quad n, m \in \mathbb{N}) \tag{5.4.77}$$

has a particular solution given by

$$y(x) = (-1)^{n+m+1} \int_x^\infty G^2_{n+m,-m}\left(\frac{x}{t}\right) f(t) \frac{dt}{t}, \tag{5.4.78}$$

where

$$G^2_{n+m,-m}(x) = \frac{(-1)^{n-1}x^n}{n!(m-1)!} \left[\log(x) + \sum_{j=0}^{n-1} \frac{1}{j-n} + \psi(m) + \gamma\right]$$

$$+ \sum_{k=0,\ (k\neq n)}^{n+m-1} \frac{(-1)^k x^k}{k!(n-k)(n+m-k-1)!}. \tag{5.4.79}$$

Example 5.33 The following ordinary differential equation of order $m+1$:

$$x^{m+1}y^{(m+1)}(x) + mx^m y^{(m)}(x) = f(x) \quad (x > 0; \quad m \in \mathbb{N}) \tag{5.4.80}$$

has a particular solution given by

$$y(x) = (-1)^{m+1} \int_x^\infty G^2_{m,-m}\left(\frac{x}{t}\right) f(t) \frac{dt}{t}, \tag{5.4.81}$$

$$G^2_{m,-m}(x) = -\frac{1}{(m-1)!}[\log(x) + \psi(m) + \gamma] - \sum_{k=1}^{m-1} \frac{(-1)^k x^k}{k!k(m-k-1)!}. \tag{5.4.82}$$

Specifically,

$$y(x) = -\int_x^\infty \log\left(\frac{x}{t}\right) f(t) \frac{dt}{t} \tag{5.4.83}$$

is a particular solution to the equation (5.4.49). It is clear that this solution coincides with the solution given by (5.4.48).

Remark 5.17 In Theorem 5.21 we obtained a particular solution to the equation (5.4.66) for all $\alpha > 0$ and $n \in \mathbb{N}_0$, except for the case when $n \in \mathbb{N}$ and $\alpha \neq 1, \cdots, n$. This is due to the fact that the function $G^2_{\alpha,n-\alpha}(x)$, given by (5.4.68), is not defined for such $\alpha = k$ $(k = 1, \cdots, n)$. This means that the Laplace transform method can not be used to obtain a particular solution to the following ordinary differential equation of order k $(k = 1, \cdots, n)$:

$$x^{k+1}y^{(k+1)}(x) + (n-k)x^k y^{(k)}(x) = f(x) \quad (x > 0; \ n \in \mathbb{N}; \ k = 1, \cdots, n). \tag{5.4.84}$$

Remark 5.18 Results presented in Sections 5.4.1-5.4.3 were proved by Kilbas and Trujillo [408]. In Sections 5.4.2 and 5.4.3 we have applied a general approach, developed in Section 5.4.1, to establish particular solutions to the equations (5.4.1) and (5.4.2) with $m = 1$. Such an approach can be also applied to the equations (5.4.1) and (5.4.2) with $m \in \mathbb{N}\backslash\{1\}$.

5.5 Fourier Transform Method for Solving Nonhomogeneous Differential Equations with Riesz Fractional Derivatives

Our investigation of the Fourier transform method involving the Riesz fractional derivatives will now be presented in the following subsections.

5.5.1 Multi-Dimensional Equations

In this section we apply the multi-dimensional Fourier transforms to derive particular solutions to linear nonhomogeneous differential equations of the form (5.1.1):

$$\sum_{k=1}^{m} A_k \left(\mathbf{D}^{\alpha_k} y\right)(x) + A_0 y(x) = f(x) \quad (x \in \mathbb{R}^n; \quad n, m \in \mathbb{N}; \quad 0 < \alpha_1 < \cdots < \alpha_m), \tag{5.5.1}$$

involving the Riesz fractional derivatives $\mathbf{D}^{\alpha_k} y$ $(k = 1, \cdots, m)$, given by (2.10.23), and constants $A_k \in \mathbb{R}$ $(k = 0, \cdots, m)$

First we present a scheme for solving such an equation by using the direct and inverse Fourier transforms $\mathcal{F}$ and $\mathcal{F}^{-1}$ given by (1.3.22) and (1.3.23), respectively. The Fourier transform method for solving the equation (5.5.1) is based on the relation (2.10.27):

$$\left(\mathcal{F}\mathbf{D}^{\alpha} y\right)(x) = |x|^{\alpha} \left(\mathcal{F}y\right)(x) \quad (x \in \mathbb{R}^n; \quad \alpha > 0). \tag{5.5.2}$$

Applying the Fourier transform $\mathcal{F}$ to both sides of (5.5.1) and using (5.5.2), we have

$$(\mathcal{F}y)(x) = \frac{1}{\left[\sum_{k=1}^{m} A_k |x|^{\alpha_k} + A_0\right]} (\mathcal{F}f)(x). \tag{5.5.3}$$

Applying the inverse Fourier transform $\mathcal{F}^{-1}$ in (5.5.3), we obtain a particular solution to the equation (5.5.1) in the form

$$y(x) = \left(\mathcal{F}^{-1}\left[\frac{1}{\left[\sum_{k=1}^{m} A_k |t|^{\alpha_k} + A_0\right]} (\mathcal{F}f)(t)\right]\right)(x). \tag{5.5.4}$$

By analogy with (5.2.81) and (5.4.9)-(5.4.10), we now introduce the *Fourier fractional analog* of the Green function:

$$G^F_{\alpha_1, \cdots, \alpha_m}(x) = \left(\mathcal{F}^{-1}\left[\frac{1}{\left[\sum_{k=1}^{m} A_k |t|^{\alpha_k} + A_0\right]}\right]\right)(x). \tag{5.5.5}$$

Applying the property (1.3.17) to the Fourier convolution (1.3.33):

$$\left(\mathcal{F}(k * f)\right)(x) := \left(\mathcal{F}\int_{\mathbb{R}^n} k(x - t) f(t) dt\right)(x) = (\mathcal{F}k)(x)(\mathcal{F}f)(x), \tag{5.5.6}$$

we present the solution (5.5.4) in the form of the Fourier convolution of $G^F_{\alpha_1,\cdots,\ \alpha_m}(x)$ and $f(x)$:

$$y(x) = \int_{\mathbb{R}^n} G^F_{\alpha_1,\cdots,\ \alpha_m}(x-t)f(t)dt. \tag{5.5.7}$$

Next we find the explicit representation for $G^F_{\alpha_1,\cdots,\ \alpha_m}(x)$. For this we need the following auxiliary assertion showing that the Fourier transform of a radial function is also a radial function and that it can be expressed in terms of the one-dimensional integral involving the Bessel function of the first kind $J_\nu(z)$ given in (1.7.1) [see Samko et al. ([729], Lemma 25.1)].

Lemma 5.6 *There holds the relation*

$$\int_{\mathbb{R}^n} e^{ix\cdot t}\varphi(|t|)dt = \frac{(2\pi)^{n/2}}{|x|^{(n-2)/2}}\int_0^\infty \varphi(\rho)\rho^{n/2}J_{(n/2)-1}(\rho|x|)d\rho \tag{5.5.8}$$

for any function $\varphi(\rho)$ such that the integral in the right-hand side of (5.5.8) *is convergent.*

Theorem 5.22 *Let $n, m \in \mathbb{N}$, $\alpha_1, \cdots, \alpha_m > 0$, and $A_0, \cdots, A_m \in \mathbb{R}$ be such that $0 < \alpha_1 < \cdots < \alpha_m$, $\alpha_m > (n-1)/2$; $A_0 \neq 0$, $A_m \neq 0$. Then the equation* (5.5.1) *is solvable, and its particular solution is given by* (5.5.7), *where*

$$G^F_{\alpha_1,\cdots,\ \alpha_m}(x) = \frac{|x|^{(2-n)/2}}{(2\pi)^{n/2}}\int_0^\infty \frac{1}{\left[\sum_{k=1}^m A_k|\rho|^{\alpha_k} + A_0\right]}\rho^{n/2}J_{(n/2)-1}(\rho|x|)d\rho. \tag{5.5.9}$$

Proof. By (5.5.4) and (1.3.23), we have

$$G^F_{\alpha_1,\cdots,\ \alpha_m}(x) = \frac{1}{(2\pi)^n}\int_{\mathbb{R}^n} \frac{1}{\left[\sum_{k=1}^m A_k|t|^{\alpha_k} + A_0\right]}e^{-ix\cdot t}dt,$$

and (5.5.9) follows from (5.5.8) with $\varphi(\rho) = \left[\sum_{k=1}^m A_k\rho^{\alpha_k} + A_0\right]^{-1}$ and with x replaced by $-x$. According to (1.7.7) and (1.7.8), for any fixed $x \in \mathbb{R}^n$, there exist the following asymptotic estimates of the integrand in (5.5.9) at zero and infinity:

$$\frac{1}{\left[\sum_{k=1}^m A_k|t|^{\alpha_k} + A_0\right]}\rho^{n/2}J_{(n/2)-1}(\rho|x|) \sim \frac{1}{\Gamma(n/2)A_0}\left(\frac{|x|}{2}\right)^{(n-2)/2}\rho^{n-1} \quad (\rho \to 0), \tag{5.5.10}$$

$$\frac{1}{\left[\sum_{k=1}^m A_k|t|^{\alpha_k} + A_0\right]}\rho^{n/2}J_{(n/2)-1}(\rho|x|) \sim$$

$$\sim \frac{1}{A_m}\left(\frac{2}{\pi|x|}\right)^{1/2}\cos\left[\rho|x| - \left(\frac{(n-1)\pi}{4}\right)\right]\rho^{[(n-1)/2]-\alpha_m} \quad (\rho \to \infty). \tag{5.5.11}$$

By (5.5.10) and (5.5.11), and the hypotheses of Theorem 5.22, the integral in the right-hand side of (5.5.9) is convergent, and thus Theorem 5.22 is proved.

Corollary 5.11 *If* $n \in \mathbb{N}$, $\alpha > (n-1)/2$, $\lambda \in \mathbb{R}$ $(\lambda \neq 0)$, *then the differential equation*

$$(\mathbf{D}^{\alpha} y)(x) + \lambda y(x) = f(x) \quad (x \in \mathbb{R}^n) \tag{5.5.12}$$

is solvable, and its particular solution has the form

$$y(x) = \int_{\mathbb{R}^n} G_{\alpha}^F(x-t) f(t) dt, \tag{5.5.13}$$

$$G_{\alpha}^F(x) = \frac{|x|^{(2-n)/2}}{(2\pi)^{n/2}} \int_0^{\infty} \frac{1}{|\rho|^{\alpha} + \lambda} \rho^{n/2} J_{(n/2)-1}(\rho|x|) d\rho. \tag{5.5.14}$$

Example 5.34 If $n \in \mathbb{N}$, then the equation

$$\left(\mathbf{D}^{3/2} y\right)(x) + \lambda y(x) = f(x) \quad (x \in \mathbb{R}^n; \quad \lambda \in \mathbb{R}) \tag{5.5.15}$$

has its solution given by

$$y(x) = \int_{\mathbb{R}^n} G_{3/2}^F(x-t) f(t) dt, \tag{5.5.16}$$

where $G_{3/2}^F(x)$ is given by (5.5.14) with $\alpha = 3/2$.

When $m \in \mathbb{N} \backslash \{1\}$, $A_0 = 0$ and $A_1 \neq 0$, the equation (5.5.1) takes the form

$$\sum_{k=1}^{m} A_k (\mathbf{D}^{\alpha_k} y)(x) = f(x) \quad (x \in \mathbb{R}^n; \quad n \in \mathbb{N}; \ m \in \mathbb{N} \backslash \{1\}; \quad 0 < \alpha_1 < \cdots < \alpha_m). \tag{5.5.17}$$

Theorem 5.23 *Let* $n \in \mathbb{N}$, $m \in \mathbb{N} \backslash \{1\}$, $\alpha_1, \cdots, \alpha_m > 0$, *and* $A_1, \cdots, A_m \in \mathbb{R}$ *be such that* $0 < \alpha_1 < \cdots < \alpha_m$, $\alpha_1 < n$, $\alpha_m > (n-1)/2$; $A_1 \neq 0$, $A_m \neq 0$. *Then the equation* (5.5.17) *is solvable, and its particular solution is given by* (5.5.7), *where*

$$G_{\alpha_1, \cdots, \alpha_m}^F(x) = \frac{|x|^{(2-n)/2}}{(2\pi)^{n/2}} \int_0^{\infty} \frac{1}{\left[\sum_{k=1}^{m} A_k |\rho|^{\alpha_k}\right]} \rho^{n/2} J_{(n/2)-1}(\rho|x|) d\rho. \tag{5.5.18}$$

Proof. The proof of Theorem 5.23 is similar to that of Theorem 5.22, using the following asymptotic estimates:

$$\frac{1}{\left[\sum_{k=1}^{m} A_k |t|^{\alpha_k}\right]} \rho^{n/2} J_{(n/2)-1}(\rho|x|) \sim \frac{1}{\Gamma(n/2) A_1} \left(\frac{|x|}{2}\right)^{(n-2)/2} \rho^{n-1-\alpha_1} \quad (\rho \to 0), \tag{5.5.19}$$

$$\frac{1}{\left[\sum_{k=1}^{m} A_k |t|^{\alpha_k}\right]} \rho^{n/2} J_{(n/2)-1}(\rho|x|) \sim$$

$$\sim \frac{1}{A_m} \left(\frac{2}{\pi|x|}\right)^{1/2} \cos\left[\rho|x| - \left(\frac{(n-1)\pi}{4}\right)\right] \rho^{[(n-1)/2]-\alpha_m} \quad (\rho \to \infty). \tag{5.5.20}$$

Corollary 5.12 *If $n \in \mathbb{N}$, $\alpha > 0$ and $\beta > 0$ are such that $\beta < \alpha$, $\alpha > (n-1)/2$, $\beta < n$ and $\lambda \in \mathbb{R}$ $(\lambda \neq 0)$, then the differential equation*

$$(\mathbf{D}^{\alpha} y)(x) + \lambda \left(\mathbf{D}^{\beta} y\right)(x) = f(x) \quad (x \in \mathbb{R}^n) \tag{5.5.21}$$

is solvable, and its particular solution has the form

$$y(x) = \int_{\mathbb{R}^n} G^F_{\alpha,\beta}(x-t) f(t) dt, \tag{5.5.22}$$

$$G^F_{\alpha,\beta}(x) = \frac{|x|^{(2-n)/2}}{(2\pi)^{n/2}} \int_0^{\infty} \frac{1}{|\rho|^{\alpha} + \lambda |\rho|^{\beta}} \rho^{n/2} J_{(n/2)-1}(\rho|x|) d\rho. \tag{5.5.23}$$

Remark 5.19 It follows from the condition $\alpha_m > (n-1)/2$ that the explicit solutions to the equations of the form (5.5.1) and (5.5.17) with the Riesz derivative of order $\alpha_m = 1/2$ can be obtained only in the one-dimensional case $n = 1$.

5.5.2 One-Dimensional Equations

Substituting $n = 1$ in the results of Section 5.5.1, we derive particular solutions to linear nonhomogeneous differential equations of the forms (5.5.1) and (5.5.17):

$$\sum_{k=1}^{m} A_k \left(\mathbf{D}^{\alpha_k} y\right)(x) + A_0 y(x) = f(x) \quad (x \in \mathbb{R}; \quad m \in \mathbb{N}; \quad 0 < \alpha_1 < \cdots < \alpha_m), \tag{5.5.24}$$

$$\sum_{k=1}^{m} A_k \left(\mathbf{D}^{\alpha_k} y\right)(x) = f(x) \quad (x \in \mathbb{R}; \quad m \in \mathbb{N}\backslash\{1\}; \quad 0 < \alpha_1 < \cdots < \alpha_m), \tag{5.5.25}$$

involving the one-dimensional Riesz fractional derivatives $\mathbf{D}^{\alpha_k} y$ given by (2.10.23) with $n = 1$:

$$\left(\mathbf{D}^{\alpha_k} y\right)(x) = \frac{1}{d_1(l, \alpha_k)} \int_{\mathbb{R}} \frac{\left(\Delta_t^l y\right)(x)}{|t|^{1+\alpha_k}} dt \quad (l > \alpha_k; \quad k = 1, \cdots, m), \tag{5.5.26}$$

where $d_1(l, \alpha_k)$ is given by (2.10.25) with $n = 1$ and $\alpha = \alpha_k$. We note that the relations (5.5.9) and (5.5.18) for the Fourier analogs of the Green function $G^F_{\alpha_1, \cdots, \alpha_m}(x)$ are simplified in view of the familiar relationship

$$J_{-1/2}(z) = \left(\frac{2}{\pi z}\right)^{1/2} \cos z. \tag{5.5.27}$$

The first result follows from Theorem 5.22 and (5.5.27).

Theorem 5.24 *Let $m \in \mathbb{N}$, $\alpha_1, \cdots, \alpha_m > 0$ and $A_0, \cdots, A_m \in \mathbb{C}$ be such that $0 < \alpha_1 < \cdots < \alpha_m$, $A_0 \neq 0$, and $A_m \neq 0$. Then the equation* (5.5.24) *is solvable, and its particular solution is given by*

$$y(x) = \int_{-\infty}^{\infty} G^F_{\alpha_1, \cdots, \alpha_m}(x-t) f(t) dt, \tag{5.5.28}$$

where

$$G^F_{\alpha_1,\cdots,\ \alpha_m}(x) = \frac{1}{\pi}\int_0^\infty \frac{1}{[\sum_{k=1}^m A_k|\rho|^{\alpha_k} + A_0]}\cos(\rho|x|)d\rho. \tag{5.5.29}$$

Corollary 5.13 *If* $\alpha > 0$, $\lambda \in \mathbb{R}$ $(\lambda \neq 0)$, *then the differential equation*

$$(\mathbf{D}^\alpha y)(x) + \lambda y(x) = f(x) \quad (x \in \mathbb{R}) \tag{5.5.30}$$

is solvable, and its particular solution has the form

$$y(x) = \int_{-\infty}^\infty G^F_\alpha(x-t)f(t)dt, \tag{5.5.31}$$

$$G^F_\alpha(x) = \frac{1}{\pi}\int_0^\infty \frac{1}{|\rho|^\alpha + \lambda}\cos(\rho|x|)d\rho. \tag{5.5.32}$$

Example 5.35 The equation

$$\left(\mathbf{D}^{1/2}y\right)(x) + \lambda y(x) = f(x) \quad (x \in \mathbb{R};\ \ \lambda \in \mathbb{R}) \tag{5.5.33}$$

has its solution given by

$$y(x) = \int_{-\infty}^\infty G^F_{1/2}(x-t)f(t)dt, \tag{5.5.34}$$

where $G^F_{1/2}(x)$ is given by (5.5.14) with $\alpha = 1/2$.

Our next result follows from Theorem 5.23 and (5.5.27).

Theorem 5.25 *Let* $m \in \mathbb{N}\backslash\{1\}$, $\alpha_1, \cdots, \alpha_m > 0$, *and* $A_1, \cdots, A_m \in \mathbb{R}$ *be such that* $0 < \alpha_1 < \cdots < \alpha_m$, $\alpha_1 < 1$; $A_1 \neq 0$, $A_m \neq 0$. *Then the equation* (5.5.25) *is solvable, and its particular solution is given by* (5.5.28), *where*

$$G^F_{\alpha_1,\cdots,\ \alpha_m}(x) = \frac{1}{\pi}\int_0^\infty \frac{1}{[\sum_{k=1}^m A_k|\rho|^{\alpha_k}]}\cos(\rho|x|)d\rho. \tag{5.5.35}$$

Corollary 5.14 *If* $\alpha > 0$ *and* $\beta > 0$ *are such that* $\beta < \alpha$ *and* $\beta < 1$ *and* $\lambda \in \mathbb{R}$ $(\lambda \neq 0)$, *then the differential equation*

$$(\mathbf{D}^\alpha y)(x) + \lambda\left(\mathbf{D}^\beta y\right)(x) = f(x) \quad (x \in \mathbb{R}) \tag{5.5.36}$$

is solvable, and its particular solution has the form

$$y(x) = \int_{-\infty}^\infty G^F_{\alpha,\beta}(x-t)f(t)dt, \tag{5.5.37}$$

$$G^F_{\alpha,\beta}(x) = \frac{1}{\pi}\int_0^\infty \frac{1}{|\rho|^\alpha + \lambda|\rho|^\beta}\cos(\rho|x|)d\rho. \tag{5.5.38}$$

Example 5.36 The equation

$$\left(\mathbf{D}^{3/2}y\right)(x)+\lambda\left(\mathbf{D}^{1/2}y\right)(x)=f(x)\quad(x\in\mathbb{R},\ \lambda\in\mathbb{R}) \tag{5.5.39}$$

has a particular solution given by

$$y(x)=\int_{-\infty}^{\infty}G_{3/2,1/2}^{F}(x-t)f(t)dt, \tag{5.5.40}$$

where $G_{3/2,1/2}^{F}(x)$ is given by (5.5.37) with $\alpha=3/2$ and $\beta=1/2$.

Remark 5.20 The Fourier and Mellin transforms were used in [370] and [372] to present a scheme for solving linear one-dimensional nonhomogeneous equations with constant coefficients of the form (5.1.1) and (5.4.1)-(5.4.2) involving the Liouville fractional derivative $D_{+}^{\alpha}y$ and $D_{-}^{\alpha}y$ given in (5.3.3)-(5.3.4), and the Hadamard fractional derivatives $\mathcal{D}_{0+}^{\alpha}y$ and $\mathcal{D}_{-}^{\alpha}y$ given in (2.7.9)-(2.7.10), respectively.

Chapter 6

PARTIAL FRACTIONAL DIFFERENTIAL EQUATIONS

The present chapter is devoted to the results for partial fractional differential equations. We give a survey of results in this field and apply Laplace and Fourier integral transforms to construct solutions in closed form of Cauchy type and Cauchy problems for fractional diffusion-wave and evolution equations.

6.1 Overview of Results

In this section we consider methods and results for partial differential equations and separately present the so-called fractional diffusion equations which arise in applications. We also present investigations of more general abstract differential equations of fractional order. More detailed information about some of these results can be found in a survey paper by Kilbas and Trujillo ([408], Sections 5-7).

6.1.1 Partial Differential Equations of Fractional Order

Gerasimov [279] derived and solved fractional-order partial differential equations for special applied problems. He studied two problems of viscoelasticity describing the motion of a viscous fluid between moving surfaces, and reduced these problems to the following two partial differential equations of fractional order:

$$\varrho\frac{\partial^2 u}{\partial t^2} = k\left({}^C D^{\alpha}_{-,t}\left[\frac{\partial^2 u}{\partial x^2}\right]\right)(x,t)\ (0<\alpha<1) \tag{6.1.1}$$

and

$$\varrho x^3\frac{\partial^2 u}{\partial t^2} = k\frac{\partial}{\partial x}\left(x^3\frac{\partial}{\partial x}({}^C D^{\alpha}_{-,t}u)\right)\ (0<\alpha<1), \tag{6.1.2}$$

with an unknown $u = u(x,t)$, the given constants ϱ and k, and a Caputo partial fractional derivative with respect to t of the form (2.4.53)

$$({}^C D_{-,t}^{\alpha} u)(x,t) = \frac{1}{\Gamma(1-\alpha)} \int_{-\infty}^{t} \frac{u_y(x,y)dy}{(t-y)^{\alpha}} \quad (0 < \alpha < 1). \tag{6.1.3}$$

Zhemukhov ([927], [928]) studied the Darboux problem for the following second-order degenerate loaded hyperbolic equations

$$y^m u_{xx} - u_{yy} = \sum_{k=1}^{n} a_k(x,y) D_{0+,x}^{\alpha_k} f\left[u(x,0), u_x(x,0), u_{xx}(x,0)\right], \tag{6.1.4}$$

$$\frac{\partial^m}{\partial y^m}\left(y^m u_{xx} - u_{yy} + a u_x + b u_y + cu + \sum_{k=1}^{n} a_k D_{0+,x}^{\alpha_k} f[\tau(x), \tau'(x)]\right) = 0, \tag{6.1.5}$$

involving the partial Riemann-Liouville fractional derivatives with respect to x defined in (2.9.9)

$$\left(D_{0+,x}^{\alpha} u\right)(x,y) = \left(\frac{\partial}{\partial x}\right)^{[\alpha]+1} \frac{1}{\Gamma(1-\{\alpha\})} \int_0^x \frac{u(t,y)dt}{(x-t)^{\{\alpha\}}} \quad (x > 0;\ y > 0;\ \alpha > 0), \tag{6.1.6}$$

$[\alpha]$ and $\{\alpha\}$ being the integral and fractional parts of α. He reduced these problems to equivalent nonlinear Volterra integral equations of the second kind and proved their solvability by the method of the contraction mapping principle, and then solved Cauchy problems for these Volterra equations in order to obtain the solutions $u(x,y)$ to the Darboux problems considered.

Conlan [145] considered the following nonlinear partial differential equation

$$\left(D_{0+,x}^{\alpha} D_{0+,y}^{\beta} u\right)(x,y) = f\left[x, y, u, \mathcal{D}_{0+,x}^{\alpha} u, \mathcal{D}_{0+,x}^{\alpha-1} \mathcal{D}_{0+,y}^{\beta} u, \mathcal{D}_{0+,y}^{\beta} u, \mathcal{D}_{0+,x}^{\alpha-1} u\right], \tag{6.1.7}$$

and used the Banach contraction mapping principle; he proved local theorems for the existence and the uniqueness of the solution $u(x,y)$ of the corresponding integral equation.

Fedosov and Yanenko [255] studied the following half-integer order partial differential equation

$$\sum_{k=0}^{n} a_k \left(D_{+,x}^{k/2} D_{+,y}^{(n-k)/2} u\right)(x,y) = f(x,y), \tag{6.1.8}$$

where a_k $(k = 1, \cdots, n)$ are constants and $D_{+,x}^{k/2}$ and $D_{+,y}^{(n-k)/2}$ are the partial Liouville fractional differentiation operators of the form (2.2.3) defined by (2.9.11) with $n = 2$ and $a_1 = -\infty$:

$$\left(D_{+,x}^{\alpha} u\right)(x,y) = \left(\frac{\partial}{\partial x}\right)^{[\alpha]+1} \frac{1}{\Gamma(1-\{\alpha\})} \int_{-\infty}^{x} \frac{u(t,y)dt}{(x-t)^{\{\alpha\}}} \quad (x \in \mathbb{R};\ y \in \mathbb{R};\ \alpha > 0), \tag{6.1.9}$$

and similarly for $\mathcal{D}^{\beta}_{+,y}u$. They proved that the operator in the left-side of (6.1.8) can be represented as the composition of n invertible operators of the form $\mathcal{D}^{1/2}_{+,x} + \lambda_j \mathcal{D}^{1/2}_{+,y}$, where λ_j are the roots of a certain characteristic polynomial. In particular, in the case $n = 1$, $a_0 = 0$ and $a_1 = 1$, they gave the relation for the general solution of the equation (6.1.8), which can contain an arbitrary function, and investigated the case $n = 2$ in detail [see Samko et al. ([729], Section 43.2, Note 42.13)]. Fedosov and Yanenko [255] also discussed a number of boundary conditions for the uniqueness of the solution $u(x, y)$ of the equation (6.1.1). They studied a similar problem in Fedosov and Yanenko [254] in connection with a mixed problem for the wave equation in cylindrical polygonal domains.

Biacino and Miseredino [83] investigated the Dirichlet problem on a rectangle $[a, b] \times [a' \times b']$ for a two-dimensional differential operator

$$Lu \equiv Eu + Pu, \tag{6.1.10}$$

where E is a uniformly strongly elliptic operator of the fourth order, while P is an operator that contains a special partial derivative $D^{\alpha}u$ of fractional order $\alpha = (\alpha_1, \alpha_2)$ $(\alpha_j > 0,\ j = 1, 2)$, $|\alpha| = \alpha_1 + \alpha_2$, and proved an alternative theorem and an existence and uniqueness result for the Dirichlet problem for the operator (6.1.10). Biacino and Miseredino [84] obtained similar statements while studying the Dirichlet problem on an interval $[a, b]$ for a one-dimensional differential operator (6.1.10).

A Cauchy type problem for the following partial fractional differential equation

$$\left(D^{\alpha}_{0+,t}u\right)(x,t) = \left(D^{\beta}_{0+,x}u\right)(x,t) \quad (1 \leqq \alpha \leqq 2; 1 \leqq \beta \leqq 2) \tag{6.1.11}$$

with

$$\left(D^{\alpha}_{0+,t}u\right)(x,t) = \left(\frac{\partial}{\partial t}\right)^{[\alpha]+1} \frac{1}{\Gamma(1-\{\alpha\})} \int_0^t \frac{u(x,\tau)}{(t-\tau)^{\{\alpha\}}} d\tau \quad (x \in \mathbf{R},\ t > 0,\ \alpha > 0), \tag{6.1.12}$$

was considered by Fujita [267]. He proved the existence and the uniqueness of a solution $u(x, t)$ and a representation for such a solution. In the case when $\beta = 2$ and $1 \leqq \alpha \leqq 2$, he proved the positivity of the fundamental solution and studied the asymptotic behavior of the solution $u(x, t)$. The results obtained can be interpreted as a phenomenon between the heat equation $(\alpha = 1; \beta = 2)$ and the wave equation $(\alpha = \beta = 2)$.

In connection with problems in fractals, the fractional partial differential equations of the forms

$$\left(D^{\alpha}_{0+,t}u\right)(x,t) = -Ax^{-\beta}\frac{\partial u(x,t)}{\partial x} \quad \left(0 < \alpha \leqq \frac{1}{2};\ A > 0;\ \beta \geqq 0\right) \tag{6.1.13}$$

and

$$\left(D^{\alpha}_{0+,t}u\right)(x,t) = -A\left[\frac{\partial u(x,t)}{\partial x} + \frac{k}{x}u(x,t)\right] \quad \left(0 < \alpha \leqq \frac{1}{2};\ A > 0;\ k \in \mathbb{R}\right) \tag{6.1.14}$$

with the Riemann-Liouville partial fractional derivative (6.1.12), were considered by Giona and Roman [282] and Roman and Giona [715], respectively. Applying the Laplace transform (1.4.1) with respect to t, they reduced these equations to certain ordinary differential equations for $U(x,p) = (\mathcal{L}_t u(x,t))(p) = \int_0^\infty u(x,t)e^{-pt}dt$, obtained explicit solutions $U(x,t)$ of these equations under a certain normalization condition, and indicated that the solutions $u(x,t)$ of (6.1.13) and (6.1.14) can be obtained by using the inverse Laplace transform $\mathcal{L}^{-1}$ with respect to t.

Kolwankar and Gangal [436] considered the following partial fractional differential equation

$$\mathbf{D}_t^\alpha u(x,t) = \frac{\Gamma(\alpha+1)}{4}\chi_\Omega(t)\frac{\partial^2 u(x,t)}{\partial x^2} \quad (0 < \alpha \leqq 1;\ x \in \mathbb{R};\ t > 0), \tag{6.1.15}$$

where use is made of a special fractional derivative $\mathbf{D}_t^\alpha f(t)$ defined by

$$\mathbf{D}_t^\alpha f(t) = \lim_{y \to x+} \frac{1}{\Gamma(n-\alpha)}\left(\frac{\partial}{\partial x}\right)^n \int_t^y \frac{f(t) - \sum_{k=0}^n f^{(k)}(t)(s-t)^k/k!}{(y-t)^{\alpha-n+1}}dt, \tag{6.1.16}$$

where $n = [\alpha] + 1$ and $\chi_\Omega(t)$ is the characteristic function of a set $\Omega \subset \mathbf{R}_+$: $\chi_\Omega(t) = 1$ if $t \in \Omega$ and $\chi_\Omega(t) = 0$ if $t \notin \Omega$. They obtained the explicit solution of the Cauchy problem for the equation (6.1.16) with the initial condition $u(x,0) = \delta(x)$, $\delta(x)$ being the Dirac delta function. Kolwankar and Gangal [436] indicated that the partial fractional differential equation (6.1.15) is an example of more general fractional differential equations of the form

$$\mathbf{D}_t^\alpha u(x,t) = L(x,t)u(x,t) \tag{6.1.17}$$

with the special operator $L(x,t)$, which are called the local fractional Fokker-Planck equations and which appear while studying a phenomenon taking place in fractal space and time.

Atanackovic and Stankovic [38] and Stankovic [801] used the Laplace transform in a certain space of distributions to solve a system of partial differential equations with fractional derivatives, and indicated that such a system may serve as a certain model for a visco-elastic rod.

Kilbas and Repin [388] studied the following equation of mixed type

$$\frac{\partial^2 u(x,y)}{\partial x^2} - \left(D_{0+,y}^\alpha u\right)(x,y) = 0\ (y > 0), \quad (-y)^m \frac{\partial^2 u(x,y)}{\partial x^2} - \frac{\partial^2 u(x,y)}{\partial^2} = 0\ (y < 0) \tag{6.1.18}$$

with $m > 0$ and the partial Riemann-Liouville derivative (6.1.12) of order $0 < \alpha < 1$ in a special domain of $\mathbb{R}^2$. They investigated a non-local problem for this equation with boundary conditions involving generalized fractional integro-differential operators with the Gauss hypergeometric function (1.6.1) in the kernel and gave conditions for the uniqueness and existence of a solution of the considered problem. Earlier, Kilbas and Repin [387] established similar results for an analog of the Bitsadze-Samarskii problem for the equation of mixed type with the partial Riemann-Liouville fractional derivative.

Nakhushev [615] and Nakhushev and Borisov [619] studied certain problems for the following degenerate integro-differential equation:

$$u_{yy} - (-y)^m u_{xx} + au_x + bu_y + \sum_{k=1}^{n} a_k D_{0+,\xi}^{\alpha_k} u(x,0) + cu = d, \tag{6.1.19}$$

and for the following so-called loaded parabolic equation:

$$u_y - ku_{xx} + \sum_{l=1}^{n} \sum_{j=1}^{m} a_l(x,y) D_{0+,y}^{\alpha_i} [k_j(x,y)u(x^j,y)] = f(x,y), \tag{6.1.20}$$

where $D_{0+,\xi}^{\alpha_k} u(x,0)$ are the partial Riemann-Liouville fractional derivatives or integrals of order α_k defined by (2.9.9) and (2.9.3), respectively, with $n = 2$.

To conclude this section, we note that the equations (6.1.13) and (6.1.14) for $\alpha = 1/2$ were first obtained by Oldham and Spanier in [641] and [642], respectively, by reducing a boundary-value problem involving Fick's second law in electroanalytical chemistry to a formulation based on the partial Riemann-Liouville fractional derivative $D_{0+,x}^{1/2}$ in (6.1.6). Oldham [640] and Oldham and Spanier ([643], Chapter 11) gave other applications of such equations for diffusion problems.

6.1.2 Fractional Partial Differential Diffusion Equations

It was indicated at the end of the previous section that Oldham and Spanier ([643], Chapter 11) first considered the partial fractional differential equations arising in diffusion problems. A series of papers was devoted to investigating such so-called fractional diffusion equations, and in most of them formal explicit solutions to certain boundary- and initial-value problems for the considered equations were obtained. Wyss [898] studied the following fractional differential equation

$$\frac{x_+^{-\alpha-1}}{\Gamma(\alpha)} * u(x,t) = \lambda^2 \frac{\partial^2 u(x,t)}{\partial x^2} \quad (x > 0;\ 0 < \alpha \leqq 1;\ \lambda > 0), \tag{6.1.21}$$

where $x_+^{-\alpha-1}/\Gamma(\alpha)$ is understood as a distribution in the space $\mathcal{D}'$ of generalized functions (see Section 1.2). He investigated two problems for the equation (6.1.21) with

$$u(x,0) = 0,\ u(0,T) = b\ (t > 0) \tag{6.1.22}$$

When $b = 0$ and $b = -1$. He sought a solution $u(x,t)$ to these problems in the form

$$u(x,t) = f(y),\ y = t^{-\alpha/2}x, \tag{6.1.23}$$

and reduced (6.1.21)-(6.1.22) to the following one-dimensional problem:

$$\lambda^2 \frac{d^2 f(y)}{dy^2} = (I_{-;2/\alpha,1}^{-\alpha} f)(y)\ (y > 0),\ f(0) = b,\ f(\infty) = 1 + b, \tag{6.1.24}$$

where $I^{-\alpha}_{-;2/\alpha,1}f$ is the Erdélyi-Kober type fractional integral of the form (2.6.4).

Applying the Mellin transform (1.4.23) to the equation in (6.1.24) and taking into account the initial conditions in (6.1.24), he arrived at the following relation:

$$(\mathcal{M}y)(s) = \pi^{-1/2}(2a)^s \frac{\Gamma(\mp s)\Gamma((1+s)/2)\Gamma(1+s/2)}{\Gamma(1\mp s)\Gamma(1+\lambda s/2)}. \tag{6.1.25}$$

Applying the inverse Mellin transform (1.4.24), Wyss [898] obtained the following solutions to the above two problems in terms of the H-function (1.12.1) in their respective forms:

$$u(x,t) = \pi^{-1/2} H^{2,1}_{2,3}\left[\frac{x}{2\lambda}t^{-\alpha/2}\,\middle|\, \begin{matrix}(1,1),(1,\alpha/2)\\ (1/2,1/2),(1,1/2),(0,1)\end{matrix}\right] \tag{6.1.26}$$

and

$$u(x,t) = -\pi^{-1/2} H^{3,0}_{2,3}\left[\frac{x}{2\lambda}t^{-\alpha/2}\,\middle|\, \begin{matrix}(1,1),(1,\alpha/2)\\ (0,1),(1/2,1/2),(1,1/2)\end{matrix}\right]. \tag{6.1.27}$$

Schneider and Wyss [746] considered the equation

$$u(x,t) = \sum_{k=0}^{l-1} f_k(x)t^k + \frac{1}{\Gamma(\alpha)}\int_0^t \frac{\Delta u(x,s)ds}{(t-s)^{1-\alpha}} \quad (l-1<\alpha \leqq l;\ l=1,2), \tag{6.1.28}$$

where $x \in \mathbb{R}^n$ ($n \in \mathbb{N}$), $t>0$, $f_k(x)$ are the initial data, i.e.,

$$\left(\frac{\partial}{\partial t}\right)^k u(x,t)|_{t=0} = f_k(x) \quad (0 \leqq k \leqq l-1;\ l=1,2), \tag{6.1.29}$$

and Δ is the Laplacian (1.3.31). (6.1.28) represents the fractional diffusion and the wave equation when $l=1$, $0<\alpha\leqq 1$ and $l=2$, $1<\alpha\leqq 2$, respectively. Applying the Laplace transform (1.4.1) with respect to t in (6.1.28), they reduced (6.1.28) to the form

$$\Delta U(x,p) - p^\alpha U(x,p) = -\sum_{k=0}^{l-1} f_k(x)p^{-k-1} \quad (l=1,2), \tag{6.1.30}$$

where $U(x,p) = (\mathcal{L}_t u(x,t))(p)$. Applying the inverse Mellin transform (1.4.24) to the general solution of (6.1.30), Schneider and Wyss [746] obtained the solution to (6.1.28) in the form

$$u(x,t) = \sum_{k=0}^{l-1} \int_{\mathbb{R}^n} G^\alpha_k(|x-y|,t) f_k(y)dy, \tag{6.1.31}$$

here $y \in \mathbb{R}^n$, $|x-y| = \left[\sum_{k=1}^n (x_k-y_k)^2\right]^{1/2}$, and the analogies of the Green functions $G^\alpha_k(r,t)$ ($0 \leqq k \leqq l-1$; $l=1,2$) are expressed via the H-function (1.12.1) by

$$G_k^\alpha(r,t) = \frac{\pi^{-n/2}}{2r^n}\left(\frac{r}{2}\right)^{2k/\alpha} H_{1,2}^{2,0}\left[\frac{r}{2}t^{-\alpha/2}\,\middle|\, \begin{matrix}(1,\alpha/2)\\ (n/2-k/\alpha,1/2),(1-k/\alpha,1/2)\end{matrix}\right]. \tag{6.1.32}$$

Schneider and Wyss [746] also obtained an explicit solution to the fractional diffusion equation (6.1.28) with $0 < \alpha \leqq 1$ in the half space $D = \mathbb{R}^{n-1} \times \mathbb{R}_+$ with the boundary $\partial D = \mathbb{R}^{n-1} \times \{0\}$, supplemented by a special boundary condition.

For $0 < \alpha \leqq 1$ and $m = 1$, Schneider [745] gave a more elegant solution of (6.1.28):

$$u(x,t) = f(x) + \frac{1}{\Gamma(\alpha)}\int_0^t \frac{\Delta u(x,s)ds}{(t-s)^{1-\alpha}} \;(0 < \alpha \leqq 1),\; u(x,0) = f(x) \tag{6.1.33}$$

for $x \in \mathbb{R}^n$ $(n \in \mathbb{N})$ and $t > 0$. Applying the Fourier transform (1.3.22) with respect to x and the Laplace transform (1.4.1) with respect to t, he reduced (6.1.33) to the form

$$(\mathcal{F}_x\mathcal{L}_t u(x,t))(y,p) = \frac{p^{\alpha-1}}{p^\alpha + y^2}(Ff)(y),\; y^2 = y\cdot y. \tag{6.1.34}$$

Taking the inverse Laplace and Fourier transforms (1.4.2) and (1.3.23) to (6.1.34), Schneider [745] obtained the solution of the problem (6.1.33) in the form (6.1.31)

$$u(x,t) = \int_{\mathbb{R}^n} G^\alpha(x-y,t)f_k(y)dy, \tag{6.1.35}$$

where

$$G^\alpha(x,t) = (4\pi t^\alpha)^{-n/2} H_{1,2}^{2,0}\left[\frac{x\cdot x}{4t^\alpha}\,\middle|\, \begin{matrix}(1-\alpha n/2,\alpha)\\ (0,1),(1-n/2,1)\end{matrix}\right]. \tag{6.1.36}$$

Wyss [898] and Schneider and Wyss [746] also investigated the properties of the fractional analogs of the Green function given by (6.1.32) and (6.1.36). Schneider and Wyss [746] indicated a physical application of (6.1.28) to special types of porous media as pointed out by Nigmatullin [623], and Schneider [745] constructed a stochastic process called grey Brownian motion, based on a probability measure called grey noise, and showed that such a process is related to fractional diffusion as ordinary Brownian motion is related to ordinary diffusion. We also mention de Arrieta and Kalla [159] who treated some fractional diffusion equations of the forms (6.1.21) and (6.1.28) by using the Laplace transform (1.4.1) and the Mellin transform (1.4.23) of the H-function (1.12.1) and of special cases of the generalized hypergeometric function ${}_pF_q(a_1,\cdots,a_p;b_1,\cdots,b_q;z)$ given by (1.6.28).

Kochubei [427] considered the following Cauchy type problem:

$$\left({}^C\!D_t^{(\alpha)}u\right)(x,t) = Lu(x,t)\;(0 < \alpha < 1;\; x \in \mathbb{R}^n;\; 0 < t \leqq T), \tag{6.1.37}$$

$$u(x,0) = f(x) \; (x \in \mathbb{R}^n), \tag{6.1.38}$$

where $\left({}^{C}D_t^{(\alpha)}u\right)(x,t)$ is the Caputo regularized partial fractional derivative with respect to t:

$$\left({}^{C}D_t^{(\alpha)}u\right)(x,t) = \frac{1}{\Gamma(1-\alpha)}\left[\frac{\partial}{\partial t}\int_0^t \frac{u(x,s)ds}{(t-s)^\alpha} - \frac{u(x,0)}{t^\alpha}\right], \tag{6.1.39}$$

and L is the second-order elliptic differential operator in n variables:

$$L = \sum_{k,j} a_{kj}\frac{\partial^{k+j}}{\partial x_k \partial x_j} + \sum_k a_k \frac{\partial^k}{\partial x_k} + a, \tag{6.1.40}$$

where the summation is taken over repeated indices. Kochubei [427] proved that the Cauchy type problem (6.1.37)-(6.1.38) has a unique solution $u(x,t)$ under the appropriate assumptions on $f(x)$ and the coefficients a_{kj}, a_l and a of the operator L, and also indicated the conditions when this problem with $f(x) = 0$ has a nonzero solution. In the case when $L = \Delta$ is the Laplacian (1.3.31), he obtained the explicit form for the fundamental solution $u(x,t)$ in terms of the H-function (1.12.1), and showed its integralibility in t over $\mathbb{R}^n$. The fundamental solution of the Cauchy problem for the equation (6.1.37), in which L is a uniformly elliptic operator, was constructed in terms of the H-function (1.12.1) by Eidelman and Kochubei [219] .

Fujita [269] studied the following two-dimensional equation (6.1.33) with $1 < \alpha < 2$

$$u(x,t) = f(x) + \frac{1}{\Gamma(\alpha)}\int_0^t \frac{\Delta u(x,\tau)d\tau}{(t-\tau)^{1-\alpha}} \quad (1 < \alpha < 2;\; x \in \mathbb{R};\; t > 0) \tag{6.1.41}$$

in a certain subspace of the Schwartz space of rapidly decreasing functions (see Section 1.2). Using the Fourier transform, he obtained its solution $u(x,t)$ as follows:

$$u(x,t) = \frac{1}{\alpha}\int_{-\infty}^{\infty} P_\alpha(|x-t|,\tau)f(\tau)d\tau, \tag{6.1.42}$$

$$P_\alpha(x,t) = \frac{1}{2\pi}\int_{-\infty}^{\infty} \exp\left[-t|\tau|^{2/\alpha}e^{-\gamma\pi sign(\tau)i/2}\right]e^{-ix\tau}d\tau, \; \gamma = 2 - \frac{2}{\alpha}. \tag{6.1.43}$$

Fujita [269] also showed that the fundamental solution of (6.1.41) takes its maximum value at $x = \pm c_\alpha t^{\alpha/2}$ for each $t > 0$, and hence these points propagate with finite speed as in the wave equation

$$\frac{\partial^2 u(x,t)}{\partial t^2} = \lambda^2 \frac{\partial^2 u(x,t)}{\partial x^2} \quad (\lambda > 0) \tag{6.1.44}$$

in the case $\alpha = 2$, and that the support of $u(\cdot, t)$ is not compact for each $t > 0$ even if f has compact support and thus, in that sense, there is an infinite propagation speed as in the case $\alpha = 1$ of the heat equation

$$\frac{\partial u(x,t)}{\partial t} = \lambda^2 \frac{\partial^2 u(x,t)}{\partial x^2} \quad (\lambda > 0). \tag{6.1.45}$$

Fujita [268] obtained the solution to the equation

$$u(x,t) = f(x) + \frac{t^{\alpha/2}}{\Gamma(1+\alpha/2)} g(x) + \frac{1}{\Gamma(\alpha)} \int_0^t \frac{\Delta u(x,\tau) d\tau}{(t-\tau)^{1-\alpha}} \tag{6.1.46}$$

with $1 < \alpha \leqq 2$, $x \in \mathbb{R}$, and $t > 0$, in the form

$$u(x,t) = \frac{E}{2} \left[f\left(x + Y_\alpha(t)\right) + f\left(x - Y_\alpha(t)\right) + \int_{x - Y_\alpha(t)}^{x + Y_\alpha(t)} g(t) dt \right], \tag{6.1.47}$$

where $Y_\alpha(t)$ is a continuous, non-decreasing and nonnegative stochastic process with Mittag-Leffler distribution of order $\alpha/2$, and E stands for the expectation. Using the Fourier transform and probability methods, Fujita [270] proved the energy inequalities for the integro-differential equations of the form (6.1.41) and (6.1.46) which correspond to the energy inequality for the wave equation (6.1.44).

The equation (6.1.21) belongs to the so-called fractional diffusion equations [see Nigmatullin [622], [623] and Westerlund [874]]. In the simplest case of the one-dimensional diffusion, such an equation is given by

$$(D_{0+,t}^\alpha u)(x,t) = \lambda^2 \frac{\partial^2 u(x,t)}{\partial x^2} \quad (\alpha > 0;\ \lambda > 0), \tag{6.1.48}$$

with the partial Riemann-Liouville fractional derivative $(D_{0+,t}^\alpha u)(x,t)$ defined in (6.1.12). When $\alpha = 1$ and $\alpha = 2$, (6.1.48) coincides with the heat (diffusion) equation (6.1.45) and the wave equation (6.1.44), respectively.

Mainardi [514], [517] [519] studied an equation of the form (6.1.48):

$$({}^C D_{0+,t}^\alpha u)(x,t) = \lambda^2 \frac{\partial^2 u(x,t)}{\partial x^2} \quad (\alpha > 0;\ \lambda > 0), \tag{6.1.49}$$

with $x \in \mathbb{R}$ and $t > 0$ for $0 < \alpha \leqq 2$. Here $({}^C D_{0+,t}^\alpha u)(x,t)$ is the Caputo partial fractional derivative with respect to t of order $\alpha > 0$ ($l-1 < \alpha < l$; $l \in \mathbb{N}$) defined, for any $x \in \mathbb{R}$ and $t > 0$, by

$$({}^C D_{0+,t}^\alpha u)(x,t) = \frac{1}{\Gamma(n-\alpha)} \int_0^t \frac{(\partial^l u(x,\tau)/\partial \tau^l) d\tau}{(t-\tau)^{\alpha-l+1}}, \tag{6.1.50}$$

$$({}^C D_{0+,t}^l u)(x,t) = \frac{\partial^l u(x,t)}{\partial t^l}. \tag{6.1.51}$$

Mainardi [514] investigated the initial-value problem

$$({}^C D_{0+,t}^\alpha u)(x,t) = \lambda^2 \frac{\partial^2 u(x,t)}{\partial x^2} \quad (x \in \mathbb{R};\ t > 0;\ 0 < \alpha < 1), \tag{6.1.52}$$

$$u(x,0) = f(x) \ (x \in \mathbb{R}), \quad \lim_{x \to \pm\infty} u(x,t) = 0 \ (t > 0). \tag{6.1.53}$$

Using the method suggested by Schneider [745] and applying the Laplace transform (1.4.1) with respect to t and the Fourier transform (1.3.1) with respect to x, he reduced the problem (6.1.52)-(6.1.53) to a form like (6.1.34). And, by applying the inverse Laplace and Fourier transforms, he found the solution $u(x,t)$ of this problem in the form (6.1.35):

$$u(x,t) = \int_{-\infty}^{\infty} G^\alpha(x-\tau,t) f(\tau) d\tau. \tag{6.1.54}$$

Here a fractional analog of the Green function $G^\alpha(x,t)$ is given by

$$G^\alpha(x,t) = \frac{1}{\pi}\int_0^\infty E_\alpha\left(-\lambda^2 y^2 t^\alpha\right)\cos(yx)dy = \frac{1}{2\lambda}t^{-\alpha/2}\varphi\left(\frac{\alpha}{2}, 1-\frac{\alpha}{2}; \frac{|x|}{\lambda}t^{-\alpha/2}\right) \tag{6.1.55}$$

in terms of the integral of the Mittag-Leffler function (1.8.1) and the Wright function (1.11.1).

Mainardi ([517], [519]) used the same approach to find the fundamental solutions of the Cauchy problem and of the so-called signaling problem for the equation (6.1.49) with $0 < \alpha \leqq 2$ in terms of the Wright function (1.11.1) with $z = |x|t^{-\alpha/2}$, which reduces to the Gaussian function in the case $\alpha = 1$ of the classical diffusion equation. For $0 < \alpha \leqq 1$, the initial conditions of the Cauchy problem are given by (6.1.53) with $f(x) = \delta(x)$, while the initial conditions of the signaling problem have the form

$$u(x,0) = \delta(x) \ (x > 0); \ u(0,t) = g(t), \ u(+\infty,t) = 0 \ \ (t > 0). \tag{6.1.56}$$

When $1 < \alpha \leqq 2$, the condition

$$\frac{\partial}{\partial t}u(x,t)|_{t=0} = 0 \tag{6.1.57}$$

must be added to the relations (6.1.53) and (6.1.56).

Podlubny [682] also applied such a technique to obtain the solution of the Cauchy type problem for the equation (6.1.48) with the following initial conditions:

$$(\mathcal{D}_{0+,t}^{\alpha-1}u)(x,t)|_{t=0} = f(x), \quad \lim_{|x|\to\infty} u(x,t) = 0, \tag{6.1.58}$$

and for the equation of Schneider-Wyss type

$$u(x,t) = f(x) + \lambda^2 \frac{1}{\Gamma(\alpha)}\frac{\partial^2}{\partial x^2}\int_0^t \frac{u(x,\tau)d\tau}{(t-\tau)^{1-\alpha}} \quad (x \in \mathbb{R}; \ 0 < \alpha \leqq 1) \tag{6.1.59}$$

with the initial condition (6.1.53), in the form (6.1.54), where

$$G^{\alpha}(x,t) = \frac{1}{2\lambda} t^{(\alpha-2)/2} \varphi \left(-\frac{\alpha}{2}, \frac{\alpha}{2}; -\frac{|x|}{\lambda} t^{-\alpha/2} \right), \tag{6.1.60}$$

which, for $\alpha = 1$, is reduced to the classical expression

$$G(x,t) = \frac{1}{2\lambda} (\pi t)^{-1/2} \exp\left(-\frac{x^2}{4\lambda^2 t} \right). \tag{6.1.61}$$

Gorenflo and Mainardi [305] considered the fractional diffusion equation (6.1.49) with $\lambda = 1$ in the quarter-plane $\mathbb{R}_{++} = \{(x,t) \in \mathbb{R}^2 : \ x > 0, \ t > 0\}$

$$({}^{C}D^{\alpha}_{0+,t}u)(x,t) = \frac{\partial^2 u(x,t)}{\partial x^2} \quad ((x,t) \in \mathbb{R}_{++}; \ 0 < \alpha \leqq 2), \tag{6.1.62}$$

with the initial and boundary conditions of the form

$$u(x,0) = 0, \ u(0,t) = \phi(t) \quad (t \geqq 0), \tag{6.1.63}$$

for $0 < \alpha \leqq 1$, while, for $1 < \alpha \leqq 2$,

$$u(x,0) = \frac{\partial}{\partial t} u(x,0) = 0, \ \frac{\partial}{\partial x} u(0,t) = \psi(t) \quad (t \geqq 0). \tag{6.1.64}$$

They applied the Laplace transform to obtain the explicit solution to these problems in the form (6.1.54), where a fractional analog of the Green function $G^{\alpha}(x,t)$ is the inverse Laplace transform with respect to t of $\exp\left(-xt^{\alpha/2}\right)$ and $-t^{-\alpha/2}\exp\left(-xt^{\alpha/2}\right)$, respectively. In this regard we indicate the papers by Mainardi [519, 520], Gorenflo and Mainardi [302], Gorenflo, Mainardi and Srivastava [310], Mainardi and Tomirotti [535] and Mainardi and Paradisi [530], where various relations of such a function $G^{\alpha}(x,t)$ with special functions and extremal stable probability densities are worked out. We also mention Luchko and Gorenflo [506] who considered scale-invariant solutions for some diffusion partial diferential equations of fractional order, and Buckwar and Luchko [109] who investigated an invariance of such equations relative to a certain Lie group.

Equations (6.1.48) and (6.1.49) are the simplest examples of the approach to "fractalization" and unification of the classical diffusion and wave equations suggested by Nonnenmacher and Nonnenmacher [636]. Another approach given by Metzler et al. [581], in the simplest case, leads to the following partial fractional differential equation:

$$(D^{\alpha}_{0+,t}u)(x,t) = Ax^{-\beta} \frac{\partial}{\partial x} \left(x^{\beta} \frac{\partial u(x,t)}{\partial x} \right) \quad (\alpha > 0; \ A > 0; \ \beta \geqq 0). \tag{6.1.65}$$

Agrawal [7] applied the direct and inverse Laplace transforms to obtain the solution, in closed form, of the Cauchy problem (6.1.53) and of the signaling problem (6.1.56) for the following fourth order fractional differential equation

$$({}^{C}D^{\alpha}_{0+,t}u)(x,t) + b^2 \frac{\partial^4 u(x,t)}{\partial x^4} = 0 \quad (x \in \mathbb{R}; \ t > 0; \ 0 < \alpha \leqq 2), \tag{6.1.66}$$

with $b \in \mathbb{R}$ and the Caputo partial fractional derivative (6.1.50).

Hilfer [344] obtained a fundamental solution of the following multi-dimensional equation of the form (6.1.48) of order $0 < \alpha < 1$

$$(D^{\alpha}_{0+,t}u)(x,t) = \lambda\,(\Delta_x u)\,(x,t) \quad (x \in \mathbb{R}^n;\ t > 0;\ \lambda > 0) \tag{6.1.67}$$

with the Laplacian (1.3.31) in terms of the $H^{2,0}_{1,2}$-function (1.12.1).

Chechkin et al. [135] studied the diffusion-like equation with the Caputo time partial fractional derivative of distributed order, proved the positivity of its solution, and established the relation of the considered equation with the continuous-time random walk theory.

Kilbas et al. [385] applied the direct and inverse Laplace and Fourier transforms with respect to $x \in \mathbb{R}$ and $t > 0$ to find the explicit solution to the Cauchy problem

$$({}^{C}D^{\alpha}_{0+,t}u)(x,t) = \lambda(D^{\beta}_{-,x}u)(x,t) \ (x \in \mathbb{R};\ t > 0;\ 0 < \alpha < 1;\ \beta > 0;\ \lambda \in \mathbb{R} \setminus \{0\}) \tag{6.1.68}$$

$$\lim_{x \to \pm\infty} u(x,t) = 0, \quad u(x,0+) = g(x) \tag{6.1.69}$$

with the Caputo and Liouville partial fractional derivatives defined for $x \in \mathbb{R}$ and $t > 0$ by

$$({}^{C}D^{\alpha}_{0+,t}u)(x,t) = \frac{1}{\Gamma(1-\alpha)}\frac{\partial}{\partial t}\int_0^t \frac{u(x,\tau)-u(x,0)}{(t-\tau)^{\alpha}}d\tau \quad (0 < \alpha < 1) \tag{6.1.70}$$

and

$$(D^{\beta}_{-,x}u)(x,t) = \frac{1}{\Gamma(1-\{\beta\})}\left(\frac{\partial}{\partial x}\right)^{[\beta]+1}\int_{-\infty}^{x}\frac{u(\tau,t)}{(x-\tau)^{\{\beta\}}} \quad (\beta > 0), \tag{6.1.71}$$

respectively. Such a method was also used in Kilbas et al. [386] to construct the explicit solution for the following Cauchy type problem

$$(D^{\alpha}_{0+,t}u)(x,t) = \lambda(D^{\beta}_{-,x}u)(x,t) \quad (x \in \mathbb{R};\ t > 0;\ 0 < \alpha \leqq 1;\ \beta > 0;\ \lambda \in \mathbb{R}\backslash\{0\}, \tag{6.1.72}$$

$$\lim_{x \to \pm\infty} u(x,t) = 0, \quad (D^{\alpha-1}_{0+,t}u)(x,0+) = g(x) \tag{6.1.73}$$

with the Riemann-Liouville partial fractional derivative $(D^{\alpha}_{0+,t}u)(x,t)$ with respect to t, given by (6.1.12). Applications were given in the above two articles to the analysis of diffusion mechanisms with internal degrees of freedom while studying the square root of the standard linear diffusion equation in one space dimension, $u_t - u_{xx} = 0$ [see Vázquez [842] and Vázquez and Vilela Mendes [845]].

Voroshilov and Kilbas ([861], [862]) used the direct and inverse Laplace and Fourier transforms to derive explicit solutions of the Cauchy type problems for the one- and multi-dimensional homogeneous fractional diffusion equations (6.1.49) and (6.1.67) with $0 < \alpha < 2$, with the Cauchy type conditions (6.1.58) in the case $0 < \alpha \leqq 1$, while, for $1 < \alpha < 2$,

$$\left(D_{0+,t}^{\alpha-k}u\right)(x,0+) = f_k(x) \quad (x \in \mathbb{R}^n;\ 1 < \alpha < 2;\ k = 1,2). \tag{6.1.74}$$

Here $\left(D_{0+,t}^{\alpha-1}u\right)(x,t)$ is the partial Riemann-Liouville fractional derivative defined by (6.1.12), while $\left(D_{0+,t}^{\alpha-2}u\right)(x,t) = \left(I_{0+,t}^{n-\alpha}u\right)(x,t)$ is the partial Riemann-Liouville fractional integral of the form (2.9.3)

$$\left(I_{0+,t}^{\alpha}u\right)(x,t) = \frac{1}{\Gamma(\alpha)}\int_0^t \frac{u(x,\tau)d\tau}{(t-\tau)^{1-\alpha}} \quad (\alpha > 0). \tag{6.1.75}$$

In particular, for $0 < \alpha < 1$, the results obtained coincide with those proved by Podlubny [682] for the Cauchy type problem (6.1.48)-(6.1.58).

Mainardi et al. [526] considered the fractional diffusion equation of the form (6.1.68) with the Caputo partial fractional derivative (6.1.50) of order $0 < \alpha \leqq 2$ in which the Liouville partial derivative $\left(D_{-,x}^{\beta}u\right)(x,t)$ is replaced by the so-called Riesz-Feller partial fractional derivative $\left(D_{\theta;x}^{\beta}u\right)(x,t)$ defined, for $0 < \beta \leqq 2$ and $|\theta| \leqq \min[\beta, 2-\beta]$ via the Fourier transform, by

$$\left(\mathcal{F}_x\ D_{\theta;x}^{\beta}u\right)(\sigma,t) = |\sigma|^{\beta}e^{i(\text{sign}(\sigma)\theta\pi/2}\left(\mathcal{F}_x u\right)(\sigma,t) \quad (\sigma \in \mathbb{R};\ t > 0). \tag{6.1.76}$$

They established the explicit solution of the form (6.1.64) for the equation considered with the initial condition (6.1.69), where $g(x) = \delta(x)$, in the case $0 < \beta \leqq 1$ and with the additional condition $\frac{\partial}{\partial t}u(x,0) = 0$ when $1 < \beta \leqq 2$. Another representation for the fractional Green function $G^{\alpha}(x.t)$ in terms of the $H_{3,3}^{1,1}$-function was given by Mainardi et al. [529].

6.1.3 Abstract Differential Equations of Fractional Order

In this section we present investigations of certain abstract fractional differential equations. First of all, we note that Berens and Westphal [79] first considered the abstract Cauchy type problem for the Riemann-Liouville fractional differentiation operator D_{0+}^{α} $(0 < \alpha < 1)$ defined in (2.1.8), with a certain domain $G(D_{\alpha};p)$ and range in $L_p(\mathbb{R}_+)$ $(1 < p < \infty)$. The problem lies is finding a functional $\omega_{\alpha}(x) = \omega_{\alpha}(x; f_0) \in G(D_{\alpha};p)$, for a given $f_0 \in L_p(\mathbb{R}_+)$, such that

$$\frac{d}{dx}\omega_{\alpha}(x) + D_{0+}^{\alpha}\omega_{\alpha}(x) = 0 \ (x > 0;\ 0 < \alpha < 1), \tag{6.1.77}$$

$$\lim_{x\to 0+} ||\omega_{\alpha}(x) - f_0||_{L_p} = 0 \ (1 \leqq p < \infty). \tag{6.1.78}$$

Berens and Westphal [79] proved that D_{0+}^{α} is a bounded linear operator whose domain is dense in $L_p(\mathbb{R}_+)$ and that $\{\alpha :\ \alpha > 0\}$ belongs to the resolvent set of D_{0+}^{α} and showed, by using the Hille-Yoshida theorem, that the Cauchy type problem (6.1.77)-(6.1.78) has a unique solution $\omega_\alpha(x)$ for any given f_0 in terms of a holomorphic contraction semigroup generated by D_{0+}^α. They constructed the solution of this problem in the form $\omega_\alpha(x) = W_\alpha^{(x)} f_0(x)$, where W_α is a semigroup of a class C_0 in $L_p(\mathbb{R}_+)$ for any $x > 0$.

Kochubei [424] studied the following Cauchy problem

$$\left({}^C\!D^{(\alpha)}y\right)(t) = Ay(t) \ (0 < t < T), \ y(0) = y_0, \tag{6.1.79}$$

where A is a closed linear operator on a Banach space X, $y_0 \in X$, and $D^{(\alpha)}$ $(0 < \alpha < 1)$ is the regularized Caputo fractional derivative of the form (6.1.39):

$$\left({}^C\!D^{(\alpha)}y\right)(t) = \frac{1}{\Gamma(1-\alpha)}\left[\frac{d}{dt}\int_0^t \frac{y(s)ds}{(t-s)^\alpha} - \frac{y(0)}{t^\alpha}\right]. \tag{6.1.80}$$

He found conditions on the resolvent $(A - \lambda E)^{-1}$ of the operator A which yield the existence and uniqueness of the solution $y(t)$ to the problem (6.1.78), and gave a formula for this solution. Kostin [443] considered the Cauchy problem for the following fractional operator equation

$$(D_{0+}^\alpha y)(t) = Ay(t) \ (t > 0), \ (D_{0+}^{\alpha-k}y)(0+) = y_k \ \ (k = 1, \cdots, n; \ n-1 < \alpha \leqq n), \tag{6.1.81}$$

with the Riemann-Liouville fractional derivative (2.1.10) of any $\alpha > 0$ and obtained the conditions for the uniform correctness on a compact set of this problem. When $0 < \alpha < 2$, he applied the obtained result to the operator $A = \Delta^{2m+1}$, Δ being the Laplacian (1.3.31), in the Banach space of bounded uniformly continuous functions on $\mathbb{R}^n$.

Using ideas related to the theory of first- and second-order abstract differential equations, Bazhlekova [75] gave necessary and sufficient conditions on an unbounded closed operator A in a Banach space X with dense domain $D \subset X$ such that the abstract Cauchy problem for the following differential equation of fractional order

$$\left({}^C\!D_{0+}^\alpha y\right)(t) = Ay(t) \ (0 \leqq t \leqq T; \ 0 < \alpha \leqq 1), \ y(0) = x \in X, \tag{6.1.82}$$

with the Caputo derivative (2.4.17), can be solved. Bazhlekova [75] gave the conditions on A obtained for the solution $y(t)$ to the problem (6.1.82) to be holomorphic, and relations between the solutions to this problem for different α $(0 < \alpha \leqq 1)$. In particular, these relations showed that (6.1.82) has a holomorphic solution whenever A generates a C_0-semigroup. Gorenflo, Luchko and Zabreiko [301] considered the Cauchy problem

$$\left({}^C\!D_{0+}^\alpha y\right)(t) = Ay(t) \ (0 \leqq t < T \leqq \infty; \ n-1 < \alpha \leqq n; \ n \in \mathbb{N}), \tag{6.1.83}$$

$$y^{(k)}(0) = x_k \in X \ (k = 0, \cdots, n-1), \tag{6.1.84}$$

which is more general than (6.1.82), with an unbounded linear operator A in a Banach space X. They described those initial data of this problem for which the solution may be represented via the Mittag-Leffler function (1.8.17) in the form

$$y(t) = \sum_{k=0}^{n-1} x_k t^k E_{\alpha,k+1}(At^\alpha). \tag{6.1.85}$$

Perturbation properties of the solution $y(t)$ to the problem (6.1.83)-(6.1.84) were investigated by Bazhlekova [76], and the results obtained generalize the known facts about C_0-semigroups and cosine operator functions.

El-Sayed in [225] and [229] studied abstract Cauchy problems for the differential equation of fractional order $(D_{0+}^\gamma y)(t) = Ay(t)$ $(0 \leqq t \leqq T;\ 0 < \gamma \leqq 2)$ with a linear closed operator A on a Banach space X. He reduced the problems considered to certain integral equations containing the abstract operator A, and proved several results for such integral equations. But he *did not prove the equivalence* between the initial-value problems and the corresponding integral equations. The same concerns his papers [227] and [228], where he studied some abstract equations with the right-sided Riemann-Liouville fractional derivative $D_{b-}^\alpha y$ and some nonlinear differential-difference and functional-differential equations of fractional order.

Kempfle and Beyer [364] studied a formal linear operator of the form

$$\mathcal{A} = \sum_{k=0}^{m} d_k D_{-}^{\alpha_k} \tag{6.1.86}$$

with $d_k \in \mathbb{R}$ $(k = 0, \cdots, m)$ and $0 \leqq \alpha_0 < \alpha_1 < \cdots < \alpha_m$. They called (6.1.86) a fractional differential operator and studied the following equation

$$\mathcal{A}x(t) = f(t) \quad (t \in \mathbb{R};\ x(t) \in L_2(\mathbb{R})), \tag{6.1.87}$$

called the related differential equation. They defined the associated symbol $p(\omega)$ of $\mathcal{A}$ by

$$p(s) = \sum_{k=0}^{m} d_k (is)^{\alpha_k} \tag{6.1.88}$$

with the special choice of the branch of the power function $(is)^{\alpha_k}$ $(k = 0, \cdots, m)$. Kempfle and Beyer [364] considered the representation of $\mathcal{A}$ as follows

$$\mathcal{A} = \mathcal{F}^{-1} p(-iD)\mathcal{F}, \quad D = \frac{d}{dx}, \tag{6.1.89}$$

in terms of the direct and inverse Fourier transforms (1.3.1) and (1.3.2). Using (6.1.89), they obtained the explicit solution $x(t) \in L_2(\mathbb{R})$ of (6.1.87) in the form

$$x(t) = (\mathcal{A}^{-1}f)(t) = (2\pi)^{-1/2} \int_{-\infty}^{\infty} \left[\mathcal{F}^{-1}\left(\frac{1}{p(s)}\right)\right](\tau) f(t-\tau)d\tau. \tag{6.1.90}$$

Physical applications were considered, and a special case of the operator (6.1.86) in the form $D^\mu + aD^\nu + b$ was studied. See also Beyer and Kempfle [82], Kempfle and Gaul [365] and Kempfle [363] in this regard. Kempfle and Schaefer [366] compared the above approach for certain fractional differential equations with the usual approach based on the Riemann-Lioville fractional derivative (2.1.10), and they concluded that the results from both approaches coincide in the case of linear equations with constant coefficients.

6.2 Solution of Cauchy Type Problems for Fractional Diffusion-Wave Equations

In this section we consider a fractional differential equation of the form

$$(D_{0+,t}^{\alpha}u)(x,t) = \lambda^2(\Delta_x u)(x,t) \quad (x \in \mathbb{R}^n;\ t > 0;\ 0 < \alpha < 2;\ \lambda > 0), \tag{6.2.1}$$

involving the partial Riemann-Liouville fractional derivative $(D_t^{\alpha}u)(x,t)$ of order $\alpha > 0$ with respect to $t > 0$ defined by (6.1.12), and the Laplacian $(\Delta_x u)(x,t)$ with respect to $x \in \mathbb{R}^n$:

$$(\Delta_x u)(x,t) = \frac{\partial^2 u(x,t)}{\partial x_1^2} + \cdots + \frac{\partial^2 u(x,t)}{\partial x_n^2} \quad (n \in \mathbb{N}). \tag{6.2.2}$$

In particular, when $n = 1$, (6.2.1) takes the form of the equation (6.1.32) known as the fractional diffusion-wave equation (see Section 6.1). Therefore, we call the equation (6.2.1) for $n \geqq 2$ the *multi-dimensional fractional diffusion-wave equation.* We apply the Fourier and Laplace transforms to obtain an explicit solution of the equation (6.2.1) with the Cauchy type initial conditions (6.1.56):

$$\left(D_{0+,t}^{\alpha-k}u\right)(x,0+) = f_k(x), \tag{6.2.3}$$

where $x \in \mathbb{R}^n$, $k = 1$ for $0 < \alpha \leqq 1$,and $k = 2$ for $1 < \alpha < 2$.

First we consider the two-dimensional case $n = 1$.

6.2.1 Cauchy Type Problems for Two-Dimensional Equations

We consider the partial differential equation (6.1.31) of order $0 < \alpha < 2$

$$(D_{0+,t}^{\alpha}u)(x,t) = \lambda^2 \frac{\partial^2 u(x,t)}{\partial x^2} \quad (x \in \mathbb{R};\ t > 0;\ \lambda > 0), \tag{6.2.4}$$

with the Cauchy type initial conditions of the form (6.2.3) for $n = 1$

$$\left(D_{0+,t}^{\alpha-k}u\right)(x,0+) = f_k(x), \tag{6.2.5}$$

where $x \in \mathbb{R}$, $k = 1$ for $0 < \alpha \leqq 1$,and $k = 2$ for $1 < \alpha < 2$.

To solve this problem, we apply the Laplace transform with respect to t:

$$(\mathcal{L}_t u)(x,s) = \int_0^{\infty} u(x,t)e^{-st}dt \quad (x \in \mathbb{R};\ s > 0) \tag{6.2.6}$$

and the Fourier transform with respect to $x \in \mathbb{R}$:

$$(\mathcal{F}_x u)(\sigma,t) = \int_{-\infty}^{\infty} u(x,t)e^{ix\sigma}dx \quad (\sigma \in \mathbb{R};\ t > 0). \tag{6.2.7}$$

Applying the Laplace transform (6.2.6) to (6.2.4), and taking into account the formula of the form (5.2.3)

$$\left(\mathcal{L}_t D_{0+,t}^{\alpha} u\right)(x,s) = s^{\alpha}\left(\mathcal{L}u\right)(x,s) - \sum_{j=1}^{l} s^{j-1}\left(D_{0+,t}^{\alpha-j}u\right)(x,0+), \tag{6.2.8}$$

with $x \in \mathbb{R}$, $l-1 < \alpha \leqq l$ and $l \in \mathbb{N}$. If $l = 1$ and $l = 2$ in the respective cases $0 < \alpha \leqq 1$ and $1 < \alpha < 2$, and the initial conditions in (6.2.5), we have

$$s^{\alpha}\left(\mathcal{L}_t u\right)(x,s) = \sum_{k=1}^{l} s^{k-1} f_k(x) + \lambda^2\left(\frac{\partial^2}{\partial x^2}\mathcal{L}_t u\right)(x,s) \quad (l = 1,2).$$

Applying the Fourier transform (6.2.7) and using the formula (1.3.11) with $k = 2$:

$$\left(\mathcal{F}_x\left[\frac{\partial^2 u(x,t)}{\partial x^2}\right]\right)(\sigma,t) = -|\sigma|^2\left(\mathcal{F}_x u\right)(\sigma,t), \tag{6.2.9}$$

we arrive at the following relation:

$$\left(\mathcal{F}_x\mathcal{L}_t u\right)(\sigma,s) = \sum_{k=1}^{l}\frac{s^{k-1}}{s^{\alpha}+\lambda^2|\sigma|^2}\left(\mathcal{F}_x f_k\right)(\sigma) \quad (\sigma \in \mathbb{R};\ t > 0;\ l = 1,2). \tag{6.2.10}$$

Now we obtain the explicit solution $u(x,t)$ by using the inverse Fourier transform with respect to σ:

$$\left(\mathcal{F}_{\sigma}^{-1}u\right)(x,t) = \frac{1}{2\pi}\int_{-\infty}^{\infty} u(\sigma,t)e^{-i\sigma x}d\sigma \quad (\sigma \in \mathbb{R};\ t > 0) \tag{6.2.11}$$

and the inverse Laplace transform with respect to s:

$$\left(\mathcal{L}_s^{-1}u\right)(x,t) = \frac{1}{2\pi}\int_{\gamma-i\infty}^{\gamma+i\infty} e^{st}u(x,s)ds \quad (x \in \mathbb{R};\ \gamma = \Re(s) > \sigma_{\varphi}). \tag{6.2.12}$$

From readily-accessible tables of the Fourier and Laplace transforms, we have

$$\left(\mathcal{F}_x e^{-c|x|}\right)(\sigma) = \frac{2c}{c^2+|\sigma|^2} \quad (c > 0;\ \sigma \in \mathbb{R}), \tag{6.2.13}$$

$$\left(\mathcal{F}e^{-\frac{|x|}{\lambda}s^{\frac{\alpha}{2}}}\right)(\sigma) = \frac{2\lambda s^{\frac{\alpha}{2}}}{s^{\alpha}+\lambda^2|\sigma|^2}.$$

Hence the relation (6.2.10) takes the form

$$\left(\mathcal{F}_x\mathcal{L}_t u\right)(\sigma,s) = \left(\mathcal{F}_x\left[\frac{1}{2\lambda}\sum_{k=1}^{l} s^{k-1-\frac{\alpha}{2}}e^{-\frac{|x|}{\lambda}s^{\frac{\alpha}{2}}}\right]\right)(\sigma)\left(\mathcal{F}_x f_k\right)(\sigma) \quad (l = 1,2),$$

or, in accordance with the convolution property (1.3.17),

$$(\mathcal{F}_x\mathcal{L}_t u)(\sigma, s) = \left(\mathcal{F}_x\left[\sum_{k=1}^{l}\frac{1}{2\lambda}s^{k-1-\frac{\alpha}{2}}e^{-\frac{|x|}{\lambda}s^{\frac{\alpha}{2}}} *_x f_k(x)\right]\right)(\sigma) \quad (l=1,2).$$

From here, applying the inverse Fourier transform (6.2.11), we derive the following relation:

$$(\mathcal{L}_t u)(x, s) = \sum_{k=1}^{l}\frac{1}{2\lambda}s^{k-1-\frac{\alpha}{2}}e^{-\frac{|x|}{\lambda}s^{\frac{\alpha}{2}}} *_x f_k(x) \quad (x\in\mathbb{R};\ s>0;\ l=1,2). \tag{6.2.14}$$

Applying the inverse Laplace transform in (6.2.14), we can obtain the explicit solution to the Cauchy type problem (6.2.4)-(6.2.5). For this we need to know the inverse Laplace transforms of the functions $s^{k-1-\frac{\alpha}{2}}e^{-\frac{|x|}{\lambda}s^{\frac{\alpha}{2}}}$ $(k=1,2)$. These functions are expressed via the Laplace transform of the Wright function (1.11.1) of the form $\phi(-\alpha/2, b; -z)$. If $0<\alpha<2$, then $\phi(-\alpha/2, b; -z)$ is an entire function of z (See section 1.11). It is easily proved by using (6.2.6) and (1.11.1) that

$$\left(\mathcal{L}_t\left[t^{\frac{\alpha}{2}-k}\phi\left(-\frac{\alpha}{2},\frac{\alpha}{2}-k+1;-\frac{|x|}{\lambda}t^{-\frac{\alpha}{2}}\right)\right]\right)(s) = s^{k-1-\frac{\alpha}{2}}e^{-\frac{|x|}{\lambda}s^{\frac{\alpha}{2}}}. \tag{6.2.15}$$

for $k=1,2$.

Applying the inverse Laplace transform in (6.2.14) and taking (6.2.15) and (1.3.13) into account, we obtain the following result.

Theorem 6.1 *If $0<\alpha<2$ and $\lambda>0$, then the Cauchy type problem* (6.2.4)-(6.2.5) *is solvable, and its solution $u(x,t)$ is given by*

$$u(x,t) = \sum_{k=1}^{l}\int_{-\infty}^{\infty}G_k^{\alpha}(x-\tau,t)f_k(\tau)d\tau \quad (l=1\ for\ 0<\alpha\leqq 1;\ l=2\ for\ 1<\alpha<2), \tag{6.2.16}$$

$$G_k^{\alpha}(x,t) = \frac{1}{2\lambda}t^{\frac{\alpha}{2}-k}\phi\left(-\frac{\alpha}{2},\frac{\alpha}{2}-k+1;-\frac{|x|}{\lambda}t^{-\frac{\alpha}{2}}\right) \quad (k=1,2), \tag{6.2.17}$$

provided that the integrals in the right-hand side of (6.2.16) *are convergent.*

Corollary 6.1 *If $0<\alpha\leqq 1$ and $\lambda>0$, then the Cauchy type problem*

$$(D_{0+,t}^{\alpha}u)(x,t) = \lambda^2\frac{\partial^2 u(x,t)}{\partial x^2}, \quad \left(D_{0+,t}^{\alpha-1}u\right)(x,0+) = f(x) \quad (x\in\mathbb{R};\ t>0) \tag{6.2.18}$$

is solvable, and its solution has the form

$$u(x,t) = \int_{-\infty}^{\infty} G_1^\alpha(x-\tau,t)f(\tau)d\tau, \tag{6.2.19}$$

$$G_1^\alpha(x,t) = \frac{1}{2\lambda}t^{\frac{\alpha}{2}-1}\phi\left(-\frac{\alpha}{2},\frac{\alpha}{2};-\frac{|x|}{\lambda}t^{-\frac{\alpha}{2}}\right), \tag{6.2.20}$$

provided that the integral in the right-hand side of (6.2.19) *is convergent.*

Corollary 6.2 *If* $1 < \alpha < 2$ *and* $\lambda > 0$, *then the Cauchy type problem*

$$(D_{0+,t}^\alpha u)(x,t) = \lambda^2 \frac{\partial^2 u(x,t)}{\partial x^2} \quad (x \in \mathbb{R};\ t > 0) \tag{6.2.21}$$

$$\left(D_{0+,t}^{\alpha-1}u\right)(x,0+) = f_1(x),\ \left(D_{0+,t}^{\alpha-2}u\right)(x,0+) = f_2(x) \quad (x \in \mathbb{R}) \tag{6.2.22}$$

is solvable, and its solution has the form

$$u(x,t) = \int_{-\infty}^{\infty} G_1^\alpha(x-\tau,t)f_1(\tau)d\tau + \int_{-\infty}^{\infty} G_2^\alpha(x-\tau,t)f_2(\tau)d\tau, \tag{6.2.23}$$

where $G_1^\alpha(x,t)$ *is given by* (6.2.20), *that is, by*

$$G_2^\alpha(x,t) = \frac{1}{2\lambda}t^{\frac{\alpha}{2}-2}\phi\left(-\frac{\alpha}{2},\frac{\alpha}{2}-1;-\frac{|x|}{\lambda}t^{-\frac{\alpha}{2}}\right), \tag{6.2.24}$$

provided that the integrals in the right-hand side of (6.2.23) *are convergent.*

Example 6.1 The following Cauchy type problem (6.2.18) with $\alpha = \frac{1}{2}$

$$(D_{0+,t}^{1/2}u)(x,t) = \lambda^2 \frac{\partial^2 u(x,t)}{\partial x^2};\quad \left(I_{0+,t}^{1/2}u\right)(x,0+) = f(x) \quad (x \in \mathbb{R};\ t > 0) \tag{6.2.25}$$

has its solution given by

$$u(x,t) = \int_{-\infty}^{\infty} G_1^{1/2}(x-\tau,t)f(\tau)d\tau, \tag{6.2.26}$$

where $G_1^{1/2}(x,t)$ is given by (6.2.20) with $\alpha = 1/2$.

Example 6.2 The following Cauchy type problem (6.2.21)-(6.2.22) with $\alpha = \frac{3}{2}$

$$(D_{0+,t}^{3/2}u)(x,t) = \lambda^2 \frac{\partial^2 u(x,t)}{\partial x^2} \quad (x \in \mathbb{R};\ t > 0) \tag{6.2.27}$$

$$\left(D_{0+,t}^{1/2}u\right)(x,0+) = f_1(x),\ \left(I_{0+,t}^{1/2}u\right)(x,0+) = f_2(x) \quad (x \in \mathbb{R}) \tag{6.2.28}$$

has its solution given by

$$u(x,t) = \int_{-\infty}^{\infty} G_1^{3/2}(x-\tau,t)f_1(\tau)d\tau + \int_{-\infty}^{\infty} G_2^{3/2}(x-\tau,t)f_2(\tau)d\tau, \qquad (6.2.29)$$

where $G_1^{3/2}(x,t)$ and $G_2^{3/2}(x,t)$ are given by (6.2.20) and (6.2.24) with $\alpha = \frac{3}{2}$.

Example 6.3 The following Cauchy problem for the heat equation (6.1.45)

$$\frac{\partial u(x,t)}{\partial t} = \lambda^2 \frac{\partial^2 u(x,t)}{\partial x^2}, \quad u(x,0) = f(x) \quad (x \in \mathbb{R};\ t > 0) \qquad (6.2.30)$$

has its solution given by

$$u(x,t) = \int_{-\infty}^{\infty} G(x-\tau,t)f(\tau)d\tau, \quad G(x,t) = \frac{1}{2\lambda\sqrt{\pi}} t^{-1/2} e^{-\frac{|x|^2}{4\lambda^2 t}}. \qquad (6.2.31)$$

This well-known result follows from Corollary 6.1, if we take into account the following relation for the Wright function

$$\phi\left(-\frac{1}{2}, \frac{1}{2}; z\right) = \frac{1}{\sqrt{\pi}} e^{-\frac{z^2}{4}}, \qquad (6.2.32)$$

which is readily verified by using (1.11.1).

Remark 6.1 The explicit solution of the Cauchy type problem (6.2.18) in the form (6.2.19) with $G_1^{\alpha}(x,t)$ given by (6.2.20) was obtained by Podlubny ([682], Example 4.4), who used a different notation.

6.2.2 Cauchy Type Problems for Multi-Dimensional Equations

In this section we extend the results of Section 6.2.1 to the multi-dimensional fractional diffusion-wave equation (6.2.1) of order $0 < \alpha < 2$ with the Cauchy type conditions (6.2.3). To solve this problem, we apply the Laplace transform with respect to t:

$$(\mathcal{L}_t u)(x,s) = \int_0^{\infty} u(x,t)e^{-st}dt \quad (x \in \mathbb{R}^n;\ s > 0) \qquad (6.2.33)$$

and the multi-dimensional Fourier transform with respect to $x \in \mathbb{R}^n$:

$$(\mathcal{F}_x u)(\sigma,t) = \int_{\mathbb{R}^n} u(x,t)e^{ix\cdot\sigma}dx \quad (\sigma \in \mathbb{R}^n;\ t > 0;\ l = 1,2). \qquad (6.2.34)$$

Applying the Laplace transform (6.2.33) to (6.2.1), taking into account the formula (6.2.8) with $x \in \mathbb{R}^n$ and the initial conditions in (6.2.3), we have

$$s^{\alpha}\left(\mathcal{L}_t u\right)(x,s)=\sum_{k=1}^{l}s^{k-1}f_k(x)+\lambda^2(\Delta_x\mathcal{L}_t u)(x,s)\quad (l=1,2).$$

Applying the Fourier transform (6.2.34) and using a formula of the form (1.3.32) given by

$$\left(\mathcal{F}_x\Delta_x u\right)(\sigma,t)=-|\sigma|^2\left(\mathcal{F}_x u\right)(\sigma,t), \tag{6.2.35}$$

we arrive at the following relation of the form (6.2.10):

$$\left(\mathcal{F}_x\mathcal{L}_t u\right)(\sigma,s)=\sum_{k=1}^{l}\frac{s^{k-1}}{s^{\alpha}+\lambda^2|\sigma|^2}\left(\mathcal{F}_x f_k\right)(\sigma)\quad (\sigma\in\mathbb{R}^n;\ t>0;\ l=1,2). \tag{6.2.36}$$

Now we obtain the explicit solution $u(x,t)$ by using the inverse Fourier transform with respect to σ:

$$\left(\mathcal{F}_\sigma^{-1}u\right)(x,t)=\frac{1}{(2\pi)^{n/2}}\int_{\mathbb{R}^n}u(\sigma,t)e^{-i\sigma\cdot x}d\sigma\quad (\sigma\in\mathbb{R}^n;\ t>0) \tag{6.2.37}$$

and the inverse Laplace transform with respect to s:

$$\left(\mathcal{L}_s^{-1}u\right)(x,t)=\frac{1}{2\pi}\int_{\gamma-i\infty}^{\gamma+i\infty}e^{st}u(x,s)ds\quad (x\in\mathbb{R}^n;\quad \gamma=\mathrm{Re}(s)>\sigma_\varphi). \tag{6.2.38}$$

To apply the inverse Fourier transform (6.2.37) to (6.2.36), we need an auxiliary assertion involving the Fourier transform of the Macdonald function $K_{(n/2)-1}(|x|)$ defined by (1.7.25).

Lemma 6.1 *For* $n\in\mathbb{N}$ *and* $c>0$, *there holds the following relation:*

$$\left(\mathcal{F}_x\left[|x|^{(2-n)/2}K_{(n-2)/2}(c|x|)\right]\right)(\sigma)=\left(\frac{2\pi}{c}\right)^{n/2}\frac{c}{c^2+|\sigma|^2}\quad (\sigma\in\mathbb{R}^n;\ n\in\mathbb{N}). \tag{6.2.39}$$

Proof. Using (1.3.32) and (5.5.8), we have

$$\left(\mathcal{F}_x\left[|x|^{(2-n)/2}K_{(n/2)-1}(c|x|)\right]\right)(\sigma)$$

$$=\frac{(2\pi)^{n/2}}{|\sigma|^{(n-2)/2}}\int_0^\infty \rho K_{(n-2)/2}(c\rho)J_{(n-2)/2}(|\sigma|\rho)d\rho. \tag{6.2.40}$$

Using a known formula in Prudnikov et al. ([689], Vol. 2, formula 2.16.21.1) and taking (1.6.8) into account, we have

$$\int_0^\infty \rho K_{(n-2)/2}(c\rho)J_{(n-2)/2}(|\sigma|\rho)d\rho=\frac{|\sigma|^{(n-2)/2}}{c^{(n/2)+1}}\ {}_2F_1\left(\frac{n}{2},1;\frac{n}{2};-\frac{|\sigma|^2}{c^2}\right)$$

$$= \left(\frac{|\sigma|}{c}\right)^{(n-2)/2} \frac{1}{c^2 + |\sigma|^2}.$$

Substituting this relation into (6.2.40) yields the result in (6.2.39).

By (6.2.39) with $c = \frac{s^{\alpha/2}}{\lambda}$, we have

$$\frac{1}{\lambda}\left(\mathcal{F}_x\left[\frac{s^{\alpha(n-2)/4}}{(2\lambda\pi)^{n/2}}|x|^{(2-n)/2}K_{(n-2)/2}\left(\frac{|x|}{\lambda}s^{\alpha/2}\right)\right]\right)(\sigma) = \frac{1}{s^\alpha + \lambda^2|\sigma|^2}. \quad (6.2.41)$$

Hence (6.2.36) takes the form

$$(\mathcal{F}_x \mathcal{L}_t u)(\sigma, s) = \left(\mathcal{F}_x\left[\sum_{k=1}^{l} \frac{s^{k-1+\alpha(n-2)/4}}{\lambda(2\lambda\pi)^{n/2}}\right.\right.$$

$$\left.\left.\cdot|x|^{(2-n)/2}K_{(n-2)/2}\left(\frac{|x|}{\lambda}s^{\alpha/2}\right)\right]\right)(\sigma)\,(\mathcal{F}_x f_k)(\sigma),$$

with $l = 1, 2$, or, in accordance with the convolution property (5.5.6),

$$(\mathcal{F}_x \mathcal{L}_t u)(\sigma, s) = \left(\mathcal{F}_x\left[\sum_{k=1}^{l} \frac{s^{k-1+\alpha(n-2)/4}}{\lambda(2\lambda\pi)^{n/2}}\right.\right.$$

$$\left.\left.\cdot|x|^{(2-n)/2}K_{(n-2)/2}\left(\frac{|x|}{\lambda}s^{\alpha/2}\right) *_x f_k(x)\right]\right)(\sigma).$$

Thus, by applying the inverse Fourier transform (6.2.37), we derive that

$$(\mathcal{L}_t u)(x, s) = \sum_{k=1}^{l} \frac{s^{k-1+\alpha(n-2)/4}}{\lambda(2\lambda\pi)^{n/2}}|x|^{(2-n)/2}K_{(n-2)/2}\left(\frac{|x|}{\lambda}s^{\alpha/2}\right) *_x f_k(x), \quad (6.2.42)$$

with $x \in \mathbb{R}^n$, $s > 0$, and $l = 1, 2$.

Applying the inverse Laplace transform in (6.2.38), we can obtain the explicit solution of the Cauchy type problem (6.2.1)-(6.2.3). For this we need to know the inverse Laplace transform of the functions

$$s^{k-1+\alpha(n-2)/4}K_{(n-2)/2}\left(\frac{|x|}{\lambda}s^{\alpha/2}\right) \quad (k = 1, 2). \quad (6.2.43)$$

We show that these functions are expressed in terms of the Laplace transform of some special H-functions (1.12.1) of the form

$$H^{2,0}_{2,2}\left[z \,\middle|\, \begin{matrix}(a, 1/2), (b, \alpha/2)\\(c, 1), (d, 1/2)\end{matrix}\right] = \frac{1}{2\pi i}\int_{\mathcal{L}} \frac{\Gamma(c+\tau)\Gamma\left(d+\frac{\tau}{2}\right)}{\Gamma\left(a+\frac{\tau}{2}\right)\Gamma\left(b+\frac{\alpha\tau}{2}\right)} z^{-\tau}d\tau, \quad (6.2.44)$$

where $a, b, c, d \in \mathbb{R}$ and an infinite contour $\mathcal{L}$ separates all poles of the gamma functions $\Gamma(c+\tau)$ and $\Gamma\left(d+\frac{\tau}{2}\right)$ to the left. By (1.11.16), $\Delta = (2-\alpha)/2$. Thus Theorem 1.4 yields the following assertion.

Lemma 6.2 *If* $0 < \alpha < 2$ *and* $a, b, c, d \in \mathbb{R}$, *then the* H*-function* (6.2.44) *is defined for* $\mathcal{L} = \mathcal{L}_{i\gamma\infty}$ *and* $|\arg(z)| < [(2-\alpha)\pi/4]$, $z \neq 0$.

The functions (6.2.43) can now be expressed in terms of the Laplace transform of the H-function (6.2.44).

Lemma 6.3 *If* $0 < \alpha \leqq 1$ $(k = 1)$ *or* $1 < \alpha < 2$ $(k = 1, 2)$, *then there exist the relations*

$$\left(\mathcal{L}_t\left[t^{-k-\frac{\alpha(n-2)}{4}} H_{2,2}^{2,0}\left[\frac{|x|}{\lambda}t^{-\frac{\alpha}{2}} \,\middle|\, \begin{matrix} \left(\frac{n}{4}, \frac{1}{2}\right), \left(1-k-\frac{\alpha(n-2)}{4}, \frac{\alpha}{2}\right) \\ \left(\frac{n}{2}-1, 1\right), \left(\frac{1}{2}-\frac{n}{4}, \frac{1}{2}\right) \end{matrix}\right]\right]\right)(s)$$

$$= 2^{n/2}\pi^{-1/2} s^{k-1+\frac{\alpha(n-2)}{4}} K_{(n-2)/2}\left(\frac{|x|}{\lambda}s^{\frac{\alpha}{2}}\right). \tag{6.2.45}$$

Proof. Use the representation (6.2.44) for the $H_{2,2}^{2,0}$-function in the integrand of (6.2.45) and choose $\mathcal{L}$ such that $\Re(\tau) > \frac{2}{\alpha} + \frac{n-2}{2}$. Then $-k - \frac{\alpha(n-2)}{4} + \frac{\alpha}{2}\Re(\tau) > -1$ for $0 < \alpha \leqq 1$ $(k = 1)$ and $1 < \alpha < 2$ $(k = 2)$, which ensure the convergence of the integrals below. By (6.2.33) and (6.2.44), we have

$$\left(\mathcal{L}_t\left[t^{-k-\frac{\alpha(n-2)}{4}} H_{2,2}^{2,0}\left[\frac{|x|}{\lambda}t^{-\frac{\alpha}{2}} \,\middle|\, \begin{matrix} \left(\frac{n}{4}, \frac{1}{2}\right), \left(1-k-\frac{\alpha(n-2)}{4}, \frac{\alpha}{2}\right) \\ \left(\frac{n}{2}-1, 1\right), \left(\frac{1}{2}-\frac{n}{4}, \frac{1}{2}\right) \end{matrix}\right]\right]\right)(s)$$

$$= \frac{1}{2\pi i}\int_{\mathcal{L}} \frac{\Gamma\left(\frac{n}{2}-1+\tau\right)\Gamma\left(\frac{1}{2}-\frac{n}{2}+\frac{\tau}{2}\right)}{\Gamma\left(\frac{n}{4}+\frac{\tau}{2}\right)\Gamma\left(1-k-\frac{\alpha(n-2)}{4}+\frac{\alpha}{2}\tau\right)}\left(\frac{|x|}{\lambda}\right)^{-\tau} d\tau$$

$$\cdot \int_0^\infty e^{-st} t^{-k-\frac{\alpha(n-2)}{4}+\frac{\alpha}{2}\tau} dt$$

$$= s^{k-1+\frac{\alpha(n-2)}{4}} \frac{1}{2\pi i}\int_{\mathcal{L}} \frac{\Gamma\left(\frac{n}{2}-1+\tau\right)\Gamma\left(\frac{1}{2}-\frac{n}{2}+\frac{\tau}{2}\right)}{\Gamma\left(\frac{n}{4}+\frac{\tau}{2}\right)}\left(\frac{|x|}{\lambda}s^{\frac{\alpha}{2}}\right)^{-\tau} d\tau$$

$$= s^{k-1+\frac{\alpha(n-2)}{4}} H_{1,2}^{2,0}\left[\frac{|x|}{\lambda}s^{\frac{\alpha}{2}} \,\middle|\, \begin{matrix} \left(\frac{n}{4}, \frac{1}{2}\right) \\ \left(\frac{n}{2}-1, 1\right), \left(\frac{1}{2}-\frac{n}{4}, \frac{1}{2}\right) \end{matrix}\right]. \tag{6.2.46}$$

By (1.12.57), we get

$$H_{1,2}^{2,0}\left[\frac{|x|}{\lambda}s^{\frac{\alpha}{2}} \,\middle|\, \begin{matrix} \left(\frac{n}{4}, \frac{1}{2}\right) \\ \left(\frac{n}{2}-1, 1\right), \left(\frac{1}{2}-\frac{n}{4}, \frac{1}{2}\right) \end{matrix}\right] = 2^{n/2}\pi^{-1/2} K_{(n-2)/2}\left(\frac{|x|}{\lambda}s^{\frac{\alpha}{2}}\right). \tag{6.2.47}$$

Hence (6.2.46) yields (6.2.45), which completes the proof of Lemma 6.3.

Using (6.2.45), rewriting (6.2.42) in the form

$$(\mathcal{L}_t u)(x, s) = \frac{2^{-n}|x|^{(2-n)/2}}{\lambda^{1+\frac{n}{2}}(\pi)^{(n-1)/2}}$$

$$\left(\mathcal{L}_t\left[\sum_{k=1}^{l} t^{-k-\frac{\alpha(n-2)}{4}} H_{2,2}^{2,0}\left[\frac{|x|}{\lambda}t^{-\frac{\alpha}{2}} \,\middle|\, \begin{matrix} \left(\frac{n}{4}, \frac{1}{2}\right), \left(1-k-\frac{\alpha(n-2)}{4}, \frac{\alpha}{2}\right) \\ \left(\frac{n}{2}-1, 1\right), \left(\frac{1}{2}-\frac{n}{4}, \frac{1}{2}\right) \end{matrix}\right] *_x f_k(x)\right]\right)(s)$$

and applying the inverse Laplace transform (6.2.38) we get the following explicit solution $u(x,t)$:

$$u(x,t)=\frac{2^{-n}|x|^{(2-n)/2}}{\lambda^{1+\frac{n}{2}}(\pi)^{(n-1)/2}}\sum_{k=1}^{l}t^{-k-\frac{\alpha(n-2)}{4}}$$

$$\cdot H^{2,0}_{2,2}\left[\frac{|x|}{\lambda}t^{-\frac{\alpha}{2}}\;\middle|\;\begin{matrix}\left(\frac{n}{4},\frac{1}{2}\right),\left(1-k-\frac{\alpha(n-2)}{4},\frac{\alpha}{2}\right)\\ \left(\frac{n}{2}-1,1\right),\left(\frac{1}{2}-\frac{n}{4},\frac{1}{2}\right)\end{matrix}\right]*_x f_k(x), \tag{6.2.48}$$

with $l=1$ and $l=2$ in the cases $0<\alpha\leqq 1$ and $1<\alpha<2$, respectively.

Thus we obtain the following result.

Theorem 6.2 *If $0<\alpha<2$ and $\lambda>0$, then the Cauchy type problem (6.2.1)-(6.2.3) is solvable, and its solution $u(x,t)$ is given by*

$$u(x,t)=\sum_{k=1}^{l}\int_{-\infty}^{\infty}G_k^{\alpha}(x-\tau,t)f_k(\tau)d\tau, \tag{6.2.49}$$

with $l=1$ for $0<\alpha\leqq 1$, $l=2$ for $1<\alpha<2$, and

$$G_k^{\alpha}(x,t)$$

$$=\frac{2^{-n}|x|^{(2-n)/2}}{\lambda^{1+\frac{n}{2}}\pi^{(n-1)/2}}t^{-k-\frac{\alpha(n-2)}{4}}H^{2,0}_{2,2}\left[\frac{|x|}{\lambda}t^{-\frac{\alpha}{2}}\;\middle|\;\begin{matrix}\left(\frac{n}{4},\frac{1}{2}\right),\left(1-k-\frac{\alpha(n-2)}{4},\frac{\alpha}{2}\right)\\ \left(\frac{n}{2}-1,1\right),\left(\frac{1}{2}-\frac{n}{4},\frac{1}{2}\right)\end{matrix}\right], \tag{6.2.50}$$

($k=1,2$), provided that the integrals in the right-hand side of (6.2.49) are convergent.

Corollary 6.3 *If $0<\alpha\leqq 1$, $n\in\mathbb{N}$ and $\lambda>0$, then the Cauchy type problem*

$$(D^{\alpha}_{0+,t}u)(x,t)=\lambda^2(\Delta_x u)(x,t);\quad \left(D^{\alpha-1}_{0+,t}u\right)(x,0+)=f(x), \tag{6.2.51}$$

with $x\in\mathbb{R}^n$, $t>0$, is solvable, and its solution has the form

$$u(x,t)=\int_{\mathbb{R}^n}G_1^{\alpha}(x-\tau,t)f(\tau)d\tau, \tag{6.2.52}$$

$$G_1^{\alpha}(x,t)=\frac{2^{-n}|x|^{(2-n)/2}}{\lambda^{1+\frac{n}{2}}\pi^{(n-1)/2}}t^{-1-\frac{\alpha(n-2)}{4}}H^{2,0}_{2,2}\left[\frac{|x|}{\lambda}t^{-\frac{\alpha}{2}}\;\middle|\;\begin{matrix}\left(\frac{n}{4},\frac{1}{2}\right),\left(-\frac{\alpha(n-2)}{4},\frac{\alpha}{2}\right)\\ \left(\frac{n}{2}-1,1\right),\left(\frac{1}{2}-\frac{n}{4},\frac{1}{2}\right)\end{matrix}\right], \tag{6.2.53}$$

provided that the integral in the right-hand side of (6.2.52) is convergent.

Corollary 6.4 *If $1<\alpha<2$ and $\lambda>0$, then the Cauchy type problem*

$$(D^{\alpha}_{0+,t}u)(x,t)=\lambda^2(\Delta_x u)(x,t)\quad (x\in\mathbb{R}^n;\ t>0) \tag{6.2.54}$$

$$\left(D^{\alpha-1}_{0+,t}u\right)(x,0+)=f_1(x),\ \left(D^{\alpha-2}_{0+,t}u\right)(x,0+)=f_2(x)\quad (x\in\mathbb{R}^n) \tag{6.2.55}$$

is solvable, and its solution has the form

$$u(x,t) = \int_{\mathbb{R}^n} G_1^\alpha(x-\tau,t) f_1(\tau) d\tau + \int_{\mathbb{R}^n} G_2^\alpha(x-\tau,t) f_2(\tau) d\tau, \tag{6.2.56}$$

where $G_1^\alpha(x,t)$ *is given by* (6.2.53) *and*

$$G_2^\alpha(x,t) = \frac{2^{-n}|x|^{(2-n)/2}}{\lambda^{1+\frac{n}{2}}\pi^{(n-1)/2}} t^{-2-\frac{\alpha(n-2)}{4}} H_{2,2}^{2,0}\left[\frac{|x|}{\lambda}t^{-\frac{\alpha}{2}} \middle| \begin{matrix} \left(\frac{n}{4},\frac{1}{2}\right), \left(-1-\frac{\alpha(n-2)}{4},\frac{\alpha}{2}\right) \\ \left(\frac{n}{2}-1,1\right), \left(\frac{1}{2}-\frac{n}{4},\frac{1}{2}\right) \end{matrix}\right], \tag{6.2.57}$$

provided that the integrals in the right-hand side of (6.2.56) *are convergent.*

Example 6.4 The following Cauchy type problem (6.2.51) with $\alpha = 1/2$, $x \in \mathbb{R}^n$ and $t > 0$:

$$(D_{0+,t}^{1/2}u)(x,t) = \lambda^2(\Delta_x u)(x,t); \quad \left(I_{0+,t}^{1/2}u\right)(x,0+) = f(x) \quad () \tag{6.2.58}$$

has solution, and it is given by

$$u(x,t) = \int_{\mathbb{R}^n} G_1^{1/2}(x-\tau,t) f(\tau) d\tau, \tag{6.2.59}$$

where $G_1^{1/2}(x,t)$ is given by (6.2.3) with $\alpha = 1/2$.

Example 6.5 The following Cauchy type problem (6.2.54)-(6.2.55) with $\alpha = 3/2$, $x \in \mathbb{R}^n$ and $t > 0$:

$$(D_{0+,t}^{3/2}u)(x,t) = \lambda^2(\Delta_x u)(x,t) \tag{6.2.60}$$

$$\left(D_{0+,t}^{1/2}u\right)(x,0+) = f_1(x), \quad \left(I_{0+,t}^{1/2}u\right)(x,0+) = f_2(x) \tag{6.2.61}$$

has solution, and it is given by

$$u(x,t) = \int_{\mathbb{R}^n} G_1^{3/2}(x-\tau,t) f(\tau) d\tau + \int_{\mathbb{R}^n} G_2^{3/2}(x-\tau,t) f_2(\tau) d\tau, \tag{6.2.62}$$

where $G_1^{3/2}(x,t)$ and $G_2^{3/2}(x,t)$ are given by (6.2.53) and (6.2.57) with $\alpha = 3/2$.

The result of Corollary 6.3 is simplified for $\alpha = 1$. This is based on the following assertion.

Lemma 6.4 *The function* $G_1^\alpha(x,t)$ *in* (6.2.53) *for* $\alpha = 1$ *has the form*

$$G(x,t) := G_1^1(x,t) = \frac{1}{(2\lambda\sqrt{\pi})^n} t^{-\frac{n}{2}} e^{-\frac{|x|^2}{4\lambda^2 t}}. \tag{6.2.63}$$

Proof. In accordance with the property (1.12.43) of the H-function, we have

$$G(x,t) := G_1^1(x,t) = \frac{2^{-n}|x|^{(2-n)/2}}{\lambda^{1+\frac{n}{2}}\pi^{(n-1)/2}} t^{-\frac{1}{2}-\frac{n}{4}} H_{1,1}^{1,0}\left[\frac{|x|}{\lambda}t^{-\frac{1}{2}} \middle| \begin{matrix} \left(\frac{n}{4},\frac{1}{2}\right) \\ \left(\frac{n}{2}-1,1\right) \end{matrix}\right]$$

$$= \frac{2^{-n}}{\lambda^n \pi^{(n-1)/2}} t^{-\frac{n}{2}} H^{1,0}_{1,1}\left[\frac{|x|}{\lambda} t^{-\frac{1}{2}} \,\middle|\, \begin{matrix} \left(\frac{1}{2}, \frac{1}{2}\right) \\ (0,1) \end{matrix}\right]. \tag{6.2.64}$$

If $0 < \alpha < 2$, then the direct evaluation of the integrand in

$$H^{1,0}_{1,1}\left[z \,\middle|\, \begin{matrix} \left(b, \frac{\alpha}{2}\right) \\ (0,1) \end{matrix}\right] = \frac{1}{2\pi i}\int_{\mathcal{L}} \frac{\Gamma(\tau)}{\Gamma\left(b + \frac{\alpha\tau}{2}\right)} z^{-\tau} d\tau, \tag{6.2.65}$$

where $\alpha > 0$, $b \in \mathbb{R}$ and an infinite contour $\mathcal{L} = \mathcal{L}_{i\gamma\infty}$ separates all poles $s = -j$ $(j \in \mathbb{N}_0)$ of the gamma function $\Gamma(s)$ to the left, yields the following relation:

$$H^{1,0}_{1,1}\left[z \,\middle|\, \begin{matrix} \left(b, \frac{\alpha}{2}\right) \\ (0,1) \end{matrix}\right] = \phi\left(-\frac{\alpha}{2}, b; -z\right). \tag{6.2.66}$$

Then (6.2.64) takes the form

$$G(x,t) := G_1^1(x,t) = \frac{2^{-n}}{\lambda^n \pi^{(n-1)/2}} t^{-\frac{n}{2}} \phi\left(-\frac{1}{2}, \frac{1}{2}; \frac{|x|}{\lambda} t^{-\frac{1}{2}}\right). \tag{6.2.67}$$

Thus the result in (6.2.63) follows from (6.2.32).

Example 6.6 The following Cauchy problem

$$\frac{\partial u(x,t)}{\partial t} = \lambda^2 (\Delta_x u)(x,t), \quad u(x,0) = f(x) \quad (x \in \mathbb{R}^n;\ t > 0) \tag{6.2.68}$$

has its well-known solution given by

$$u(x,t) = \int_{\mathbb{R}^n} G(x-\tau, t) f(\tau) d\tau, \quad G(x,t) = \frac{1}{(2\lambda\sqrt{\pi})^{n/2}} t^{-\frac{n}{2}} e^{-\frac{|x|^2}{4\lambda^2 t}}. \tag{6.2.69}$$

Remark 6.2 If $n = 1$, then (6.2.39) coincides with (6.2.13), because (see (1.7.27))

$$K_{-1/2}(z) = \left(\frac{\pi}{2z}\right)^{1/2} e^{-z}, \tag{6.2.70}$$

while the relations (6.2.45) coincide with (6.2.15). Indeed, by the property (1.12.43) of the H-function, we have

$$H^{2,0}_{2,2}\left[\frac{|x|}{\lambda} t^{-\frac{\alpha}{2}} \,\middle|\, \begin{matrix} \left(\frac{1}{4}, \frac{1}{2}\right), \left(1-k+\frac{\alpha}{4}, \frac{\alpha}{2}\right) \\ \left(-\frac{1}{2}, 1\right), \left(\frac{1}{4}, \frac{1}{2}\right) \end{matrix}\right] = H^{1,0}_{1,1}\left[\frac{|x|}{\lambda} t^{-\frac{\alpha}{2}} \,\middle|\, \begin{matrix} \left(1-k+\frac{\alpha}{4}, \frac{\alpha}{2}\right) \\ \left(-\frac{1}{2}, 1\right) \end{matrix}\right], \tag{6.2.71}$$

and, in accordance with (6.2.70), the formula (6.2.45) takes the form

$$\left(\frac{|x|}{\lambda}\right)^{1/2} \left(\mathcal{L}_t\left[t^{\frac{\alpha}{4}-k} H^{1,0}_{1,1}\left[\frac{|x|}{\lambda} t^{-\frac{\alpha}{2}} \,\middle|\, \begin{matrix} \left(1-k+\frac{\alpha}{4}, \frac{\alpha}{2}\right) \\ \left(-\frac{1}{2}, 1\right) \end{matrix}\right]\right]\right)(s) = s^{k-1-\frac{\alpha}{2}} e^{-\frac{|x|}{\lambda} s^{\frac{\alpha}{2}}}. \tag{6.2.72}$$

By the property (1.12.45) of the H-function with $\sigma = 1/2$, we have

$$\left(\frac{|x|}{\lambda} t^{-\frac{\alpha}{2}}\right)^{1/2} H^{1,0}_{1,1}\left[\frac{|x|}{\lambda} t^{-\frac{\alpha}{2}} \,\middle|\, \begin{matrix} \left(1-k+\frac{\alpha}{4}, \frac{\alpha}{2}\right) \\ \left(-\frac{1}{2}, 1\right) \end{matrix}\right]$$

$$= H^{1,0}_{1,1}\left[\frac{|x|}{\lambda} t^{-\frac{\alpha}{2}} \,\middle|\, \begin{matrix} \left(1-k+\frac{\alpha}{2}, \frac{\alpha}{2}\right) \\ (0,1) \end{matrix}\right]. \tag{6.2.73}$$

Then, in accordance with (6.2.66) and (6.2.70), (6.2.45) with $n = 1$ yields (6.2.15).

Remark 6.3 When $n = 1$, then, in accordance with (6.2.71),(6.2.73) and (6.2.66), the functions $G_k^\alpha(x,t)$ $(k = 1, 2)$ in (6.2.50) coincide with those in (6.2.17). Thus Theorem 6.1 and Corollaries 6.1-6.2 follow from Theorem 6.3 and Corollaries 6.3-6.4, respectively. This also means that Examples 6.1-6.3 are particular cases of Examples 6.4-6.6 for $n = 1$.

Remark 6.4 By analogy with Green functions for partial differential equations, functions $G_1^\alpha(x,t)$ and $G_2^\alpha(x,t)$, given in (6.2.20), (6.2.53) and (6.2.24), (6.2.57) can be called *fractional Green functions* (see Section 6.1). In particular, the functions $G_1^1(x,t)$ yield the usual Green functions $G(x,t)$ in (6.2.31) and (6.2.69).

Remark 6.5 The fundamental solution of the Cauchy type problem (6.2.51) with $0 < \alpha < 1$ was obtained by Hilfer [345] in terms of the $H_{1,2}^{2,0}$-function (1.12.1).

Remark 6.6 Results presented in this section were given in Voroshilov and Kilbas [861], [862] and Kilbas at al. [413].

6.3 Solution of Cauchy Problems for Fractional Diffusion-Wave Equations

In Section 6.2 we established explicit solutions of Cauchy type problems for fractional diffusion-wave partial differential equations involving the Riemann-Liouville partial fractional derivatives $(D_{0+,t}^\alpha u)(x,t)$ of order $\alpha > 0$. In this section we consider the fractional differential equation of the form (6.2.1)

$$({}^CD_{t,0+}^\alpha u)(x,t) = \lambda^2(\Delta_x u)(x,t) \quad (x \in \mathbb{R}^n;\ t > 0;\ 0 < \alpha < 2;\ \lambda > 0), \tag{6.3.1}$$

involving the partial Caputo fractional derivative $({}^CD_{t,0+}^\alpha u)(x,t)$ with respect to $t > 0$, and the Laplacian $(\Delta_x u)(x,t)$ with respect to $x \in \mathbb{R}^n$ given in (6.2.2). Just as in (2.4.1), the above Caputo derivative of order $\alpha > 0$ and $l-1 < \alpha < l$ $(l \in \mathbb{N})$ is defined in terms of the Riemann-Liouville partial fractional derivative (6.1.12) by

$$({}^CD_{0+,t}^\alpha u)(x,t) = \left(D_{0+,t}^\alpha\left[u(x,\tau) - \sum_{k=0}^{l-1}\frac{\partial^k u(x,0)}{\partial t^l}\frac{\tau^k}{k!}\right]\right)(x,t). \tag{6.3.2}$$

When, for any fixed $x \in \mathbb{R}$, $u(x,t) \in C^l(\mathbb{R}_+)$ as a function of $t > 0$, then $({}^CD_{0+,t}^\alpha u)(x,t)$ has the representation (6.1.50). Following an approach developed in Section 6.2, we apply the Fourier and Laplace transforms to obtain an explicit solution to the equation (6.2.1) with the Cauchy initial conditions

$$\frac{\partial^k u(x,0)}{\partial t^k} = f_k(x) \quad (x \in \mathbb{R}^n;\ k = 0 \text{ for } 0 < \alpha \leqq 1;\ k = 1 \text{ for } 1 < \alpha < 2). \tag{6.3.3}$$

$$\frac{\partial^0 u(x,0)}{\partial t^0} := u(x,0). \tag{6.3.4}$$

As in Section 6.2, we begin from the two-dimensional (case $n = 1$).

6.3.1 Cauchy Problems for Two-Dimensional Equations

We consider the following partial differential equation (6.2.1) of order $0 < \alpha < 2$

$$({}^C D^{\alpha}_{0+,t} u)(x,t) = \lambda^2 \frac{\partial^2 u(x,t)}{\partial x^2} \quad (x \in \mathbb{R};\ t > 0;\ \lambda > 0), \tag{6.3.5}$$

with the Cauchy conditions (6.3.3).

Applying the Laplace transform (6.2.6) with respect to t and the Fourier transform (6.2.7) with respect to $x \in \mathbb{R}$ and using the following relation of the form (5.3.3)

$$\left(\mathcal{L}_t\ {}^C D^{\alpha}_{0+,t} u\right)(x,s) = s^{\alpha}\left(\mathcal{L}u\right)(x,s) - \sum_{k=0}^{l-1} s^{\alpha-k-1} \frac{\partial^k u(x,0)}{\partial t^k} \tag{6.3.6}$$

$$(x \in \mathbb{R};\ l-1 < \alpha \leqq l;\ l \in \mathbb{N}),$$

where $x \in \mathbb{R}$, $l-1 < \alpha \leqq l$, $l \in \mathbb{N}$, and with $l = 1$ and $l = 2$ in the respective cases $0 < \alpha \leqq 1$ and $1 < \alpha < 2$, the initial conditions in (6.3.3) and the formula (6.2.9), we arrive at the relation of the form (6.2.10)

$$\left(\mathcal{F}_x\mathcal{L}_t u\right)(\sigma,s) = \sum_{k=0}^{l-1} \frac{s^{\alpha-k-1}}{s^{\alpha}+\lambda^2|\sigma|^2} \left(\mathcal{F}_x f_k\right)(\sigma) \quad (\sigma \in \mathbb{R};\ t > 0;\ l = 1,2). \tag{6.3.7}$$

By (6.2.13), we have

$$\left(\mathcal{F}_x\left[\frac{1}{2\lambda} s^{\frac{\alpha}{2}-k-1} e^{-\frac{|x|}{\lambda}s^{\frac{\alpha}{2}}}\right]\right)(\sigma) = \frac{s^{\alpha-k-1}}{s^{\alpha}+\lambda^2|\sigma|^2} \quad (k = 0,1),$$

and hence (6.3.7) takes the following form:

$$\left(\mathcal{F}_x\mathcal{L}_t u\right)(\sigma,s) = \left(\mathcal{F}_x\left[\sum_{k=0}^{l-1} \frac{1}{2\lambda} s^{\frac{\alpha}{2}-k-1} e^{-\frac{|x|}{\lambda}s^{\frac{\alpha}{2}}}\right]\right)(\sigma)\left(\mathcal{F}_x f_k\right)(\sigma) \quad (l = 1,2),$$

or, by the convolution property (1.3.17),

$$\left(\mathcal{F}_x\mathcal{L}_t u\right)(\sigma,s) = \left(\mathcal{F}_x\left[\sum_{k=0}^{l-1} \frac{1}{2\lambda} s^{\frac{\alpha}{2}-k-1} e^{-\frac{|x|}{\lambda}s^{\frac{\alpha}{2}}} *_x f_k\right]\right)(\sigma) \quad (l = 1,2).$$

Applying the inverse Fourier transform (6.2.11) to this formula, we obtain the relation

$$\left(\mathcal{L}_t u\right)(x,s) = \sum_{k=0}^{l-1} \frac{1}{2\lambda} s^{\frac{\alpha}{2}-k-1} e^{-\frac{|x|}{\lambda}s^{\frac{\alpha}{2}}} *_x f_k(x) \quad (l = 1,2). \tag{6.3.8}$$

When $0 < \alpha < 2$, then the functions $s^{\frac{\alpha}{2}-k-1} e^{-\frac{|x|}{\lambda}s^{\frac{\alpha}{2}}}$ $(k = 0,1)$ are expressed via the Laplace transform of the Wright function $\phi(-\alpha/2, b; z)$ as follows:

$$\left(\mathcal{L}_t\left[t^{k-\frac{\alpha}{2}}\phi\left(-\frac{\alpha}{2},k+1-\frac{\alpha}{2};-\frac{|x|}{\lambda}t^{-\frac{\alpha}{2}}\right)\right]\right)(s)=s^{\frac{\alpha}{2}-k-1}e^{-\frac{|x|}{\lambda}s^{\frac{\alpha}{2}}}\quad(k=0,1). \tag{6.3.9}$$

Applying the inverse Laplace transform to (6.3.8) and taking (6.3.9) and (1.3.13) into account, we obtain the following result.

Theorem 6.3 *If $0<\alpha<2$ and $\lambda>0$, then the Cauchy problem* (6.3.5), (6.3.3) *is solvable, and its solution $u(x,t)$ is given by*

$$u(x,t)=\sum_{k=0}^{l-1}\int_{-\infty}^{\infty}G_k^{\alpha}(x-\tau,t)f_k(\tau)d\tau\quad(l=0\ for\ 0<\alpha\leqq 1,\ l=1\ for\ 1<\alpha<2), \tag{6.3.10}$$

$$G_k^{\alpha}(x,t)=\frac{1}{2\lambda}t^{k-\frac{\alpha}{2}}\psi\left(-\frac{\alpha}{2},k+1-\frac{\alpha}{2};\frac{|x|}{\lambda}t^{-\frac{\alpha}{2}}\right)\quad(k=0,1), \tag{6.3.11}$$

provided that the integrals in the right-hand side of (6.3.10) *are convergent.*

Corollary 6.5 *If $0<\alpha<1$ and $\lambda>0$, then the Cauchy problem*

$$({}^C D_{0+,t}^{\alpha}u)(x,t)=\lambda^2\frac{\partial^2 u(x,t)}{\partial x^2},\quad u(x,0)=f(x)\quad(x\in\mathbb{R};\ t>0) \tag{6.3.12}$$

is solvable, and its solution has the form

$$u(x,t)=\int_{-\infty}^{\infty}G_1^{\alpha}(x-\tau,t)f(\tau)d\tau, \tag{6.3.13}$$

$$G_1^{\alpha}(x,t)=\frac{1}{2\lambda}t^{-\frac{\alpha}{2}}\phi\left(-\frac{\alpha}{2},1-\frac{\alpha}{2};-\frac{|x|}{\lambda}t^{-\frac{\alpha}{2}}\right), \tag{6.3.14}$$

provided that the integral in the right-hand side of (6.3.13) *is convergent.*

Corollary 6.6 *If $1<\alpha<2$ and $\lambda>0$, then the Cauchy problem*

$$({}^C D_{0+,t}^{\alpha}u)(x,t)=\lambda^2\frac{\partial^2 u(x,t)}{\partial x^2}\quad(x\in\mathbb{R};\ t>0) \tag{6.3.15}$$

$$u(x,0)=f_0(x),\quad\frac{\partial u(x,0)}{\partial t}=f_1(x)\quad(x\in\mathbb{R}) \tag{6.3.16}$$

is solvable, and its solution has the form

$$u(x,t)=\int_{-\infty}^{\infty}G_1^{\alpha}(x-\tau,t)f_0(\tau)d\tau+\int_{-\infty}^{\infty}G_2^{\alpha}(x-\tau,t)f_1(\tau)d\tau, \tag{6.3.17}$$

where $G_1^{\alpha}(x,t)$ is given by (6.3.14) *and*

$$G_2^{\alpha}(x,t)=\frac{1}{2\lambda}t^{1-\frac{\alpha}{2}}\phi\left(-\frac{\alpha}{2},2-\frac{\alpha}{2};-\frac{|x|}{\lambda}t^{-\frac{\alpha}{2}}\right), \tag{6.3.18}$$

provided that the integrals in the right-hand side of (6.3.17) *are convergent.*

Example 6.7 The following Cauchy problem (6.3.12) with $\alpha = \frac{1}{2}$

$$({}^C D_{0+,t}^{1/2} u)(x,t) = \lambda^2 \frac{\partial^2 u(x,t)}{\partial x^2}; \quad u(x,0) = f(x) \quad (x \in \mathbb{R};\ t > 0) \tag{6.3.19}$$

has its solution given by

$$u(x,t) = \int_{-\infty}^{\infty} G_1^{1/2}(x-\tau,t) f(\tau) d\tau, \tag{6.3.20}$$

where $G_1^{1/2}(x,t)$ is given by (6.3.14) with $\alpha = \frac{1}{2}$.

Example 6.8 The following Cauchy problem (6.3.15)-(6.3.16) with $\alpha = \frac{3}{2}$

$$({}^C D_{0+,t}^{3/2} u)(x,t) = \lambda^2 \frac{\partial^2 u(x,t)}{\partial x^2} \quad (x \in \mathbb{R};\ t > 0) \tag{6.3.21}$$

$$u(x,0) = f_0(x), \quad \frac{\partial u(x,0)}{\partial t} = f_1(x) \tag{6.3.22}$$

has its solution given by

$$u(x,t) = \int_{-\infty}^{\infty} G_1^{3/2}(x-\tau,t) f_0(\tau) d\tau + \int_{-\infty}^{\infty} G_2^{3/2}(x-\tau,t) f_1(\tau) d\tau, \tag{6.3.23}$$

where $G_1^{3/2}(x,t)$ and $G_2^{3/2}(x,t)$ are given by (6.3.14) and (6.3.18) with $\alpha = \frac{3}{2}$.

Remark 6.7 The explicit solution of the Cauchy problem (6.3.12) in the form (6.3.13) with $G_1^{\alpha}(x,t)$ given by (6.3.14) was constructed by Mainardi [515] [see also Gorenflo et al.[310] and Podlubny ([682], Example 4.5)].

6.3.2 Cauchy Problems for Multi-Dimensional Equations

In this section we extend the results of Section 6.3.1 to the multi-dimensional fractional diffusion-wave equation (6.3.1) of order $0 < \alpha < 2$ with the Cauchy conditions (6.3.3). Applying the Laplace transform (6.2.33) with respect to $t > 0$ and the Fourier transform (6.2.34) with respect to $x \in \mathbb{R}^n$, and using the relation (6.3.6), the initial conditions (6.3.3) and the formula (6.2.35), we arrive at the following relation of the form (6.3.7):

$$\left(\mathcal{F}_x \mathcal{L}_t u\right)(\sigma,s) = \sum_{k=0}^{l-1} \frac{s^{\alpha-k-1}}{s^{\alpha} + \lambda^2 |\sigma|^2} \left(\mathcal{F}_x f_k\right)(\sigma) \quad (\sigma \in \mathbb{R}^n;\ t > 0;\ l = 1, 2). \tag{6.3.24}$$

According to (6.2.39) with $c = \frac{s^{\frac{\alpha}{2}}}{\lambda}$, this formula takes the form

$$\left(\mathcal{F}_x \mathcal{L}_t u\right)(\sigma,s)$$

$$= \left(\mathcal{F}_x \left[\sum_{k=0}^{l-1} \frac{s^{\frac{\alpha(n+2)}{4}-k-1}}{\lambda(2\lambda\pi)^{n/2}} |x|^{(2-n)/2} K_{(n-2)/2}\left(\frac{|x|}{\lambda} s^{\alpha/2}\right)\right]\right)(\sigma) \left(\mathcal{F}_x f_k\right)(\sigma),$$

with $l = 1, 2$, or, in accordance with the convolution property (5.5.6),

$$(\mathcal{F}_x \mathcal{L}_t u)(\sigma, s)$$

$$= \left(\mathcal{F}_x \left[\sum_{k=0}^{l-1} \frac{s^{\frac{\alpha(n+2)}{4} - k - 1}}{\lambda(2\lambda\pi)^{n/2}} |x|^{(2-n)/2} K_{(n-2)/2}\left(\frac{|x|}{\lambda} s^{\alpha/2} \right) *_x f_k(x) \right] \right)(\sigma).$$

Now, by applying the inverse Fourier transform (6.2.37), we derive that

$$(\mathcal{L}_t u)(x, s) = \sum_{k=0}^{l-1} \frac{s^{\frac{\alpha(n+2)}{4} - k - 1}}{\lambda(2\lambda\pi)^{n/2}} |x|^{(2-n)/2} K_{(n-2)/2}\left(\frac{|x|}{\lambda} s^{\alpha/2} \right) *_x f_k(x) \quad (6.3.25)$$

$$(x \in \mathbb{R}^n;\ s > 0;\ l = 1, 2).$$

The functions

$$s^{\frac{\alpha(n+2)}{4} - k - 1} K_{(n-2)/2}\left(\frac{|x|}{\lambda} s^{\alpha/2} \right) \quad (k = 0, 1) \quad (6.3.26)$$

are expressed via the Laplace transform of the $H^{2,0}_{2,2}$-function (6.2.44). There exists the following assertion, which can be proved as Lemma 6.3.

Lemma 6.5 *If* $0 < \alpha \leqq 1$ $(k = 0, 1)$ *or* $1 < \alpha < 2$ $(k = 1)$, *then there exist the following relations:*

$$\left(\mathcal{L}_t \left[t^{k - \frac{\alpha(n+2)}{4}} H^{2,0}_{2,2} \left[\frac{|x|}{\lambda} t^{-\frac{\alpha}{2}} \;\middle|\; \begin{matrix} \left(\frac{n}{4}, \frac{1}{2}\right), \left(1 + k - \frac{\alpha(n+2)}{4}, \frac{\alpha}{2}\right) \\ \left(\frac{n}{2} - 1, 1\right), \left(\frac{1}{2} - \frac{n}{4}, \frac{1}{2}\right) \end{matrix} \right] \right] \right)(s)$$

$$= 2^{n/2} \pi^{-1/2} s^{\frac{\alpha(n+2)}{4} - k - 1} K_{(n-2)/2}\left(\frac{|x|}{\lambda} s^{\frac{\alpha}{2}} \right). \quad (6.3.27)$$

Using (6.3.27), rewriting (6.3.25) in the form

$$(\mathcal{L}_t u)(x, s) = \frac{2^{-n} |x|^{(2-n)/2}}{\lambda^{1+\frac{n}{2}} (\pi)^{(n-1)/2}}$$

$$\cdot \left(\mathcal{L}_t \left[\sum_{k=0}^{l-1} t^{k - \frac{\alpha(n+2)}{4}} H^{2,0}_{2,2} \left[\frac{|x|}{\lambda} t^{-\frac{\alpha}{2}} \;\middle|\; \begin{matrix} \left(\frac{n}{4}, \frac{1}{2}\right), \left(1 + k - \frac{\alpha(n+2)}{4}, \frac{\alpha}{2}\right) \\ \left(\frac{n}{2} - 1, 1\right), \left(\frac{1}{2} - \frac{n}{4}, \frac{1}{2}\right) \end{matrix} \right] *_x f_k(x) \right] \right)(s),$$

and applying the inverse Laplace transform (6.2.38), we are led to the explicit solution $u(x, t)$:

$$u(x, t) = \frac{2^{-n} |x|^{(2-n)/2}}{\lambda^{1+\frac{n}{2}} (\pi)^{(n-1)/2}} \sum_{k=1}^{l} t^{k - \frac{\alpha(n+2)}{4}}$$

$$\cdot H^{2,0}_{2,2} \left[\frac{|x|}{\lambda} t^{-\frac{\alpha}{2}} \;\middle|\; \begin{matrix} \left(\frac{n}{4}, \frac{1}{2}\right), \left(1 + k - \frac{\alpha(n+2)}{4}, \frac{\alpha}{2}\right) \\ \left(\frac{n}{2} - 1, 1\right), \left(\frac{1}{2} - \frac{n}{4}, \frac{1}{2}\right) \end{matrix} \right] *_x f_k(x), \quad (6.3.28)$$

with $l = 1$ and $l = 2$ in the respective cases $0 < \alpha \leqq 1$ and $1 < \alpha < 2$.

Thus we obtain the following result.

Theorem 6.4 *If $0 < \alpha < 2$ and $\lambda > 0$, then the Cauchy problem* (6.3.1), (6.3.3) *is solvable, and its solution $u(x,t)$ is given by*

$$u(x,t) = \sum_{k=0}^{l-1} \int_{-\infty}^{\infty} G_k^\alpha(x-\tau,t) f_k(\tau) d\tau \ \ (l = 0 \ for \ 0 < \alpha \leqq 1; \ l = 1 \ for \ 1 < \alpha < 2), \tag{6.3.29}$$

$$G_k^\alpha(x,t)$$
$$= \frac{2^{-n}|x|^{(2-n)/2}}{\lambda^{1+\frac{n}{2}}\pi^{(n-1)/2}} t^{k-\frac{\alpha(n+2)}{4}} H_{2,2}^{2,0}\left[\frac{|x|}{\lambda} t^{-\frac{\alpha}{2}} \left| \begin{matrix} \left(\frac{n}{4},\frac{1}{2}\right), \left(1+k-\frac{\alpha(n+2)}{4},\frac{\alpha}{2}\right) \\ \left(\frac{n}{2}-1,1\right), \left(\frac{1}{2}-\frac{n}{4},\frac{1}{2}\right) \end{matrix} \right. \right], \tag{6.3.30}$$

where $k = 0, 1$, provided that the integrals in the right-hand side of (6.3.29) *are convergent.*

Corollary 6.7 *If $0 < \alpha < 1$, $n \in \mathbb{N}$ and $\lambda > 0$, then the Cauchy problem*

$$({}^C D_{0+,t}^\alpha u)(x,t) = \lambda^2 (\Delta_x u)(x,t), \ \ u(x,0) = f(x) \ \ (x \in \mathbb{R}^n; \ t > 0) \tag{6.3.31}$$

is solvable, and its solution has the form

$$u(x,t) = \int_{\mathbb{R}^n} G_1^\alpha(x-\tau,t) f(\tau) d\tau, \tag{6.3.32}$$

$$G_1^\alpha(x,t) = \frac{2^{-n}|x|^{(2-n)/2}}{\lambda^{1+\frac{n}{2}}\pi^{(n-1)/2}} t^{-\frac{\alpha(n+2)}{4}} H_{2,2}^{2,0}\left[\frac{|x|}{\lambda} t^{-\frac{\alpha}{2}} \left| \begin{matrix} \left(\frac{n}{4},\frac{1}{2}\right), \left(1-\frac{\alpha(n+2)}{4},\frac{\alpha}{2}\right) \\ \left(\frac{n}{2}-1,1\right), \left(\frac{1}{2}-\frac{n}{4},\frac{1}{2}\right) \end{matrix} \right. \right], \tag{6.3.33}$$

provided that the integral in the right-hand side of (6.3.32) *is convergent.*

Corollary 6.8 *If $1 < \alpha < 2$ and $\lambda > 0$, then the Cauchy problem*

$$({}^C D_{0+,t}^\alpha u)(x,t) = \lambda^2 (\Delta_x u)(x,t) \ \ (x \in \mathbb{R}^n; \ t > 0) \tag{6.3.34}$$

$$u(x,0) = f_0(x), \ \ \frac{\partial u(x,0)}{\partial t} = f_1(x) \ \ (x \in \mathbb{R}^n) \tag{6.3.35}$$

is solvable, and its solution has the form

$$u(x,t) = \int_{\mathbb{R}^n} G_1^\alpha(x-\tau,t) f_0(\tau) d\tau + \int_{\mathbb{R}^n} G_2^\alpha(x-\tau,t) f_1(\tau) d\tau, \tag{6.3.36}$$

where $G_1^\alpha(x,t)$ is given by (6.3.33) *and*

$$G_2^\alpha(x,t) = \frac{2^{-n}|x|^{(2-n)/2}}{\lambda^{1+\frac{n}{2}}\pi^{(n-1)/2}} t^{1-\frac{\alpha(n+2)}{4}} H_{2,2}^{2,0}\left[\frac{|x|}{\lambda} t^{-\frac{\alpha}{2}} \left| \begin{matrix} \left(\frac{n}{4},\frac{1}{2}\right), \left(2-\frac{\alpha(n+2)}{4},\frac{\alpha}{2}\right) \\ \left(\frac{n}{2}-1,1\right), \left(\frac{1}{2}-\frac{n}{4},\frac{1}{2}\right) \end{matrix} \right. \right], \tag{6.3.37}$$

provided that the integrals in the right-hand side of (6.3.36) *are convergent.*

Example 6.9 The following Cauchy problem (6.3.31) with $\alpha = 1/2$

$$({}^C D_{0+,t}^{1/2} u)(x,t) = \lambda^2 (\Delta_x u)(x,t), \quad u(x,0) = f(x) \quad (x \in \mathbb{R}^n, \ t > 0) \tag{6.3.38}$$

has its solution given by

$$u(x,t) = \int_{\mathbb{R}^n} G_1^{1/2}(x - \tau, t) f(\tau) d\tau, \tag{6.3.39}$$

where $G_1^{1/2}(x,t)$ is given by (6.3.33) with $\alpha = 1/2$.

Example 6.10 The following Cauchy problem (6.3.34)-(6.3.35) with $\alpha = 3/2$

$$({}^C D_{0+,t}^{3/2} u)(x,t) = \lambda^2 (\Delta_x u)(x,t) \quad (x \in \mathbb{R}^n; \ t > 0) \tag{6.3.40}$$

$$u(x,0) = f_0(x), \quad \frac{\partial u(x,0)}{\partial t} = f_1(x) \quad (x \in \mathbb{R}^n) \tag{6.3.41}$$

has its solution given by

$$u(x,t) = \int_{\mathbb{R}^n} G_1^{3/2}(x - \tau, t) f(\tau) d\tau + \int_{\mathbb{R}^n} G_2^{3/2}(x - \tau, t) f_2(\tau) d\tau, \tag{6.3.42}$$

where $G_1^{3/2}(x,t)$ and $G_2^{3/2}(x,t)$ are given by (6.3.33) and (6.3.37) with $\alpha = 3/2$.

Remark 6.8 It can be verified, by using the properties (1.12.43) and (1.12.45) of the H-function (1.12.1) and the relation (6.2.73), that the formula (6.3.27) for $n = 1$ coincides with (6.3.9).

Remark 6.9 When $n = 1$, then it is directly verified, by using the properties (1.12.43) and (1.12.45) of the H-function and the formula (6.2.73), that the functions $G_k^{\alpha}(x,t)$ $(k = 0, 1)$ in (6.3.30) coincide with those in (6.3.11). Thus Theorem 6.5 and Corollaries 6.5-6.6 follow from Theorem 6.7 and Corollaries 6.7-6.8, respectively. This also means that Examples 6.7-6.8 are particular cases of Examples 6.9-6.10 for $n = 1$.

Remark 6.10 The functions $G_1^{\alpha}(x,t)$ and $G_2^{\alpha}(x,t)$, given in (6.3.14), (6.3.33) and (6.3.18), (6.3.37) provide two- and multi-dimensional fractional Green functions.

Remark 6.11 In Sections 6.2 and 6.3 we constructed explicit solutions to the Cauchy type and Cauchy problems for the fractional diffusion-wave equations (6.2.1) and (6.3.1) of the same form involving the Riemann-Liouville and Caputo partial fractional derivatives, respectively. The results obtained show that these problems have different solutions in the case of non-integer $\alpha \in \mathbb{R}$ $(\alpha \notin \mathbb{N})$. This is caused by the fact that the Riemann-Liouville and Caputo partial fractional derivatives differ by a quasi-polynomial term (see definitions in (6.1.12) and (6.3.2)).

6.4 Solution of Cauchy Problems for Fractional Evolution Equations

In this section we consider the fractional differential equation (6.1.68):

$$({}^{C}D^{\alpha}_{0+,t}u)(x,t) = \lambda(D^{\beta}_{-,x}u)(x,t), \tag{6.4.1}$$

with $x \in \mathbb{R}$, $t > 0$, $\alpha > 0$, $\beta > 0$, $\lambda \in \mathbb{R} \setminus \{0\}$, and involving the Caputo partial fractional derivative $({}^{C}D^{\alpha}_{0+,t}u)(x,t)$ with respect to $t > 0$ of order $\alpha > 0$ $(l-1 < \alpha \leqq l;\ l \in \mathbb{N})$ and the Liouville partial fractional derivative $(D^{\beta}_{-,x}u)(x,t)$ with respect to $x \in \mathbb{R}$ of order $\beta > 0$ defined by (6.3.2) and (6.1.71), respectively. We apply the Fourier and Laplace transforms to establish the explicit solution of the equation (6.4.1) with the following Cauchy conditions of the form (6.3.3):

$$u(x,0) = f_0(x), \quad \frac{\partial^k u(x,0)}{\partial t^k} = f_k(x), \quad (k = 1, \cdots, l-1;\ x \in \mathbb{R}). \tag{6.4.2}$$

As in Section 6.3, we begin from the simplest case of the equation (6.4.1) with $\beta = 1$.

$$({}^{C}D^{\alpha}_{0+,t}u)(x,t) = \lambda\frac{\partial u(x,t)}{\partial x} \quad (x \in \mathbb{R};\ t > 0;\ \ \alpha > 0). \tag{6.4.3}$$

6.4.1 Solution of the Simplest Problem

In this section we consider the Cauchy type problem (6.4.3)-(6.4.2). Applying the Laplace transform (6.2.6) with respect to $t > 0$ and taking into account the formula (6.3.6) and the initial conditions in (6.4.2), we have

$$s^{\alpha}(\mathcal{L}_t u)(x,s) - \sum_{k=0}^{l-1} s^{\alpha-k-1} f_k(x) = \lambda\frac{\partial}{\partial x}(\mathcal{L}_t u)(x,s). \tag{6.4.4}$$

Applying the Fourier transform (6.2.7) with respect to $x \in \mathbb{R}$ and taking into account the following formula for the Fourier transform of derivatives

$$\left(\mathcal{F}_x D^m_x u\right)(\sigma,s) = (-ik)^m(\mathcal{F}_x u)(\sigma,s) \quad \left(D_x = \frac{\partial}{\partial x};\ m \in \mathbb{N}\right) \tag{6.4.5}$$

with $m = 1$, we arrive at the relation for $\left(\mathcal{F}_x\mathcal{L}_t u\right)(\sigma,s)$:

$$\left(\mathcal{F}_x\mathcal{L}_t u\right)(\sigma,s) = \sum_{k=0}^{l-1} \frac{s^{\alpha-k-1}}{s^{\alpha} + i\lambda\sigma}\left(\mathcal{F}_x f_k\right)(\sigma). \tag{6.4.6}$$

Applying the inverse Fourier transform (6.2.11) with respect to $\sigma \in \mathbb{R}$ and the inverse Laplace transform (6.2.12) with respect to s $(\Re(s) > \sigma_\phi)$ and taking

(6.4.11) and (6.4.12) into account, we obtain the solution of the problem (6.4.2)-(6.4.3) in the form

$$u(x,t)=\sum_{k=0}^{l-1}\frac{1}{2\pi i}\int_{\gamma-i\infty}^{\gamma+i\infty}s^{\alpha-k-1}e^{st}ds\frac{1}{2\pi}\int_{-\infty}^{\infty}\frac{(\mathcal{F}_x f_k)(\sigma)}{s^\alpha+i\lambda\sigma}e^{-i\sigma x}d\sigma\quad(\gamma\in\mathbb{R}).\tag{6.4.7}$$

If $(\mathcal{F}f_k)(\sigma)$ $(k=1,\cdots,l)$ satisfy some additional conditions, the inner integrals in (6.4.7) can be evaluated by using the residue theory [see, for example, Ditkin and Prudnikov ([195], Chapter 2)].

On the basis of (6.4.6), we can give another representation for the solution $u(x,t)$ in terms of the

Mittag-Leffler function $E_{\alpha,\beta}(z)$ defined in (1.8.17). According to (1.10.9) with $\beta=k+1$, we thus have

$$\left(\mathcal{L}\left[t^k E_{\alpha,k+1}\left(-i\lambda\sigma t^\alpha\right)\right]\right)=\frac{s^{\alpha-k-1}}{s^\alpha+i\lambda\sigma}\quad\left(|\lambda\sigma s^{-\alpha}|<1;\ \ k=1,\cdots,l\right).\tag{6.4.8}$$

Applying the inverse Laplace transform (6.2.12) in (6.4.6) and using (6.4.8), we obtain

$$(\mathcal{F}_x u)(\sigma,t)=\sum_{k=0}^{l-1}t^k E_{\alpha,k+1}\left(-i\lambda\sigma t^\alpha\right)\left(\mathcal{F}_x f_k\right)(\sigma).\tag{6.4.9}$$

Then the inverse Fourier transform (6.2.11) yields the following solution $u(x,t)$:

$$u(x,t)=\sum_{k=0}^{l-1}\frac{t^k}{2\pi}\int_{-\infty}^{+\infty}E_{\alpha,k+1}\left(-i\lambda\sigma t^\alpha\right)\left(\mathcal{F}_x f_k\right)(\sigma)e^{-i\sigma x}d\sigma.\tag{6.4.10}$$

Thus we have established the following result.

Theorem 6.5 *Let $\alpha>0$ be such that $l-1<\alpha\leqq l$ $(l\in\mathbb{N})$. Then the Cauchy problem* (6.4.3), (6.4.2) *is solvable, and its explicit solution $u(x,t)$ is given by* (6.4.7) *or* (6.4.10), *provided that the integrals on the right-hand sides of* (6.4.7) *and* (6.4.10) *exist.*

Corollary 6.9 *If $0<\alpha\leqq 1$, then the Cauchy problem*

$$({}^C D_{0+,t}^\alpha u)(x,t)=\lambda\frac{\partial u(x,t)}{\partial x},\quad u(x,0)=f(x)\ (x\in\mathbb{R};\ t>0)\tag{6.4.11}$$

is solvable, and its solution has the form

$$u(x,t)=\frac{1}{2\pi i}\int_{\gamma-i\infty}^{\gamma+i\infty}s^{\alpha-1}e^{st}ds\frac{1}{2\pi}\int_{-\infty}^{\infty}\frac{(\mathcal{F}_x f)(\sigma)}{s^\alpha+i\lambda\sigma}e^{-i\sigma x}d\sigma\quad(\gamma\in\mathbb{R})\tag{6.4.12}$$

or, equivalently,

$$u(x,t) = \frac{1}{2\pi}\int_{-\infty}^{+\infty} E_\alpha\left(-i\lambda\sigma t^\alpha\right)\left(\mathcal{F}_x f\right)(\sigma)e^{-i\sigma x}d\sigma, \tag{6.4.13}$$

provided that the integrals in the right-hand sides of (6.4.12) *and* (6.4.13) *exist.*

Corollary 6.10 *If* $1 < \alpha \leqq 2$, *then the Cauchy problem*

$$({}^C D_{0+,t}^{\alpha} u)(x,t) = \lambda \frac{\partial u(x,t)}{\partial x} \quad (x \in \mathbb{R};\ t > 0) \tag{6.4.14}$$

$$u(x,0) = f_0(x), \quad \frac{\partial u(x,0)}{\partial t} = f_1(x) \tag{6.4.15}$$

is solvable, and its solution has the form

$$u(x,t) = \frac{1}{2\pi i}\int_{\gamma-i\infty}^{\gamma+i\infty} s^{\alpha-1}e^{st}ds\frac{1}{2\pi}\int_{-\infty}^{\infty}\frac{(\mathcal{F}_x f_0)(\sigma)}{s^\alpha + i\lambda\sigma}e^{-i\sigma x}d\sigma$$

$$+\frac{1}{2\pi i}\int_{\gamma-i\infty}^{\gamma+i\infty} s^{\alpha-2}e^{st}ds\frac{1}{2\pi}\int_{-\infty}^{\infty}\frac{(\mathcal{F}_x f_1)(\sigma)}{s^\alpha + i\lambda\sigma}e^{-i\sigma x}d\sigma \quad (\gamma \in \mathbb{R}) \tag{6.4.16}$$

or, equivalently,

$$u(x,t) = \frac{1}{2\pi}\int_{-\infty}^{+\infty} E_\alpha\left(-i\lambda\sigma t^\alpha\right)\left(\mathcal{F}_x f_0\right)(\sigma)e^{-i\sigma x}d\sigma$$

$$+\frac{t}{2\pi}\int_{-\infty}^{+\infty} E_{\alpha,2}\left(-i\lambda\sigma t^\alpha\right)\left(\mathcal{F}_x f_1\right)(\sigma)e^{-i\sigma x}d\sigma, \tag{6.4.17}$$

provided that the integrals in the right-hand sides of (6.4.16) *and* (6.4.17) *exist.*

Example 6.11 The Cauchy problem

$$({}^C D_{0+,t}^{1/2} u)(x,t) = \lambda \frac{\partial u(x,t)}{\partial x}, \quad u(x,0) = f(x)\ (x \in \mathbb{R};\ t > 0) \tag{6.4.18}$$

has its solution given by

$$u(x,t) = \frac{1}{2\pi}\int_{-\infty}^{+\infty} E_{1/2}\left(-i\lambda\sigma t^{1/2}\right)\left(\mathcal{F}_x f\right)(\sigma)e^{-i\sigma x}d\sigma. \tag{6.4.19}$$

Example 6.12 The Cauchy problem

$$({}^C D_{0+,t}^{3/2} u)(x,t) = \lambda \frac{\partial u(x,t)}{\partial x} \quad (x \in \mathbb{R};\ t > 0) \tag{6.4.20}$$

$$u(x,0) = f_0(x), \quad \frac{\partial u(x,0)}{\partial t} = f_1(x) \tag{6.4.21}$$

has its solution given by

$$u(x,t) = \frac{1}{2\pi}\int_{-\infty}^{+\infty} E_{3/2}\left(-i\lambda\sigma t^{3/2}\right)(\mathcal{F}_x f_1)(\sigma)e^{-i\sigma x}d\sigma$$
$$+\frac{t}{2\pi}\int_{-\infty}^{+\infty} E_{3/2,2}\left(-i\lambda\sigma t^{3/2}\right)(\mathcal{F}_x f_2)(\sigma)e^{-i\sigma x}d\sigma. \tag{6.4.22}$$

The solution $u(x,t)$ of the Cauchy problem (6.4.2)-(6.4.3) can be represented in a form different from that in (6.4.7) and (6.4.10), if the given functions $f_k(x)$ $(k = 1, \cdots, l)$ are analytic.

Theorem 6.6 *Let* $\alpha > 0$ *be such that* $l - 1 < \alpha \leqq l$ $(l \in \mathbb{N})$, *and let* $f_k(x)$ $(k = 1, \cdots, l)$ *be analytic functions of the real variable* $x \in \mathbb{R}$ *such that*

$$\lim_{|x|\to\infty} f_k^{(j)}(x) = 0 \quad (k = 1, \cdots, l;\ j \in \mathbb{N}_0). \tag{6.4.23}$$

Then the solution of the Cauchy problem (6.4.3), (6.4.2) *is given by*

$$u(x,t) = \sum_{k=0}^{l-1} t^k \sum_{j=0}^{\infty} \frac{(\lambda t^\alpha)^j}{\Gamma(\alpha j + k + 1)} f_k^{(j)}(x), \tag{6.4.24}$$

provided that the series in the right-hand side of (6.4.24) *converges for any* $x \in \mathbb{R}$ *and any* $t > 0$.

Using (6.4.10) and interchanging the order of integration and summation (which is permissible by the uniform convergence of the series represented by the entire Mittag-Leffler function), we have

$$u(x,t) = \sum_{k=1}^{l} \frac{t^k}{2\pi}\int_{-\infty}^{+\infty}\left[\sum_{j=0}^{\infty}\frac{(-i\lambda\sigma t^\alpha)^j}{\Gamma(\alpha j + k + 1)}\right](\mathcal{F}_x f_k)(\sigma)e^{-i\sigma x}d\sigma$$

$$= \sum_{k=1}^{l} t^k \sum_{j=0}^{\infty}\frac{(\lambda t^\alpha)^j}{\Gamma(\alpha j + k + 1)}\frac{1}{2\pi}\int_{-\infty}^{+\infty}(-i\sigma)^j(\mathcal{F}_x f_k)(\sigma)e^{-ikx}dk. \tag{6.4.25}$$

By means of the formula (1.3.11), we have, for the inverse Fourier transform,

$$\frac{1}{2\pi}\int_{-\infty}^{+\infty}(-i\sigma)^j e^{-i\sigma x}d\sigma - (\mathcal{F}_\sigma^{-1}\mathcal{F}_x f_k)^{(j)}(x) = f_k^{(j)}(x) \quad (k = 1, \cdots, l;\ j \in \mathbb{N}_0). \tag{6.4.26}$$

Using (6.4.26), from (6.4.25) we derive the representation for the solution $u(x,t)$ of the problem (6.4.2)-(6.4.3) in the form (6.4.24).

Example 6.13 The following Cauchy problem

$$({}^C D_{0+,t}^\alpha u)(x,t) = \lambda\frac{\partial u(x,t)}{\partial x} \quad (x \in \mathbb{R},;\ t > 0;\ \ l - 1 < \alpha \leqq l;\ l \in \mathbb{N}), \tag{6.4.27}$$

$$\frac{\partial^k u(x,0)}{\partial t^k} = b_k e^{-\mu_k|x|} \quad (b_k \in \mathbb{R};\ \mu_k > 0;\quad k = 0, \cdots, l-1) \tag{6.4.28}$$

has its solution given by

$$u(x,t) = \sum_{k=0}^{l-1} b_k t^k e^{-\mu_k|x|} E_{\alpha,k+1}\left(-\text{sign}(x)\lambda\mu_k t^\alpha\right). \tag{6.4.29}$$

In particular, if $0 < \alpha \leqq 1$, then

$$u(x,t) = be^{-\mu|x|} E_\alpha\left(-\text{sign}(x)\lambda\mu t^\alpha\right) \tag{6.4.30}$$

is the solution to the following Cauchy problem:

$$({}^C D_{0+,t}^\alpha u)(x,t) = \lambda \frac{\partial u(x,t)}{\partial x}, \quad u(x,0) = be^{-\mu|x|} \quad (x \in \mathbb{R};\ t > 0). \tag{6.4.31}$$

6.4.2 Solution to the General Problem

In this section we generalize the results of Section 6.4.1 for the fractional differential equation (6.4.1) with the Liouville fractional derivative $\left(D_{-,x}^\beta u\right)(x,t)$ with respect to $x \in \mathbb{R}$ of any order $\beta > 0$. For this we need the following formula for the Fourier transform of such a derivative [see ([729], formula (7.4))]:

$$(\mathcal{F}_x(D_{-,x}^\beta u))(\sigma,t) = (-i\sigma)^\beta (\mathcal{F}u)(\sigma,t) \quad (\beta > 0), \tag{6.4.32}$$

where the function $(-i\sigma)^\beta$ is understood as

$$(-i\sigma)^\beta := |\sigma|^\beta e^{-\frac{\beta\pi i}{2} sign(\sigma)} \quad (\sigma \in \mathbb{R};\ \beta > 0). \tag{6.4.33}$$

Following Section 6.4.1, we apply the Laplace transform (6.2.6) with respect to $t > 0$ and the Fourier transform (6.2.7) with respect to $x \in \mathbb{R}$ to the equation (6.4.1). Using the relation (6.2.8) and (6.4.32), we arrive at the following relation of the form (6.4.6):

$$(\mathcal{F}_x \mathcal{L}_t u)(\sigma, s) = \sum_{k=0}^{l-1} \frac{s^{\alpha-k-1}}{s^\alpha - \lambda(-i\sigma)^\beta} (\mathcal{F}_x f_k)(\sigma). \tag{6.4.34}$$

Using the inverse Laplace and Fourier transforms and applying the same arguments as in Section 6.4.1, we obtain the solution $u(x,t)$ to the problem (6.4.1)-(6.4.2) in the form (6.4.7) and (6.4.10) as follows:

$$u(x,t) = \sum_{k=0}^{l-1} \frac{1}{2\pi i} \int_{\gamma-i\infty}^{\gamma+i\infty} s^{\alpha-k-1} e^{st} ds \frac{1}{2\pi} \int_{-\infty}^{\infty} \frac{(\mathcal{F}_x f_k)(\sigma)}{s^\alpha - \lambda(-i\sigma)^\beta} e^{-i\sigma x} d\sigma \quad (\gamma \in \mathbb{R}) \tag{6.4.35}$$

and

$$u(x,t) = \sum_{k=0}^{l-1} \frac{t^k}{2\pi} \int_{-\infty}^{+\infty} E_{\alpha,k+1}\left(\lambda(-i\sigma)^\beta t^\alpha\right) (\mathcal{F}_x f_k)(\sigma) e^{-i\sigma x} d\sigma. \tag{6.4.36}$$

Thus we have proved the following result.

Theorem 6.7 *Let $\alpha > 0$ be such that $l - 1 < \alpha \leqq l$ $(l \in \mathbb{N})$ and let $\beta > 0$. Then the Cauchy problem (6.4.1)-(6.4.2) is solvable, and its explicit solution $u(x,t)$ is given by (6.4.35) or (6.4.36), provided that the integrals on the right-hand sides of (6.4.35) and (6.4.36) exist.*

Corollary 6.11 *If $0 < \alpha \leqq 1$ and $\beta > 0$, then the Cauchy problem*

$$({}^{C}D^{\alpha}_{0+,t}u)(x,t) = \lambda\left(D^{\beta}_{-,x}u\right)(x,t), \quad u(x,0) = f(x) \ (x \in \mathbb{R};\ t > 0) \tag{6.4.37}$$

is solvable, and its solution has the form

$$u(x,t) = \frac{1}{2\pi i}\int_{\gamma-i\infty}^{\gamma+i\infty} s^{\alpha-1}e^{st}ds\frac{1}{2\pi}\int_{-\infty}^{\infty}\frac{(\mathcal{F}_x f)(\sigma)}{s^{\alpha}-\lambda(-i\sigma)^{\beta}}e^{-i\sigma x}d\sigma \quad (\gamma \in \mathbb{R}) \tag{6.4.38}$$

or, equivalently,

$$u(x,t) = \frac{1}{2\pi}\int_{-\infty}^{+\infty} E_{\alpha}\left(\lambda(-i\sigma)^{\beta}t^{\alpha}\right)(\mathcal{F}_x f)(\sigma)e^{-i\sigma x}d\sigma, \tag{6.4.39}$$

provided that the integrals in the right-hand sides of (6.4.38)and (6.4.39) exist.

Corollary 6.12 *If $1 < \alpha \leqq 2$ and $\beta > 0$, then the Cauchy problem*

$$({}^{C}D^{\alpha}_{0+,t}u)(x,t) = \lambda\left(D^{\beta}_{-,x}u\right)(x,t) \quad (x \in \mathbb{R};\ t > 0), \tag{6.4.40}$$

$$u(x,0) = f_0(x), \quad \frac{\partial u(x,0)}{\partial t} = f_1(x) \quad (x \in \mathbb{R}) \tag{6.4.41}$$

is solvable, and its solution has the form

$$u(x,t) = \frac{1}{2\pi i}\int_{\gamma-i\infty}^{\gamma+i\infty} s^{\alpha-1}e^{st}ds\frac{1}{2\pi}\int_{-\infty}^{\infty}\frac{(\mathcal{F}_x f_0)(\sigma)}{s^{\alpha}-\lambda(-i\sigma)^{\beta}}e^{-i\sigma x}d\sigma$$

$$+\frac{t}{2\pi i}\int_{\gamma-i\infty}^{\gamma+i\infty} s^{\alpha-2}e^{st}ds\frac{1}{2\pi}\int_{-\infty}^{\infty}\frac{(\mathcal{F}_x f_1)(\sigma)}{s^{\alpha}-\lambda(-i\sigma)^{\beta}}e^{-i\sigma x}d\sigma \quad (\gamma \in \mathbb{R}) \tag{6.4.42}$$

or, equivalently,

$$u(x,t) = \frac{1}{2\pi}\int_{-\infty}^{+\infty} E_{\alpha}\left(\lambda(-i\sigma)^{\beta}t^{\alpha}\right)(\mathcal{F}_x f_0)(\sigma)e^{-i\sigma x}d\sigma$$

$$+\frac{t}{2\pi}\int_{-\infty}^{+\infty} E_{\alpha,2}\left(\lambda(-i\sigma)^{\beta}t^{\alpha}\right)(\mathcal{F}_x f_1)(\sigma)e^{-i\sigma x}d\sigma, \tag{6.4.43}$$

provided that the integrals in the right-hand sides of (6.4.42) and (6.4.43) exist.

Corollary 6.13 *If $\alpha > 0$ ($l-1 < \alpha \leqq l$, $l \in \mathbb{N}$), then the Cauchy problem*

$$({}^CD^{\alpha}_{0+,t}u)(x,t) = \lambda \frac{\partial^2 u(x,t)}{\partial x^2} \quad (x \in \mathbb{R};\ t > 0) \tag{6.4.44}$$

$$u(x,0) = f_0(x), \quad \frac{\partial^k u(x,0)}{\partial t^k} = f_k(x) \ (k = 1, \cdots, l-1;\ x \in \mathbb{R}) \tag{6.4.45}$$

is solvable, and its solution has the form

$$u(x,t) = \sum_{k=0}^{l-1} \frac{t^k}{2\pi i} \int_{\gamma - i\infty}^{\gamma + i\infty} s^{\alpha-k-1} e^{st} ds \frac{1}{2\pi} \int_{-\infty}^{\infty} \frac{(\mathcal{F}_x f_k)(\sigma)}{s^{\alpha} + \lambda\sigma^2} e^{-i\sigma x} d\sigma \ (\gamma \in \mathbb{R}) \tag{6.4.46}$$

or, equivalently,

$$u(x,t) = \sum_{k=0}^{l-1} \frac{t^k}{2\pi} \int_{-\infty}^{+\infty} E_{\alpha,k+1}\left(-\lambda\sigma^2 t^{\alpha}\right) \left(\mathcal{F}_x f_k\right)(\sigma) e^{-i\sigma x} d\sigma, \tag{6.4.47}$$

provided that the integrals in the right-hand sides of (6.4.46) *and* (6.4.47) *exist.*

Example 6.14 The following Cauchy problem

$$({}^CD^{1/2}_{0+,t}u)(x,t) = \lambda \left(D^{\beta}_{-,x}u\right)(x,t) \ \ u(x,0) = f(x) \ (x \in \mathbb{R};\ t > 0;\ \beta > 0) \tag{6.4.48}$$

has its solution given by

$$u(x,t) = \frac{1}{2\pi} \int_{-\infty}^{+\infty} E_{1/2}\left(\lambda(-i\sigma)^{\beta} t^{1/2}\right) \left(\mathcal{F}_x f\right)(\sigma) e^{-i\sigma x} d\sigma. \tag{6.4.49}$$

Example 6.15 The following Cauchy problem

$$({}^CD^{3/2}_{0+,t}u)(x,t) = \lambda \left(D^{\beta}_{-,x}u\right)(x,t) \quad (x \in \mathbb{R};\ t > 0;\ \beta > 0) \tag{6.4.50}$$

$$u(x,0) = f_0(x), \quad \frac{\partial u(x,0)}{\partial x} = f_1(x) \tag{6.4.51}$$

has its solution given by

$$u(x,t) = \frac{1}{2\pi} \int_{-\infty}^{+\infty} E_{3/2}\left(\lambda(-i\sigma)^{\beta} t^{3/2}\right) \left(\mathcal{F}_x f_0\right)(\sigma) e^{-i\sigma x} d\sigma$$

$$+ \frac{t}{2\pi} \int_{-\infty}^{+\infty} E_{3/2,2}\left(\lambda(-i\sigma)^{\beta} t^{3/2}\right) \left(\mathcal{F}_x f_1\right)(\sigma) e^{-i\sigma x} d\sigma. \tag{6.4.52}$$

Example 6.16 The following Cauchy problem

$$\frac{\partial^l u(x,t)}{\partial t^l} = \lambda \frac{\partial^m u(x,t)}{\partial x^m} \quad (x \in \mathbb{R};\ t > 0;\ \ l, m \in \mathbb{N}), \tag{6.4.53}$$

$$u(x,0) = f_0(x), \quad \frac{\partial^k u(x,0)}{\partial t^k} = f_k(x) \quad (k = 1, \cdots, l-1;\ x \in \mathbb{R}) \tag{6.4.54}$$

has its solution given by

$$u(x,t) = \sum_{k=0}^{l-1} \frac{t^k}{2\pi} \int_{-\infty}^{+\infty} E_{l,k+1}\left(\lambda(-i\sigma)^m t^l\right) (\mathcal{F}_x f_k)(\sigma) e^{-i\sigma x} d\sigma. \tag{6.4.55}$$

Example 6.17 The following Cauchy problem

$$\frac{\partial^l u(x,t)}{\partial t^l} = \lambda \frac{\partial^{2m} u(x,t)}{\partial x^{2m}} \quad (x \in \mathbb{R};\ t > 0;\ \ l, m \in \mathbb{N}), \tag{6.4.56}$$

$$u(x,0) = f_0(x), \quad \frac{\partial^k u(x,0)}{\partial t^k} = f_k(x) \quad (k = 1, \cdots, l-1;\ x \in \mathbb{R}) \tag{6.4.57}$$

has its solution given by

$$u(x,t) = \sum_{k=0}^{l-1} \frac{t^k}{2\pi} \int_{-\infty}^{+\infty} E_{l,k+1}\left[\lambda(-1)^m \sigma^{2m} t^l\right] (\mathcal{F}_x f_k)(\sigma) e^{-i\sigma x} d\sigma. \tag{6.4.58}$$

In particular,

$$u(x,t) = \frac{1}{2\pi} \int_{-\infty}^{+\infty} E_2\left(-\lambda\sigma^2 t^2\right) (\mathcal{F}_x f_0)(\sigma) e^{-i\sigma x} d\sigma$$

$$+ \frac{t}{2\pi} \int_{-\infty}^{+\infty} E_{2,2}\left(-\lambda\sigma^2 t^2\right) (\mathcal{F}_x f_1)(\sigma) e^{-i\sigma x} d\sigma \tag{6.4.59}$$

is the solution to the following Cauchy problem:

$$\frac{\partial^2 u(x,t)}{\partial t^2} = \lambda \frac{\partial^2 u(x,t)}{\partial x^2} \quad (x \in \mathbb{R};\ t > 0), \tag{6.4.60}$$

$$u(x,0) = f_0(x), \quad \frac{\partial u(x,0)}{\partial t} = f_1(x) \quad (x \in \mathbb{R}). \tag{6.4.61}$$

Remark 6.12 If $\lambda > 0$ and λ is replaced by λ^2, then the relations (6.4.46), (6.4.47) with $l = 1$ and $l = 2$ yield, in different form than (6.3.13) and (6.3.17), explicit solutions of the Cauchy problems, (6.3.12) and (6.3.15)-(6.3.16). It is easily verified that these solutions coincide for "suitable" functions $f_0(x)$ and $f_1(x)$.

Remark 6.13 The results of Corollaries 6.9 and 6.11 with $0 < \alpha < 1$ were given by Kilbas et al., [385]. Similar results were established in Kilbas et al. [386] for the Cauchy type problem for equations of the form (6.4.11) and (6.4.37), with the Caputo derivative replaced by the Riemann-Liouville derivative, and with the initial condition (6.1.58).

6.4.3 Solutions of Cauchy Problems via the H-Functions

We show that, in the case $l \neq 2m$ ($l, m \in \mathbb{N}$), the solution (6.4.58) can be expressed in terms of the H-functions (1.12.1) of the form

$$H_{3,3}^{1,2}\left[z \left| \begin{matrix} (0,1),(0,2m),(0,m) \\ (0,1),(0,m),(-k,l) \end{matrix} \right.\right]$$

$$= \frac{1}{2\pi i}\int_{\mathcal{L}} \frac{\Gamma(s)\Gamma(1-s)\Gamma(1-2ms)}{\Gamma(ms)\Gamma(1-ms)\Gamma(1+k-ls)} z^{-s} ds \ (l, m \in \mathbb{N};\ l \neq 2m;\ k = 0, \cdots, l-1), \tag{6.4.62}$$

where $a, b, c, d \in \mathbb{R}$ and an infinite contour $\mathcal{L}$ separates all poles of the gamma functions $\Gamma(s)$ to the left and poles of gamma functions $\Gamma(1-s)$ and $\Gamma(1-2ms)$ to the right. By (1.11.16), $\Delta = l - 2m$. Thus Theorem 1.6 yields the following assertion.

Lemma 6.6 *If $l, m \in \mathbb{N}$ ($l \neq 2m$) and $k = 0, \cdots, l-1$, then the H-functions (6.4.62) are defined in the following cases:*

$$l > 2m, \quad \mathcal{L} = \mathcal{L}_{-\infty}, \tag{6.4.63}$$

$$l < 2m, \quad \mathcal{L} = \mathcal{L}_{+\infty}. \tag{6.4.64}$$

Now we prove that the functions

$$E_{l,k+1}\left(\lambda(-1)^m \sigma^{2m} t^l\right) \quad (k = 0, \cdots, l-1) \tag{6.4.65}$$

can be expressed in terms of the Fourier transform of the H-functions (6.4.62).

Lemma 6.7 *If $l, m \in \mathbb{N}$ are such that $l \neq 2m$, then, for $k = 0, \cdots, l-1$, there hold the following formulas:*

$$\left(\mathcal{F}_x\left[\frac{1}{|x|} H_{3,3}^{1,2}\left[(-1)^m \frac{\lambda t^l}{|x|^{2m}} \left| \begin{matrix} (0,1),(0,2m),(0,m) \\ (0,1),(0,m),(-k,l) \end{matrix} \right.\right]\right]\right)(\sigma)$$

$$= E_{l,k+1}\left[(-1)^m \lambda \sigma^{2m} t^l\right]. \tag{6.4.66}$$

Proof. By (6.2.7) and (6.4.62), we have

$$\left(\mathcal{F}_x\left[\frac{1}{|x|}H^{1,2}_{3,3}\left[(-1)^m\frac{\lambda t^l}{|x|^{2m}}\,\middle|\,\begin{matrix}(0,1),(0,2m),(0,m)\\(0,1),(0,m),(-k,l)\end{matrix}\right]\right]\right)(\sigma)$$

$$=\frac{1}{2\pi i}\int_{\mathcal{L}}\frac{\Gamma(s)\Gamma(1-s)\Gamma(1-2ms)}{\Gamma(ms)\Gamma(1-ms)\Gamma(1+k-ls)}\left[-\lambda(-1)^m t^l\right]^{-s}ds\int_{-\infty}^{\infty}\frac{e^{ix\sigma}}{|x|^{1-2ms}}dx. \tag{6.4.67}$$

Using (5.5.8) and (5.5.27), we find for the inner integral in (6.4.67) that

$$\int_{-\infty}^{\infty}\frac{e^{ix\sigma}}{|x|^{1-2ms}}dx=2\int_0^{\infty}\frac{\cos(\rho|\sigma|)}{\rho^{1-2ms}}d\rho. \tag{6.4.68}$$

We can choose $s\in\mathbb{C}$ in (6.4.67) such that $0<\Re(s)<\frac{1}{2m}$. Then, using the following known formula [see, for example, Brychkov and Prudnikov ([108], formula 7.12)]

$$\int_0^{\infty}\frac{\cos(\lambda\rho)}{\rho^{\gamma}}d\rho=\sin\left(\frac{\gamma\pi}{2}\right)\Gamma(1-\gamma)\lambda^{\gamma-1}\quad(0<\Re(\gamma)<1) \tag{6.4.69}$$

and the functional equation (1.5.8) for the gamma function (1.5.1), we evaluate the integral in (6.4.68) as follows:

$$\int_{-\infty}^{\infty}\frac{e^{ix\sigma}}{|x|^{1-2ms}}dx=2\pi\frac{\Gamma(2ms)}{\Gamma\left(\frac{1}{2}-ms\right)\Gamma\left(\frac{1}{2}+ms\right)}|\sigma|^{-2m}. \tag{6.4.70}$$

Substituting (6.4.70) into (6.4.67), we have

$$\left(\mathcal{F}_x\left[\frac{1}{|x|}H^{1,2}_{3,3}\left[(-1)^m\frac{\lambda t^l}{|x|^{2m}}\,\middle|\,\begin{matrix}(0,1),(0,2m),(0,m)\\(0,1),(0,m),(-k,l)\end{matrix}\right]\right]\right)(\sigma)$$

$$= \frac{1}{2\pi i}\int_{\mathcal{L}} \frac{\Gamma(s)\Gamma(1-s)}{\Gamma(1+k-ls)} \left[\frac{2\pi\Gamma(1-2ms)\Gamma(2ms)}{\Gamma(ms)\Gamma(1-ms)\Gamma\left(\frac{1}{2}-ms\right)\Gamma\left(\frac{1}{2}+ms\right)}\right]$$

$$\cdot \left[-\lambda(-1)^m t^l |\sigma|^{2m}\right]^{-s} ds.$$

In accordance with the Legendre duplication formula (1.5.9), we find that

$$\frac{2\pi\Gamma(1-2ms)\Gamma(2ms)}{\Gamma(ms)\Gamma(1-ms)\Gamma\left(\frac{1}{2}-ms\right)\Gamma\left(\frac{1}{2}+ms\right)} = 1$$

and, therefore, that

$$\left(\mathcal{F}_x\left[\frac{1}{|x|}H^{1,2}_{3,3}\left[(-1)^m\frac{\lambda t^l}{|x|^{2m}}\,\middle|\,\begin{matrix}(0,1),(0,2m),(0,m)\\(0,1),(0,m),(-k,l)\end{matrix}\right]\right]\right)(\sigma)$$

$$= \frac{1}{2\pi i}\int_{\mathcal{L}} \frac{\Gamma(s)\Gamma(1-s)}{\Gamma(1+k-ls)}\left[-\lambda(-1)^m t^l|\sigma|^{2m}\right]^{-s} ds. \tag{6.4.71}$$

By (1.8.32) with $\alpha = l$, $\beta = k+1$ and $z = \lambda(-1)^m t^l |\sigma|^{2m}$, (6.4.71) yields the result in (6.4.66), and thus Lemma 6.7 is proved.

By the convolution property (1.3.17) and (6.4.66), we obtain following result analogous to (6.4.58).

Theorem 6.8 *If $l, m \in \mathbb{N}$ are such that $l \neq 2m$, then the Cauchy problem* (6.4.56)-(6.4.57) *has its solution given by*

$$u(x,t) = \sum_{k=1}^{l}\int_{-\infty}^{\infty} G^l_{mk}(x-\tau,t)f_k(\tau)d\tau, \tag{6.4.72}$$

where

$$G^l_{mk}(x,t) = \frac{t^k}{|x|}H^{1,2}_{3,3}\left[(-1)^m\frac{\lambda t^l}{|x|^{2m}}\,\middle|\,\begin{matrix}(0,1),(0,2m),(0,m)\\(0,1),(0,m),(-k,l)\end{matrix}\right] \quad (k = 0,\cdots,l-1). \tag{6.4.73}$$

Corollary 6.14 *If $l \in \mathbb{N}$ $(l \neq 2)$, then the solution to the Cauchy problem*

$$\frac{\partial^l u(x,t)}{\partial t^l} = \lambda\frac{\partial^2 u(x,t)}{\partial x^2} \quad (x \in \mathbb{R};\ t>0;\ \ l \in \mathbb{N}\backslash\{2\}), \tag{6.4.74}$$

$$u(x,0) = f_0(x), \quad \frac{\partial^k u(x,0)}{\partial t^k} = f_k(x) \quad (k = 1,\cdots,l-1;\ x \in \mathbb{R}) \tag{6.4.75}$$

is given by

$$u(x,t) = \sum_{k=0}^{l-1} \int_{-\infty}^{\infty} G_{2k}^{l}(x-\tau,t) f(\tau) d\tau, \tag{6.4.76}$$

$$G_{2k}^{l}(x,t) = \frac{t^k}{|x|} H_{1,1}^{0,1} \left[-\frac{\lambda t^l}{|x|^2} \middle| \begin{matrix} (0,2) \\ (-k,l) \end{matrix} \right] \quad (k = 0, \cdots, l-1). \tag{6.4.77}$$

Example 6.18 The following Cauchy problem

$$\frac{\partial u(x,t)}{\partial t} = \lambda \frac{\partial^{2m} u(x,t)}{\partial x^{2m}}, \quad u(x,0) = f(x) \quad (x \in \mathbb{R};\ t > 0;\ \ m \in \mathbb{N}), \tag{6.4.78}$$

has its solution given by

$$u(x,t) = \int_{-\infty}^{\infty} G_m(x-\tau,t) f(\tau) d\tau, \tag{6.4.79}$$

$$G_m(x,t) = \frac{1}{|x|} H_{2,2}^{1,1} \left[(-1)^m \frac{\lambda t}{|x|^{2m}} \middle| \begin{matrix} (0,2m),(0,m) \\ (0,1),(0,m) \end{matrix} \right]. \tag{6.4.80}$$

In particular, the solution to the Cauchy problem

$$\frac{\partial u(x,t)}{\partial t} = \lambda \frac{\partial^2 u(x,t)}{\partial x^2}, \quad u(x,0) = f(x) \quad (x \in \mathbb{R};\ t > 0), \tag{6.4.81}$$

is given by

$$u(x,t) = \int_{-\infty}^{\infty} G_2(x-\tau,t) f(\tau) d\tau, \quad G_2(x,t) = \frac{1}{|x|} H_{1,1}^{0,1} \left[(-1)^m \frac{\lambda t}{|x|^2} \middle| \begin{matrix} (0,2) \\ (0,1) \end{matrix} \right]. \tag{6.4.82}$$

Remark 6.14 The functions $G_{mk}^{l}(x,t)$ ($l, m \in \mathbb{N}$; $k = 0, \cdots, l-1$) and $G_m(x,t)$ ($m \in \mathbb{N}$), given in (6.4.73) and (6.4.80), yield the Green functions for the ordinary partial equations (6.4.56) and (6.4.78).

Remark 6.15 The explicit solution (6.4.72) of the Cauchy problem (6.4.56)-(6.4.57), presented in Theorem 6.8 via the Green functions (6.4.73), is valid for $l, m \in \mathbb{N}$ such that $l \neq 2m$. In particular, such Green functions cannot be constructed for the classical Cauchy problem for the wave (diffusion) equation:

$$\frac{\partial^2 u(x,t)}{\partial t^2} = \frac{\partial^2 u(x,t)}{\partial x^2}, \quad \frac{\partial u(x,0)}{\partial t} = f_1(x), \ u(x,0) = f_2(x). \tag{6.4.83}$$

Therefore, the problem of constructing Green functions for the problem (6.4.56)-(6.4.57) for the case $l = 2m$ (in particular, for the problem (6.4.83)) is still open.

Chapter 7

SEQUENTIAL LINEAR DIFFERENTIAL EQUATIONS OF FRACTIONAL ORDER

As mentioned in Chapter 3, the problem of studying linear differential equations and systems of linear differential equations of fractional order has been dealt with by numerous authors throughout history, particularly in recent years. We also noted that many of the questions related to this topic have only been partially resolved, and in a less than systematic fashion.

As for the problem of obtaining explicit solutions to linear differential equations and to systems of linear differential equations of fractional order with constant coefficients, Miller, Ross, Gorenflo, Mainardi , and Podlubny, mainly, have provided in recent years some methods based primarily on the use of Laplace transforms for certain equations associated with the Riemann-Liouville D^{α}_{0+} or the Caputo ${}^{C}D^{\alpha}_{0+}$ derivatives. This leaves certain problems with initial values involving the fractional derivatives D^{α}_{a+} or ${}^{C}D^{\alpha}_{a+}$ $(a \in \mathbb{R})$, or those associated with boundary conditions, for example, beyond their scope.

This chapter presents a general theory for sequential linear fractional differential equations, and systems of these equations with Riemann-Liouville and Caputo derivatives which provides a transparent and systematic approach for obtaining a general solution for the corresponding cases with constant coefficients. The contents of this chapter are based essentially on the works of Bonilla et al. [97, 98]. See also Dattoli et al. [155] and Vázquez [842], and Vázquez and Vilela Mendez Vázquez[845].

Specifically, we will give the explicit solution, in the case of constant coefficients, for both the homogeneous and the non-homogeneous problem, using a fractional Green function which generalizes the corresponding ordinary function.

Also in this chapter, we apply series of fractional order to investigate the solutions of certain linear fractional differential equations with variable coefficients, including a generalization of the well-known Frobenius theory.

In general, in this chapter it will be understood that $0 < \alpha \leqq 1$, $a \in \mathbb{R}$, $\mathbf{A} \in M_{n\times n}(\mathbb{R})$ and $\bar{\mathbf{B}} \in M_{n\times 1}(\mathbb{R})$, $M_{m\times n}(\mathbb{R})$ being the set of real matrices of order $m \times n$. A real matrix function of order $n \times n$ with domain in $\mathbb{R}$ will be represented by $\mathbf{A}(x)$.

We will use frequently the Mittag-Leffler type functions $e_{\alpha}^{\lambda t}$ and $e_{\alpha,m}^{\lambda t}$, which were introduced in Chapter 1; see (1.10.11) and (1.10.29) in Section 1.10.

7.1 Sequential Linear Differential Equations of Fractional Order

This section introduces a general theory for the fractional case, analogous to that of the normal case, applicable to Riemann-Liouville (R-L) sequential linear fractional differential equations (LFDE). General methods, independent of the integral transforms, are developed for obtaining the general solution to a LFDE with constant coefficients, by using the roots of the characteristic polynomial of the corresponding homogeneous equation, or for finding a certain Green function in the non-homogeneous case.

It must be pointed out here that a theory, similar to that developed in this chapter for the R-L fractional differential operator D_{a+}^{α}, is feasible for the case of LFDE with the differential operator of the Caputo ${}^{C}D_{a+}^{\alpha}$ type, keeping in mind properties of the Mittag-Leffler functions presented in Chapter 1. In any case, it is also possible to connect this theory for the Riemann-Liouville derivative with the corresponding theory for the Caputo derivative operator if we use the following close connection between $D_{a+}^{\alpha}y$ and ${}^{C}D_{a+}^{\alpha}y$ (see (2.4.4) in Section 2.4):

$$({}^{C}D_{a+}^{\alpha}y)(x) = \left(D_{a+}^{\alpha}[y(t) - y(a)]\right)(x) \quad (0 < \Re(\alpha) < 1). \tag{7.1.1}$$

We begin this section with the following definitions.

Definition 7.1 *Let $n \in \mathbb{N}$. We shall call linear sequential fractional differential equation of order $n\alpha$ the equations of the form*

$$\sum_{k=0}^{n} b_k(x) y^{(k\alpha}(x) = f(x) \quad (a < x < b), \tag{7.1.2}$$

where $b_k(x)$ are given real functions, $y^{(0}(x) := y(x)$, and $y^{(k\alpha}(x) := \left(\mathcal{D}_{a+}^{k\alpha}y\right)(x)$ $(k = 1, ..., n)$ represents a fractional sequential derivative. For example, for the R-L derivative, $\mathcal{D}_{a+}^{\alpha}y$ is

$$\begin{aligned} \mathcal{D}_{a+}^{\alpha}y &:= D_{a+}^{\alpha}y \\ \mathcal{D}_{a+}^{k\alpha}y &:= \mathcal{D}_{a+}^{\alpha}\mathcal{D}_{a+}^{(k-1)\alpha}y \quad (k = 2, 3, \cdots). \end{aligned} \tag{7.1.3}$$

For the case $k = 2$ and $0 < \alpha < 1/2$, the relationship between $\mathcal{D}^{k\alpha}_{a+}y$ and $D^{k\alpha}_{a+}y$ is given by

$$(\mathcal{D}^{2\alpha}_{a+}y)(x) = \left(D^{2\alpha}_{a+} \left[y(t) - (I^{1-\alpha}_{a+}y)(a+)\frac{(t-a)^{\alpha-1}}{\Gamma(\alpha)} \right] \right)(x), \tag{7.1.4}$$

by applying Property 2.4.

If $\forall x \in [a,b]$, $b_n(x) \neq 0$, equation (7.1.2) may be expressed in its normal form for the R-L derivative as follows:

$$[\mathbf{L_{n\alpha}}(y)](x)$$

$$:= \left(\mathcal{D}^{n\alpha}_{a+}y\right)(x) + \sum_{k-0}^{n-1} a_k(x) \left(\mathcal{D}^{k\alpha}_{a+}y\right)(x) \equiv y^{(n\alpha}(x) + \sum_{k=0}^{n-1} a_k(x) y^{(k\alpha}(x) = f(x). \tag{7.1.5}$$

Note that the equation (7.1.5) reduces to the system

$$\left(D^{\alpha}_{a+}\bar{Y}\right)(x) = \mathbf{A}(x)\bar{Y}(x) + \bar{\mathbf{B}}(x), \tag{7.1.6}$$

with

$$\mathbf{A}(x) = \begin{pmatrix} 0 & 1 & 0 & . & . & . & 0 \\ 0 & 0 & 1 & . & . & . & 0 \\ \cdots & . & . & . & . & . & . \\ \cdots & . & . & . & . & . & . \\ \cdots & . & . & . & . & . & 1 \\ -a_0 & -a_1 & . & . & . & . & -a_{n-1} \end{pmatrix} \tag{7.1.7}$$

$$\bar{\mathbf{B}}(x) = \begin{pmatrix} 0 \\ 0 \\ \cdots \\ \cdots \\ \cdots \\ f(x) \end{pmatrix}; \quad \bar{Y}(x) = \begin{pmatrix} y_1(x) \\ y_2(x) \\ \cdots \\ \cdots \\ \cdots \\ y_n(x) \end{pmatrix}, \tag{7.1.8}$$

just by changing the variables

$$y_1(x) = y(x); \quad \left(D^{\alpha}_{a+}y_j\right)(x) = y_{j+1}(x) \quad (j = 1, ..., n-1). \tag{7.1.9}$$

Definition 7.2 *For $\alpha \in \mathbb{R}$, we call α-Wronskian of n functions $u_j(x)$ $(j = 1, ..., n)$, having fractional sequential derivatives up to order $(n-1)\alpha$ in the interval $V \subset (a,b]$, the following determinant:*

$$|\mathbf{W}_\alpha(u_1, ..., u_n)(x)| = \begin{vmatrix} u_1(x) & u_2(x) & . & . & . & u_n(x) \\ u_1^{(\alpha}(x) & u_2^{(\alpha}(x) & . & . & . & u_n^{(\alpha}(x) \\ u_1^{(2\alpha}(x) & u_2^{(2\alpha}(x) & . & . & . & u_n^{(2\alpha}(x) \\ \cdots & . & . & . & . & . \\ \cdots & . & . & . & . & . \\ u_1^{((n-1)\alpha}(x) & u_2^{((n-1)\alpha}(x) & \cdots & . & . & u_n^{((n-1)\alpha}(x) \end{vmatrix}, \tag{7.1.10}$$

To simplify the notation, this will be represented by $|\mathbf{W}_\alpha(x)|$.

We will use $\mathbf{W}_\alpha(x)$ for the corresponding Wronskian matrix.

Definition 7.3 *We shall call fundamental system of solutions of the equation* $\mathbf{L_{n\alpha}}(y) = 0$ *in* $V \subset [a,b]$ *a family of* n *functions linearly independent in* V*, which are solutions of this equation.*

Definition 7.4 *The term general solution of the fractional differential equation* $\mathbf{L_{n\alpha}}(y) = f(x)$ *refers to any solution to this equation, which depends on* n *independent constants.*

Next we will prove two theorems on the existence and uniqueness of global solutions to the equation $\mathbf{L_{n\alpha}}(y) = f(x)$ with certain initial conditions.

Theorem 7.1 *Let* $x_0 > a$ *, let* $y_0^k \in \mathbb{R}$ $(k = 0, 1, ..., n-1)$ *be given real numbers, and let* $a_j(x) \in C[a,b]$ $(j = 0, 1, .., n-1)$ *and* $f(x) \in C[a,b]$ *be given continuous functions on* $[a,b]$*. There exists a unique continuous solution* $y(x) \in C(a,b]$ *of the Cauchy problem*

$$[\mathbf{L_{n\alpha}}(y)](x) = f(x) \tag{7.1.11}$$

$$\left(\mathcal{D}_{a+}^{k\alpha} y\right)(x_0) := y^{(k\alpha}(x_0) = y_0^k \quad (k = 0, 1, ..., n-1), \tag{7.1.12}$$

In addition, this solution $y(x)$ *is a-singular of order* α*, that is,*

$$\lim_{x \to a+} (x-a)^{1-\alpha} y(x) < \infty, \tag{7.1.13}$$

and satisfies the inequality $\left(I_{a+}^{1-\alpha} y\right)(x) < \infty$.

Proof. Theorem 7.1 follows immediately upon application of the corresponding theory developed in Chapter 3 and the solution prolongation method until the maximal solution is found.

Theorem 7.2 *Let* $a_j(x) \in C([a,b])$ $(j = 0, 1, .., n-1)$ *and* $f(x) \in C_{1-\alpha}([a,b])$ *be given functions on* $[a,b]$*. Then there exists a unique continuous solution* $y(x) \in C(a,b]$ *for the equation*

$$[\mathbf{L_{n\alpha}}(y)](x) = f(x), \tag{7.1.14}$$

such that, for $k = 0, 1, ..., n-1$,

$$\lim_{x \to a+} (x-a)^{1-\alpha} (\mathcal{D}_{a+}^{k\alpha} y)(x) = b_k \quad (b_k \in \mathbb{R}) \tag{7.1.15}$$

or

$$\left(I_{a+}^{1-\alpha} \mathcal{D}_{a+}^{k\alpha} y\right)(a+) = b_k \quad (b_k \in \mathbb{R}). \tag{7.1.16}$$

Proof. The proof of Theorem 7.2 is analogous to that of Theorem 7.1.

Proposition 7.1 *Any linear combination of solutions of the homogeneous equation*

$$[\mathbf{L_{n\alpha}}(y)](x) = 0 \tag{7.1.17}$$

is also a solution to this equation.

Proof. This property is immediately evident keeping in mind the linearity of $\mathbf{L}_{n\alpha}$.

Proposition 7.2 *Let* $x_0 \in (a, b]$ *(or* $x_0 = a$*) and let* $a_j(x) \in C((a, b])$ $(j = 0, 1, .., n-1)$ *be such that* $(x-a)^{1-\alpha} a_j(x)|_{x=a} < \infty$ $(j = 1, ..., n)$*. Then the homogeneous differential equation* (7.1.17) *with the initial conditions*

$$y^{(j\alpha}(x_0) = 0 \quad (or\ [(x-a)^{1-\alpha} y^{(j\alpha}(x)]_{x=a+} = 0) \quad (j = 0, 1, ..., n-1). \qquad (7.1.18)$$

has only the trivial continuous solution $y(x) = 0$.

Proof. Proposition 7.2 follows clearly from Theorems 7.1 and 7.2.

Proposition 7.3 *Let* $\{u_j(x)\}_{j=1}^{n}$ *be a family of functions which admit fractional sequential derivatives up to order* $(n-1)\alpha$ *in* $(a, b]$*, satisfying for* $j = 1, 2, ..., n$ *and* $k = 0, ..., n-1$*, the condition*

$$\lim_{x \to a+} [(x-a)^{1-\alpha} u_j^{(k\alpha}(x)] < \infty, \qquad (7.1.19)$$

that is, the functions $\{u_j^{(k\alpha}\}_{\substack{j=1,2,...,n \\ k=0,1,...,n-1}}$ *are a-singular of order* α.

If functions $(x-a)^{1-\alpha} u_j(x)$ $(j = 1, \cdots n)$ *are linearly dependent in* $[a, b]$*, then for all* $x \in [a, b]$*,*

$$(x-a)^{n-n\alpha} |\mathbf{W}_\alpha(x)| = 0, \qquad (7.1.20)$$

where $|\mathbf{W}_\alpha(x)|$ *is defined by* (7.1.10).

Proof. Since $\{(x-a)^{1-\alpha} u_j(x)\}_{j=1}^{n}$ are linearly dependent in $[a, b]$, there exist n constants $\{c_j\}_{j=1}^{n}$, not all zero, such that, $\forall x \in [a, b]$,

$$\sum_{j=1}^{n} c_j (x-a)^{1-\alpha} u_j(x) = 0, \qquad (7.1.21)$$

and therefore, $\forall x \in (a, b]$,

$$\sum_{k=1}^{n} c_k u_k(x) = 0. \qquad (7.1.22)$$

Successive application of the sequential derivatives $\mathcal{D}_{a+}^{k\alpha}$ $(k = 1, ..., n-1)$ to (7.1.22) lead to the following relation:

$$\mathbf{W}_\alpha(x) \bar{\mathbf{C}} - \bar{0} \qquad (7.1.23)$$

with

$$\bar{\mathbf{C}} = \begin{pmatrix} c_1 \\ c_2 \\ \cdots \\ \cdots \\ \cdots \\ c_n \end{pmatrix} \neq \bar{0}, \qquad (7.1.24)$$

and therefore, $|\mathbf{W}_\alpha(x)| = 0 \ \forall x \in (a, b]$.

In addition, by the hypothesis (7.1.19), we can also conclude that

$$\lim_{x\to a+} (x-a)^{1-\alpha}\mathbf{W}_\alpha(x)\bar{\mathbf{C}} = 0, \tag{7.1.25}$$

and hence

$$\lim_{x\to a+} (x-a)^{n-n\alpha}|\mathbf{W}_\alpha(x)| = 0, \tag{7.1.26}$$

which proves Proposition 7.3.

Proposition 7.4 *If the conditions of Proposition 7.3 are valid and there exists an $x_0 \in [a, b]$ such that $[(x-a)^{n-n\alpha}|\mathbf{W}_\alpha(x)|]_{x=x_0} \neq 0$, then the functions $\{(x-a)^{1-\alpha}u_j(x)\}_{j=1}^n$ are linearly independent.*

Proof. By Proposition 7.3, Proposition 7.4 is proved by *reductio ad absurdum.*

Proposition 7.5 *Let solutions $u_j(x)$ $(j = 1, \cdots, n)$ of equation $[\mathbf{L}_{\mathbf{n}\alpha}(y)](x) = 0$ in $(a, b]$ be a-singular of order α. Then the functions*

$$\{(x-a)^{1-\alpha}u_j(x)\}_{j=1}^n \tag{7.1.27}$$

are linearly dependent in $[a, b]$ if, and only if, there exists an $x_0 \in [a, b]$ such that

$$(x-a)^{n-n\alpha}|\mathbf{W}_\alpha(x)|]_{x=x_0} = 0. \tag{7.1.28}$$

Proof. If $\{(x-a)^{1-\alpha}u_j(x)\}_{j=1}^n$ are linearly dependent, (7.1.28) follows as a consequence of Proposition 7.3.

To demonstrate the converse, we will first consider the case $x_0 \neq a$ and consider the system of equations

$$\mathbf{W}_\alpha(x_0)\bar{\mathbf{C}} = \bar{0}, \tag{7.1.29}$$

where $\bar{\mathbf{C}}$ is given by (7.1.24) and $\{c_j^0\}_{j=1}^n$ are independent constants, which yield the solution $\bar{\mathbf{C}}_0 \neq \bar{0}$.

For the function

$$y(x) = \sum_{j=1}^n c_j^0 u_j(x), \tag{7.1.30}$$

defined in $(a, b]$, which is a solution to $\mathbf{L}_{\mathbf{n}\alpha}(y) = 0$ in $(a, b]$ (because it is a linear combination of solutions) it holds that

$$y^{(k\alpha}(x_0) = 0 \quad (\forall k = 0, 1, \ldots, n-1) \tag{7.1.31}$$

Therefore, by Proposition 7.2, $y(x) = 0$ $(\forall x \in (a, b])$, or, equivalently, the functions $\{u_j(x)\}_{j=1}^n$ are linearly dependent in $(a, b]$ and consequently $\{(x-a)^{1-\alpha}u_j(x)\}_{j=1}^n$ are linearly dependent in $[a, b]$.

For the case $x_0 = a$, the result is similarly obtained by considering the system $[(x-a)^{1-\alpha}\mathbf{W}_\alpha(x)]_{x=x_0}\bar{\mathbf{C}} = \bar{0}$.

Proposition 7.6 *If the conditions of Proposition 7.5 are satisfied, then* $\forall x \in [a, b]$,

$$(x-a)^{n-n\alpha}|\mathbf{W}_\alpha(x)| = 0 \quad or \quad (x-a)^{n-n\alpha}|\mathbf{W}_\alpha(x)| \neq 0, \tag{7.1.32}$$

where $\mathbf{W}_\alpha(x) = \mathbf{W}_\alpha(u_1, ..., u_n)(x)$.

Proof. Proposition 7.6 follows from Propositions 7.3 and 7.5.

Proposition 7.7 *If* $\{u_j(x)\}_{j=1}^n$ *is a fundamental system of solutions to the equation* $[\mathbf{L}_{n\alpha}(y)](x) = 0$ *in a certain interval* $V \subset (a, b]$, *then the general solution to this differential equation in* V *is given by*

$$y_g(x) = \sum_{k=1}^{n} c_k u_k(x), \tag{7.1.33}$$

where $\{c_k\}_{k=1}^n$ *are arbitrary constants.*

Proof. Since $y_g(x)$ is a solution to (7.1.17), we need only show that any other solution is a special case of $y_g(x)$. Let $u(x)$ be a solution to the equation (7.1.17) which satisfies, for a given $x_0 \in (a, b]$, the following conditions:

$$u^{(k\alpha}(x_0) = u_0^k \in \mathbb{R} \quad (k = 0, 1, ..., n-1). \tag{7.1.34}$$

Then, applying Theorem 7.1, there exist certain particular constants $\{c_j\}_{j=1}^n$ such that

$$u(x) = \sum_{k=1}^{n} c_k u_k(x), \tag{7.1.35}$$

which proves the proposition.

Proposition 7.8 *The set of all solutions of the differential equation* $[\mathbf{L}_{n\alpha}(y)](x) = 0$, *in a certain interval* $V \subset (a, b]$, *is a vector space of* n *dimensions.*

Proof. Proposition 7.8 follows directly from Proposition 7.7.

Proposition 7.9 *There exist a fundamental system of solutions of the differential equation of fractional order* $[\mathbf{L}_{n\alpha}(y)](x) = 0$ *in an interval* $V \subset (a, b]$, *such that*

$$[(x-a)^{n-n\alpha}|\mathbf{W}_\alpha(x)|]_{x=a} = \frac{1}{[\Gamma(\alpha)]^n}. \tag{7.1.36}$$

Proof. Proposition 7.9 follows directly from Theorem 7.2.

Proposition 7.10 *If* $y_p(x)$ *is a particular solution to the equation* $[\mathbf{L}_{n\alpha}(y)](x) = f(x)$, *then the general solution to this equation is given by*

$$y_g(x) = y_h(x) + y_p(x), \tag{7.1.37}$$

where $y_h(x)$ *is the general solution to the associated homogeneous equation* $[\mathbf{L}_{n\alpha}(y)](x) = 0$.

Proof. Proposition 7.10 is evident.

7.2 Solution of Linear Differential Equations with Constant Coefficients

In this section we introduce a method independent of the Laplace transform, analogous to that for the ordinary case, for obtaining a fundamental system of solutions to the equation $\mathbf{L}_{n\alpha}(y) = 0$, which also yields an explicit expression for the general solution to the non-homogeneous equation $\mathbf{L}_{n\alpha}(y) = f(x)$.

7.2.1 General Solution in the Homogeneous Case

Let

$$[\mathbf{L}_{n\alpha}(y)](x) := \left(\mathcal{D}_{a+}^{n\alpha}\right)(x) + \sum_{k=0}^{n-1} a_k \left(\mathcal{D}_{a+}^{k\alpha}\right) y(x) = 0, \tag{7.2.1}$$

where the coefficients $\{a_j\}_{j=1}^{n-1}$ are real constants.

As in the usual case, we shall seek the solution of (7.2.1) in the form $y(x) = e_\alpha^{\lambda(x-a)}$, where $e_\alpha^{\lambda(x-a)}$ is defined by (1.10.11). It follows from (7.2.1) and (1.10.54) that

$$\left[\mathbf{L}_{n\alpha}\left(e_\alpha^{\lambda(t-a)}\right)\right](x) = P_n(\lambda) e_\alpha^{\lambda(x-a)}, \tag{7.2.2}$$

where

$$P_n(\lambda) = \lambda^n + \sum_{k=1}^{n-1} a_k \lambda^k \tag{7.2.3}$$

is the characteristic polynomial associated with the equation $[\mathbf{L}_{n\alpha}(y)](x) = 0$.

The following assertion is true for complex $\lambda \in \mathbb{C}$.

Lemma 7.1 *If $\lambda \in \mathbf{C}$ is a root of the characteristic polynomial* (7.2.3), *then*

$$\frac{\partial}{\partial \lambda}\left[\mathbf{L}_{n\alpha}\left(e_\alpha^{\lambda(t-a)}\right)\right](x) = \left[\mathbf{L}_{n\alpha}\left(\frac{\partial}{\partial \lambda} e_\alpha^{\lambda(t-a)}\right)\right](x) \tag{7.2.4}$$

and

$$\frac{\partial^l}{\partial \lambda^l} e_\alpha^{\lambda(x-a)} = (x-a)^{l\alpha} e_{\alpha,l}^{\lambda(x-a)}. \tag{7.2.5}$$

Proof. Lemma 7.1 follows from the linearity of the operators $\frac{\partial}{\partial \lambda}$ and $\mathbf{L}_{n\alpha}$ and from (1.10.28).

Proposition 7.11 *If λ_1 is a root of multiplicity μ_1 of the characteristic polynomial* (7.2.3), *then the functions $\{y_{1,l}(x)\}_{l=0}^{\mu_1-1}$:*

$$y_{1,l}(x) = (x-a)^{l\alpha} e_{\alpha,l}^{\lambda_1(x-a)}, \tag{7.2.6}$$

where $e_{\alpha,l}^{\lambda_1(x-a)}$ is defined by (1.10.31), *are solutions of the equation* $[\mathbf{L}_{n\alpha}(y)](x) = 0$.

Proof. Taking (7.2.2), (7.2.4) and (1.10.28) into account and using the classical Leibniz rule, we have

$$\left\{\left[\mathbf{L}_{\mathbf{n}\alpha}\left(\frac{\partial^l}{\partial\lambda^l}e_\alpha^{\lambda(t-a)}\right)\right](x)\right\}_{\lambda=\lambda_1} = \left\{\frac{\partial^l}{\partial\lambda^l}\left[\mathbf{L}_{\mathbf{n}\alpha}\left(e_\alpha^{\lambda(t-a)}\right)\right](x)\right\}_{\lambda=\lambda_1} \tag{7.2.7}$$

$$= \sum_{j=0}^{l}\binom{l}{j}\left[\frac{\partial^{l-j}}{\partial\lambda^{l-j}}\left(e_\alpha^{\lambda(x-a)}\right)\right]_{\lambda=\lambda_1}\frac{\partial^j}{\partial\lambda^j}\left[P_n(\lambda)\right]_{\lambda=\lambda_1} = 0. \tag{7.2.8}$$

Since

$$\left[\frac{\partial^j}{\partial\lambda^j}P_n(\lambda)\right]_{\lambda=\lambda_1} = 0 \quad (j = 0, 1, ..., \mu_1 - 1), \tag{7.2.9}$$

by the hypothesis, then $[\mathbf{L}_{\mathbf{n}\alpha}(y_{1,l})](x) = 0$, and the proposition is proved.

Corollary 7.1 *Let* $\{\lambda_j\}_{j=1}^k$ *be k distinct roots of multiplicity* $\{\mu_j\}_{j=1}^k$ *of the characteristic polynomial* (7.2.3). *Then the functions*

$$\bigcup_{m=1}^{k}\left\{(x-a)^{l\alpha}e_{\alpha,l}^{\lambda_m(x-a)}\right\}_{l=0}^{\mu_m-1} \tag{7.2.10}$$

are linearly independent solutions of the equation (7.2.1).

Proof. Corollary 7.1 follows from Propositions 7.11 and 7.4.

Proposition 7.12 *If* λ_1 *and* $\overline{\lambda_1}$ $(\lambda_1 = b + ic,\ c \neq 0)$ *are two complex solutions of multiplicity* σ_1 *of the characteristic polynomial* (7.2.3), *then the functions*

$$\left\{\sum_{j=0}^{\infty}(-1)^j\frac{c^{2j}}{(2j)!}(x-a)^{(2j+l)\alpha}e_{\alpha,l+2j}^{b(x-a)}\right\}_{l=0}^{\sigma_1-1} \tag{7.2.11}$$

and

$$\left\{\sum_{j=0}^{\infty}(-1)^j\frac{c^{2j+1}}{(2j+1)!}(x-a)^{(2j+l+1)\alpha}e_{\alpha,l+2j+1}^{b(x-a)}\right\}_{l=0}^{\sigma_1-1} \tag{7.2.12}$$

form $2\sigma_1$ *linearly independent real solutions of the equation* $[\mathbf{L}_{\mathbf{n}\alpha}(y)](x) = 0$.

Proof. Since $\lambda_1 = b + ic$ is a root of multiplicity order σ_1 of the polynomial (7.2.3), it follows from Proposition 7.11 that $\forall l = 0, 1, ..., \sigma_1 - 1$,

$$y_{1,l}(x) = (x-a)^{l\alpha}e_{\alpha,l}^{\lambda_1(x-a)} \tag{7.2.13}$$

is a complex solution to (7.2.1). Using similar arguments for $\overline{\lambda_1}$, and taking into account (1.10.33), (1.10.34), and Proposition 7.4, we complete the proof of Proposition 7.12.

Corollary 7.2 *Let $\{\lambda_m, \overline{\lambda}_m\}_{m=1}^{p}$, $\lambda_m = b_m + ic_m$ $(c_m \neq 0)$, be all distinct pairs of complex conjugate solutions of multiplicity $\{\sigma_m\}_{m=1}^{p}$ of the characteristic polynomial* (7.2.3) *for the fractional differential equation* (7.2.1). *Then the functions*

$$\bigcup_{m=1}^{p}\left\{\sum_{j=0}^{\infty}(-1)^j\frac{c_m^{2j}}{(2j)!}(x-a)^{(2j+l)\alpha}e_{\alpha,l+2j}^{b_m(x-a)}\right\}_{l=0}^{\sigma_m-1} \tag{7.2.14}$$

and

$$\bigcup_{m=1}^{p}\left\{\sum_{j=0}^{\infty}(-1)^j\frac{c_m^{2j+1}}{(2j+1)!}(x-a)^{(2j+l+1)\alpha}e_{\alpha,l+2j+1}^{b_m(x-a)}\right\}_{l=0}^{\sigma_m-1} \tag{7.2.15}$$

determine a linearly independent set of solutions to the equation (7.2.1).

Proof. To derive Corollary 7.2, it is sufficient to apply Propositions 7.4 and 7.12.

Theorem 7.3 *Let $P_n(\lambda)$, given by* (7.2.3), *be the characteristic polynomial for the linear fractional differential equation* (7.2.1). *Let $\{\lambda_j\}_{j=1}^{k}$ be all real and distinct roots of* (7.2.3) *of the multiplicity $\{\mu_j\}_{j=1}^{k}$, and let $\{r_j, \overline{r}_j\}_{j=1}^{p}$ $(r_j = b_j + ic_j)$ be the set of all distinct pairs of complex conjugate roots of* (7.2.3) *of the multiplicity $\{\sigma_j\}_{j=1}^{p}$ such that $\sum_{j=1}^{k}\mu_j + 2\sum_{j=1}^{p}\sigma_j = n$. Then the functions*

$$\bigcup_{m=1}^{k}\left\{(x-a)^{l\alpha}e_{\alpha,l}^{\lambda_m(x-a)}\right\}_{l=1}^{\mu_m-1}, \tag{7.2.16}$$

$$\bigcup_{m=1}^{p}\left\{\sum_{j=0}^{\infty}(-1)^j\frac{c_m^{2j}}{(2j)!}(x-a)^{(2j+l)\alpha}e_{\alpha,l+2j}^{b_m(x-a)}\right\}_{l=1}^{\sigma_m-1} \tag{7.2.17}$$

and

$$\bigcup_{m=1}^{p}\left\{\sum_{j=0}^{\infty}(-1)^j\frac{c_m^{2j+1}}{(2j+1)!}(x-a)^{(2j+l+1)\alpha}e_{\alpha,l+2j+1}^{b_m(x-a)}\right\}_{l=1}^{\sigma_m-1} \tag{7.2.18}$$

form the fundamental system of solutions of the differential equation (7.2.1).

Proof. Theorem 7.3 follows from 7.1-7.2 and Proposition 7.4.

Note that in the case $a = 0$, operational methods such as the Laplace transform may be also applied to solve the problem involving constant coefficients.

Example 7.1

1. Consider the fractional differential equation $y^{(2\alpha}(x) + \mu^2 y(x) = 0$ $(\mu > 0)$, with $y^{(2\alpha}(x) = \left(\mathcal{D}_{a+}^{2\alpha}y\right)(x)$. The characteristic polynomial (7.2.3) for this equation has

the form $P_2(\lambda) := \lambda^2 + \mu^2 = 0$, and $\lambda_1 = \mu i$ and $\lambda_2 = -\mu i$ are its roots. Therefore from Corollary 7.2 we obtain the following fundamental system of solutions:

$$\{\cos_\alpha[\mu(\mathrm{x}-\mathrm{a})],\ \sin_\alpha[\mu(\mathrm{x}-\mathrm{a})]\}, \tag{7.2.19}$$

where

$$\cos_\alpha[\mu(x-a)] = \sum_{j=0}^{\infty}(-1)^j\mu^{(2j+1)}\frac{(x-a)^{(2j+1)\alpha}}{(2j+1)!}e_{\alpha,2j+1}^{(x-a)} \tag{7.2.20}$$

$$= \sum_{j=0}^{\infty}(-1)^j\mu^{(2j+1)}\frac{(x-a)^{(j+1)2\alpha-1}}{\Gamma[(j+1)2\alpha]} \tag{7.2.21}$$

and

$$\sin_\alpha[\mu(x-a)] = \sum_{j=0}^{\infty}(-1)^j\mu^{(2j)}\frac{(x-a)^{(2j)\alpha}}{(2j)!}e_{\alpha,2j}^{(x-a)} \tag{7.2.22}$$

$$= \sum_{j=0}^{\infty}(-1)^j\mu^{(2j)}\frac{(x-a)^{(2j+1)\alpha-1}}{\Gamma[(2j+1)\alpha]} \tag{7.2.23}$$

2. The fractional differential equation $y^{(2\alpha}(x) - y(x) = 0$ has the following fundamental system of solutions:

$$\{e_\alpha^{(x-a)}, e_\alpha^{-(x-a)}\}. \tag{7.2.24}$$

This result follows from Corollary 7.1 if we take into account that $\lambda_1 = 1$ and $\lambda_2 = -1$ are the roots of the characteristic polynomial $P_2(\lambda) := \lambda^2 - 1 = 0$.

3. The fractional differential equation $y^{(2\alpha}(x) - 2y^{(\alpha}(x) + y(x) = 0$ has the following fundamental system of solutions:

$$\{e_\alpha^{(x-a)}, (x-a)^\alpha e_{\alpha,1}^{(x-a)}\}. \tag{7.2.25}$$

Since $\lambda_1 = 1$ is the root of multiplicity two of the characteristic polynomial $P_2(\lambda) := \lambda^2 - 2\lambda - 1 = 0$, Corollary 7.1 yields the above result.

7.2.2 General Solution in the Non-Homogeneous Case. Fractional Green Function

Consider the non-homogeneous linear differential equation of fractional order

$$[\mathbf{L}_{\mathbf{n}\alpha}(y)](x) = f(x) \tag{7.2.26}$$

defined by (7.1.5). As in the case of ordinary differential equations, the general solution to this equation is a sum of a particular solution to the equation (7.2.26) and a general solution to the corresponding homogeneous equation (7.2.1).

In this section we apply the operational method and the Laplace transform method to derive a particular solution $y_p(x)$ of (7.2.26). Note that the Laplace transform method is valid only when $a = 0$, since methods similar to that of the variation of parameters are not applicable due to a noticeable lack of basic properties in the R-L derivative. Note that a particular solution to the equation (7.2.26) may be also obtained by reducing this equation to a non-homogeneous linear system.

A. Operational method.

Proposition 7.13 *Let $\{\lambda_j\}_{j=1}^k$ be k different roots of the multiplicity $\{\mu_j\}_{j=1}^k$ ($\sum_{j=1}^k \mu_j = n$) of the characteristic polynomial $P_n(\lambda) = 0$ associated with the equation $[\mathbf{L}_{n\alpha}(y)](x) = 0$, where $y^{(k\alpha}(x) = (\mathcal{D}_{a+}^{k\alpha} y)(x)$. Then the particular solution to the equation* (7.2.26) *with a given function $f(x)$ on $(a, b]$ is given by*

$$y_p(x) = \left[\prod_{j=1}^{k}\left(\sum_{n=0}^{\infty}(-\lambda_j)^n I_{a+}^{(n+1)\alpha}\right)^{\mu_j} f\right](x) \tag{7.2.27}$$

provided that the series in the right-hand side of (7.2.27) *are uniformly convergent.*

Proof. We can consider the inverse operator to the left for $\mathbf{L}_{n\alpha}$ in the form

$$\mathbf{L}_{n\alpha}^{-1} = \prod_{j=1}^{k}\left[(D_{a+}^{\alpha} - \lambda_j)^{-1}\right]^{\mu_j} \tag{7.2.28}$$

$$= \prod_{j=1}^{k}\left\{[D_{a+}^{\alpha}(1 - \lambda_j I_{a+}^{\alpha})]^{-1}\right\}^{\mu_j} = \prod_{j=1}^{k}\left(\sum_{n=0}^{\infty}(-\lambda_j)^n I_{a+}^{(n+1)\alpha}\right)^{\mu_j}, \tag{7.2.29}$$

which yields the formal solution to the equation (7.2.26) in (7.2.27). Applying the operator

$$\mathbf{L}_{n\alpha} = \prod_{j=1}^{k}\left(D_{a+}^{\alpha} - \lambda_j\right)^{\mu_j} \tag{7.2.30}$$

to the function y_p in (7.2.27) and using the uniform convergence of the series considered, we derive that $y_p(x)$ is the solution to the equation $[\mathbf{L}_{n\alpha}(y)](x) = f(x)$.

B. Laplace Transform Method

The Laplace transform method is only useful for the case $a = 0$, and has been used by various authors as already discussed. We will apply this method to explicitly derive a particular solution for the non-homogeneous equation

$$[\mathbf{L}_{n\alpha}(y)](x) = f(x). \tag{7.2.31}$$

By (2.2.36), the Laplace transform $\mathcal{L}$ of $D_{0+}^{\alpha} f(x)$, for a suitable $f(x)$, is given by

$$\left(\mathcal{L}D_{0+}^{\alpha} f\right)(s) = s^{\alpha}(\mathcal{L}f)(s) \tag{7.2.32}$$

and, therefore,

$$\left(\mathcal{L}\mathcal{D}_{0+}^{n\alpha} f\right)(s)\} = s^{n\alpha}(\mathcal{L}f)(s). \tag{7.2.33}$$

Proposition 7.14 *A particular solution to the non-homogeneous LFDE* (7.2.31) *with* $y^{(k\alpha}(x) = (\mathcal{D}_{0+}^{k\alpha} y)(x)$ *is given by*

$$y_p(x) = \sum_{j=0}^{k} \sum_{m=1}^{\mu_j} c_{j,m} (g_{j,m} * f)(x), \tag{7.2.34}$$

where

$$g_{j,m}(x) = \mathcal{L}^{-1}\left[\frac{1}{(s^\alpha - \lambda_j)^m}\right] = \frac{x^{\alpha(m-1)}}{(m-1)!}\mathcal{E}_{\alpha,m-1}^{\lambda_j x} = \frac{1}{(m-1)!}\frac{\partial^{m-1}}{\partial \lambda_j^{m-1}} e_\alpha^{\lambda_j x}, \tag{7.2.35}$$

$c_{j,m}$ $(j = 1, 2, ..., k;\ m = 1, 2, ..., \mu_j)$ *are constants defined by the following decomposition into simple fractions:*

$$\frac{1}{P_n(\lambda)} = \frac{1}{\prod_{j=1}^{k} (\lambda - \lambda_j)^{\mu_j}} = \sum_{j=1}^{k} \sum_{m=1}^{\mu_m} \frac{c_{j,m}}{(\lambda - \lambda_j)^m}, \tag{7.2.36}$$

and $(g * f)(x)$ *represents the Laplace convolution of* $g(x)$ *and* $f(x)$ *defined by* (1.4.10).

Proof. Applying the Laplace transform (1.4.1) to the equation (7.2.31), using (7.2.36) and taking the inverse Laplace transform (1.4.2), we find the particular solution to (7.2.31) in the form

$$y_p(x) = \left(\mathcal{L}^{-1}\left[\sum_{j=1}^{k} \sum_{m=1}^{\mu_m} c_{j,m} \frac{(\mathcal{L}f)(s)\}}{(s^\alpha - \lambda_j)^m}\right]\right)(x). \tag{7.2.37}$$

This yields the result in (7.2.34), if we take into account (7.2.35) and the Laplace convolution formula (1.4.12).

C. Operational decomposition method

In this section we extend the result obtained in Proposition 7.14 for the equation (7.2.31) with $y^{(k\alpha}(x) = (\mathcal{D}_{0+}^{k\alpha} y)(x)$ to the case $y^{(k\alpha}(x) = (\mathcal{D}_{a+}^{k\alpha} y)(x)$ $\forall a \in \mathbb{R}$. First we consider the simplest equation.

Proposition 7.15 *Let* $f(x) \in L_1(a, b) \cap C[(a, b)]$. *Then the LFDE*

$$y^{(\alpha}(x) - \lambda y(x) = f(x) \quad (x > a), \tag{7.2.38}$$

where $y^{(\alpha}(x) = \left(D_{a+}^{\alpha} y\right)(x)$, *has the general solution*

$$y_g(x) = c e_\alpha^{\lambda(x-a)} + y_p(x), \tag{7.2.39}$$

where

$$y_p(x) = (e_\alpha^{\lambda t} *^a f)(x) \tag{7.2.40}$$

is a particular solution to (7.2.38), $*^a$ *being the following convolution:*

$$(g *^a f)(x) = \int_a^x g(x-t)f(t)dt. \tag{7.2.41}$$

In addition, if $f(x)$ *is continuous in* $[a,b]$, *then* $y_p(a+) = 0$, *while if* $f(x) \in C_{1-\alpha}([a,b])$, *then* $\left(I_{a+}^{1-\alpha}y_p\right)(a+) = 0$.

Proof. By Proposition 7.7, $y(x) = ce_\alpha^{\lambda(x-a)}$ is the general solution to the corresponding homogeneous equation (7.2.38) $y^{(\alpha}(x) - \lambda y(x) = 0$. Therefore it is sufficient to verify that $y_p(x)$ is a particular solution to (7.2.38). Applying Lemma 2.10 and taking (2.1.56) into account, we have

$$\left(D_{a+}^\alpha y_p\right)(x) = \left(D_{a+}^\alpha \int_a^t e_\alpha^{\lambda(t-\tau)} f(\tau)d\tau\right)(x) \tag{7.2.42}$$

$$= \int_a^x \{D_{a+}^\alpha e_\alpha^{\lambda(x-a)}\}(t)f(x-t+a)dt + f(x) \lim_{t\to a+} \{I_{a+}^{1-\alpha} e_\alpha^{\lambda(x-a)}\}(t) \tag{7.2.43}$$

$$= \lambda \int_a^x e_\alpha^{\lambda(t-a)} f(x+a-t)dt + f(x) \tag{7.2.44}$$

$$= \lambda \int_a^x e_\alpha^{\lambda(x-\xi)} f(\xi)d\xi + f(x) = \lambda y(x) + f(x). \tag{7.2.45}$$

This concludes the proof of Proposition 7.15,

Theorem 7.4 *Let* $\{\lambda_j\}_{j=1}^k$ *be the* k *distinct complex roots of the multiplicity* $\{\sigma_j\}_{j=1}^k$ *of the characteristic polynomial* (7.2.3) *for the following non-homogeneous LFDE:*

$$\left[\mathbf{L}_{\mathbf{n}\alpha}(y)\right](x) = \left(\prod_{j=1}^k (D_{a+}^\alpha - \lambda_j)^{\sigma_j} y\right)(x) = f(x) \quad (x > a). \tag{7.2.46}$$

Then the particular solution to (7.2.1) *is given by*

$$y_p = (G_\alpha *^a f)(x), \tag{7.2.47}$$

where $G_\alpha(x)$ *is given by*

$$G_\alpha(x) = \left(\prod_{j=1}^{k}{}^{*^a} \left(\prod_{l=1}^{\sigma_j}{}^{*^a} e_\alpha^{\lambda_j t}\right)\right)(x). \tag{7.2.48}$$

Furthermore, if $f(x) \in C_{1-\alpha}([a,b])$, *then* $\left(I_{a+}^{1-\alpha}y_p\right)(a+) = 0$, *while* $y_p(a+) = 0$ *when* $f(x)$ *is continuous in* $[a,b]$.

Additionally, $\left(I_{a+}^{1-\alpha}G_\alpha\right)(a+) = 0$.

Proof. It is sufficient to apply successively the result of Proposition 7.15 while taking into account the weak singularity of the function $e_\alpha^{\lambda x}$.

The function $G_\alpha(x-\xi)$ plays the role of the fractional Green function associated with the non-homogeneous LFDE $[\mathbf{L}_{\mathbf{n}\alpha}(y)](x) = f(x)$, analogous to the case of ordinary differential equations with constant coefficients.

The result (7.2.47) is given, in general, by the function of a complex variable. Its real part will be a particular solution to (7.2.46) and its imaginary part will be a solution to the corresponding homogeneous equation $[\mathbf{L}_{\mathbf{n}\alpha}(y)](x) = 0$.

7.3 Non-Sequential Linear Differential Equations with Constant Coefficients

The results derived in Sections 7.1 and 7.2 may be applied to certain cases of non-sequential fractional differential equations of the Riemann-Liouville type. Without going into specifics, we consider the following case of order 2α:

$$\left(D_{a+}^{2\alpha}y\right)(x) + a_1(x)\left(D_{a+}^{\alpha}y\right)(x) + a_0(x)y(x) = f(x) \ (0 < \alpha < \frac{1}{2}). \tag{7.3.1}$$

Recall that the relationship between $\left(D_{a+}^{2\alpha}y\right)(x)$ and $\left(\mathcal{D}_{a+}^{2\alpha}y\right)(x)$ for $0 < \alpha < \frac{1}{2}$ is given by (7.1.4):

$$\left(\mathcal{D}_{a+}^{2\alpha}y\right)(x) = \left(D_{a+}^{2\alpha}\left[y(t) - \frac{\left(I_{a+}^{1-\alpha}y\right)(a+)}{\Gamma(\alpha)}(t-a)^{\alpha-1}\right]\right)(x) \tag{7.3.2}$$

and that

$$\left(D_{a+}^{\alpha}(t-a)^{\alpha-1}\right)(x) = 0. \tag{7.3.3}$$

Corollary 7.3 *Let* $0 < \alpha < \frac{1}{2}$, $a_k(x) \in C([a,b])$ $(k = 0,1)$ *and* $f(x) \in C_{1-\alpha}([a,b])$. *Then the Cauchy type problem for the non-homogeneous LFDE* (7.3.1)

$$D_{a+}^{2\alpha}y + a_1(x)D_{a+}^{\alpha}y + a_0(x)y = f(x) \tag{7.3.4}$$

$$\begin{aligned} &\lim_{x\to a+}(x-a)^{1-\alpha}y(x) = b_0 \\ &\lim_{x\to a+}(x-a)^{1-\alpha}\left(D_{a+}^{\alpha}y\right)(x) = b_1 \end{aligned} \tag{7.3.5}$$

has a unique continuous solution $y(x) \in C(a,b]$.

Proof. The above result follows from Theorem 7.2 and Property 2.4, because the Cauchy type problem in (7.3.4)-(7.3.5) is equivalent to

$$\left(\mathcal{D}_{a+}^{2\alpha}y\right)(x) + a_1(x)\left(\mathcal{D}_{a+}^{\alpha}y\right)(x) + a_0(x)y(x) = f(x) \tag{7.3.6}$$

$$\lim_{x\to a+}(x-a)^{1-\alpha}\left(\mathcal{D}_{a+}^{k\alpha}y\right)(x) = b_k \ (k = 0,1). \tag{7.3.7}$$

Corollary 7.4 *Let $0 < \alpha < \frac{1}{2}$ and $f \in C_{1-\alpha}([a,b])$. Then the LFDE (7.3.4) with real constant coefficients a_0 and a_1 has the solution given by*

$$y(x) = C_1 z_1(x) + C_2 z_2(x) + z_p(x) - \frac{C}{\Gamma(\alpha)}(x-a)^{\alpha-1}. \tag{7.3.8}$$

Here z_j $(j = 1, 2)$ are two linearly independent solutions of the homogeneous sequential LFDE

$$\left(\mathcal{D}_{a+}^{2\alpha} z\right)(x) + a_1 \left(\mathcal{D}_{a+}^{\alpha} z\right)(x) + a_0 z(x) = 0, \tag{7.3.9}$$

and

$$z_p(x) = \left(z_1 *^a z_2 *^a \left[f(t) + a_0 C(t-a)^{\alpha-1}\right]\right)(x) \tag{7.3.10}$$

is a particular solution to the non-homogeneous equation

$$\left(\mathcal{D}_{a+}^{2\alpha} z\right)(x) + a_1 \left(\mathcal{D}_{a+}^{\alpha} z\right)(x) + a_0 z(x) = f(x) + a_0 C(x-a)^{\alpha-1}, \tag{7.3.11}$$

where C, C_1 and C_2 are real constants satisfying the following relationships:

a) *$C_1 + C_2 = C$ for the case when the roots of the characteristic polynomial associated with (7.3.9) are distinct;*

b) *$C_1 = C$ for any other case.*

Proof. The result in the corollary follows from Theorems 7.3 and 7.4.

Example 7.2 Let $0 < \alpha < \frac{1}{2}$ and $f \in C_{1-\alpha}([a,b])$. The equation

$$\left(D_{a+}^{2\alpha} y\right)(x) - y(x) = f(x) \quad (x > a) \tag{7.3.12}$$

has the general solution

$$y_g(x) = C_1 e_\alpha^{(x-a)} + (C - C_1) e_\alpha^{-(x-a)} + h(x) - \frac{C}{\Gamma(\alpha)}(x-a)^{\alpha-1}, \tag{7.3.13}$$

where

$$h(x) = \left(e_\alpha^{(t-a)} *^a e_\alpha^{-(t-a)} *^a \left[f(t) - \frac{C}{\Gamma(\alpha)}(t-a)^{\alpha-1}\right]\right)(x), \tag{7.3.14}$$

C and C_1 being two arbitrary real constants.

Example 7.3 Let $0 < \alpha < \frac{1}{2}$ and $f \in C_{1-\alpha}([a,b])$. The equation

$$\left(D_{a+}^{2\alpha} y\right)(x) - 2\left(D_{a+}^{\alpha} y\right)(x) + y(x) = f(x) \quad (x > a) \tag{7.3.15}$$

has the general solution

$$y_g(x) = C e_\alpha^{(x-a)} + C_2 e_{\alpha,1}^{(x-a)} + u(x) - \frac{C}{\Gamma(\alpha)}(x-a)^{\alpha-1}, \tag{7.3.16}$$

where

$$u(x) = \left(e_\alpha^{(t-a)} *^a e_{\alpha,1}^{(t-a)} *^a \left[f(t) + \frac{C}{\Gamma(\alpha)}(t-a)^{\alpha-1}\right]\right)(x), \tag{7.3.17}$$

C_2 and C being two arbitrary real constants.

7.4 Systems of Equations Associated with Riemann-Liouville and Caputo Derivatives

This section deals with the study of the following system of linear fractional differential equations (SLFDE) associated with R-L and Caputo derivatives

$$\left(D^{\alpha}\bar{Y}\right)(x) := \left(\bar{Y}^{(\alpha)}\right)(x) = \mathbf{A}(x)\bar{Y}(x) + \bar{\mathbf{B}}(x), \tag{7.4.1}$$

where

$$\mathbf{A}(x) = \begin{pmatrix} a_{11}(x) & . & . & . & a_{1n}(x) \\ \cdots & . & . & . & . \\ \cdots & . & . & . & . \\ \cdots & . & . & . & . \\ a_{n1}(x) & . & . & . & a_{nn}(x) \end{pmatrix}; \quad \bar{\mathbf{B}}(x) = \begin{pmatrix} b_1(x) \\ \cdots \\ \cdots \\ \cdots \\ b_n(x) \end{pmatrix} \tag{7.4.2}$$

are matrices of known real functions and D^{α} represents the R-L or Caputo fractional derivative.

We give explicit solutions for the homogeneous equation with variable coefficients and for the non-homogeneous equation with constant coefficients. In the last case we use a fractional Green function generalizing the corresponding ordinary function.

7.4.1 General Theory

First we prove two theorems which establish the existence and uniqueness of global solutions for the system (7.4.1), where $D^{\alpha} = D^{\alpha}_{a+}$, with certain initial conditions. An analogous theory could be developed for the Caputo case.

Theorem 7.5 *Let* $0 < \alpha < 1$, *let* x_0 *and* $\bar{Y}_0$ *be fixed real numbers, and let* $\mathbf{A}(x)$ *and* $\bar{\mathbf{B}}(x)$ *be continuous matrices on* $[a, b]$. *Then the SLFDE* (7.4.1) *where* $D^{\alpha} = D^{\alpha}_{a+}$, *has a unique continuous solution* $\bar{Y}(x)$, *defined in* $(a, b]$, *which satisfies*

$$\bar{Y}(x_0) = \bar{Y}_0. \tag{7.4.3}$$

This solution is a-singular of order α, *that is,*

$$\lim_{x \to a+} \left[(x-a)^{1-\alpha}\bar{Y}(x)\right] < \infty \tag{7.4.4}$$

and, in addition,

$$\left(I^{1-\alpha}_{a+}\bar{Y}\right)(a+) < \infty. \tag{7.4.5}$$

Proof. Theorem 7.5 follows by applying the theory developed in Chapter 3.

Theorem 7.6 *Let* $0 < \alpha < 1$, *let* $\mathbf{A}(x)$ *be continuous in* $[a, b]$, *and let* $\mathbf{B}(x) \in \bar{C}_{1-\alpha}([a, b])$. *Then the SLFDE* (7.4.1) *has a unique continuous solution on* $(a, b]$ *such that*

$$\lim_{x \to a+} \left[(x-a)^{1-\alpha}\bar{Y}(x)\right] = \bar{b} \ \left(\text{or } \left(I^{1-\alpha}_{a+}\bar{Y}\right)(a+) = \bar{b}\right), \tag{7.4.6}$$

where $\bar{b} \in \mathbb{R}^n$.

Proof. The proof of Proposition 7.6 is analogous to that of Theorem 7.5.

Proposition 7.16 *Any linear combination of solutions of the homogeneous system*

$$\bar{Y}^{(\alpha}(x) = \mathbf{A}(x)\bar{Y}(x) \tag{7.4.7}$$

is also a solution to the system (7.4.7).

Proof. The proof of Proposition 7.16 is evident.

Definition 7.5 *We shall call the fundamental system of solutions of the homogeneous sequential LFDE* (7.4.7) *in* $V \subset (a, b]$ *a family of linearly independent real vectorial functions* $\{\bar{U}_i\}_{i=1}^n$, *which are solutions to* (7.4.7) *in* V.

Definition 7.6 *We shall call the fundamental solution-matrix of the system* (7.4.7) *a square matrix* $\mathbf{X}(x)$ *of order* n, *the columns of which form a fundamental system in* (7.4.7), *and which thus satisfies the matrix equation*

$$\mathbf{X}^{(\alpha}(x) = \mathbf{A}(x)\mathbf{X}(x). \tag{7.4.8}$$

Definition 7.7 *If* $\mathbf{X}$ *is a fundamental matrix solution for the system* (7.4.7), *then we shall call the general solution to* (7.4.7) *the following relation:*

$$\bar{Y}_h(x) = \mathbf{X}(x)\bar{\mathbf{C}}, \tag{7.4.9}$$

where

$$\bar{\mathbf{C}} = \begin{pmatrix} c_1 \\ \cdots \\ \cdots \\ \cdots \\ c_n \end{pmatrix} \tag{7.4.10}$$

is a vector whose components are arbitrary constants.

Theorem 7.7 *The general solution to the non-homogeneous sequential LFDE* (7.4.1) *has the form*

$$\bar{Y}_g(x) = \mathbf{X}(x)\bar{\mathbf{C}} + \bar{Y}_p(x), \tag{7.4.11}$$

where $\mathbf{X}(x)$ *is a fundamental matrix of* (7.4.7) *and* $\bar{Y}_p(x)$ *is a particular solution to* (7.4.1).

Proof. The proof of Theorem 7.7 is evident.

Theorem 7.8 *The matrix*

$$\mathbf{X}(x) = e_\alpha^{\mathbf{A}(x-a)} \quad (\mathbf{A} \in M_{n\times n}(\mathbb{R})) \tag{7.4.12}$$

is a fundamental solution-matrix for the system

$$\bar{Y}^{(\alpha}(x) = \mathbf{A}\bar{Y}(x). \tag{7.4.13}$$

Proof. It is sufficient to recall that

$$\left(D_{a+}^{\alpha} e_{\alpha}^{\mathbf{A}(t-a)}\right)(x) = \mathbf{A} e_{\alpha}^{\mathbf{A}(x-a)} \tag{7.4.14}$$

and to note that

$$\lim_{x \to a+} \left[(x-a)^{n-n\alpha} |e_{\alpha}^{\mathbf{A}(x-a)}|\right] \neq 0. \tag{7.4.15}$$

Therefore, the general solution to (7.4.7) is given by

$$\bar{Y}_h(x) = e_{\alpha}^{\mathbf{A}(x-a)} \bar{\mathbf{C}} \quad (\bar{\mathbf{C}} \in \mathbb{R}^n). \tag{7.4.16}$$

Corollary 7.5 *The following initial-value problem*

$$\bar{Y}^{(\alpha}(x) = \mathbf{A}(x)\bar{Y}(x) \tag{7.4.17}$$

$$\bar{Y}(x_0) = \bar{0} \; (x_0 > a) \tag{7.4.18}$$

has only the trivial solution $\bar{Y}(x) = \bar{0}$.

Proof. Corollary 7.5 follow from Theorem 7.5.

Theorem 7.9 *The following initial-value problem*

$$\bar{Y}^{(\alpha}(x) = \mathbf{A}\bar{Y}(x) \tag{7.4.19}$$

$$\bar{Y}(x_0) = \bar{Y}_0 \; (x_0 > a), \tag{7.4.20}$$

where $\mathbf{A} \in M_n(\mathbb{R})$ *and* $D^{\alpha} = D_{a+}^{\alpha}$, *has a unique continuous global solution* $\bar{Y}(x)$ *on* $(a, \infty) \subset \mathbb{R}$, *which is a-singular of order* α, *that is,*

$$\lim_{x \to a+} \left[(x-a)^{1-\alpha} \bar{Y}(x)\right] < \infty. \tag{7.4.21}$$

This solution is given by

$$\bar{Y}(x) = e_{\alpha}^{\mathbf{A}(x-a)} \left(e^{\mathbf{A}(x_0-a)}\right)^{-1} \bar{Y}_0. \tag{7.4.22}$$

Proof. Theorem 7.9 is proved by an application of Theorems 7.5 and 7.7.

Theorem 7.10 *The following initial-value problem*

$$\bar{Y}^{(\alpha}(x) = \mathbf{A}\bar{Y}(x) \tag{7.4.23}$$

$$\lim_{x \to a+} \left[(x-a)^{1-\alpha} \bar{Y}(x)\right] = \bar{Y}_0 \; (\bar{Y}_0 \in \mathbb{R}^n), \tag{7.4.24}$$

where $\mathbf{A} \in M_n(\mathbb{R})$ *and* $D^{\alpha} = D_{a+}^{\alpha}$, *has its unique continuous global solution defined in* $(a, \infty) \subset \mathbb{R}$, *given by*

$$\bar{Y}(x) = e_{\alpha}^{\mathbf{A}(x-a)} \bar{Y}_0. \tag{7.4.25}$$

Proof. Theorem 7.10 is deduced by an application of Theorems 7.6 and 7.7.

Example 7.4 Consider the following equation:

$$\left(\mathcal{D}_{0+}^{2\alpha}y\right)(x) + y(x) = 0. \tag{7.4.26}$$

Making the substitutions

$$\begin{aligned} y(x) &= y_1(x) \\ y_1^{(\alpha}(x) &= y_2(x) \end{aligned} \tag{7.4.27}$$

we reduce equation (7.4.26) to the following system of equations with constant coefficients:

$$\bar{Y}^{(\alpha}(x) = \mathbf{A}(x)\bar{Y}(x) \tag{7.4.28}$$

with

$$\mathbf{A} = \begin{pmatrix} 0 & 1 \\ -1 & 0 \end{pmatrix}; \quad \bar{Y} = \begin{pmatrix} y_1 \\ y_2 \end{pmatrix}. \tag{7.4.29}$$

According to (7.2.19), its general solution has the form

$$Y_g(x) = e_\alpha^{\mathbf{A}x}\bar{C} = \begin{pmatrix} \cos_\alpha x & \sin_\alpha x \\ -\sin_\alpha x & \cos_\alpha x \end{pmatrix}\bar{\mathbf{C}}. \tag{7.4.30}$$

Then the general solution to (7.4.26) is given by

$$y(x) = y_1(x) = c_1 \cos_\alpha x + c_2 \sin_\alpha x. \tag{7.4.31}$$

7.4.2 General Solution for the Case of Constant Coefficients. Fractional Green Vectorial Function

We analyze the general solution to the following non-homogeneous sequential LFDE

$$\left(D^\alpha \bar{Y}\right)(x) = \mathbf{A}\bar{Y} + \bar{\mathbf{B}}(x). \tag{7.4.32}$$

with constant matrix $\mathbf{A}(x)$: $\mathbf{A} \in M_{n\times n}(\mathbb{R})$:

Theorem 7.11 *The system*

$$\left(D_{a+}^\alpha \bar{Y}\right)(x) = \mathbf{A}\bar{Y}(x) + \bar{\mathbf{B}}(x), \tag{7.4.33}$$

where $\mathbf{A} \in M_{n\times n}(\mathbb{R}), \bar{\mathbf{B}} \in \bar{C}_{1-\alpha}((a,b])$ *, has the general solution given by*

$$\bar{Y}(x) = e_\alpha^{\mathbf{A}(x-a)}\bar{C} + \int_a^x e_\alpha^{\mathbf{A}(x-\xi)}\bar{\mathbf{B}}(\xi)d\xi, \tag{7.4.34}$$

where

$$G_\alpha(x-\xi) = e_\alpha^{\mathbf{A}(x-\xi)} \tag{7.4.35}$$

is the Green function associated with (7.4.33).

In addition, if $\bar{\mathbf{B}}(x)$ *is continuous in* $[a,b]$, *, then the particular solution*

$$\bar{Y}_p(x) = \left(G_\alpha *^a \bar{\mathbf{B}}\right)(x) = \int_a^x G_\alpha(x-\xi)\bar{\mathbf{B}}(\xi)d\xi \tag{7.4.36}$$

satisfies the relation $\bar{Y}_p(a+) = 0$, *while* $\left(I_{a+}^{1-\alpha}\bar{Y}_p\right)(a+) = 0$ *if* $\bar{\mathbf{B}}(x) \in \bar{C}_{1-\alpha}([a,b])$.

Proof. First we show that $\bar{Y}_p(x)$ is a particular solution to (7.4.33). For that we use the following operational method:

$$\left(D_{a+}^{\alpha}[1 - \mathbf{A}I_{a+}^{\alpha}]\bar{Y}\right)(x) = \bar{\mathbf{B}}(x) \tag{7.4.37}$$

from where

$$\bar{Y}_p(x) = \left(\left\{D_{a+}^{\alpha}[1 - \mathbf{A}I_{a+}^{\alpha}]\right\}^{-1}\bar{\mathbf{B}}\right)(x) \tag{7.4.38}$$

$$= \left([1 - \mathbf{A}I_{a+}^{\alpha}]^{-1}\left(I_{a+}^{\alpha}\bar{\mathbf{B}}\right)(t) + K[1 - \mathbf{A}I_{a+}^{\alpha}]^{-1}\{(t-a)^{\alpha-1}\}\right)(x). \tag{7.4.39}$$

For the case $K = 0$, it follows that $\bar{Y}_p(x)$ is equal to

$$\sum_{i=0}^{\infty}\left(I_{a+}^{(i+1)\alpha}\mathbf{A}^i\bar{\mathbf{B}}\right)(x) = \sum_{i=0}^{\infty}\frac{1}{\Gamma[(i+1)\alpha]}\int_a^x (x-\xi)^{(i+1)\alpha-1}\mathbf{A}^i\bar{\mathbf{B}}(\xi)d\xi \tag{7.4.40}$$

$$= \int_a^x \sum_{i=0}^{\infty}\mathbf{A}^i(x-\xi)^{(i+1)\alpha-1}\frac{1}{\Gamma[(i+1)\alpha]}\bar{\mathbf{B}}(\xi)d\xi = \int_a^x e_{\alpha}^{\mathbf{A}(x-\xi)}\bar{\mathbf{B}}(\xi)d\xi \tag{7.4.41}$$

$$= \left(G_{\alpha} *^{a} \bar{\mathbf{B}}\right)(x). \tag{7.4.42}$$

The above arguments are formal. Now we show that, in fact, $\bar{Y}_p(x)$ is a particular solution to (7.4.33). For this we will use Lemma 2.10, and the relations

$$\left(D_{a+}^{\alpha}e_{\alpha}^{\mathbf{A}(t-a)}\right)(x) = \mathbf{A}e_{\alpha}^{\mathbf{A}(x-a)}, \tag{7.4.43}$$

and

$$\lim_{x\to a+}\left(I_{a+}^{1-\alpha}e_{\alpha}^{\mathbf{A}(t-a)}\right)(x) = \mathbf{E}. \tag{7.4.44}$$

Then we have

$$\left(D_{a+}^{\alpha}\bar{Y}_p\right)(x) = \left(D_{a+}^{\alpha}\int_a^t e_{\alpha}^{\mathbf{A}(t-\xi)}\bar{\mathbf{B}}(\xi)d\xi\right)(x) \tag{7.4.45}$$

$$= \int_a^x \left(D_{a+}^{\alpha}e_{\alpha}^{\mathbf{A}(x-a)}\right)(\xi)\bar{\mathbf{B}}(x+a-\xi)d\xi + \bar{\mathbf{B}}(x)\lim_{x\to a+}\left(I_{a+}^{1-\alpha}e_{\alpha}^{\mathbf{A}(t-a)}\right)(x) \tag{7.4.46}$$

$$= \mathbf{A}\int_a^x e_{\alpha}^{\mathbf{A}(\xi-a)}\bar{\mathbf{B}}(x+a-\xi)d\xi + \bar{\mathbf{B}}(x) \tag{7.4.47}$$

$$= \mathbf{A}\int_a^x e_{\alpha}^{\mathbf{A}(x-\xi)}\bar{\mathbf{B}}(\xi)d\xi + \bar{\mathbf{B}}(x). \tag{7.4.48}$$

This concludes the proof of Theorem 7.11.

Example 7.5 The general solution to the following equation

$$\bar{Y}^{(\alpha}(x) = \begin{pmatrix} 0 & 1 \\ -1 & 0 \end{pmatrix}\bar{Y}(x) + \begin{pmatrix} t \\ 0 \end{pmatrix}. \tag{7.4.49}$$

is given by

$$\bar{Y}_g(x) = \bar{Y}_h(x) + \bar{Y}_p(x), \tag{7.4.50}$$

where

$$\bar{Y}_h(x) = e_\alpha^{\mathbf{A}(x)}\bar{C} = \begin{pmatrix} \cos_\alpha x & \sin_\alpha x \\ -\sin_\alpha x & \cos_\alpha x \end{pmatrix} \bar{C} \tag{7.4.51}$$

and

$$\bar{Y}_p(x) = \int_a^x e_\alpha^{\mathbf{A}(x-\xi)} \begin{pmatrix} \xi \\ 0 \end{pmatrix} d\xi = \int_a^x \begin{pmatrix} \xi\cos_\alpha(x-\xi) \\ -\xi\sin_\alpha(x-\xi) \end{pmatrix} d\xi. \tag{7.4.52}$$

For the system of equations with constant coefficients involving the Caputo derivatives, the following statement holds.

Theorem 7.12 *The initial-value problem*

$$\left({}^C D_{a+}^\alpha \bar{Y}\right)(x) = \mathbf{A}\bar{Y}(x), \tag{7.4.53}$$

$$\bar{Y}(a) = \bar{b} \quad (\bar{b} \in \mathbb{R}^n), \tag{7.4.54}$$

where $\mathbf{A} \in M_{n\times n}(\mathbb{R})$*, has a unique continuous global solution* $\bar{Y}(x)$ *in* $[a,\infty) \subset \mathbb{R}$ *given by*

$$\bar{Y}(x) = \int_a^x e_\alpha^{\mathbf{A}(x-\xi)}\mathbf{A}\bar{b}d\xi + \bar{b}. \tag{7.4.55}$$

Proof. Since $\left({}^C D_{a+}^\alpha \bar{Y}\right)(x) = \left(D_{a+}^\alpha[\bar{Y}(t) - \bar{Y}(a)]\right)(x)$, the system (7.4.53)-(7.4.54) is equivalent to

$$\left(D_{a+}^\alpha \bar{Z}\right)(x) = \mathbf{A}\bar{Z}(x) + \mathbf{A}\bar{b}, \tag{7.4.56}$$

$$\bar{Z}(a) = \bar{0}, \tag{7.4.57}$$

where $\bar{Z}(x) = \bar{Y}(x) - \bar{b}$.

Theorem 7.13 *The differential equation*

$$\left({}^C D_{a+}^\alpha \bar{Y}\right)(x) = \mathbf{A}\bar{Y}(x) + \bar{\mathbf{B}}(x), \tag{7.4.58}$$

where $\mathbf{A} \in M_n(\mathbb{R})$ *and* $\bar{\mathbf{B}} \in \bar{C}_{1-\alpha}([a,b])$*, has its general solution given by*

$$\bar{Y}(x) = \int_a^x e_\alpha^{\mathbf{A}(x-\xi)}[\bar{\mathbf{B}}(\xi) + \mathbf{A}\bar{Y}(a)]d\xi + \bar{Y}(a) \tag{7.4.59}$$

Proof. In order to establish Theorem 7.13 it is sufficient to use the definition of a Caputo derivative (1.4.1), the inequality $\bar{y}(a) < \infty$, and Theorem 7.7.

Theorem 7.14 *The following initial-value problem*

$$\left({}^C D_{a+}^\alpha \bar{Y}\right)(x) = \mathbf{A}\bar{Y}(x) + \bar{\mathbf{B}}(x), \tag{7.4.60}$$

$$\bar{Y}(a) = \bar{b} \ (\bar{b} \in \mathbb{R}^n), \tag{7.4.61}$$

where $\mathbf{A} \in M_n(\mathbb{R})$ *and* $\bar{\mathbf{B}} \in \bar{C}_{1-\alpha}([a,b])$*, has its unique solution given by*

$$\bar{Y}(x) = \int_a^x e_\alpha^{\mathbf{A}(x-\xi)}[\bar{\mathbf{B}}(\xi) + \mathbf{A}\bar{b}]d\xi + \bar{b}. \tag{7.4.62}$$

Proof. Theorem 7.14 is an immediate consequence of Theorem 7.13.

7.5 Solution of Fractional Differential Equations with Variable Coefficients. Generalized Method of Frobenius

In this section we will consider solutions around a point $x_0 \in [a,b]$ for differential equations of the form

$$[\mathbf{L_{n\alpha}}(y)](x) = g(x), \tag{7.5.1}$$

where

$$[\mathbf{L_{n\alpha}}(y)](x) = y^{(n\alpha}(x) + \sum_{k=0}^{n-1} a_k(x) y^{(k\alpha}(x), \tag{7.5.2}$$

where $\alpha \in (0,1]$; $n \in \mathbb{N}$, real functions $g(x)$ and $a_k(x)$ $(k = 0, 1, \cdots, n-1)$ are defined on the interval $[a,b]$, and $y^{(k\alpha}(x)$ represent sequential fractional derivatives of order $k\alpha$ of the function $y(x)$ (see (7.1.3)).

We note that the series method, based on the expansion of the unknown solution $y(x)$ in the fractional power series to obtain solutions of fractional differential equations, was first suggested by Al-Bassam [17]. Using that idea, he derived the formal solution to the Cauchy type problem (4.1.34) with $a = 0$. Al-Bassam in [18], [19] [see also [21], [22] and [23]] considered more general LFDE of the form (7.5.16) and (7.5.34) with series coefficients $p(x)$ and $q(x)$ given by (7.5.17) and (7.5.36), respectively. Such a method for obtaining solutions for certain particular cases of the equation (7.5.16) was also used formally by Oldham and Spanier ([643], p. 103) and by Hadid and Grzaslewicz [322]. In this regard see also Miller and Ross ([602], Chapter VI, Section 3), Podlubny ([682], Example 3.2 and Sections 6.2.1-6.2.3.) and Kilbas and Trujillo ([408], Section 4 and [377]).

On the other hand, we must indicate that, so far we know, the study of solutions around an environment of a singular point of linear fractional differential equations has been no development till now, see [709]. Here, we point out that the generalization of the power fractional series and the Frobenius methods could be very interesting as a natural tool to apply the separation method to obtain the solution to many fractional models connected with Complex Systems.

7.5.1 Introduction

In this section we analyze solutions of the homogeneous equation $[\mathbf{L_{n\alpha}}(y)](x) = 0$, which are expressed as a series of powers of x^α. For this we introduce the concept of α-analyticity which generalizes the concept of an analytic function in the ordinary case; [in this regard see Bonilla et al.[94]].

Definition 7.8 *Let $\alpha \in (0,1]$, $f(x)$ be a real function defined on the interval $[a,b]$, and $x_0 \in [a,b]$. Then $f(x)$ is said to be α-analytic at x_0 if there exists an interval $N(x_0)$ such that, for all $x \in N(x_0)$, $f(x)$ can be expressed as a series of natural powers of $(x-x_0)^\alpha$. That is, $f(x)$ can be expressed as* $\sum_{n=0}^{\infty} c_n (x-x_o)^{n\alpha}$ $(c_n \in R)$, *this series being absolutely convergent for* $|x - x_0| < \rho$ $(\rho > 0)$. *The radius of convergence of the series is* ρ.

Example 7.6 a) The Mittag-Leffler function (1.8.1)

$$E_\alpha((x-a)^\alpha) = \sum_{i=0}^{\infty} \frac{(x-a)^{i\alpha}}{\Gamma(i\alpha+1)} \tag{7.5.3}$$

is an α-analytic function around the point $x_0 = a$ with radius of convergence $\rho = \infty$.

It follows from Example 4.9, that $y(x) = CE_\alpha(x^\alpha)$ is the general solution to the following FDE of order $\alpha \in (0,1)$:

$$\left({}^C D_{a+}^\alpha y\right)(x) = y(x) \quad (x > a;\ a \in R) \tag{7.5.4}$$

b) The function

$$z(x) = x^{1-\alpha} e_\alpha^x = \sum_{i=0}^{\infty} \frac{x^\alpha}{\Gamma(\alpha x + \alpha)} \tag{7.5.5}$$

is α-analytic around the point $x_0 = 0$ with radius of convergence $\rho = \infty$ (see (1.10.12)).

By Corollary 4.6, $y(x) = C\ e_\alpha^x$ is the general solution to the following FDE:

$$\left(D_{0+}^\alpha y\right)(x) = y(x); \quad (x > 0;\ 0 < \alpha < 1). \tag{7.5.6}$$

Definition 7.9 *A point* $x_0 \in [a,b]$ *is said to be an* α*-ordinary point of the equation* $[\mathbf{L}_{n\alpha}(y)](x) = 0$, *if the functions* $a_k(x)$ $(k = 0, 1, \cdots, n-1)$ *are* α*-analytic in* x_0. *A point* $x_0 \in [a,b]$ *which is not* α*-ordinary will be called* α*-singular.*

By analogy with the ordinary case, we classify the α-singular points as regular α-singular and essential α-singular as follows:

Definition 7.10 *Let* $x_0 \in [a,b]$ *be an* α*-singular point of the equation* $[\mathbf{L}_{n\alpha}(y)](x) = 0$. *Then* x_0 *is said to be a regular* α*-singular point of this equation if the functions* $(x - x_0)^{(n-k)\alpha} a_k(\alpha)$ *are* α*-analytic in* x_0 $(k = 0, 1, \cdots, n-1)$. *Otherwise,* x_0 *is said to be an essential* α*-singular point.*

If x_0 is a regular α-singular point, then the equation $\mathbf{L}_{n\alpha}(y) = 0$ can be expressed as follows:

$$(x - x_0)^{n\alpha} y^{(n\alpha}(x) + \sum_{k=0}^{n-1} (x - x_0)^{k\alpha} b_k(x) y^{(k\alpha}(x) = 0, \tag{7.5.7}$$

where the functions $b_k(x)$ $(k = 0, 1, \cdots, n-1)$ are α-analytic around x_0.

Example 7.7 a) Any point $x = x_0 > 0$ is an ordinary point for the following equations:

$$y^{(\alpha}(x) - x^{\alpha} y(x) = 0 \tag{7.5.8}$$

$$y^{(2\alpha}(x) - x^{\alpha} y(x) = 0 \tag{7.5.9}$$

$$x^{2\alpha}\, y^{(2\alpha}(x) - x^{\alpha} y^{(\alpha}(x) + x^{2\alpha}\, y(x) = 0. \tag{7.5.10}$$

b) Any point $x = x_0 > 1$ is an ordinary point for the following equations:

$$(x-1)^{\alpha}\, y^{(\alpha}(x) - y(x) = 0 \tag{7.5.11}$$

$$(x-1)^{2\alpha}\, y^{(2\alpha}(x) + (x-1)^{\alpha}\, y^{(\alpha}(x) + (x-1)^{2\alpha}\, y(x) = 0 \tag{7.5.12}$$

c) The point $x = 0$ is a regular α-singular point for the equation (7.5.10), while the point $x = 1$ is a regular α-singular point for the equations (7.5.11) and (7.5.12).

In the previous examples, $y^{(\alpha}$ represents the R-L fractional derivative D^{α}_{a+} as well as the Caputo fractional derivative ${}^{C}D^{\alpha}_{a+}$.

Next for $n = 1, 2$, we develop a generalization of the usual method used for obtaining the power series solutions for a linear equation with variable coefficients around ordinary points, as in the equation (7.5.6), when the fractional derivatives are of the Riemann-Liouville or Caputo type. We also extend the well-known Frobenius method so as to obtain solutions to the equation (7.5.7) around regular α-singular points.

Remark 7.1 We should point out that we consider the case $n = 1$, since for linear differential equations with variable coefficients of order α, $\alpha \in (0, 1)$, there is no direct method for obtaining a general solution as there is for the case when $\alpha = 1$.

We shall make use of the following property, whose general proof is analogous to that in the case $\alpha = 1$.

Property 7.1 *Let $\alpha \in (0, 1]$. If $f(x)$ is α-analytic at x_0, with convergence radius ρ $(\rho > 0)$, then*

$$\left(D^{\alpha}_{a+} f\right)(x) = \left(D^{\alpha}_{a+}\left(\sum_{n=0}^{\infty} c_n (t - x_0)^{n\alpha}\right)\right)(x) = \sum_{n=0}^{\infty} c_n \ \left(D^{\alpha}_{a+} (t - x_0)^{n\alpha}\right)(x) \tag{7.5.13}$$

$$(x \in [x_0, x_0 + \rho)).$$

Next, on the basis of Property 2.1, we give the following definition:

Definition 7.11 *Let $\alpha \in (0,1]$, $a \in \mathbb{R}$ and $\Re(\beta) \in \mathbb{R}\backslash\mathbb{Z}^-$. Then the derivative of order α of $(x-a)^\beta$ is defined as follows:*

$$\left(D_{a+}^\alpha (t-a)^\beta\right)(x) := \frac{\Gamma(\beta+1)}{\Gamma(\beta-\alpha+1)}(x-a)^{\beta-\alpha} \ (x > a). \tag{7.5.14}$$

In accordance with this definition, we have the following property:

Property 7.2 *Let $\alpha \in (0,1]$, $a \in \mathbb{R}$ and $x_0 \geqq a$. If $\Re(\beta) \notin \mathbb{Z}$, then*

$$\left(D_{a+}^\alpha (t-x_0)^\beta\right)(x) = \frac{\Gamma(\beta+1)}{\Gamma(\beta-\alpha+1)}(x-x_0)^{\beta-\alpha} \ (x > x_0). \tag{7.5.15}$$

Proof. For $x_0 = a$, (7.5.14) holds. If $x_0 > a$, notice that

$$(1-x)^\beta = \sum_{n=0}^{\infty} \binom{\beta}{n} (-x)^n,$$

this series being uniformly convergent for $|x| < 1$. Therefore,

$$\begin{aligned}
\left(D_{a+}^\alpha (t-x_0)^\beta\right)(x) &= \left(D_{a+}^\alpha \left[(t-a)^\beta \left(1-\frac{x_0-a}{t-a}\right)^\beta\right]\right)(x) \\
&= \left(D_{a+}^\alpha \left[(t-a)^\beta \sum_{n=0}^{\infty} \binom{\beta}{n} (-1)^n \left(\frac{x_0-a}{t-a}\right)^n\right]\right)(x) \\
&= \left(D_{a+}^\alpha \left[\sum_{n=0}^{\infty} \binom{\beta}{n} (-1)^n (x_0-a)^n (t-a)^{\beta-n}\right]\right)(x) \\
&= \sum_{n=0}^{\infty} \binom{\beta}{n} (-1)^n (x_0-a)^n \left(D_{a+}^\alpha (t-a)^{\beta-n}\right)(x).
\end{aligned}$$

Since $\Re(\beta-n) \in \mathbb{R}\backslash\mathbb{Z}^-$ ($n \in \mathbb{N}_0$), applying (7.5.14) and taking the same arguments as above we obtain (7.5.15).

Remark 7.2 Properties 7.1 and 7.2 stay valid if the derivative D_{a+}^α is replaced by the Caputo derivative ${}^C D_{a+}^\alpha$.

7.5.2 Solutions Around an Ordinary Point of a Fractional Differential Equation of Order α

In this section we analyze the existence of solutions for the equation

$$[\mathbf{L}_\alpha(y)](x) := y^{(\alpha}(x) + p(x)y(x) = 0 \tag{7.5.16}$$

around an α-ordinary point $x_0 \in [a,b]$, with $p(x)$ defined in the interval $[a,b]$ and $\alpha \in (0,1)$. We consider separately the cases where $y^{(\alpha}$ represents the Riemann-Liouville and Caputo fractional derivatives of order α of the function $y(x)$. Since x_0 is an α-ordinary point, $p(x)$ can be expressed as follows:

$$p(x) = \sum_{n=0}^{\infty} p_n (x - x_0)^{n\alpha}, \tag{7.5.17}$$

this series being convergent for $x \in [x_0, x_0 + \rho]$, with $\rho > 0$.

Theorem 7.15 *Let $\alpha \in (0,1]$ and $a_0 \in \mathbb{R}$, and let $x_0 \in [a,b]$ be an α-ordinary point for the equation*

$$[\mathbf{L}_\alpha(y)](x) := \left(D_{a+}^{\alpha} y\right)(x) + p(x)y(x) = 0. \tag{7.5.18}$$

Then there exists a unique function

$$y(x) = (x - x_0)^{\alpha-1} \sum_{n=0}^{\infty} a_n (x - x_0)^{n\alpha},$$

which is the solution to the equation (7.5.18) for $x \in (x_0, x_0+\rho)$ and which satisfies the initial condition $a_0 = \lim_{x \to x_0} (x - x_0)^{1-\alpha} y(x)$.

Proof. We shall seek a solutions of (7.5.18) as follows:

$$y(x) = (x - x_0)^{\alpha-1} \sum_{n=0}^{\infty} a_n (x - x_0)^{n\alpha} = \sum_{n=0}^{\infty} a_n (x - x_0)^{(n+1)\alpha-1}. \tag{7.5.19}$$

Substituting (7.5.19) in (7.5.18) and using (7.5.15), we obtain the following recurrence formula which allows us to express a_n $(n > 0)$ in terms of a_0:

$$\frac{\Gamma[(n+2)\alpha]}{\Gamma[(n+1)\alpha]} a_{n+1} = -\sum_{k=0}^{n} p_{n-k} a_k \quad (k = 0, 1, 2, \cdots) \tag{7.5.20}$$

We show that, for $x \in (x_0, x_0 + \rho)$, the series in (7.5.19) converges.

Let $r < \rho$. Since (7.5.17) converges, there exists a constant $M > 0$ such that

$$|p_{n-k}| \leqq \frac{M r^{k\alpha}}{r^{n\alpha}}$$

and, therefore,

$$\frac{\Gamma[(n+2)\alpha]}{\Gamma[(n+1)\alpha]} |a_{n+1}| \leqq \frac{M}{r^{n\alpha}} \sum_{k=0}^{n} a_k r^{n\alpha}.$$

Denoting $b_0 = |a_0|$, by recurrence we define b_n $(n \in \mathbb{N})$ as follows:

$$\frac{\Gamma[(n+2)\alpha]}{\Gamma[(n+1)\alpha]} b_{n+1} = \frac{M}{r^{n\alpha}} \sum_{k=0}^{n} b_k r^{n\alpha} \ (n \in \mathbb{N}_0).$$

It is clear that $0 \leqq |a_n| \leqq b_n$ for $n \in \mathbb{N}_0$. The series

$$g_r(x) \sum_{n=0}^{\infty} b_n (x - x_0)^{n\alpha+\alpha-1} = (x - x_0) \sum_{n=0}^{\infty} b_n (x - x_0)^{n\alpha}$$

is convergent for $|x - x_0| < r$, because in accordance with the asymptotic representation (1.5.15), there holds the following estimate:

$$\lim_{n\to\infty}\left|\frac{b_{n+1}(x-x_0)^{(n+1)\alpha}}{b_n(x-x_0)^{n\alpha}}\right| = \left(\frac{|x-x_0|}{r}\right)^{\alpha} < 1.$$

This proves the convergence of (7.5.19) for $x \in (x_0, x_0+\rho)$. The uniqueness of the above solution follows from the corresponding existence and uniqueness theorem in Chapter 3.

Theorem 7.16 *Let $\alpha \in (0,1]$ and $a_0 \in \mathbb{R}$. Let $x_0 \in [a,b]$ be an α-ordinary point for the equation*

$$[\mathbf{L}_\alpha(y)](x) := \left({}^C D^{\alpha}_{a+}y\right)(x) + p(x)y(x) = 0 \tag{7.5.21}$$

with $p(x)$ defined in $[a,b]$. Then there exists a unique α-analytic function $y(x)$ in x_0 as a solution to (7.5.21) for $x \in [x_0, x_0+\rho]$, such that the initial condition $y(x_0) = a_0$ is satisfied.

Proof. The proof is analogous to that of Theorem 7.15, by seeking the series representation for the solution in the form $y(x) = \sum_{n=0}^{\infty} a_n(x-x_0)^{n\alpha}$ and using the following recurrence formula for coefficients a_n:

$$a_{n+1}\frac{\Gamma[(n+1)\alpha+1]}{\Gamma(n\alpha+1)} = -\sum_{k=0}^{n} p_{n-k}a_k. \tag{7.5.22}$$

Example 7.8 Consider the following FDE of order α $(0 < \alpha < 1)$:

$$\left(D^{\alpha}_{a+}y\right)(x) + (x+1)^{\alpha}y(x) = 0 \quad (a \leqq -1). \tag{7.5.23}$$

The point $x_0 = -1$ is an α-ordinary point for this equation. Let us find the general solution to (7.5.23) around the point $x_0 = -1$. We seek solutions of the form

$$y(x) = (x+1)^{\alpha-1}\sum_{n=0}^{\infty} a_n(x+1)^{n\alpha}. \tag{7.5.24}$$

Substituting (7.5.24) in (7.5.23), we have

$$a_{2k+1} = 0; \ (k \in \mathbb{N}), \tag{7.5.25}$$

$$a_{2k} = (-1)^k\, Q(k)\cdot a_0, \tag{7.5.26}$$

with

$$Q(k) = \prod_{j=1}^{k}\frac{\Gamma(2j\alpha)}{\Gamma((2j+1)\alpha)}. \tag{7.5.27}$$

Therefore, the general solution to (7.5.23) has the form

$$y(x) = a_0(x+1)^{\alpha-1}\left[1+\sum_{k=1}^{\infty}(-1)^k Q(k)(x+1)^{2k\alpha}\right]. \tag{7.5.28}$$

The particular solution to (7.5.23) which satisfies $y(0) = 1$ is given by (7.5.28), with a_0 given by:

$$a_0 = \frac{1}{1+\sum_{k=1}^{\infty}(-1)^k Q(k)}. \tag{7.5.29}$$

Example 7.9 The general solution to the following FDE of order α $(0 < \alpha < 1)$

$$\left({}^C D_a^\alpha y\right)(x) + (x+1)^\alpha y(x) = 0 \quad (a \leq -1) \tag{7.5.30}$$

is given by

$$y(x) = a_0\left[1+\sum_{k=1}^{\infty}(-1)^k Q^*(k)(x+1)^{2k\alpha}\right], \tag{7.5.31}$$

with

$$Q^*(k) = \prod_{j=1}^{k}\frac{\Gamma((2j-1)\alpha+1)}{\Gamma(2j\alpha+1)}. \tag{7.5.32}$$

The particular solution to (7.5.30), which satisfies $y(0) = 1$, is given by (7.5.31) with

$$a_0 = \frac{1}{1+\sum_{k=1}^{\infty}(-1)^k Q^*(k)}. \tag{7.5.33}$$

7.5.3 Solutions Around an Ordinary Point of a Fractional Differential Equation of Order 2α

In this section we shall consider the solutions around an α-ordinary point $x_0 \in [a, b]$ to the equation

$$[\mathbf{L}_{2\alpha}(y)](x) := y^{(2\alpha}(x) + p(x)y^{(\alpha}(x) + q(x)y(x) = 0 \tag{7.5.34}$$

where $\alpha \in (0, 1]$, $p(x)$ and $q(x)$ are defined on an interval $[a, b]$, and $y^{2\alpha}$ and y^α represent the Riemann-Liouville or Caputo sequential derivatives of order 2α and α, respectively, of the function $y(x)$.

Since x_0 is an α-ordinary point of (7.5.34), then by definitions 7.8 and 7.9,

$$p(x) = \sum_{n=0}^{\infty} p_n(x-x_0)^{n\alpha} \quad (x \in [x_0, x_0+\rho_1]; \quad \rho_1 > 0) \tag{7.5.35}$$

and

$$q(x) = \sum_{n=0}^{\infty} q_n(x - x_0)^{n\alpha} \quad (x \in [x_0, x_0 + \rho_2];\ \rho_2 > 0). \tag{7.5.36}$$

Theorem 7.17 *Let $\alpha \in (0,1]$, let $a_0, a_1 \in \mathbb{R}$, and let $x_0 \in [a,b]$ be an α-ordinary point of the equation*

$$[\mathbf{L}_{2\alpha}(y)](x) := \left(\mathcal{D}_{a+}^{2\alpha} y\right)(x) + p(x)\left(D_{a+}^{\alpha} y\right)(x) + q(x)y(x) = 0. \tag{7.5.37}$$

Then there exists a unique solution to the equation (7.5.37) given by

$$y(x) = (x - x_0)^{\alpha-1} \sum_{n=0}^{\infty} a_n(x - x_0)^{n\alpha}$$

for $x \in (x_0, x_0 + \rho)$ with $\rho = \min\{\varrho_1, \varrho_2\}$, which satisfies the following initial conditions:

$$\lim_{x \to x_0} \left[(x - x_0)^{1-\alpha} y(x)\right] = a_0$$

and

$$\frac{\Gamma(\alpha)}{\Gamma(2\alpha)} \lim_{x \to x_0} \left[(x - x_0)^{1-\alpha} (D_{a+}^{\alpha} y)(x)\right] = a_1$$

Proof. We seek the solution to the equation (7.5.37) in the form

$$y(x) = (x - x_0)^{\alpha-1} \sum_{n=0}^{\infty} a_n(x - x_0)^{n\alpha} = \sum_{n=0}^{\infty} a_n(x - x_0)^{(n+1)\alpha-1}. \tag{7.5.38}$$

Calculating $\left(D_{a+}^{\alpha} y\right)(x)$ and $\left(D_{a+}^{2\alpha} y\right)(x)$, taking (7.5.15) into account and substituting these values in equation (7.5.37), we arrive at the following recurrence formula for the coefficients a_n:

$$\frac{\Gamma[(n+3)\alpha]}{\Gamma[(n+1)\alpha]} a_{n+2} = -\sum_{k=0}^{n} \frac{\Gamma[(k+2)\alpha]}{\Gamma[(k+1)\alpha]} p_{n-k} a_{k+1} + q_{n-k} a_k. \tag{7.5.39}$$

Therefore the coefficients a_n $(n \geqq 2)$ are expressed in terms of a_0 and a_1.

We show that the series in (7.5.38) converges for $x \in (x_0, x_0 + \rho)$ with $\rho = \min\{\rho_1, \rho_2\}$. Let us fix r $(0 < r < \rho)$. Since the series (7.5.35) and (7.5.36) are convergent for $x \in [x_0, x_0 + r]$, there exists $M > 0$ such that

$$|p_{n-k}| \leqq \frac{M r^{k\alpha}}{r^{n\alpha}} \quad (n \in \mathbb{N}_0;\ 0 \leqq k \leqq n) \tag{7.5.40}$$

and

$$|q_{n-k}| \leqq \frac{M r^{k\alpha}}{r^{n\alpha}} \quad (n \in \mathbb{N}_0;\ 0 \leqq k \leqq n) \tag{7.5.41}$$

According to (7.5.39), (7.5.40) and (7.5.41), there holds the following estimate:

$$\frac{\Gamma[(n+3)\alpha]}{\Gamma[(n+1)\alpha]}|a_{n+2}| \leqq \frac{M}{r^{n\alpha}} \sum \left[\frac{\Gamma[(n+2)\alpha]}{\Gamma[(n+1)\alpha]}|a_{k+1}| + |a_k|\right] r^{k\alpha} + M|a_{n+1}|r^{\alpha}.$$

Let $b_0 = |a_0|$, $b_1 = |a_1|$, and let b_k $(k > 1)$ be defined by the following recurrence formula:

$$\frac{\Gamma[(n+3)\alpha]}{\Gamma[(n+1)\alpha]} b_{n+2} = \frac{M}{r^{n\alpha}} \sum \left[\frac{\Gamma[(n+2)\alpha]}{\Gamma[(n+1)\alpha]} b_{k+1} + b_k \right] r^{k\alpha} + M b_{n+1} r^{\alpha}.$$

Since $0 \leqq |a_n| \leqq b_n$ $(n \in \mathbb{N}_0)$, the series (7.5.38) converges for $x \in (x_0, x_0 + \rho)$ if the series

$$g_r(x) = \sum_{n=0}^{\infty} b_n (x - x_0)^{(n+1)\alpha - 1} = (x - x_0)^{\alpha - 1} \sum_{n=0}^{\infty} b_n (x - x_0)^{n\alpha}$$

converges. This series is convergent because the estimate

$$\lim_{n \to \infty} \left| \frac{b_{n+1}(x - x_0)^{(n+1)\alpha}}{b_n (x - x_0)^{n\alpha}} \right| = \left(\frac{|x - x_0|}{r} \right)^{\alpha} < 1,$$

follows from the recurrence relation for b_n, using the asymptotic relation (1.5.15).

The uniqueness of the obtained solution follows from the corresponding existence and uniqueness theorems in Chapter 3.

Theorem 7.18 *Let* $\alpha \in (0, 1]$, *and* $a_0, a_1 \in \mathbb{R}$, *and let* $x_0 \in [a, b]$ *be an* α*-ordinary point of the equation*

$$[\mathbf{L}_{2\alpha} y](x) := \left({}^{C}\!D_{a+}^{2\alpha} y\right)(x) + p(x) \left({}^{C}\!D_{a+}^{\alpha} y\right)(x) + q(x) y(x) = 0. \tag{7.5.42}$$

Then there exists a unique solution $y(x)$ *of* (7.5.42), *for* $x \in (x_0, x_0 + \rho)$ *with* $\rho = \min\{\rho_1, \rho_2\}$. *This solution is* α*-analytic function in* x_0 *and satisfies the following initial conditions:*

$$\lim_{x \to x_0} y(x) = a_0$$

and

$$lim_{x \to 0} \left({}^{C}\!D_{a+}^{\alpha} y\right)(x) = a_1.$$

Proof. Seeking a solution to equation (7.5.42) in the form

$$y(x) = \sum_{n=0}^{\infty} a_n (x - x_0)^{n\alpha},$$

we arrive at the following recurrence formula for the coefficients a_n:

$$a_{n+2} \frac{\Gamma(n+2)\alpha + 1)}{\Gamma(n\alpha + 1)} = - \sum_{k=0}^{n} \left[p_{n-k} \frac{\Gamma[(k+1)\alpha + 1]}{\Gamma(k\alpha + 1)} a_{k+1} + q_{n-k} \cdot a_k \right] \tag{7.5.43}$$

The rest of the proof is analogous to that in Theorem 7.17.

Example 7.10 Find the general solution to the following FDE of order 2α $(0 < \alpha < 1)$, which could be called the fractional Ayre equation:

$$\left(\mathcal{D}^{2\alpha}y\right)(x) - x^{\alpha}y(x) = 0 \tag{7.5.44}$$

with $\mathcal{D}^{(2\alpha}y = D^{\alpha}D^{\alpha}y$, for $D^{\alpha} := D^{\alpha}_{0+}$ (case (a)), as well as for $D^{\alpha}y := {}^{C}D^{\alpha}_{0+}y$ (case (b)).

a) As shown earlier, we seek a solution in the form (7.5.38)

$$y(x) = x^{\alpha-1}\sum_{n=0}^{\infty} a_n x^{n\alpha}. \tag{7.5.45}$$

Substituting (7.5.45) in (7.5.44), we have

$$a_{n+2} = \frac{\Gamma((n+1)\alpha)}{\Gamma((n+3)\alpha)}\, a_{n-1} \quad (n \in \mathbb{N}) \tag{7.5.46}$$

It follows from here that $a_{2k-1} = 0$ $(k \in \mathbb{N})$. Supposing that a_0 and a_1 are arbitrary constants, we obtain the following general solution to (7.5.44):

$$y(x) = a_0 x^{\alpha-1}\left[1 + \sum_{k=1}^{\infty} a_k x^{3k\alpha}\right] + a_1 x^{\alpha-1}\left[x^{\alpha} + \sum_{k=1}^{\infty} b_k x^{(3k+1)\alpha}\right] \tag{7.5.47}$$

where

$$a_k = \prod_{j=1}^{k} \frac{\Gamma(3j-1)\alpha)}{\Gamma((3j+1)\alpha)}; \qquad b_k = \prod_{j=1}^{k} \frac{\Gamma(3j\alpha)}{\Gamma((3j+2)\alpha)} \tag{7.5.48}$$

b) The general solution to (7.5.44) with $D^{\alpha}y := {}^{C}D^{\alpha}_{0+}y$ is obtained similarly to the previous case, and this solution is given by

$$y(x) = a_0\left[1 + \sum_{k=1}^{\infty} a_k x^{3k\alpha}\right] + a_1\left[x^{\alpha} + \sum_{k=1}^{\infty} b_k x^{(3k+1)\alpha}\right] \tag{7.5.49}$$

with

$$a_k = \prod_{j=1}^{k} \frac{\Gamma(3j-2)\alpha+1)}{\Gamma(3j\alpha+1)}; \qquad b_k = \prod_{j=1}^{k} \frac{\Gamma((3j-1)\alpha+1)}{\Gamma((3j+1)\alpha+1)}. \tag{7.5.50}$$

7.5.4 Solution Around an α-Singular Point of a Fractional Differential Equation of Order α

In this section we obtain the fractional power series solutions of a homogeneous LFDE of order α $(0 < \alpha \leqq 1)$, around a regular α-singular point x_0 of the equation. It will be more convenient, for our purposes, to consider the differential equation written as

$$[\mathbf{L}_{\alpha}(y)](x) := (x - x_0)^{\alpha} y^{(\alpha}(x) + q(x)y(x) = 0, \tag{7.5.51}$$

where $q(x)$, due to the regular α-singular character of x_0, is an α-analytic function around x_0.

Theorem 7.19 *Let $x_0 \geqq a$ be a regular α-singular point of the differential equation (7.5.51) of order α, and let*

$$q(x) = \sum_{n=0}^{\infty} q_n (x - x_0)^{n\alpha} \tag{7.5.52}$$

be the power series expansion of the α-analytic function $q(x)$. Then there exists the solution

$$y(x; \alpha, s_1) = (x - x_0)^{s_1} \sum_{n=0}^{\infty} a_n (x - x_0)^{n\alpha} \tag{7.5.53}$$

of equation (7.5.51) on a certain interval to the right of x_0. Here a_0 is a non-zero arbitrary constant, $s_1 > -1$ is the real solution to the equation

$$\frac{\Gamma(s+1)}{\Gamma(s-\alpha+1)} + q_0 = 0$$

and the coefficients a_n $(n \geqq 1)$ are given by the following recurrence formula:

$$a_n = -\frac{\Gamma(n\alpha + s - \alpha + 1)}{\Gamma(n\alpha + s + 1)} \sum_{l=0}^{n-1} a_l \, q_{n-l}.$$

Moreover, if the series (7.5.52) converges for all x in a semi-interval $0 < x - x_0 < R$ $(R > 0)$, then the series solution (7.5.53) of equation (7.5.51) is also convergent in the same interval.

Proof. Seeking a solution to equation (7.5.51) in the form (7.5.53), taking fractional differentiation of $y(x; \alpha, s_1)$ with using (7.5.14) and substituting the result and (7.5.52) into (7.5.51), we get

$$[\mathbf{L}_\alpha(y(x; \alpha, s))](x) := \sum_{n=1}^{\infty} \left[a_n \left[\frac{\Gamma(n\alpha + s + 1)}{\Gamma(n\alpha + s - \alpha + 1)}\right] + \sum_{l=0}^{n} a_l q_{n-l} \right] (x - x_0)^{n\alpha + s}$$

$$+ a_0 \left[\frac{\Gamma(s+1)}{\Gamma(s-\alpha+1)} + q_0 \right] (x - x_0)^s = 0. \tag{7.5.54}$$

If we now put $f_0(s) = \dfrac{\Gamma(s+1)}{\Gamma(s-\alpha+1)} + q_0$, then, from (7.5.54), we obtain

$$a_0 f_0(s) = 0 \tag{7.5.55}$$

and

$$a_n = -\frac{\sum_{l=0}^{n-1} a_l \, q_{n-l}}{f_0(n\alpha + s)}. \tag{7.5.56}$$

Suppose that $a_0 \neq 0$, then $f_0(s) = 0$. Thus, if s_1 is the only real root of the equation $f_0(s) = 0$, then the expression (7.5.56) provides, by recurrence, the coefficients a_n of (7.5.53) in terms of a_0.

Now we prove the convergence of the series. Let $0 < r < R$. Since the series (7.5.52) is convergent, there exists a constant $M > 0$, such that

$$|q_{n-l}| \leqq \frac{Mr^{l\alpha}}{r^{n\alpha}} \ (n \in \mathbb{N}) \tag{7.5.57}$$

Applying now (7.5.57) to the relation (7.5.53), we have

$$|a_n| \leqq \frac{M}{|f_0(n\alpha + s)|} \sum_{k=1}^{n} \frac{|a_{n-k}|}{r^{k\alpha}}.$$

Now we define $c_0 = |a_0|$ and $c_n = \dfrac{M}{|f_0(n\alpha + s)|} \displaystyle\sum_{k=1}^{n} \frac{|a_{n-k}|}{r^{k\alpha}}$ for $n \geqq 1$. Then

$$\left|\frac{c_{n+1}}{c_n}\right| = \left[\frac{M}{|f_0((n+1)\alpha + s)|} + \frac{f_0(n\alpha + s)}{f_0((n+1)\alpha + s)}\right] \frac{1}{r^\alpha},$$

and using the asymptotic representation (1.5.15), we get

$$\left|\frac{c_{n+1}(x - x_0)^{(n+1)\alpha}}{c_n(x - x_0)^{n\alpha}}\right| \longrightarrow \left(\frac{|x - x_0|}{r}\right)^\alpha$$

when $n \to \infty$. Therefore the series $\displaystyle\sum_{n=0}^{\infty} c_n(x - x_0)^{n\alpha}$ converges for all x such that $0 < |x - x_0| < r$.

From this we conclude that (7.5.53) converges for $0 < x - x_0 < R$.

Example 7.11 Consider the FDE

$$(x - 1)^\alpha y^{(\alpha}(x) - y(x) = 0, \tag{7.5.58}$$

where $y^{(\alpha}(x)$ represents either the Caputo or the Riemann-Liouville fractional derivative.

Since the point $x = 1$ is a regular α-singular point of (7.5.58), we shall seek a solution to this equation around the point $x = 1$ in the form

$$y(x) = (x - 1)^s \sum_{n=0}^{\infty} a_n(x - 1)^{n\alpha} \tag{7.5.59}$$

As we have seen earlier, s must be a real solution (which always exists) of the fractional index equation associated with (7.5.58); namely, of the following equation:

$$\frac{\Gamma(s+1)}{\Gamma(s - \alpha + 1)} = 1 \ \ (s > -1) \tag{7.5.60}$$

Suppose that the above-mentioned solution is $s = \beta$, $(-1 < \beta < 0)$.

It is directly verified that $a_n = 0$ $(n \in \mathbb{N})$. Then the general solution to equation (7.5.58) has the following form:

$$y(x) = a_0(x-1)^{\beta} \quad (a_0 \neq 0). \tag{7.5.61}$$

Naturally, the result would be the same if we directly seek a solution as

$$y(x) = c(x-1)^{s}. \tag{7.5.62}$$

Lastly, let us see the general solution to equation (7.5.58) for $\alpha = 0.01,\ 0.1,\ 0.3,\ 0.5,\ 0.6,\ 0.9,\ 0.95,\ 0.99$. The solution is given by

$$y(x) = C(x-1)^{\beta}, \tag{7.5.63}$$

where the β with the corresponding α are given in the following table:

α	β
0.01	0.46664
0.1	0.51201
0.3	0.61506
0.5	0.72118
0.6	0.77539
0.9	0.94267
0.95	0.97124
0.99	0.99423

(7.5.64)

7.5.5 Solution Around an α-Singular Point of a Fractional Differential Equation of Order 2α

We now focus our attention on the following homogeneous sequential linear fractional differential equation of order 2α $(0 < \alpha \leqq 1)$:

$$[\mathbf{L}_{2\alpha}(y)](x) := (x-x_0)^{2\alpha} y^{(2\alpha}(x) + (x-x_0)^{\alpha} p(x) y^{(\alpha}(x) + q(x)y(x) = 0. \tag{7.5.65}$$

If $x_0 \geqq a$ is a regular α-singular point of equation (7.5.65), then the functions $p(x)$ and $q(x)$ are α-analytic around x_0 and, therefore, they have the following series expansions:

$$p(x) = \sum_{n=0}^{\infty} p_n (x-x_0)^{n\alpha} \tag{7.5.66}$$

and

$$q(x) = \sum_{n=0}^{\infty} q_n (x-x_0)^{n\alpha}, \tag{7.5.67}$$

valid on a semi-interval $0 < x - x_0 < R$ for some $R > 0$. Our goal is to find a solution to (7.5.65) in the form

$$y(x;\alpha,s) = (x-x_0)^{s} \sum_{n=0}^{\infty} a_n (x-x_0)^{n\alpha} \tag{7.5.68}$$

with $a_0 \neq 0$, s being a number to be determined.

Differentiating (7.5.68) we get for $\Re(s) > \alpha - 1$ and $s \notin Z^-$,

$$y^{(\alpha}(x;\alpha,s) = \sum_{n=0}^{\infty} a_n \frac{\Gamma(n\alpha+s+1)}{\Gamma(n\alpha+s-\alpha+1)} (x-x_0)^{n\alpha+s-\alpha}$$

and

$$y^{(2\alpha}(x;\alpha,s) = \sum_{n=0}^{\infty} a_n \frac{\Gamma(n\alpha+s+1)}{\Gamma(n\alpha+s-2\alpha+1)} (x-x_0)^{n\alpha+s-2\alpha}.$$

Substitution of these expressions into the equation (7.5.65) yields

$$[\mathbf{L}_{2\alpha}(y(t;\alpha,s))](x) = a_0 f_0(s)(x-x_0)^s$$

$$+\sum_{n=1}^{\infty}\left[a_n f_0(n\alpha+s) + \sum_{l=0}^{n-1} a_l f_{n-l}(n\alpha+s)\right](x-x_0)^{n\alpha+s} = 0, \tag{7.5.69}$$

where

$$f_0(s) = \frac{\Gamma(s+1)}{\Gamma(s-2\alpha+1)} + \frac{\Gamma(s+1)}{\Gamma(s-\alpha+1)} p_0 + q_0 \tag{7.5.70}$$

and

$$f_k(s) = p_k \frac{\Gamma(s+1)}{\Gamma(s-\alpha+1)} + q_k. \tag{7.5.71}$$

Therefore

$$a_0 f_0(s) = 0 \tag{7.5.72}$$

and

$$a_n = -\frac{\sum_{l=0}^{n-1} a_l f_{n-l}(n\alpha+s)}{f_0(n\alpha+s)}. \tag{7.5.73}$$

Since we assumed that $a_0 \neq 0$, then from (7.5.72) we get

$$f_0(s) = \frac{\Gamma(s+1)}{\Gamma(s-2\alpha+1)} + \frac{\Gamma(s+1)}{\Gamma(s-\alpha+1)} p_0 + q_0 = 0. \tag{7.5.74}$$

Equation (7.5.74) is the so-called "fractional indicial equation" of the differential equation (7.5.65). We will assume that the equation (7.5.74), for values s with $\Re(s) > -1$, has two roots which, if complex, must be complex conjugates (because $\Gamma(\overline{z}) = \overline{\Gamma(z)}$ for all $z \in \mathbb{C}$).

If s_1 is the larger real root of $f_0(s)$, then we could get a solution to (7.5.65) in the form (7.5.68), which we denote by $y(x;\alpha,s_1)$. The coefficients a_n, determined by the recurrence formula (7.5.73), for any $n \geqq 1$ have the following explicit representations:

$$a_n(s_1) = \frac{(-1)^n \begin{vmatrix} f_1(s_1) & f_0(s_1+\alpha) & 0 & \cdots & 0 \\ f_2(s_1) & f_1(s_1+\alpha) & f_0(s_1+2\alpha) & \cdots & 0 \\ \cdots & \cdots & \cdots & \cdots & \cdots \\ f_{n-1}(s_1) & f_{n-2}(s_1+\alpha) & f_{n-3}(s_1+2\alpha) & \cdots & f_0(s_1+(n-1)\alpha) \\ f_n(s_1) & f_{n-1}(s_1+\alpha) & f_{n-2}(s_1+2\alpha) & \ldots & f_1(s_1+(n-1)\alpha) \end{vmatrix}}{f_0(s_1+\alpha)\, f_0(s_1+2\alpha) \cdots f_0(s_1+n\alpha) a_0}. \tag{7.5.75}$$

If s_2 is the other root of the indicial equation $f_0(s) = 0$, then the expression (7.5.75) with s_1 replaced by s_2 yield a second solution to equation (7.5.65), except for the case $s_1 = s_2 + n\alpha$ for all $n \geqq 0$.

The above arguments can be summarized as follows:

Theorem 7.20 *Let $x_0 \geqq a$ be a regular α-singular point of the equation (7.5.65) and let the series (7.5.66) and (7.5.67) be convergent on a semi-interval $0 < x - x_0 < R$ with $R > 0$. Let s_1, $s_2 > \alpha - 1$ be two real roots of the fractional indicial equation (7.5.74) with $s_1 > s_2$ and $s_1 - s_2 \neq n\alpha$ with $n \in \mathbb{N}_0$. Then, in the interval $0 < x - x_0 < R$, the equation (7.5.65) has two linearly independent solutions:*

$$y_1(s;\alpha,s_1) = (x-x_0)^{s_1} \sum_{n=0}^{\infty} a_n(s_1)(x-x_0)^{n\alpha}, \quad a_0(s_1) \neq 0 \tag{7.5.76}$$

and

$$y_2(s;\alpha,s_2) = (x-x_0)^{s_2} \sum_{n=0}^{\infty} a_n(s_2)(x-x_0)^{n\alpha}, \quad a_0(s_2) \neq 0. \tag{7.5.77}$$

Here the coefficients $a_n(s_1)$ and $a_n(s_2)$ are given in terms of $a_0(s_1)$ and $a_0(s_2)$ by the recurrence formula (7.5.73) with s replaced by s_1 and s_2, respectively. Moreover, the series

$$\sum_{n=0}^{\infty} a_n(s_1)(x-x_0)^{n\alpha} \tag{7.5.78}$$

and

$$\sum_{n=0}^{\infty} a_n(s_2)(x-x_0)^{n\alpha} \tag{7.5.79}$$

converge for $0 < x - x_0 < R$.

Proof. By the above arguments, to complete the proof of this theorem we would show the convergence of the series (7.5.77) and (7.5.78). The details of this proof are analogous to those used to prove the convergence of the Frobenius series solution near a regular singular point of an ordinary differential equation of second order. In addition, we must take into account the above-mentioned asymptotic relation (1.5.15).

There still remains the problem of how to find a second solution to equation (7.5.65) in the case when $s_1 - s_2 = n\alpha$ $(n \in \mathbb{N}_0)$. The following theorem gives an answer to these special cases.

Theorem 7.21 *Let $x_0 \geqq a$ be a regular α-singular point of the equation (7.5.65), and let the series (7.5.66) and (7.5.67) be convergent on a semi-interval $0 < x - x_0 < R$ with $R > 0$. Let s_1 and s_2 be two real roots of the fractional indicial equation (7.5.74) with $s_1, s_2 > \alpha - 1$ and $s_1 \geqq s_2$. Then, in the interval $0 < x - x_0 < R$, the equation (7.5.65) has one solution of the form (7.5.68):*

$$y_1(x;\alpha,s_1) = (x-x_0)^{s_1} \sum_{n=0}^{\infty} a_n(s_1)(x-x_0)^{n\alpha} \quad (a_0(s_1) \neq 0), \tag{7.5.80}$$

where the coefficients $a_n(s_1)$ are given in terms of $a_0(s_1)$ by the formula (7.5.73) with s replaced by s_1. The following assertions also hold:

a) *If $s_1 = s_2$, then a second solution to (7.5.65) has the following form:*

$$y_2(x;\alpha,s_1) = y_1(\alpha,s_1)\ lg(x-x_0) + \sum_{n=0}^{\infty} b_n(x-x_0)^{n\alpha+s_1}, \tag{7.5.81}$$

where

$$b_n = \frac{\partial}{\partial s}(a_n(s))\Big|_{s=s_1} \tag{7.5.82}$$

and $a_n(s)$ is given by (7.5.73).

b) *If $s_1 - s_2 = n\alpha$ with $n \in \mathbb{N}$, then a second solution to the equation (7.5.65) is given by*

$$y_2(x;\alpha,s_2) = y_1(\alpha,s_1)\cdot A(x)\cdot lg(x-x_0) + \sum_{n=0}^{\infty} c_n(s_2)(x-x_0)^{n\alpha+s_2}, \tag{7.5.83}$$

where $A(x)$ is a function obtained by evaluating the derivative

$$A(x) = \frac{\partial}{\partial s}\Big[(s-s_2)\ y(x;\alpha,s)\Big]\Big|_{s=s_2} \tag{7.5.84}$$

Moreover, the series (7.5.80)), (7.5.81), and (7.5.82) are convergent for the values of x on the semi-interval $0 < x - x_0 < R$.

Proof. First consider the case a). It is clear that $y_1(x;\alpha,s_1)$ in (7.5.66) is the solution to the equation (7.5.65). To get the second solution we make arguments analogous to the case of ordinary differential equations.

Consider s as a continuous variable and determine the coefficients $a_n(s)$ of $y(x;\alpha,s)$ as functions of s, using formulas (7.5.73). Then

$$[\mathbf{L}_{2\alpha}(y(t;\alpha,s))](x) = a_0 f_0(s)(x-x_0)^s. \tag{7.5.85}$$

Since $f_0(s) = (s-s_1)^2 g(s)$ (with analytic $g(s)$), then

$$\left[\mathbf{L}_{2\alpha}\left[\frac{\partial}{\partial s}\ y(t;\alpha,s)\Big|_{s=s_1}\right]\right](x) = 0, \tag{7.5.86}$$

and so the second solution $y_2(x;\alpha,s_1)$ is obtained from specifying the derivative

$$\frac{\partial}{\partial s}\, y(x;\alpha,s)\Big|_{s=s_1} \tag{7.5.87}$$

In the case b) the proof is also analogous to that of the ordinary case. Consider

$$y^*(x;\alpha,s) = (s-s_2)\; y(x;\alpha,s), \tag{7.5.88}$$

where we again consider the coefficients $a_n(s)$ of $y(x;\alpha,s)$ as functions of s given by (7.5.73). Then there exists a function $A(x)$ such that

$$y^*(x;\alpha,s_2) = A(x)\; y(s;\alpha,s_1). \tag{7.5.89}$$

On the other hand, by using (7.5.85), there also holds that

$$\left[\mathbf{L}_{2\alpha}\left[\frac{\partial}{\partial s}\, y^*(t;\alpha,s)\right]_{s=s_2}\right](x) = 0, \tag{7.5.90}$$

and, therefore,

$$\frac{\partial}{\partial s}\left[(s-s_2)\; y(x;\alpha,s)\right]\Big|_{s=s_2} \tag{7.5.91}$$

will be the second solution. Finally, evaluating the derivative (7.5.91) and taking (7.5.88) into account, we obtain the desired result (7.5.83).

Remark 7.3 Let us emphasize the fact that, for the case when the Riemann-Liouville fractional derivative is replaced by the Caputo derivative, the results obtained for solutions around regular α-singular points coincide exactly with those in the Riemann-Liouville case.

Example 7.12 Consider the following sequential FDE of order 2α $(0<\alpha<1)$:

$$(x-x_0)^{2\alpha}y^{(2\alpha}(x) + P_0(x-x_0)^{\alpha}y^{(\alpha}(x) + Q_0 y(x) = 0, \tag{7.5.92}$$

where $y^{(\alpha}(x)$ is the Riemann-Liouville fractional derivative $\left(D_{a+}^{\alpha}y\right)(x)$ or the Caputo derivative $\left({}^{C}D_{a+}^{\alpha}y\right)(x)$, and P_0 and Q_0 are constants. We shall seek at least one solution to the equation (7.5.92) around the regular α-singular point $x = x_0 > a$ $(a \in \mathbb{R})$. Let us look for a solution in the form:

$$y = (x-x_0)^s, \tag{7.5.93}$$

This expression with $s = \beta_1$, is the solution to the equation (7.5.92), if β_1 is a solution to the fractional indicial equation associated with (7.5.92):

$$\frac{\Gamma(s+1)}{\Gamma(s-2\alpha+1)} + P_0\frac{\Gamma(s+1)}{\Gamma(s-\alpha+1)} + Q_0 = 0. \tag{7.5.94}$$

Depending on the values of P_0, Q_0, and α, the equation (7.5.94) will have one real solution, two distinct real solutions, two double real solutions, or two complex conjugate solutions. If there are two real solutions, we take the greater one.

Next we present a table where, for given values of P_0, Q_0 and α, the corresponding roots β_1 and β_2 of (7.5.94) are given.

α	P_0	Q_0	β_1	β_2
0.95	1	-1	-0.82399	0.93479
0.65	1	-1	$-.097981$	0.50594
0.3	1	-1	-0.14703	$-$
0.1	1	-1	-0.68953	$-$
10^{-3}	1	-1	-0.99699	$-$
0.7	1	0.7	$-0.69651+0.23147\,i$	$-0.69651-0.23147\,i$
0.6	1.9	1.2529	0.757083	0.75783

(7.5.95)

So, for example:

a) If $\alpha=0.95$, $P_0=1$ and $Q_0=-1$, then the following linearly independent solutions of (7.5.92) around $x=x_0$ exist:

$$y_1(x)=(x-x_0)^{-0.82399} \tag{7.5.96}$$

$$y_2(x)=(x-x_0)^{0.93479} \tag{7.5.97}$$

b) If $\alpha=0.3$, $P_0=1$ and $Q_0=-1$, then only one solution exists of (7.5.92) around $x=x_0$, given by:

$$y_1(x)=(x-x_0)^{-0.14703} \tag{7.5.98}$$

c) If $\alpha=0.7$, $P_0=1$ and $Q_0=0.7$, then two linearly independent solutions of (7.5.92) around $x=x_0$ exist:

$$y_1=Re[(x-x_0)^{c+di}] \tag{7.5.99}$$

$$y_2=Im[(x-x_0)^{c+di}] \tag{7.5.100}$$

where $c=-0.69651$ and $d=0.23147$.

d) If $\alpha=0.6$, $P_0=1.9$ and $Q_0=1.2529$, then the equation (7.5.92) has the following linearly independent solutions around $x=x_0$:

$$y_1=(x-x_0)^{0.757083} \tag{7.5.101}$$

$$y_2(x)=\frac{\partial}{\partial s}\Big[(x-x_0)^s\Big]_{s=0.757083}=(x-x_0)^{0.757083}L(x-x_0) \tag{7.5.102}$$

Example 7.13 Consider the following generalized Bessel equation of order $\nu=0$:

$$x^{2\alpha}y^{(2\alpha}(x)+x^{\alpha}y^{(\alpha}(x)+x^{2\alpha}y(x)=0\ (0<\alpha<1) \tag{7.5.103}$$

Let us find two linearly independent solutions of (7.5.103) around the regular α-singular point $x=0$.

The relations (7.5.66) and (7.5.67) with the corresponding constants p_n and q_n $(n \geqq 0)$ take the following forms:

$$p(x)=1 \text{ and } p_0=1;\ p_n=0\quad(n\geqq 1),$$

$$q(x) = x^{2\alpha} \text{ and } q_0 = 0;\ q_1 = 0;\ q_2 = 1;\ q_n = 0 \quad (n \geqq 3).$$

Therefore, the indicial equation associated with (7.5.103) is given by

$$\frac{\Gamma(s+1)}{\Gamma(s-2\alpha+1)} + P_0 \frac{\Gamma(\beta+1)}{\Gamma(\beta-\alpha+1)} + q_0 = 0. \tag{7.5.104}$$

This yields the following solutions for different values of α:

α	β_1	β_2
0.99	−0.459	0.08194
0.8	−.6784	0.0535
0.65	−0.9629	−0.1051
0.628	−0.1312	–
10^{-4}	−0.9999	–

(7.5.105)

This table lets us obtain a solution around $x = 0$ of (7.5.103) for $\alpha = 0.628$ and for $\alpha = 10^{-4}$, and two linearly independent solutions in the cases $\alpha = 0.99$, $\alpha = 0.8$, and $\alpha = 0.65$.

Remark 7.4 As in the case of LFDE, the theory presented above can be applied to cases including non-sequential derivatives of Riemman-Liouville by the use of the relation (7.1.4).

7.6 Some Applications of Linear Ordinary Fractional Differential Equations

As we have already indicated, there are various applications of fractional differential equations in practically all areas of science and engineering.

It must be noted, however, that although there are many applications, there is still no clear interpretation of the different concepts in the context of fractional derivatives.

We will not try to present here an exhaustive resource of these applications. For that we refer the reader to the works of Carpinteri and Mainardi [132], Podlubny [682], Hilfer [340], Metzler and Klatfer [586], Schumer et al. [747], Zaslavsky [911], Turchetti et al. [834], Sokolov et al [781], and Kilbas et al. [404], where an important collection of the applications of fractional differential equations may be found.

In this section we will solve some generalized differential models related to the dynamics of a sphere immersed in an incompressible viscous fluid with oscillatory processes that have fractional damping terms. It should be pointed out that the use of damping terms also provides a very important tool in the study of fluid dynamics across a porous boundary. Likewise, we will resolve the four-parameter model proposed by Bagley and Torvik and Caputo and Mainardi (with different fractional derivatives) for modeling the mechanical behavior, from a rheologic standpoint, of certain new materials, especially polymers, which have viscoelastic properties.

Lastly, we will formally consider two fractional models associated with slow diffusion which are resolved through the method of separation of variables.

We note here the following property which will be of great use in some of the applications that will be covered in this section.

$$(D^n y)(t) = (\mathcal{D}^{p\alpha}_{a+} y)(t) \quad (\forall n, p \in \mathbb{N};\ \alpha = \frac{n}{p};\ p > n;\ t > a). \tag{7.6.1}$$

This is an immediate remembering that, if $y \in C([a,b])$, then

$$(I^{\beta}_{a+} y)(a+) = 0 \quad (\forall \beta \in \mathbb{R}^+). \tag{7.6.2}$$

From this we conclude that any ordinary differential equation can be expressed as a sequential fractional differential equation.

For example, the following differential equation

$$x'(t) - a^2 x(t) = 0, \tag{7.6.3}$$

where $x'(t) = (D^{1/2}_{0+} D^{1/2}_{0+} x)(t)$ for $t > 0$, can be expressed as follows:

$$(\mathcal{D}^{2\alpha}_{0+} x)(t) - a^2 x(t) = 0 \quad (\alpha = 1/2). \tag{7.6.4}$$

By Corollary 7.1, its general solution is given by

$$x(t) = C_1 e^{at}_{\alpha} + C_2 e^{-at}_{\alpha} \tag{7.6.5}$$

and since $x(0) < \infty$, we have $C_2 = -C_1$. Therefore,

$$x(t) = C_1 \sum_{j=1}^{\infty} \frac{[1 - (-1)^j] a^j t^{j\alpha + \alpha - 1}}{\Gamma[(j+1)\alpha]}; \tag{7.6.6}$$

and since $[1 - (-1)^j] = 0$ for $j = 2k$ $(k \in \mathbb{N})$, the solution (7.6.6) takes the form

$$x(t) = 2C_1 a \sum_{j=1}^{\infty} \frac{a^{2j} t^j}{j!} = Ce^{a^2 t}, \tag{7.6.7}$$

which is the general solution to (7.6.3), as expected.

7.6.1 Dynamics of a Sphere Immersed in an Incompressible Viscous Fluid. Basset's Problem

The dynamics of a sphere immersed in an incompressible viscous fluid is a classical problem with numerous applications in engineering and in the study of geophysical flows. A particularly important problem is the study of a sphere subjected to gravity, which was considered by Basset first in [70], and subsequently in [71], who introduced a special hydraulic force, generally known as "Basset's force". This force was interpreted by Mainardi [518] in terms of a fractional derivative of order 1/2 of the velocity of the particle relative to the fluid.

Mainardi et al. [533] introduced a new formulation of that force, called the generalized Basset's force, based on the fractional Caputo derivative ${}^C D_{0+}^{\alpha}$ $(0 < \alpha < 1)$ and on a generalized model of the one originally considered by Basset. That is, when the fluid is at rest and the particle moves vertically under the effect of gravity with a certain initial velocity $V(0+) = V_0$:

$$V'(t) + a\left({}^C D_{0+}^{\alpha} V\right)(t) + V(t) = 1; \; V(0+) = V_0 \;\; (a > 0;\; 0 < \alpha < 1). \qquad (7.6.8)$$

This problem arises, in particular, from the following fractional relaxation equation, also studied by Mainardi [518], using the Laplace transform [see also Gorenflo et al. [310]]:

$$U'(t) + a\left({}^C D_{0+}^{\alpha} U\right)(t) + U(t) = g(t); \; U(0+) = C_0 \; (0 < \alpha < 1). \qquad (7.6.9)$$

Equation (7.6.9) can be solved explicitly for $\alpha \in \mathbb{Q}$, without using integral transforms, with the theory developed in this text and keeping in mind (7.6.1), similar to the method we will use here to solve Basset's original problem:

$$U'(t) + a(D_{0+}^{1/2} U)(t) + U(t) = 1; \; (a > 0;\; t > 0) \qquad (7.6.10)$$

$$U(0) = 0. \qquad (7.6.11)$$

If we get $U'(t) = (\mathcal{D}^{2\alpha} U)(t)$ $(\alpha = 1/2)$, the equation (7.6.10) reduces to the form

$$[\mathbf{L}_{2\alpha}(U)](xt) = (\mathcal{D}_{0+}^{2\alpha} U)(t) + a(D_{0+}^{\alpha} U)(t) + U(t) = 1. \qquad (7.6.12)$$

This is a non-homogeneous sequential LFDE with constant coefficients, its solution dependent on the roots of the characteristic polynomial $m^2 + am + 1 = 0$ associated with the corresponding homogeneous equation. These roots are the following:

1. $\lambda_{1,2} = \frac{-a \pm \sqrt{a^2-4}}{2}$ $(\lambda_1 > \lambda_2)$, if $a > 2$.
2. $\lambda_1 = \lambda_2 = -\frac{a}{2} \in \mathbb{R}$, if $a = 2$.
3. $\lambda_2 = \bar{\lambda}_1$, with $\lambda_1 = \frac{-a + i\sqrt{4 - a^2}}{2}$, if $a < 2$.

Therefore, by applying Theorem 7.4, the solution $U(t)$ of (7.6.12) with the initial condition $U(0) = 0$ is given as follows:

1. If $a > 2$,

$$U(t) = \int_0^t e_{\alpha}^{\lambda_1(t-\eta)} \int_0^{\eta} e_{\alpha}^{\lambda_2(\eta-\xi)} d\xi d\eta$$

$$= \int_0^t \int_{\xi}^t e_{\alpha}^{\lambda_1(t-\eta)} e_{\alpha}^{\lambda_2(\eta-\xi)} d\eta d\xi = \lambda_1^{-1} \int_0^t e_{\alpha}^{\lambda_2(t-\eta)} \left[E_{\alpha}(\lambda_1 \eta^{\alpha}) - 1\right] d\eta$$

$$= t^{\alpha} \sum_{j=0}^{\infty} \sum_{k=0}^{\infty} (\lambda_2)^j (\lambda_1)^k \frac{t^{(j+k)\alpha}}{\Gamma[(j+k+1)\alpha]}$$

$$= t e_{\alpha}^{\lambda_1 t} \sum_{j=0}^{\infty} \left(\frac{\lambda_2}{\lambda_1}\right)^j - t^{\alpha} \sum_{j=0}^{\infty} \sum_{l=0}^{j-1} (\lambda_1)^l \frac{t^{l\alpha}}{\Gamma[(l+1)\alpha]}. \qquad (7.6.13)$$

2. If $a = 2$,

$$U(t) = \int_0^t e_\alpha^{\lambda_1(t-\eta)} \int_0^\eta (\eta - \xi)^\alpha e_{\alpha,1}^{\lambda_1(\eta-\xi)} d\xi d\eta. \tag{7.6.14}$$

3. If $a < 2$

$$U(t) = \int_0^t R_e \left[e_\alpha^{\lambda_1(t-\eta)}\right] \int_0^\eta I_m \left[e_\alpha^{\lambda_1(\eta-\xi)}\right] d\xi d\eta. \tag{7.6.15}$$

7.6.2 Oscillatory Processes with Fractional Damping

In 1984, Bagley and Torvik [52] considered the following initial-value problem:

$$\begin{aligned} &y''(t) + a\left(D_{0+}^{3/2} y\right)(t) + by(t) = f(t) \quad (t > 0), \\ &y(0) = 0; \; y'(0) = 0, \end{aligned} \tag{7.6.16}$$

with $a = \frac{2s\sqrt{\mu\rho}}{m}$ and $b = \frac{k}{m}$, and solved this problem by using the Laplace transform. This model is associated with the following problem.

Consider a thin, rigid plate of mass m and area s, immersed in a Newtonian fluid of infinite extension with density ρ and viscoelastic constant μ, and connected to a fixed point via a spring with spring constant k. In addition suppose that the surface of the plate is sufficiently large so that the fluid slows the movement of the plate.

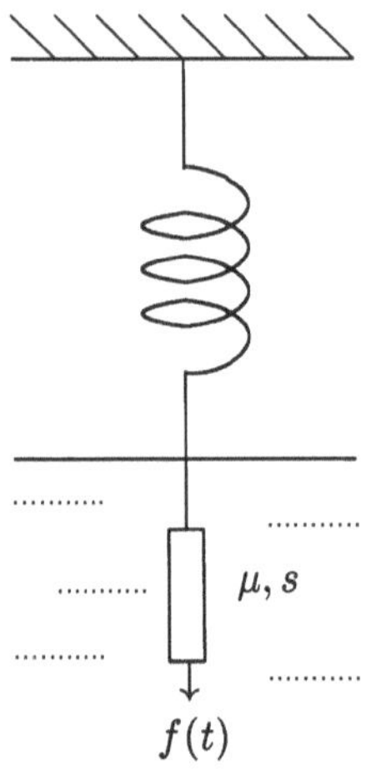

Newtonian Viscous Fluid
$\varphi = density$
$\mu = viscosity$
$f(t) = displacement\ of\ the\ plate$

Figure 1

Another approach, using the Laplace transform, was studied by Mainardi [515] for the following initial-value problem:

$$\begin{aligned} &V''(t) + a\left({}^C D_{0+}^\alpha V\right)(t) + bV(t) = g(t) \quad (a > 0, \; 0 < \alpha < 2), \\ &V(0) = C_0; \; V'(0) = C_1, \end{aligned} \tag{7.6.17}$$

as a model associated with damping processes with a fractional damping term.

Both models can be solved directly by applying the fractional differential equation theory developed in Chapter 7 of this book.

Nigmatullin and Ryabov [625], Kempfle and Gaul [365], and Beyer and Kempfle [82] also studied the model (7.6.17), but they considered the fractional damping term $D_+^\alpha V$ or $D_{0+}^\alpha V$ instead of ${}^C D_{0+}^\alpha V$, in order to use the Fourier transform.

Here we solve a special case of (7.6.16).

Consider the function

$$f(t) = \begin{cases} 8, & \text{if } 0 \leqq t \leqq 1, \\ 0, & \text{if } t > 1, \end{cases} \tag{7.6.18}$$

and the differential equation

$$y''(t) + 3\left(D_{0+}^{3/2} y\right)(t) + y(t) = f(t) \quad (t > 0), \tag{7.6.19}$$

with the initial conditions

$$y(0) = 0; \; y'(0) = 0. \tag{7.6.20}$$

We set $y''(t) = \left(\mathcal{D}_{0+}^{4\frac{1}{2}} y\right)(t)$, taking into account that y is differentiable, and the following Property:

Let $\eta > 0$ and let the functional series $\{f_n\}_{n=0}^\infty$ be such that, for $t \in [a,b]$ $(D_{a+}^\eta f_n)(t) < \infty$ $(\forall n \in \mathbb{N}_0)$. If the series $\sum_{j=0}^\infty f_j(t)$ and $\sum_{j=0}^\infty (D_{a+}^\eta f_j)(t)$ are uniformly convergent on any $[a+\epsilon, b]$, $\epsilon > 0$, then

$$(D_{a+}^\eta \sum_{j=0}^\infty f_j)(t) = \sum_{j=0}^\infty (D_{a+}^\eta f_j)(t) \quad (a < t < b). \tag{7.6.21}$$

Then the equation (7.6.19) reduces to the sequential LFDE given by

$$[\mathbf{L}_{4\alpha}(y)](t) := \left(\mathcal{D}_{0+}^{4\alpha} y\right)(t) + 3\left(\mathcal{D}_{0+}^{3\alpha} y\right)(t) + y(t) = f(t) \; (\alpha = 1/2), \tag{7.6.22}$$

which can be written in vectorial form as follows:

$$\left(D_{0+}^\alpha \bar{Z}\right)(t) = \mathbf{A}\bar{Z}(t) + \bar{\mathbf{B}}, \tag{7.6.23}$$

with the changes $\left(\mathcal{D}_{0+}^{j\alpha} y\right)(t) = z_{j+1}$ $(j = 0,1,2,3)$, in which

$$\mathbf{A} = \begin{pmatrix} 0 & 1 & 0 & 0 \\ 0 & 0 & 1 & 0 \\ 0 & 0 & 0 & 1 \\ -1 & 0 & 0 & -3 \end{pmatrix}; \; \bar{\mathbf{B}}(t) = \begin{pmatrix} 0 \\ 0 \\ 0 \\ f(t) \end{pmatrix}; \; \bar{Z}(t) = \begin{pmatrix} z_1(t) \\ z_2(t) \\ z_3(t) \\ z_4(t) \end{pmatrix}. \tag{7.6.24}$$

In addition, the initial conditions $y(0) = y'(0) = 0$, knowing that $y(t)$ must be differentiable in $[0,t]$, guarantee that

$$y(0) = (D_{0+}^\alpha y)(0) = (D_{0+}^{2\alpha} y)(0) = (D_{0+}^{3\alpha} y)(0) = 0 \tag{7.6.25}$$

and therefore, the desired solution to (7.6.23) must satisfy $\bar{Z}(0) = 0$.

Let $\mathbf{P}$ be the similarity matrix for the diagonalization of $\mathbf{A}$, that is, it satisfies the relationship

$$\mathbf{P}^{-1}\mathbf{A}\mathbf{P} = \mathbf{J} = \begin{pmatrix} \lambda_1 & 0 & 0 & 0 \\ 0 & \lambda_2 & 0 & 0 \\ 0 & 0 & \lambda_3 & 0 \\ 0 & 0 & 0 & \lambda_4 \end{pmatrix}, \tag{7.6.26}$$

where λ_j $(j = 1, 2, 3, 4)$ are the eigenvalues of $\mathbf{A}$. It is directly verified that

$$\mathbf{P} = \begin{pmatrix} -3.363 + 0.556i & -3.363 - 0.556i & -0.038 & -2.235 \\ -0.911 + 2.073i & -0.911 - 2.073i & 0.114 & 1.709 \\ 0.822 + 1.260i & 0.822 - 1.260i & -0.338 & -1.307 \\ 1 & 1 & 1 & 1 \end{pmatrix}, \tag{7.6.27}$$

and that its inverse matrix $\mathbf{P}^{-1}$ is equal to

$$\begin{pmatrix} -0.092 - 0.019i & -0.050 - 0.131i & 0.123 - 0.172i & 0.044 - 0.044i \\ -0.092 + 0.019i & -0.050 + 0.131i & 0.123 + 0.172i & 0.044 + 0.044i \\ 0.351 & -0.119 & 0.040 & 1.040 \\ -0.168 & 0.220 & -0.287 & -0.128 \end{pmatrix}, \tag{7.6.28}$$

and also that the eigenvalues of $\mathbf{A}$ are given by

$$\begin{aligned} \lambda_1 &= 0.363 - 0.556i \\ \lambda_2 &= 0.363 + 0.556i \\ \lambda_3 &= -2.962 \\ \lambda_4 &= -0.765. \end{aligned} \tag{7.6.29}$$

If the transform $\bar{Z}(t) = \mathbf{P}\bar{W}(t)$ is applied to the equation (7.6.23), it reduces to

$$(D_{0+}^{\alpha}\bar{W})(t) = \mathbf{J}\bar{W}(t) + \hat{\mathbf{B}}(t), \tag{7.6.30}$$

where

$$\hat{\mathbf{B}}(t) = \mathbf{P}^{-1}\bar{\mathbf{B}}(t) = \bar{b}f(t), \tag{7.6.31}$$

with

$$\bar{b} = (b_i) = \begin{pmatrix} 0.044 - 0.044i \\ 0.044 + 0.044i \\ 1.040 \\ -0.128 \end{pmatrix} \tag{7.6.32}$$

and with the initial condition

$$\bar{W}(0) = \bar{0}. \tag{7.6.33}$$

The only solution to (7.6.30) which satisfies (7.6.33), is obtained by applying Theorem 7.11 and this solution is given by

$$\bar{W}(t) = 8\int_0^t e_{\alpha}^{\mathbf{J}(t-\xi)}\bar{b}d\xi = 8\begin{pmatrix} b_1\int_0^t e_{\alpha}^{\lambda_1(t-\xi)}d\xi \\ b_2\int_0^t e_{\alpha}^{\lambda_2(t-\xi)}d\xi \\ b_3\int_0^t e_{\alpha}^{\lambda_3(t-\xi)}d\xi \\ b_4\int_0^t e_{\alpha}^{\lambda_4(t-\xi)}d\xi \end{pmatrix}. \tag{7.6.34}$$

Therefore,

$$\bar{Z}(t) = 8\mathbf{P}\int_0^t e_\alpha^{\mathbf{J}(t-\xi)}\bar{b}d\xi = 8\mathbf{P}\mathbf{J}^{-1}\begin{pmatrix} b_1E_\alpha(\lambda_1 t^\alpha) \\ b_2E_\alpha(\lambda_2 t^\alpha) \\ b_3E_\alpha(\lambda_3 t^\alpha) \\ b_4E_\alpha(\lambda_4 t^\alpha) \end{pmatrix}, \tag{7.6.35}$$

where $E_\alpha(z)$ is the Mittag-Leffler function (1.8.1), and hence the real part of this solution has the following form:

$$y(t) = Re[z_1(t)] = 10^{-2}Re[(-4.8+1.4i)E_\alpha(\lambda_1 t^\alpha)$$

$$+(0.9-7.4i)E_\alpha(\lambda_2 t^\alpha) + 0.348E_\alpha(\lambda_3 t^\alpha) - 0.59E_\alpha(\lambda_4 t^\alpha)]. \tag{7.6.36}$$

7.6.3 Study of the Tension-Deformation Relationship of Viscoelastic Materials

The appearance and proliferation of the so-called "new materials" beginning in the middle of the 19th century, such as plastics, rubbers, fibers, resins, and polymers, have provided a set of materials whose mechanical behavior cannot be explained via the classical rheologic means, such as those contained in the theories of Kelvin or Maxwell.

It should be noted that the field of rheology studies, among other things, the flow and deformation of materials with unusual behavior, particularly the flow of non-Newtonian fluids, the behavior of solids which can flow, or the flow of substances across porous media with fractal geometry.

Viscoelasticity is an intrinsic property of many of the "new materials", and it also constitutes an element of study within rheology. This property describes varying degrees of behavior between elasticity and viscosity, depending on the material (see, for example, González [290], Podlubny [682] Mainardi [518, 520], Rossikhin and Shitikova [718], Caputo and Mainardi [129, 130] and Christensen [139]).

Depending on the temperature of the materials, these will achieve, in general, four different states:

1. Vitreous state (in the vitreous region): this region contains organic crystals which only allow small deformations;

2. Transition state (in the transition region): this region contains, among others, linear thermoplastic and reticular polymers (cellulosic materials, polyamides, polyesters ...);

3. Elastic state (in the elastic region): this region contains elastomeric materials (rubbers, plastics...);

4. Fluid state (in the influence region).

At ambient temperature, a given polymer may be found in one of these four regions.

In this book we will only consider the models which attempt to explain the viscoelastic deformation of certain materials.

Many rheologic models have been developed to study this behavior. In general, we have the following model:

$$\boxed{\sum_{k=0}^{n} a_k \frac{d^k \sigma}{dt^k} = \sum_{k=0}^{m} b_k \frac{d^k \epsilon}{dt^k}}$$

Next, as an example, we will describe Burger's model.

a) If a sample of an amorphous polymer is subjected to a constant tension, an instantaneous deformation will be observed when the force is applied. If the force is immediately relaxed, the deformation disappears. This state is called the "elastic domain". Mechanically, this behavior is represented with a spring of modulus E_1 which describes the instantaneous elastic response:

$$\sigma(t) = E_1 \varepsilon(t) \quad \text{(Hooke's Law)}, \tag{7.6.37}$$

where $\sigma(t)$ represents the tension (force/surface) and $\varepsilon(t)$ is the unit deformation $(\Delta L/L)$.

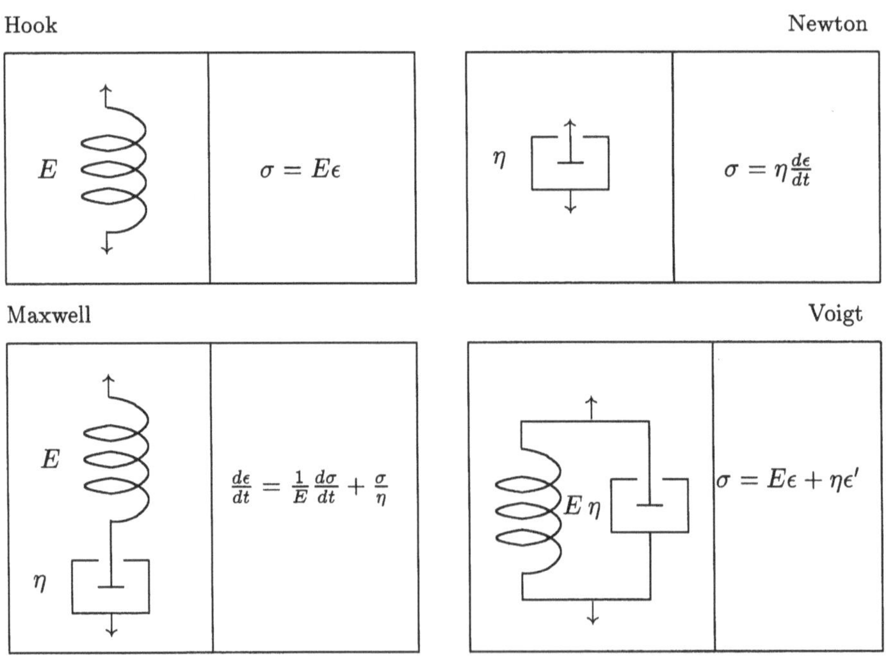

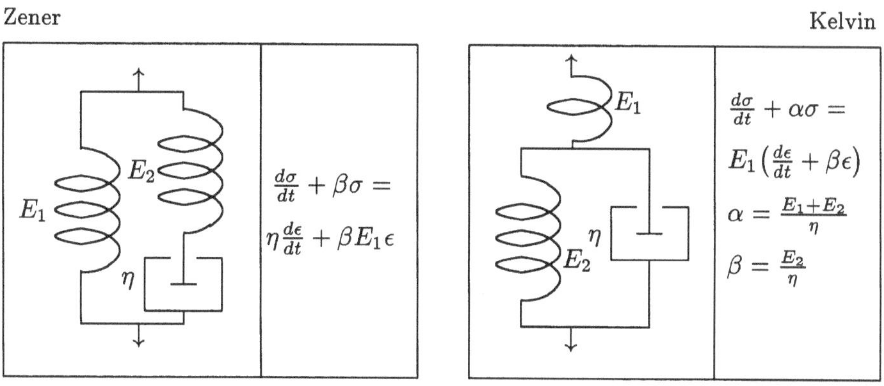

b) As the tension is prolonged, we witness a transition to a "deformation as a function of time". The deformation slows until it progresses at a constant speed. This mechanism is represented as being made up of a Voigt type element (E_2, h_2) and which corresponds to a "delayed elasticity". When the tension is removed, the material begins to return to its original shape but the recovery slows and finally disappears before it is completed. Mechanically, this behavior is represented by the parallel grouping of a spring and a piston

$$\sigma(t) = E_2\varepsilon(t) + \eta\frac{d\varepsilon(t)}{dt} \quad \text{(Voigt's model).} \tag{7.6.38}$$

c) Lastly, a piston h_1 represents the viscous fluid

$$\sigma(t) = \eta\frac{d\varepsilon(t)}{dt} \quad \text{(Law of Newtonian Fluids).} \tag{7.6.39}$$

The corresponding mechanical representation of Burger's model is shown in the following figure:

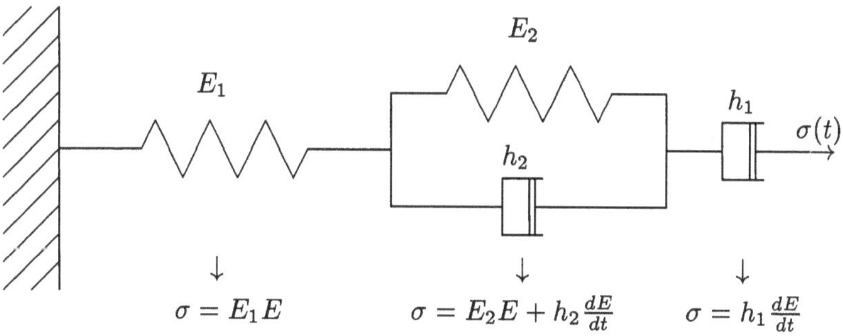

Figure 2

Therefore, this model has the following mathematical representation associated with it:

$$\sigma''(t) + a_1\sigma'(t) + a_0\sigma(t) = b_2\varepsilon''(t) + b_1\varepsilon'(t) + b_0\varepsilon(t), \tag{7.6.40}$$

where

$$a_1 = \frac{E_1 + E_2}{h_2} + \frac{E_1}{h_1}; \quad a_0 = \frac{E_1 \cdot E_2}{h_1 h_2} \tag{7.6.41}$$

$$b_2 = E_1; \quad b_1 = \frac{E_1 \cdot E_2}{h_2}; \quad b_0 = 0. \tag{7.6.42}$$

Thus the ordinary linear model was widely studied. For example, Caputo-Mainardi [130] and Christensen [139]) have used the following model:

$$\left[1 + \sum_{k=1}^{p} a_k D^k\right] \sigma(t) = \left[b_0 + \sum_{k=1}^{q} b_k D^k\right] \varepsilon(t), \tag{7.6.43}$$

where $p = q$ or $p = q + 1$.

This model represents a series or parallel combination of basic Zener or Kelvin type units.

These models, however, did not adjust themselves well to the behavior demonstrated by many viscoelastic materials. Gerasimov [279] was probably the first to propose, in 1948, a study of models on the anomalous dynamics of the mechanical behavior of these viscoelastic materials. He used a generalization of the classical models which implied that the tension could be proportional to the R-L fractional derivative to the right of D_+^α of the material deformation, that is,

$$\sigma(t) = E(D_+^\alpha \varepsilon)(t) \quad (0 < \alpha < 1). \tag{7.6.44}$$

In 1947, Blair [752] suggested an analogous generalization using the R-L fractional derivative D_{0+}^α, while Caputo and Mainardi [130, 129] in 1971 and Gorenflo and Mainardi [304] in 1997 suggested an analogous generalization based on the Caputo derivative to the right ${}^C D_+^\alpha$ or that of Caputo ${}^C D_{0+}^\alpha$ considering the deformation and tension to be negligible for $t \leqq 0$.

These observations were made based on the behavior, experimentally verified, that the modulus of a viscoelastic material in relation to its frequency, that is, the modulus

$$G(is) = \frac{\hat{\sigma}(is)}{\hat{\varepsilon}(is)}, \tag{7.6.45}$$

where $\hat{\sigma}(is)$ and $\hat{\varepsilon}(is)$ represent the Fourier transform (or the Laplace transform, as the case may be) of $\sigma(t)$ and $\varepsilon(t)$ respectively, adjusts itself better to real powers of order α $(0 < \alpha < 1)$ of the frequency s than to natural powers, as was classically thought.

This drove many specialists, most notably: Nutting [639], Gemant [277, 278] Blair [752, 753], Blair and Caffyn [754], Rabotnov [692, 693] and Meshkov and Rossikhin [573] before 1950; between 1966 and 1969 Caputo [120, 121, 122], Caputo and Mainardi [130, 129](who, in addition, suggested the use of the fractional derivative in seismological and metallurgical dissipation models); and, in the last 25 years, Lokshin and Suvorova [497], Caputo [123, 124, 125], Mainardi [516], Mainardi and Bonetti [523], Smith and Vries [772], Scarpi [736], Stiassnie [803], Bagley and Torvik [50, 51, 52, 53], Rogers [714], Koeller [431, 432] Koh and Kelly

[433], Friedrich [266], Nonnenmacher and Glöckle [635], Glöckle and Nomenmacher [285], Makris and Constantinou [539], Fernander [257], Pritz [688], Rossikhin and Shitikova [718], and Lion [490], to propose the following fractional model which generalizes (7.6.43)

$$\sigma^{\alpha_n}(t) + \sum_{k=1}^{p} a_k D^{\alpha_k}\sigma(t) = \sum_{k=0}^{p} b_k D^{\beta_k}\varepsilon(t) \tag{7.6.46}$$

with $a_k, b_k, \alpha_k, \beta_k \in \mathbb{R}$ $(k = 0, 1, ..., p)$ and where D^α represents the appropriate fractional derivative such that the Fourier transform of (7.6.46) becomes

$$\left[(is)^{\alpha_n} + \sum_{k=1}^{p} a_k (is)^{\alpha_k}\right] \hat{\sigma}(is) = \left[\sum_{k=0}^{p} b_k (is)^{\beta_k}\right] \hat{\varepsilon}(is). \tag{7.6.47}$$

Along these lines, Caputo and Mainardi proposed in [130] the following model, based on four parameters, as the most adequate for representing the viscoelastic behavior of certain materials from a rheologic point of view:

$$[1 + bD^\alpha]\,\sigma(t) = [E_0 + E_1 D^\alpha]\,\varepsilon(t) \quad (0 < \alpha < 1). \tag{7.6.48}$$

Between 1979 and 1986, Bagley and Torvik [50, 51, 52, 53, 54] and Bagley [45] made a series of fundamental contributions which would consolidate this as the definitive model showing in 1986 [54] that, with the following restrictions

$$E_0 \geqq 0, E_1 > 0, b \geqq 0; E_1 \geqq bE_0, \tag{7.6.49}$$

the model satisfies the First and Second Laws of Thermodynamics.

In [54], Bagley and Torvic also showed experimentally that this model is in close agreement with the behavior of over 150 viscoelastic materials.

It must be noted that although Bagley and Torvik suggested the R-L fractional derivative D^α_{0+}, their work does not make clear which D^α fractional derivative is best suited to their model, since their only consideration was that the Fourier transform should satisfy the relation

$$\mathcal{F}(D^\alpha f(t))(s) = (is)^\alpha \hat{f}(is), \tag{7.6.50}$$

where $\hat{f}(is) = \mathcal{F}(f(t))(s)$, and with this sole condition there are several definitions of a fractional derivative which can be of use, such as that of Grünwald-Letnikov ${}^G D^\alpha_+$, that of Liouville D^α_+ or that of Caputo-Liouville ${}^C D^\alpha_+$, over infinite intervals, or that of R-L $D^\alpha_{0+} f$ (if $f(t) = 0$ for $t < 0$) or $D^\alpha_{a+} f$ (if $f(t) = 0$ for $t < a$), or those corresponding to Caputo ${}^C D^\alpha_{0+}$ and ${}^C D^\alpha_{a+}$.

Here we present, as an example of the application of the theory developed in the previous Chapters, the explicit solutions of diverse problems based on the four-parameter model (7.6.48). We propose a generalization of the same type which is perhaps more representative of the behavior of certain viscoelastic materials and whose explicit solution follows from the application of the theory presented previously.

It must be pointed out that the methods used for obtaining solutions do not require the use of an integral transform, which in a way provides a certain degree of flexibility.

Proceeding from the four-parameter model (7.6.48), generally accepted,

$$\sigma(t) + b(D^{\alpha}\sigma)(t) = E_0\varepsilon(t) + E_1(D^{\alpha}\varepsilon)(t) \quad (0 < \alpha \leqq 1), \tag{7.6.51}$$

with $b \geqq 0, E_0 \geqq 0, E_1 > 0, b \leqq \frac{E_1}{E_0}$, and where D^{α} represents an adequate fractional derivative, we find that (7.6.51) can be represented as follows:

a) For a known tension $\sigma(t)$,

$$(D^{\alpha}z)(t) + \lambda z(t) = f(t), \tag{7.6.52}$$

with $z(t) = \frac{b}{E_1}\sigma(t) - \varepsilon(t)$, $\lambda = \frac{E_0}{E_1}$, $A = \frac{bE_0-1}{E_1^2}$, and $f(t) = A\sigma(t)$.

The general solution to (7.6.52), when $D^{\alpha} \equiv D^{\alpha}_{a+}$, if σ is continuous and integrable in (a,t), is given by

$$z(t) = Ce_{\alpha}^{-\lambda(t-a)} + A\int_a^t e_{\alpha}^{-\lambda(t-\xi)}\sigma(\xi)d\xi, \tag{7.6.53}$$

where C is an arbitrary real constant.

Therefore the general solution to the equation (7.6.51) has the form

$$\varepsilon(t) = \frac{b}{E_1}\sigma(t) - Ce_{\alpha}^{-\frac{E_0}{E_1}(t-a)} - \frac{bE_0-1}{E_1^2}\int_a^t e_{\alpha}^{-\frac{E_0}{E_1}(t-\xi)}\sigma(\xi)d\xi. \tag{7.6.54}$$

Moreover, if $C = 0$ and $\sigma \in C([a,t])$, then $\varepsilon(a) = \frac{b}{E_1}\sigma(a)$. In particular, $\varepsilon(a) = \sigma(a) = 0$, if $C = 0$ and $\sigma \in C([a,t])$, while for the case when

$$\begin{array}{l} \lim_{t\to a+}(t-a)^{1-\alpha}\sigma(t) = k_1, \\ \lim_{t\to a+}(t-a)^{1-\alpha}\varepsilon(t) = k_2, \end{array} \tag{7.6.55}$$

the solution to the equation (7.6.51) is given by (7.6.54), where

$$C = \Gamma(\alpha)\Big[\frac{b}{E_1}k_1 - k_2\Big]. \tag{7.6.56}$$

In addition, by application of Theorem 7.13, we derive the general solution to the equation (7.6.52) with $D^{\alpha}y \equiv {}^{C}D^{\alpha}_{a+}y$:

$$z(t) = \int_a^t e_{\alpha}^{-\lambda(t-\xi)}[A\sigma(\xi) - \lambda k]d\xi + k. \tag{7.6.57}$$

If $z(a) = k$, then the corresponding solution to the equation (7.6.51) has the form

$$\varepsilon(t) = \frac{b}{E_1}\sigma(t) - \int_a^t e_{\alpha}^{-\lambda(t-\xi)}\left(\frac{bE_0-1}{E_1^2}\sigma(\xi) - \lambda k\right)d\xi + k. \tag{7.6.58}$$

b) For a known unit deformation $\varepsilon(t)$,

$$(D^\alpha y)(t) + \gamma y(t) = g(t), \tag{7.6.59}$$

with $y(t) = \frac{E_1}{b}\varepsilon(t) - \sigma(t)$, $\gamma = \frac{1}{b}$, $B = \frac{E_1 - E_0^2}{b^2}$, and $g(t) = B\varepsilon(t)$.

If $\varepsilon(t)$ is continuous and integrable in (a,t), then the general solution to the equation (7.6.59) with $D^\alpha y \equiv D^\alpha_{a+}y$ is given by

$$y(t) = Ce_\alpha^{-\gamma(t-a)} + B\int_a^t e_\alpha^{-\gamma(t-\xi)}\varepsilon(\xi)d\xi. \tag{7.6.60}$$

Therefore the general solution to the equation (7.6.51) can be represented as follows:

$$\sigma(t) = \frac{E_1}{b}\varepsilon(t) - Ce_\alpha^{-\frac{1}{b}(t-a)} - \frac{E_1 - E_0^2}{b^2}\int_a^x e_\alpha^{-\frac{1}{b}(x-\xi)}\varepsilon(\xi)d\xi. \tag{7.6.61}$$

In addition, if $C = 0$ and $\varepsilon \in C([a,t])$, then $\sigma(a) = \frac{E_1}{b}\varepsilon(a)$. In particular, $\varepsilon(a) = \sigma(a) = 0$, if $C = 0$ and $\varepsilon \in C([a,t])$, as expected.

In the case when the conditions in (7.6.55) are satisfied, the solution to the equation (7.6.51) is given by (7.6.61) with

$$C = \Gamma(\alpha)\left(\frac{E_1}{b}k_2 - k_1\right). \tag{7.6.62}$$

Moreover, applying Theorem 7.13 to the equation (7.6.59) with $D^\alpha y = {}^CD^\alpha_{a+}y$, we obtain the following general solution:

$$y(t) = \int_a^t e_\alpha^{\gamma(t-\xi)}\left[B\varepsilon(\xi) - \gamma k\right]d\xi + k, \tag{7.6.63}$$

if $y(a) = k$, and the corresponding solution to the equation (7.6.51) has the form:

$$\sigma(t) = \frac{E_1}{b}\varepsilon(t) - \int_a^t e_\alpha^{\gamma(t-\xi)}\left[\left(\frac{E_1 - E_0^2}{b^2}\right)\varepsilon(\xi) - \gamma k\right]d\xi + k. \tag{7.6.64}$$

As we have already noted, the model (7.6.51) is the one proposed by Torvik-Bagley with $D^\alpha \equiv D^\alpha_{a+}$. Therefore in the case $a = 0$, (7.6.54) or (7.6.58) represent the solutions of this model when the tension $\sigma(t)$ or the deformation $\varepsilon(t)$, respectively, are known.

It seems natural, therefore, to propose the following more general "new rheologic models" along the lines of (7.6.46) based on the R-L sequential fractional derivative D^α_{a+} and on that of Caputo ${}^CD^\alpha_{a+}$.

Let $\mathcal{D}^{j\alpha}y$ $(0 < \alpha < 1)$ be the R-L or Caputo sequential fractional derivative.

Model A.

$$(\mathcal{D}^{n\alpha}z)(t)+\sum_{j=0}^{n-1}\lambda_j(\mathcal{D}^{j\alpha}z)(t)=K\sigma(t), \tag{7.6.65}$$

if $\sigma(t)$ is known, with

$$z(t)=\beta\sigma(t)-\varepsilon(t) \tag{7.6.66}$$

and with the parameters a, K, β, and $\lambda_j \in \mathbb{R}$ $(j=1,2,...,n-1)$, or

$$(\mathcal{D}^{n\alpha}z)(t)+\sum_{j=0}^{n-1}\eta_j(\mathcal{D}^{j\alpha}z)(t)=E\varepsilon(t), \tag{7.6.67}$$

if $\varepsilon(t)$ is known, with

$$z(t)=\gamma\varepsilon(t)-\sigma(t) \tag{7.6.68}$$

and with the parameters a, E, γ and $\eta_j \in \mathbb{R}$ $(j=1,2,...,n-1)$.

The general solution to these models could be explicitly obtained without difficulty by applying the theory developed above. For example, the model given by (7.6.65) can be expressed as follows:

$$(D^{\alpha}\bar{Z})(t)=\mathbf{A}\bar{Z}(t)+\bar{\mathbf{B}}(t), \tag{7.6.69}$$

where $D^{\alpha}y$ is the Riemann-Liouville derivative $D^{\alpha}_{a+}y$ or the Caputo derivative ${}^{C}D^{\alpha}_{a+}y$ and

$$\mathbf{A}=\begin{pmatrix} 0 & 1 & 0 & \cdot\ \cdot\ \cdot & 0 & 0\\ 0 & 0 & 1 & \cdot\ \cdot\ \cdot & 0 & 0\\ \cdots & \cdot & \cdot & \cdot\ \cdot\ \cdot & \cdot & \cdot\\ \cdots & \cdot & \cdot & \cdot\ \cdot\ \cdot & \cdot & \cdot\\ \cdots & \cdot & \cdot & \cdot\ \cdot\ \cdot & \cdot & \cdot\\ 0 & 0 & 0 & \cdot\ \cdot\ \cdot & 0 & 1\\ -\lambda_0 & -\lambda_1 & -\lambda_2 & \cdot\ \cdot\ \cdot & -\lambda_{n-2} & -\lambda_{n-1} \end{pmatrix}; \tag{7.6.70}$$

$$\bar{\mathbf{B}}(t)=K\begin{pmatrix} 0\\ 0\\ 0\\ \sigma(t) \end{pmatrix} \quad \text{and} \quad \bar{Z}(t)=\begin{pmatrix} z_1(t)\\ \cdots\\ \cdots\\ \cdots\\ z_n(t) \end{pmatrix}, \tag{7.6.71}$$

with $z_j(t)=(\mathcal{D}^{(j-1)\alpha}_{a+}z)(t)$ $(j=1,...,n)$ or $\left(z_j(t)=({}^{C}D^{(j-1)\alpha}_{a+}z)(t)\right)$. According to Theorem 7.11, the general solution to the system (7.6.69) is represented as follows:

$$\bar{Z}(t)=e_{\alpha}^{\mathbf{A}(t-a)}\bar{C}+\int_a^t e_{\alpha}^{\mathbf{A}(t-\xi)}\bar{\mathbf{B}}(\xi)d\xi, \tag{7.6.72}$$

where $z_1(t)=z(t)=\beta\sigma(t)-\varepsilon(t)$ $(\beta\in\mathbb{R})$ and $\bar{C}$ is an arbitrary constant vector.

This representation suggests a new generalization of the model presented in **Model A.**

Model B.

$$(D_{a+}^{\alpha}\bar{Z})(t) = \mathbf{A}\bar{Z}(t) + \bar{\mathbf{B}}(t), \tag{7.6.73}$$

where

$$\mathbf{A} = (a_{ij})_{i,j=1,\dots,n}, \tag{7.6.74}$$

$$\bar{\mathbf{B}}(t) = \begin{pmatrix} \sigma_1(t) \\ \cdots \\ \cdots \\ \cdots \\ \sigma_n(t) \end{pmatrix} \quad \text{and} \quad \bar{Z}(t) = \begin{pmatrix} z_1(t) \\ \cdots \\ \cdots \\ \cdots \\ z_n(t) \end{pmatrix}, \tag{7.6.75}$$

and $z_j(t) = \left(\mathcal{D}_{a+}^{(j-1)\alpha} z\right)(t)$ $(j = 1, 2, ..., n-1)$. The general solution to the system (7.6.73) can also be easily expressed by

$$\bar{Z}(t) = e_{\alpha}^{\mathbf{A}(t-a)}\bar{C} + \int_a^t e_{\alpha}^{\mathbf{A}(t-\xi)}\bar{\mathbf{B}}(\xi)d\xi. \tag{7.6.76}$$

The function (7.6.76) is also the solution to the system of the form (7.6.73), with the Caputo fractional derivative ${}^{C}D_{a+}^{\alpha}\bar{Z}$ instead of the Riemann-Liouville one $D_{a+}^{\alpha}\bar{Z}$.

Model C.

$$({}^{C}D_{a+}^{\alpha}\bar{Z})(t) = \mathbf{A}\bar{Z}(t) + \bar{\mathbf{B}}(t), \tag{7.6.77}$$

where $\mathbf{A}$, $\bar{\mathbf{B}}(t)$, and $\bar{Z}(t)$ are given in (7.6.74)-(7.6.75). The general solution to the equation (7.6.77), also obtained in Theorem 7.13, has the form

$$\bar{Z}(t) = \int_a^t e_{\alpha}^{\mathbf{A}(t-\xi)}\left[\bar{\mathbf{B}}(\xi) + \mathbf{A}\bar{Y}(a)\right]d\xi + \bar{Y}(a), \tag{7.6.78}$$

In particular, the solution which satisfies $\bar{Y}(a) = \bar{b}$ is given by

$$\bar{Z}(t) = \int_a^t e_{\alpha}^{\mathbf{A}(t-\xi)}\left[\bar{\mathbf{B}}(\xi) + \mathbf{A}\bar{b}\right]d\xi + \bar{b}. \tag{7.6.79}$$

Chapter 8

FURTHER APPLICATIONS OF FRACTIONAL MODELS

This chapter is devoted to some aspects of differential equations of fractional order and their applications. The first part supports and explains the emergence of fractional differential equations as a useful tool for the modeling of many anomalous phenomena in nature and in the theory of complex systems. This conclusion is already well known to many applied researchers, but it is not so evident to many mathematical analysts and researchers. The main reason for the utility of such fractional derivative operators is the strong relationship between these operators and fractional Brownian motion, the continuous time random walk (CTRW) method, the Lévi stable distributions, and the generalized central limit theorem. Moreover, the fractional derivative operators allow for the representation of the long-memory and non-local dependence of many anomalous processes.

In the second part of this chapter, some useful properties of various fractional integral and fractional derivative operators are presented. We give examples of simple fractional models with boundary conditions in one dimension in connection with anomalous diffusion, and explain why several specific fractional derivative operators are suitable in the sub-diffusive case. We introduce here a new model useful for simulating both sub-diffusion and super-fast diffusion processes. Such a model involves a generalized Liouville fractional derivative operator over the time variable.

8.1 Preliminary Review

In the following subsections we present the reader with a historical review of the applications of the fractional calculus to complex systems, which is not intended to be exhaustive. In the last subsection we introduce the properties of certain fractional operators which will be useful in later sections.

8.1.1 Historical Overview

Questions as to what we mean by, and where we could apply, the fractional calculus operators have fascinated us all ever since 1695, when the so-called fractional calculus was conceptualized in connection with the infinitesimal calculus [717] [643]. We should like to recall the famous correspondence between Leibniz and L'Hopital. Many great mathematicians have been interested in these questions during the last 170 years, and some remarkable applications of fractional calculus have emerged as a result. A reasonably adequate study of fractional calculus can be found in [729] and some of its applications in [643] and [682].

The interest in the study of the fractional differential equations (FDE) as a separate topic arose some 40 years ago. Questions about the existence of solutions to Cauchy type problems involving the Riemann-Liouville operators D^{α}_{a+} ($\alpha \in \mathbb{C}$ and $a \in \mathbb{R}$) and some other problems have attracted the attention of several mathematicians. A systematic and rigorous study of some problems of this kind involving FDE can be found in recent literature [see [407] and [408] for a review].

Although numerous theoretical applications of the fractional calculus operators have been found during their long history, many mathematicians and applied researchers have tried to model real processes using FDE. However, except possibly in a handful of applications, the literature on fractional calculus does not seem to contain results of far-reaching consequences. This may be due, in part, to the fact that many of the useful properties of the ordinary derivative D are not known to carry over analogously for the case of the fractional derivative operator D^{α}, such as, for example, a clear geometric or physical meaning, product rules, chain rules, and so on.

So then what are the useful properties of these fractional calculus operators which help in the modeling of so many anomalous processes? From the point of view of the authors and from known experimental results, most of the processes associated with complex systems have non-local dynamics involving long-memory in time, and the fractional integral and fractional derivative operators do have some of those characteristics. Perhaps this is one of the reasons why these fractional calculus operators lose the above-mentioned useful properties of the ordinary derivative D.

It is known that the classical calculus provides a powerful tool for explaining and modeling important dynamic processes in many areas of the applied sciences. But experiments and reality teach us that there are many complex systems in nature with anomalous dynamics, such as the transport of chemical contaminants through water around rocks, the dynamics of viscoelastic materials as polymers, atmospheric diffusion of pollution, cellular diffusion processes, network traffic, signal transmission through strong magnetic fields, the effect of speculation on the profitability of stocks in financial markets, and many more.

In most of the above-mentioned cases, these kinds of anomalous processes have a macroscopic complex behavior, and their dynamics can not be characterized by classical derivative models. Nevertheless, the heuristic explanation of the corresponding models of some of these processes are often not so complicated when tools from statistical physics are applied. For such an explanation, one must use some generalized concepts from classical physics such as fractional Brownian motion, the continuous time random walk (CTRW) method involving Lévi stable distributions (instead of Gaussian distributions), the generalized central limit theorem (instead of the classical limit theorem), and non-Markovian (and thus non-local) distributions (instead of the classical Markovian ones). From this approach it is also important to note that the anomalous behavior of many complex processes includes multi-scaling in the time and space variables, and therefore also in the fractal characteristics of the media.

The above-mentioned tools have been used extensively during the last 30 years. But the connections between these statistical models and certain FDE involving the fractional integral and derivative operators (Riemann-Liouville, Caputo, Liouville or Weyl and Riesz) had not been formally established until the last 15 years and, more intensively, during the last 5 years, as we will show later. The mentioned connections have been obtained in several ways. One of the most used approaches is based on generalizing the propagator of the model through the use of Fourier and/or Laplace transforms. In this approach the characteristic function of the corresponding probability density function (pdf), which contains the exponential function e^x in the classical case, is replaced by the Mittag-Leffler, Wright, or Fox functions which satisfy some of the main properties of e^x. Using this concept, many researchers have obtained useful fractional dynamic models where fractional derivative operators in the time and space variables assume the long-memory and non-local dependence characteristics of the anomalous processes [67], [66], [534], [586], [429], [512], [135] and [873]. At present, the more commonly studied models are connected with the anomalous diffusion processes. These models have generated the fractional versions of the diffusion and advection-diffusion equations, the Fokker-Planck and Kramer equations, and some others [see [586] for a review].

Here we should point out that the above-mentioned fractional integral and fractional derivative operators have allowed for the modeling of some special situations. But there are many other types of fractional calculus operators, fractional pseudo-differential operators, and perhaps other singular operators that could help us to remodel some of the anomalous processes. In this regard, we shall use a generalized Liouville operator to obtain a new model for super-diffusion processes where the fractional derivative operator only works over the time variable.

We may conclude from the foregoing presentation that it is important to work out a rigorous theory of FDE analogous to that for ordinary differential equations (ODE). In fact, there are many similar and new open questions in this subject, the answers to which can help in the development of the applied sciences. Let us mention here, as an illustrative historical example, that the delta function, introduced by Dirac at the end of 19th century, has been used formally by physicists in many of their ordinary derivative models for a long time. And their investigations became rigorous only about 50 years later after using the distribution theory presented by Laurent Schwartz in 1944 (see, for example, [750]).

This section is organized as follows. The first part given in Subsection 8.1.2 is dedicated to an explanation of FDE as an emergent topic in the development of complex system modeling. In Subsection 8.1.3 some properties of several The second part includes a mathematical justification of new models for super-diffusion processes in terms of Liouville-type fractional derivatives in the time variable. In Section 3 some properties of several fractional integrals and derivatives, considering i Chapter 2, are presented. Subsection 8.2 includes a mathematical justification of a new models for superdiffusion processes in terms of Louiville-type fractional derivatives in the time variable. Subsection 8.3 deals with the study of diffusion mechanisms with internal degrees of freedom. Finally, in Section 4 examples of simple partial FDE with boundary conditions are discussed in connection with anomalous diffusion equations. Here, a new model useful in simulating both sub-diffusion and super-fast diffusion processes is introduced. Such a model involves a generalized Liouville fractional derivative operator in the time variable. In this context we note also that some earlier authors have studied only the models of super-diffusion processes involving the Riesz fractional derivative in the space variables.

8.1.2 Complex Systems

The asymptotic approach to equilibrium of physical systems is captured by the second law of thermodynamics, which states that the entropy of an isolated system either remains fixed or increases over time. All physical states in equilibrium are statistically equivalent and this is the endpoint of all physical activity. There are many systems in nature which apparently contradict the second law of thermodynamics. This apparent conflict with physical law arises mainly because there are no systems which are completely isolated [see [868], [58], [78], [869], [771], and [873]].

Schrödinger discussed the idea that macroscopic laws are the consequences of the interactions between a large number of particles. Therefore, such laws can only be described by means of statistical physics. For this reason, the Gaussian or normal probability distribution, Brownian motion, and the classical central limit theorem are important and useful. These tools have permitted important advances in explaining many natural processes stemming from the classical physical models which are based upon the use of the infinitesimal calculus, and which involve the ordinary and local derivative operator D. One relevant example of such utility is classical diffusion and Fokker-Planck equations [66], [586], and [62].

Complex systems are very familiar and occur often in nature. Examples of this kind of dynamic process appear throughout science. In the majority of such systems, classical models and analytical functions are often not sufficient to describe their basic features. A familiar case is the so-called Brownian motion, which describes the erratic path of a particle through space.

This Brownian motion in nature could be modeled as a random walk on a two-dimensional plane, where the size of the jumps and the wait time before each jump are random and not uniform. This is the background of the so-called "continuous time random walk" (CTRW) method. Such a method is the basis for a heuristic explanation of the physical behavior of normal and anomalous diffusion processes.

For example, we can see macroscopic dynamic processes of normal diffusion that are modeled by a diffusion equation or by the Fokker-Planck equation, with fixed parameters, by using the CTRW approach. That is, the above processes can be seen as the mean line of the CTRW method applied to a set of microscopic particles whose motions are random in the time-wait and in the space-jump, and these random motions are governed by characteristic distributions. When both random variables are independent and their pdf's are Gaussian and Markovian (local process in the sense that the entire past of the system does not influence the future and only the present determines the immediate future), we can apply the classical central limit theorem. But when, for instance, the wait-time follows a Lévi stable distribution instead of a Gaussian one, we can describe more complex processes such as sub-diffusion and super-diffusion by using the CTRW method and the generalized central limit theorem.

The CTRW method can be characterized by the moments of the random mean motion. If the process is non-local or has a memory wait-time, then the second moment of the random variable X of the jumps $< X^2 >$ is proportional to a power t^α of order α of the time when the time is sufficiently large. This type of stochastic model lets us characterize the sub-diffusion, normal-diffusion, and super-diffusion processes as the cases where $0 < \alpha < 1$, $\alpha = 1$, and $\alpha > 1$, respectively.

On the other hand, when the random variables of the jumps in space do not follow a Gaussian distribution, then a certain relationship with the so-called fractal media behavior of many complex phenomena can be obtained.

We should point out that close connections exist between fractional differential equations and the dynamics of many complex systems, including anomalous processes, CTRW method, fractional Brownian motion, fractional diffusion, and fractal media.

Anomalous diffusion is perhaps the most frequently studied complex problem. Richardson [705] was probably the first to consider the existence of such anomalous kinetics in connection with turbulent diffusion in the atmosphere. Later on, other authors observed the same phenomena in other areas, such as chemical and plasma physics [835] and [445]. The characterization of the Lévi stable distributions in terms of the Fourier transform of their probability density functions (pdf), the generalized central limit theorem, and fractional Brownian motion were introduced by Lévi [479] and [480]. The last concept was developed by Mandelbrot and Van Ness [548]. Subsequently, Mandelbrot [546] found some connections between fractals and the geometry of complex systems in nature.

The CTRW method was probably first introduced by Gnedenko and Kolmogorov in 1949 in the Russian version of their classical book [288]. Later on, such an approach was applied by Montroll and Weiss [608] to study the mechanical properties of lattices. A good introduction to the above statistical tools can be found in the book by Feller [256].

Oldham and Spanier [643] wrote the first book dedicated specifically to the subject of fractional calculus and its applications. This book contains a very good history written by B. Ross of the developments of fractional integration and fractional differentiation and their applications from 1695 to 1974, and also some of their applications to the chemical and physical sciences. In 1987 Samko et al. published a book in Russian [729], where one can find an encyclopedic, detailed, and rigorous study of the fractional calculus operators. The book by Miller and Ross [603] presents the first introduction to fractional differential equations [see also a more recent book by Podlubny [682]].

The first Ph.D. thesis to apply fractional differential equations in modeling the behavior of viscoelastic materials was written by Bagley, advised by Torvik [45], at the Air Force Institute of Technology and Material Laboratory of Ohio. After that many researchers paid attention to the application of fractional differential equations to viscoelastic materials and other complex processes, including anomalous diffusion phenomena.

We shall classify a short selection of relevant applications in selected fields. Additionally, we will provide some survey-type references on the topic of this presentation which will give the reader an idea of the evolution of the mathematical modeling of some *Complex Systems.* We begin with the articles published in the period 1970-1990. Perhaps [641] [see also [643]] was the first to introduce the fractional differential equation as a model for certain chemical processes.

The *CTRW method* was applied to some physical problems in [764], [767], [422], and [866]; the last paper contains a survey of the results in this approach. *Anomalous Processes in Amorphous and Disordered Media* were studied in [739], [672], and [673]. *Models of Viscoelastic Materials* in terms of fractional differential equations were discussed in [803], [51], [52], [431], and [54]. Problems involving the *Kinetics of Anomalous Diffusion* were considered using random walk methods in [332], [333], [420], [765], and [99]. A generalization of the entropy concept was presented in [824] and [825], and this concept has a close connection with the behavior of some complex systems. Fractional diffusion and fractional wave equations, generalizing the classical ones, were investigated in [746] and [267].

Here, we present some of the many articles devoted to fractional models in different applied sciences from 1991, *Material Theory:* [284], [537], [538], [775], [240], [761], [92], [770][594], and [39]; *Transport Processes, Fluid Flow Phenomena, Earthquakes, Solute Transport:* [433], [358], [61], [80], [660], [540], [2], [165], [659], [747] [748], [458], [570], [201], [59], and [351]; *Chemistry, Wave Propagation and Signal Theory:* [326], [730], [348], [329], [806], [203], and [203]; *Biology:* [287], [334], [33], [336], [813], [465], [838], [337], [854], and [468] *Electromagnetic Theory:* [259], [241], [242], [244], [245], [247], [86], and [840]; *Thermodynamics:* [142], [575], [776], [780], and [452]; *Mechanics:* [708], [718], and [676]; *Geology and Astrophysics:* [536], [460], and [135]; *Economics:* [58], [534], [873], [457], and [414]; *Control Theory:* [653], [678], [814], [668], [691],[815],[11], [204], and [598]; *Chaos and Fractals:* [911], [648], [918], [360], [463], [464], [649], [826] [922], [564], and [915].

Models of some *anomalous processes in terms of fractional differential equations*

Table 8.1: **Evolution in the Number of Publications in Connection with Models of Complex Systems Considered by JCR**

Descriptive Phrases	45-80	81-90	91-96	97-03	Total
Fractional Brownian Motion	3	6	158	507	674
Anomalous Diffusion	116	108	520	1047	1791
Anomalous Relaxation	10	14	23	51	98
Super-Diffusion or Sub-Diffusion	0	18	30	202	250
Anomalous Dynamics or Anomalous Processes or Fractional Models or Fractional Relaxation or Fractional Kinetics	18	28	79	105	230
Fractional Diffusion Equation or Fractional Fokker-Planck Equation	0	2	19	129	150

were discussed in [270], [260], [286], [581], [140], [723], [67], [305], [310], [711], [65], [66], [582], [512], [778], [834], [8], [327], [328], [327], [355], and [593].

Note also that the applications of the described fractional calculus methods to modeling anomalous processes in the above and other sciences can be found in the following books and survey papers: [666], [901], [620], [865], [352], [626], [766], [352], [444], [100], [132], [67], [917], [547], [665], [751], [77], [340], [586], [638], [658], [338], [345], [871], and [916]. See also [740], [768] and [421], where one may find an introduction to anomalous phenomena.

We have mentioned a series of papers and books devoted to the modeling of some complex systems through the use of different tools, including fractional differential equations. We chose only some of the many publications in this field, which appeared basically during the last few years. In Table 8.1 we show the evolution in the number of these publications. We have used for this only those papers in scientific journals which are included in the lists of the Science Citation Index and Social Science Citation Index of the Institute for Scientific Information (JCR) from 1945 to 2001.

We have not included here those papers that are devoted to applications of the CTRW method, although they have close links with the fractional derivative models we have considered in this table.

We can conclude from this study that the number of publications devoted to the modeling of anomalous processes has considerably increased during the last 10 years. We also note that the use of the fractional derivative operators plays an important role in obtaining the corresponding derivative models of such processes. These models include non-local and long-memory properties and are strongly related with the CTRW method.

We note that, during the last few years, there has been a special interest in the modeling of anomalous diffusion processes involving ultra-slow diffusion. But the anomalous phenomena, such as super-diffusion, super-conductivity, wave propagation, and others, have not been studied thoroughly.

8.1.3 Fractional Integral and Fractional Derivative Operators

In this section, we recall some properties of the known fractional integral and fractional derivative operators from Chapter 2. Let f be a real function, $\alpha \in \mathbb{C}$ and $n \in \mathbb{N} = \{1, 2, 3, \cdots\}$. The classical Liouville (or Weyl) fractional integral $I_{-}^{\alpha} f$ of order α with $\Re(\alpha) > 0$ is defined by

$$(I_{-}^{\alpha} f)(x) = \frac{1}{\Gamma(\alpha)} \int_{x}^{\infty} \frac{f(t)dt}{(t-x)^{1-\alpha}} \quad (x > a \geqq -\infty), \tag{8.1.1}$$

where $\Gamma(\alpha)$ is the Euler Gamma function. The corresponding fractional derivative $D_{-}^{\alpha} f$ of order α ($\Re(\alpha) \geqq 0;\ \alpha \neq 0$) has the form

$$(D_{-}^{\alpha} f)(x) = (-D)^{n} (I_{-}^{n-\alpha} f)(x) \quad (x > a \geqq -\infty), \tag{8.1.2}$$

with $D^n = (d/dx)^n$ and $n = [\Re(\alpha)] + 1$, where $[\Re(\alpha)]$ means the integer part of $\Re(\alpha)$.

The Liouville fractional calculus operators I_{-}^{α} and D_{-}^{α} of exponential function yield the same exponential function, both apart from a constant multiplication factor.

Lemma 8.1 *Let* $\alpha, \lambda \in \mathbb{C}$ ($\Re(\alpha) > 0;\ \Re(\lambda) > 0$). *Then*

$$(I_{-}^{\alpha} e^{-\lambda t})(x) = \lambda^{-\alpha} e^{-\lambda x} \quad \textit{and} \quad (D_{-}^{\alpha} e^{-\lambda t})(x) = \lambda^{\alpha} e^{-\lambda x}. \tag{8.1.3}$$

Now we present the so-called fractional integrals and fractional derivatives of a function f with respect to another function g. Such operators play the main role in the new model of sub- and super-diffusion processes. We shall introduce these processes in the next section.

Let $g(x)$ be an increasing and positive monotone function on (a, ∞), having a continuous derivative $g'(x)$ on (a, ∞). We define the general fractional integral operator $I_{-;g}^{\alpha} f$ of a function f with respect to a function g, of order α ($\Re(\alpha) > 0$) by

$$(I_{-;g}^{\alpha} f)(x) = \frac{1}{\Gamma(\alpha)} \int_{x}^{\infty} \frac{g'(t) f(t)dt}{[g(t) - g(x)]^{1-\alpha}} \quad (x > a \geqq -\infty). \tag{8.1.4}$$

If $g'(x) \neq 0$ ($a < x < \infty$), then the operator $I_{-;g}^{\alpha}$ can be expressed *via* the fractional integral I_{-}^{α} by

$$I_{-;g}^{\alpha} f = Q_g I_{-}^{\alpha} Q_g^{-1} f, \tag{8.1.5}$$

where Q_g is the substitution operator $(Q_g f)(x) = f[g(x)]$ and Q_g^{-1} is its inverse operator. When $g'(x) \neq 0$ ($a < x < b$), the corresponding fractional derivative $D_{-;g}^{\alpha} f$ of a function f with respect to g of order α ($\Re(\alpha) \geqq 0;\ \alpha \neq 0$) has the form

$$(D_{-;g}^{\alpha} f)(x) = \left(-\frac{1}{g'(x)} D \right)^{n} (I_{-;g}^{n-\alpha} f)(x) \tag{8.1.6}$$

$$= \frac{1}{\Gamma(n-\alpha)}\left(-\frac{1}{g'(x)}D\right)^n \int_x^\infty \frac{g'(t)f(t)dt}{[g(t)-g(x)]^{\alpha-n+1}} \quad (x > a \geqq -\infty), \tag{8.1.7}$$

where $D = d/dx$ and $n = [\Re(\alpha)] + 1$.

When $\alpha = n \in \mathbb{N}$, the Liouville derivative (8.1.2) and the generalized derivative (8.1.6) have the following forms:

$$(D_-^n f)(x) = (-1)^n f^{(n)}(x) \text{ and } (D_{-;g}^n f)(x) = \left(-\frac{1}{g'(x)}\frac{d}{dx}\right)^n f(x). \tag{8.1.8}$$

In particular, $(D_-^1 f)(x) = -f'(x)$ and $(D_{-;g}^1 f)(x) = -f'(x)/g'(x)$.

The following properties will be important in the development of new dynamic models of super-diffusion and sub-diffusion (see Properties 2.19 and 2.21 in Subsection 2.5):

$$(I_{-;g}^\alpha e^{-\lambda g(t)})(x) = \lambda^{-\alpha} e^{-\lambda g(x)} \tag{8.1.9}$$

and

$$(D_{-;g}^\alpha e^{-\lambda g(t)})(x) = \lambda^\alpha e^{-\lambda g(x)}. \tag{8.1.10}$$

where $\alpha \in \mathbb{C}$ ($\Re(\alpha) > 0$) and $\lambda > 0$.

In particular, when $a = 0$ and $g(x) = x^\sigma$ ($\sigma > 0$), (8.1.4) and (8.1.6) yield the following special cases of the fractional integral and fractional derivative operators of order α with respect to the function x^σ:

$$(I_{-;x^\sigma}^\alpha f)(x) = \frac{\sigma}{\Gamma(\alpha)} \int_x^\infty \frac{t^{\sigma-1} f(t)dt}{(t^\sigma - x^\sigma)^{1-\alpha}} \quad (x > 0), \tag{8.1.11}$$

when $\Re(\alpha) > 0$, and

$$(D_{-;x^\sigma}^\alpha f)(x) = \frac{\sigma^{1-n}}{\Gamma(n-\alpha)} \left(-x^{1-\sigma} D\right)^n \int_x^\infty \frac{t^{\sigma-1} f(t)dt}{(t^\sigma - x^\sigma)^{\alpha-n+1}} \quad (x > 0), \tag{8.1.12}$$

when $\Re(\alpha) \geqq 0$ ($\alpha \neq 0$) *with* $n = [\Re(\alpha)] + 1$.

We note that $I_{-;x^\sigma}^\alpha f$ is related to the fractional integral $I_-^\alpha f$ by

$$(I_{-;x^\sigma}^\alpha f)(x) = \left(I_-^\alpha f(t^{1/\sigma})\right)(x^\sigma) \quad (x > 0), \tag{8.1.13}$$

and also that it is connected with the so-called Erdélyi-Kober type operator apart from a power multiplier factor (see Section 2.6 for details).

Lemma 8.2 *Let* $\sigma > 0$, $\lambda > 0$, *and* $\alpha \in \mathbb{C}$ ($\Re(\alpha) > 0$). *Then*

$$(I_{-;x^\sigma}^\alpha e^{-\lambda t^\sigma})(x) = \lambda^{-\alpha} e^{-\lambda x^\sigma} \quad \textit{and} \quad (D_{-;x^\sigma}^\alpha e^{-\lambda t^\sigma})(x) = \lambda^\alpha e^{-\lambda x^\sigma}. \tag{8.1.14}$$

Remark 8.1 The formula (8.1.10) and the second relations in (8.1.3) and (8.1.14) are valid for $\alpha \in \mathbb{C}$ with $\Re(\alpha) = 0$ and $\alpha \neq 0$.

The Riesz integral and Riesz derivative operators I^α and D^α of order $\alpha \in \mathbb{C}$ ($\Re(\alpha) > 0$) in the n-dimensional Euclidean space $\mathbb{R}^n$ deserve special attention. These operators are defined as negative $(-\Delta)^{-\alpha/2}$ and positive $(-\Delta)^{\alpha/2}$ powers of the Laplace operator

$$\Delta = \frac{\partial^2}{\partial x_1^2} + \frac{\partial^2}{\partial x_2^2} + \cdots + \frac{\partial^2}{\partial x_n^2}. \tag{8.1.15}$$

They can be represented in terms of the direct Fourier $\mathcal{F}$ and inverse Fourier $\mathcal{F}^{-1}$ transforms by

$$I^\alpha f \equiv (-\Delta)^{-\alpha/2} f = \mathcal{F}^{-1} |\bar{x}|^{-\alpha} \mathcal{F} f; \quad D^\alpha f \equiv (-\Delta)^{\alpha/2} f = \mathcal{F}^{-1} |\bar{x}|^{\alpha} \mathcal{F} f, \tag{8.1.16}$$

where $\bar{x} = (x_1, \cdots, x_n) \in \mathbb{R}^n$ and $|\bar{x}| = \left(x_1^2 + \cdots + x_n^2\right)^{1/2}$.

When $0 < \alpha < n$, the Riesz fractional integration I^α can be represented for *sufficiently good* functions f by the Riesz potential, given (for $\bar{x} \in \mathbb{R}^n$) by

$$(I^\alpha f)(\bar{x}) = \gamma(n, \alpha) \int_{\mathbb{R}^n} \frac{f(\bar{y})}{|\bar{x} - \bar{y}|^{n-\alpha}} d\bar{y} \quad \left(\gamma(n, \alpha) = \frac{\Gamma[(n-\alpha)/2]}{2^\alpha \pi^{n/2} \Gamma(\alpha/2)}\right). \tag{8.1.17}$$

We note that the Riesz potential (8.1.17) and the Riesz derivative operator D^α, expressed explicitly as a hypersingular integral by E. M. Stein in 1961, have been sufficiently studied (see, for example [707], [728, 727] and Sections 25-27 in [729]). In connection with the fractional diffusion processes such a multidimensional Riesz fractional differentiation was used by Kochubei in [424], [425], [427], [428] and [429]. Although such a Riesz derivative operator was used by many applied researchers, they only used its characterization in terms of Fourier transforms.

We conclude this section by noting that a great number of other fractional calculus operators of one and several variables exist which preserve different memory properties. Some fractional calculus operators, such as the Riemann-Liouville, Grünwald-Letnikov, Marchaud, Erdélyi-Kober, Hadamard, etc., can be found in Chapter 2. We also mention another smooth fractional calculus operator, known as the Caputo derivative (see, for example [305] and [310]), which is often used in many applied fractional calculus models, explicitly or implicitly (see for details, Section 2.4). Furthermore, it is possible to introduce or define new singular operators with long memory and non-local properties.

8.2 Fractional Model for the Super-Diffusion Processes

As mentioned in the preceding section, many papers published during the last ten years presented fractional calculus models for the kinetics of natural anomalous processes in complex systems, which maintain the long-memory and non-local properties of the corresponding dynamics. Special interest has been paid to the anomalous diffusion processes, that is, super-slow diffusion (or sub-diffusion) and super-fast diffusion (or super-diffusion) processes. Nigmatullin in [622] and [621] was probably the first who considered the following *diffusion equation with memory*:

$$\frac{\partial U(\bar{x},t)}{\partial t} = \int_a^t K(t-\tau)\Delta U(\bar{x},\tau)d\tau \quad (t > 0;\ \bar{x} \in \mathbb{R}^n), \tag{8.2.1}$$

where Δ is the Laplace operator (8.1.15). In the particular case when $K(t) = \rho^2\delta(t-a)$, $\delta(t-a)$ being the Dirac delta function, this relation represents the classical diffusion equation

$$\frac{\partial U(\bar{x},t)}{\partial t} = \rho^2 \Delta U(\bar{x},t) \quad (t > 0;\ \bar{x} \in \mathbb{R}^n). \tag{8.2.2}$$

As we have seen in the first part of this presentation, there are many fractional calculus models for the sub-diffusion processes, without and with an external force field, such as the *Fractional Diffusion Equation* and the *Advection-Diffusion* or the *Fokker-Planck Equations*, respectively. All these models include the Riemann-Liouville and Caputo fractional derivative operators acting on the time variable, and the Liouville (or Weyl) $D^{\alpha}_{\pm}$ and Riesz D^{α} fractional derivative operators acting on the space variable.

The super-diffusion processes, however, are studied less frequently. Furthermore, only a few papers include fractional calculus models, where the authors introduce the Riesz derivative operator D^{α} acting on the space variables by using its definition, which is given in (8.1.16) in terms of the direct and inverse Fourier transforms (see, for example, [250], [872], [474], and [135]).

We propose here a (presumably new) fractional derivative model, including the generalized Liouville fractional derivative operator $D^{\alpha}_{-;g}$, which acts on the time variable. This model allows us to have strong control of both the sub- and super-diffusion processes. It also has great flexibility due to the free function g, as well as many other advantages from a purely technical point of view in the sense that we can use the same methods as in the ordinary case (that is, those based upon integral transforms and separation of variables) to solve boundary-value problems associated with the fractional calculus models.

First we consider our fractional calculus model in the one-dimensional case. We begin with a scheme for the solution of the classical problem associated with the one-dimensional heat equation

$$\frac{\partial U(x,t)}{\partial t} = \rho^2 \frac{\partial^2}{\partial x^2} U(x,t) \quad (\rho > 0;\ \ t > 0;\ \ 0 < x < l), \tag{8.2.3}$$

with the boundary conditions

$$U(0,t) = U(l,t) = 0 \quad (t > 0), \tag{8.2.4}$$

(f being a *sufficiently smooth* function on $[0,l]$) and with the initial condition

$$U(x,0) = f(x) \quad (0 < x < l). \tag{8.2.5}$$

We note that the parameter ρ^2 can characterize the thermal or diffusion constant of the media.

Using the method of separation of variables, we seek the solution to the problem (8.2.3)-(8.2.5) in the form $U(x,t) = X(x)T(t)$. Then

$$\frac{X''(x)}{X(x)} = \frac{T'(t)}{\rho^2 T(t)} = -\lambda^2. \tag{8.2.6}$$

Therefore, $X(x)$ is a solution of the Sturm-Liouville problem

$$X''(x) + \lambda X(x) = 0, \quad X(0) = X(l) = 0, \tag{8.2.7}$$

for which the eigenvalues and the corresponding eigenfunctions are given by

$$\lambda_n = \mathbf{a}^2; \; X_n(x) = \sin(\mathbf{a}x); \; \mathbf{a} = \frac{n\pi}{l} \quad (n \in \mathbb{N}). \tag{8.2.8}$$

On the other hand, $T(t)$ is a solution of the differential equation

$$T'(t) + (\mathbf{a}\rho)^2 T(t) = 0, \tag{8.2.9}$$

given by

$$T_n(t) = \exp\left[-(\mathbf{a}\rho)^2 t\right] \quad (n \in \mathbb{N}). \tag{8.2.10}$$

Finally, the solution of the problem modeled by (8.2.3) to (8.2.5) is obtained as a series solution whose coefficients are found from the following Fourier development of the function f involved in the initial condition (8.2.5):

$$U(x,t) = \sum_{n=1}^{\infty} c_n X_n(x) T_n(t), \tag{8.2.11}$$

where

$$c_n = \frac{2}{l} \int_0^l \sin(\mathbf{a}x) f(x) dx \quad (n \in \mathbb{N}), \tag{8.2.12}$$

and $X_n(x)$ and $T_n(t)$ are given in (8.2.8) and (8.2.10), respectively.

Thus we need only show the convergence on $[0,l] \times [0,\infty)$ of the series for $U(x,t)$ in (8.2.11) and of the series for $U_{xx}(x,t)$ and $U_t(x,t)$ obtained from (8.2.11) by term-by-term differentiation with respect to x and t, respectively.

Now we can see more clearly that a sub-diffusive process, associated with the above problem, can be obtained by replacing the factor $T_n(t)$ in the solution (8.2.11) by some other function $T_n(t,\alpha)$ depending on the time t and a new parameter α. Such a new parameter gives us the possibility of controlling the decay of the process in the time variable; for instance, we have

$$T_n(t,\alpha) = \exp\left[-(\mathbf{a}\rho)^{2/\alpha} g(t,\alpha)\right] \quad (n \in \mathbb{N}). \tag{8.2.13}$$

In this case, (8.2.11) is extended to the form

$$U(x,t) = \sum_{n=1}^{\infty} c_n X_n(x) T_n(t,\alpha), \tag{8.2.14}$$

where $X_n(t)$ and $T_n(t,\alpha)$ are given in (8.2.8) and (8.2.13), respectively. Thus a sub- or super-diffusive process can be controlled via the function $g(t,\alpha)$.

For example, if $g(t,\alpha) = t^\alpha$, we have sub-diffusion kinetics when $0 < \alpha < 1$ and super-diffusion kinetics when $\alpha > 1$. This is in agreement with the CTRW method.

In the above-mentioned way, by choosing a function $g(t,\alpha)$, we can obtain other suitable solutions to our problem so as to control a very strong super-diffusion process. For example, if we take $g(t,\alpha) = \exp(t^\alpha)$ with $\alpha > 1$, perhaps (8.2.14) would yield a new model for the dynamics of the properties of super-conductivity of some materials under certain boundary conditions.

Now we can express the corresponding differential model which has the solution (8.2.14). The problem then becomes finding some derivative operator $\mathbf{D}^\alpha$ (fractional, singular with a memory, or others) with the following property:

$$\mathbf{D}^\alpha T(t,\alpha) = (\mathbf{a}\rho)^2 T(t,\alpha), \tag{8.2.15}$$

where

$$T(t,\alpha) \equiv T_n(t,\alpha) = \exp\left[-(\mathbf{a}\rho)^{2/\alpha} g(t,\alpha)\right] \quad (n \in \mathbb{N}). \tag{8.2.16}$$

Such a problem is solved by taking $\mathbf{D}^\alpha = D^\alpha_{-,g(t,\alpha)}$, where $D^\alpha_{-;g(t,\alpha)} f$, as a function of t, is the fractional derivative of a function f with respect to $g(t,\alpha)$ defined in (8.1.6). Indeed, if we suppose that $g(t,\alpha)$ (for any fixed α) is an increasing and monotonic function of $t \in \mathbb{R}_+ = (0,\infty)$ having the continuous derivative $\frac{\partial}{\partial t} g(t,\alpha) \neq 0$, then, in accordance with (8.1.10),

$$\left(D^\alpha_{-;g(t,\alpha)} \exp\left[-(\mathbf{a}\rho)^{2/\alpha} g(\tau,\alpha)\right]\right)(t) = (\mathbf{a}\rho)^2 \exp\left[-(\mathbf{a}\rho)^{2/\alpha} g(t,\alpha)\right], \tag{8.2.17}$$

and hence (8.2.15) holds true for $\mathbf{D}^\alpha = D^\alpha_{-;g(t,\alpha)}$. Therefore, the new fractional diffusion model can be presented by the fractional differential equation

$$D^\alpha_{-;g(t,\alpha)} U(x,t) = \rho^2 \frac{\partial^2}{\partial x^2} U(x,t) \quad (\alpha > 0;\ t > 0;\ x \in \mathbb{R}). \tag{8.2.18}$$

If we additionally suppose that $\lim_{t\to 0+} g(t,\alpha) = k(\alpha) \in \mathbb{R}$ for any $\alpha > 0$, then the explicit solution to the problem modeled by (8.2.18), (8.2.4), and (8.2.5) has the form (8.2.14)

$$U(x,t) = \frac{1}{k(\alpha)} \sum_{n=1}^{\infty} c_n \sin(\mathbf{a}x) \exp\left[-(\mathbf{a}\rho)^{2/\alpha} g(t,\alpha)\right], \tag{8.2.19}$$

provided that $g(t,\alpha)$ is chosen so that $U(x,t)$, $D^\alpha_{-;g(t,\alpha)} U(x,t)$, and $U_{xx}(x,t)$ are convergent series on $[0,l] \times [0,\infty)$, c_n being given in (8.2.12).

For example, when $g(t,\alpha) = t^\alpha$ $(\alpha > 0)$, we can prove the above convergence just as in the ordinary case when $\alpha = 1$. In this case, $D^\alpha_{-;g(t,\alpha)} = D^\alpha_{-;t^\alpha}$ is the fractional derivative (8.1.12), and the equation (8.2.18) takes the form

$$D^\alpha_{-;t^\alpha} U(x,t) = \rho^2 \frac{\partial^2}{\partial x^2} U(x,t) \quad (\alpha > 0;\ t > 0;\ x \in \mathbb{R}), \tag{8.2.20}$$

and the explicit solution of the problem modeled by (8.2.20), (8.2.4), and (8.2.5) is represented by

$$U(x,t)=\sum_{n=1}^{\infty} c_n \sin(\mathrm{a}x)\exp\left[-(\mathrm{a}\rho)^{2/\alpha}t^{\alpha}\right]. \tag{8.2.21}$$

In a similar manner, we can consider the case when $g(t,\alpha)=\exp(t^{\alpha})$.

The model (8.2.18) can be extended to the space variable $\bar{x}\in\mathbb{R}^n$ if we replace the partial derivative $\partial^2/\partial x^2$ by the Laplace operator (8.1.15) with respect to $\bar{x}$:

$$D^{\alpha}_{-;g(t,\alpha)}U(\bar{x},t)=\rho^2\Delta_{\bar{x}}U(\bar{x},t) \quad (\alpha>0;\ t>0;\ \bar{x}\in\mathbb{R}^n). \tag{8.2.22}$$

In particular, the model (8.2.20) is extended to the form

$$D^{\alpha}_{-;t^{\alpha}}U(\bar{x},t)=\rho^2\Delta_{\bar{x}}U(\bar{x},t) \quad (\alpha>0;\ t>0;\ \bar{x}\in\mathbb{R}^n). \tag{8.2.23}$$

Analogous to (8.2.19) and (8.2.21), the explicit solutions to these equations under the corresponding boundary and initial conditions can be also derived.

It seems that the above approach, based on the use of the generalized fractional derivative (8.1.6), can also be applied in order to construct the corresponding generalized fractional calculus models for the Fokker-Planck and advection-diffusion equations. We think that these models will be compatible with those for the corresponding generalizations including, for example, the Riesz operator D^{α} acting on the space variable.

We note that there are many more possibilities for controlling the exponential-type decay of the fundamental solutions of the generalized fractional diffusion equations, and that (in this way) the same mathematical technique may be used to study new fractional calculus models for the control of anomalous wave processes.

8.3 Dirac Equations for the Ordinary Diffusion Equation

In this section we will apply the results obtained in Section 6.4 to the study of diffusion mechanisms with internal degrees of freedom.

If, in Example 6.11, we set $\lambda=-1$ and $\lambda=1$, we obtain the following solutions:

$$u(x,t)=\frac{1}{2\pi}\int_{-\infty}^{+\infty} E_{1/2}\left(i\sigma t^{1/2}\right)(\mathcal{F}_x f)(\sigma)e^{-i\sigma x}d\sigma \tag{8.3.1}$$

and

$$u(x,t)=\frac{1}{2\pi}\int_{-\infty}^{+\infty} E_{1/2}\left(-i\sigma t^{1/2}\right)(\mathcal{F}_x f)(\sigma)e^{-i\sigma x}d\sigma, \tag{8.3.2}$$

to the respective Cauchy type problems

$$({}^{C}D^{1/2}_{0+,t}u)(x,t)+\frac{\partial u(x,t)}{\partial x}=0, \quad u(x,0)=f(x) \ (x\in\mathbb{R};\ t>0) \tag{8.3.3}$$

and

$$({}^{C}D_{0+,t}^{1/2}u)(x,t) - \frac{\partial u(x,t)}{\partial x} = 0, \quad u(x,0) = f(x) \ (x \in \mathbb{R};\ t > 0). \tag{8.3.4}$$

Finally, we will apply the above results to equations of the forms (8.3.3) and (8.3.4) which arise in the study of diffusion mechanisms with internal degrees of freedom related to the square root of the one-dimensional diffusion equation $u_t - u_{xx} = 0$ [see Vázquez [842] and Vázquez and Vilela Mendez [845]].

If we use the property $\partial_t^{1/2}\partial_t^{1/2}u = \partial_t u$, which holds for "suitable" functions $u(x,t)$, then $u_t - u_{xx} = 0$ can be rewritten in the following form [see Vázquez [842]]:

$$\left(A\partial_t^{1/2} + B\frac{\partial}{\partial x}\right)\psi(x,t) = 0, \quad \psi(x,t) = \begin{pmatrix} u(x,t) \\ v(x,t) \end{pmatrix}, \tag{8.3.5}$$

where A and B are 2×2 matrices satisfying the conditions

$$A^2 = I, \quad B^2 = -I, \quad AB + BA = 0, \tag{8.3.6}$$

and I is the identity operator. One of the possible choices, according to *Pauli's algebra*, is $A = \begin{pmatrix} 0 & 1 \\ 1 & 0 \end{pmatrix}$ and $B = \begin{pmatrix} 0 & 1 \\ -1 & 0 \end{pmatrix}$, and, therefore, the system (8.3.6) is reduced to the following Dirac type equations:

$$\partial_t^{1/2}v(x,t) + \frac{\partial v(x,t)}{\partial x} = 0, \tag{8.3.7}$$

$$\partial_t^{1/2}u(x,t) - \frac{\partial u(x,t)}{\partial x} = 0. \tag{8.3.8}$$

Now, if we substitute $\partial_t^{1/2} = {}^{C}D_{0+,t}^{1/2}$, then the equations (8.3.7) and (8.3.8) coincide with the equations (8.3.3) and (8.3.4), respectively. Hence (8.3.1) and (8.3.2) are solutions of the equations (8.3.7) and (8.3.8) with the initial condition $u(x,0) = f(x)$.

8.4 Applications Describing Carrier Transport in Amorphous Semiconductors with Multiple Trapping

The physical analysis of dispersive transport was pioneered in the 1970s by Scher and Montroll [739] and Pfister and Scher [673]. At that time, the main interest of research was the operation of photocopy machines. In order to create the printing image, carriers photogenerated by a flash of light in a sheet of organic conductor must be separated by drift under a strong applied electrical field. It was observed that the photocurrents departed markedly from that expected of gaussian transport; instead, the photocurrents displayed a power law time dependence both at short and at long times with respect to the typical transit time of a carrier through the film. Such behavior was attributed to the strong disorder of the amorphous organic conductor, and the formalism of the CTRW was developed to account for it [739] and [673], assuming a wide distribution of wait times for an individual carrier to hop from one site to another. Such a distribution was interpreted in terms of a significant number of traps for the carriers in the material [see [744] and [637]]. In the early 1980s, the formalism of multiple trapping (MT) was developed and applied in a variety of transport experiments [see [818], [817], [646], and [596]]. The main assumption of MT is the distinction between a band of transport states, where carriers can move freely, and a distribution of bandgap (trap) states, where carriers remain immobile. Both kinds of states are able to exchange carriers, so that these can be trapped from or released to the transport states [see [818], [817], [646]]. Obviously, the rate of transport is reduced when the fraction of trapped carriers increases. A formal connection was also established between the MT model and the CTRW model [see [744] and [637]]. The CTRW can be formulated rigorously in terms of a fractional diffusion equation (FDE)[344] and [343]:

$$\frac{\partial}{\partial t} f(x,t) = K_\alpha \left(D^{1-\alpha}{}_{0+,t} \frac{\partial^2}{\partial x^2} \right)(x,t) \tag{8.4.1}$$

This FDE has been amply studied in the literature ([344], [347], [140], [575], [586], [336] and [62]). Another possible FDE that generalizes the ordinary Fick's law is based on the replacement of the time derivative in the ordinary diffusion equation with a derivative of non-integer order:

$$D^{\alpha}_{0+,t} f(x,t) = C_\alpha \frac{\partial^2 f(x,t)}{\partial x^2}. \tag{8.4.2}$$

Here, f is a probability distribution and C_α is the fractional diffusion coefficient, which takes the form

$$C_\alpha = K_0 \ \tau_\alpha^{1-\alpha} \tag{8.4.3}$$

with respect to the ordinary diffusion coefficient K_0 (in cm2/s) and a time constant τ_α. The fractional Riemann-Liouville derivative operator of order $\alpha (0 < \alpha \leqq 1)$, ${}_tD^{\alpha}_{0+}$, is defined (see (2.9.3) and (2.9.9)) as follows:

$$\left({}_tD^{\alpha}_{a+} f\right)(x,t) := \frac{d}{dx} \ {}_tI^{1-\alpha}_{a+}, \tag{8.4.4}$$

where

$$\left(I^{\alpha}_{a+,t} f\right)(x,t) = \frac{1}{\Gamma(\alpha)} \int_a^t (t-y)^{\alpha-1} f(x,y) \, dy \tag{8.4.5}$$

is the partial Riemann-Liouville fractional integral operator of order α. The fractional time derivative can be written by

$$\left(D^{\alpha}_{0+,t} f\right)(x,t) = \frac{1}{\Gamma(1-\alpha)} \frac{d}{dt} \int_0^t \frac{f(x,y)}{(t-y)^\alpha} dy, \tag{8.4.6}$$

and its Laplace transform is

$$\left(\mathcal{L}_t D_{0+}^{\alpha} f\right)(s) = s^{\alpha} f(x,s) - f_0 \tag{8.4.7}$$

where [see [389]]

$$f_0 = (I_{0+,t}^{1-\alpha} f)(x,t)\Big|_{t=0} = \lim_{t\to 0} t^{1-\alpha} f(x,t) \tag{8.4.8}$$

The equation (8.4.2) was discussed as a natural generalized diffusion equation, suggesting that it could describe anomalous diffusion processes in [344], [575], [62], [360], [342], [143] and [65]. However, those possible applications have been scarcely developed because of a problem of physical interpretation, owing to the fact that the function $f(x,t)$ is not a normalized function ([575] and [62]) and the initial condition could not be the usual one. Indeed, it was found in [344] that $f(x,t)$ is divergent for $t \to 0$, as described in the context of the Cauchy problem by Kilbas et al. in ([389], Lemma 9). Hilfer related the nonlocal form of the initial condition in [344] to the fractional stationarity concept [see [342] and [65]] which establishes, in addition to the conventional constants, a second class of stationary states that obey a power law time dependence. Ryabov in [720] discussed the solution and initial conditions of (8.4.2) in different dimensions and concluded that "virtual sources" of the diffusion agent are required at the origin, meaning that some mass is injected at the origin. However, this injection is not defined by the boundary conditions of the problem. Thus, Ryabov in [720] concluded that the presence of such "virtual sources" in solutions is physically meaningless and that this contradiction needed a resolution. The first interpretation of (8.4.2) in terms of carrier transport was provided by Bisquert, who showed [86] that the FDE (8.4.2) describes the diffusion of free carriers in MT with an exponential distribution of gap states. This seems contradictory, because earlier papers had demonstrated an equivalence of MT with CTRW [see [744] and [637]] as already mentioned, while (8.4.1) and (8.4.2) are not equivalent. However, the earlier results [744], [637], referred to the total injected charge in MT, which indeed undergoes a CTRW, as confirmed by the analysis of Bisquert in [86]. In contrast, the equation (8.4.2) describes only the diffusion of carriers in transport states [see [86]]. The motion of the total charge is an important magnitude for those where the measured current is a displacement current, corresponding to standard time-of-flight measurements. However, in the 1990s another kind of system became highly visible: nanostructured semiconductors permeated with an electrolyte, particularly in connection with dye-sensitized solar cells [see [645] and [89]]. In later systems, electrical fields were shielded at the nanoscale in the transport layer, so that measured photocurrents corresponded with the arrival of diffusing carriers at an extracting contact [see [160], [87] and [840]]. In this case, the magnitude of physical interest is the diffusive flux of free carriers. Hence both equations (8.4.1) and (8.4.2) are legitimate interpretations of MT transport, but the one that applies depends on the kind of measurement. In further work [88], Bisquert clarified the question of the non-conserved probability density of equation (8.4.2). This is a natural feature, because the equation (8.4.2) refers to the transport of a fraction of the injected carriers, those in transport states, which may decay to trap states. Hence, the density of carriers described by (8.4.2) is not conserved. Indeed, considering the expression of the FDE (8.4.2) in Fourier-Laplace space (with $x \to u$ variables and $t \to s$), then

$$f(u,s) = \frac{f_{0,\alpha}}{\tau_\alpha^{\alpha-1} s^\alpha + K_0 u^2} \tag{8.4.9}$$

where $f_{0,\alpha}$ is a constant. The case $q \to 0$ describes a situation in which the carrier density is homogenous, i.e., there is no diffusion at all, which gives

$$f(u,0) = f_{0,\alpha}\ \tau_\alpha^{\alpha-1}\ u^{-\alpha}. \tag{8.4.10}$$

Therefore, the time decay of the probability in spatially homogeneous conditions is given by

$$f(0,t) = \frac{f_{0,\alpha}\ \tau_\alpha^{-\alpha+1}}{\Gamma(\alpha)}\ t^{\alpha-1}. \tag{8.4.11}$$

The last result, the decay law for (8.4.2), shows that the number of particles decreases with time. The decay law in (8.4.11) can be explained physically by the heuristic arguments for the evolution of free carrier density in MT provided by Tiedje and Rose in [818] and [817], and by Orenstein and Kastner in [646]. In brief [for details see [88] and [90]], photogenerated carriers in a semiconductor with an exponential distribution of traps will be rapidly trapped. Subsequently, the carriers in shallow traps (close to the transport level) will be released and retrapped, so that a time-dependent demarcation level exists which separates the frozen charge below it from the thermalized charge above it, as indicated in Fig. 1. This demarcation level plays the role of a Fermi level, so that in effect the density of free carriers decreases with time. These arguments explain and give a simple interpretation for the temporal decay of the probability of the FDE, and indeed (8.4.11) is confirmed by the experimental results of transient photocurrents in conditions of homogeneous light absorption [see [646] and [289]].

It has been stressed in the literature [see [344] and [389]] that the equation (8.4.2) requires an initial condition for the Green function of the form

$${}_tI_{0+}^{1-\alpha}[f(x,0+)] = f_{0,\alpha}\ \delta(x), \tag{8.4.12}$$

where $\delta(x)$ is the Dirac measure at the origin and $f_{\alpha,0}$ is a constant. Indeed, in the study of the Cauchy problem for (8.4.1), it has been shown rigorously in [389] that the initial condition $f(t)$ satisfies the condition

$$\lim_{x \to 0+} t^{1-\alpha} f(t) < \infty. \tag{8.4.13}$$

It is clear that the temporal dynamics in the FDE imply the divergence of $f(t)$ as $t \to 0$ [see [344] and [389]]. However, it must be recognized that in the MT interpretation discussed above, the decay law cannot be extrapolated to $t = 0$, because the equation (8.4.11) makes no sense without a minimal time for initial thermalization such that $E_d < E_c$ (see Fig. 1).

In summary, the equation (8.4.2) describes a dissipative dynamics in which the total energy (determined by the electrochemical potential of the electrons $E_d(t)$) decreases with time while the number density is conserved. The relaxation in the full energy space is not considered explicitly in (8.4.2), which only contains the resulting evolution of the carriers in extended states, in correspondence with the requirements of the experimental techniques that monitor carrier transport by detecting only the free carriers.

Example 8.1 Let us find the solution to the following model [see [90]]):

$$D^{\alpha}_{0+,t}\; u(x,t) = \sigma \frac{\partial^2 \mu(x,t)}{\partial x^2} \quad (t>0;\ 0<x<l) \tag{8.4.14}$$

$$\begin{cases} u_x(0,t) = 0 \\ u(l,t) = K \end{cases} \tag{8.4.15}$$

$$\lim_{t\to 0} t^{1-\alpha} u(x,t) = h_{\epsilon}(x) - \frac{1}{2l} + \frac{Kl}{\Gamma(\alpha+1)}, \tag{8.4.16}$$

where $\epsilon << l$, y

$$h_{\epsilon}(x) = \begin{cases} 1/\epsilon;\ 0<x<\epsilon \\ 0;\ x>\epsilon \end{cases} \tag{8.4.17}$$

To solve it we will not use the usual and powerful technique of integral transforms; instead, we will reduce the problem (8.4.14)-(8.4.15)-(8.4.16) to one with homogeneous boundary conditions via the usual change

$$u(x,t) = \frac{Kx^2}{l^2} + V(x,t). \tag{8.4.18}$$

Our problem thus becomes the following:

$${}_tD^{\alpha}_{0+} V(x,t) = \sigma V_{xx} + \frac{2\sigma K}{l^2} \tag{8.4.19}$$

$$V_x(0,t) = 0;\quad V(0,l) = 0 \tag{8.4.20}$$

$$\lim_{t\to 0} t^{1-\alpha} V(x,t) = h_C(x) - \frac{1}{2l} + \frac{Kl}{\Gamma(\alpha+1)}. \tag{8.4.21}$$

Now we look for a solution V of (8.4.19) in the form

$$V(x,t) = \sum_{n=1}^{\infty} T_n(t)\ cos\left(\frac{n\tau}{2l}x\right) + T_0(t) \tag{8.4.22}$$

which satisfies (8.4.20) if $T_0(l) = 0$.

Substituting (??) into (8.4.19) we have

$$\left[T_0^{(\alpha}(t) - \frac{2\sigma K}{l^2}\right] + \sum_{n=1}^{\infty}\left[T_n^{\alpha}(t) + \sigma\left(\frac{n\pi}{2l}T_n(t)\right)^2 cos\left(\frac{n\pi}{2l}x\right) = 0 \tag{8.4.23}$$

and, therefore, since $T_0^{(\alpha} = \frac{2\sigma K}{l^2}$, it holds that

$$T_0(t) = \frac{Kt^{\alpha}}{\Gamma(\alpha+1)} + l\ t^{\alpha-1} \tag{8.4.24}$$

Since $T_0(l) = 0$, then

$$T_0(t) = \frac{Kt^\alpha}{\Gamma(\alpha+1)} - \frac{Kl}{\Gamma(\alpha+1)}\, t^{\alpha-1} \tag{8.4.25}$$

Additionally, (8.4.23) requires that the following relation must be satisfied:

$$T_n^{(\alpha}(t) + \sigma\Big(\frac{n\pi}{2l}\Big)^2 T_n(t) = 0, \tag{8.4.26}$$

and imposing the initial condition (8.4.21) it holds that

$$h_\epsilon(x) = \frac{1}{2l} + \sum_{n=1}^{\infty} \lim_{t\to 0}[t^{1-\alpha}T_n(t)]\; cos\Big(\frac{n\pi}{2l}\, x\Big) \tag{8.4.27}$$

Therefore, any function $T_n(t)$ is established as a solution of the Cauchy type problem

$$T_n^{(\alpha}(t) + \sigma\Big(\frac{n\pi}{2l}\Big)^2 T_n(t) = 0 \tag{8.4.28}$$

$$\Big(I_{0+}^{1-\alpha}T_n\Big)(0) = \lim_{t\to 0} t^{1-\alpha}T_n(t) = \frac{1}{l}, \tag{8.4.29}$$

By Example 4.2, the unique solution of this problem is given by

$$T_n(t) = \frac{\Gamma(\alpha)}{l}\; l_\alpha^{-\sigma}\Big(\frac{n\pi}{2l}\Big)^2 t; \quad (n \in \mathbb{N}) \tag{8.4.30}$$

Hence,in accordance with (8.4.18), (8.4.22), (8.4.25) and (??), the solution to of the initial problem (8.4.14)-(8.4.16) has the form:

$$u(x,t) = \frac{Kx^2}{l^2} + \frac{K[t^\alpha - l\; t^{\alpha-1}]}{\Gamma(\alpha+1)} + \frac{\Gamma(\alpha)}{l}\sum_{n=1}^{\infty} l_\alpha^{-\sigma(\frac{n\pi}{2l})^2 t}\; cos\Big(\frac{n\pi}{2l}\, x\Big) \tag{8.4.31}$$

Bibliography

[1] ABRAMOWITZ, M. and STEGUN, I. A., *Handbook of Mathematical Functions with Formulas, Graphics and Mathematical Tables*, Dover, New York, 1972.

[2] ADDISON, P. S., QU, B., NISBET, A., and PENDER, G., A non-Fickian particle-tracking diffusion model based on fractional Brownian motion, *Int. J. Numer. Methods Fluids*, **25**(12), (1997) 1373–1384.

[3] ADOMIAN, G., A review of the decomposition method applied mathematics, *J. Math. Anal. Appl.*, **135**, (1988), 501–544.

[4] ADOMIAN, G., *Solving Frontier Problems of Physics: The Decomposition Method* , Kluwer Acad. Publ., Boston, 1994.

[5] AGARWAL, R. P., A propos d'une note de M. Pierre Humbert, *C. R. Acad. Sci. Paris*, **236**(21), (1953) 2031–2032.

[6] AGARWAL, R. and O'REGAN, D.(Eds.), *Integral and Integro-Differential Equations*, in *Nonlinear Analysis and Applications*, Gordon and Breach, 1999.

[7] AGRAWAL, O. P., A general solution for the fourth-order fractional diffusion-wave equation, *Fract. Calc. Appl. Anal.*, **3**(1), (2000) 1–12.

[8] AGRAWAL, O. P., A new Lagrangian and a new Lagrangian equation of motion for fractionally damped systems, *Trans. ASME J. Appl. Mech.*, **68**(2), (2001) 339–341.

[9] AGRAWAL, O. P., Solution for a fractional diffusion-wave equation defined in a bounded domain, in: TENREIRO-MACHADO, J. A. (Ed.), *Special Issue of Fractional Order Calculus and Its Applications*, Nonlinear Dynamics, **29**(1-4), 2002, 145–155.

[10] AGRAWAL, O. P., Response of a diffusion-wave system subjected to deterministic and stochastic fields, *Z. Angew. Math. Mech.*, **83**(4), (2003) 265–274.

[11] AGRAWAL, O. P., A general formulation and solution scheme for fractional optimal control problems, in: DUARTE, M.and TENREIRO-MACHADO, J. A.(Eds.), *Special Issue of Fractional Order Derivatives and their Applications*, Nonlinear Dynamics, **38**(1-2), 2004, 323–337.

[12] AL-ABEDEEN, A. Z., Existence theorem on differential equation of generalized order, *Al-Rafidain J. Sci. Mosul University Iraq*, **1**, (1976) 95–104.

[13] AL-ABEDEEN, A. Z. and ARORA, H. L., A global existence and uniqueness theorem for ordinary differential equation of generalized order, *Canad. Math. Bull.*, **21**(3), (1978) 267–271.

[14] AL-BASSAM, M. A., Concerning Holmgren-Riesz transform equations of Gauss-Riemann type, *Rend. del Circolo Mat. di Palermo, Serie II*, **9**, (1962) 1–20.

[15] AL-BASSAM, M. A., On functional representation by a certain type of generalized power series, *Ann. della Scoula Norm. Sup. di Pisa, Ser. III*, **16**(4), (1962) 351–366.

[16] AL-BASSAM, M. A., Some properties of Holmgren-Riesz transform in two dimensions, *Ann. della Scoula Norm. Sup. di Pisa, Ser. III*, **16**(1), (1962) 75–90.

[17] AL-BASSAM, M. A., Some existence theorems on differential equations of generalized order, *J. Reine Angew. Math.*, **218**(1), (1965) 70–78.

[18] AL-BASSAM, M. A., Existence of series solution of a type of differential equation of generalized order, *Bull. College Sci. Baghdad*, **9**(1), (1966) 175–180.

[19] AL-BASSAM, M. A., On the existence of series solution of differential equations of generalized order, *Portugal. Math.*, **29**(1), (1970) 5–11.

[20] AL-BASSAM, M. A., Fractional calculus and its applications, in: ROSS, B. (Ed.), *International Conference (New Haven, 1974)*, Springer-Verlag, Berlin, 1975, 91–105.

[21] AL-BASSAM, M. A., On fractional calculus and its applications to the theory of ordinary differential equations of generalized order, in: *Nonlinear Analysis and Applications*, vol. 80 of Lect. Notes Pure Appl. Math., Dekker, New York, 1981, pp. 305–331.

[22] AL-BASSAM, M. A., On fractional analysis and its applications, in: MANOCHA, H. L. (Ed.), *Modern Analysis and its Applications*, Prentice-Hall, New Delhi, 1983, 269–307.

[23] AL-BASSAM, M. A., *On Generalized Power Series and Generalized Operational Calculus and its Applications*, World Sci. Publishing, Singapore, 1987, 51–88.

[24] AL-BASSAM, M. A. and LUCHKO, Y. F., On generalized fractional calculus and its application to the solution of integro-differential equations, *J. Fract. Calc.*, **7**, (1995) 69–88.

[25] AL-SAQABI, B. N., Solution of a class of differintegral equations by means of Riemann-Liouville operator, *J. Fract. Calc.*, **8**, (1995) 95–102.

[26] AL-SAQABI, B. N. and KIRYAKOVA, V. S., Explicit solutions of fractional integral and differential equations involving Erdélyi-Kober operators, *Appl. Math. Comp.*, **95**(1), (1998) 1–13.

[27] AL-SAQABI, B. N. and TUAN, V. K., Solution of a fractional differintegral equation, *Integral Transform. Spec. Funct.*, **4**(4), (1996) 321–326.

[28] AL-SHAMMERY, A. H., KALLA, S. L., and KHAJAH, H. G., A fractional generalization of the free electron laser equation, *Fract. Calc. Appl. Anal.*, **2**(4), (1999) 501–508.

[29] ALEROEV, T. S., The Sturm-Liouville Problem for a second-order differential equation with fractional derivatives in the lower terms, *Differentsial'nye Uravneniya*, **18**(2), (1982) 341–342.

[30] ALEROEV, T. S., Spectral analysis of a class of nonselfadjoint operators, *Differentsial'nye Uravneniya*, **20**(1), (1984) 171–172.

[31] ALFIMOV, G. L., USERO, D., and VÁZQUEZ, L., On complex singularities of solutions of the equation $HUx - U + U^p = 0$, *Physica A*, **33**(38), (2000) 6707–6720.

[32] ALI, I., KIRYAKOVA, V., and KALLA, S. L., Solutions of fractional multi-order integral and differential equations using a Poisson-type transform, *J. Math. Anal. Appl.*, **269**(1), (2002) 172–199.

[33] ALLEGRINI, P., BUIATTI, M., GRINOLINI, P., and WEST, B. J., Fractional Brownian motion as a nonstationary process: Analternative paradigm for DNA sequences, *Phys. Rev. E*, **57**(4), (1998) 558–567.

[34] ALONSO, J., On differential equations of fractional order, *University Cincinnati, Diss. Abstrs.*, **26**(4), (1964) 947–1017.

[35] ANASTASIO, T. J., The fractional-order dynamics of Brainstem Vestibulo-Oculomotor neurons, *Biol. Cybernet*, **72**(1), (1994), 69–79.

[36] ANNABY, M. and BUTZER, P. L., Mellin-type differential equations and associated sampling expansions, in: *Proceedings of International Conference on Fourier Analysis and Applications* (Kuwait, 1998), vol. 21, New Delhi, 2000, 1–24.

[37] ARORA, H. L. and ALSHAMANI, J. G., Stability of differential equation of noninteger order through fixed point in the large, *Indian J. Pure Appl. Math.*, **11**(3), (1980) 307–313.

[38] ATANACKOVIC, T. M. and STANKOVIC, B., Dynamics of a viscoelastic rod of fractional derivative type, *Z. Angew. Math. Mech.*, **82**(6), (2002) 377–386.

[39] ATANACKOVIC, T. M. and STANKOVIC, B., Dynamics of a viscoelastic rod of fractional derivative type, *Z. Angew. Math. Mech.*, **82**(6), (2004) 377–386.

[40] ATANACKOVIC, T. M. and STANKOVIC, B., On a system of differential equations with fractional derivative arising in Rod theory, *J. Phys. A: Math Gen.*, **37**(4), (2004) 1241–1250.

[41] AYDINER, E., Anomalous rotational relaxation: A fractional Fokker-Planck equation approach, *Phys. Rev. E*, **71**(4), (2005) 6103- 6107.

[42] BABAKHANI, A. and DAFTADAR-GEJJI, V., Existence of positive solutions of nonlinear fractional differential equations, *J. Math. Anal. Appl.*, **252**(2), (2000) 804–812.

[43] BABAKHANI, A. and DAFTARDAR-GEJJI, V., Existence of positive solutions of nonlinear fractional differential equations, *J. Math. Anal. Appl.*, **278**(2), (2003) 434–442.

[44] BABENKO, Y. I., *Heat and Mass Transfer. The Method of Calculation for the Heat and Diffusion Flows*, Himaya, Moscow, 1986.

[45] BAGLEY, R. L., *Applications of Generalized Derivatives to Viscoelasticity*, Ph.D. thesis, Air Force Institute of Technology, 1979.

[46] BAGLEY, R. L., The initial value problem for fractional order differential equations with constant coefficients, *Air Force Instit. of Tech.*, **AFIT-TR-88-1**.

[47] BAGLEY, R.L., Power law and fractional calculus model of viscoelasticity, *AIAA J.*, **27**, (1989) 1414–1417.

[48] BAGLEY, R. L., On the fractional order initial value problem and its engineering applications, in: NISHIMOTO, K. (Ed.), *International Conf. on Fractional Calculus and its Applications*, College Engrg. Nihon University, Japan, Tokyo, 1990, 12–20.

[49] BAGLEY, R. L. and CALIPO, R. A., Fractional order state equations for the control of viscoelasticity damped structures, *J. Guid. Control. Dynam.*, **14**(2), (1991) 304–311.

[50] BAGLEY, R. L. and TORVIK, P. J., A generalized derivative model an elastomer damper, *Shock Vibr. Bull.*, **49**(2), (1979) 135–143.

[51] BAGLEY, R. L. and TORVIK, P. J., A theorical basis for the application of fractional calculus to viscoelasticity, *J. Rheol.*, **27**(3), (1983) 201–210.

[52] BAGLEY, R. L. and TORVIK, P. J., On the appearence of the fractional derivative in the behavior of real materials, *J. Appl. Mech.*, **51**(4), (1984) 294–298.

[53] BAGLEY, R. L. and TORVIK, P. J., Fractional calculus in the transient analysis of viscoelastically damped strutures, *AIAA J.*, **23**(6), (1985) 918–925.

[54] BAGLEY, R. L. and TORVIK, P. J., On the fractional calculus model of viscoelastic behavior, *J. Rheol.*, **30**(1), (1986) 133–155.

[55] BAGLEY, R. L. and TORVIK, P. J., On the existence of the order domain and the solution of distributed order equations., *Int. J. Appl. Math.*, **2**(7), (2000) 865–882.

[56] BAGLEY, R. L. and TORVIK, P. J., On the existence of the order domain and the solution of distributed order equations. II, *Int. J. Appl. Math.*, **2**(8), (2000) 965–987.

[57] BAI, C. Z. and FANG, J. X., The existence of a positive solution for a singular coupled system of nonlinear fractional differential equations, *Appl. Math. Comp.*, **150**(3), (2004) 611–621.

[58] BAILLIE, R. T. and KING, M. L., Fractional differencing and long memory processes, *J. Econometrics*, **73**(1), (1996) 1–3.

[59] BAKUNIN, O. G., Quasi-diffusion and correlations in models of anisotropic transport, *Physica A: Stat. Mech. Appl.*, **337**(1-2), (2004) 27–35.

[60] BALAKRISHNAN, A. V., Fractional powers of closed operators and the semigroups generated by them, *Pacific J. Math.*, **10**, (1960), 419–437.

[61] BALESCU, R., Anomalous transport in turbulent plasmas and continuous time random walks, *Phys. Rev. E*, **51**(5), (1995) 807–822.

[62] BARKAI, E., Fractional Fokker-Planck equation, solution and application, *Phys. Rev. E*, **63**(4), (2001) 6111.

[63] BARKAI, E., CTRW pathways to the fractional diffusion equation, *J. Chem. Phys.*, **284**(1-2), (2002) 13–27.

[64] BARKAI, E. and CHENG, Y. C., Aging continuous time random walks, *J. Chem. Phys.*, **118**(14), (2003) 6167–6178.

[65] BARKAI, E., METZLER, R., and KLAFTER, J., From continuous time random walk to the fractional Fokker-Planck equation, *Phys. Rev. E*, **61**(1), (2000) 132–138.

[66] BARKAI, E. and SILBEY, R. J., Fractional Kramers equation, *J. Phys. Chem. B*, **104**(16), (2000) 3866–3874.

[67] BARLOW, M. T., Diffusions on fractals, in: *Lectures on Probability Theory and Statistics*, vol. 1690 of Lect. Notes in Math., Springer-Verlag, Berlin, 1995, 1–121.

[68] BARRETT, J. H., Differential equations of non-integer order, *Canad. J. Math.*, **6**(4), (1954) 529–541.

[69] BARSEGOV, V. and MUKAMEL, S., Multipoint fluorescence quenching-time statistics for single molecules with anomalous diffusion, *J. Phys. Chem. A*, **108**(1), (2004) 15–24.

[70] BASSET, A. B., *A Treatise on Hydrodynamics*, vol. 2, Cambridge University Press., England, 1888.

[71] BASSET, A. B., On the descent of a sphere in a viscous liquid, *Quart. J. Math.*, **41**, (1910) 369–381.

[72] BHATIA, D. and KUMAR, P., On multiobjective fractional control problem,*Asia-Pacific J. Oper. Res.*, **13**(2), 1996, 115-132.

[73] BHATT, S. K., An existence theorem for a fractional control problem ,*J. Optimization Theory Appl.*, **11**, (1973) 378–385.

[74] BAUMANN, G., SUDLAND, N., and NONNENMACHER, T. F., Anomalous relaxation and diffusion processes in complex systems, *Trans. Th. Stat. Phys*, **29**(1-2), (2000) 157–171.

[75] BAZHLEKOVA, E., The abstract Cauchy problem for the fractional evolution equation, *Fract. Calc. Appl. Anal.*, **1**(3), (1998) 255–270.

[76] BAZHLEKOVA, E., Perturbation properties for abstract evaluation equations of fractional order, *Fract. Calc. Appl. Anal.*, **2**(4), (1999) 359–366.

[77] BEN-AVRAHAM, D. and HAVLIN, S., *Diffusion and Reactions in Fractals and Disordered Systems.*, Cambridge University Press, Cambridge, 2000.

[78] BENSON, D. A., *The Fractional Advection-Dispersion Equation: Development and Applications*, Ph.D. thesis, University of Nevada, Nevada, 1998.

[79] BERENS, H. and WESTPHAL, U., A Cauchy problem for a generalized wave equation, *Acta Sci. Math. (Szeged)*, **29**(1-2), (1968) 93–106.

[80] BERKOWITS, B. and SCHER, H., Theory of anomalous chemical transport in ran dom fracture networks, *Phys. Rev. E*, **57**(5), (1998) 858–869.

[81] BERKOWITZ, B., KLAFTER, J., METZLER, R., and SCHER, H., Physical pictures of transport in heterogeneous media: Advection-dispersion, random-walk, and fractional derivative formulations, *Wat. Res. Res.*, **38**(10), (2002) 1191–11203.

[82] BEYER, H. and KEMPFLE, S., Definition of physically consistent damping laws with fractional derivatives, *Z. Angew. Math. Mech.*, **75**(8), (1995) 623–635.

[83] BIACINO, L. and MISERENDINO, D., Perturbazioni di operatori elliptici mediante operatori contenenti derivate di ordine frazionario, *Matematiche (Catania)*,**34**(1-2), (1979) 166–187.

[84] BIACINO, L. and MISERENDINO, D., Derivatives of fractional order for functions in W^r (Italian), *Boll. University Mat. Ital. C*, **6**(1), (1982) 235–278.

[85] BIRKOFF, G. and ROTA, G., *Ordinary Differential Equations*, Wiley and Sons, New York, 1989.

[86] BISQUERT, J., Fractional diffusion in the multiple-trapping regime and revision of the equivalence with the continuous time random walk, *Phys. Rev. Lett.*, **91**(1), (2003) 1602.

[87] BISQUERT, J., Chemical diffusion coefficient in nanostructured semiconductor electrodes and dye-sensitized solar cells, *J. Phys. Chem. B*, **108**, (2004) 2323–2332.

[88] BISQUERT, J., Interpretation of fractional difussion equation with nonconserved probability density in terms of experimental systems with trapping or recombination, *Phys. Rev. Lett.*, **91**(1), (2005) 10602–10606.

[89] BISQUERT, J., CAHEN, D., RÜHLE, S., HODES, G., and ZABAN, A., Physical chemical principles of photovoltaic conversion with nanoparticulate, mesoporous dye-sensitized solar cells., *J. Phys. Chem. B*, **108**, (2004) 8106–8118.

[90] BISQUERT, J. and TRUJILLO, J. J., Transient photocurrents in anonmalous diffusion (Preprint), *Depto. Análisis Matemático* (Universidad de La Laguna), (2005).

[91] BLANK, L., Numerical treatment of differential equations of fractional order, *Nonlinear World*, **4**(4), (1997) 473–491.

[92] BLUMEN, A., GURTOVENKO, A. A., and JESPERSEN, S., Anomalous dynamics of model polymer systems, *J. Lumin.*, **94-95**, (2001) 437–440.

[93] BONILLA, B., KILBAS, A. A., RIVERO, M., RODRÍGUEZ-GERMÁ, L., and TRUJILLO, J. J., Modified Bessel-type function and solution of differential and integral equations, *Indian J. Pure Appl. Math.*, **31**(1), (2000) 93–109.

[94] BONILLA, B., KILBAS, A. A., and TRUJILLO, J. J., Fractional order continuity and some properties about integrability and differentiability of real functions, *J. Math. Anal. Appl.*, **231**(1), (1999) 205–212.

[95] BONILLA, B., KILBAS, A. A., and TRUJILLO, J. J., Systems of nonlinear fractional differential equations in the space of summable functions, *Tr. Inst. Mat. Minsk*, **6**, (2000) 38–46.

[96] BONILLA, B., KILBAS, A. A., and TRUJILLO, J. J., *Cálculo Fraccionario y Ecuaciones Diferenciales Fraccionarias*, Uned, Madrid, 2003.

[97] BONILLA, B., RIVERO, M., and TRUJILLO, J. J., Theory of secuencial linear differential equations. Applications (Preprint), *Depto. Análisis Matemático* (Universidad de La Laguna), (2005).

[98] BONILLA, B., RIVERO, M., and TRUJILLO, J. J., Theory of systems of linear differential equations of fractional order. Applications (Preprint), *Depto. Análisis Matemático* (Universidad de La Laguna), (2005).

[99] BOUCHAUD, J. P. and GEORGES, A., Anomalous diffusion in disordered media: statistical, mechanics, models and physical applications, *Phys. Rep.*, **195**(4-5), (1999) 127–293.

[100] BOUCHAUD, J. P. and POTTERS, M., *Theory of Financial Risks*, Cambridge University Press, Cambridge, 2000.

[101] BOYADJIEV, L. and DOBNER, H. J., Analytical and numerical treatment of multi-dimensional fractional FEL equation, *Appl. Anal.*, **70**(1-2), (1998) 1–18.

[102] BOYADJIEV, L. and DOBNER, H. J., On the solution of a fractional free electron laser equation, *Fract. Calc. Appl. Anal.*, **1**(4), (1998) 385–400.

[103] BOYADJIEV, L. and DOBNER, H. J., Fractional free electron laser equations, *Integral Transform. Spec. Funct.*, **11**(2), (2001) 113–136.

[104] BOYADJIEV, L., DOBNER, H. J., and KALLA, S. M., A fractional integro-differential equation of Volterra type, *Math. Comput. Modelling*, **10**(28), (1998) 103–113.

[105] BOYADJIEV, L. and KALLA, S. L., Series representations of analytic functions and applications, *Fract. Calc. Appl. Anal.*, **4**(3), (2001) 379–408.

[106] BOYADJIEV, L., KALLA, S. L., and KHAJAH, H. G., Analytical and numerical treatment of a fractional integro-differentiation equation of Volterra-type, *Math. Comput. Modelling*, **25**(12), (1997) 1–9.

[107] BRYCHKOV, Y. A., GLAESKE, H.-J., PRUDNIKOV, A. P., and TUAN, V. K., *Multidimensional Integral Transformations*, Gordon and Breach, Philadelphia, 1992.

[108] BRYCHKOV, Y. A. and PRUDNIKOV, A. P., *Integral Transforms of Generalized Functions*, Gordon and Breach, New York, 1989.

[109] Buckwar, E. and Luchko, Y. F., Invariance of a partial differential equation of fractional order under the Lie group of scaling transformations, *J. Math. Anal. Appl.*, **227**(1), (1998) 81–97.

[110] Bultheel, A. and Martínez-Sulbaran, H., Computation of the Fractional Fourier Transform, *Appl. Comp. Harm. Anal.*, **16** (3), (2004) 182–202.

[111] Bultheel, A. and Martínez-Sulbaran, H.,Recent developments in the theory of the fractional Fourier transforms and linear canonical transforms, *Bull. Belgian Math. Soc. Simon Stevin* (to appear), 2006.

[112] Butzer, P. L., Kilbas, A. A., and Trujillo, J. J., Compositions of Hadamard-type fractional integration operators and the semigroup property, *J. Math. Anal. Appl.*, **269**(2), (2002) 387–400.

[113] Butzer, P. L., Kilbas, A. A., and Trujillo, J. J., Fractional calculus in the Mellin setting and Hadamard-type fractional integrals, *J. Math. Anal. Appl.*, **269**(1), (2002) 1–27.

[114] Butzer, P. L., Kilbas, A. A., and Trujillo, J. J., Mellin transform and integration by parts for Hadamard-type fractional integrals, *J. Math. Anal. Appl.*, **270**(1), (2002) 1–15.

[115] Butzer, P. L., Kilbas, A. A., and Trujillo, J. J., Generalized Stirling functions of second kind and representations of fractional order differences via derivatives, *J. Diff. Equat. Appl.*, **9**(5), (2003) 503–533.

[116] Butzer, P. L. and Westphal, U., An access to fractional differentiation via fractional difference quotients, in: *Fractional Calculus and Its Applications*, vol. 457 of Lect. Notes in Math., Springer-Verlag, Berlin, 1974, 116–145.

[117] Butzer, P. L. and Westphal, U., *An introduction to fractional calculus.*, World Sci. Publishing, 2000, 1–85.

[118] Campos, L. M., On the solution of some simple fractional differential equations, *Int. J. Math. Math. Sci.*, **13**(3), (1990) 481–496.

[119] Candan, C., Kutay, M. A., and Ozaktas, H. M., The discrete fractional Fourier transform *IEEE Trans. Sig. Proc.*, **48**, 2000, 1329–1337

[120] Caputo, M., Linear models of dissipation whose Q is almost frequency independent, *Anna. di Geofis.*, **19**, (1966) 3–393.

[121] Caputo, M., Linear models of dissipation whose Q is almost frequency independent II, *Geophys. J. Royal Astronom. Soc.*, **13**, (1967) 529–539.

[122] Caputo, M., *Elasticita e Dissipazione*, Zanichelli, Bologna, 1969.

[123] Caputo, M., Vibrations of an infinite viscoelastic layer with a dissipative memory, *J. Acoust. Soc. Amer.*, **56**, (1974) 897–904.

[124] Caputo, M., A model for the fatigue in elastic materials with frequency independent Q, *J. Acoust. Soc. Amer.*, **66**, (1979) 176–179.

[125] Caputo, M., Generalized rheology and geophysical consequences, *Tectonophysics*, **116**, (1985) 163–172.

[126] Caputo, M., Modern rheology and electric induction: multivalued index of refraction, splitting of eigenvalues and fatigues, *Anna. di Geofis.*, **39**(5), (1996) 941–966.

[127] Caputo, M., Mean fractional-order-derivatives differential equations and filters, *Ann. University Ferrara, Sez. VII*, **41**, (1997) 73–84.

[128] CAPUTO, M., Distributed order differential equations modelling dielectric induction and diffusion, *Fract. Calc. Appl. Anal.*, **4**(4), (2001) 421–442.

[129] CAPUTO, M. and MAINARDI, F., Linear models of dissipation in anelastic solids, *Riv. Nuovo Cimento*, **1**(2), (1971) 161–198.

[130] CAPUTO, M. and MAINARDI, F., A new dissipation model based on memory mechanism, *Pure Appl. Geophys.*, **91**(8), (1971) 134–147.

[131] CARPINTERI, A., CHIAIA, B., and CORNETTI, P., A disordered microstructure material model based on fractal geometry and fractional calculus, *Z. Angewandte Math. Mech.*, **84**(2), (2004) 128–135.

[132] CARPINTERI, A. and MAINARDI, F. (Eds.), *Fractals and Fractional Calculus in Continuum Mechanics*, vol. 378 of CISM Courses and Lectures, Springer-Verlag, Berlin, 1997.

[133] CARRACEDO, C. M. and ALIX, M. S., *The Theory of Fractional Powers of Operators*, vol. 187 of North-Holland Mathematical Studies, Elsevier Science Publishers, Amsterdam, London and New York, 2001.

[134] CHECHKIN, A., GONCHAR, V., KLAFTER, J., METZLER, R., and TANATAROV, L., Stationary states of non-linear oscillators driven by Levy noise, *J. Chem. Phys.*, **284**(1-2), (2002) 233–251.

[135] CHECHKIN, A. V., GONCHAR, V. Y., and SZYDLOWSKI, M., Fractional kinetics for relaxation and superdiffusion in a magnetic field, *Phys. Plasma*, **9**(1), (2002) 78–88.

[136] CHECHKIN, A. V., KLAFTER, J., and SOKOLOV, I. M., Fractional Fokker-Planck equation for ultraslow kinetics, *Europhys. Lett.*, **63**(3), (2003) 326–332.

[137] CHEN, Y. Q., VINAGRE, B. M., and PODLUBNY, I., Fractional order disturbance observer for Robust vibration suppression, in: DUARTE, M.and TENREIRO-MACHADO, J. A.(Eds.), *Special Issue of Fractional Order Derivatives and their Applications*, Nonlinear Dynamics, **38**(1-2), 2004, 355–367.

[138] CHOW, C. W. and LIU, K. L., Fokker-Planck equation and subdiffusive fractional Fokker-Planck equation of bistable systems with sinks, *Physica A: Stat. Mech. Appl.*, **341**, (2004) 87–106.

[139] CHRISTENSEN, R. M., *Theory of Viscoelasticity*, Academic Press, New York, 1982.

[140] COMPTE, A., Stochastic foundations of fractional dynamics, *Phys. Rev. E*, **53**(4), (1996) 191–193.

[141] COMPTE, A., Continuous time random walks on moving fluids, *Phys. Rev. E*, **55**(6), (1997) 6821–6831.

[142] COMPTE, A. and JOU, D., Non-equilibrium thermodynamics and anomalous diffusion, *J. Phys. A: Math. Gen.*, **29**(15), (1996) 4321–4329.

[143] COMPTE, A. and METZLER, R., The generalized Cattaneo equation for the description of anomalous transport processes, *J. Phys. A: Math. Gen.*, **30**(21), (1997) 7277–7289.

[144] COMPTE, A., METZLER, R., and CAMACHO, J., Biased continuous time random walks between parallel plates, *Phys. Rev. E*, **56**(2), (1997) 1445–1454.

[145] CONLAN, J., Hyperbolic differential equations of generalized order, *Appl. Anal.*, **14**(3), (1983) 167–177.

[146] COPSON, E. T., Some applications of Marcel Riesz's integral of fractional order, *Proc. Roy. Soc. Edinburg*, **A61**, (1943) 260–272.

[147] CUESTA, E., , *Métodos lineales multipaso para ecuaciones integro-diferenciales de orden fraccionario en espacios de Banach* , Ph.D. thesis, Universidad de Valladolid, 2001.

[148] CUESTA, E. and PALENCIA, C., A numerical method for an integro-differential equation with memory in Banach spaces: Qualitative properties , *SIAM J. Numer. Anal.*, **41**, 2003, 1232–1241.

[149] CUESTA, E. and PALENCIA, C., Propiedades cualitativas de una discretización de orden 1 para ecuaciones integro-diferenciales de orden fraccionario (Spanish), in: JORNET, J. M., LÓPEZ, J. M., OLIVÉ, C., RAMÍREZ, R. (Eds.) *Proceeding of XVIII CEDYA and VIII CMA*, Univ. Rovira i Virgili, Tarragona, 2003, 40–52

[150] CUESTA, E. and PALENCIA, C., A fractional trapezoidal rule for integro-differential equations of fractional order in Banach spaces, *Appl. Numer. Math.*, **45** (2-3), 2003, 139–159.

[151] CUNHA, M. D., KONOTOP, V. V., and VÁZQUEZ, L., Small-amplitude solitons in a nonlocal Sine-Gordon model, *Phys. Lett. A*, **221**(5), (1996) 317–322.

[152] DAFTARDAR-GEJJI, V., Positive solution of as system of non-autonomous fractional differential equations, *J. Math. Anal. Appl.*, **302**(1), (2005) 56–64.

[153] DAFTARDAR-GEJJI, V. and BABAKHANI, A., Analysis of a system of fractional differential equations, *J. Math. Anal. Appl.*, **293**(2), (2004) 511–522.

[154] DAFTARDAR-GEJJI, V. and JAFARI, A., Adomian descomposition: A tool for solving a system of fractional differential equations, *J. Math. Anal. Appl.*, **301**(2), (2005) 508–518.

[155] DATTOLI, G., CESARANO, C., RICCI, P. E., and VÁZQUEZ, L., Special polynomials and fractional calculus, *Math. Comput. Modelling*, **37**(7-8), (2003) 729–733.

[156] DATTOLI, G., GIANESSI, L., MEZI, L., TOCCI, D., and COLAI, R., FEL time-evolution operator, *Nucl. Instr. Methods*, **304**(1-3), (1991) 541–544.

[157] DATTOLI, G., LORENZUTTA, S., MAINO, G., and TORRE, A., Analytic treatment of the high-gain free electron laser equation, *Radiat. Phys. Chem.*, **48**(1), (1996) 29–40.

[158] DAVIS, H. T., The applications of fractional operators to functional equations, *Amer. J. Math.*, **49**(1), (1927) 123–142.

[159] DE ARRIETA, U. X. and KALLA, S. L., Some results on the fractional diffusion equation, *Rev. Técn. Fac. Ingr. University Zulia*, **16**(2), (1993) 177–184.

[160] DE JONGH, P. E. and VANMAEKELBERGH, D., Trap-limited electronic transport in assemblies of nanometer-size *TiO*2 particles, *Phys. Rev. Lett.*, **77**(3), (1996) 427–430.

[161] DEBNATH, L., Fractional integral and fractional differential equations in fluid mechanics, *Fract. Calc. Appl. Anal.*, **6**(2), (2003) 119–156.

[162] DEBNATH, L., Recents applications of fractional calculus to Science and Engineering, *Int. J. Math. Appl. Sci.*, **54**, (2003) 3413–3442.

[163] DELBOSCO, D., Fractional calculus and function spaces, *J. Fract. Calc.*, **6**, (1994) 45–53.

[164] DELBOSCO, D. and RODINO, L., Existence and uniqueness for a nonlinear fractional differential equation, *J. Math. Anal. Appl.*, **204**(2), (1996) 609–625.

[165] DEL CASTILLO-NEGRETE, D., Chaotic transport in zonal flows in analogous geophysical and plasma systems, *Phys. Plasma*, **7**(5), (2000) 1702–1711.

[166] DEL CASTILLO-NEGRETE, D.,CARRERAS, B.A., and LYNCH, V. E., Front dynamics in reaction-diffusion systems with Levy flights: A fractional diffusion approach *Phys. Rev. Lett.*, **91**(1), 2003, 8302–8309.

[167] DIDENKO, A. V., On the solution of systems of some convolution type equations in the space of generalized functions, *VINITI (Minsk, Bel. St. University)*, **11.09.84**(6170), (1984) 1–10.

[168] DIDENKO, A. V., On the solution of the composite equation of Abel type in generalized functions, *VINITI (Odessa, Dep. in Ukrain.)*, **15.08.84**(1464), (1984) 1–11.

[169] DIDENKO, A. V., On the solution of a class of the integro-differential equations of fractional order, *VINITI (Odessa, Dep. in Ukrain.)*, **01.04.85**(662), (1985) 1–13.

[170] DIETHELM, K., An algorithm for the numerical solution of differential equations of fractional order, *Electron. Trans. Numer. Anal.*, **5**, (1997) 1–6.

[171] DIETHELM, K., Generalized compound quadrature formulae for finite-part integrals, *IMA J. Numer. Anal.*, **17**(3), (1997) 479–493.

[172] DIETHELM, K., Estimation of quadrature errors in terms of Caputo-type fractional derivatives, *Fract. Calc. Appl. Anal.*, **2**, 1999, 313–327.

[173] DIETHELM, K., Fractional differential equations. Theory and numerical treatment, 2000, (Preprint).

[174] DIETHELM, K., Predictor-corrector strategies for single- and multi-term fractional differential equations, in: LIPITAKIS, E. A. (Ed.) *Proceedings of the 5th Hellenic-European Conference on Computer Mathematics and its Applications*,LEA Press, Athens (2002), 117-122.

[175] DIETHELM, K., Efficient solution of multi-term fractional differential equations using P(EC)E-m methods, *Computing*, **71**(4), (2003) 305–319.

[176] DIETHELM, K. and FORD, N. J., Numerical solution methods for distributed order differential equations, *Fract. Calc. Appl. Anal.*, **4**(4), (2001) 531–542.

[177] DIETHELM, K. and FORD, N. J., Analysis of fractional differential equations, *J. Math. Anal. Appl.*, **265**(2), (2002) 229–248.

[178] DIETHELM, K. and FORD, N. J., Numerical Solution of the Bagley-Torvik equation, *BIT*, **42**(3), (2002) 490–507.

[179] DIETHELM, K. and FORD, N. J., Multi-order fractional differential equations and their numerical solution, *Appl. Math. Comp.*, **3**(154), (2004) 621–640.

[180] DIETHELM, K., FORD, J.M., FORD, N. J., and WEILBEER, A comparison of backward differentiation approaches for ordinary and partial differential equations of fractional order, in: LE MÉHAUTÉ, A., TENREIRO-MACHADO, J. A., TRIGEASSOU, J. C., and SABATIER, J. (Eds.) *Proceedings of the 1st IFAC Workshop on Fractional Differentiation and its Applications*, ENSEIRB, Bordeaux, (2004), 353-358.

[181] DIETHELM, K., FORD, J.M., FORD, N. J., and WEILBEER, M., Pitfalls in fast numerical solvers for fractional differential equations, *J. Comput. Appl. Math.*, **83**(11) (to appear), 2005.

[182] DIETHELM, K., FORD, N. J., and FREED, A. D., A predictor-corrector approach for the numerical solution of fractional differential equations. Fractional order calculus and its applications, in: TENREIRO-MACHADO, J. A. (Ed.), *Special Issue of Fractional Order Calculus and Its Applications*, Nonlinear Dynamics, **29**(1-4), (2002) 3–22.

[183] DIETHELM, K., FORD, N. J., and FREED, A. D., Detailed error analysis for a fractional Adams method, *Numer. Algorit.*, **1**(36), (2004) 31–52.

[184] DIETHELM, K., FORD, N. J., FREED, A. D., and LUCHKO, Y. F., Algorithms for the fractional calculus: A selection of numerical methods, *Computer Meth. Appl. Mech. Engin.*, **194**(6-8), (2005) 743–773.

[185] DIETHELM, K. and FREED, A. D., The FracPECE. Subroutine for the numerical solutions of differential equations of fractional order, in: HEINZEL, S. and PLESSER, T. (Eds.), *Forschung und wissenschaftliches Rechnen: Beiträge zum Heinz-Billing-Preis 1998*. Gesellschaft für wissenschaftliche Datenverarbeitung, Gottingen, 1999, 57–71.

[186] DIETHELM, K. and FREED, A. D., On the solution of nonlinear fractional differential equations used in the modelling of viscoelasticity, in: KEIL, F., MACKENS, W., VOB, H., and WERTHER, J. (Eds.), *Scientific Computing in Chemical Engineering II. Computational Fluid Dynamics Reaction Engineering, and Molecular Properties*, Springer-Verlag, Heidlberg, 1999, 217–224.

[187] DIETHELM, K. and LUCHKO, Y., Numerical solution of linear multi-term initial value problems of fractional order, *Fract. Calc. Appl. Anal.* (to appear), 2005.

[188] DIETHELM, K., NEVILLE, J., FORD, A. D., and FREED, A., Predictor-corrector approach for the numerical solution of fractional differential equations, in: TENREIRO-MACHADO, J. A. (Ed.), *Special Issue of Fractional Order Calculus and Its Applications*, Nonlinear Dynamics, **29**(1-4), 2002, 3–22.

[189] DIETHELM, K. and WALZ, G., Numerical solution of fractional order differential equations by extrapolation, *Numer. Algorit.*, **16**(3-4), (1997) 231–253.

[190] DIETHELM, K. and WEILBEER, Initial-boundary value problems for time-fractional diffusion-wave equations and their numerical solution, in: LE MÉHAUTÉ, A., TENREIRO-MACHADO, J. A., TRIGEASSOU, J. C., and SABATIER, J. (Eds.) *Proceedings of the 1st IFAC Workshop on Fractional Differentiation and its Applications*, ENSEIRB, Bordeaux, (2004), 551–557.

[191] DIMOVSKI, I. H., Operational calculus for a class of differential operators, *C.R. Acad. Bulgar. Sci.*, **19**(12), (1966) 1111–1114.

[192] DIMOVSKI, I. H., *Convolutional Calculus*, vol. 43 of Mathematics and Its Applications, Kluwer, Dordrecht, 1990.

[193] DITKIN, V. A., The theory of operator calculus, *Dokl. Acad. Nauk. SSSR*, **116**, (1957) 15–17.

[194] DITKIN, V. A. and PRUDNIKOV, A. P., On the theory of operational calculus Stemming from the Bessel equation, *Z. Vychisl. Mat. i Mat. Fiz*, **3**, (1963) 223–238.

[195] DITKIN, V. A. and PRUDNIKOV, A. P., *Integral Transforms and Operational Calculus*, Pergamon Press, Oxford, 1965.

[196] DITKIN, V. A. and PRUDNIKOV, A. P., *Operational Calculus*, Vysshaja Shkola, Moscow, 1966.

[197] DITKIN, V. A. and PRUDNIKOV, A. P., *Calcul Operationnel*, Mir, Moscow, 1979.

[198] DOETSCH, G., *Handbuch der Laplace-Transformation*, Birkhauser-Verlag, Basel and Stuttgart, 1955.

[199] DOETSCH, G., *Introduction to the Theory and Application of the Laplace Transformation*, Springer-Verlag, Berlin, 1974.

[200] DORTA, J. A., Sobre ecuaciones diferenciales de orden no entero, *Depto. Análisis Matemático.* (Universidad de La Laguna).

[201] DRANIKOV, I. L., KONDRATENKO, P. S., and MATVEEV, L. V., Anomalous transport regimes in a stochastic advection-diffusion model, *J. Exper. Theoret. Phys.*, **98**(5), (2004) 945–952.

[202] DUARTE, F. and TENREIRO-MACHADO, J. A., Chaotic phenomena and fractional-order dynamics in the trajectory control of redundant manipulators, in: TENREIRO-MACHADO, J. A.(Ed.), *Special Issue of Fractional Order Calculus and Its Applications*, Nonlinear Dynamics, **29**(1-4), 2002, 315–342.

[203] DUARTE, M. and TENREIRO-MACHADO, J. A.(Eds.), *Special Issue of Fractional Processing and Applications*, Signal Proces. J., **83**(11), (2003).

[204] DUARTE, M. and TENREIRO-MACHADO, J. A.(Eds.), *Special Issue of Fractional Order Derivatives and their Applications*, Nonlinear Dynamics, **38**(1-2), 2004.

[205] DZHERBASHYAN, M. M., *Integral Transforms and Representations of Functions in Complex Domain*, Nauka, Moskow, 1966.

[206] DZHERBASHYAN, M. M., A boundary value problem for a Sturm-Liouville type differential operator of fractional order, *Izv. Akad. Nauk Armajan. SSR, Ser. Mat.*, **5**(2), (1970) 71–96.

[207] DZHERBASHYAN, M. M., Differential operators of fractional order and boundary value problems in the complex domain, in: *The Gohberg Anniversary Collection, Vol II*, vol. 41 of Oper. Theory Adv. Appl., Birkhauser-Verlag, Calgary, 1988, 153–172.

[208] DZHERBASHYAN, M. M., *Harmonic Analysis and Boundary Value Problems in the Complex Domain*, Oper. Theory Adv. Appl., Birkhauser-Verlag, Berlin, 1993.

[209] DZHERBASHYAN, M. M. and NERSESYAN, A. B., Criteria of expansibility of functions in Dirichlet series, *Izv. Akad. Nauk Arm. SSR, Ser. Fiz-Mat. Nauk*, **11**(5), (1958) 85–106.

[210] DZHERBASHYAN, M. M. and NERSESYAN, A. B., Fractional derivatives and the Cauchy problem for differential equations of fractional order, *Izv. Akad. Nauk Armajan. SSR, Ser. Mat.*, **3**(1), (1968) 3–29.

[211] DZHERBASHYAN, M. M. and RAFAELYAN, S. G., Singular boundary value problems that generate systems of Fourier type, *Izv. Akad. Nauk Armyan. SSR, Ser. Mat.*, **23**(2), (1988) 109–122.

[212] DZHRBASHYAN, M. M., Interpolation and spectral expansions associated with differential operators of fractional order, *Izv. Akad. Nauk Armyan. SSR, Ser. Mat.*, **19**(2), (1984) 81–181.

[213] DZHRBASHYAN, M. M., Interpolation and spectral expansions associated with differential operators of fractional order, *Colloq. Math. Soc. Janos Bolyai, Budapest*, **49**, (1987) 303–322.

[214] DZHRBASHYAN, M. M., The basis property of systems of eigenfunctions that are generated by a boundary value problem in a complex domain, *Soviet J. Contemp. Math. Anal.*, **23**(5), (1988) 1–14.

[215] DZHRBASHYAN, M. M. and NERSESYAN, A. B., Expansions in special biorthogonal systems and boundary value problems for differential equations of fractional order, *Dokl. Akad. Nauk SSSR*, **132**(4), (1960) 747–750.

[216] DZHRBASHYAN, M. M. and NERSESYAN, A. B., Expansions in certain biorthogonal systems and boundary value problems for differential equations of fractional order, *Trudy Moskov. Mat. Obsc.*, **10**, (1961) 89–179.

[217] DZHRBASHYAN, M. M. and NERSESYAN, A. B., Fractional derivatives and the Cauchy problem for differential equations of fractional order. (Russian), *Izv. Akad. Nauk Armyan. SSR, Ser. Mat.*, **3**(1), (1968) 3–29.

[218] EDWARDS, J. T., FORD, N. J., and SIMPSON, A. C., The numerical solution of linear multi-term fractional differential equations; systems of equations, *J. Comp. Math. Appl.*, **148**(2), (2002) 401–418.

[219] EIDELMAN, S. D. and KOCHUBEI, A. N., Cauchy problem for fractional diffusion equation, *J. Diff. Equat.*, **199**(2), (2004) 211–255.

[220] EL-MESIRY, A. E. M., EL-SAYED, A. M. A., and EL-SAKA, H. A. A., Numerical methods for multi-term fractional (arbitrary) orders differential equations, *Applied Mathematics and Computation*, **160**(3), (2005) 683–699.

[221] EL-RAHEEM, Z. F. A., Modification of the application of a contraction mapping method on a class of fractional differential equation, *Appl. Math. Comp.*, **137**(2-3), (2003) 371–374.

[222] EL-SAYED, A. M. A., Fractional differential equations, *Kyungpook Math. J.*, **28**(2), (1988) 119–122.

[223] EL-SAYED, A. M. A., On the fractional differential equations, *Appl. Math. Comp.*, **49**(2-3), (1992) 205–213.

[224] EL-SAYED, A. M. A., Linear differential equations of fractional order, *Appl. Math. Comp.*, **55**(1), (1993) 1–12.

[225] EL-SAYED, A. M. A., Fractional order evolution equations, *J. Fract. Calc.*, **7**, (1995) 89–100.

[226] EL-SAYED, A. M. A., Finite Weyl fractional calculus and abstract fractional differential equation, *J. Fract. Calc.*, **9**, (1996) 59–68.

[227] EL-SAYED, A. M. A., Fractional differential-difference equations, *J. Fract. Calc.*, **10**, (1996) 101–107.

[228] EL-SAYED, A. M. A., Fractional order diffusion-wave equation, *Int. J. Theoret. Phys.*, **35**(2), (1996) 311–322.

[229] EL-SAYED, A. M. A., Nonlinear functional-differential equations of arbitrary order, *Nonlinear Anal.*, **33**(2), (1998) 181–186.

[230] EL-SAYED, A. M. A. and GAAFAR, F. M., Exact solution of linear differential equations of fractional order, *Kyungpook Math. J.*, **37**(2), (1997) 181–189.

[231] EL-SAYED, A. M. A. and GAAFAR, F. M., Sturm-Liouville equation of fractional order, *Kyungpook Math. J.*, **37**(1), (1997) 9–17.

[232] EL-SAYED, A. M. A. and IBRAHIM, A. G., Multivalued fractional differential equations, *Appl. Math. Comp.*, **68**(1), (1995) 15–25.

[233] El-Wakil, S. A. and Zahran, M. A., Fractional Fokker-Planck equation, *Chaos Solit. Fract.*, **11**(5), (2000) 791–798.

[234] El-Wakil, S. A. and Zahran, M. A., Fractional representation of Fokker-Planck equation, *Chaos Solitons and Fractals*, **12**(10), (2001) 1929–1935.

[235] El-Wakil, S. A., Zahran, M. A., and Abulwafa, E. M., The diffusion-drift equation on comb-like structure, *Physica A: Stat. Mech. Appl.*, **303**(1-2), (2002) 27–34.

[236] Eldred, L. B.,Baker, W. P., and Palazotto, A. N., Numerical application of fractional derivative model constitutive relations for viscoelastic materials, *Computers Struct.*, **60**, 1996, 875–882.

[237] Elizarraraz, D. and Verde-Star, L., Fractional divided differences and the solution of differential equations of fractional order, *Adv. Appl. Math.*, **24**(3), (2000) 260–283.

[238] Elwakil, S. A. and Zahran, M. A., Fractional integral representation of master equation, *Chaos Solit. Fract.*, **10**(9), (1999) 1545–1548.

[239] Elwakil, S. A., Zahran, M. A., and Abdou, M. A., The operator method for solving the fractional Fokker-Planck equation, *J. Quantit. Spectros. and Radiat. Transfer*, **77**(3), (2003) 317–327.

[240] Enelund, M. and Olsson, P., Damping described by fading memory: analysis and application to fractional derivative models., *Int. J. Solids Structures*, **36**(7), (1999) 939–970.

[241] Engheta, N., On fractional calculus and fractional multipoles in electromagnetism, *IEEE Trans. Antennas and Propagation*, **44**(4), (1996) 554–566.

[242] Engheta, N., On the role of fractional calculus in electromagnetic theory, *IEEE Antenn. Propag.*, **39**(4), (1997) 3546.

[243] Engheta, N., Fractional curl operator in electromagnetics, *Microwave Opt. Tech. Lett.*, **17**(2), (1998) 86–91.

[244] Engheta, N., On Fractional paradigm and intermediate zones in electromagnetism, I: Planar observation, *Microw. Optim. Tech. Lett.*, **22**(4), (1999) 236–241.

[245] Engheta, N., On Fractional paradigm and intermediate zones in electromagnetism. II: Cylindrical and spherical observations, *Microw. Optim. Tech. Lett.*, **23**(2), (1999) 100–103.

[246] Engheta, N.,Phase and amplitude of the fractional-order intermediate wave, *Microwave Opt. Tech. Lett.*, **21**(5), (1999) 338–343.

[247] Engheta, N., Fractional paradigm in electromagnetic theory, in: Werner, H. and Mittra, R. (Eds.), *Frontiers in Electromagnetics*, IEEE Pres, 2000, 523–552.

[248] Erdélyi, A., Fractional integrals of generalized functions, *J. Austral. Math. Soc.*, **14**(1), (1972) 30–37.

[249] Erdélyi, A., Magnus, W., Oberhettinger, F., and Tricomi, F., *Higher Transcendental Functions*, vol. I-III, Krieger Pub., Melbourne, Florida, 1981.

[250] Ertas, D. and Kardar, M., Dynamic roughening od directed lines, *Phys. Rev. Lett.*, **69**(6), (1992) 929–932.

[251] Erugin, N. P., *Reading Book on General Course of Differential Equations*, Nauka i Tekhnika, Minsk, 1970.

[252] EULER, L., De progressionibus transcentibus, sev Quarum Termini Algebraice Dari Nequeunt, *Comment. Acad. Sci. Imperiales Petropolitanae*, **1**(5), (1738) 38–57.

[253] FATTORINI, H. O., *The Cauchy problem*, vol.18 of Encyclopedia of Mathematics and its Applications, Addison-Wesley, Massachussets, 1983.

[254] FEDOSOV, V. P. and YANENKO, N. N., Solution of a mixed problem for the wave equation, *Dokl. Acad. Nauk SSSR*, **262**(3), (1982) 549–553.

[255] FEDOSOV, V. P. and YANENKO, N. N., Equations with partial derivatives of half-integral degree, *Dokl. Akad. Nauk SSSR*, **276**(4), (1984) 804–808.

[256] FELLER, W., *An Introduction to Probability Theory and Its Applications*, Wiley and Sons, New York, 1971.

[257] FERNANDER, A. A., Modal synthesis when modelling damping by use of fractional derivatives, *AIAA J.*, **34**(5), (1996) 1051–1058.

[258] FERNÁNDEZ-CARA, E. and ZUAZUA, E., Control theory: History, mathematical achievements and perspectives, *Journal Systems Analysis Modelling Simulation*, **26**, (2003) 79–140.

[259] FLORIANI, E., TREFAN, G., GRIGOLINI, P., and WEST, B. J., A dynamical approach to anomalous conductivity, *J. Statist. Phys.*, **84**, (1996) 1043–1066.

[260] FOGEDBY, H., Langevin-equations for continuous time levy flights, *Phys. Rev. E*, **50**(2), (1994) 1657–1660.

[261] FORD, N. J. and SIMPSON, A. C., The numerical solution of fractional differential equations: speed versus accuracy, *Numer. Algorit.*, **26**(4), (2001) 333–346.

[262] FOURIER, J. B., *Theorie Analytic de la Chaleur*, vol. 1 of Oeuvres de Fourier, Gautier-Villars, Paris, 1888.

[263] FOX, C., The asymptotic expansion of generalized hypergeometric functions, *Proc. London Math. Soc. (Ser. 2)*, **27**, (1928) 389–400.

[264] FOX, C., The *G* and *H* functions as symmetrical Fourier kernels, *Trans. Amer. Math. Soc.*, **98**, (1961) 395–429.

[265] FREMBERG, N. E., A study of generalized hyperbolic potentials, in: *Dissertation Meddelanden Fran Lunds Universitets Matematiska Seminarium, Band 7*, 1946.

[266] FRIEDRICH, C., Mechanical stress relaxation in polymers: fractional integral versus fractional differential model, *J. Non-Newtton. Fluid Mech.*, **46**(2-3), (1993) 307–314.

[267] FUJITA, Y., Cauchy problems of fractional order and stable processes, *Japan J. Appl. Math.*, **7**(3), (1990) 459–476.

[268] FUJITA, Y., Integro-differential equations which interpolate the heat equation and the wave equation. II, *Osaka J. Math.*, **27**(4), (1990) 797–804.

[269] FUJITA, Y., Integro-differential equations which interpolates the heat equation and the wave equation, *Osaka J. Math.*, **27**(2), (1990) 309–321.

[270] FUJITA, Y., Energy inequalities for integro-partial differential equations with Riemann-Liouville integrals, *SIAM J. Math. Anal.*, **23**(5), (1992) 1182–1188.

[271] FUJIWARA, M., On the integration and differentiation of an arbitrary order, *Tôhoku Math. J.*, **37**, (1933) 110–121.

[272] FUKUNAGA, M., On uniqueness of the solutionn of initial value problems for ordinary fractional differential equations, *Int. J. Appl. Math*, **2**(10), (2002) 177–189.

[273] FUKUNAGA, M., On uniqueness of the solutionn of initial value problems for ordinary fractional differential equations II, *Int. J. Appl. Math*, **3**(10), (2002) 269–283.

[274] FUKUNAGA, M. and SHIMIZU, N., Role of prehistories in the initial value problems of fractional viscoelastic equations, in: DUARTE, M.and TENREIRO-MACHADO, J. A.(Eds.), *Special Issue of Fractional Order Derivatives and their Applications*, Nonlinear Dynamics, **38**(1-2), 2004, 207–220.

[275] GAUL, L., KLEIN, P., and KEMPFLE, S., Damping description involving fractional operators, *Mech. Syst. Signal Process.*, **5**(2), (1991) 81–88.

[276] GEL'FAND, I. M. and SHILOV, G. E., *Generalized Functions*, vol. 1, Academic Press, London, 1964.

[277] GEMANT, A., A Method of analyzing experimental results obtained from elasto-viscous bodies, *Physics*, **7**(8), (1936) 311–317.

[278] GEMANT, A., On Fractional Differentials, *Philos. Mag. Lett.*, **25**, (1938) 540–549.

[279] GERASIMOV, A. N., A generalization of linear laws of deformation and its application to problems of internal friction, *Akad. Nauk SSSR, Prikl. Mat. Mekh.*, **12**, (1948) 251–259.

[280] GIANNANTONI, C., The problem of the initial conditions and their physical meaning in linear differential equations of fractional order, *Appl. Math. Comp.*, **141**(1), (2003) 87–102.

[281] GILBERT, P. P., *Function Theoric Methods in Partial Differential Equations*, vol. 54 of *Mathematics in Science and Engineering*, Academic Press, New York, 1969.

[282] GIONA, M. and ROMAN, H. E., Fractional diffusion equation on fractals: One-dimensional case and asymptotic behaviour, *J. Phys. A: Math. Gen.*, **25**(8), (1992) 2107–2117.

[283] GLAESKE, H.-J., KILBAS, A. A., and SAIGO, M., Modified Bessel type integral transform and its compositions with fractional calculus operators on spaces $F_{p,\mu}$ and $F'_{p,\mu}$, *J. Comp. Math. Appl.*, **118**(1-2), (2000) 151–168.

[284] GLOCKLE, W. G. and NONNEMACHER, T. F., Fractional integral operators and Fox functions in theory of viscoelasticity, *Macromolecules*, **24**(24), (1991) 6426–6434.

[285] GLOCKLE, W. G. and NONNEMACHER, T. F., Fox function representation of non-Debye relaxation processes, *J. Statist. Phys.*, **71**(3-4), (1993) 741–757.

[286] GLOCKLE, W. G. and NONNENMACHER, T. F., Fractional relaxation and the time temperature superposition principle, *Rheol. Acta*, **33**(4), (1994) 337–343.

[287] GLOCKLE, W. G. and NONNENMACHER, T. F., A fractional calculus approach to self-similar protein dynamics, *Biophys. J.*, **68**(1), (1995) 46–53.

[288] GNEDENKO, B. V. and KOLMOGOROV, A. N., *Limit Distributions for Sums of Independent Random Variables*, Reading, Addison-Wesley, Massachusetts, 1954.

[289] GÖEL, E. O. and GRAUDSZUS, W., Optical detection of multiple-trapping relaxation in disordered crystalline semiconductors, *Phys. Rev. Lett.*, **48**(1), (1982) 277–280.

[290] GONZALEZ, E., *Estudio Dimensional de la Teoria Viscoelastica*, Ph.D. thesis, Univ. Politecnica de Madrid, 1999.

[291] GORENFLO, R., Fractional calculus: Some numerical methods, in: CARPINTERI, A. and MAINARDI, F.(Eds.), *Fractals and Fractional Calculus in Continuum Mechanics*, vol. 378 of CISM Courses and Lectures, Springer-Verlag, 1997, 277–290.

[292] GORENFLO, R., DE FABRITIS, G., and MAINARDI, F., Discrete random walk models for symmetric Levy-Feller diffusion process, *Physica A*, **269**(1), (1999) 79–89.

[293] GORENFLO, R., ISKENDEROV, A., and LUCHKO, Y. F., Mapping between solutions of fractional diffusion-wave equations, *Fract. Calc. Appl. Anal.*, **3**(1), (2000) 75–86.

[294] GORENFLO, R., KILBAS, A. A., and ROGOSIN, S. V., On the generalized Mittag-Leffler type function, *Integral Transform. Spec. Funct.*, **7**(3-4), (1998) 215–224.

[295] GORENFLO, R., KILBAS, A. A., and ROGOSIN, S. V., On the properties of a generalized Mittag-Leffler type function, *Dokl. Nats. Akad. Nauk Belarusi*, **42**(5), (1998) 34–39.

[296] GORENFLO, R., LUCHKO, Y. F., and MAINARDI, F., Analytical properties and applications of the Wright functions, *Fract. Calc. Appl. Anal.*, **2**(4), (1999) 383–414.

[297] GORENFLO, R., LUCHKO, Y. F., and MAINARDI, F., Wright functions as scale-invariant solutions of the diffusion-wave equation. Higher transcendental functions and their applications, *J. Comput. Appl. Math.*, **118**(1-2), (2000) 175-191.

[298] GORENFLO, R., LUCHKO, Y. F., and SRIVASTAVA, H. M., Operational method for solving integral equations with Gauss's hypergeometric function as a kernel, *Internat. J. Math. Statist. Sci.*, **6**(2), (1997) 179–200.

[299] GORENFLO, R., LUCHKO, Y. F., and UMAROV, S. R., On some boundary value problems for pseudo-differential equations with boundary operators of fractional order, *Fract. Calc. Appl. Anal.*, **2**(2), (2000) 163–174.

[300] GORENFLO, R., LUCHKO, Y. F., and UMAROV, S. R., On the Cauchy and multi-point problems for partial pseudo-differential equations of fractional order, *Fract. Calc. Appl. Anal.*, **4**(3), (2000) 453–468.

[301] GORENFLO, R., LUCHKO, Y. F., and ZADREIKO, P. P., On solvability of linear fractional differential equations in Banach spaces, *Fract. Calc. Appl. Anal.*, **2**(2), (1999) 163–176.

[302] GORENFLO, R. and MAINARDI, F., Signalling problem and Dirichlet-Neumann map for time-fractional diffusion-wave equations, in: NAVARRO, M. P. and GORENFLO, R. (Eds.), *International Conference on Inverse Problems and Applications*, Philippines University, Manila, Dilima (Philippines), 1988, 23–27.

[303] GORENFLO, R. and MAINARDI, F., Fractional oscillations and Mittag-Leffler functions, in: UNIVERSITY KUWAIT, D. M. C. S. (Ed.), *International Workshop on the Recent Advances in Applied Mathematics (Kuwait, RAAM'96)*, Kuwait, 1996, 193–208.

[304] GORENFLO, R. and MAINARDI, F., *Fractional Calculus: Integral and Differential Equations of Fractional Order*, vol. 378 of CISM courses and lectures, Springer-Verlag, Berlin, 1997, 223–276.

[305] GORENFLO, R. and MAINARDI, F., Fractional calculus and stable probability distributions, *Arch. Mech. (Basel)*, **50**(3), (1998) 377–388.

[306] GORENFLO, R. and MAINARDI, F., Random walk models for space-fractional diffusion processes, *Fract. Calc. Appl. Anal.*, **1**(2), (1998) 167–191.

[307] GORENFLO, R. and MAINARDI, F., Random walk models approximating symmetric space-fractional diffusion processes, in: ELSCHNER, J., GOHBERG, I. and SILBERMANN, B. (Eds.), *Problems in Mathematical Physics*, vol. 121, Birkhauser-Verlag, Berlin, 2001, 120–145.

[308] GORENFLO, R., MAINARDI, F., MORETTI, D., PAGNINI, P., and PARADISI, P., Fractional diffusion: Probability distributions and random walk models, non extensive thermodynamics and physic applications, *Physica A*, **305**(1-2), (2002) 106–112.

[309] GORENFLO, R., MAINARDI, F., MORETTI, D., and PARADISI, P., Time fractional diffusion: A discrete random walk approach. Fractional order calculus and its applications, in: TENREIRO-MACHADO, J. A. (Ed.), *Special Issue of Fractional Order Calculus and Its Applications*, Nonlinear Dynamics, **29**(1-4), 2002, 129–143.

[310] GORENFLO, R., MAINARDI, F., and SRIVASTAVA, H. M., Special functions in fractional relaxation-oscillation and fractional diffusion-wave phenomena, in: *The Eighth International Colloquium on Differential Equations (Plovdiv, 1997)*, VSP Publising Company, Utrecht, 1998, 195–202.

[311] GORENFLO, R. and RUTMAN, R., On ultraslow and intermediate processes:, in: RUSEV, I. P., DIMOSKI, I., and KIRYAKOVA, V. (Eds.), *International Workshop on Transforms Methods and Special Functions*, Bulgarian Acad. Sci., Sofia, 1994, 61–81.

[312] GORENFLO, R. and VESSELA, S., *Abel Integral Equations*, vol. 1461 of Lect. Notes in Math., Springer-Verlag, Berlin, 1991.

[313] GORENFLO, R., VIVOLI, A., and MAINARDI, F., Discrete and continuous random walk models for space-time fractional diffusion, *Nonlinear Dynam.*, **38**(1-4), (2004) 101–116.

[314] GOTO, M. and ISHII, D., Semidifferential electroanalysis, *J. Electroanal. Chem. and Interfacial Electrochem.*, **61**, (1975), 361–365.

[315] GRIN'KO, A. P., Solution of a non-linear differential equation with a generalized fractional derivative, *Dokl. Akad. Nauk BSSR*, **35**(1), (1991) 27–31.

[316] GRIPENBERG, G., LONDEN, S.-O., and STAFFANS, O., *Volterra Integral and Functional Equations*, Cambridge University Press, Cambridge, 1990.

[317] GRÜWALD, A. K., Uber "begrenzte" derivationen und deren anwendung, *Z. Angew. Math. Phys.*, **12**, (1867) 441–480.

[318] GUELLAL, S., CHERRUAULT, Y.,PUJOL, M. J. and GRIMALT, P. Decomposition method applied to Hydrology, *Kybernetes*, **29**(4), (2000), 499–504.

[319] HADAMARD, J., Essai sur létude des fonctions données par leur développement de Taylor, *J. Math. Pures Appl.*, **8**(Ser. 4), (1892) 101–186.

[320] HADID, S. B., Local and global existence theorems on differential equations of non-integer order, *J. Fract. Calc.*, **7**, (1995) 101–105.

[321] HADID, S. B. and ALSHAMANI, J. G., Liapunov stability of differential equations of non-integer order, *Arab J. Math. Sci.*, **7**(1-2), (1986) 7–17.

[322] HADID, S. B. and GRZASLEWICZ, R., On the solution of certain type of differential equations with fractional derivatives, in: NISHIMOTO, K. (Ed.), *International Conference on Fractional Calculus and Its Applications (Tokyo, 1989)*, College Engrg. Nihon University, Tokyo, 1989, 52–55.

[323] HADID, S. B. and LUCHKO, Y. F., An operational method for solving fractional differential equations of an arbitrary real order, *PanAmer. Math. J.*, **6**(1), (1996) 57–73.

[324] HADID, S. B., MASAEDEH, B., and MOMANI, S., On the existence of maximal and minimal solutions of differential equations of non-integer order, *J. Fract. Calc.*, **9**, (1996) 41–44.

[325] HADID, S. B., TA'ANI, A. A., and MOMANI, S. M., Some existence theorems on differential equation of generalized order through a fixed-point theorem, *J. Fract. Calc.*, **9**, (1996) 45–49.

[326] HANYGA, A., Wave propagation in media with singular memory, *Math. Comput. Modelling*, **34**(12-13), (2001) 1399–1421.

[327] HANYGA, A., Multidimensional solutions of space-time-fractional equations, *Proc. Roy. Soc. London, Ser. A*, **458**(2018), (2002) 429–450.

[328] HANYGA, A., Multidimensional solutions of time-fractional diffusion-wave equations, *Proc. Roy. Soc. London, Ser. A*, **458**(2020), (2002) 933–957.

[329] HANYGA, A., Wave propagation in poroelasticity: Equations and solutions, in: SHANG, E. C., LI, Q., and GAO, T. F. (Eds.), *Theoretical and Computational Acoustics*, World Scientific, River Edge, 2002, 310–318.

[330] HARDY, G. H., Riemann's form of Taylor's series, *J. London Math. Soc.*, **20**(10), (1945) 48–57.

[331] HARDY, G. H. and LITTLEWOOD, J. E., Some properties of fractional integrals, *Math. Z.*, **27**(4), (1928) 565–606.

[332] HAUS, J. W. and KEHR, K. W., Diffusion in regular and disordered lattices, *Phys. Rep.*, **150**(5-6), (1987) 263–406.

[333] HAVLIN, S. and BEN-AVRAHAM, D., Diffusion in disordered media, *Adv. Phys.*, **51**(1), (2002) 187–292.

[334] HAVLIN, S., BULDYREV, S. V., GOLDBERGER, A. L., MANTEGNA, R. N., OSSADNIK, S. M., PENG, C. K., SIMON, M., and STANLEY, H., Fractals in biology and medicine, *Chaos Solit. Fract.*, **6**(1), (1995) 171–201.

[335] HAYEK, N., TRUJILLO, J., RIVERO, M., BONILLA, B., and MORENO, J. C., An extension of Picard-Lindeloff theorem to fractional differential equations, *Appl. Anal.*, **70**(3-4), (1999) 347–361.

[336] HENRY, B. I. and WEARNE, S. L., Fractional reaction-diffusion, *Physica A*, **276**(3-4), (2000) 448–455.

[337] HENRY, B. I. and WEARNE, S. L., Existence of tuning instabilities in two-species fractional reaction-diffusion system, *SIAM J. Appl. Math.*, **62**(3), (2002) 870–887.

[338] HERRCHEN, M. P., *Stochastic Modelling of Dispersive Diffusion by Non-Gaussian Noise*, Ph.D. thesis, Swiss Federal Institute of Technology, 2001.

[339] HEYMANS, N. and BAUWENS, J. C., Fractal rheological models and fractional differential-equations for viscoelastic behavior, *Rheol. Acta*, **33**(3), (1994) 210–219.

[340] HILFER, E. (Ed.), *Applications of Fractional Calculus in Physics*, World Sci. Publishing, New York, 2000.

[341] HILFER, R., An extension of dynamical foundation for the statistical equilibrium concept, *Physica A*, **221**(1-3), (1995) 89–96.

[342] HILFER, R., Foundations of fractional dynamics, *Fractals*, **3**(3), (1995) 549–555.

[343] HILFER, R., On fractional diffusion and its relation with continuous time random walks, in: KUTNER, R., PEKALSKI, A., and SZNAIJ-WERON, K. (Eds.), *Anomalous Diffusion: From Basics to Applications*, Springer-Verlag, Berlin, 1999, 77–82.

[344] HILFER, R., Fractional diffusion based on Riemann-Liouville fractional derivatives, *J. Phys. Chem. B*, **104**(3), (2000) 914–924.

[345] HILFER, R., Experimental evidence for fractional time evolution in glass forming materials, *J. Chem. Phys.*, **284**(1-2), (2002) 399–408.

[346] HILFER, R., On fractional diffusion and continuous time random walks, *Physica. A*, **329**(1-2), (2003) 35–40.

[347] HILFER, R. and ANTON, L., Fractional master equations and fractal time random walks, *Phys. Rev. E*, **51**(2), (1995) 848–851.

[348] HILFER, R., METZLER, R., BLUMEN, A. and KLAFTER, J.(Eds.), *Special Issue on Strange Kinetics*, Chemical Physics, **284**(1-2), 2002.

[349] HILLE, E. and TAMARKIN, J. D., On the theory of linear integral equations, *Ann. Math.*, **31**(3), (1930) 479–528.

[350] HOFFMANN, K. H., ESSEX, C., and SCHULZKY, C., Fractional diffusion and entropy production, *J. Non-Equilibrium Thermod.*, **23**(2), (1998) 166–175.

[351] HUANG, F. and LIU, F., The time fractional diffusion equation and the advection-dispersion equation, *ANZIAM J.*, **46**, (2005) 317–330.

[352] HUGHES, B. D., *Random Walks and Random Environments, vol. I*, Oxford University Press, Oxford, 1995.

[353] HUMBERT, P., Quelques resultas relatifs a la fonction de Mittag-Leffler, *C. R. Acad. Sci. Paris*, **236**, (1953) 1467–1468.

[354] HUMBERT, P. and AGARWAL, R. P., Sur la fonction de Mittag-Leffler et quelques-unes de ses généralisations, *Bull. Sci. Math.*, **77**(2), (1953) 180–185.

[355] HWANG, C., LEU, J. F., and TSAY, S. Y., A note on time-domain simulation of feedback fractional-order systems, *IEEE Trans. Automat. Control*, **47**(4), (2002) 625–631.

[356] IMANALIEV, M. I. and VEBER, V. K., Generalisation and applications of functions of Mittag-Leffler type, *Estud. IntegroDiff. Equat.*, **13**, (1980) 49–59, 378.

[357] INGMAN, D. and SUZDALNITSKY, J., Control of damping oscillations by fractional differential operator with time-dependent order, *Computer Meth. Appl. Mech. Engin.*, **193**(52), (2004) 5585–5595.

[358] ISICHENKO, M. B., Percolation, statistical topography and transport in random media, *Rev. Modern Phys.*, **64**(4), (1992) 961–1043.

[359] JIMÉNEZ, S., PASCUAL, P., AGUIRRE, C., and VÁZQUEZ, L., A panoramic view of some perturbed nonlinear wave equations, *Int. J. Bifur. Chaos*, **14**(1), (2004) 1–40.

[360] JUMARIE, G., A new approach to complex-valued fractional brownian motion via rotating white noise, *Chaos, Solitons and Fractals*, **9**(6), (1998) 881–893.

[361] JUMARIE, G., New results on Fokker-Planck equations of fractional order, *Chaos Solit. Fract.*, **12**(10), (2001) 1873–1886.

[362] KARAPETYANTS, N. K., KILBAS, A. A., and SAIGO, M., On the solution of nonlinear Volterra convolution equation with power nonlinearity, *J. Int. Equat. Appl.*, **8**(4), (1996) 429–445.

[363] KEMPFLE, S., Causality criteria for solutions of linear fractional differential equations, *Fract. Calc. Appl. Anal.*, **1**(4), (1998) 351–364.

[364] KEMPFLE, S. and BEYER, H., Global and causal solutions of fractional linear differential equations, in: RUSEV, P., DIMOSKI, I., and KIRYAKOVA, V. (Eds.), *International Workshop on Transform Methods and Special Functions (Varna'96)*, Bulgarian Acad. Sci., Sofía, 1996, 210–226.

[365] KEMPFLE, S. and GAUL, L., Global solutions of fractional linear differential equations, in: *ICIAM 95*, Z. Angew Math. Mech., **76** (2), (1996) 571–572.

[366] KEMPFLE, S. and SCHAEFER, I., Functional calculus method versus Riemann-Liouville Approach, *Fract. Calc. Appl. Anal.*, **2**(4), (1999) 415–427.

[367] KEMPFLE, S. and SCHAEFER, I., Fractional differential equations and initial conditions, *Fract. Calc. Appl. Anal.*, **3**(4), (2000) 387–400.

[368] KEMPFLE, S., SCHAEFER, I., and BEYER, H., Fractional calculus via funtional calculus: theory and applications, in: TENREIRO-MACHADO, J. A. (Ed.), *Special Issue of Fractional Order Calculus and Its Applications*, Nonlinear Dynamics, **29**(1-4), 2002, 99–127.

[369] KEMPFLE, S., SCHAEFER, I., and BEYER, H., Functional calculus and a link to fractional calculus, *Fract. Calc. Appl. Anal.*, **5**(4), (2002) 411–426.

[370] KILBAS, A. A., Hadamard-type fractional calculus, *J. Korean Math. Soc.*, **38**(6), (2001) 1191–1204.

[371] KILBAS, A. A., Some problems in the theory of integral and differential equations of fractional order, in: SAMKO, S., LEBRE, A. and DOS SANTOS, A.F. (Eds.), *Factorization, Singular Operators and Related Problems*, Kluwer Acad. Publ., London, 2003, 131–149.

[372] KILBAS, A. A., Some aspects of differential equations of fractional order, *Rev. R. Acad. Cienc. Exactas, Fis. Nat.*, **98**(1), (2004) 27–38.

[373] KILBAS, A., BONILLA, B., RODRÍGUEZ, J., TRUJILLO, J. J., and RIVERO, M., Compositions of fractional integrals and derivatives with a Bessel-type function and the solution of differential and integral equations, *Dokl. Nats. Akad. Nauk Belarusi*, **42**(2), (1998) 25–29.

[374] KILBAS, A. A., BONILLA, B., and TRUJILLO, J. J., Existence and uniqueness theorems for nonlinear fractional differential equations, *Demonstr. Math.*, **33**(3), (2000) 583–602.

[375] KILBAS, A. A., BONILLA, B., and TRUJILLO, J. J., Fractional integrals and derivatives, and diferential equations of fractional order in weighted spaces of continuous functions (Russian), *Dokl. Nats. Akad. Nauk Belarusi*, **44**(6), (2000) 18–22.

[376] KILBAS, A. A., BONILLA, B., and TRUJILLO, J. J., Nonlinear differential equations of fractional order in space of integrable functions (Russian), *Dokl. Akad. Nauk*, **374**(4), (2000) 445–449. Translated in *Dokl. Math.*, **62**(2) (2000), 222–226.

[377] KILBAS, A. A., RIVERO, M., RODRÍGUEZ-GERMÁ, L., and TRUJILLO, J. J., On solutions in series of linear fractional differential equations with variable coefficients (Preprint), *Depto. Análisis Matemático* (Universidad de La Laguna), (2005).

[378] KILBAS, A. A. and MARZAN, S. A., Nonlinear differential equations of fractional order in weighted spaces of continuous functions (Russian), *Dokl. Nats. Akad. Nauk Belarusi*, **47**(1), (2003) 29–35.

[379] KILBAS, A. A. and MARZAN, S. A., Approximate method for solution of a class of differential equations of fractional order (Rusian), *Vescī Nats. Akad. Navuk Belarusi, Ser. Fiz.-Mat. Navuk*, **no. 3**, (2004) 5–9.

[380] KILBAS, A. A. and MARZAN, S. A., Cauchy problem for differential equations with Caputo derivative (Russian), *Dokl. Akad. Nauk*, **399**(1), (2004) 7–11.

[381] KILBAS, A. A. and MARZAN, S. A., Cauchy Type problem for the differential equation of fractional order in the weighted space of continuous functions (Russian), *Dokl. Nats. Akad. Navuk Belarusi*, **48**(5), (2005) 20–24.

[382] KILBAS, A. A. and MARZAN, S. A., Nonlinear Differential equations with the Caputo fractional derivative in the space of continously differentiable functions (Russian), *Differentsial'nye Uravneniya*, **41**(1), (2005) 82–86. Translated in *Diff. Equat.*, **41**(1) (2005), 84–89.

[383] KILBAS, A. A., MARZAN, S. A., and TITYURA, A. A., Hadamard-type fractional integrals and derivatives and differential equations of fractional order (Russian), *Dokl. Akad. Nauk*, **389**(6), (2003) 734–738. Translated in *Dokl. Math.*, **67**(2) (2003), 263–267.

[384] KILBAS, A. A., PIERANTOZZI, T., TRUJILLO, J. J., and VÁZQUEZ, L., On the initial value problem for the fractional diffusion-wave equation, in: *Proceedings of the 1st IFAC Workshop on Fractional Differentiation and its Applications*, ENSEIRB, Bordeaux, (2004), 128–131.

[385] KILBAS, A. A., PIERANTOZZI, T., TRUJILLO, J. J., and VÁZQUEZ, L., On the solution of fractional evolution equations, *J. Phys. A: Math. Gen.*, **37**(9), (2004) 3271–3282.

[386] KILBAS, A. A., PIERANTOZZI, T., TRUJILLO, J. J., and VÁZQUEZ, L., On generalized fractional evolution-diffusion equation, in: LE MEAUTÉ, A., TENREIRO, J.A., TRIGEASSOU, J.C., and SABATIER, J. (Eds.), *Fractionnal Derivatives an their Applications: Mathematical tools, Geometrical and Physical aspects*, Ubooks, Germany, 2005, 135–150.

[387] KILBAS, A. A. and REPIN, O. A., Analog of the Bitsadze-Samarskii problem for a mixed type equation with a fractional derivative (Russian), *Differentsial'nye Uravneniya*, **5**(39), (2003) 638–644, translation in *Differential Equations*, **5**(39), (2003) 674–680 .

[388] KILBAS, A. A. and REPIN, O. A., A Non-Local Problem for the equation of mixed type with the partial Riemann-Liouville derivative and operators of generalized integration in boundary condition (Russian), *Tr. Inst. Mat., Minsk*, **2**(12), (2004) 76–82.

[389] KILBAS, A. A., RIVERO, M., and TRUJILLO, J. J., Existence and uniqueness theorems for differential equations of fractional order in weighted spaces of continuous functions, *Fract. Calc. Appl. Anal.*, **6**(4), (2003) 363–399.

[390] KILBAS, A. A., RODRÍGUEZ-GERMÁ, L., and TRUJILLO, J. J., Asymptotic representations for Hypergeometric-Bessel type function and fractional integrals, *J. Comp. Math. Appl.*, **149**(2), (2002) 469–487.

[391] KILBAS, A. A. and SAIGO, M., Fractional integrals and derivatives of functions of Mittag-Leffler type, *Dokl. Akad. Nauk Belarusi*, **39**(4), (1995) 22–26.

[392] KILBAS, A. A. and SAIGO, M., On solution of integral equation of Abel-Volterra type, *Diff. Integr. Equat.*, **8**(5), (1995) 993–1011.

[393] KILBAS, A. A. and SAIGO, M., Solutions of Abel integral equations of the second kind and of differential equations of fractional order, *Dokl. Nats. Akad. Nauk Belarusi*, **39**(5), (1995) 29–34.

[394] KILBAS, A. A. and SAIGO, M., Application of fractional calculus to solve Abel-Volterra nonlinear and linear integral equations, in: RUSEV, P., DIMOSKI, I., and KIRYAKOVA, V. (Eds.), *International Workshop on Transform Methods and Special Functions, Varna'96*, Bulgarian Acad. Sci., Sofia, 1996, 236–263.

[395] KILBAS, A. A. and SAIGO, M., On Mittag-Leffler type function, fractional calculus operators and solutions of integral equations, *Integral Transform. Spec. Funct.*, **4**(4), (1996) 355–370.

[396] KILBAS, A. A. and SAIGO, M., *On solution in closed form of nonlinear integral and differential equations of fractional order*, Surikaisekikenkyusho Kokyuroku 963, Kyoto, vol. 963, 1996, 39–50.

[397] KILBAS, A. A. and SAIGO, M., Solution in closed form of a class of linear differential equations of fractional order, *Differntial'nye. Uravnenija.*, **33**(2), (1997) 195–204.

[398] KILBAS, A. A. and SAIGO, M., *H-Transforms. Theory and Applications*, Chapman and Hall, Boca Raton, Florida, 2004.

[399] KILBAS, A. A., SAIGO, M., and SAXENA, R. K., Solution of Volterra integro-differential equations with generalized Mittag-Leffler functions in the kernels, *J. Int. Equat. Appl.*, **14**(4), (2002) 377-396.

[400] KILBAS, A. A., SAIGO, M., and SAXENA, R. K., Generalized Mittag-Leffler function and generalized fractional calculus operators, *Integral Transform. Spec. Funct.*, **15**(1), (2004) 31–49.

[401] KILBAS, A. A., SAIGO, M., SAXENA, R. K., and TRUJILLO, J. J., Asymptotic behavior of the Kräzel function and evaluation of integrals (Preprint), *Depart. Math. and Mech.* (Belarusin State University), (2005).

[402] KILBAS, A. A., SAIGO, M., and TRUJILLO, J. J., On the generalized Wright function, *Fract. Calc. Appl. Anal.*, **5**(4), (2002) 437–460.

[403] KILBAS, A. A. and SHLAPAKOV, S. A., On the composition of a Bessel-type integral operator with operators of fractional integro-differentiation and the solution of differential equations, *Differ. Equat.*, **30**(2), (1994) 325–338.

[404] KILBAS, A. A., SRIVASTAVA, H. M., and TRUJILLO, J. J., Fractional differential equations: An emergent field in applied and mathematical sciences, in: SAMKO, S., LEBRE, A. and DOS SANTOS, A.F. (Eds.) *Factorization, Singular Operators and Related Problems*, Kluwer Acad. Publ., London, 2003, 151–173.

[405] KILBAS, A. A. and TITJURA, A. A., Hadamard-type fractional integrals and derivatives, *Trudy Inst. Mat. Minsk*, **11**, (2002) 79–87.

[406] KILBAS, A. A. and TRUJILLO, J. J., Computation of fractional integrals via functions of hypergeometric and Bessel type, *J. Comp. Math. Appl.*, **118**(1-2), (2000) 223–239.

[407] KILBAS, A. A. and TRUJILLO, J. J., Differential equation of fractional order: methods, results and problems. I, *Appl. Anal.*, **78**(1-2), (2001) 153–192.

[408] KILBAS, A. A. and TRUJILLO, J. J., Differential equation of fractional order: methods, results and problems. II, *Appl. Anal.*, **81**(2), (2002) 435–493.

[409] KILBAS, A. A. and TRUJILLO, J. J., On solutions of a homogeneous fractional differential equations with nonconstant coefficients (Preprint), *Depto. Análisis Matemático* (Universidad de La Laguna), (2005).

[410] KILBAS, A. A. and TRUJILLO, J. J., Solutions of homogeneous linear differential equations with Caputo derivatives and constant coefficients (Preprint), *Depto. Análisis Matemático* (Universidad de La Laguna), (2005).

[411] KILBAS, A. A. and TRUJILLO, J. J., Solutions of homogeneous linear fractional differential equations with constant coefficients (Preprint), *Depto. Análisis Matemático* (Universidad de La Laguna), (2005).

[412] KILBAS, A. A. and TRUJILLO, J. J., Solutions of nonhomogeneous linear fractional differential equations with Caputo derivatives and constant coefficients (Preprint), *Depto. Análisis Matemático* (Universidad de La Laguna), (2005).

[413] KILBAS, A. A., TRUJILLO, J. J., and VOROSHILOV, Cauchy type problem for the diffusion-wave equation with the Riemann-Liouville partial derivative, *Fract. Calc. Appl. Anal.* (to appear), (2005).

[414] KIM, K., CHOI, J. S., YOON, S. M., and CHANG, K. H., Volatility and returns in Korean financial markets, *J. Korean Phys. Soc.*, **44**(4), (2004) 1015–1018.

[415] KIM, K. and KONG, Y. S., Anomalous behaviors in fractional Fokker-Planck equation, *J. Korean Phys. Soc.*, **40**(6), (2002) 979–982.

[416] KIRANE, M., A wave equation with fractional damping, *Zeit. Anal. Anw.*, **22**(3), (2003) 609–617.

[417] KIRYAKOVA, V. S., *Generalized Fractional Calculus and Applications*, vol. 301 of Pitman Res. Notes in Math., Wiley and Sons, New York, 1994.

[418] KIRYAKOVA, V. S. and AL-SAQABI, B. N., Solutions of Erdélyi-Kober fractional integral, differential and differintegral equations of second type, *C.R. Acad. Bulgare Sci.*, **50**(1), (1997) 27–30.

[419] KIRYAKOVA, V. S. and AL-SAQABI, B. N., Transmutation method for solving Erdelyi-Kober fractional differintegral equations, *J. Math. Anal. Appl.*, **211**(1), (1997) 347–364.

[420] KLAFTER, J., BLUMEN, A., and SCHLESINGER, M. F., Stochastic pathway to anomalous diffusion, *Phys. Rev. A*, **35**(3), (1987) 81–85.

[421] KLAFTER, J., SHLESINGER, M. F., and ZUMOFEN, G., Beyond Brownian motion, *Phys. Today*, **49**, (1996) 33–39.

[422] KLAFTER, J. and SILBEY, R., Derivation of the continuous time random walk equation, *Phys. Rev. Lett.*, **44**(2), (1980) 55–58.

[423] KLIMEK, M., Stationarity-conservation laws for fractional differential equations with variable coefficients, *Physics A: Math. Gen.*, **35**(31), (2002) 6675–6693.

[424] KOCHUBEI, A. N., Cauchy problem for equations of fractional order, *Differentsial'nye Uravnenija*, **25**(8), (1989) 1359–1368.

[425] KOCHUBEI, A. N., The Cauchy Pproblem for evolution equations of fractional order, *Differ. Equat.*, **25**(8), (1989) 967–974.

[426] KOCHUBEI, A. N., Parabolic pseudo-differential equations, hypersingular integrals and Markov processes, *Math. USSR-Izv.*, **33**(2), (1989) 233–259.

[427] KOCHUBEI, A. N., Diffusion of fractional order, *Differ. Equat.*, **26**(4), (1990) 485–492.

[428] KOCHUBEI, A. N., *Fractional Differentiation Operator over an Infinite Extension of Local Field*, vol. 207 of Lect. Notes in Pure and Appl. Math., Marcel Dekker, New York, 1999, 167–178.

[429] KOCHUBEI, A. N., *Pseudo-Differential Equations and Stochastic over non-Archimedean Fields*, vol. 244 of Monographs and Textbooks in Pure and Appl. Math., Marcel Dekker, New York, 2001.

[430] KOCHURA, A. I. and DIDENKO, A. V., On the solution of the differential equation of fractional order with piecewise constant coefficients, *NIINTI, Dep. in Ukrain., Odessa*, **855**, (1985) 1–13.

[431] KOELLER, R. C., Applications of fractional calculus to the theory of viscoelasticity, *Trans. ASME J. Appl. Mech.*, **51**(2), (1984) 299–307.

[432] KOELLER, R. C., Polynomial operators, Stieltjes convolution and fractional calculus in hereditary mechanics, *Acta Mech.*, **58**(3-4), (1986) 251–264.

[433] KOH, C. G. and KELLY, J. M., Application of fractional derivatives to seismic analysis of base-isolated models, *Earthq. Engrg. Struct. Dynam.*, **19**, (1990) 229–241.

[434] KOLMOGOROV, A. N. and FOMIN, S. V., *Fundamentals of the Theory of Functions and Functional Analysis*, Nauka, Moscow, 1968.

[435] KOLMOGOROV, A. N. and FOMIN, S. V., *Elements of Theory of Functions and Functional Analysis*, Bibfismat, Dover, New York, 1984.

[436] KOLWANKAR, K. M. and GANGAL, A. D., Local fractional Fokker-Planck equation, *Phys. Rev. Lett.*, **80**(2), (1998) 14–17.

[437] KOMATSU, H., Fractional powers of operators, *Pacific J. Math.*, **19**, (1966), 285–346.

[438] KOMATSU, H., Interpolation spaces, *Pacific J. Math.*, **21**, (1967), 89–111.

[439] KOMATSU, H., Fractional powers of operators. Negative powers, *J. Math. Soc. Japan*, **21**, (1969),205–220.

[440] KOMATSU, H., Fractional powers of operators. Potential operators, *J. Math. Soc. Japan*, **21**, (1969),221–228.

[441] KOMATSU, H., Fractional powers of operators. Dual operators, *J. Fac. Sci. Univ. Tokyo Sect. IA Math.*, **17**, (1970),373–396.

[442] KOMATSU, H., Fractional powers of operators. Interpolation of non-negative operators and imbedding theorems operators, *J. Fac. Sci. Univ. Tokyo Sect. IA Math.*, **19**, (1972),1–63.

[443] KOSTIN, V. A., The Cauchy problem for an abstract differential equation with fractional derivatives, *Dokl. Akad. Nauk*, **326**(4), (1992) 597–600.

[444] KOTOMIN, E. and KUZOVKOV, V., *Modern Aspects of Diffusion-Controlled Reactions: Cooperative Phenomena in Bimolecular Processes*, vol. 34 of Comprehensive Chemical Kinetics, Elsevier, Amsterdam, 1996.

[445] KRALL, N. and TRIVELPIECE, A. W., *Principles of Plasma Physics*, Academic Press, New York and London, 1973.

[446] KRASNOV, M. I., KISELEV, A. I., and MAKARENKO, G. I., *Integral Equations*, Nauka, Moscow, 1976.

[447] KRÄTZEL, E., Bemerkungen zur Meijer-transformation und anwendungen, *Math. Nachr.*, **30**, (1965) 327–334.

[448] KRÄTZEL, E., Die faltung der L-transformation, *Wiss. Z. Friedrich-Schiller-University Jena*, **14**, (1965) 383–390.

[449] KRÄTZEL, E., Eine Verallgemeinerung der Laplace- und Meijer-transformation, *Wiss. Z. Friedrich-Schiller-University Jena*, **14**, (1965) 369–381.

[450] KRÄTZEL, E., Differentiationssätze der L-transformation und differentialgleichungen nach dem operator $(d/dt)t^{1/n-\nu}(t^{1-1/n}d/dt)^{n-1}t^{\nu+1-2/n}$, *Math. Nachr.*, **35**, (1967) 105–114.

[451] KRÄTZEL, E., Integral transformations of Bessel-type., in: RUSEV, P., DIMOSKI, I., and KIRYAKOVA, V. (Eds.), *Generalised Functions and Operational Calculus*, Bulgarian Acad. Sci., Sofia, 1975, 148–155.

[452] KUPFERMAN, R., Fractional kinetics in Kac-Zwanzig heat bath models, *J. Stat. Phys.*, **114**(1-2), (2004) 291–326.

[453] KUTNER, R., Hierarchical spatio-temporal coupling in fractional wanderings. (I) -Continuous-time Weierstrass flights, *Physica A*, **264**(1-2), (1999) 84–106.

[454] KUTNER, R., Spatio-temporal coupling in the continuous-time Levy flights, *Solid State Ionics*, **119**(1-4), (1999) 323–329.

[455] KUTNER, R. and REGULSKI, M., Hierarchical spatio-temporal coupling in fractional wanderings. (II). Diffusion phase diagram for Weierstrass walks, *Physica A*, **264**(1-2), (1999) 107–133.

[456] KUTNER, R. and SWITALA, F., Study of the non-linear autocorrelations within the Gaussian regime, *Europhys. Lett.*, **33**(4), (2003) 495–503.

[457] KUTNER, R. and SWITALA, F., Remarks on the possible universal mechanism of the non-linear long-term autocorrelations in financial time-series, *Physica A: Stat. Mech. Appl.*, **344**(1-2), (2004) 244–251.

[458] KUN, Y., LI, L., and JIAXIANG, T., Stochastic seismic response of structures with added viscoelastic dampers modeled by fractional derivative, . *Earthquake Engineering and Engineering Vibration*, **2**(1), 2003, 133–140.

[459] LACROIX, S. F., *Traite du Calcul Differential et du Calcul Integral*, Courcier, Paris, 1819.

[460] LAGUTIN, A. A., NIKULIN, Y. A., and UCHAIKIN, W., The "knee" in the primary cosmic ray spectrum as consequence of the anomalous diffusion of the particles in the fractal interstellar medium, *Nuclear Phys. B (Proc. Suppl.)*, **97**(1), (2001) 267–270.

[461] LANGLANDS, T. A. M. and HENRY, B. I., The accuracy and stability of an implicit solution method for the fractional diffusion equation *J. Comput. Phys.*, **205**(2), 2005, 79–736.

[462] LAPLACE, P. S., *Theorie Analytique des Probabilites*, Courcier, Paris, 1812.

[463] LATORA, V., RAPISARDA, A., and RUFFO, S., Superdiffusion and out-of-equilibrium chaotic dynamics with many degrees of freedom, *Phys. Rev. Lett.*, **83**(2), (1999) 104–107.

[464] LATORA, V., RAPISARDA, A., and RUFFO, S., Chaotic dynamics and superdiffusion in a Hamiltonian system with many degrees of freedom, *Physica A*, **280**(1-2), (2000) 81–86.

[465] LEE, C., HOOPES, M. F., DIEHL, J., GILLILAND, W., HUXEL, G., LEAVER, E. V., MCCANN, K., UMBANHOWAR, J., and MOGILNER, A., Non-local concepts and models in biology, *J. Theoret. Biol.*, **210**(2), (2001) 201–219.

[466] LEE, C.-Y., SRIVASTAVA, H. M., and YUEH, W.-C., Explicit solutions of some linear ordinary and partial fractional differintegral equations, *Appl. Math. Comp.*, **144**(1), (2003) 11–25.

[467] LEIBNIZ, G. W., *Leibniz Mathematische Schriften*, Olms-Verlag, Hildesheim, 1962.

[468] LEITH, J. R., Fractal scaling of fractional diffusion processes, *Signal Process.*, **83**(11), (2003) 2397–2409.

[469] LU, J.-F. and HANYGA, A., Wave field simulation for heterogeneous porous media with singular memory drag forc, *J. Comput. Phys.*, **208**(2), 2005, 651–674.

[470] LE MEHAUTE, A. and CREPY, G., Introduction to transfer and motion in fractal media: the geometry of kinetics, *Solid State Ionics*, **9-10**, (1983), 17–30.

[471] LENORMAND, R., Flow through porous media: limits of fractal patterns, *Proc. Roy. Soc. London Ser. A*, **423**, (1989), 159–168.

[472] LENZI, E. K., MALACARNE, L. C., MENDES, R. S., and PEDRON, I. T., Anomalous diffusion, nonlinear fractional Fokker-Planck equation and solutions, *Physica A: Stat. Mech. Appl.*, **319**, (2003) 245–252.

[473] LENZI, E. K., MENDES, R. S., FA, K. S., MALACARNE, L. C., and DA SILVA, L. R., Anomalous diffusion: Fractional Fokker-Planck equation and its solutions, *J. Math. Phys.*, **44**(5), (2003) 2179–2185.

[474] LEONCINI, X., KUZNETSOV, L., and ZASLAVSKY, G. M., Chaotic advection near a three-vortex collapse, *Phys. Rev. E*, **63**(3), (2001) 6224.

[475] LEONCINI, X., KUZNETSOV, L., and ZASLAVSKY, G. M., Evidence of fractional transport in point vortex flow, *Chaos Solit. Fract.*, **19**(2), (2004) 259–273.

[476] LESKOVSKIJ, I. P., Theory of fractional differentiation, *Differ. Equations*, **13**, (1977) 118–120.

[477] LESKOVSKIJ, I. P., On the solution of the linear homogeneous differential equations with fractional derivatives and constant coefficients, *P. Tul'sk Politech. Inst.*, Tula, 1980, 85–88.

[478] LETNIKOV, A. V., Theory of differentiation with an arbitrary index, *Mat. Sb.*, **3**, (1868) 1–66.

[479] LÉVI, P., *Theorie de l'Addition des Variables Aleatories*, Gauthier-Villars, Paris, 1937.

[480] LÉVI, P., *Processus Stochastiques et Mouvement Brownien*, Gauthier-Villars, Paris, 1948.

[481] LIN, S.-D., LING, W.-C., NISHIMOTO, K., and SRIVASTAVA, H. M., A simple fractional-calculus approach to the solutions of the Bessel differential equation of general order and some of its applications, *Comput. Math. Appl.*, **49**(49), (2005) 1487–1498.

[482] Lin, S.-D., Shyu, J.-C., Nishimoto, K., and Srivastava, H. M., Explicit solutions of some general families of ordinary and partial differential equations associated with the Bessel equation by means of fractional calculus, *J. Fract. Calc.*, **25**, (2004) 33–45.

[483] Lin, S.-D., Srivastava, H. M., Tu, S.-T., and Wang, P.-Y., Some families of linear ordinary and partial differential equations solvable by means of fractional calculus, *Internat. J. Differential Equations Appl.*, **4**(4), (2002) 405–421.

[484] Lin, S.-D., Tu, S.-T., Chen, I-C., and Srivastava, H. M., Explicit solutions of certain family of fractional differintegral equations, *Hyperion Sci. J. Ser. A Math. Phys. Electr. Engrg.*, **2**, (2001) 85–90.

[485] Lin, S.-D., Tu, S.-T., and Srivastava, H. M., Explicit solutions of certain ordinary differential equations by means of fractional calculus, *J. Fract. Calc.*, **20**, (2001) 35–43.

[486] Lin, S.-D., Tu, S.-T., and Srivastava, H. M., Certain classes of ordinary and partial differential equations solvable by means of fractional calculus, *Appl. Math. Comput.*, **131**(2-3), (2002) 223–233.

[487] Lin, S.-D., Tu, S.-T., and Srivastava, H. M., Explicit solutions of some classes of non-Fuchsian differential equations by means of fractional calculus, *J. Fract. Calc.*, **21**, (2002) 49–60.

[488] Lin, S.-D., Tu, S.-T., and Srivastava, H. M., A unified presentation of certain families of non-Fuchsian differential equations via fractional calculus operators, *Comput. Math. Appl.*, **45**(12), (2003) 1861–1870.

[489] Lin, S.-D., Tu, S.-T., Srivastava, H. M., and Wang, P.-Y., Certain operators of fractional calculus and their applications to differential equations, *Comput. Math. Appl.*, **44**(12), (2002) 1557–1565.

[490] Lion, A., On the thermodynamics of fractional damping elements, *Contin. Mech. Thermodyn.*, **9**(2), (1997) 83–96.

[491] Lion, A. and Kardelky, C., The Payne effect in finite viscoelasticity: constitutive modelling based on fractional derivatives and intrinsic time scales, *Internat. J. Plasticity*, **20**(7), (2004) 1313–1345.

[492] Liouville, J., Memoire sur quelques question de geometrie et de mecanique et sur un nouveau genre de calcul pour resudre ces question, *J. Ecole Polytech.*, **13**(21), (1832) 1–69.

[493] Liouville, J., Memoire sur le changemant de variables dans calcul des differentielles d'indices quelconques, *J. Ecole Polytech.*, **15**(24), (1835) 17–54.

[494] Liu, F., Anh, V., and Turner, I., Numerical solution of the space fractional Fokker-Planck equation, *J. Comput. Appl. Math.*, **166**(1), (2004) 209–219.

[495] Liu, G. and Han, R. S., Dynamic behaviour of vortex matter, memory effect and Mittag-Leffler relaxation, *Chinese Phys. Lett.*, **18**(2), (2001) 269–271.

[496] Liu, G., Li, B. G., Han, R. S., and Yang, S. D., The motion of repulsive Brownian particles in quenched disorder, *Commun. in Theoretical Phys.*, **35**(2), (2001) 241–244.

[497] Lokshin, A. A. and Suvorova, Y. V., *The Mathematical Theory of Wave Propagation in Media with Memory*, Moskov. Gos. University, Moscow, 1982.

[498] LU, S. L., MOLZ, F. J., and FIX, G. J., Possible problems of scale dependency in applications of the three-dimensional fractional advection-dispersion equation to natural porous media, *Water Resources Research*, **38**(9), (2002) 1165.

[499] LUBICH, C., Fractional linear multisep methods for Abel-Volterra integral equations of the second kind, *Math. Comput.*, **45**(172), (1985) 463–469.

[500] LUBICH, C., Discretized fractional calculus, *SIAM J. Math. Anal.*, **17**(3), (1986) 704–719.

[501] LUBICH, C., Fractional linear multistep methods for Abel-Volterra integral equations of first kind, *IMA J. Numer. Anal.*, **1**(7), (1987) 97–106.

[502] LUCHKO, Y. F., Operational method in fractional calculus, *Fract. Calc. Appl. Anal.*, **2**(4), (1999) 463–488.

[503] LUCHKO, Y. F., Asymptotics of zeros of the Wright function, *Z. Anal. Anwendungen*, **19**(2), (2000) 583–595.

[504] LUCHKO, Y. F., On the distribution of zeros of the Wright function, *Integral Transform. Spec. Funct.*, **11**(2), (2001) 195–200.

[505] LUCHKO, Y. F. and GORENFLO, R., The initial value problem for some fractional differential equations with the Caputo derivatives, *Fachbereich Math. und Inf. A*, **8**(98), (1998) 1–23.

[506] LUCHKO, Y. F. and GORENFLO, R., Scale-invariant solutions of a partial differential equation of fractional order, *Fract. Calc. Appl. Anal.*, **1**(1), (1998) 63–78.

[507] LUCHKO, Y. F. and GORENFLO, R., An operational method for solving fractional differential equations with the Caputo derivatives, *Acta Math. Vietnam.*, **24**(2), (1999) 207–233.

[508] LUCHKO, Y. F. and SRIVASTAVA, H. M., The exact solution of certain differential equations of fractional order by using operational calculus, *Comput. Math. Appl.*, **29**(8), (1995) 73–85.

[509] LUCHKO, Y. F. and YAKUBOVICH, S. B., Operational calculus for the generalized fractional differential operator and applications, *Math. Balkanica*, **7**(2), (1993) 119–130.

[510] LUCHKO, Y. F. and YAKUBOVICH, S. B., An operational method for solving some classes of integro-differential equations, *Differ. Equat.*, **30**(2), (1994) 247–256.

[511] LUSZCZKI, Z. and RZEPECKI, B., An existence theorem for ordinary differential equations of order $\alpha \in (0, 1]$, *Demonstr. Math.*, **20**(3-4), (1987) 471–475.

[512] LUTZ, E., Fractional Langevin equation, *Phys. Rev. E*, **64**(5), (2001) 1106.

[513] LYNCH, V. E., CARRERAS, B.A., DEL CASTILLO-NEGRETE, D., FERREIRA, K.M., and HICKS, H. R., Numerical methods for the solution of partial differential equations of fractional order, *J. Comput. Phys.*, **192**(2), (2003) 406–421.

[514] MAINARDI, F., On the initial value problem for the fractional diffusion-wave equation, in: RUGGERI, S. and RIONERO, T. (Eds.), *Waves and Stability in Continuous Media*, World Sci. Publishing, Singapure, Bologna, 1993, 246–251.

[515] MAINARDI, F., Fractional relaxation and fractional diffusion equations of fractional order, in: AMES, W. F. (Ed.), *12th IMACS World Congress*, vol. 1, Atlanta, 1994, 329–333.

[516] MAINARDI, F., Fractional relaxation in anaelastic solids, *J. Alloys and Compounds*, **211-212**, (1994) 534–538.

[517] MAINARDI, F., The time fractional diffusion-wave equation, *Izv. Vyssh. Uchebn. Zaved. Radiofiz.*, **38**(1-2), (1995) 20–36.

[518] MAINARDI, F., The fractional relaxation-oscillation and fractional diffusion-wave phenomena, *Chaos, Solitons and Fractals*, **7**(9), (1996) 1461–1477.

[519] MAINARDI, F., The fundamental solutions for the fractional diffusion-wave equation, *Appl. Math. Lett.*, **9**(6), (1996) 23–28.

[520] MAINARDI, F., Fractional calculus: Some basic problems in continuum and statistical mechanics, in: *CISM courses and lectures*, vol. 378, Springer-Verlag, Berlin, 1997, 291–348.

[521] MAINARDI, F., Applications of fractional calculus in mechanics, in: RUSEV, P., DIMOVSKI, I. and KIRYAKOVA, V. (Eds.) *Transform Method and Special Functions, Varna'96*, Bulgarian Academy of Sciences, Sofia, 1998, pp. 309–334.

[522] MAINARDI, F., Applications of integral transforms in fractional diffusion processes, *Integral Transform. Spec. Funct.*, **15**(6), (2004) 477–484.

[523] MAINARDI, F. and BONETTI, E., The application of real-order derivatives in linear viscoelasticity, *Rheol. Acta*, **26**(suppl), (1988) 64–67.

[524] MAINARDI, F. and GORENFLO, R., Fractional calculus: special functions and applications, in: D., C., DATTOLI, G., and SRIVASTAVA, H. M. (Eds.), *Advanced Special Functions and Applications (Melfi, 1999)*, Aracne Edictrice, Rome, 2000, 165–188.

[525] MAINARDI, F. and GORENFLO, R., On Mittag-Leffler type functions in fractional evolution processes, *J. Comp. Math. Appl.*, **118**(1-2), (2000) 293–299.

[526] MAINARDI, F., LUCHKO, Y. F., and PAGNINI, G., The fundamental solution of the space-time fractional diffusion equation, *Fract. Calc. Appl. Anal.*, **4**(2), (2001) 153–192.

[527] MAINARDI, F. and PAGNINI, G., Space-time fractional diffusion: exact solutions and probability interpretation, in: *"WASCOM 2001"-11th Conference on Waves and Stability in Continuous Media*, World Sci. Publishing, River Edge, New Jersey, 2002, 296–301.

[528] MAINARDI, F. and PAGNINI, G., The Wright functions as solutions of the time-fractional diffusion equation, in : SRIVASTAVA, H. M., DATTOLI, G., and RICCI, P. E. (Eds), *Advanced Special Functions and Related Topics in Differential Equations*, Appl. Math. Comp., **141**(1), (2003) 51–62.

[529] MAINARDI, F., PAGNINI, G., and SAXENA, R. K., Fox *H* functions in fractional diffusion, *J. Comp. Math. Appl.*, **178**(1-2), (2005) 321–331.

[530] MAINARDI, F. and PARADISI, P., Fractional diffusive waves, *J. Comp. Acoust.*, **9**(4), (2001) 1417–1436.

[531] MAINARDI, F., PARADISI, P., and GORENFLO, R., Probability distributions generated by the fractional diffusion equation, in: KERTESZ, J. and KONDOR, I. (Eds.), *International Workshop on Econophysics*, Kluwer Acad Publishers., Dordrecht, 1998.

[532] MAINARDI, F., PIRONI, P., and TAMPIERI, F., A numerical approach to the generalized Basset problem for a sphere accelerating in a viscous fluid, in: P.A.THIBAULT and BERGERON, D. M. (Eds.), *3rd Annual Conference of the Computational Fluid Dynamics Society of Canada*, vol. 2, Alberta, Canada, 1995, 105–112.

[533] MAINARDI, F., PIRONI, P., and TAMPIERI, F., On a generalization of the Basset problem via fractional calculus, in: TABARROK, B. and DOST, S. (Eds.), *5th Canadian Congress of Applied Mechanics*, vol. 2, Victoria, Canada, 1995, 836–837.

[534] MAINARDI, F., RABERTO, M., GORENFLO, R., and SCALAS, E., Fractional calculus and continuous-time finance II: the waiting-time distribution, *Physica A*, **287**(3-4), (2000) 468–481.

[535] MAINARDI, F. and TOMIROTTI, M., On a special function arising in the time fractional diffusion-wave equation, in: *Transform Methods and Special Functions (Sofia 1994)*, SCTP, Singapore, 1995, 171–183.

[536] MAINARDI, F. and TOMIROTTI, M., Seismic pulse propagation with constant Q and stable probability distributions, *Ann. Geofis.*, **40**(5), (1997) 1311–1328.

[537] MAKRIS, N. and CONSTANTINOU, M. C., Fractional derivative Maxwell model for viscous dampers, *J. Struct. Engrg. ASCE*, **117**(9), (1991) 2708–2724.

[538] MAKRIS, N. and CONSTANTINOU, M. C., Spring-viscous damper systems for combined seismic and vibration isolation, *Earthq. Engrg. Struct. Dynam.*, **21**(8), (1992) 649–664.

[539] MAKRIS, N. and CONSTANTINOU, M. C., Models of viscoelasticity with complex-order derivatives, *J. Engrg. Mech. ASCE*, **119**(7), (1993) 1453–343.

[540] MAKRIS, N., DARGUSH, G. F., and CONSTANTINOU, M. C., Dynamic analysis of generalized viscoelastic fluids, *ASCE J. Engrg. Mech.*, **119**(8), (1993) 1663–1679.

[541] MALAKHOVSKAYA, R. M. and SHIKHMANTER, E. D., Certain formulae for the realization of operators, and their applications to the solution of integro-diferential equations, *Trudy Tomsk. Gos. University*, **220**, (1975) 46–56.

[542] MALLAMACE, F. and STANLEY, H. E. (Eds.), *The Physics of Complex Systems*, vol. 134 of International School of Physics Enrico Fermi, IOS Press, Amsterdam, 1997.

[543] MAMBRIANI, A., The derivative of any order and the solution of hyper-geometric equation, *Boll. University Mat. Ital.*, **2**(3), (1940) 9–10.

[544] MAMBRIANI, A., On the partial derivative of any order and the solution of Euler and Poisson equation, *Ann. Scoula Norm. Sup. Pisa*, **2**(11), (1942) 79–97.

[545] MANDELBROJT, S., Sulla generalizzazione del calcolo delle variazione, *Atti Accad. Naz. Lincei. Rend. Cl. Sci. Fis. Mat.-Natur. Ser. 6*, **1**, (1925) 151–156.

[546] MANDELBROJT, S., *The Fractal Geometry of Nature*, Freeman, New York, 1982.

[547] MANDELBROT, B. B., *Fractals and Scaling in Finance*, Springer-Verlag, Berlin, 1997.

[548] MANDELBROT, B. B. and VAN NESS, J. W., Fractional Brownian motions, fractional noises and applications, *SIAM Rev.*, **10**(4), (1968) 422–437.

[549] MARAVALL, D., Nuevos tipos de ecuaciones integrales e integrodiferenciales. Nuevos fenómenos de oscilaciones (Spanish), *Rev. R. Acad. Cienc. Exactas, Fis. Nat.*, (1956) 1–151.

[550] MARAVALL, D., *Ingeniería de las Oscilaciones* (Spanish), Dossat, Madrid, 1959.

[551] MARAVALL, D., *Teoria e Aplicaçaõ das Oscilações* (Portuguese), Editora Globo Porto Alegre, Brasil, 1964.

[552] MARAVALL, D., *Mecánica y Cálculo Tensorial* (Spanish), Dossat, Madrid, 1965.

[553] MARAVALL, D.,Ecuaciones diferenciales lineales no enteras y oscilaciones fraccionarias (Spanish), *Rev. R. Acad. Cienc. Exactas, Fis. Nat.*, **65**(1), (1971) 245–258.

[554] MARAVALL, D., Las soluciones exactas de las ecuaciones integrales de la rheología y de la viscoelasticidad y sus consecuencioas físicas (Spanish), *Rev. R. Acad. Cienc. Exactas, Fis. Nat.*, (1978).

[555] MARAVALL, D., Ecuaciones diferenciales e integrales fraccionarias. Relaciones entre el cálculo simbólico y el fraccionario (Spanish), *Rev. R. Acad. Cienc. Exactas, Fis. Nat.*, **98**(1), (2004) 3–16.

[556] MARAVALL, D., Sistemas de ecuaciones diferenciales fraccionarias. Ecuaciones diferenciales e integrales de orden infinito. Nuevas funciones trascendentes (Spanish), *Rev. R. Acad. Cienc. Exactas, Fis. Nat.* (to appear), (2005).

[557] MARGOLIN, G. and BERKOWITZ, B., Continuous time random walks revisited: first passage time and spatial distributions, *Physica A: Stat. Mech. Appl.*, **334**(1-2), (2004) 46–66.

[558] MARTÍNEZ, C. and SANZ, M. A., *The Theory of Fractional Powers of Operators*, vol. 187 of North-Holland Mathematical Studies, Elsevier Science Publishers, Amsterdam, London and New York, 2001.

[559] MARZAN, S. A., Aproximate methods to solutions of fractional differential equations, *Vesti Nats. Akad. Nauk Belarusi*, **48**.

[560] MATHAI, A. M. and SAXENA, R. K., *The H- Function with Applications in Statistics and other Disciplines*, Wiley and Sons, New York, 1978.

[561] MATIGNON, D., *Représentations en variables d'état de modèles de guides d'ondes avec dérivation fractionnaire*, Ph.D. thesis, Université Paris XI, Orsay, 1994.

[562] MATIGNON, D., Stability results on fractional differential equations with applications to control processing, in: *Computational Engineering in Systems and Applications Multiconference*, vol. 2, IMACS, IEEE-SMC, Lille, 1996, 963–968.

[563] MATIGNON, D. and MONTSENY, G., *Fractional Differential Systems: Models, Methods, and Applications*, vol. 5 of *ESAIM Proceedings*, SMAI, Paris, 1998.

[564] MATSUZAKI, T. and NAKAGAWA, M., A chaos neuron model with fractional differential equation, *J. Phys. Soc. Japan*, **72**(10), (2003) 2678–2684.

[565] MBODJE, B. and MONTSENY, G., Boundary fractional derivative control of the wave equation, *IEEE Trans. Automat. Control*, **40**(2), (1995) 378–382.

[566] MCBRIDE, A. C., *Fractional Calculus and Integral Transforms of Generalized Functions*, vol. 31 of *Research Notes in Math.*, Pitman, London, 1979.

[567] MCBRIDE, A. and ROACH, G. (Eds.), *Fractional Calculusand its applications*, vol. 138 of Research Notes in Math., Pitman, London, (1985).

[568] MEERSCHAERT, M. M., BENSON, D. A., and BAUMER, B., Multidimensional advection and fractional dispersion, *Phys. Rev. E*, **59**(5), (1999) 5026–5028.

[569] MEERSCHAERT, M. M., BENSON, D. A., SCHEFFLER, H.-P., and BAEUMER, B., Stochastic solution of space-time fractional diffusion equation, *Phys. Rev. E*, **65**(4), (2002) 1103.

[570] MEERSCHAERT, M. M. and TADJERAN, C., Finite difference approximations for fractional advection-dispersion flow equations, *J. Comp. Appl. Math.*, **172**(1), (2004) 65–77.

[571] MELCHIOR, P., POTY, A., and OUSTALOUP, A., Motion control by ZV shaper synthesis extended for fractional systems and its application to CRONE control, in: DUARTE, M.and TENREIRO-MACHADO, J. A. (Eds.), *Special Issue of Fractional Order Derivatives and their Applications*, Nonlinear Dynamics, **38**(1-2), 2004, 401–416.

[572] MELLER, N. A., Some applications of operational calculus to problems of analysis (Rusian), *J. Vychisl. Mat. i Mat. Fiz.*, **1**(3), (1963) 71–78.

[573] MESHKOV, S. and ROSSIKHIN, Y. A., Sound wave propagation in a viscoelastic medium whose hereditary properties are determined by weakly singular kernels, in: RABOTNOV, Y. (Ed.), *Waves in Inelastic Media*, Technical Kishinev University, Moldova, Kishinev, Moldova, 1970, 162–172.

[574] METZLER, R., BARKAI, E., and KLAFTER, J., Anomalous diffusion and relaxation close to thermal equilibrium: A fractional Fokker-Planck equation approach, *Phys. Rev. Lett.*, **82**(18), (1999) 3563–3567.

[575] METZLER, R., BARKAI, E., and KLAFTER, J., Anomalous diffusion and relaxation close to thermal equilibrium: A fractional Fokker-Plank equation approach, *Phys. Rev. Lett.*, **82**(3), (1999) 563–567.

[576] METZLER, R., BARKAI, E., and KLAFTER, J., Anomalous transport in disordered systems under the influence of external fields, *Physica A*, **266**(1-4), (1999) 343–350.

[577] METZLER, R., BARKAI, E., and KLAFTER, J., Deriving fractional Fokker-Planck equations from a generalised master equation, *Europhys. Lett.*, **46**(4), (1999) 431–436.

[578] METZLER, R. and COMPTE, A., Stochastic foundation of normal and anomalous Cattaneo-type transport, *Physica A: Stat. Mech. Appl.*, **268**(3-4), (1999) 454–468.

[579] METZLER, R. and COMPTE, A., Generalized diffusion-advection schemes and dispersive sedimentation: A fractional approach, *J. Phys. Chem. B*, **104**(16), (2000) 3858–3865.

[580] METZLER, R., GLOCKLE, W. G., NONNENMACHER, T. F., and WEST, B. J., Fractional tuning of the Riccati equation, *Fractals Interdiscipl. J. Complex Geometry of Nature*, **5**(4), (1997) 597–601.

[581] METZLER, R., GLÖKLE, W. G., and NONNENMACHER, T. F., Fractional model equation for anomalous diffusion, *Physica A: Stat. Mech. Appl.*, **211**(1), (1994) 13–24.

[582] METZLER, R. and KLAFTER, J., Accelerating Brownian motion: A fractional dynamics approach to fast diffusion, *Europhys. Lett.*, **51**(5), (2000) 492–498.

[583] METZLER, R. and KLAFTER, J., Boundary value problems for fractional diffusion equations, *Physica A: Stat. Mech. Appl.*, **278**(1-2), (2000) 107–125.

[584] METZLER, R. and KLAFTER, J., The fractional Fokker-Planck equation: Dispersive transport in an external force field, *J. Molec. Liquids*, **86**(1-3), (2000) 219–228.

[585] METZLER, R. and KLAFTER, J., From a generalized Chapman-Kolmogorov equation to the fractional Klein-Kramers equation, *J. Phys. Chem. B*, **104**(16), (2000) 3851–3857.

[586] METZLER, R. and KLAFTER, J., The random walk's guide to anomalous diffusion: A fractional dynamics approach, *Phys. Rep.*, **339**(1), (2000) 1–77.

[587] METZLER, R. and KLAFTER, J., Levy meets Boltzmann: strange initial conditions for Brownian and fractional Fokker-Planck equations, *Physica A: Stat. Mech. Appl.*, **302**(1-4), (2001) 290–296.

[588] METZLER, R. and KLAFTER, J., From stretched exponential to inverse power-law: fractional dynamics, Cole-Cole relaxation processes, and beyond, *J. Non-Crystalline Solids*, **305**(1-3), (2002) 81–87.

[589] METZLER, R. and KLAFTER, J., The restaurant at the end of the random walk: recent developments in the description of anomalous transport by fractional dynamics, *Physics A: Math. Gen.*, **37**(31), (2004) 161–208.

[590] METZLER, R., KLAFTER, J., and SOKOLOV, I. M., Anomalous transport in external fields: Continuous time random walks and fractional diffusion equations extended, *Phys. Rev. E*, **58**(2), (1998) 1621–1633.

[591] METZLER, R. and NONNENMACHER, T. F., Fractional diffusion: exact representations of spectral functions, *J. Phys. A: Math. Gen.*, **30**(4), (1997) 1089–1093.

[592] METZLER, R. and NONNENMACHER, T. F., Fractional diffusion, waiting-time distributions, and Cattaneo-type equations, *Phys. Rev. E*, **57**(6), (1998) 409–414.

[593] METZLER, R. and NONNENMACHER, T. F., Space- and time-fractional diffusion and wave equations, fractional Fokker-Planck equations, and physical motivation, *J. Chem. Phys.*, **284**(1), (2002) 67–90.

[594] METZLER, R. and NONNENMACHER, T. F., Fractional relaxation processes and fractional rheological models for the description of a class of viscoelastic materials, *Int. J. Plasticity*, **19**(7), (2003) 941–959.

[595] METZLER, R. and SOKOLOV, I. M., Superdiffusive Klein-Kramers equation: Normal and anomalous time evolution and Levy walk moments, *Europhys. Lett.*, **58**(4), (2002) 482–488.

[596] MICHIEL, H., ADRIAENSSENS, G. J., and DAVIS, E. A., Extended-state mobility and its relation to the tail-state distribution in a-$Si : H$, *Phys. Rev. B*, **34**(2), (1986) 486–499.

[597] MICU, S. and ZUAZUA, E., On the control of a 1-d fractional order parabolic equation, in: *Procceding of XVIII CEDYA, Tarragona*, (2003), 41–43.

[598] MICU, S. and ZUAZUA, E., On the controllability of a fractional order parabolic equation, *SIAM J. Cont. Optim* (to appear), 2005.

[599] MIKUSINSKI, J., *Operational Calculus*, Pergamon Press, New York, 1959.

[600] MILLER, K. S., Fractional differential equations, *J. Fract. Calc.*, **3**, (1993) 49–57.

[601] MILLER, K. S., Derivatives of non-integer order, *Math. Mag.*, **68**(3), (1995) 183–192.

[602] MILLER, K. S. and ROSS, B., Fractional Green's functions, *Indian J. Pure Appl. Math.*, **22**(9), (1991) 763–767.

[603] MILLER, K. S. and ROSS, B., *An Introduction to the Fractional Calculus and Fractional Differential Equations*, Wiley and Sons, New York, 1993.

[604] MITTAG-LEFFLER, G. M., Sur la nouvelle function E_α, *C.R. Acad. Sci. Paris*, **137**, (1903) 554–558.

[605] MITTAG-LEFFLER, G. M., Sopra la funzione $E_\alpha(x)$, *Rend. Accad. Lincei, ser. 5*, **13**, (1904) 3–5.

[606] MITTAG-LEFFLER, G. M., Sur la representation analytique d'une branche uniforme d'une fonction monogene, *Acta Math.*, **29**, (1905) 101–182.

[607] MOMANI, S. and AL-KHALED, K., Numerical solutions for systems of fractional differential equations by the decomposition method, *Appl. Math. Comp.*, **162**(3), (2005) 1351–1365.

[608] MONTROLL, E. W. and WEISS, G. H., Random walk on lattices. II, *J. Math. Phys.*, **6**(2), (1965) 167–181.

[609] MONTSENY, G., AUDOUNET, J., and MATIGNON, D., Diffusive representation for pseudo-differentially damped nonlinear systems, in: ISIDORI, A., LAMNABHI-LAGARRIGUE, F., and RESPONDEK, W. (Eds.) *Nonlinear control*, vol. 2, Springer-Verlag, Berlin, 2000, 163–182.

[610] MOREAU, X., RAMUS-SERMENT, C., and OUSTALOUP, A., Fractional differentiation in passive vibration control, in: TENREIRO MACHADO, J. A. (Ed.), *Special Issue of Fractional Order Calculus and Its Applications*, Nonlinear Dynamics, **29**(1-4), 2002, 343–362.

[611] MORITA, T., Solution for the initial-value problem of a fractional differential equation, *J. Korean Phys. Soc.*, **46**(3), (2005) 580–584.

[612] NAHAK, C. and NANDA, S., Duality for multiobjective fractional control problems with generalized invexity *Korean J. Comput. Appl. Math.*, **5**(2), (1998) 433–446.

[613] NAKHUSHEV, A. M., Inverse problems for degenerate equations and Volterra integral equations of the third kind, *Differentsial'nye Uravnenija*, **10**(1), (1974) 100–111.

[614] NAKHUSHEV, A. M., A certain mixed problem for degenerate elliptic equations, *Differentsial'nye Uravnenija*, **11**(1), (1975) 192–195.

[615] NAKHUSHEV, A. M., The Darboux problem for a certain degenerate second order loaded integro-differential equation, *Differentsial'nye Uravnenija*, **12**(1), (1976) 103–108.

[616] NAKHUSHEV, A. M., The Sturm-Liouville Problem for a second order ordinary differential equation with fractional derivatives in the lower terms, *Dokl. Akad. Nauk SSSR*, **234**(2), (1977) 308–311.

[617] NAKHUSHEV, A. M., Continuous differential equation and their difference analogues, *Dokl. Akad. Nauk SSSR*, **300**(4), (1988) 796–799.

[618] NAKHUSHEV, A. M., *Mathematical Problems in Biology*, Nauka, Moscow, 1996.

[619] NAKHUSHEV, A. M. and BORISOV, V. N., Boundary value problems for loaded parabolic equations and their applications to the prediction of ground water level, *Differ. Equat.*, **13**(1), (1977) 105–110.

[620] NAUMKIN, P. I. and SHISHMAREV, I. A., *Nonlinear Nonlocal Equations in the Theory of Waves*, vol. 133 of *Translations of Mathematical Monographs*, A.M.S., Providence, Rhode Island, 1994.

[621] NIGMATULLIN, R. R., On the theory of relaxation with "remnant" memory, *Phys. Status Solidi B*, **124**(1), (1984) 389–393.

[622] NIGMATULLIN, R. R., To the theoretical explanation of the "universal" response, *Phys. Status Solidi B*, **123**(2), (1984) 739–745.

[623] NIGMATULLIN, R. R., The realization of the generalized transfer equation in a medium with fractal geometry, *Phys. Status Solidi B*, **133**, (1986) 425–430.

[624] NIGMATULLIN, R. R., A fractional integral ant its physical interpretation, *J. Theoret. Math. Phys.*, **90**(3), (1992) 242–251.

[625] NIGMATULLIN, R. R. and RYABOV, Y. E., Cole-Davidson dielectric relaxation as a self-similar relaxation process, *Phys. Solid State*, **39**(1), (1997) 87–99.

[626] NIKIAS, C. L. and SHAO, M., *Signal Processing with Alpha-Stable Distributions and Applications*, Wiley and Sons, New York, 1995.

[627] NIKOL'SKII, S. M., *The Approximation of Functions of Several Variables and the Imbedding Theorems* (Russian), Nauka, Moscow, 1977.

[628] NIKOL'SKII, S. M., *Course of Mathematical Analysis (Russian)*, vol. 1-2, Nauka, Moscow, 1983.

[629] NISHIMOTO, K., *Fractional Calculus: Integration and Differentiation of Arbitrary Order,* Vols. I-V, Descartes Press, Koriyama, 1984, 1989, 1991, and 1996.

[630] NISHIMOTO, K., *An Essence of Nishimoto's Fractional Calculus*, Descartes Press, Koriyama, 1991.

[631] NISHIMOTO, K., OWA, S., and SRIVASTAVA, H. M., Solutions to a new class of fractional differintegral equations, *J. College Engrg. Nihon Univ. Ser. B*, **25**, (1984) 75–78.

[632] NISHIMOTO, K., SRIVASTAVA, H. M., and TU, S.-T., Application of fractional calculus in solving certain classes of Fuchsian differential equations, *J. College Engrg. Nihon Univ. Ser. B*, **32**, (1991) 119–126.

[633] NISHIMOTO, K., SRIVASTAVA, H. M., and TU, S.-T., Solutions of some second-order linear differential equations by means of fractional calculus, *J. College Engrg. Nihon Univ. Ser. B*, **33**, (1992) 15–25.

[634] NONNENMACHER, T. F., *Fractional relaxation equations for viscoelasticity and related phenomena*, in: CASAS-VÁZQUEZ, J. and JOU, D. (Eds.), *Rheological Modelling:Thermodynamical and Statistical Aproches*, vol. 381 of Lect. Notes in Phys., Springer-Verlag, Berlin, 1991, 309–320.

[635] NONNENMACHER, T. F. and GLOKLE, W. G., A fractional model for mechanical stress relaxation, *Philos. Mag. Lett.*, **64**(2), (1991) 89–93.

[636] NONNENMACHER, T. F. and NONNENMACHER, D. J. F., Towards the formulation of a non-linear fractional extended irreversible thermodynamics, *Acta Phys. Hungar.*, **66**(1-4), (1989) 145–154.

[637] NOOLANDI, J., Equivalence of multiple-trapping model and time-dependent random walk, *Phys. Rev. B*, **16**(4), (1977) 474–479.

[638] NOVAK, M. M., *Paradigms of Complexity*, World Scientific Publishing Company, Singapore and London, 2000.

[639] NUTTING, P. G., A new generalized law of deformation, *J. Franklin Inst.*, **191**, (1921) 679–685.

[640] OLDHAM, K. B., Diffusive transport to planar, cylindrical, and spherical electrodes, *J. Electroanal. Chem.*, **41**, (1973) 351–360.

[641] OLDHAM, K. B. and SPANIER, J., The replacement of Fick's law by a formulation involving semi-differentiation, *J. Electroanal. Chem.*, **26**, (1970) 331–341.

[642] OLDHAM, K. B. and SPANIER, J., A general solution of the diffusion equation for semi-infinite geometries, *J. Math. Anal. Appl.*, **39**, (1972) 655–669.

[643] OLDHAM, K. B. and SPANIER, J., *The Fractional Calculus*, Academic Press, New York, 1974.

[644] OLDHAM, K. B. and SPANIER, J., Fractional calculus and its applications, *Bull. Inst. Politehn. Iasi. Sect. I*, **24 (28)**(3-4), (1978) 29–34.

[645] O'REGAN, B. and GRÄTZEL, M., A Low-Cost Hight-efficiency solar cell based on dye-sensitized colloidal $TiO2$ films, *Nature*, **353**, (1991) 737.

[646] ORENSTEIN, J. and KASTNER, M., Photocurrent transient spectroscopy: Mesurament of the density of localized states in a-$As2Se3$, *Phys. Rev. Lett.*, **46**(1), (1981) 421–424.

[647] ORSONI, B., MELCHIOR, P., OUSTALOUP, A., BADIE, T., and ROBIN, G., Fractional motion control: Application to an XY cutting table, in: TENREIRO-MACHADO, J. A. (Ed.), *Special Issue of Fractional Order Calculus and Its Applications*, Nonlinear Dynamics, **29**(1-4), 2002, 297–314.

[648] OSADA, H., Self-similar diffusions on a class of infinitely ramified fractals, *J. Math. Soc. Japan*, **47**(4), (1995) 591–616.

[649] OSADA, H., A family of diffusion processes on Sierpinski carpets, *Prob. Theory Related Fields*, **119**(2), (2001) 275–310.

[650] O'SHAUGHNESSY, L., Problem ♯ 433, *Amer. Math. Monthly*, **25**, (1918) 172–173.

[651] ÖTÜK, V. Z. and SHKHANEKOV, M. H., Difference method for the differential equation with fractional derivative, *Indian J. Pure Appl. Math.*, **30**(5), (1999) 517–523.

[652] OUSTALOUP, A., *Systèmes Asservis Linéaires d'Ordre Fractionnaire* (French), Masson, Paris, 1983 .

[653] OUSTALOUP, A., *La Commande CRONE. Comande Robuste D'ordre Non Entiere*, Hermes Sci., Paris, 1991.

[654] OUSTALOUP, A., *La Derivation non Entiére, Theorie Synthese et Applications*, Hermes Sci., Paris, 1995.

[655] OUSTALOUP, A. and BENHAFSIA, M., Active fractional order filters of any kind and interest of their incorporation in feedback-control systems, *Onde Electrique*, **61**(3), (1981) 31–37.

[656] OUSTALOUP, A., POMMIER, V., and LANUSSE, P., Design of a fractional control using performance contours. Application to an electromechanical system *Fract. Calc. Appl. Anal.*, **6**(1), 2003, 1–24.

[657] OZAKTAS, H. M., KUTAY, M. A., and BOZDAGI, G.,Digital computation of the fractional Fourier transform , *IEEE Trans. Sig. Proc.*, **44**, 1996, 2141–2150.

[658] OZAKTAS, H. M., ZALEVSKY, Z., and KUTAY, M. A., *The Fractional Fourier Transform with Applications in Optics and Signal Processing*, John Wiley and Sons, New York, 2000.

[659] PACHEPSKY, Y., BENSON, D., and RAWLS, W., Simulating scale-dependent solute transport in soils with the fractional advective-dispersive equation, *Soil Sci. Soc. Amer. J.*, **64**(4), (2000) 1234–1243.

[660] PARADISI, P., CESARI, R., MAINARDI, F., and TAMPIERI, F., The Fractional Fick's law for non-local transport processes, *Physica A*, **293**(1-2), (2001) 130–142.

[661] PAZY, A., *Semigroups of Linear Operators and Applications to Partial Differential Equations*, Springer-Verlag, 1983.

[662] PEČARIĆ, J. E., PERIĆ, I., and SRIVASTAVA, H. M., A family of the Cauchy type mean-value theorems, *J. Math. Anal. Appl.*, **306**(2), (2005) 730–739.

[663] PEI, S.C. and YEH, M.H., The discrete fractional cosine and sine transforms, *IEEE Trans. Sig. Proc.*, **49**(6), (2001) 1198–1207.

[664] PEI, S.C., YEH, M.H., and TSENG, C.-C., Discrete fractional Fourier-transform based on orthogonal projections, *IEEE Trans. Sig. Proc.*, **47**(5), (1999) 1335-1348.

[665] PEKALSKI, A. and SZNAJD-WERON, K., *Anomalous Diffusion from Basics to Applications*, Springer-Verlag, Berlin, 1999.

[666] PELITI, L. and VULPIANI, A. (Eds.), *Measures of Complexity*, Springer-Verlag, Berlin, 1988.

[667] PETRAS, I.,The fractional-order controllers: Methods for their synthesis and application, *J. Electrical Engineering*, **50**(9-10), (1999) 284-288.

[668] PETRAS, I., *Fractional Calculus in Control*, Ph.D. thesis, Department of Informatics and Process Control, Faculty BERG, Technical University of Kosice, 2000.

[669] PETRAS, I.,Control of fractional-order Chua's system, *J. Electrical Engineering*, **53**(7-8), (2002) 219-222.

[670] PETRAS, I., DORCAK, L., and KOSTIAL, I., Control quality enhancement by fractional order controllers, *Acta Montanistica Slovaca*, **2**, (1998) 143–148.

[671] PETRAS, I. and DORCAK, L., Fractional-order control systems: Modelling and simulation, *Dynamics and Control*, **1**, (2000) 1–8.

[672] PFISTER, G. and SCHER, H., Time dependent electrical transport in amorphous solids-$AS2SE$, *Phys. Rev. B*, **15**(2), (1977) 62–83.

[673] PFISTER, G. and SCHER, H., Dispersive (non-Gaussian) transient transport in disordered solids, *Adv. Phys.*, **27**(5), (1978) 747–798.

[674] PICOZZI, S. and WEST, B. J., Fractional Langevin model of memory in financial markets, *Phys. Rev. E*, **66**(4-5), (2002) 6118.

[675] PITCHER, E. and SEWELL, W. E., Existence theorems for solutions of differential equations of non-integral order, *Bull. Amer. Math. Soc.*, **44**(2), (1938) 100–107.

[676] PLONKA, A., Recent developments in dispersive kinetics, *Progr. React. Kinet. Mech.*, **25**(2), (2000) 109–127.

[677] PODLUBNY, I., Fractional derivatives: a new stage in process modelling and control, in: *4th International DAAAM Symposium*, Brno, Czech Republic, 1993, 263–264.

[678] PODLUBNY, I., Fractional-order systems and fractional-order controllers, *Inst. Exper. Phys. Slovak Acad. Sci.*, **3**(94).

[679] PODLUBNY, I., The Laplace transform method for linear differential equations of the fractional order, *Inst. Exper. Phys. Slovak Acad. Sci.*, **2**, (1994) 1–35.

[680] PODLUBNY, I., Solution of linear fractional differential equations with constant coefficients, in: RUSEV, P., DIMOSKI, I., and KIRYAKOVA, V. (Eds.), *1st International Workshop on Transforms Methods and Special Functions*, Bulgarian Acad. Sci., Sofia, 1994, 239–254.

[681] PODLUBNY, I., Numerical solution of ordinary fractional differential equations by the fractional difference method, in: *Advances in Differential Equations*, Gordon and Breach, Amsterdam, 1995, 507–515.

[682] PODLUBNY, I., *Fractional Differential Equations*, Academic Press, San Diego, 1999.

[683] PODLUBNY, I., Fractional-order systems and $PI^{\lambda}D^{\mu}$-controllers, *Ieee Trans. Automat. Control*, **44**(1), (1999) 208–214.

[684] PODLUBNY, I., PETRAS, I., O'LEARY, P., DORCAK, P. and VINAGRE, B., Analogue realization of fractional order controllers, in: TENREIRO-MACHADO, J. A. (Ed.), *Special Issue of Fractional Order Calculus and Its Applications*, Nonlinear Dynamics, **29**(1-4), 2002, 281-296.

[685] PODLUBNY, I., PETRAS, I., VINAGRE, B. M., CHEN, Y. Q., O'LEARY, P., and DORCAK, L., Realization of Fractional Order Controllers , *Acta Montanistica Slovaca* , **8**(4), 2003, 233–235.

[686] POST, E. L., Discussion of the solution of $(d/dx)^{1/2}y = y/x$ (♯ problem 433), *Amer. Math. Monthly*, **26**, (1919) 37–39.

[687] PRABHAKAR, T. R., A singular integral equation with a generalized Mittag-Leffler function in the kernel, *Yokohama Math. J.*, **19**, (1971) 7 15.

[688] PRITZ, T., Analysis of four-parameter fractional derivative model of real solid materials, *J. Sound Vibrat.*, **195**(1), (1996) 103–115.

[689] PRUDNIKOV, A. P., BRYCHKOV, Y. A., and MARICHEV, O. I., *Integrals and Series, vols. I, II and III*, Elementary Functions, Gordon and Breach, New York, 1986.

[690] PRUDNIKOV, A. P., BRYCHOV, Y. A., and MARICHEV, O. I., *Integrals and Series*, vols. 2 and 3, Fiziko-Matematicheskaya Literatura, Moscow, 2003.

[691] QUEVEDO, J. and ESCOBET, T.(Eds.), *Digital Control 2000: Past, Present and Future of PID Control*, Elsevier Science, Holland, 2000.

[692] RABOTNOV, Y. N., The Equilibrium of an elastic medium with after effect, *Akad. Nauk SSSR Prikl. Mat. Mekh.*, **12**, (1948) 81-91.

[693] RABOTNOV, Y. N., *Elements of Hereditary Solid Mechanics*, MIR, Moscow, 1980.

[694] RAFAELYAN, S. G., The basis property of certain biorthogonal systems in $L^2(-\sigma,\sigma)$ with weights, *Izv. Akad. Nauk Armyan. SSR, Ser. Mat.*, **19**(3), (1984) 207–218.

[695] RAINA, R. K., SRIVASTAVA, H. M., KILBAS, A. A., and SAIGO, M., Solvability of some Abel-type integral equations involving the Gauss hypergeometric function as kernels in the spaces of summable functions, *ANZIAM J.*, **43**(2), (2001) 291–320.

[696] RAMUS-SERMENT, C., MOREAU, X., NOUILLANT, M., OUSTALOUP, A., and LEVRON, F., Generalised approach on fractional response of fractal networks, *Chaos Solit. Fract.*, **14**(3), (2002) 479–488.

[697] RANGARAJAN, G. and DING, M. Z., Anomalous diffusion and the first passage time problem, *Phys. Rev. E*, **62**(1), (2000) 120–133.

[698] RANGARAJAN, G. and DING, M. Z., First passage time distribution for anomalous diffusion, *Phys. Lett. A*, **273**(5-6), (2000) 322–330.

[699] REN, F. Y., LIANG, J. R., QIU, W. Y., WANG, X. T., XU, Y., and NIGMATULLIN, R. R., An anomalous diffusion model in an external force fields on fractals, *Phys. Lett. A*, **312**(3-4), (2003) 187–197.

[700] REN, F. Y., LIANG, J. R., QIU, W. Y., and XU, Y., Fractional Fokker-Planck equation on heterogeneous fractal structures in external force fields and its solutions, *Physics A: Stat. Mech. Appl.*, **326**(3-4), (2003) 430–440.

[701] REN, F. Y., LIANG, J. R., QIU, W. Y., and XU, Y., Universality of stretched Gaussian asymptotic behaviour for the fractional Fokker-Planck equation in external force fields, *Physics A: Math. Gen.*, **36**(27), (2003) 7533–7543.

[702] REN, F. Y., QIU, W. Y., XU, Y., and LIANG, J. R., Answer to an open problem proposed by E Barkai and J Klafter, *Physics A: Math. Gen.*, **37**(42), (2004) 9919–9922.

[703] REN, F. Y., XU, Y., QIU, W. Y., and LIANG, J. R., Universality of stretched Gaussian asymptotic diffusion behavior on biased heterogeneous fractal structure in external force fields, *Chaos Solit. Fract.*, **24**(1), (2005) 273–278.

[704] REN, F. Y., YU, Z. G., ZHOU, J., LE MEHAUTE, A., and NIGMATULLIN, R. R., The relationship between the fractional integral and the fractal structure of a memory set, *Physica A*, **246**(3-4), (1997) 419–429.

[705] RICHARDSON, L. F., Atmosphere diffusion shown on a distance neighbor graph, *Proc. Roy. Soc. London, Ser. A*, **110**, (1926) 709–737.

[706] RIEMANN, B., Versuch einer allgemeinen auffasung der integration und differentiation, *Gesammelte Werke*, (1876) pp. 62.

[707] RIESZ, M., L'integral de Riemann-Liouville et le probleme de Cauchy, *Acta Math.*, **81**, (1949) 1–223.

[708] RIEWE, F., Mechanics with fractional derivative, *Phys. Rev. E*, **55**(3), (1997) 581–592.

[709] RIVERO, M., RODRÍGUEZ-GERMÁ, L., and SRIVASTAVA, H. M., and TRUJILLO, J. J., Generalized method of Frobenius for linear fractional differential equations (Preprint), *Depto. Análisis Matemático* (Universidad de La Laguna), (2005).

[710] ROBINSON, D. A., The use of control systems analysis in neurophysiology of eye movements, *Ann. Rev. Neurosci*, **4**, (1981), 462–503.

[711] ROCCO, A. and WEST, B. J., Fractional calculus and the evolution of fractal phenomena, *Physica A*, **265**(3-4), (1999) 535–546.

[712] RODRÍGUEZ, J., Operational calculus for the generalized Bessel operator, *Serdica*, **15**(3), (1989) 179–186.

[713] RODRÍGUEZ, J., TRUJILLO, J. J., and RIVERO, M., Operational fractional calculus of Kräzel integral transformation, in: *Differential Equations*, vol. 118 of Lect. Notes in Pure and Appl. Math, Xanthi (Greece), 1987, 613–620.

[714] ROGERS, L., Operators and fractional derivatives for viscoelastic constitutive equations, *J. Rheol.*, **27**, (1983) 351–372.

[715] ROMAN, H. E. and GIONA, M., Fractional diffusion equation on fractals: three-dimensional case and scattering function, *J. Phys. A: Math. Gen.*, **25**(8), (1992) 2107–2117.

[716] ROSS, B.(Ed.), *Fractional Calculus and Its Applications: Proceedings of the International Conference, New Haven, June 1974*, Springer Verlag, New York, 1974.

[717] ROSS, B., The development of fractional calculus: 1695-1900, *Historia Math.*, **4**, (1977) 75–89.

[718] ROSSIKHIN, Y. A. and SHITIKOVA, M. V., Applications of fractional calculus to dynamic problems of linear and non linear hereditary mechanics of solids, *Appl. Mech. Rev.*, **50**(1), (1997) 15–67.

[719] RUBIN, B., *Fractional Integrals and Potentials*, Addison Wesley Longman Ltd., 1996.

[720] RYABOV, Y. E., Behaviour of fractional diffusion at the origin, *Phys. Rev. E*, **68**(13), (2003) 102.

[721] SABATIER, J., OUSTALOUP, A., GARCIA, A., and LANUSSE, P., CRONE control: Principles and extension to time-variant plants with asymptotically constant coefficients. in: TENREIRO-MACHADO, J. A. (Ed.), *Special Issue of Fractional Order Calculus and Its Applications*, Nonlinear Dynamics, **29**(1-4), 2002, 363–385.

[722] SAHA-RAY, S. and BERA, R. K. Solution of an extraordinary differential equation by Adomian decomposition method, *J. Appl. Math.*, **4**, (2004), 4331–338.

[723] SAICHEV, A. and ZASLAVSKY, G. M., Fractional kinetic equations: solutions and applications, *Chaos*, **7**(4), (1997) 753–764.

[724] SAIGO, M. and KILBAS, A. A., On Mittag-Leffler type function and applications, *Integral Transform. Spec. Funct.*, **7**(1-2), (1998) 97–112.

[725] SAIGO, M. and KILBAS, A. A., The solution of a class of linear differential equations in terms of functions of the Mittag-Leffler type, *Differ. Equat.*, **36**(2), (2000) 193–202.

[726] SAKAKIBARA, S., Properties of vibration with fractional derivative damping of order 1/2, *JSME Internat. J. C*, **40**(3), (1997) 393–399.

[727] SAMKO, S. G., Spaces of Riesz potentials, *Izv. Akad. Nauk SSSR, Ser. Math.*, **40**(5), (1976) 1143–1172.

[728] SAMKO, S. G., *Hypersingular Integrals and their Applications*, Taylor and Francis, London, 2002.

[729] SAMKO, S. G., KILBAS, A. A., and MARICHEV, O. I., *Fractional Integrals and Derivatives: Theory and Applications*, Gordon and Breach Science Publishers, Switzerland, 1993.

[730] SÁNCHEZ, A. and VÁZQUEZ, L. M., Nonlinear wave propagation in disordered media, *Int. J. Modern Phys. B*, **5**(18), (1991) 2825–2882.

[731] SANZ, M. A. and MARTÍNEZ, C., Potencias fraccionarias de operadores y cálculo funcional (Spanish), *Rev. R. Acad. Cienc. Exactas, Fis. Nat.*, **98**(1), (2004) 39–53.

[732] SANZ-SERNA, J. M., A numerical method for a partial integro-differential equation, *SIAM J. Numer. Anal.*, **25**(2), (1988) 319–327.

[733] SAXENA, R. K., MATHAI, A. M., and HAUBOLD, H. J., Unified fractional kinetic equation and a fractional diffusion equation, *Astrophys. Space Science*, **290**(3-4), (2004) 299–310.

[734] SCALAS, E., GORENFLO, R., and MAINARDI, F., Uncoupled continuous-time random walks: Solution and limiting behavior of the master equation, *Phys. Rev. E*, **69**(1), (2004) 1107.

[735] SCALAS, E., GORENFLO, R., MAINARDI, F., and RABERTO, M., Revisiting the derivation of the fractional order diffusion equation. Scalling and disordered systems, *Fractals*, **11**(suppl.), (2003) 281–289.

[736] SCARPI, G. B., Sui modelli reologici intermedi per liquidi viscoelatici, *Atti Accad. Sci. Torino Cl. Sci. Fis. Mat.-Natur.*, **107**, (1973) 239–243.

[737] SCHER, H., MARGOLIN, G., and BERKOWITZ, B., Towards a unified framework for anomalous transport in heterogeneous media, *J. Chem. Phys.*, **284**(1-2), (2002) 349–359.

[738] SCHER, H., MARGOLIN, G., METZLER, R., KLAFTER, J., and BERKOWITZ, B., The dynamical foundation of fractal stream chemistry: The origin of extremely long retention times, *Geophys. Research Lett.*, **29**(5), (2002) 1029.

[739] SCHER, H. and MONTROLL, E. W., Anomalous transit-time dispersion in amorphous solids, *Phys. Rev. B*, **12**(2), (1975) 455–477.

[740] SCHER, H., SHLESINGER, M., and BENDLER, J., Time-scale invariance in transport and relaxation, *Phys. Today*, **44**(1), (1991) 26–34.

[741] SCHERTZER, D., LARCHEVEQUE, M., DUAN, J., YANOVSKY, V. V., and LOVEJOY, S., Fractional Fokker-Planck equation for nonlinear stochastic differential equations driven by non-Gaussian Levy stable noises, *J. Math. Phys.*, **42**(1), (2001) 200–212.

[742] SCHIESSEL, H., METZLER, R., BLUMEN, A., and NONNENMACHER, T. F., Generalized viscoelastic models: Their fractional equations with solutions, *Physics A: Math. Gen.*, **28**(23), (1995) 6567–6584.

[743] SCHIESSEL, H., OSHANIN, G., and BLUMEN, A., Dynamics and conformational properties of polyampholytes in external electrical fields, *J. Chem. Phys.*, **103**(12), (1995) 5070–5074.

[744] SCHMIDLIN, F. W., Theory of trap-controlled transient photoconduction, *Phys. Rev. B*, **16**(4), (1978) 474–479.

[745] SCHNEIDER, W. R., Fractional diffusion, in: *Dynamics and Stochastic Processes*, vol. 355 of Lect. Notes in Phys., Springer-Verlag, Lisbon, 1988, 276–286.

[746] SCHNEIDER, W. R. and WYSS, W., Fractional diffusion and wave equation, *J. Math. Phys.*, **30**(1), (1989) 134–144.

[747] SCHUMER, R., BENSON, D. A., MEERSCAERT, M. M., and WHEATCRAFT, S. W., Eulerian derivation of fractional advection-dispertion equation, *J. Contam. Hydrol.*, **48**(1-2), (2001) 69–88.

[748] SCHUMER, R., BENSON, D. A., MEERSCHAERT, M. M., and BAEUMER, B., Fractal mobile/immobile solute transport, *Wat. Res. Res.*, **39**(10), (2003) 1296.

[749] SCHUMER, R., BENSON, D. A., MEERSCHAERT, M. M., and BAEUMER, B., Multiscaling fractional advection-dispersion equations and their solutions, *Wat. Res. Res.*, **39**(1), (2003) 1022.

[750] SCHWARTZ, L., *Theorie des Distributions*, Hermann and Cie., Paris, 1950-1951.

[751] SCOTT, A., *Nonlinear Science: Emergence and Dynamics of Coherent Structures*, Oxford University Press, Oxford, 1999.

[752] SCOTT BLAIR, G. W., The role of psychophysics in rheology, *J. Colloid Sci.*, **2**, (1947) 21–32.

[753] SCOTT BLAIR, G. W., *Survey of General and Applied Rheology*, Pitman, London, 1949.

[754] SCOTT BLAIR, G. W. and CAFFYN, J. E., An application of the theory of quasi-properties to the treatment of anomalous strain-stress relations, *Philos. Mag. Lett.*, **40**, (1949) 80–94.

[755] SHLESINGER, M. F., ZALAVSKY, G. M., and KLAFTER, J., Strange kinetics, *Nature*, **363**, (1993) 31–37.

[756] SEITKAZIEVA, A., Integro-differential equations with a kernel which has a weak singularity, *Stud. IntegroDiffer. Equat.*, **13**, (1980) 132–135.

[757] SEMENCHUK, N. P., A class of differential equations of nonintegral order, *Differentsial'nye Uravnenija*, **18**(10), (1982) 1831–1833.

[758] SEREDYNSKA, M. and HANYGA, A., Nonlinear Hamiltonian equations with fractional damping, *J. Math. Phys.*, **4**(14), (2000) 1291–2156.

[759] SEREDYNSKA, M. and HANYGA, A., A nonlinear differential equation of fractional order with chaotic properties, *Int. J. Bifur. Chaos Appl. Sci. Engrg.*, **4**(14), (2004) 1291–1304.

[760] SHAWAGFEH, N. T., Analytical approximate solutions for nonlinear fractional differential equations, *Appl. Math. Comp.*, **131**(2-3), (2002) 517–529.

[761] SHIMIZU, N. and ZHANG, W., Fractional calculus approach to dynamic problems of viscoelastic materials, *JSME Internat. J. C*, **42**(4), (1999) 825–837.

[762] SHKHANUKOV, M. K., On the convergence of difference schemes for differential equations with a fractional derivative, *Dokl. Akad. Nauk*, **348**(6), (1996) 746–748.

[763] SHKHANUKOV, M. K., KEREFOV, A. A., and BEREZOVSKII, A. A., Boundary value problems for the heat equation with a fractional derivative in the boundary conditions, and difference methods for their numerical realization, *Ukrain. Math. J.*, **45**(9), (1993) 1445–1455.

[764] SHLESINGER, M. F., Asymptotic solutions of continuous time random walks, *J. Statist. Phys.*, **10**(5), (1974) 421–433.

[765] SHLESINGER, M. F., WEST, B. J., and KLAFTER, J., Levi dynamics of enhanced diffusion: Application to turbulence, *Phys. Rev. Lett.*, **58**(1), (1987) 100–103.

[766] SHLESINGER, M. F., ZASLAVSKY, G. M., and FRISCH, U., *Levi Flight and Related Topics in Physics*, Springer-Verlag, Berlin, 1995.

[767] SHLESINGER, M. F., ZASLAVSKY, G. M., and KLAFTER, J., Asymptotic solutions of continuous time random walks, *J. Statist. Phys.*, **10**(5), (1974) 421–434.

[768] SHLESINGER, M. F., ZASLAVSKY, G. M., and KLAFTER, J., Strange kinetics, *Nature*, **363**(6424), (1993) 31–37.

[769] SHUSHIN, A. I., Kinetics of subdiffusion-assisted reactions: Non-Markovian stochastic Liouville equation approach, *New J. of Physics*, **7**, (2005) 21–38.

[770] SJOBERG, M. and KARI, L., Non-linear behavior of a rubber isolator system using fractional derivatives, *Vehicle Syst. Dynam.*, **37**(3), (2002) 217–236.

[771] SMIRNOV, I. P., VIROVLYANSKY, A. L., and ZASLAVSKY, G. M., Theory and applications of ray chaos to underwater acoustics, *Phys. Rev. E*, **64**(3), (2001) 6221.

[772] SMITH, W. and VRIES, H., Rheological models containing fractional derivatives, *Rheol. Acta.*, **9**, (1970) 525–534.

[773] SNEDDON, I. N., *Fourier Transforms*, Dover Publications, New York, 1995.

[774] SO, F. and LIU, K. L., A study of the subdiffusive fractional Fokker-Planck equation of bistable systems, *Physica A: Stat. Mech. Appl.*, **331**(3-4), (2004) 378–390.

[775] SODDEMANN, T., SCHIESSEL, H., and BLUMEN, A., Molecular-dynamics simulations of polyampholytes: Instabilities due to excess charges and external fields, *Phys. Rev. E*, **57**(2), (1998) 81–90.

[776] SOKOLOV, I. M., Thermodynamics and fractional Fokker-Planck equations, *Phys. Rev. E*, **63**(5), (2001) 6111.

[777] SOKOLOV, I. M., BLUMEN, A., and KLAFTER, J., Dynamics of annealed systems under external fields: CTRW and the fractional Fokker-Planck equations, *Europhys. Lett.*, **56**(2), (2001) 175–180.

[778] SOKOLOV, I. M., BLUMEN, A., and KLAFTER, J., Linear response in complex systems: CTRW and the fractional Fokker-Planck equations, *Physica A*, **302**(1-4), (2001) 268–278.

[779] SOKOLOV, I. M., CHECHKIN, A. V., and KLAFTER, J., Distributed-order fractional kinetics, *Acta Phys. Polonica B*, **35**(4), (2004) 1323–1341.

[780] SOKOLOV, I. M., KLAFTER, J., and BLUMEN, A., Does strange kinetics imply unusual thermodynamics?, *Phys. Rev. E*, **64**(2), (2001) 1107.

[781] SOKOLOV, I. M., KLAFTER, J., and BLUMEN, A., Fractional kinetics, *Phys. Today*, **55**(11), (2002) 48–54.

[782] SOKOLOV, I. M. and METZLER, R., Towards deterministic equations for Levy walks: The fractional material derivative, *Phys. Rev. E*, **67**(1), (2003) 101.

[783] SOKOLOV, I. M. and METZLER, R., Non-uniqueness of the first passage time density of Levy random processes, *Physics A: Math. Gen.*, **37**(46), (2004) L609–L615.

[784] SOLTEIRO, E. J., TENREIRO-MACHADO, J. A., and MOURA, P. B., Fractional order dynamics in a GA planner, *J. Signal Proces.*, **83**(11), 2003, 2377-2386.

[785] SRIVASTAVA, H. M., On an extension of the Mittag-Leffler function, *Yokohama Math. J.*, **16**, (1968) 77–88.

[786] SRIVASTAVA, H. M., Fractional calculus and its applications, *Cubo Math. Ed.*, **5**(1), (2003) 33–48.

[787] SRIVASTAVA, H. M. and BUSCHMAN, R. G., *Convolution Integral Equations with Special Function Kernels*, John Wiley and Sons, New York, 1977.

[788] SRIVASTAVA, H. M. and BUSCHMAN, R. G., *Theory and Applications of Convolution Integral Equations*, vol. 79 of Kluwer Series on Mathematics, Kluwer Academic Publishers, Dordrecht, Boston and London, 1992.

[789] SRIVASTAVA, H. M., GUPTA, K. C., and GOYAL, S. P., *The H-Functions of One and Two Variables with Applications*, South Asian Publishers, New Delhi, 1982.

[790] SRIVASTAVA, H. M. and KARLSSON, P. W., *Multiple Gaussian Hypergeometric Series*, Halsted Press (Ellis Horwood Limited, Chichester), John Wiley and Sons, New York, Chichester, Brisbane and Toronto, 1985.

[791] SRIVASTAVA, H. M. and OWA, S. (Eds.), *Univalent Functions, Fractional Calculus, and Their Applications*, Ellis Horwood Ltd., John Wiley and Sons, New York, 1989.

[792] SRIVASTAVA, H. M., OWA, S., and NISHIMOTO, K., A note on a certain class of fractional differintegral equations, *J. College Engrg. Nihon Univ. Ser. B*, **25**, (1984) 69–73.

[793] SRIVASTAVA, H. M., OWA, S., and NISHIMOTO, K., Some fractional differintegral equations, *J. Math. Anal. Appl.*, **106**(2), (1985) 360–366.

[794] SRIVASTAVA, H. M. and SAIGO, M., Multiplication of fractional calculus operators and boundary value problems involving the Euler-Darboux equation, *J. Math. Anal. Appl.*, **121**(2), (1987) 325–369.

[795] SRIVASTAVA, H. M. and SAXENA, R. K., Operators of fractional integration and their applications, *Appl. Math. Comput.*, **118**, (2001) 1–52.

[796] SRIVASTAVA, H. M. and SAXENA, R. K., Some Volterra-type fractional integro-differential equations with a multivariable confluent hypergeometric function in the kernel, *J. Integral Equations Appl.*, **17**(2), (2005) 199–217.

[797] STANISLAVSKY, A. A., Fractional dynamics from the ordinary Langevin equation, *Phys. Rev. E*, **67**(2), (2003) 1111-1122.

[798] STANISLAVSKY, A. A., Subordinated brownian motion and its fractional Fokker-Planck equation, *Phys. Scripta*, **67**(4), (2003) 265–268.

[799] STANESCU, D., KIM, D., and WOYCZYNSKI, W. A., Numerical study of interacting particles approximation for integro-differential equations , *J. Comput. Phys.*, **206**(2), 2005, 706–726.

[800] STANKOVIC, B., Differential equations with fractional derivatives and nonconstant coefficients, *Integral Transform. Spec. Funct.*, **6**(13), (2002) 489–496.

[801] STANKOVIC, B., A system of partial differential equations with fractional derivatives, *Math. Vesnik*, **3-4**(54), (2002) 187–194.

[802] STEIN, E. M. and WEISS, G., *Introduction to Fourier Analysis on Euclidean Spaces*, vol. 32 of *Princeton Mathematical Series*, Princeton University Press, Princeton, New Jersey, 1971.

[803] STIASSNIE, M., On the application of fractional calculus for the formulation of viscoelastic models, *Appl. Math. Modell.*, **3**, (1979) 300–302.

[804] SUÁREZ, J. I., VINAGRE, B. M., CALDERÓN, A. J., MONJE, C. A., and CHEN, Y. Q., Using fractional calculus for lateral and longitudinal control of autonomous vehicles, in: PICHLER, F. R.and MORENO, R.(Eds.) *Autonomous and Control System* , vol. 2809 of Lect. Notes in Comput. Sci., 2003, 337–348.

[805] SUN, J. P. and ZHAO, Y. H., Multiplicity of positive solutions of a class of nonlinear fractional differential equations, *Computers. Math. Appl.*, **49**(1), (2005) 73–80.

[806] SUNG, Y. M., BARKAI, E., SILBEY, R. J., and LEE, S., Fractional dynamics approach to diffusion-assisted reactions in disordered media, *J. Chem. Phys.*, **116**(6), (2002) 2338–2341.

[807] SUZUKI, N. and BIYAJIMA, M., Fractional Fokker-Planck equation and oscillatory behavior of cumulant moments, *Phys. Rev. E*, **65**(1), (2002).

[808] TANABE, H., *Equations of evolution*, Pitman Press, London,1979.

[809] TATAR, N. E., The decay rate for a fractional differential equation, *J. Math. Anal. Appl.*, **295**(2), (2004) 303–314.

[810] TATAR, N. and KIRANE, M., Exponential growth for a fractionally damped wave equation, *Zeit. Anal. Anw.*, **22**(1), (2003) 167–177.

[811] TAZALI, A. Z., Local existence theorems for ordinary differential equations of fractional order, in: *Ordinary and Partial Differential Equations*, vol.964 of Lecture Notes in Math., Springer-Verlag, Berlin, 1982, 652–665.

[812] TAZALI, A. Z. and KARIM, I. F., Continuous dependence on a parameter and continuation of solutions in the extended sense in ordinary differential equations of fractional order, *Pure Appl. Sci. B*, **21**(1), (1994) 270–280.

[813] TELBAN, R. J., CARDULLO, F. M., and GUO, L., Investigation of mathematical models of otolith organs for human centered motion cueing algorithms, *J. Amer. Inst. Aeronaut. Astronaut*, (No. AIAA-2000-4291).

[814] TENREIRO-MACHADO, J. A., Analysis and design of fractional-order digital control systems, *J. Systems Anal. Model. Simul.*, **27**, 1997, 107–122.

[815] TENREIRO-MACHADO, J. A.(Ed.), *Special Issue of Fractional Order Calculus and Its Applications*, Nonlinear Dynamics, **29**(1-4), 2002.

[816] TENREIRO-MACHADO, J. A., Implementing discrete-time fractional-order controllers, *J. Adv. Comput. Intelligence*, **5**(5), 2001, 279-285.

[817] TIEDJE, T., CEBULKA, J. M., MOREL, D. L., and ABELES, B., Evidence for dispersive, *Phys. Rev. Lett.*, **46**(1), (1981) 425.

[818] TIEDJE, T. and ROSE, A., A physical interpretation of dispersive transport in disordered semiconductors, *Solid State Commun.*, **37**, (1981) 49.

[819] TITCHMARSH, E. C., *Introduction to the Theory of Fourier Integrals*, Chelsea Publishing Company, New York, 1986 (First Edition: Oxford University Press, Oxford, 1937).

[820] TORVIK, P. J. and BAGLEY, R. L., On the appearance of the fractional derivative in the behavior of real materials, *J. Appl. Mech.*, **51**, (1984) 294–298.

[821] TRIEBEL, H., *Interpolation Theory, Function Spaces and Differential Operators*, North-Holland, Amsterdam-New York, 1978.

[822] TRINKS, C. and RUGE, P., Treatment of dynamic systems with fractional derivatives without evaluating memory-integrals, *Computat. Mech.*, **29**(6), (2002) 471–476.

[823] TRUJILLO, J., RIVERO, M., and BONILLA, B., On a Riemann-Liouville Generalized Taylor's Formula, *J. Math. Anal. Appl.*, **231**(1), (1999) 255–265.

[824] TSALLIS, C., General statistical mechanics: connection with thermodynamics, *J. Phys. A: Math. Gen.*, **24**(2), (1991) 69–72.

[825] TSALLIS, C., Generalized entropy-based criterion for consistent testing, *Phys. Rev. E*, **58**(2), (1998) 1442–1445.

[826] TSALLIS, C., Entropic nonextensivity: A possible measure of complexity chaos, *Chaos Solit. Fract.*, **13**(3), (2002) 371–391.

[827] TSALLIS, C. and LENZI, E. K., Anomalous diffusion: Nonlinear fractional Fokker-Planck equation, *J. Chem. Phys.*, **284**(1-2), (2002) 341–347.

[828] TSALLIS, C. and LENZI, E. K., Anomalous diffusion: Nonlinear fractional Fokker-Planck equation (vol 284, pg 341, 2002), *Chemical Physics*, **287**(1-2), (2003) 295–295.

[829] TU, S.-T., CHYAN, D.-K., and SRIVASTAVA, H. M., Some families of ordinary and partial fractional differintegral equations, *Integral Transform. Spec. Funct.*, **11**, (2001) 291–302.

[830] TU, S.-T., HUANG, Y.-T., CHEN, I-C., and SRIVASTAVA, H. M., A certain family of fractional differintegral equations, *Taiwanese J. Math.*, **4**, (2000) 417–426.

[831] TU, S.-T., LIN, S.-D., HUANG, Y.-T., and SRIVASTAVA, H. M., Solutions of a certain class of fractional differintegral equations, *Appl. Math. Lett.*, **14**(2), (2001), 223–229.

[832] TU, S.-T., LIN, S.-D., and SRIVASTAVA, H. M., Solutions of a class of ordinary and partial differential equations via fractional calculus, *J. Fract. Calc.*, **18**, (2000) 103–110.

[833] TUAN, V. K. and GORENFLO, R., Extrapolation to the limit for numerical fractional differentiation, *Z. Angew. Math. Mech.*, **75**(8), (1995) 646–648.

[834] TURCHETTI, G., USERO, D., and VÁZQUEZ, L. M., Hamiltonian systems with fractional time derivative, *Tamsui Oxford J. Math. Sci.*, **18**(1), (2002) 31–44.

[835] TURING, A. M., The chemical basis of morphogenesis, *Philos. Trans. Roy. Soc. London B*, **237**, (1952) 37–72.

[836] UCHAIKIN, V. V., Multidimensional symmetric anomalous diffusion, *J. Chem. Phys.*, **284**(1-2), (2002) 507–520.

[837] UCHAIKIN, V. V., Relaxation processes and fractional differential equations, *Int. J. Theor. Phys.*, **42**(1), (2003) 121–134.

[838] UPADHYAYA, A., RIEU, J. P., SAWADA, Y., and GLAZIER, J. A., Anomalous diffusion and non-Gaussian velocity distribution of Hydra cells in cellular aggregates, *Physica A*, **293**(3-4), (2001) 549–558.

[839] USERO, D. and VÁZQUEZ, L., Fractional derivative. A new formulation for damped systems, in: *Localization and Energy Transfer in Nonlinear System*, World Scientific, 2003, 296–303.

[840] VAN DE LAGEMAAT, J., KOPIDAKIS, N., NEALE, N. R., and FRANK, A. J., Effect of nonideal statistics on electron diffusion in sensitized nanocrystalline $TiO2$, *Phys. Rev. B*, **71**(3), (2005) 5304.

[841] VÁZQUEZ, L., Singularity analysis of a nonlinear fractional differential equation, *Rev. R. Acad. Cienc. Serie A. Mat.* (to appear), (2005).

[842] VÁZQUEZ, L., Fractional diffusion equations with internal degrees of freedom, *J. Comp. Math.*, **21**(4), (2003) 491–494.

[843] VÁZQUEZ, L., Panoramic view of fractional calculus and its applications (Spanish), *Rev. R. Acad. Cienc. Exactas, Fis. Nat.*, **98**(1), (2004) 17–25.

[844] VÁZQUEZ, L., EVANS, W. A. B., and RICKAYZEN, G., Numerical investigation of non-local Sine-Gordon model, *Phys. Lett. A*, **189**(6), (1994) 454–459.

[845] VÁZQUEZ, L. and VILELA MENDES, R., Fractionally coupled solutions of the diffusion equation, *Appl. Math. Comp.*, **141**(1), (2003) 125–130.

[846] VEBER, V. K., On a differential equation of non-integer order, *Sb. Trudov Aspir. i Soiskat. Kirgiz. University Ser. Mat. Nauk*, **10**, (1973) 7–14.

[847] VEBER, V. K., A space of fundamental functions in the theory of Liouville differentiation, *Trudy Kirgiz. Gos. University, Ser. Mat. Nauk*, **9**, (1974) 164–168.

[848] VEBER, V. K., The structure of general solution of the system $y^{(\alpha)} = Ay$; $0 < \alpha \leq 1$, *Trudy Kirgiz. Gos. University Ser. Mat. Nauk, Vyp.*, **11**, (1976) 26–32.

[849] VEBER, V. K., Asymptotic behavior of solutions of a linear system of differential equations of fractional order, *Stud. IntegroDiff. Equat.*, **16**, (1983) 19–125.

[850] VEBER, V. K., Passivity of linear systems of differential equations with fractional derivative and quasiasymptotic behaviour of solutions, *Stud. in IntegroDiff. Equat.*, **16**, (1983) 349–356.

[851] VEBER, V. K., Linear equations with fractional derivatives and constant coefficients in spaces of generalized functions, *Stud. IntegroDiff. Equat.*, **18**, (1985) 306–312.

[852] VEBER, V. K., On the general theory of linear systems with fractional derivatives, *Stud. IntegroDiff. Equat.*, **18**, (1985) 301–305.

[853] VEBER, V. K., Solution of problem of Sturm-Liouville type, *Stud. IntegroDiff. Equat.*, **21**, (1988) 245–249.

[854] VERKMAN, A. S., Solute and macromolecule diffusion in cellular aqueous compartments of living cells, *Trends Biochem. Sci.*, **27**(1), (2002) 27–33.

[855] VILELA MENDES, R. and VÁZQUEZ, L., The dynamical nature of a backlash system with and without fluid friction, *Nonlinear Dynam.* (to appear), (2005).

[856] VINAGRE, B. M., PETRAS, I., PODLUBNY, I., and CHEN, Y. Q., Using fractional order adjustment rules and fractional order reference models in model-reference adaptive control, in: TENREIRO-MACHADO, J.A. (Ed.) *Special Issue of Fractional Order Calculus and Its Applications*, Nonlinear Dynam., **29**(1-4), 2002, 269–279.

[857] VLAD, M. O., ROSS, J., and SCHNEIDER, F. W., Levy diffusion in a force field, Huber relaxation kinetics, and nonequilibrium thermodynamics: H theorem for enhanced diffusion with Levy white noise, *Phys. Rev. E*, **62**(2), (2000) 1743–1763.

[858] VLAD, M. O., VELARDE, M. G., and ROSS, J., Analogies between colored Levy noise and random channel approach to disordered kinetics, *J. Math. Phys.*, **45**(2), (2004) 736–760.

[859] VLADIMIROV, V. S., *Generalized Functions in Mathematical Physics*, Mir Publishing, Moscow, 1979.

[860] VOROSHILOV, A. A. and KILBAS, A. A., Cauchy type problem for the diffusion equation of fractional order (Russian), *Dokl. Nats. Acad. Belarusi* **49**(3), (2005) 14-18.

[861] VOROSHILOV, A. A. and KILBAS, A. A., Cauchy type problem for the diffusion-wave equation with partial Riemann-Liouville derivative (Russian), *Dokl. Akad. Nauk Belarusi* (to appear), (2005).

[862] VOROSHILOV, A. A. and KILBAS, A. A., Cauchy type problem to multi-dimensional fractional diffusion-wave equation, *Vestnik Beloruss. Gos. University Ser. 1 Fiz.-Mat. Inform.* (to appear), (2005).

[863] WANG, P.-Y., LIN, S.-D., and SRIVASTAVA, H. M., Remarks on a simple fractional-calculus approach to the solutions of Bessel differential equation of general order and its applications, *Comput. Math. Appl.* (to appear), (2005).

[864] WATANABE, Y., On some properties of fractional powers of linear operators, *Proc. Japan Acad.*, **37**(6), (1961) 273–275.

[865] WEISS, G. H., *Aspects and Applications of the Random Walk*, North-Holland, Amsterdam, 1994.

[866] WEISS, G. H. and RUBIN, R. J., Random walks theory and selected applications, *Adv. Chem. Phys.*, **52**, (1983) 363–505.

[867] WERNER, D. H. and MITTRA, R., *Frontiers in Electromagnetic*, IEEE Press and John Wiley and Sons, New York, 1999.

[868] WEST, B. J., *Fractal Physiology and Chaos in Medicine*, World Scientific Publishing Company, Singapore and London, 1990.

[869] WEST, B. J., *Physiology, Promiscuity and Prophecy at the Millennium. A Tale of Tails*, World Scientific Publishing Company, Singapore and London, 1999.

[870] WEST, B. J., Comments on the renormalization group, scaling and measures of complexity, *Chaos Solit. Fract.*, **20**(1), (2004) 33–44.

[871] WEST, B. J., BOLOGNA, M., and GRIGOLINI, P., *Physics of Fractal Operators*, Springer-Verlag, New York and Berlin, 2003.

[872] WEST, B. J., GRIGOLINI, P., METZLER, R., and NONNENMACHER, T. F., Fractional diffusion and Levy stable process, *Phys. Rev. E*, **55**(1), (1997) 99–106.

[873] WEST, B. J. and PICOZZI, S., Fractional Langevin model of memory in financial time series, *Phys. Rev. E*, **65**(3), (2002) 7106.

[874] WESTERLUND, S., Dead matter has memory!, *Phys. Scripta*, **43**(2), (1991) 174–179.

[875] WESTPHAL, U., Gebrochene potenzen abgeschlossener opertoren, definiert mit Hilfe gebrochener differenzen. Linear operators and approximations II (German), in: BIRKHÄSER, B. (Ed.), *Conf. Oberwolfach Marh. Res. Inst. (Oberwolfach, 1974)*, vol. 25 of Internat. Ser. Numer. Math., 1974, 23–27 .

[876] WESTPHAL, U., Fractional powers of infinitesimal generators of semigroups, in: HILFER, R. (Eds.) *Applications of Fractional Calculus in Physics*, World Scientific Publishing Co., 2000, 131-170.

[877] WEYL, H., Bemerkungen zum begriff des differentialquotienten gebrochener ordnung (German), *Zurich Naturf. Ges.*, **62**, (1917), 296–302.

[878] WHITTAKER, E. T. and WATSON, G. N., *A Course of Modern Analysis*, Cambridge University Press, Cambridge, 1969.

[879] WIENER, K., On solutions of boundary value problems for differential equations of noninteger order, *Math. Nachr.*, **88**, (1979) 181–190.

[880] WIENER, K., On a system of differential equations with derivatives of nonintegral order, *Beitrage Anal.*, **18**, (1981) 105–112.

[881] WIENER, K., On solutions of a differential equation of nonintegral order that occurs in the theory of polarography, *Wiss. Z. Martin-Luther-University Halle Wittenberg Math.-Natur. Reihe*, **32**(1), (1983) 41–46.

[882] WIENER, K., On the asymptotic behavior of the solutions of differential equation of nonintegral order from polarography, *Wiss. Z. Martin-Luther-University Halle-Wittenberg Math.-Natur. Reihe*, **32**(5), (1983) 75–86.

[883] WIENER, K., On the number of solutions for differential equations of nonintegral order, *Wiss. Z. Martin-Luther-University Halle-Wittenberg Math.-Natur. Reihe*, **34**(3), (1985) 112–120.

[884] WIENER, K., On the solutions of homogeneous differential equations of nonintegral order, *Wiss. Z. Martin-Luther-University Halle Wittenberg Math.-Natur. Reihe*, **34**(4), (1985) 54–58.

[885] WIENER, K., Solutions of inhomogeneous differential equations with derivatives of nonintegral order. I, *Wiss. Z. Martin-Luther-University Halle-Wittenberg Math.-Natur. Reihe*, **36**(3), (1987) 22–29.

[886] WIENER, K., Solutions of inhomogeneous differential equations with derivatives of nonintegral order. III, *Wiss. Z. Martin-Luther-University Halle-Wittenberg Math.-Natur. Reihe*, **37**(3), (1988) 127–131.

[887] WIENER, K., Solutions of systems of differential equations with derivatives of noninteger order, *Wiss. Z. Martin-Luther-University Halle Wittenberg Math.-Natur. Reihe*, **38**(5), (1989) 13–19.

[888] WIENER, K., Solutions of inhomogeneous differential equations with derivatives of nonintegral order. IV, *Wiss. Z. Martin-Luther-University Halle-Wittenberg Math.-Natur. Reihe*, **40**(2), (1991) 69–72.

[889] WILMES, G., On Riesz-type inequalities and K-functionals related Riesz potential in R^n, *Numer. Funct. Anal. Optim.*, **1**(1), (1979) 57–77.

[890] WIMAN, A., Überden fundamentalsatz in der theorie der funktionen $E_\alpha(x)$, *Acta Math.*, **29**, (1905) 191–201.

[891] WRIGHT, E. M., On the coefficients of power series having essential singularities, *J. London Math. Soc.*, **8**, (1933) 71–80.

[892] WRIGHT, E. M., The asymptotic expansion of the generalized Bessel function, *Proc. London Math. Soc.*, **38**(2), (1934) 257–270.

[893] WRIGHT, E. M., The asymptotic expansion of the generalized hypergeometric function, *J. London Math. Soc.*, **10**, (1935) 286–293.

[894] WRIGHT, E. M., The asymptotic expansion of integral functions defined by Taylor Series, *Philos. Trans. Roy. Soc. London A*, **238**, (1940) 423–451.

[895] WRIGHT, E. M., The asymptotic expansion of the generalized hypergeometric function II, *Proc. London Math. Soc.*, **46**(2), (1940) 389–408.

[896] WRIGHT, E. M., The generalized Bessel function of order greater than one, *Quart. J. Math. Oxford Ser.*, **11**, (1940) 36–48.

[897] WU, J. L. and CHEN, C. H., A new operational approach for solving fractional calculus and fractional differential equations numerically, *Ieice Transact. Fundament. Electr. Commun. Computer Scienc.*, **E87A**(5), (2004) 1077–1082.

[898] WYSS, W., The fractional diffusion equation, *J. Math. Phys.*, **27**(11), (1986) 2782–2785.

[899] XU, Y., REN, F. Y., LIANG, J. R., and QIU, W. Y., Stretched Gaussian asymptotic behavior for fractional Fokker-Planck equation on fractal structure in external force fields, *Chaos Solit. Fract.*, **20**(3), (2004) 581–586.

[900] XUE, D. and CHEN, Y. Q., A comparative introduction of four fractional order controllers, in: *Proceedings of the 4th World Congress on Intelligent Control and Automation*, vol.4, 2002, 3228–3235.

[901] YAKUSHEVICH, L. V., *Nonlinear Physics of DNA*, Wiley and Sons, New York, 1990.

[902] YANG, B. and DEBNATH, L., A strengthened Hardy-Hilbert's inequality, *Proc. Jangjeon Math. Soc.*, **6**(2), (2003) 119–124.

[903] YANOVSKY, V. V., CHECHKIN, A. V., SCHERTZER, D., and TUR, A. V., Levy anomalous diffusion and fractional Fokker-Planck equation, *Physica A: Stat. Mech. Appl.*, **282**(1-2), (2000) 13–34.

[904] YEH, M.H. and PEI, S.C., A method for the discrete fractional Fourier transform computation, *IEEE Trans. Sig. Proc.*, **51**(3), (2003) 889–891.

[905] YUAN, L. X. and AGRAWAL, O. P., A numerical scheme for dynamic systems containing fractional derivatives, in: *Proceedings of DETC' 98, ASME Design Engineering Technical Conferences*, Atlanta, 1998, 1–4.

[906] ZABURDAEV, V. Y. and CHUKBAR, K. V., Enhanced superdiffusion and finite velocity of Levy flights, *J. Exper. Theoret. Phys.*, **94**(2), (2002) 252–259.

[907] ZAHRAN, M. A., 1/2-order fractional Fokker-Planck equation on comb-like model, *J. Stat. Phys.*, **109**(5-6), (2002) 1005–1016.

[908] ZAHRAN, M. A., ABULWAFA, E. M., and ELWAKIL, S. A., The fractional Fokker-Planck equation on comb-like model, *Physica A: Stat. Mech. Appl.*, **323**, (2003) 237–248.

[909] ZAIN, A. and TAZALI, A., A global existence and uniqueness theorem for ordinary differential equations of generalized order, *Canad. Math. Bull.*, **21**(3), (1978) 267–271.

[910] ZAIN, A. and TAZALI, A. Z., Local existence theorems for ordinary differential equations of fractional order, in: *7th Conference on Ordinary and Partial Differential Equations*, vol. 964 of Lect. Notes Math., Springer-Verlag, New York, 1982, 652–665.

[911] ZASLAVSKY, G. M., Fractional kinetic equation for Hamiltonian chaos, *Physica D*, **76**(1-3), (1994) 110–122.

[912] ZALAVSKY, G. M., *Physics of Chaos in Hamiltonian Dynamics*, Imperial College Press, London, 1998.

[913] ZALAVSKY, G. M., Chaotic dynamics and the original of statistical laws, *Phys. Today*, **8**(1), (1999) 39–45.

[914] ZALAVSKY, G. M., Chaos, fractional kinetics, and anomalous transport, *Phys. Rep.*, **371**, (2002) 461–580.

[915] ZASLAVSKY, G. M., *Chaos and Fractional Dynamics*, vol. 511 of Lect. Notes in Phys., Oxford University Press, Oxford, 2005.

[916] ZASLAVSKY, G. M., *Hamiltonian Chaos and Fractional Dynamics*, Oxford University Press, Oxford, 2005.

[917] ZASLAVSKY, G. M. and BENKADDA, M., *Chaos. Kinetics and Nonlinear Dynamics in Fluids and Plasmas*, vol. 511 of Lect. Notes in Phys., Springer-Verlag, Berlin, 1998.

[918] ZASLAVSKY, G. M., EDELMAN, M., and NIYAZOV, B. A., Self-similarity, renormalization and phase space nonuniformity of Hamiltonian chaotic dynamics, *Chaos*, **7**(1), (1997) 159–181.

[919] ZASLAVSKY, G. M., EDELMAN, M., WEITZNER, H., CARRERAS, B., MCKEE, G., BRAVENEC, R., and FONCK, R., Large-scale behavior of the tokamak density fluctuations, *Phys. Plasmas*, **7**(9), (2000) 3691–3698.

[920] ZASLAVSKY, G. M. and NIYAZOV, B. A., Fractional kinetics and accelerator modes, *Physics Re.*, **283**(1-4), (1997) 73–93.

[921] ZEMANIAN, A. H., *Generalized Integral Transformations*, vol. 18 of Pure Appl. Math., Wiley and Sons, New York, 1968.

[922] ZHANG, S., Existence of positive solutions for some class of nonlinear fractional differential equations, *J. Math. Anal. App.*, **278**(1), (2003) 136–148.

[923] ZHANG, S. Q., The existence of a positive solution for a nonlinear fractional differential equation, *J. Math. Anal. Appl.*, **252**(2), (2000) 804–812.

[924] ZHANG, W. and SHIMIZU, N., Damping properties of the viscoelastic material described by fractional Kelvin-Voigt model, *JSME Internat. J. C*, **42**(1), (1999) 1–9.

[925] ZHELUDEV, V., The well-posedness of the certain class of convolution equations, *Zh. Vychisl. Mat. i Mat. Fiz.*, **14**(3), (1974) 610–630.

[926] ZHELUDEV, V. A., Derivatives of fractional order and the numerical solution of a class of convolution equations, *Differentsial'nye Uravneniya*, **18**(11), (1982) 1950–1960.

[927] ZHEMUKHOV, K. K., *An Analogue of the Darboux Problem for a Higher-Order Loaded Hyperbolic Equation*, Acad. Nauk Ukrain. SSR (Inst. Mat.), Kiev, 1986, 97–101.

[928] ZHEMUKHOV, K. K., The Darboux problem for a second-order degenerate loaded integro-differential equation, ẽmphDifferentsial'nye Uravneniya, **22**(12), (1986) 2174–2176.

Subject Index

FSC
www.fsc.org
MIX
Papier | Fördert
gute Waldnutzung
FSC® C083411

Kolibri 360 GmbH
Ferdinand-Jühlke-Straße 7
99095 Erfurt, Deutschland
produktsicherheit@kolibri360.de